T0390570

Circular Economy and SUSTAINABILITY

VOLUME 1

Circular Economy and SUSTAINABILITY

VOLUME 1

Management and Policy

Edited by

ALEXANDROS STEFANAKIS
Assistant Professor, Environmental Engineering and Management Laboratory, School of Chemical and Environmental Engineering, Technical University of Crete, Greece

IOANNIS NIKOLAOU
Associate Professor, Business Economics and Environmental Technology Laboratory, Department of Environmental Engineering, Democritus University of Thrace, Greece

ELSEVIER

Elsevier
Radarweg 29, PO Box 211, 1000 AE Amsterdam, Netherlands
The Boulevard, Langford Lane, Kidlington, Oxford OX5 1GB, United Kingdom
50 Hampshire Street, 5th Floor, Cambridge, MA 02139, United States

Library of Congress Cataloging-in-Publication Data
A catalog record for this book is available from the Library of Congress

British Library Cataloguing-in-Publication Data
A catalogue record for this book is available from the British Library

ISBN 978-0-12-819817-9

For information on all Elsevier publications
visit our website at https://www.elsevier.com/books-and-journals

Publisher: Candice Janco
Acquisitions Editor: Marisa LaFleur
Editorial Project Manager: Andrea Dulberger
Production Project Manager: Paul Prasad Chandramohan
Cover designer: Matthew Limbert

Typeset by STRAIVE, India

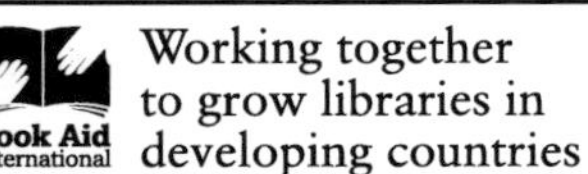

Contents

Contributors

Sami G. Al-Ghamdi
Division of Sustainable Development, College of Science and Engineering, Hamad Bin Khalifa University, Qatar Foundation, Doha, Qatar

Ibtisam Sulaiman Alhosni
Civil Engineering Department, Middle East College, Knowledge Oasis Muscat, Al Rusayl, Oman

Omar Amoudi
College of Engineering, National University of Science and Technology, Muscat, Oman

Fenna Blomsma
Universität Hamburg, Faculty of Business, Economics and Social Sciences, Hamburg, Germany

Geraldine Brennan
Imperial College London, Centre for Environmental Policy, Faculty of Natural Sciences, London, United Kingdom; Irish Manufacturing Research Co., Dublin, Ireland

Anton Brummelhuis
Head of Sustainable Innovation & Sustainable Design, Signify, Eindhoven, The Netherlands

Nicola Callaghan
School of Computing, Engineering and Built Environment, Glasgow Caledonian University, Glasgow, Scotland, United Kingdom

Joana Costa
DEGEIT, Universidade de Aveiro, GOVCOPP and INESCTEC, Aveiro, Portugal

Marco Cucchi
Gruppo Ceramiche Gresmalt, Sassuolo, Italy

Dimo Dimov
School of Management, University of Bath, Bath, United Kingdom

Azadeh Dindarian
Co-founder, NA, Women AI Academy & Consulting GmbH, Berlin, Germany

Karine Van Doorsselaer
Department Productdevelopment, University of Antwerp, Antwerp, Belgium

Hilde Engels
Department of Strategy, Institute for Management Research, Radboud University Nijmegen, Nijmegen, The Netherlands

K.I. Evangelinos
Department of Environment, University of the Aegean, Mytilini, Greece

M.A. Fernández López
Universidad Camilo José Cela (UCJC), School of Technology and Science, Madrid, Spain

Anna Maria Ferrari
Department of Sciences and Methods for Engineering, University of Modena and Reggio Emilia, Reggio Emilia, Italy

Karen Fortuin
Environmental Systems Analysis Group, Wageningen University, Wageningen, Netherlands

Robin Foster
HSSMI Limited, London, United Kingdom

G. Fountoulakis
Department of Environment, University of the Aegean, Mytilini, Greece

Tânia Freitas
FEP—Universidade do Porto, Porto, Portugal

Victor Fukumoto
Federal University of Mato Grosso do Sul, Brazil

Fernando E. García-Muiña
Department of Business Administration (ADO), Applied Economics II and Fundamentals of Economic Analysis, Rey-Juan-Carlos University, Madrid, Spain

Darya Gerasimenko
Professor of Sustainable Development at Samara National Research University (SSAU), Russia; Lecturer in Economics at St Gallen University (HSG); Co-facilitator, co-creator, *Science and Mindfulness Lead* (as a Senior Researcher at École polytechnique fédérale de Lausanne (EPFL)) of "*Beyond Waste: Circular Resources Lab 2018*" in canton Vaud, Switzerland

Helena I. Gomes
Food, Water, Waste Research Group, Faculty of Engineering, University of Nottingham, Nottingham, United Kingdom

João Pedro Moreira Gonçalves
Finca Equilibrium, Montijo, Portugal

Emma Goosey
Envaqua Research Limited, Warwick, United Kingdom

Rafael Hernández López
Universidad Camilo José Cela (UCJC), School of Technology and Science, Madrid, Spain

Alistair Ho
HSSMI Limited, London, United Kingdom

Holger Hoppe
Sustainability Management Department, Linde Material Handling GmbH, Aschaffenburg, Germany

Beijia Huang
School of Environment and Architecture, University of Shanghai for Science and Technology, Shanghai, China

M. Hughes
Wood Material Technology Research Group, Department of Bioproducts and Biosystems, School of Chemical Technology, Aalto University, Espoo, Finland; Faculty of Health Science, Department of Human Care Engineering, Nihon Fukushi University, Aichi, Japan

Susan L. Hull
Department of Biological and Marine Sciences, University of Hull, Hull, United Kingdom

R. Husgafvel
Wood Material Technology Research Group, Department of Bioproducts and Biosystems, School of Chemical Technology, Aalto University, Espoo, Finland; Faculty of Health Science, Department of Human Care Engineering, Nihon Fukushi University, Aichi, Japan

Konstantinos Iatridis
School of Management, University of Bath, Bath, United Kingdom

Jan Jonker
Department of Strategy, Institute for Management Research, Radboud University Nijmegen, Nijmegen, The Netherlands

René Kemp
United Nations University-MERIT, Maastricht University, Maastricht, The Netherlands

A. Kipritsis
Business Economics and Environmental Technology Laboratory, Department of Environmental Engineering, Democritus University of Thrace, Greece

Nermin Kişi
Zonguldak Bülent Ecevit University, Zonguldak, Turkey

G. Lanaras-Mamounis
Business Economics and Environmental Technology Laboratory, Department of Environmental Engineering, Democritus University of Thrace, Greece

L. Linkosalmi
Clean Technologies Research Group, Wood Material Technology Research Group, Department of Bioproducts and Biosystems, School of Chemical Technology, Aalto University, Espoo, Finland

Cembalo Luigi
University of Naples Federico II, Department of Agricultural Sciences, AgEcon and Policy Group, Italy

Warren E. Mabee
Department of Geography and Planning/School of Policy Studies, Queen's University, Kingston, ON, Canada

Monica Mihaela Maer-Matei
National Research Institute for Labour and Social Protection; Bucharest University of Economic Studies, Bucharest, Romania

Sergey Makaryan
Sustainability Management Department, Linde Material Handling GmbH, Aschaffenburg, Germany; Environmental Systems Analysis Group, Wageningen University, Wageningen, Netherlands

Mehzabeen Mannan
Division of Sustainable Development, College of Science and Engineering, Hamad Bin Khalifa University, Qatar Foundation, Doha, Qatar

Thomas Marinelli
Head of Sustainable Innovation & Sustainable Design, Signify, Eindhoven, The Netherlands

Borrello Massimiliano
University of Naples Federico II, Department of Agricultural Sciences, AgEcon and Policy Group, Italy

Deepak John Mathew
Department of Design, Indian Institute of Technology Hyderabad, India

William M. Mayes
Department of Geography, Geology and Environment, University of Hull, Hull, United Kingdom

Erica Mazerolle-Castillo
Social Innovation & Communities of Practice Lead at Impact Hub Lausanne/Geneva (IHL/IHG); Co-facilitator, co-creator of "Beyond Waste: Circular Resources Lab 2018", Switzerland

M. Sonia Medina-Salgado
Department of Business Administration (ADO), Applied Economics II and Fundamentals of Economic Analysis, Rey-Juan-Carlos University, Madrid, Spain

O. Mikroni
Department of Environment, University of the Aegean, Mytilini, Greece

Eva Militaru
National Research Institute for Labour and Social Protection, Bucharest, Romania

Alberto Fernandez Minguela
HSSMI Limited, London, United Kingdom

Irina Mkrtchyan
"ISSD - Innovative Solutions for Sustainable Development of Communities" NGO, Yerevan, Armenia

Cristina Mocanu
National Research Institute for Labour and Social Protection; University of Bucharest, Bucharest, Romania

Ioannis E. Nikolaou
Business Economics and Environmental Technology Laboratory, Department of Environmental Engineering, Democritus University of Thrace, Greece

Piotr Nowakowski
Faculty of Transport, Silesian University of Technology, Katowice, Poland

Satoshi Ohnishi
Tokyo University of Science, Faculty of Science and Technology, Department of Industrial Administration, Noda-shi, Chiba-ken, Japan

Juyeon Park
National Physical Laboratory (NPL), Electrochemistry Group, Teddington, United Kingdom

Seigo Robinson
University College London, Central House, London, United Kingdom

Indranil Saha
Department of Design, Indian Institute of Technology Hyderabad, India

D. Sakaguchi
Faculty of Health Science, Department of Human Care Engineering, Nihon Fukushi University, Aichi, Japan; Clean Technologies Research Group, Department of Bioproducts and Biosystems, Aalto University, School of Chemical Technology, Espoo, Finland

Daniel Rolph Schneider
Department of Energy, Power Engineering and Environment, Faculty of Mechanical Engineering and Naval Architecture, University of Zagreb, Zagreb, Croatia

Anastasios Sepetis
Department of Business Administration, University of West Attica and Hellenic Open University, Athens, Greece

Davide Settembre-Blundo
Gruppo Ceramiche Gresmalt, Sassuolo, Italy

Manuela Castro Silva
FEP—Universidade do Porto, Porto, Portugal

Keith R. Skene
Biosphere Research Institute, Angus, United Kingdom

Alexandros I. Stefanakis
Environmental Engineering and Management Laboratory, School of Chemical and Environmental Engineering, Technical University of Crete, Greece

Krzysztof Szwarc
Institute of Computer Science, University of Silesia in Katowice, Sosnowiec, Poland

Mike Tennant
Imperial College London, Centre for Environmental Policy, Faculty of Natural Sciences, London, United Kingdom

Tihomir Tomić
Department of Energy, Power Engineering and Environment, Faculty of Mechanical Engineering and Naval Architecture, University of Zagreb, Zagreb, Croatia

Thomas A. Tsalis
Business Economics and Environmental Technology Laboratory, Department of Environmental Engineering, Democritus University of Thrace, Greece

Alisha Tuladhar
School of Management, University of Bath, Bath, United Kingdom

Serdar Türkeli
United Nations University-MERIT, Maastricht University, Maastricht, The Netherlands

Elena Turrado Domínguez
Universidad Camilo José Cela (UCJC), School of Technology and Science, Madrid, Spain

Alexandre Meira de Vasconcelos
Federal University of Mato Grosso do Sul, Brazil

Konstantinos I. Vatalis
Department of Mineral Resources Engineering, University of Western Macedonia; Department of Environmental Engineering and Antipollution Control, Technological Educational Institute of Western Macedonia (TEI), Kozani, Greece

P. Vouros
Department of Environment, University of the Aegean, Mytilini, Greece

Mariusz Wala
PST Transgór S.A, Rybnik, Poland

K.S. Winans
UCD Industrial Ecology Program, Davis, CA, United States

Ana Maria Zamfir
National Research Institute for Labour and Social Protection, Bucharest, Romania

Preface

Circular economy is today a rising and widely discussed topic among professionals, academics, and the public. The trigger for this increasing attraction to this relatively new concept is the gradual realization that the way our economies and societies have grown over the last 50 years has caused global environmental problems. We have experienced an unforeseen and continuous increase in materials extraction and consumption in our societies, doubled the production of goods, and quadrupled our economic development, reaching the smallest ever recorded percentage of people living under the poverty threshold. However, this much-desired growth and improvement of our global living standards has also resulted in tremendous biodiversity loss, increasing water stress and greenhouse gas emissions, and ever-progressing climate change.

While the necessity to tackle these global environmental issues effectively is now recognized, we are still trying to identify the means to do that, i.e., to reach the goal of a truly sustainable society. While the discussion on sustainable society and economy is ongoing, it is clear that we need a new narrative, i.e., a new direction and scientific and epistemological theoretical frameworks. Circular economy is essentially an upcoming concept that could create the conditions to overcome the limitations of the existing linear economic model (take–produce–dispose) and focus on more efficient use of materials and flow optimization through engineering advances to preserve natural resources.

However, to achieve feasible sustainable solutions, the interface between economy, policy, and engineering, and implementation and production should be viewed from a different angle. A single environmental or technical approach is not sufficient, and a wider view is needed of the way we deal with environmental issues and industrial processes, to include goals such as social justice, poverty alleviation, and global and local connections. Hence, to get the answers and the solutions we seek, we see circular economy as a necessary vehicle to achieve the concept of sustainability, i.e., we believe that these two terms go hand-in-hand, thanks to social equity, economic growth, and environmental protection.

The new concept of circular economy and sustainability should be built on three main pillars: (i) environmental engineering, (ii) business, management, and economy, and (iii) society. The challenge now is to investigate the interconnections, interrelations, interactions, and synergies between these three pillars and how these relations are or can be realized in the real world. This is exactly the main motivation and the rationale behind this book. With this book we wanted to explore this new approach of circular economy and sustainability through chapters that would touch on and discuss different solutions and concepts to implement circular economy, viewed from multiple different angles.

The response was excellent, with more than 100 chapter proposals received. Therefore, we consulted Elsevier and they happily agreed to publish a second volume of this book, so that more chapters could be accepted. Thus, after the initial screening of the chapter proposals, the overall outcome is impressive: 63 chapters written by 198 authors from 33 different countries from all continents. The separation into two volumes was based on the contents: the first volume of the book includes chapters that focus more on the management and policy aspects, while the second volume deals mostly with the engineering and technology aspect. This two-volume book is the first to present an integrated approach of circular economy in various fields and disciplines, and to relate it to the goal of a sustainable society.

We hope that the readers, including professionals, academics, engineers, consultants, researchers, scientists, government agency employees, industrial stakeholders/entities, and students, will find the information provided here valuable for their work and tasks and useful for the fulfillment of their goals. We also hope that this book will contribute to a better understanding of the circular economy concept and to a clear definition of its contents.

Alexandros Stefanakis
Ioannis Nikolaou

CHAPTER 1

A review of circular economy literature through a threefold level framework and engineering-management approach

Ioannis E. Nikolaou[a] **and Alexandros I. Stefanakis**[b]
[a]Business Economics and Environmental Technology Laboratory, Department of Environmental Engineering, Democritus University of Thrace, Greece
[b]Environmental Engineering and Management Laboratory, School of Chemical and Environmental Engineering, Technical University of Crete, Greece

1. Introduction

Today, circular economy (CE) is considered to be a new concept, which has gained great momentum among scholars and practitioners (Su et al., 2013; Hart and Pomponi, 2021; Lewandowski, 2016; Nikolaou et al., 2021; Webster, 2021). Essentially, the term CE consists of two general components, i.e., "circular" and "economy." The former component implies that products and services should be organized in a way that slows, narrows, and closes the loop of materials and resources within organizations' production processes (Bocken et al., 2016). Thus, several strategies have been advocated to enable the shift from the conventional linear and behavioral thinking of organizations, consumers, and decision-makers to a CE concept (Michelini et al., 2017). The later component is focused on economic aspects such as production procedures and financial outcomes (Bocken et al., 2021; Genovese et al., 2017).

In essence, the majority of existing studies have focused on the circular side of the concept and primarily on engineering processes. Korhonen et al. (2004) pointed out that, despite the potential for engineering and nature-based sciences to support and facilitate modern societies in introducing the principles of CE into their day-to-day operation, they have failed to make comprehensible systems by drawing insights from management science. Two very significant (economic and management) aspects to achieve the principles of CE in modern societies are through strengthening supply and demand sides.

This distinction between engineering/nature-based sciences and management/economic sciences has been mainly examined at a threefold level of analysis. At the first level, a single firm is examined, which adopts various practices to change its conventional model, e.g., adopting a circular business model through reducing, reusing, recycling, and redesigning materials and waste (Franco, 2017). The next level implies collective

Circular Economy and Sustainability, Volume 1
https://doi.org/10.1016/B978-0-12-819817-9.00001-6

collaborations among firms to exchange by-products as raw materials and succeeding either involuntarily or voluntarily to close the loop of materials, products, and packaging by adopting an industrial ecology and symbiosis paradigm (Gómez et al., 2018). The final level entails the integration of CE principles (e.g., reduce, reuse, recycle, and remanufacture) into an overall economy, a city, a region, or a municipality (Ferronato et al., 2019).

The methodology of this chapter is a bibliometric analysis that focuses on the examination of the CE concept from engineering/nature-based sciences and management/economic-based sciences through a threefold-level analysis. Six areas of literature have been recognized through engineering/nature-based sciences, management/economic-based analysis, a single firm, among joint corporate endeavors, and overall prefectures and countries. This study aims at finding the topics that have been overemphasized and those that have been less examined in the current literature. It also aims at describing an overall agenda suitable for future research.

The rest of this chapter is organized in four sections. The first section describes the necessary theoretical background, where a number of existing literature review studies are analyzed to determine the key findings and results. The second section develops the methodology of this study, and the third section analyzes the most significant findings. Finally, a conclusion section is presented.

2. Theoretical background

Today, CE is an overused concept in the current literature but without an uncontested and a clear definition. Many scholars and organizations have conducted studies to collect and analyze the bulk of current definitions and establish a more comprehend definition for CE. Having analyzed the current literature, Kirchherr et al. (2017) identified 114 CE definitions, 3% of which promote a hierarchy of waste and only 40% of the 114 definitions have focused on a system analysis. Similarly, Korhonen et al. (2018a) have classified CE definitions, according to the Ellen MacArthur Foundation definition (EMF_D), into two categories. The former category of definitions utilizes some key traits of EMF_D such as "restorative," "regenerative," "reuse," "recycle," and "recovery." The latter category of definitions analyzes key concepts such as "reduce the use of virgin materials," "restoration," "recycling," "sustainable economy," and "eco-efficiency." However, Korhonen et al. (2018b) have developed a CE definition that emphasizes shifting the behavior of consumers and producers from a linear to a circular state.

By studying the existing literature, Homrich et al. (2018) identified that most current definitions place emphasis on the effective exploitation of existing materials and energy, as well as on the use of less virgin natural resources, by introducing longevity-thinking into the design procedures of products. They have also settled on a definition for CE that is focused on combining economic and engineering aspects by closing materials loops, with simultaneous economic benefits (win-win solutions). Having analyzed many definitions of CE, Prieto-Sandoval et al. (2018) identified that the key points of most of the existing definitions put emphasis on closing the loop through transformation, distribution, use, and recovery of materials. By basing it on EMF_D, Geissdoerfer et al. (2017) defined CE by highlighting the "regenerative" concept of use of materials through lowering, closing, and narrowing the loops of materials and energy.

Furthermore, many scholars have endeavored to explore these definitions in a more operational way by discussing them in tandem with the principles of sustainable development (Alonso-Almeida et al., 2020). Correspondingly, several models have been advocated as suitable to quantify different principles of CE (e.g., reuse, recycle, and reduce). A well-known and complete model has been provided by Potting et al. (2017), where three phases are considered vital to move an economy from the linear to the circular form. These phases are the efficient use of materials (e.g., recover, recycle), the extended longevity of products (e.g., repurpose, remanufacture, refurbish, repair, and reuse), and smart products (e.g., reduce, rethink, and refuse).

To study CE subjects, many literature review studies have been conducted that examine strategies, practices, and frameworks suitable for implementing the principles of CE (Merli et al., 2018; Sassanelli et al., 2019). By conducting a literature review at the micro-level, Lüdeke-Freund et al. (2019) recognized six types of circular business models found in everyday business operation, including the recycling type, business to business (B2B), business to customer (B2C), business model (BM), customer to customer (C2C), and circular economy business model. Similarly, Urbinati et al. (2017) proposed a taxonomy concerning circular business models, which are based on internal incentives for businesses to make a value proposition to their customers by adopting CE practices, and external incentives for businesses to adjust their operation in an institutional context in relation to CE. Merli et al. (2018) identified that many of the existing studies regarding CE pay more attention to the institutional effects of production and consumption processes. By reviewing current techniques and methods of CE, Sassanelli et al. (2019) emphasized the move from an end-of-life to closed-loop thinking for product design. They also recorded all the current measurement and assessment techniques regarding circularity performance of products such as life cycle analysis, multicriteria decision-making methods, and design for the circular economy. In the same way, a taxonomy of CE measurement performance frameworks has been presented by Saidani et al. (2019) including 10 categories of indicators classified in a three-level context, i.e., micro-, meso-, and macro-level.

Most of the current literature review studies mainly focus on analyzing only one level of CE literature (e.g., micro-level), one side of the economy (e.g., consumption), or one principle of CE (e.g., remanufacturing). Although the focus on only one level or one topic seems to be methodically correct, it nevertheless alienates the findings from the leading objectives of a concept like CE. Indeed, the description of each level itself could ultimately help reduce the consumption of virgin natural resources without taking into account the greater picture of the global environmental problems and certainly the overall solution to such problems. It is likely that the existing experience of the other levels would form a good groundwork for designing adequate CE frameworks to solve environmental problems through a systemic and systematic approach. One further substantial weakness of the current literature reviews is their emphasis on the analysis of the CE concept through one scientific principle such as environmental, management sciences, or engineering sciences. However, the concept of CE is also associated with topics of different academic fields, which are necessary and useful to solve the environmental problems. Thus, several studies to date promote the interdisciplinary nature of these issues (Korhonen et al., 2018a,b).

3. Methodology

This study conducts a systematic review of the current literature regarding CE from engineering/nature-based and management/economic-based perspectives (Nikolaou et al., 2021). This review is based on quantitative and qualitative approaches. Particularly, a bibliometric methodology is utilized to assist in conducting quantitative and qualitative research.

3.1 Research structure

The structure of the suggested methodology is based on five interrelated steps (Fig. 1). The first step designates the basic research questions of this study, in keeping with the classical approach, before conducting a study, of determining suitable research questions that should be answered through the suggested methodology. The second step describes the necessary processes of gathering the appropriate data in order to answer the defined

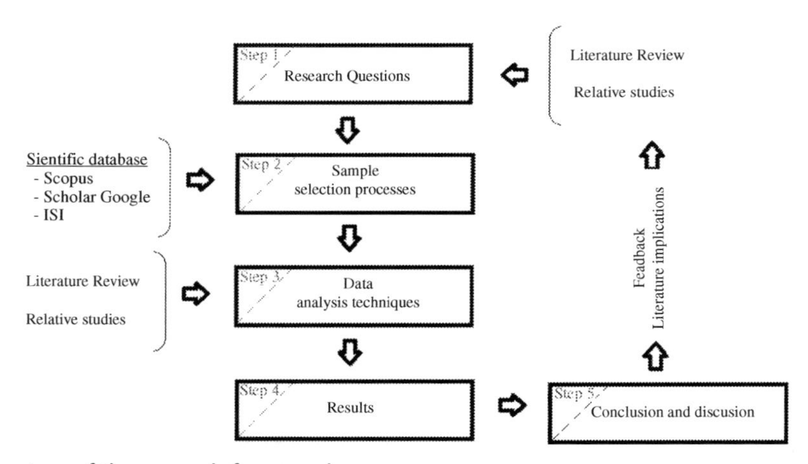

Fig. 1 Structure of the research framework.

research questions. This section is considered very substantial since it provides the essential information to examine the concept of CE from two basic academic fields: engineering/nature-based and management/economic-based sciences. The next step contributes to the analysis of the most common techniques and methods used to describe the most crucial topics of the current literature. The fourth step presents the results of this research and the final step provides the implications for the engineering and management literature.

3.2 Research questions

Nowadays, the term CE is very popular, and many scholars have conducted studies to make its content clearer (Lewandowski, 2016; Cecchin et al., 2021; Hart and Pomponi, 2021; Kirchherr et al., 2018; Nikolaou et al., 2021; Webster, 2021). Similarly, several principles have been suggested as fundamental CE concepts, arising from different academic fields (Bocken et al., 2021; Ghisellini et al., 2021; Ozili, 2021; Stefanakis et al., 2021). One significant stream of literature, which is associated with the engineering and nature-based academic fields, puts greater emphasis on pinpointing applied solutions to solve large environmental problems, and closing materials loops (Molina-Moreno et al., 2017; Avraamidou et al., 2020). Their first task is to study the potential technical and biological cycles in order to preserve materials over a longer period in both the production and consumption procedures and obviously to avoid the use of virgin natural resources. The second stream of literature emerges from the management and economic sciences, and focuses on identifying certain ways to incorporate CE ideas into organizations in operationally effective and cost-efficient ways.

Essentially, these scientific fields are necessary to make CE useful, functional, and efficient for contemporary and future economies and societies. However, a mapping of the CE territory is necessary to show the key topics of each field, the common topics, and the potential connections among such fields. To examine these ideas, the research questions of this study are as follows:

Research Question 1 (RQ_1): Which scientific fields have covered the greatest part of CE literature?

Research Question 2 (RQ_2): What has been the evolution of CE studies over time?

Research Question 3 (RQ_3): What level of analysis attracts the greatest emphasis regarding the concept of CE?

3.3 Data selection processes

One significant subject in conducting a study is to create suitable conditions to collect all necessary data. To do so, a robust context is needed to collect necessary information from the existing literature. As aforementioned, this study utilizes a bibliometric methodology to gather data by analyzing the existing literature through specific steps. Table 1 shows the steps that are adopted for the purpose of this chapter.

Table 1 Steps and criteria for data selection.

Steps	Description	Measurement scale	Clarifications
Scientific databases selection	The most popular and transparent databases.	ISI Scopus Scholar Google	ISI is a large publisher citation index Scopus is a citation database Google Scholar is a Web search engine for scholarly literature.
Criteria selection (**C*i***)	**C_1:** *Title of the study*	1–10 (1 low relative, 10 high relative)	The relevance of the title of selected study with CE
	C_2: *Year of publication*	1–4 1 before 20 years, 4 last five years	The (old or contemporary) publication year of the study
	C_3: *Scientific field of the journal*	1–10 (1 low relative, 10 high relative)	Engineering/nature-based science or management/economic-based science
	C_4: *Keywords of the study*	1–10 (1 low relative, 10 high relative)	The relevance of keywords with CE
	C_5: *Citations of the study*	1–4 (1 under 50, 4 over 500)	The citations of a study by other studies
	C_6: *Content of the study*	1–10 (1 low relative, 10 high relative)	The relevance of a study with CE
	C_7: *Impact factor (IF) of the journal*	0–4 (0 without IF, 4 over 5 IF)	The impact of the Journal in which the selected study has been published.
	C_8: *SCimago journal classification*	1–4 (1 for Q_4, 4 for Q_1)	Q_1 implies the top 25% of the Journals, and Q4 the lowest 25% of the journals. Q2 and Q3 imply 50% and 75%, respectively.
Final selection index (FDI)	$FDI = \sum_{i=1}^{8} \frac{C_i}{n}$	$0 \le FDI \le 46$	Finally, analysis of studies with the highest score

As a first step, the appropriate scientific databases are selected, considering where the majority of the relative studies are available. For the scope of this study, the best-known and most scientifically transparent databases for finding journals are Scopus, Scholar Google, and the ISI Web of Knowledge. In particular, the choice of such databases is based on certain criteria such as the quantity and the quality of scientific studies that are published in them. The second step focuses on designing a methodology to include or exclude the necessary studies from the overall studies gathered in the previous step. Some representative excluding or including criteria are: the title of the study, the year of publication, the impact factor of the journal, the academic field that the journal focuses on, and so on. As a third step, a mathematical formula is suggested to calculate the final score, which ranges from 0 to 46 points. The final selection of the studies has been made based on them achieving a higher final score.

Another task in the first step is the selection of suitable keywords on which the search procedures are based. Some significant keywords are "circular economy in engineering," "circular economy in environmental science," "circular economy in natural-based science," "circular economy in economic-based science," "circular economy in management science," "circular economy in micro-level," "circular economy in meso-level," and "circular economy in macro-level."

3.4 Data analysis techniques

To analyze the content of selected studies, a framework is designed that is based primarily on the classical threefold level approach that includes micro-, meso-, and macro-levels (Fig. 2). The first (micro) level encompasses studies with emphasis on CE strategies, methodologies, and practices in relation to the production, operation, and products processes. The second (meso) level includes studies regarding the industrial ecology and symbiosis that advocate cooperative actions among firms to exchange the waste materials of some firms as raw materials for others. Finally, the last (macro) level includes studies regarding the principles and policies of CE at a national level. These levels also include

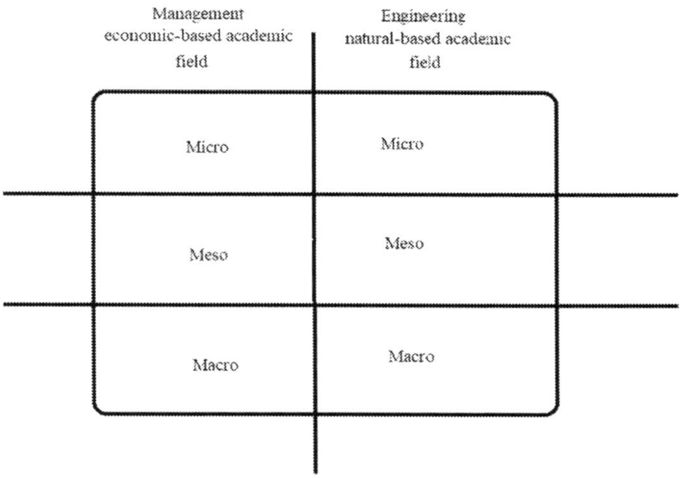

Fig. 2 Methodological framework structure.

studies from two general academic fields: engineering/nature-based sciences and management/economic-based sciences (Fig. 2).

4. Results

To answer the first research question (RQ_1) of this study, a classification of selected papers by academic field is made. This implies that selected papers could be classified in engineering/nature-based science and in management/economic-based fields. The selected studies have an engineering/nature-based or management/economic-based orientation, and the highest score achieved in FDI. Table 2 shows some indicative studies for both fields.

Fig. 3 shows a total of 11,565 studies regarding the circular economy from different academic fields. The majority of such studies are associated with engineering and nature-based sciences, i.e., 11,926 studies, and 2718 studies are associated with management and economic-based sciences. Finally, 76 of the selected studies use multidisciplinary scientific fields.

It should be noted that the field of engineering and nature-based sciences includes studies related to environmental science, engineering, energy, materials, chemical engineering, agriculture and biological sciences, and earth and planetary sciences. The majority of the studies have emerged from environmental science (35%), engineering science (25%), and energy science (18%) (Fig. 4).

Similarly, Fig. 5 shows the allocation of studies in management/economic-based sciences, i.e., business, management and accounting, economics, econometrics, finance, and decision sciences. The majority of the selected studies arise from the business management and accounting field (57%) and the economics, econometrics, and finance fields (31%). The final category of decision science includes 12% of the selected studies.

The second research question (RQ_2) requires an analysis of the studies of CE over time. To this end, the numbers of examined studies regarding CE seem to have grown rapidly over the last decade. The relevant studies have quadrupled in this decade; specifically, from 320 in 2003, they reached 4600 in 2020. A gradual evolution during the last decade (Fig. 6) is identified. Many of these papers focused on reviewing existing studies for a variety of topics on the CE concept.

The last research question (RQ_3) focuses on examining the allocation of such studies into three levels. Fig. 7 shows the allocation of CE studies in relation to the first (micro) level. The majority of studies emerge from environmental science (720) and engineering science (510). Overall, engineering and nature-based sciences (e.g., environmental science, earth and planetary sciences, engineering, and energy material sciences) represent the major part of these studies, totaling 1749 studies, while management and economic-based sciences (e.g., business, management and accounting, social sciences, economics, econometrics, finance, and decision sciences) total 1178 studies.

Table 2 Some indicative studies in the engineering/nature-based and management/economic-based fields.

Engineering/nature-based fields	Key topics
Avraamidou et al., 2020; Awasthi et al., 2019; Bao et al., 2019; Bekchanov and Mirzabaev, 2018; Belaud et al., 2019; Bocken et al., 2016; Bruel et al., 2019; Confente et al., 2020; Deutz et al., 2017; Dominguez et al., 2018; Elia et al., 2017; Esa et al., 2017; Esposito et al., 2018; Ferronato et al., 2019; Fischer and Pascucci, 2017; Gallagher et al., 2019; Garmulewicz et al., 2018; Geissdoerfer et al., 2018, Geissdoerfer et al., 2017; Geng and Doberstein, 2008; Genovese et al., 2017; Ghisellini et al., 2016, Ghisellini et al., 2018; Han et al., 2018; Hazen et al., 2017; Huang et al., 2018; Jakhar et al., 2019; Keijer et al., 2019; Korhonen et al., 2018a,b; Lausselet et al., 2017; Li et al., 2010; Lieder and Rashid, 2016; Loizia et al., 2019; Lüdeke-Freund et al., 2019; Ma et al., 2014; Mahpour, 2018; Mikulčić et al., 2016; Molina-Moreno et al., 2017; O'Connor et al., 2016; Palafox-Alcantar et al., 2020; Pan et al., 2015; Quina et al., 2018; Raheem et al., 2018; Schetters et al., 2015; Sgroi et al., 2018; Sica et al., 2018; Smol et al., 2015; Stiles et al., 2018; Voulvoulis, 2018; Wall et al., 2017.	Circular economy in process systems; electronic waste; circular economy of construction and demolition waste; composting; circular economy and industrial ecology; product design; bio-plastics products for a circular economy; resource recovery and remediation; LCA of greywater; measuring circular economy strategies waste valorization; material use in the textile industry; renewable energy technologies; disruptive technology; supply chains for the circular economy; a new sustainability paradigm; leapfrog development; sustainable supply chain management; ecological and health risks assessment; remanufacturing for the circular economy; stakeholder pressure drives the circular economy; circular chemistry; circular economy limitations; waste-to-energy; energy conservation and circular economy; circular economy in food waste management; circular economy in steel industry; greenhouse gasses emissions; material supply chain sustainability and circular economy; MSW incineration ashes; reuse of sewage sludge; reuse of end-of-life photovoltaic panels; microalgae in the circular economy
Management/economic-based field	**Key topics**
Abu-Ghunmi et al., 2016; Andersen, 2007; Aranda-Usón et al., 2019; Cardoso de Oliveira et al., 2019; Oliveira et al., 2021; D'Amato et al., 2019; De Jesus and Mendonça, 2018; Dubey et al., 2019; Esposito et al., 2017; Ferreira Gregorio et al., 2018; Geng and Doberstein, 2008; Gusmerotti et al., 2019; Hussain and Malik, 2020; Jabbour et al., 2019; Kalmykova et al., 2018; Khan et al., 2020; Liu et al., 2018; Pieroni et al., 2019; Reh, 2013; Popescu, 2018; Rajput and Singh, 2019; Reike et al., 2018; Saavedra et al., 2018; Schroeder et al., 2019; Smol et al., 2018; Stewart and Niero, 2018; Ünal and Shao, 2019; Urbinati et al., 2017; Wysokińska, 2016; Zeng et al., 2017; Zhijun and Nailing, 2007; Zhu et al., 2010; Zink and Geyer, 2017.	Circular economy and the opportunity cost; environmental economics of the circular economy; circular economy and supply chains; circular, green, and bio economy; eco-innovation; supplier relationship management for circular economy; circular business management; leapfrog development; rebound effect; circular economy practices for manufacturers; institutional pressures; environmental policy for circular economy; cradle-to-cradle products; circular economy in corporate sustainability strategies; public awareness of circular economy; circular economy and sustainable development goals; industrial ecology to circular economy; refurbish; circular economy and industry 4.0; social responsibility and business ethics; business model innovation for circular economy; microfoundations of dynamic capabilities and circular economy.

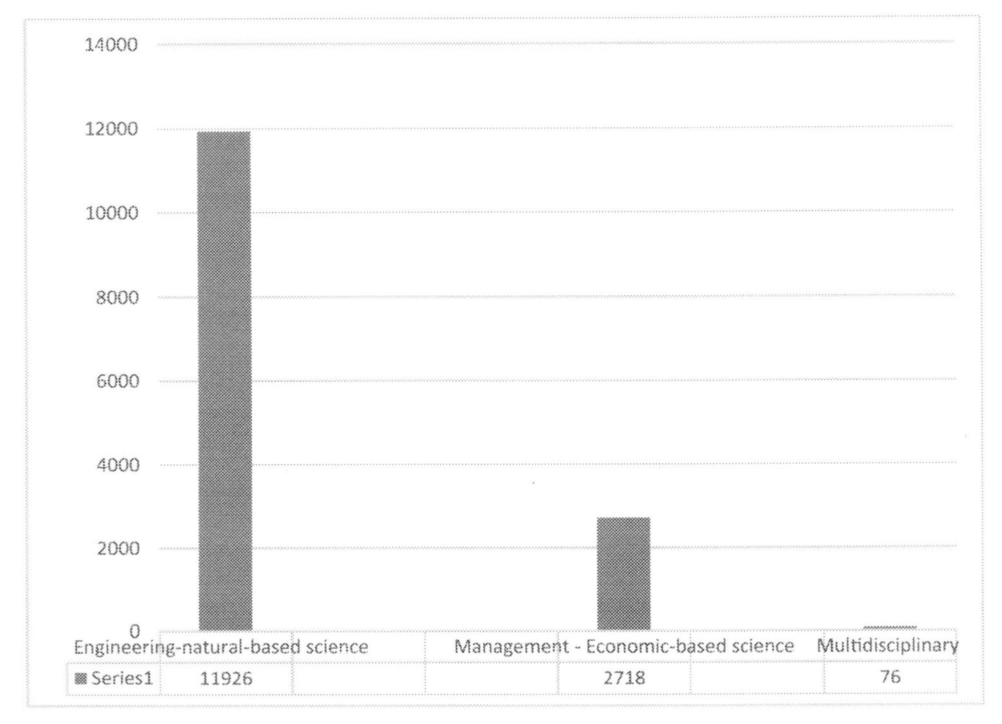

Fig. 3 Allocation of studies based on engineering and management sciences.

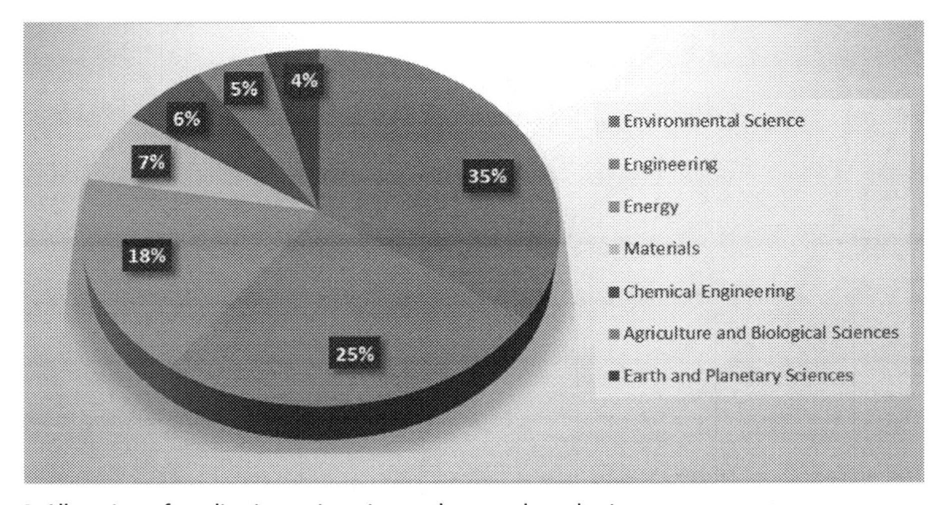

Fig. 4 Allocation of studies in engineering and nature-based sciences.

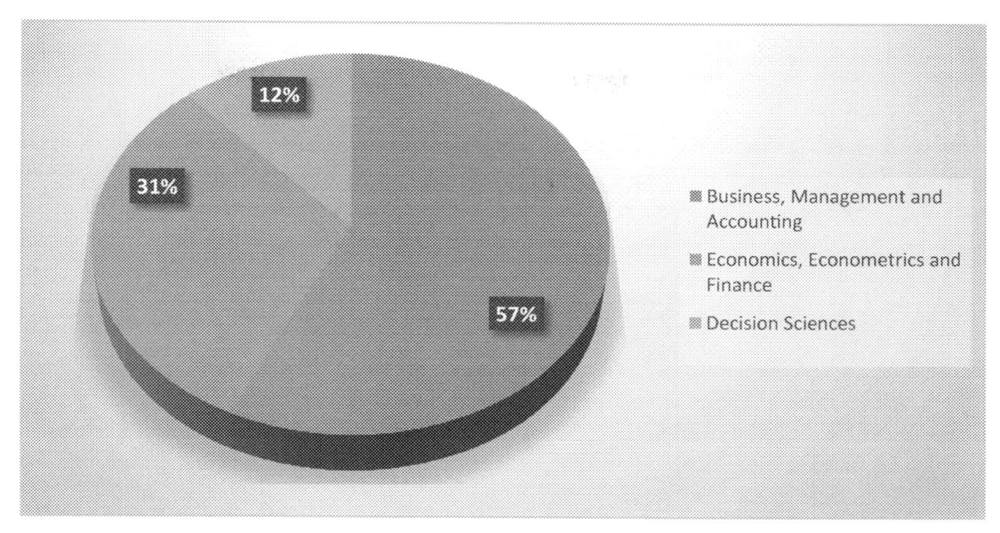

Fig. 5 Allocation of studies in management/economic-based sciences.

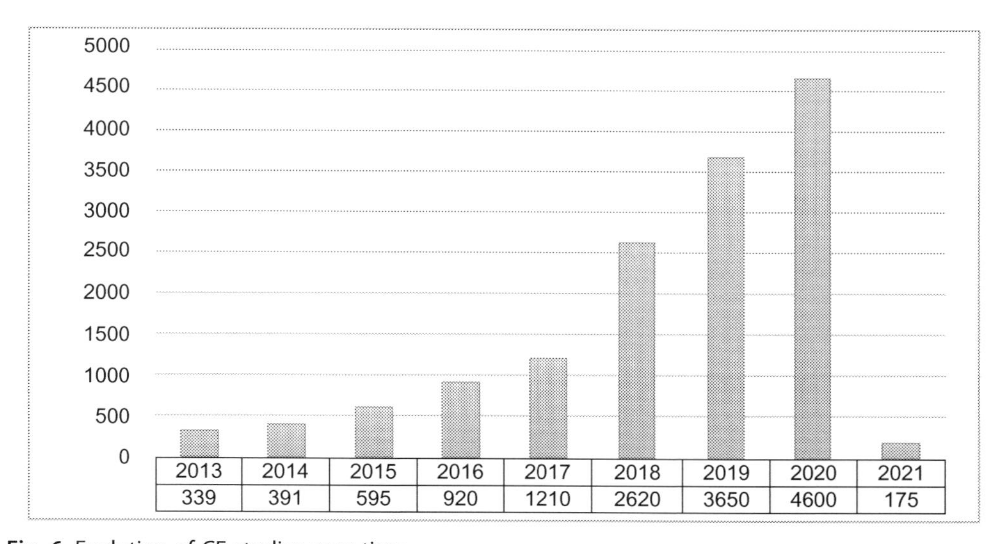

2013	2014	2015	2016	2017	2018	2019	2020	2021
339	391	595	920	1210	2620	3650	4600	175

Fig. 6 Evolution of CE studies over time.

In this level, the subjects of studies coming from the field of engineering and nature-based sciences are mainly focused on sharing resources, substituting nonrenewable mineral sources, proposing product-service systems, converting CO_2 to CO_2-based polymers, reprocessing fly ash, carbide slag, and iron oxide, among others. Actually, the main idea of this literature is to provide certain technological solutions to reuse or recycle by-products in order to transform them to raw materials for various production and

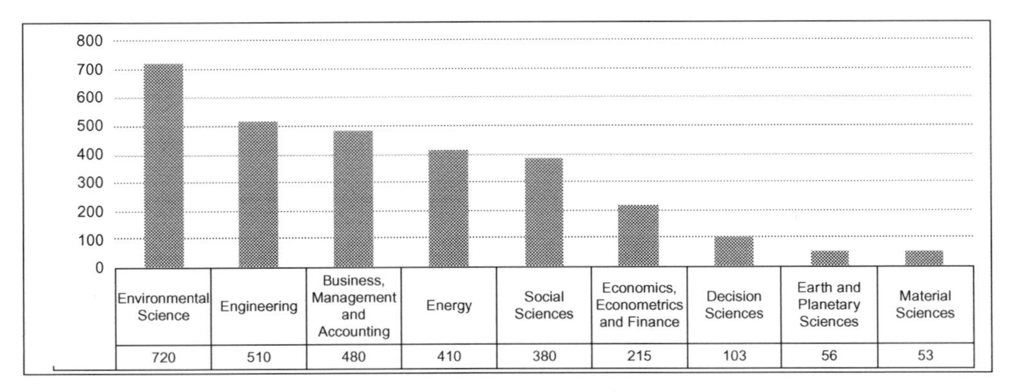

Fig. 7 Allocation of micro-level studies in various academic fields.

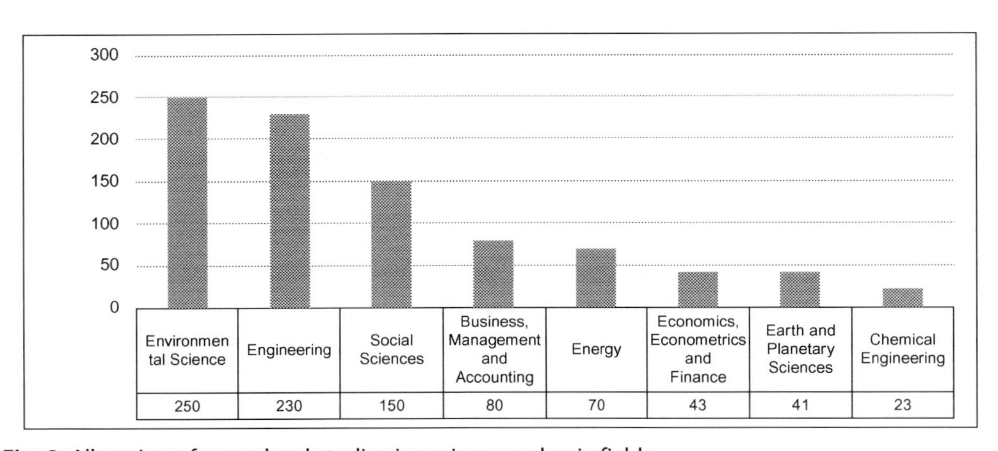

Fig. 8 Allocation of meso-level studies in various academic fields.

consumptions uses. Similarly, management and economic-based sciences put more emphasis on circular business models, accounting systems, circular design principles, and cost-effectiveness solutions.

Fig. 8 describes some characteristic studies that could be classified into the meso-level from a range of academic fields. Similar to the micro-level, the majority of the meso-level studies are associated with engineering and nature-based sciences (e.g., environmental science, earth and planetary sciences, engineering, energy, and chemical engineering), totaling 614 studies; the rest of the studies, from management and economic-based sciences (e.g., business, management and accounting, social sciences, economics, econometrics and finance, and social sciences), total only 273. The topics of these fields focus on industrial symbiosis, industrial metabolism, industrial ecology, and eco-industrial parks.

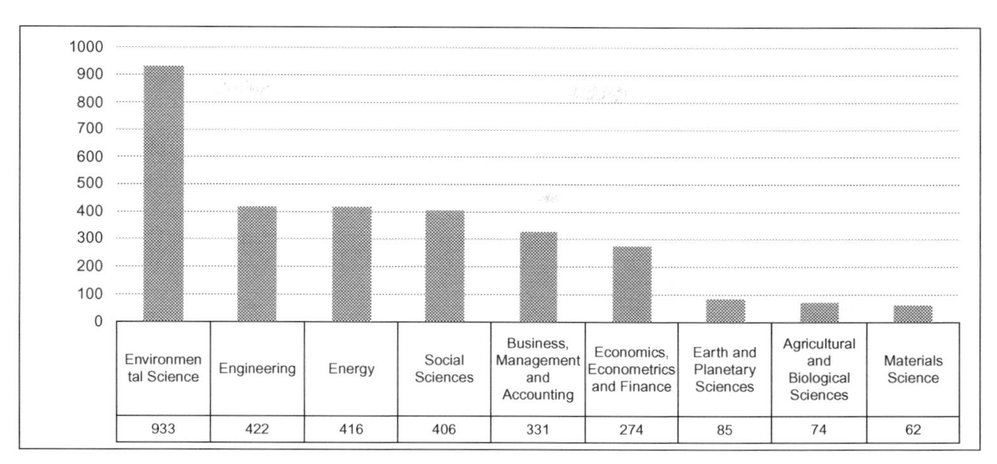

Fig. 9 Allocation of macro-level studies in various academic fields.

About 3000 studies are selected from macro-level studies to be analyzed (Fig. 9). The majority of such studies are associated with nature-based science and engineering. Specifically, the fields of environmental science, earth and planetary sciences, and agricultural and biological sciences provide 36% of the total macro-level studies. The rest of this field consists of engineering (30%) and management- and economic-based sciences (33%). The topics of such methodologies are based on policy tools (e.g., command and control, market-based, and voluntary instruments) and complete decision-making tools for local or national public organization regarding CE issues.

Finally, Fig. 10 shows the allocation of the total number of studies (see Fig. 3) per level and academic field. This shows that the macro-level includes the greatest number of studies in both scientific fields: engineering/nature-based sciences and management/

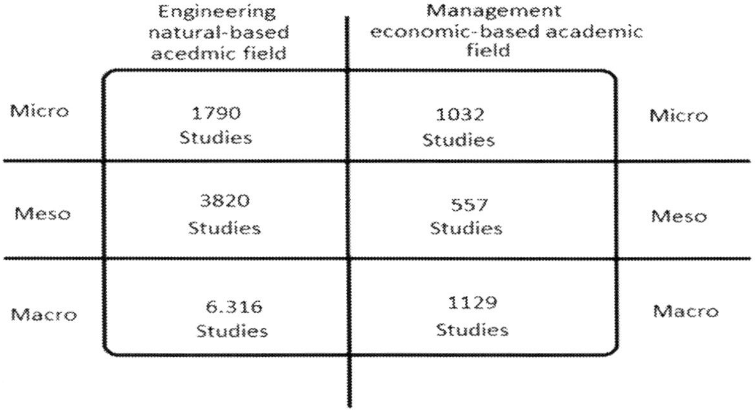

	Engineering natural-based acedmic field	Management economic-based academic field	
Micro	1790 Studies	1032 Studies	Micro
Meso	3820 Studies	557 Studies	Meso
Macro	6.316 Studies	1129 Studies	Macro

Fig. 10 Allocation of studies per level and academic field.

economic-based sciences. As expected, the meso-level of the CE represents a more advanced part in the literature in the field of engineering and environmental sciences. This is because the focus of this literature is on industrial ecology issues that date back to the 1970s, which has contributed to their faster transformation into a circular economy. The field of management/economic-based sciences pays more attention to the micro- and macro-levels by examining business circular models.

5. Conclusion and discussion

This chapter presents a bibliometric analysis regarding CE studies. Specifically, a methodological framework has been developed to analyze the literature. It is based on the classical threefold structure of micro-, meso-, and macro-levels. Another significant point of the suggested framework is its intention to identify the emphasis of current studies between engineering/nature-based and management/economic-based sciences. The threefold level and science fields are the main pillars that the framework of analysis was built on.

The results have shown that the macro-level includes the majority of CE studies. This is expected, since the majority of studies have only lately paid more attention to waste management and general policies at the municipal and international levels, with many studies suggesting complete plans to reuse and recycle materials from the building sector, from food waste, and from waste management. Indeed, the majority of existing studies focus on end-of-life strategies.

Another important section of literature has focused on the meso-level. In engineering/nature-based sciences, it seems that the meso-level includes over 3000 studies that focus on industrial ecology and symbiosis. These studies examine how by-products of firms or sectors could be raw materials for other firms and sectors. The idea is based on ecosystems functions. Finally, the micro-level also plays a critical role by covering a number of interesting themes such as reuse, recycle, remanufacture, and redesign.

It is worth noting that the majority of the current studies focus on literature review and normative models. Actually, the majority of the current literature focuses on recording the current situation regarding CE and not on proposing specific technologies and successful examples to solve current and future problems. This is a possible and necessary future research field for CE. Another absence in the current research is the supply side (production), while very little work focuses on examining the demand side (consumers). For this, further studies to explore the intentions and awareness of consumers regarding CE are necessary.

References

Abu-Ghunmi, D., Abu-Ghunmi, L., Kayal, B., Bino, A., 2016. Circular economy and the opportunity cost of not 'closing the loop' of water industry: the case of Jordan. J. Clean. Prod. 131, 228–236.

Alonso-Almeida, M.D.M., Rodríguez-Antón, J.M., Bagur-Femenías, L., Perramon, J., 2020. Sustainable development and circular economy: the role of institutional promotion on circular consumption and

market competitiveness from a multistakeholder engagement approach. Bus. Strateg. Environ. 29 (6), 2803–2814.
Andersen, M.S., 2007. An introductory note on the environmental economics of the circular economy. Sustain. Sci. 2 (1), 133–140.
Aranda-Usón, A., Portillo-Tarragona, P., Marín-Vinuesa, L.M., Scarpellini, S., 2019. Financial resources for the circular economy: a perspective from businesses. Sustainability 11 (3), 888.
Avraamidou, S., Baratsas, S.G., Tian, Y., Pistikopoulos, E.N., 2020. Circular economy—a challenge and an opportunity for process systems engineering. Comput. Chem. Eng. 133, 106629.
Awasthi, A.K., Li, J., Koh, L., Ogunseitan, O.A., 2019. Circular economy and electronic waste. Nat. Electron. 2 (3), 86–89.
Bao, Z., Lu, W., Chi, B., Yuan, H., Hao, J., 2019. Procurement innovation for a circular economy of construction and demolition waste: lessons learnt from Suzhou, China. Waste Manag. 99, 12–21.
Bekchanov, M., Mirzabaev, A., 2018. Circular economy of composting in Sri Lanka: opportunities and challenges for reducing waste related pollution and improving soil health. J. Clean. Prod. 202, 1107–1119.
Belaud, J.P., Adoue, C., Vialle, C., Chorro, A., Sablayrolles, C., 2019. A circular economy and industrial ecology toolbox for developing an eco-industrial park: perspectives from French policy. Clean Techn. Environ. Policy 21 (5), 967–985.
Bocken, N.M., De Pauw, I., Bakker, C., Van Der Grinten, B., 2016. Product design and business model strategies for a circular economy. J. Ind. Prod. Eng. 33 (5), 308–320.
Bocken, N.M.P., Weissbrod, I., Antikainen, M., 2021. Business model experimentation for the circular economy: definition and approaches. Circ. Econ. Sust. https://doi.org/10.1007/s43615-021-00026-z.
Bruel, A., Kronenberg, J., Troussier, N., Guillaume, B., 2019. Linking industrial ecology and ecological economics: a theoretical and empirical foundation for the circular economy. J. Ind. Ecol. 23 (1), 12–21.
Cardoso de Oliveira, M.C., Machado, M.C., Chiappetta Jabbour, C.J., Lopes de Sousa Jabbour, A.B., 2019. Paving the way for the circular economy and more sustainable supply chains: shedding light on formal and informal governance instruments used to induce green networks. Manag. Environ. Qual. 30 (5), 1095–1113.
Cecchin, A., Salomone, R., Deutz, P., et al., 2021. What Is in a name? The rising star of the circular economy as a resource-related concept for sustainable development. Circ. Econ. Sust. https://doi.org/10.1007/s43615-021-00021-4.
Confente, I., Scarpi, D., Russo, I., 2020. Marketing a new generation of bio-plastics products for a circular economy: the role of green self-identity, self-congruity, and perceived value. J. Bus. Res. 112, 431–439.
D'Amato, D., Korhonen, J., Toppinen, A., 2019. Circular, green, and bio economy: how do companies in land-use intensive sectors align with sustainability concepts? Ecol. Econ. 158, 116–133.
De Jesus, A., Mendonça, S., 2018. Lost in transition? Drivers and barriers in the eco-innovation road to the circular economy. Ecol. Econ. 145, 75–89.
Deutz, P., Baxter, H., Gibbs, D., Mayes, W.M., Gomes, H.I., 2017. Resource recovery and remediation of highly alkaline residues: a political-industrial ecology approach to building a circular economy. Geoforum 85, 336–344.
Dominguez, S., Laso, J., Margallo, M., Aldaco, R., Rivero, M.J., Irabien, Á., Ortiz, I., 2018. LCA of greywater management within a water circular economy restorative thinking framework. Sci. Total Environ. 621, 1047–1056.
Dubey, R., Gunasekaran, A., Childe, S.J., Papadopoulos, T., Helo, P., 2019. Supplier relationship management for circular economy: influence of external pressures and top management commitment. Manag. Decis. 57 (4), 767–790.
Elia, V., Gnoni, M.G., Tornese, F., 2017. Measuring circular economy strategies through index methods: a critical analysis. J. Clean. Prod. 142, 2741–2751.
Esa, M.R., Halog, A., Rigamonti, L., 2017. Developing strategies for managing construction and demolition wastes in Malaysia based on the concept of circular economy. J. Mater. Cycles Waste Manag. 19 (3), 1144–1154.
Esposito, M., Tse, T., Soufani, K., 2017. Is the circular economy a new fast-expanding market? Thunderbird Int. Bus. Rev. 59 (1), 9–14.
Esposito, M., Tse, T., Soufani, K., 2018. Introducing a circular economy: new thinking with new managerial and policy implications. Calif. Manag. Rev. 60 (3), 5–19.

Ferreira Gregorio, V., Pié, L., Terceño, A., 2018. A systematic literature review of bio, green and circular economy trends in publications in the field of economics and business management. Sustainability 10 (11), 4232.
Ferronato, N., Rada, E.C., Portillo, M.A.G., Cioca, L.I., Ragazzi, M., Torretta, V., 2019. Introduction of the circular economy within developing regions: a comparative analysis of advantages and opportunities for waste valorization. J. Environ. Manag. 230, 366–378.
Fischer, A., Pascucci, S., 2017. Institutional incentives in circular economy transition: the case of material use in the Dutch textile industry. J. Clean. Prod. 155, 17–32.
Franco, M.A., 2017. Circular economy at the micro level: a dynamic view of incumbents' struggles and challenges in the textile industry. J. Clean. Prod. 168, 833–845.
Gallagher, J., Basu, B., Browne, M., Kenna, A., McCormack, S., Pilla, F., Styles, D., 2019. Adapting stand-alone renewable energy technologies for the circular economy through eco-design and recycling. J. Ind. Ecol. 23 (1), 133–140.
Garmulewicz, A., Holweg, M., Veldhuis, H., Yang, A., 2018. Disruptive technology as an enabler of the circular economy: what potential does 3D printing hold? Calif. Manag. Rev. 60 (3), 112–132.
Geissdoerfer, M., Savaget, P., Bocken, N.M., Hultink, E.J., 2017. The circular economy—a new sustainability paradigm? J. Clean. Prod. 143, 757–768.
Geissdoerfer, M., Morioka, S.N., de Carvalho, M.M., Evans, S., 2018. Business models and supply chains for the circular economy. J. Clean. Prod. 190, 712–721.
Geng, Y., Doberstein, B., 2008. Developing the circular economy in China: challenges and opportunities for achieving 'leapfrog development'. Int. J. Sustain. Dev. World Ecol. 15 (3), 231–239.
Genovese, A., Acquaye, A.A., Figueroa, A., Koh, S.L., 2017. Sustainable supply chain management and the transition towards a circular economy: evidence and some applications. Omega 66, 344–357.
Ghisellini, P., Cialani, C., Ulgiati, S., 2016. A review on circular economy: the expected transition to a balanced interplay of environmental and economic systems. J. Clean. Prod. 114, 11–32.
Ghisellini, P., Passaro, R., Ulgiati, S., 2021. Revisiting Keynes in the light of the transition to circular economy. Circ. Econ. Sust. https://doi.org/10.1007/s43615-021-00016-1.
Ghisellini, P., Ripa, M., Ulgiati, S., 2018. Exploring environmental and economic costs and benefits of a circular economy approach to the construction and demolition sector. A literature review. J. Clean. Prod. 178, 618–643.
Gómez, A.M.M., González, F.A., Bárcena, M.M., 2018. Smart eco-industrial parks: a circular economy implementation based on industrial metabolism. Resour. Conserv. Recycl. 135, 58–69.
Gusmerotti, N.M., Testa, F., Corsini, F., Pretner, G., Iraldo, F., 2019. Drivers and approaches to the circular economy in manufacturing firms. J. Clean. Prod. 230, 314–327.
Han, W., Gao, G., Geng, J., Li, Y., Wang, Y., 2018. Ecological and health risks assessment and spatial distribution of residual heavy metals in the soil of an e-waste circular economy park in Tianjin, China. Chemosphere 197, 325–335.
Hart, J., Pomponi, F., 2021. A circular economy: where will it take us? Circ. Econ. Sust. https://doi.org/10.1007/s43615-021-00013-4.
Hazen, B.T., Mollenkopf, D.A., Wang, Y., 2017. Remanufacturing for the circular economy: an examination of consumer switching behavior. Bus. Strateg. Environ. 26 (4), 451–464.
Homrich, A.S., Galvao, G., Abadia, L.G., Carvalho, M.M., 2018. The circular economy umbrella: trends and gaps on integrating pathways. J. Clean. Prod. 175, 525–543.
Huang, B., Wang, X., Kua, H., Geng, Y., Bleischwitz, R., Ren, J., 2018. Construction and demolition waste management in China through the 3R principle. Resour. Conserv. Recycl. 129, 36–44.
Hussain, M., Malik, M., 2020. Organizational enablers for circular economy in the context of sustainable supply chain management. J. Clean. Prod. 256, 120–375.
Jabbour, C.J.C., Sarkis, J., de Sousa Jabbour, A.B.L., Renwick, D.W.S., Singh, S.K., Grebinevych, O., Kruglianskas, I., Godinho Filho, M., 2019. Who is in charge? A review and a research agenda on the 'human side' of the circular economy. J. Clean. Prod. 222, 793–801.
Jakhar, S.K., Mangla, S.K., Luthra, S., Kusi-Sarpong, S., 2019. When stakeholder pressure drives the circular economy: measuring the mediating role of innovation capabilities. Manag. Decis. 57 (4), 904–920.
Kalmykova, Y., Sadagopan, M., Rosado, L., 2018. Circular economy—from review of theories and practices to development of implementation tools. Resour. Conserv. Recycl. 135, 190–201.

Keijer, T., Bakker, V., Slootweg, J.C., 2019. Circular chemistry to enable a circular economy. Nat. Chem. 11 (3), 190–195.
Khan, O., Daddi, T., Iraldo, F., 2020. Microfoundations of dynamic capabilities: insights from circular economy business cases. Bus. Strateg. Environ. 29 (3), 1479–1493.
Kirchherr, J., Reike, D., Hekkert, M., 2017. Conceptualizing the circular economy: an analysis of 114 definitions. Resour. Conserv. Recycl. 127, 221–232.
Kirchherr, J., Piscicelli, L., Bour, R., Kostense-Smit, E., Muller, J., Huibrechtse-Truijens, A., Hekkert, M., 2018. Barriers to the circular economy: evidence from the European Union (EU). Ecol. Econ. 150, 264–272.
Korhonen, J., Von Malmborg, F., Strachan, P.A., Ehrenfeld, J.R., 2004. Management and policy aspects of industrial ecology: an emerging research agenda. Bus. Strateg. Environ. 13 (5), 289–305.
Korhonen, J., Honkasalo, A., Seppälä, J., 2018a. Circular economy: the concept and its limitations. Ecol. Econ. 143, 37–46.
Korhonen, J., Nuur, C., Feldmann, A., Birkie, S.E., 2018b. Circular economy as an essentially contested concept. J. Clean. Prod. 175, 544–552.
Lausselet, C., Cherubini, F., Oreggioni, G.D., del Alamo Serrano, G., Becidan, M., Hu, X., Rørstad, P.K., Strømman, A.H., 2017. Norwegian waste-to-energy: climate change, circular economy and carbon capture and storage. Resour. Conserv. Recycl. 126, 50–61.
Lewandowski, M., 2016. Designing the business models for circular economy—towards the conceptual framework. Sustainability 8 (1), 43, 1–28.
Li, H., Bao, W., Xiu, C., Zhang, Y., Xu, H., 2010. Energy conservation and circular economy in China's process industries. Energy 35 (11), 4273–4281.
Lieder, M., Rashid, A., 2016. Towards circular economy implementation: a comprehensive review in context of manufacturing industry. J. Clean. Prod. 115, 36–51.
Liu, J., Feng, Y., Zhu, Q., Sarkis, J., 2018. Green supply chain management and the circular economy. Int. J. Phys. Distrib. Logist. Manag. 48 (8), 794–817.
Loizia, P., Neofytou, N., Zorpas, A.A., 2019. The concept of circular economy strategy in food waste management for the optimization of energy production through anaerobic digestion. Environ. Sci. Pollut. Res. 26 (15), 14766–14773.
Lüdeke-Freund, F., Gold, S., Bocken, N.M., 2019. A review and typology of circular economy business model patterns. J. Ind. Ecol. 23 (1), 36–61.
Ma, S.H., Wen, Z.G., Chen, J.N., Wen, Z.C., 2014. Mode of circular economy in China's iron and steel industry: a case study in Wu'an city. J. Clean. Prod. 64, 505–512.
Mahpour, A., 2018. Prioritizing barriers to adopt circular economy in construction and demolition waste management. Resour. Conserv. Recycl. 134, 216–227.
Merli, R., Preziosi, M., Acampora, A., 2018. How do scholars approach the circular economy? A systematic literature review. J. Clean. Prod. 178, 703–722.
Michelini, G., Moraes, R.N., Cunha, R.N., Costa, J.M., Ometto, A.R., 2017. From linear to circular economy: PSS conducting the transition. Proc. CIRP 64 (2017), 2–6.
Mikulčić, H., Klemeš, J.J., Vujanović, M., Urbaniec, K., Duić, N., 2016. Reducing greenhouse gasses emissions by fostering the deployment of alternative raw materials and energy sources in the cleaner cement manufacturing process. J. Clean. Prod. 136, 119–132.
Molina-Moreno, V., Leyva-Díaz, J.C., Sánchez-Molina, J., Peña-García, A., 2017. Proposal to foster sustainability through circular economy-based engineering: a profitable chain from waste management to tunnel lighting. Sustainability 9 (12), 2229.
Nikolaou, I.E., Jones, N., Stefanakis, A., 2021. Circular economy and sustainability: the past, the present and the future directions. Circ. Econ. Sust. https://doi.org/10.1007/s43615-021-00030-3.
O'Connor, M.P., Zimmerman, J.B., Anastas, P.T., Plata, D.L., 2016. A strategy for material supply chain sustainability: enabling a circular economy in the electronics industry through green engineering. ACS Sustain. Chem. Eng. 4 (11), 5879–5888.
Oliveira, M., Miguel, M., van Langen, S.K., et al., 2021. Circular economy and the transition to a sustainable society: integrated assessment methods for a new paradigm. Circ. Econ. Sust. https://doi.org/10.1007/s43615-021-00019-y.

Ozili, P.K., 2021. Circular economy, banks, and other financial institutions: what's in it for them? Circ. Econ. Sust. https://doi.org/10.1007/s43615-021-00043-y.

Palafox-Alcantar, P.G., Hunt, D.V.L., Rogers, C.D.F., 2020. The complementary use of game theory for the circular economy: a review of waste management decision-making methods in civil engineering. Waste Manag. 102, 598–612.

Pan, S.Y., Du, M.A., Huang, I.T., Liu, I.H., Chang, E.E., Chiang, P.C., 2015. Strategies on implementation of waste-to-energy (WTE) supply chain for circular economy system: a review. J. Clean. Prod. 108, 409–421.

Pieroni, M.P., McAloone, T.C., Pigosso, D.C., 2019. Business model innovation for circular economy and sustainability: a review of approaches. J. Clean. Prod. 215, 198–216.

Popescu, D.I., 2018. Social responsibility and business ethics: VII. Circular economy and the role of corporate social marketing. Calitatea 19 (163), 118–121.

Potting, J., Hekke, M., Worrell, E., Hanemaaijer, A., 2017. Circular economy: measuring innovation in the product chain. Available from: http://www.pbl.nl/sites/default/files/cms/publicaties/pbl-2016-circular-economy-measuring-innovation-in-product-chains-2544.pdf.

Prieto-Sandoval, V., Jaca, C., Ormazabal, M., 2018. Towards a consensus on the circular economy. J. Clean. Prod. 179, 605–615.

Quina, M.J., Bontempi, E., Bogush, A., Schlumberger, S., Weibel, G., Braga, R., Funari, V., Hyks, J., Rasmussen, E., Lederer, J., 2018. Technologies for the management of MSW incineration ashes from gas cleaning: new perspectives on recovery of secondary raw materials and circular economy. Sci. Total Environ. 635, 526–542.

Raheem, A., Sikarwar, V.S., He, J., Dastyar, W., Dionysiou, D.D., Wang, W., Zhao, M., 2018. Opportunities and challenges in sustainable treatment and resource reuse of sewage sludge: a review. Chem. Eng. J. 337, 616–641.

Rajput, S., Singh, S.P., 2019. Connecting circular economy and industry 4.0. Int. J. Inf. Manag. 49, 98–113.

Reh, L., 2013. Process engineering in circular economy. Particuology 11 (2), 119–133.

Reike, D., Vermeulen, W.J., Witjes, S., 2018. The circular economy: new or refurbished as CE 3.0?—exploring controversies in the conceptualization of the circular economy through a focus on history and resource value retention options. Resour. Conserv. Recycl. 135, 246–264.

Saavedra, Y.M., Iritani, D.R., Pavan, A.L., Ometto, A.R., 2018. Theoretical contribution of industrial ecology to circular economy. J. Clean. Prod. 170, 1514–1522.

Saidani, M., Yannou, B., Leroy, Y., Cluzel, F., Kendall, A., 2019. A taxonomy of circular economy indicators. J. Clean. Prod. 207, 542–559.

Sassanelli, C., Rosa, P., Rocca, R., Terzi, S., 2019. Circular economy performance assessment methods: a systematic literature review. J. Clean. Prod. 229, 440–453.

Schetters, M.J.A., Van Der Hoek, J.P., Kramer, O.J.I., Kors, L.J., Palmen, L.J., Hofs, B., Koppers, H., 2015. Circular economy in drinking water treatment: reuse of ground pellets as seeding material in the pellet softening process. Water Sci. Technol. 71 (4), 479–486.

Schroeder, P., Anggraeni, K., Weber, U., 2019. The relevance of circular economy practices to the sustainable development goals. J. Ind. Ecol. 23 (1), 77–95.

Sgroi, M., Vagliasindi, F.G., Roccaro, P., 2018. Feasibility, sustainability and circular economy concepts in water reuse. Curr. Opin. Environ. Sci. Health 2, 20–25.

Sica, D., Malandrino, O., Supino, S., Testa, M., Lucchetti, M.C., 2018. Management of end-of-life photovoltaic panels as a step towards a circular economy. Renew. Sust. Energ. Rev. 82, 2934–2945.

Smol, M., Kulczycka, J., Henclik, A., Gorazda, K., Wzorek, Z., 2015. The possible use of sewage sludge ash (SSA) in the construction industry as a way towards a circular economy. J. Clean. Prod. 95, 45–54.

Smol, M., Avdiushchenko, A., Kulczycka, J., Nowaczek, A., 2018. Public awareness of circular economy in southern Poland: case of the Malopolska region. J. Clean. Prod. 197, 1035–1045.

Stefanakis, A.I., Calheiros, C.S., Nikolaou, I., 2021. Nature-based solutions as a tool in the new circular economic model for climate change adaptation. Circ. Econ. Sust. https://doi.org/10.1007/s43615-021-00022-3.

Stewart, R., Niero, M., 2018. Circular economy in corporate sustainability strategies: a review of corporate sustainability reports in the fast-moving consumer goods sector. Bus. Strateg. Environ. 27 (7), 1005–1022.

Stiles, W.A., Styles, D., Chapman, S.P., Esteves, S., Bywater, A., Melville, L., Silkina, A., Lupatsch, I., Grünewald, C.F., Lovitt, R., Bull, A., Morris, C., Llewellyn, C.A., Chaloner, T., 2018. Using microalgae in the circular economy to valorise anaerobic digestate: challenges and opportunities. Bioresour. Technol. 267, 732–742.

Su, B., Heshmati, A., Geng, Y., Yu, X., 2013. A review of the circular economy in China: moving from rhetoric to implementation. J. Clean. Prod. 42, 215–227.

Ünal, E., Shao, J., 2019. A taxonomy of circular economy implementation strategies for manufacturing firms: analysis of 391 cradle-to-cradle products. J. Clean. Prod. 212, 754–765.

Urbinati, A., Chiaroni, D., Chiesa, V., 2017. Towards a new taxonomy of circular economy business models. J. Clean. Prod. 168, 487–498.

Voulvoulis, N., 2018. Water reuse from a circular economy perspective and potential risks from an unregulated approach. Curr. Opin. Environ. Sci. Health 2, 32–45.

Wall, D.M., McDonagh, S., Murphy, J.D., 2017. Cascading biomethane energy systems for sustainable green gas production in a circular economy. Bioresour. Technol. 243, 1207–1215.

Webster, K., 2021. A circular economy is about the economy. Circ. Econ. Sust. https://doi.org/10.1007/s43615-021-00034-z.

Wysokińska, Z., 2016. The "new" environmental policy of the European Union: a path to development of a circular economy and mitigation of the negative effects of climate change. Comp. Econ. Res. 19 (2), 57–73.

Zeng, H., Chen, X., Xiao, X., Zhou, Z., 2017. Institutional pressures, sustainable supply chain management, and circular economy capability: empirical evidence from Chinese eco-industrial park firms. J. Clean. Prod. 155, 54–65.

Zhijun, F., Nailing, Y., 2007. Putting a circular economy into practice in China. Sustain. Sci. 2 (1), 95–101.

Zhu, Q., Geng, Y., Lai, K.H., 2010. Circular economy practices among Chinese manufacturers varying in environmental-oriented supply chain cooperation and the performance implications. J. Environ. Manag. 91 (6), 1324–1331.

Zink, T., Geyer, R., 2017. Circular economy rebound. J. Ind. Ecol. 21 (3), 593–602.

CHAPTER 2

Steering the circular economy: A new role for Adam Smith's invisible hand

Keith R. Skene
Biosphere Research Institute, Angus, United Kingdom

1. Introduction

1.1 Defining economics

Robbins (1935) defined economics as "the science which studies human behaviour as a relationship between ends and scarce means which have alternative uses." What is interesting about this definition is that it positions economics as a social or behavioral science, rather than a mathematical subject. Robbins was a great admirer of the work of Adam Smith, who underpinned his economic theory with social theory. Indeed, Smith's work in economics, as espoused in his second book, *An Inquiry into the Nature and Causes of the Wealth of Nations* (Smith, 1776) was built upon his first book, *The Theory of Moral Sentiments* (Smith, 1759), wherein he emphasized the emergent morality of a functional society as underpinning the "invisible hand" that would steer free trade in a positive direction. This in turn would strengthen that society. He wrote: "[The rich] consume little more than the poor, and in spite of their natural selfishness and rapacity . . . . they divide with the poor the produce of all their improvements. They are led by an invisible hand to make nearly the same distribution of the necessaries of life, which would have been made, had the earth been divided into equal portions among all its inhabitants, and thus without intending it, without knowing it, advance the interest of the society, and afford means to the multiplication of the species" (Smith, 1759).

Unfortunately, the interconnected concepts within these two books, society and economics, became separated and the outcome has been an increase in inequality that has damaged society, resulting in a withered hand and the death of societal feedback. Zucman (2019) reports that wealth inequality has increased dramatically since the 1980s. In the USA, the share of the national wealth owned by the richest 0.00025% of the population (amounting to around 400 individuals) has increased fourfold since the early 1980s.

In this chapter, we explore the significance of the invisible hand and its relevance to the circular economy. We then examine the honey economy of the Ogiek people, an indigenous tribe from Kenya, and introduce the concept of the invisible tripartite

Circular Economy and Sustainability, Volume 1
https://doi.org/10.1016/B978-0-12-819817-9.00028-4

embrace, a more expansive version of the invisible hand, which interconnects the three arenas of human activity: economics, society and the environment. It is suggested that only such connectivity can steer the circular economy in such a way as to integrate our economic activities within the Earth system, thus delivering meaningful sustainability. We conclude by realigning the circular economy within a truly sustainable context.

1.2 The circular economy

The circular economy is a broad church with a global outreach. The Chinese version, ensconced within the 5-year plans of recent times, has a very different political context in comparison with that proclaimed by the EU (Skene and Murray, 2017). The underpinning theory was in place many years ago. Waste in food (Simmonds, 1862) and industrial symbiosis (Devas, 1901; Parkins, 1934) can be traced back over a century, while Desrochers (2001, 2002, 2008) argues that the concepts of recycling and resource-use efficiency can be found in ancient times, driven by a scarcity of resources due to technological shortcomings in terms of extraction, rather than the current drivers of excessive extraction and the profligacy of waste.

Kirchherr et al. (2017) encountered 114 different definitions of the circular economy, and summarized these with the following working definition: "a regenerative system in which resource input and waste, emission, and energy leakage are minimised by slowing, closing, and narrowing material and energy loops. This can be achieved through long-lasting design, maintenance, repair, reuse, remanufacturing, refurbishing and recycling."

Quite clearly, the circular economy is defined strictly within the limits of resource use and waste. Obviously, this is important, but it misses out on two other essential elements of sustainability: society and the environment. Economics is not an isolated realm, wherein greening of supply and waste chains will deliver a perfect world. While the circular economy relates to the economics-environment nexus, the restoration of ecosystem function requires a much deeper response than this; and ecosystem function is an essential component of sustainability. Adam Smith recognized the importance of the society-economy nexus, and his foundation was not economics, but society.

In order to understand the environment-society-economy nexus, we need to ask what sustainability actually means.

1.3 What do we mean by sustainability?

Sustainability is often interpreted as a form of dynamic equilibrium, where losses and gains balance each other, resulting in the maintenance of some status quo (Giampietro and Mayumi, 1997; Lozano, 2007; Sakuragawa and Hosono, 2010). In terms of sustainable resource use, for example, some form of circular flow is envisaged, where materials are used but then recycled in such a way that the stock is not diminished. There are three forms of sustainability recognized today: economic, social and environmental. These are often referred to as the three arenas and are frequently represented as three overlapping circles.

The history of humankind in many ways reflects changes in the emphasis between these arenas. For around 95% of our existence as a subspecies on Earth, *Homo sapiens sapiens*, akin to the rest of nature, found ourselves within the environmental arena. Our evolution, ongoing existence and societal structure were emergent from and contingent upon this arena and represented our natural ecology, wherein we interacted with each other and the landscape, which in turn formed the context of our survival.

Some 12,000 years ago, near the conclusion of the last ice age, we began to settle, to farm, and to trade. The economic arena was formed and steadily grew, coming to dominate our behavior and our relationships both with each other and with our environment. Nature was no longer recognized as the designer, director, and arbitrator, but merely as a sink and source, a utility. The surplus theory of stratification became a reality, wherein surplus food allowed the population to proliferate, leading to specialization in work, with increasing complexity in social organization, ownership, inheritance and exchange. Inequality also increased, as did the emphasis on individualization as opposed to the collective. In many ways, the rest of our history has merely been an intensification of all of these characteristics, through industrial and technological development.

Nature's value became unilateral, wherein the benefits to humans were not counterbalanced by the costs to ecosystem functioning (Costanza et al., 2014; Seják et al., 2018). Since the end of the 19th century, it has been acknowledged that any objective valuation of goods and services cannot be derived from benefits to humans alone. Marshall (1920) attempted to reconcile the classical (cost) supply-side and neoclassical demand-side (marginal utility) theories of economic value in his "two blades of scissors" analogy. Yet today, the unilateral value concept has regained traction, where valuation of ecosystem services, also referred to as nature's contributions to people (Diaz et al., 2018), is seen as representing natural value. One of the reasons for this is that our thinking is dominated by empirical, reductionist philosophy, wherein interventionist strategies are legitimized by a comprehension of the planet as being built of small blocks, which can be rearranged and shifted at will, allowing us to replace natural capital with man-made capital. To understand the meaning of this, we need to discuss the two schools of sustainability: weak and strong sustainability.

2. Weak and strong sustainability

Weak sustainability is defined as development that meets the needs of the present generation without compromising the ability of future generations to meet their own needs (WCED, 1987). The focus is on our needs, and the three arenas are interchangeable, meaning that it doesn't matter if the environmental arena diminishes, as long as the economic and societal arenas can replace it. Technology is seen as central to this exchange of capital.

It is argued that technology can replicate nature in providing ecosystem services for our future survival, whether it be through genetic engineering (Gates, 2018), cloud seeding (Rasch et al., 2009), iron enrichment of the oceans (to move CO_2 from the atmosphere into the hydrosphere) (Zhang et al., 2015) or biomimicry, where we borrow ideas from nature and implant them within technology (Benyus, 1997). Weak sustainability allows for almost unlimited substitution between man-made and natural capital (Pearce and Turner, 1990).

In a weak sustainability paradigm, provided that mean global wealth and welfare increase, those countries doing the best (i.e., the developed nations) can compensate the less successful. Not only can each arena compensate for the others, paying our way out of trouble, but the inequalities, while maintained, can be ironed out too. It is a morally contestable position, but lies at the heart of this form of thinking.

On the other side of the sustainability argument lies strong sustainability, which advocates that nature cannot be replaced by technology, but rather, each pool of capital (economic, social, and natural) should be maintained independently (Brekke, 1997; Daly and Cobb, 1989). Strong sustainability can be defined as development that allows future generations to access the same amount of natural resources and the same economic and social capital as the current generation. Ott (2003) argues that: "Natural capital is characterized by internal and dynamic complexity. Its components form a network of relationships. In principle, they are mutually non-substitutable."

There is a problem at the heart of both schools of thought, in that, given the damage that already exists, these definitions translate as the maintenance of the current damage as well as the current capital, be it total capital or in separate pools, without including any measure of restoration. The circular economy suffers from a similar weakness, allowing no pathway to recovery, but, rather, halting the increasing depletion. However, the Earth system is seriously damaged already, and halting the damage will not be enough.

Thus, we see that any position on sustainability will depend upon which school is advocated, and can have a very different meaning, according to this choice. These differences depend on your philosophical foundations. The reasoning underpinning strong sustainability arguments lies in the reality that the Earth system is an emergent system. Weak sustainability rests upon reductionist thinking, where we can build structures that substitute for the Earth system, in a form of terraforming. To understand the importance of this we need to examine systems theory.

3. Systems theory

Any complex system, such as the Earth system, is composed of multiple parts, which are connected to and interdependent upon each other and their environment (Nicolis and Prigogine, 1989). Systems are self-assembling and self-organizing. The Earth system has self-assembled, self-organized, reassembled, and reorganized many times over the last 3.4 billion years, recovering from mass extinctions along the way. It has a creative force within it that allows it to restructure itself without human intervention. Nature has no need for the wisdom of humankind and doesn't require the formation of an organizing committee nor an action plan to repair itself.

Systems have a number of key characteristics. They are nonlinear, meaning that they do not display cause and effect, but are asymmetrical. Folke et al. (2010) highlight the point that "Causation is often non-linear in complex adaptive systems with the potential for chaotic dynamics, multiple basins of attraction, and shifts between pathways or regimes, some of which may be irreversible." This is important, because systems can undergo dramatic change with little warning, switching to a new state. Unintended consequences can result, wherein outcomes are unpredictable and rapid in nature. Such outcomes are an expected result of the dynamic nature of complex systems (Aoi et al., 2007).

Systems are emergent, meaning that they exhibit properties that belong to the whole, rather than the parts (Bedau and Humphreys, 2008). This means that they cannot be understood using reductionist thinking. Emergent characteristics are both autonomous from the underlying components and consequent upon them (Bedau, 1997). The whole is not unrelated to the parts that make up the whole, but it is the interaction of these parts that adds complexity to the whole, which then behaves and responds in ways that cannot necessarily be deduced through a reductionist approach.

Systems are also suboptimal at each level of organization, a prerequisite, given the necessity of trade-offs in a multichallenge solution space. One cannot optimize for any single level, as this would prevent functionality at other levels. Farnsworth and Niklas (1995) concluded that, as the number of challenges increase upon a process, only solutions that are increasingly suboptimal for each challenge will work. This is a classic design reality in natural and human-driven design. Trade-offs exist everywhere. For example, a car with large storage capacity, such as a family saloon, will not have the aerodynamic properties of a Lamborghini sports car, but will have much greater space for offspring and a pram. Orchid seeds are so small that they have insufficient food stores to allow germination, but can spread huge distances (some weigh only one millionth of a gram; Arditti, 1967). However, they require specific fungi to scavenge food for them and this places limitations upon where they can germinate successfully (Batty et al., 2001; Bernard, 1906).

Finally, systems rely on real-time feedback, an essential element in self-organization. Feedback lies at the heart of a system, conveying information between different levels of organization and within the one level. Feedback is what leads to dynamism, wherein change is constantly occurring, impacting on functionality at all levels. It is like an electric current running through the system, and a glue that binds the entirety together. The Earth system is continuously providing feedback, but we humans have so distanced ourselves from our environment that we do not hear it.

The Earth system displays these key characteristics throughout all of its levels of organization. As the fundamental basis of life on Earth, with its inbuilt, complex interactivity and resilience, the question of sustainability is within its domain, not our own. If we are to continue to thrive on the planet, we can only do so if our activities resonate with the Earth system. It is not a matter of human conservation programs, or of replacing parts of the natural system with technology, because this is not a reductionist challenge. Emergence rules, and such is the complexity that our interference is unlikely to result in what we expect. Furthermore, nonlinearity can result in rapid and irreversible transition.

This doesn't mean that technology is without its place. The internet of things, with its many billion smart devices (not including smartphones, tablets, and laptops) scattered across the world, provides a powerful array of ears and eyes to deliver the feedback that we need (Skene, 2020). Remote sensing can give us insights into planetary health, while artificial intelligence offers powerful analytical approaches to interpret the vast sea of data that now flows through the infosphere. This information can act as a portal, allowing us to reconnect to the natural world in ways never before available. While indigenous people have never lost their ecological intelligence, the rest of us have become isolated.

This isolation is important in terms of a meaningful circular economy for two reasons. Firstly, it has blinded us to the damage we are doing to the Earth system, by impacting on our decision-making. Evidence is accumulating that economic and cultural globalization is decoupling social and ecological systems, through innovation, increased technological connectivity and the speed and scale of linkages that drive social system change (Young et al., 2006). This has resulted in a loss of what Curry (2006) refers to as ecological ethics, wherein nature is the ultimate source of all value. Without such a value system, decisions lack systemic context, further exacerbating the problem.

We suggest that this resembles what has happened to Adam Smith's invisible hand. Once free-market capitalism, espoused by Smith, became isolated from the nudging and nurdling of a functional society whose members practiced virtuous self-interest, then it became a different beast. What had been set out as an economy that would work alongside an ever more enlightened humanity instead became the conveyer of inequality, greed, and injustice. We further suggest that these two breaches, between society and environment and between society and economics, lie at the heart of the current environmental crisis.

4. The tripartite invisible embrace

True sustainability can only be delivered by restoring these relationships, and thus the circular economy must fundamentally facilitate this restoration if it is to work as a path to a sustainable future. Returning to Robbins' (1935) definition of economics as a behavioral science, it is society that takes center stage here, responding to a reawakened environmental awareness, while ameliorating its actions in terms of economic activity for the good of the Earth system. This we call the tripartite invisible embrace: "tripartite" because it reaches across the three arenas, "invisible" because it is, ultimately, a resonance, and "embrace" because it is seen to bring these arenas together, allowing a "sympathy," as Smith would put it, across these formerly counter-positioned areas.

Stretching across all three arenas, it is an emergent property, wherein the resonance between ourselves and our landscape informs our activities. Marie-Jean-Antoine-Nicolas de Caritat, Marquis de Condorcet, the great French revolutionary and educationist, wrote that "Nature has fixed no limits to our hopes" (Condorcet, 1955). We may have been the product of its tormented and mindless struggle, but we were better than this. Émile Durkheim, the sociologist, claimed that "it is civilization that has made man what he is: it is what distinguishes him from the animal: man is man only because he is civilized" (Durkheim, 1973). Individualism became the clarion call of our age, whether it be individual empowerment (Staples, 1990), individual actualization (Rogers, 1959), or personalization (Needham, 2014). Lord and Hutchison (2009) define empowerment as "processes whereby individuals achieve increasing control of various aspects of their lives and participate in the community with dignity."

The neoliberal dogma emphasizes the centrality of the individual (Clements, 2008), leading to a move away from society as a meaningful reference point (Rose, 1999). The neoliberal tradition, particularly in Britain, has sought to rebuild society through economic means, or as Margaret Thatcher put it:

> *What's irritated me about the whole direction of politics in the last 30 years is that it's always been towards the collectivist society. People have forgotten about the personal society. And they say: do I count, do I matter? To which the short answer is, yes. And therefore, it isn't that I set out on economic policies; it's that I set out really to change the approach, and changing the economics is the means of changing that approach. If you change the approach you really are after the heart and soul of the nation. Economics are the method; the object is to change the heart and soul (Butt, 1981).*

This was a reductionist approach, where values were to be internalized by all, one individual at a time, in response to economics (Sutcliffe-Braithwaite, 2012)—quite the polar opposite of the invisible hand of Smith.

However, this concept of the individual as the unit of social currency has been challenged. In addition to the work of Smith, who espoused the functioning society as essential to progress, Husband (1995) writes "In non-European cultures, the self-evident primacy of the individual in relation to the collective cannot be assumed." Mbiti (1969) states "I am what we are." Ubuntu, a sub-Saharan African philosophy, can be summarized as the concept that no one can be self-sufficient and that interdependence is a reality for all (Nussbaum, 2003). Other philosophical positions broaden the interactive net still further. Buen Vivir, the Andean philosophy, stresses that wellbeing can only exist within a community, and the concept of community is expanded to include nature (Gudynas, 2011). Friluftsliv, meaning "free air life," is a Scandinavian philosophy.

Gelter (2000) defines it as a "philosophical lifestyle based on experiences of freedom in nature and spiritual connectedness with the landscape."

MacIntyre (1999) concluded that we do not have individual rights at our foundation, but that we are irreducibly social animals. He discussed the virtues of acknowledged dependence, contrasting this with the virtues of independence of Friedrich Nietzsche, who sought freedom from the imprisoning power of relationships in his novel, *Thus Spoke Zarathustra* (MacIntyre, 1999). Wilks (2005) suggests that "feminist ethicists have argued that our moral identities are located in and constructed through our caring relations with others." Ecological ethicists would argue that this duty of care extends to nature.

How then could such a tripartite embrace work? How can the circular economy concept be steered in order to deliver true sustainability? To explore this, we need to examine cultures as contextualized entities, whose societies function within their landscapes and whose economies are in resonance with the society-environment relationship. One excellent example is the honey economy of the Ogiek people of the Mau forest in Kenya.

5. The Ogiek people and the honey economy

The Mau forest complex grows on the edge of the Mau Escarpment, rising 1000 m above the floor of the African Rift Valley. Occupying 900 km^2, the forest supplies 40% of the freshwater of Kenya. Within this forest live the Ogiek people, some of the last indigenous forest hunter gatherers on the planet. The name Ogiek means caretakers of the plants and animals (Sang, 2002). They are some of the earliest known inhabitants of East Africa, originating in the Nile Valley until the expansion of the Sahara forced them south (National Archives, 1913).

Because their territory occurs between 2000 and 3000 m above sea level, it contains at least five ecotypes, each with different plant and animal species. This habitat complexity offers food all year round. Thus, the Ogiek are seminomadic, moving between ecotypes depending on the season.

Central to this people is honey. Honey is the most important commodity in Ogiek life. In order to understand the Ogiek, you must understand the bee. Honey is central to trade, ritual, social communication, medicine, and nutrition. A working member of the Ogiek will eat up to 2 kg of honey each day. Prior to colonialism, the Ogiek had the "honey barrel rule," which stated that wherever they hung their barrel-shaped beehives was their land (Kitching, 1980).

For the Ogiek, the three arenas of the environment, society, and economics are permeated by honey. The Ogiek are embedded within the ecology of the bee. There is no word for beekeeping in the Ogiek language, emphasizing the collaborative nature of the relationship between humans and the bees.

The council of elders, made up of all senior male and female members of sound mind and good character, determines how many beehives should be built (Blackburn, 1970). In the Mau forest, the main sources of nectar are tree flowers. There are always trees in flower all year round at different altitudes, and the Ogiek follow the bees and the flowering seasons. No tree may be cut down unless the elders agree. A single tribesman may gather around 400 kg of honey each year. Honey is shared within the community, allowing families that have suffered hardship to be cared for. In addition, excess honey is traded with neighboring tribes such as the Maasai. This trade also incubates peace between the tribes. The Ogiek alone have the ecological intelligence and skill sets to gather honey, while ensuring that the bees are unaffected by this activity, and so have secured an important position amongst the other tribes. Honey is also used to preserve meat.

Socially, honey water and honey wine play central roles in important rituals such as births, purification, marriages, new homes, and communicating with the deceased, while honey is also important as a dowry and for legal compensation (Kimaiyo, 2004; Nganga, 2006; Nightingale, 1983). The use of honey in such rituals would seem counterproductive in terms of economic profitability, meaning there is less honey for trade. However, the Ogiek recognize the importance of ritual in culture, tying together society and ecology in deeper ways.

Thus, the three arenas of economics, society and the environment are tied together through honey. It is honey that forms the invisible embrace metaphysically and physically, binding the activities across each arena. Ecological intelligence and ecological ethics inform decision-making. By protecting the trees, the bees have access to their flowers. By the noneconomic use of honey in society, the entirety is bound together. The integrity of the forest is central to the success of the bees.

Meanwhile, a healthy, functioning forest is important in terms of soil stability, water cycling, and as a habitat for wild birds and animals that provide meat for the tribe. Hunting and gathering are also carefully controlled by the elders. Animals with young offspring are not allowed to be killed for any purpose, nor are those in their late gestation period (Ronoh and Barasa, 2012). No cultivation is permitted either side of streams and rivers, protecting the water from soil erosion and pollution (Ottenberg and Ottenberg, 1960).

The land is divided up amongst the people so that each clan has ample trees for honey production. This territorial approach ensures accountability and resource management at the level of the family *in situ*, avoiding the tragedy of the commons as portrayed by Hardin (1968): if you chop down a fruit tree to burn the wood, you won't have any fruit. Because they are not exporting the consequences of their supply chains, the Ogiek are fully aware of the repercussions of their actions based on direct feedback.

Fundamentally, their indigenous economic system has had a very low impact on biological diversity (Towett, 2004). The Ogiek focus on managing their own behavior, in tune with the ecological feedback and ecological intelligence gathered over years, rather than increased honey production efficiency and maximum beehive yields. By working with the bees, this ensures that pollination within the forest continues, allowing the ecosystem to function well (Agera, 2011). Rituals are important, as is participatory education (Kratz, 1989). This gives us a unique insight into the invisible embrace, wherein intergenerational knowledge transfer, married to ecological ethics derived from a value system emergent from the landscape-society interaction, provides the perfect basis for decision-making. The oral tradition of storytelling also plays an important role, as it does in many indigenous cultures (Deloria, 1979).

A recent IPBES report (IPBES, 2019) demonstrates that lands occupied by indigenous peoples such as the Ogiek are declining less rapidly than the rest of the planet. This is because these people are rooted within their ecology, not separated from it.

6. Conclusions

Commoner (1971) explored circularity in terms of closing the circle between economics, technology, environment and government. Here we set out the idea of the circularity of a functional, sustainable economy relating to the resonance between landscape, society and economics, wherein sustainability can only be delivered within the framework of the Earth system. This doesn't mean that we must all head for the hills and place our honey barrels in the trees. Rather, we can beat our technological swords into ploughshares, utilizing the power of the internet of things and artificial intelligence to open ourselves to natural feedback, reintegrating ourselves and our activities within the Earth system (Skene, 2020).

For the circular economy to truly aspire to contributing to a sustainable role, it must be integrated within the Earth system, in terms of both the social and environmental contexts. Merely addressing resource use and waste is insufficient. Important elements, such as suboptimality, feedback, accountability and restoration must play central roles, wherein our economic activities are actively addressing ecological and societal empowerment, actualization and recovery. The tripartite embrace, much like the invisible hand, can only work if society is functional and demonstrates relatedness to the environment, and thus the circular economy must facilitate this if it is to be part of a positive future for humanity.

Given that the Earth system is unquestionably the only system that matters, then our economics must resonate with it (strong sustainability) rather than attempt to replicate it (weak sustainability). However, technology has an important part to play, not in mimicking nature, but as a portal of feedback, allowing our activities to dynamically respond to the emergent, nonlinear properties of the Earth system, while tracking the impacts of our decision-making. In conjunction, technology and the Earth system can reconnect us with our ecology, in a powerful, albeit very different approach to that used by indigenous people, who dwell within their ecology already. In order to be fit for purpose, the economy must become an emergent property, serving the societal and environmental arenas rather than destroying them. This is a truly circular economy.

Acknowledgment

This chapter is dedicated to the memory of Professor Klement Rejšek.

References

Agera, S.I.N., 2011. Role of beekeeping in the conservation of forests. Int. J. Agric. Sci. 10 (1), 27–32.
Aoi, C., de Coning, C.H., Thakur, R., 2007. The Unintended Consequences of Peacekeeping Operations. United Nations University Press, Tokyo.
Arditti, J., 1967. Factors affecting the germination of orchid seeds. Bot. Rev. 33 (1), 1–97.
Batty, A.L., Dixon, K.W., Brundrett, M., Sivasithamparam, K., 2001. Constraints to symbiotic germination of terrestrial orchid seed in a mediterranean bushland. New Phytol. 152 (3), 511–520.
Bedau, M.A., 1997. Weak emergence. Noûs 31, 375–399.
Bedau, M.A., Humphreys, P.E., 2008. Emergence: Contemporary Readings in Philosophy and Science. MIT Press, Boston.
Benyus, J.M., 1997. Biomimicry: Innovation Inspired by Nature. William Morrow & Co., New York.
Bernard, N., 1906. Fungus cooperation in orchid roots. Orchid Rev. 14 (163), 201–203.
Blackburn, R., 1970. A preliminary report of research on the Ogiek Tribe of Kenya. Discussion Paper 89. Institute for Development Studies, University College, Nairobi.
Brekke, K.A., 1997. Economic Growth and the Environment: On the Measurement of Income and Welfare. Edward Elgar, Cheltenham.
Butt, R., 1981. Mrs Thatcher: the first two years. Sunday Times, 1 May.
Clements, L., 2008. Individual budgets and irrational exuberance. CCLR 11, 413–430.
Commoner, B., 1971. The Closing Circle: Nature, Man, Technology. Alfred A. Knopf, Inc., New York.
Condorcet, M.J.A.d., 1955. Sketch for a Historical Picture of the Progress of the Human Mind (J. Barraclough, Trans.). Weidenfeld and Nicolson, London (Originally 1779).
Costanza, R., de Groot, R., Sutton, P., van der Ploeg, S., Anderson, S.J., Kubiszewski, I., Farber, S., Turner, R.K., 2014. Changes in the global value of ecosystem services. Glob. Environ. Chang. 26, 152–158.
Curry, P., 2006. Ecological Ethics: An Introduction. Polity Press, Cambridge.
Daly, H.E., Cobb Jr., J.B., 1989. For the Common Good: Redirecting the Economy Toward Community—The Environment and a Sustainable Future. Beacon Press, Boston.
Deloria Jr., V., 1979. The Metaphysics of Modern Existence. Harper and Row, San Francisco.
Desrochers, P., 2001. Cities and industrial symbiosis: some historical perspectives and policy implications. J. Ind. Ecol. 5 (4), 29–44.
Desrochers, P., 2002. Industrial ecology and the rediscovery of inter-firm recycling linkages: historical evidence and policy implications. Ind. Corp. Chang. 11 (5), 1031–1057.
Desrochers, P., 2008. Did the invisible hand need a regulatory glove to develop a green thumb? Some historical perspective on market incentives, win-win innovations and the Porter hypothesis. Environ. Resour. Econ. 41 (4), 519–539.
Devas, C.S., 1901. Monopolies and fair dealing. Int. J. Ethics 12 (1), 59–68.
Diaz, S., Pascual, U., Stenseke, M., Martin-López, B., Watson, R.T., Molnár, Z., Hill, R., Chan, K.M.A., Baste, I.A., Brauman, K.A., Polasky, S., 2018. Assessing nature's contributions to people. Science 359, 270–272.
Durkheim, É., 1973. The dualism of human nature and its social conditions. In: Bellah, R.N. (Ed.), Émile Durkheim on Morality and Society: Selected Writings. Chicago University Press, Chicago, pp. 149–163.
Farnsworth, K.D., Niklas, K.J., 1995. Theories of optimization, form and function in branching architecture in plants. Funct. Ecol. 9, 355–363.

Folke, C., Carpenter, S., Walker, B., Scheffer, M., Chapin, T., Rockström, R., 2010. Resilience thinking: integrating resilience, adaptability and transformability. Ecol. Soc. 15 (4), 1–20.
Gates, B., 2018. Gene editing for good: how CRISPR could transform global development. Foreign Aff. 97, 166.
Gelter, H., 2000. Friluftsliv: the Scandinavian philosophy of outdoor life. Can. J. Environ. Educ. 5, 77–92.
Giampietro, M., Mayumi, K., 1997. A dynamic model of socioeconomic systems based on hierarchy theory and its application to sustainability. Struct. Chang. Econ. Dyn. 8 (4), 453–469.
Gudynas, E., 2011. Buen Vivir: today's tomorrow. Development 54, 441–447.
Hardin, G., 1968. The tragedy of the commons. Science 162 (3859), 1243–1248.
Husband, C., 1995. The morally active practitioner and the ethics of anti-racist social work. In: Hugman, R., Smith, D. (Eds.), Ethical Issues in Social Work. Routledge, London, pp. 96–115.
IPBES, 2019. IPBES Global Assessment summary for policymakers. https://www.ipbes.net/sites/default/files/downloads/spm_unedited_advance_for_posting_htn.pdf. (Accessed 29 October 2019).
Kimaiyo, J.T., 2004. Ogiek Land Cases and Historical Injustices. Ogiek Welfare Council, Egerton, Nakuru.
Kirchherr, J., Reike, D., Hekkert, M., 2017. Conceptualizing the circular economy: an analysis of 114 definitions. Resour. Conserv. Recycl. 127, 221–232.
Kitching, G., 1980. Class and Economic Change in Kenya: The Making of an African Petit Bourgeoisie 1905–1970. Yale University Press, New Haven.
Kratz, C., 1989. Okiek potters and their wares. In: Barbuour, L., Wandibba, S. (Eds.), Kenyan Pots and Potters. Oxford University Press, Nairobi, pp. 24–60.
Lord, J., Hutchison, P., 2009. The process of empowerment: implications for theory and practice. Can. J. Commun. Ment. Health. 12, 5–22.
Lozano, R., 2007. Collaboration as a pathway for sustainability. Sustain. Dev. 15 (6), 370–381.
MacIntyre, A., 1999. Dependent Rational Animals: Why Human Beings Need the Virtues. Open Court, Chicago/La Salle.
Marshall, A., 1920. Principles of Economics. Macmillan, London.
Mbiti, J.S., 1969. African Religions and Philosophy. Heinemann, London.
National Archives, 1913. Land Tenure in Kiambu in Central Province of Kenya. A Case Study of the Ogiek. Ministry of Culture and National Heritage, Republic of Kenya.
Needham, C., 2014. Personalization: from day centres to community hubs? Crit. Soc. Policy 34 (1), 90–108.
Nganga, W., 2006. Kenya's Ethnic Communities: Foundation of the Nation. Gatũndũ Publishers, Nairobi.
Nicolis, G., Prigogine, I., 1989. Exploring Complexity. W.H. Freeman, New York.
Nightingale, J., 1983. A Lifetime's Recollections of Kenya Tribal Beekeeping. International Bee Research Association, London.
Nussbaum, B., 2003. African culture and Ubuntu. Perspectives 17, 1–12.
Ott, K., 2003. The case for strong sustainability. In: Ott, K., Thapa, P.P. (Eds.), Greifswald's Environmental Ethics: From the Work of the Michael Otto Professorship at Ernst Moritz Arndt University, 1997–2002. Steinbecker Verlag Ulrich Rose, Greifswald, pp. 59–64.
Ottenberg, S., Ottenberg, P., 1960. Cultures and Societies of Africa. Open University Press, New York, pp. 22–35.
Parkins, E., 1934. The geography of American geographers. Geogr. J. 33 (9), 221–230.
Pearce, D.W., Turner, R.K., 1990. Economics of Natural Resources and the Environment. Harvester Wheatsheaf, Hemel Hempstead.
Rasch, P.J., Latham, J., Chen, C.C.J., 2009. Geoengineering by cloud seeding: influence on sea ice and climate system. Environ. Res. Lett. 4 (4), 045112.
Robbins, L., 1935. An Essay on the Nature and Significance of Economic Science. MacMillan and Company, London.
Rogers, C.R., 1959. A Theory of Therapy, Personality and Interpersonal Relationships: As Developed in the Client-Centered Framework. McGraw-Hill, New York.
Ronoh, T.K., Barasa, F.S., 2012. Integrating indigenous knowledge in adult environmental education through the ODL strategies for sustainable conservation of Mau Forest, Kenya: the Ogiek experience. Huria 13 (2), 373–386.
Rose, N., 1999. Powers of Freedom: Reframing Political Thought. Cambridge University Press, Cambridge.

Sakuragawa, M., Hosono, K., 2010. Fiscal sustainability of Japan: a dynamic stochastic general equilibrium approach. Jpn. Econ. Rev. 61 (4), 517–537.
Sang, J.K., 2002. The Ogiek Land Question. Paper submitted at Indigenous Rights in the Commonwealth Project Africa Regional Expert Meeting, Cape Town, South Africa, 16th October.
Seják, J., Pokorný, J., Seeley, K., 2018. Achieving sustainable valuations of biotopes and ecosystem services. Sustainability 10 (11), 4251–4265.
Simmonds, P.L., 1862. Waste Products and Undeveloped Substances. R. Hardwicke, London.
Skene, K.R., 2020. Artificial Intelligence and the Environmental Crisis: Can Technology Really Save the World? CRC Press, Boca Raton.
Skene, K., Murray, A., 2017. sustainable Economics: Context, Challenges and Opportunities for the 21st-Century Practitioner. Routledge, London.
Smith, A., 1759. The Theory of Moral Sentiments. Printed for Andrew Millar, in the Strand; and Alexander Kincaid and J. Bell, Edinburgh.
Smith, A., 1776. An Inquiry into the Nature and Causes of the Wealth of Nations. W. Strahan and T. Cadell, London.
Staples, L.H., 1990. Powerful ideas about empowerment. Adm. Soc. Work. 14 (2), 29–42.
Sutcliffe-Braithwaite, F., 2012. Neo-liberalism and morality in the making of Thatcherite social policy. Hist. J. 55, 497–520.
Towett, J.K., 2004. Ogiek Land Cases and Historical Injustices, 1902–2004. Ogiek Welfare Council, Nairobi.
WCED, 1987. Our Common Future. Oxford University Press, Oxford.
Wilks, T., 2005. Social work and narrative ethics. Br. J. Sociol. 35, 1249–1264.
Young, O.R., Berkhout, F., Gallopin, G.C., Janssen, M.A., Ostrom, E., Leeuw, S.V.D., 2006. The globalization of socio-ecological systems: an agenda for scientific research. Glob. Environ. Chang. 16, 304–316.
Zhang, Z., Moore, J.C., Huisingh, D., Zhao, Y., 2015. Review of geoengineering approaches to mitigating climate change. J. Clean. Prod. 103, 898–907.
Zucman, G., 2019. Global wealth inequality. Annu. Rev. Econ. 11, 109–138.

CHAPTER 3

A systems thinking perspective for the circular economy

Seigo Robinson
University College London, Central House, London, United Kingdom

1. Introduction

While the circular economy concept has received wide interest and is much heralded, it is struggling to scale-up without a systemic approach (Geng et al., 2019) with just 6% of material inputs recycled (Haas et al., 2015) and circular business model innovation stalling, resulting in a "design-implementation gap" (Geissdoerfer et al., 2018). While there are a wide range of circular economy definitions and interpretations, they typically have the purpose of delivering economic and environmental benefit while omitting societal impact. A review of 114 definitions highlights the three core concepts that are most repeated: (i) economic benefit; (ii) the "3Rs," i.e., variations and expansions on the familiar "reduce, reuse, recycle" refrain, aligned to delivering environmental benefits; and (iii) a systems perspective (Kirchherr et al., 2017).

It is into this largely overlooked latter dimension that this chapter seeks to delve, with the aim of setting the context for the type of interventions that may help to bring about a circular economy. The following sections cover an overview of systems thinking, their characteristics and leverage points. These perspectives are applied to the circular economy context and the implications for the innovation of circular business models by considering them as subsystems of the wider circular economy system.

1.1 Systems thinking overview

Systems, typically described as being more than the sum of their parts, comprise elements, relationships, structure, and a purpose (Fig. 1). Systems thinking is a holistic approach to analyzing a system's constituent components, how they are interconnected to form a structure, and how systems work over time, including within the context of larger systems. For example, a person and bicycle can transform into a moving system with the aim of transportation, which individually they do not have. *Elements* are typically the easiest parts to identify, often tangible and performing a specific activity, e.g., the wheels on a bicycle. *Relationships*, or interconnections, allow flows that can be physical or informational, the latter being less easy to ascertain. Relationships underpin feedbacks, critical to

Circular Economy and Sustainability, Volume 1
https://doi.org/10.1016/B978-0-12-819817-9.00034-X

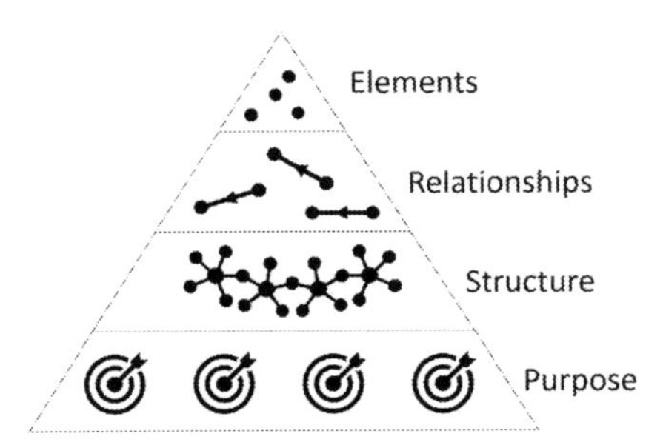

Fig. 1 Framework of categories for systems thinking.

how the system functions, e.g., the pedals are physically connected via chain and gears to move the wheel, or the rider seeing a pothole (information) corrects their steering. The structure is the design of the elements and relationships that yield the behavior, e.g., coordinated mechanical connections and body movements that propel the person-bicycle system. Finally, the *purpose* of a system can be even more difficult to discern. The stated intention of transportation to the shops to buy a newspaper may be understood differently with observation; a smiling rider, taking the long route via the park, stopping to converse with friends, and returning with bread but no newspaper, for example. Here, leisure rather than transport is revealed as the dominant purpose. Thus, the stated purpose of a system is often revealed to be different from demonstrated behavior.

The economy can be considered as a system where participants use resources to perform value-adding activities, exchanging goods and services with each other in the markets to maximize utility for society. Yet, deducing from its behavior, a key function of the economy is to grow, often being desired to do so at a constant (or even accelerating) growth rate, i.e., exponentially. This behavior is manifested in the gross domestic product, (GDP), the economy's foremost metric (Fig. 2).

2. Structure as a driver of behavior

It is possible to replicate the historic dynamic behavior of GDP (Fig. 2) with the "engine of growth" structure (Fig. 3) resulting in exponential output growth (Meadows et al., 1974); indeed industrial stock and GDP have been shown to be strongly coupled (Krausmann et al., 2017). This engine of growth model uses a key tenet of systems thinking and the simulation thereof through system dynamics, modeling systems as alternate sequences of stocks and flows. Here the flow rate of new capital stock is determined by the amount of investment. This stock then generates economic output, a fixed proportion of which is reinvested to build new capital. This generates a virtuous cycle, termed a reinforcing or positive feedback loop, resulting in stock growth. Analogously, there is the case of stabilizing or negative feedback loops. Consider the case of the thermostat that senses the temperature, checks against the desired set point, and switches the boiler

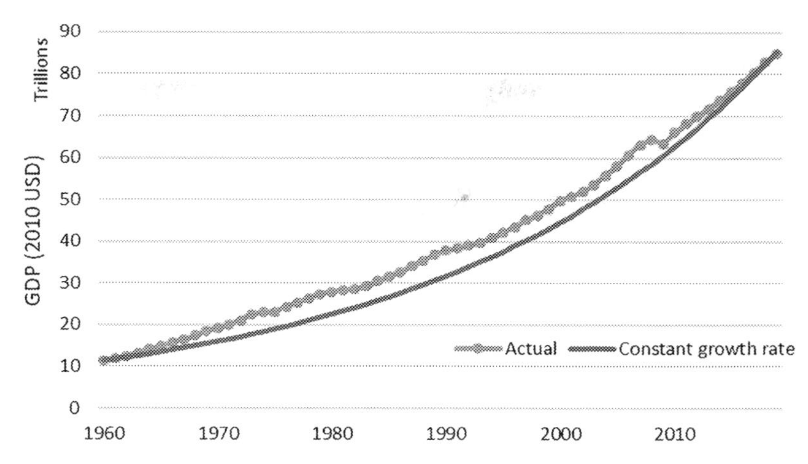

Fig. 2 Global GDP 1960–2017 versus constant growth. *(Data source: World Bank.)*

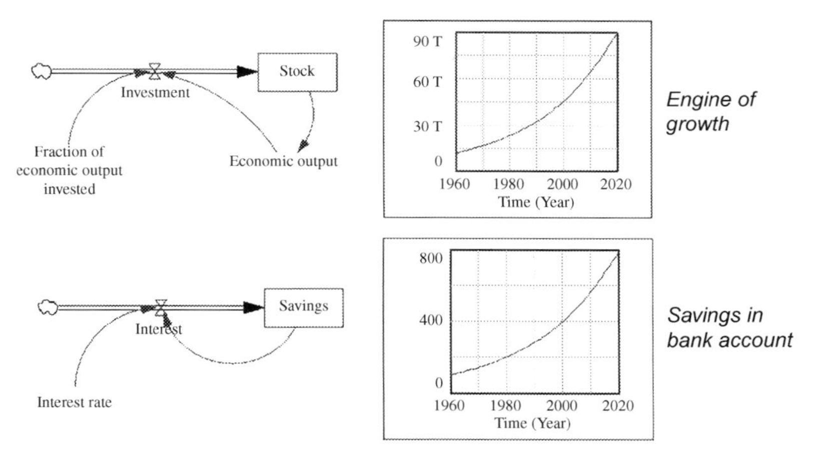

Fig. 3 Engine of growth structure generating GDP profile. Savings in bank account model generating similar exponential growth.

on or off to add heat or not. Complex systems are the combination of reinforcing and stabilizing feedback loops acting at different timescales, delays, and amplitudes. The more distantly in time and space the constituent relationships act, the more often the system yields surprising results. While subsequent behaviors can be complex, nonlinear, and highly sensitive to initial conditions, systems often fit into several patterns (Fig. 4).

1. Exponential patterns are based on constant growth rates determined by the size of the underlying stock—as stock increases, the amount added increases, in a reinforcing feedback cycle, e.g., the GDP and bank account examples (Fig. 3). Similarly, this works in the opposite form, e.g., exponential radioactive decay.

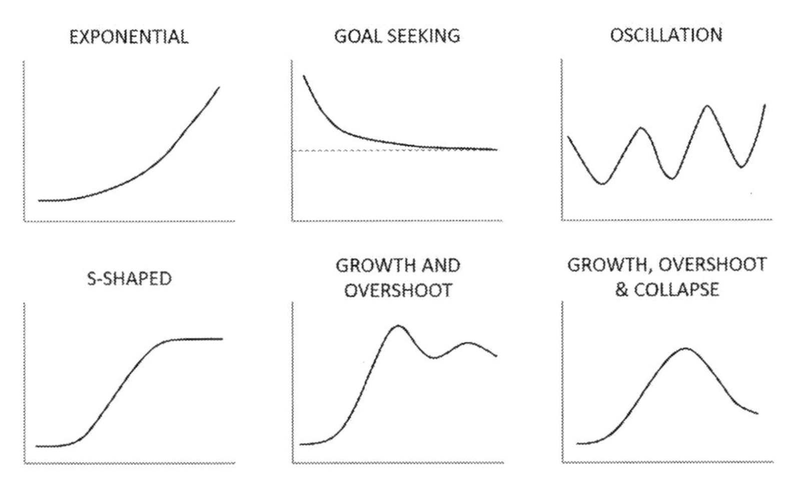

Fig. 4 Common behavior modes of dynamic systems.

2. Goal-seeking behavior is produced by stabilizing loops where the change is based on the discrepancy between a goal and the state of the system, with an appropriate adaptation time, e.g., reordering to maintain a set inventory level.
3. Oscillation occurs when there is a delay in the response of a system that is trying to stabilize, e.g., a hot water system that takes time to respond to the shower lever position, meaning the temperature lurches from too hot to too cold, iterating until the right position and thus temperature is found.
4. S-shaped patterns are best known in how innovations are adopted over time. At its simplest, S-shaped patterns are based on a structure linking together a positive and a negative loop. Initially the positive reinforcing loop dominates, generating an exponential growth phase, giving way to the stabilizing loop, which drives an asymptotic growth phase, and finally resulting in a stable steady state.
5. Growth and overshoot is similar to the S-shaped pattern, with the addition of a delay in the system that provides an oscillatory component.
6. A growth, overshoot, and collapse pattern is where a third loop is introduced—a negative feedback loop, which reduces the strength of the positive feedback loop. This is seen in nature when humans reduce predator numbers, resulting in prey populations growing exponentially until food sources become scarce and damaged. This damage reduces the food regeneration rate and thus the prey numbers that can be sustained, resulting in a collapse of the prey population.

Systems thinking can provide theoretical explanation of the underlying structures that generate the events and patterns that are observed in real life. It also allows us to investigate what systemic structures are needed to realize desired behaviors. Furthermore, it challenges the mental models that sustain these systemic structures. For example, it represents the difference between debating the merits of 1.6% or 1.9% third quarter GDP growth, and asking whether growth itself is desirable; or pondering the increase of municipal recycling rate targets from 45% to 50%, compared to asking whether waste creation is acceptable and if consumers or producers should be responsible. Thus, there are levels of systems thinking: from mental models and paradigms that form structures; structures generate patterns of behavior; and behavior is manifested through events.

2.1 System structure applied to the circular economy

Effectively, all definitions of the circular economy consider a varying number of "Rs:" not only reduce, reuse, and recycle but also remanufacture, refurbish, refuse, and rethink, i.e., eliminating the use of the product. Indeed Reike et al. (2018) identified 38 "R" imperatives, which can be rounded up to 40 by adding "rebuild" and "refill." Most "Rs" can be considered "loops", at least of information (e.g., a business model gathering data to reduce food waste), but often of material resources (e.g., collecting used automotive components to remanufacture, or donating furniture for charities to reuse/resell). This differs qualitatively in structure to the traditional linear economy. At its core, the circular economy has feedback cycles inherent in its conception, meaning a systems perspective is particularly apt.

2.1.1 Macro-level

The structure of material resource consumption of the current state of the economy can be seen in Fig. 5. A key trait of this system is the net addition to stock, i.e., far less comes out of use each year than goes in. This dynamic is relatively little discussed, with the emphasis on the vision of achieving a "perfect" cycling of material outflows. It may be appropriate then to consider what is meant by the circular economy, or specifically the intended or implicit goal, since a stock perspective is not often taken. For some, we may have "arrived" at a circular economy when we achieve 100% recycling rates, when all outflows are diverted to become inflows. However, if we expect to continue to make large net additions to stock each year, we must continue to extract resources. If,

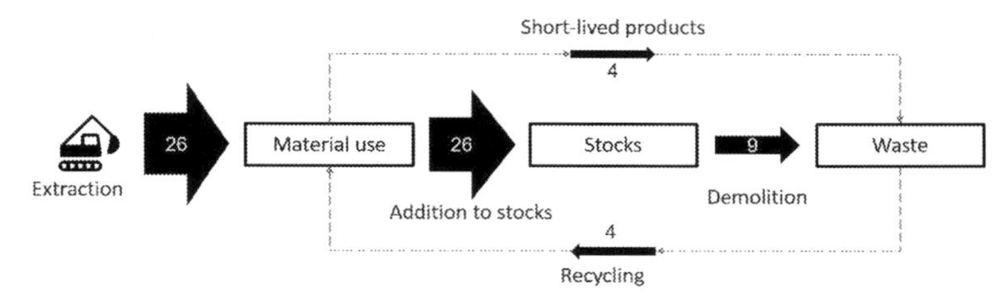

Fig. 5 Structure of material stocks and flows. *(Data from Haas, W., Krausmann, F., Wiedenhofer, D., Heinz, M., 2015. How circular is the global economy?: an assessment of material flows, waste production, and recycling in the European union and the world in 2005, J. Ind. Ecol. 19 (5), 765–777, https://doi.org/10.1111/jiec.12244.)*

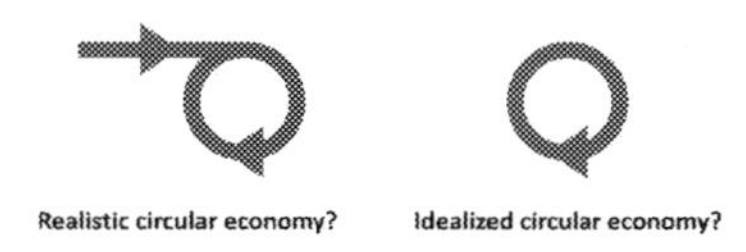

Fig. 6 100% Cycling of materials with and without net additions to stock: realistic vs idealized circular economy?

instead, we envision no longer extracting *any* nonrenewable resources and living off the stock-in-use by endlessly recovering materials, then metrics based on inflows are not particularly useful. So, while recycling rates are clearly an important metric for the circular economy, it must be with an understanding that net additions to stock are at least as important (Fig. 6). Indeed, net additions to stock represent more than half the inputs, are larger than end of life wastes, and more than quadruple recycled amounts.

Linking this idea to policy, the two most utilized metrics are resource productivity and recycling rates—both captured in Sustainable Development Goal 12 and the European Union's Circular Economy Action Plan. Here, resource productivity is the GDP generated in the economy per kg of annual material inputs into an economy; recycling rates are how much we recycle back into the economy of what we dispose. Therefore, these describe the flows in and out of the stock-in-use but do not provide a sense for the size or utility of the stock. While an idealized circular economy would have a 100% recycling rate, it would be underpinned by scaled reuse and remanufacturing, i.e., resources are kept as stock-in-use and tend not to become waste to be recycled. It may, therefore, be apt to question the helpfulness of the current macro-level policy indicators: is it more appropriate to focus on metrics based on stock-in-use?

Organization for Economic Co-operation and Development (OECD) forecasts on material consumption and GDP through to 2060 yield resource productivity improvements of ~1% per year; superficially this sounds rather positive, yet it means that absolute resource inflow still *grows* at ~1.5% per year, more than doubling from 2011. If we want an idealized circular economy with no nonrenewable materials extraction, a rough proxy may be to assume that the total material footprint needs to fall to equal 2060 forecast biomass (i.e., renewable) inputs. In this crude example, material use *reductions* of ~1.5% per year are needed, yielding resource productivity of over seven-times current levels. Referring to the behavior modes, the former resource use trajectory is an exponential process, albeit less steep than for GDP forecasts, while the latter is a goal-seeking process. These behaviors are fundamentally different; the current relationship between GDP and resource use is positive: more GDP means more resource use. For the latter it must "flip sign": more GDP means *less* resource use (Fig. 7A). Between these two end states, to maintain current levels of absolute resource consumption, resource productivity gains (i.e., innovation) must equal GDP growth. This idea is illustrated in Fig. 7B, where there is now a feedback relationship from the growth mechanism to ensure resource

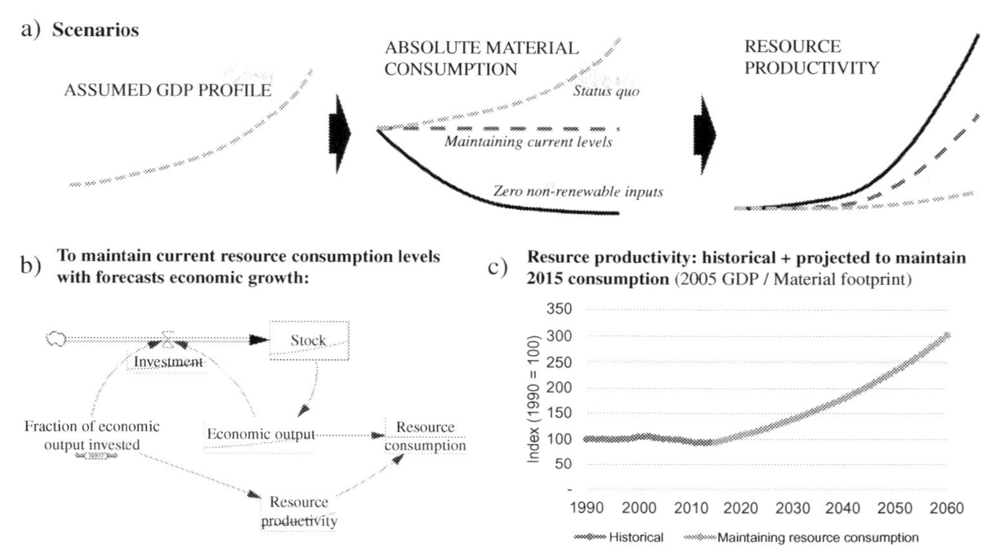

Fig. 7 (A) Three scenarios based on forecast GDP showing illustrative absolute material consumption and resource productivity profiles for: status quo trajectory; maintaining current resource consumption levels; zero nonrenewable inputs. (B) Feedback relationship to ensure resource productivity keeps pace with economic growth to maintain current material consumption. (C) Resultant resource productivity trajectory to maintain material consumption is at odds with historical trends. *(Data source: OECD, International Resource Panel.)*

productivity rises commensurately to maintain absolute resource consumption. The resultant resource productivity profile is unprecedented; since 1990 resource productivity has been effectively constant at the global level (Fig. 7C). While sustained reductions in energy use and waste-to-landfill have been observed at micro level by leading innovative firms, there is no clear evidence that it is feasible for material inputs into products. To deliver this behavior mode requires tremendous innovation and change on the part of businesses and consumers.

2.1.2 Micro-level

To transition to a circular economy, how actors create, deliver, and capture value in the economy needs to change by moving from linear business models (BMs) to circular business models (CBMs), specifically through CBM innovation. Like the circular economy, a wide range of definitions have been proposed for the wider BM concept; they fall across three levels of abstraction. The first are narratives and archetypes, focused on a singular concept; the second understands the BM as comprising elements, often presented in graphical frameworks such as the Business Model Canvas (Osterwalder and Pigneur, 2010); the third conceptualizes them as dynamic systems with interconnections and an architecture—a practical manifestation of this is that of activity systems, which views

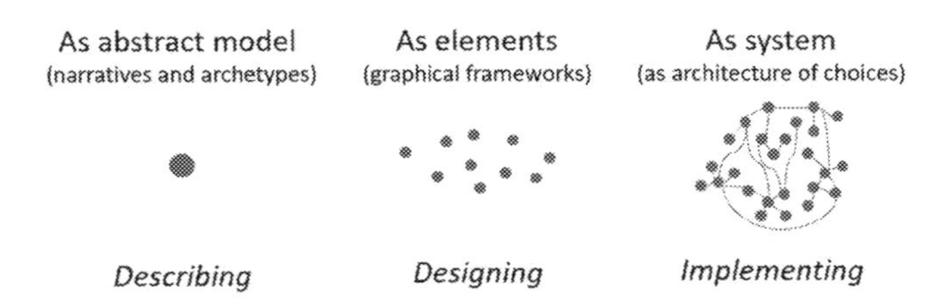

Fig. 8 Business model conceptualization: levels of abstraction.

the BM as a system of activities based on a structure with interconnections and a governance of people and rules, going beyond the focal firm (Zott and Amit, 2011). In the ecosystem literature, the ecosystem-as-structure perspective (Adner, 2017) presents a similar framing, with elements of structure being activities, actors, positions, and links. Both these models emphasize a system designed around delivering a focal value proposition. Thus, rather than one "correct" perspective, we can consider the BM as we use other models, i.e., context-driven and fit for purpose. Therefore, a BM is an abstract entity when *describing* a BM; made up of key components when *designing* a BM; and conceptualized as a dynamic system for *implementing* a BM (Fig. 8).

To date, CBM research has followed the same evolution as BM research, by developing case studies and taxonomies to describe and using graphical tools and frameworks to design CBMs. Yet CBM research is only tentatively approaching the third level of abstraction, aligned to run CBMs. This coincides with the observations in CBM innovation research, where CBMs are readily designed but rarely implemented, suggesting researchers and practitioners may benefit from understanding CBMs as systems. With the qualitatively feedback-rich nature of CBMs, this approach is particularly congruous, representing useful units of analysis, and considered as subsystems within a macro circular economy system.

Both BM and CBM innovation is typically framed as a phased process, e.g., preparation, ideation, detailed design, experimentation, piloting, and scaling. It is a dynamic process that evolves a concept through to a fledgling system that is then allowed to self-organize into a full, resilient system. Indeed, there may be several interim states; progress in one area may regress as delayed feedbacks return the system to a previous state. This tallies with empirical findings that (C)BM innovation process steps are blurry and often iterative, rather than a ratcheted stage-gate process, and thus needing a flexible organizational structure that understands where it is along the innovation progression. This notion of unstable structures and a sense for past and future transitions is reflected in the idea of "semistructures" and "rhythmic transitions" exhibited by continually changing innovation organizations (Brown and Eisenhardt, 1997). Semistructures describe a state between order and disorder—sufficiently organized for change to happen but not rigid enough to prevent it. Rhythmic transitions allow an understanding of the actors, structures, and interconnections that are required for different BM innovation phases. The innovation process is enabled by dynamic capabilities: the skills of sensing (identifying BMs and generating ideas), seizing (designing and testing BMs), and transforming (building new competences and implementation). Dynamic capabilities help to integrate, build, and reconfigure internal and external competences, so-called "ordinary capabilities," to address changing environments and increase a firm's resilience (Teece et al., 1997). The most successful firms pursue low-cost experiments to gain a better understanding of feasible pathways. When experimentation yields positive results then the structure is reinforced with policies and thus self-organizes around the improved system, i.e., the next evolutionary phase. It does mean that negative results also occur, meaning a consequent tolerance for failed experiments is essential. This is held up by research that highlights firms that tolerate failure innovate better (Farson and Keyes, 2002).

Thus, a systems perspective allows us to shift our mental model of CBMs to one of a dynamic system of interconnected elements, purposely structured, and governed by certain rules; one better suited to implementing CBMs and thus helping to address the design-implementation gap. Furthermore, a systems perspective can help to comprehend the messy innovation process itself—one founded on dynamic capabilities, deliberately maintaining a semistructure that encourages experimentation, allowing for failures, and appreciating the haphazard path to scaled implementation.

3. System characteristics

3.1 Background

Systems tend to exhibit three characteristics: hierarchies, resilience, and self-organization.

Hierarchies evolve from the bottom up with the upper levels serving the goals of lower ones. For example, a government exists to serve its citizens; human organs exist to serve the feeding, cleaning, and renewal needs of cells. As hierarchies form, systems become subsystems of larger systems. Subsystems operate relatively autonomously, regulating and maintaining themselves while contributing to the wider system. The larger system supports the coordination of subsystems, ensuring the smooth running of the whole, relying on one subsystem more if another needs to ease off.

Resilience describes the extent to which a system can survive under changing conditions. It arises when systems have stabilizing feedback loops meaning perturbations can be weathered. The loops act in different pathways to counter shocks, providing cover through redundancy, and working at various time scales. The resilience of systems is eroded when focusing on constancy and efficiency, as other parts of the system become neglected or lost—when a disruption outside the normal operating range hits, there are no correcting feedback loops to provide a restorative function.

Self-organization reflects the ability for systems to increase the complexity of their own structure over time, and thus eventually form hierarchies. It is an unpredictable process that requires experimentation and produces heterogeneity, yet is often based on simple rules. It is seen particularly frequently in nature, from the shapes of snowflakes to seeds growing into trees. In the real world, similarly to resilience, it can be neglected or even suppressed in favor of immediate productivity or to protect the establishment.

3.2 Circular economy system characteristics

3.2.1 Hierarchy

Even though a systems perspective is one of the few repeated dimensions across circular economy definitions, most references to it are observations on the need for large scale transformation and the hierarchies through which it is practiced. These fall into three levels: the aforementioned macro (global/national) and micro (business models, consumers) levels, along with the midway *meso* level (industrial cities/networks). Many focus on the macro, concentrating on economy-wide policies and what adjustments may be necessary to the economy. The meso level discusses eco industrial parks and industrial symbiosis, with significant focus on China since its policy drive in 2008. The micro level focuses on CBMs and innovation thereof; there is relatively little on consumers. While there is an understanding of this hierarchical nature, few have discussed how the levels should interconnect, the appropriate feedback mechanisms needed, or how they should be coordinated and changed together in tandem. Reflecting that higher levels of system hierarchy should serve lower ones, the lack of linkage between the levels can be considered a gap. The framing thus moves from somewhat abstract national targets to how policy instruments are deployed at specific leverage points to foster subsystems—both the development of meso level "networks" and at the micro level, helping to enable CBMs, easing their transition from design to implementation.

3.2.2 Resilience

In the traditional linear economy, the pursuit of the single least cost option to enhance short run economic output tends to sacrifice resilience. The focus is on efficiency over effectiveness with the calculation that resultant savings will be greater than any losses due to the increased susceptibility to occasional sourcing price shocks or production downtime. Since generally all firms make this calculus, global supply chains tend toward fragility, resulting in increased resource price volatility in recent years, exacerbated by speculative trading. Examples of reduced resilience include Volkswagen stopping production on 10,000 vehicles when a component supplier refused delivery, and Samsung's Galaxy Note 7 smartphone requiring a global recall, having sourced most of its batteries from a single supplier that suffered a manufacturing issue (Felgendreher, 2020).

With the increasing complexity of products, ecosystems of firms arise to deliver a value proposition. Each partner is dependent on the success of the ecosystem, yet without resilience, one partner's inability to deliver jeopardizes the whole. This applies perhaps more to CBMs that also deal with a reverse logistics process which may require a different set of actors to the traditional forward supply chain, meaning two potentially fragile pathways rather than just one. In the circular economy context then, resilience may look like a network of localized partners to avoid single pathways. Networks would not only deliver products and services but also recover them to valorize and deploy again—the system is able to rely on some partners more if others are incapacitated. While this builds resilience, contracting costs must be low to ensure that the increased complexity is financially viable.

3.2.3 Self-organization

In the context of the globalized economy, manufacturing has self-organized in low cost countries. There are three corollary effects of note for the circular economy. Firstly, richer nations have increasingly imported more finished goods, meaning they are able to relatively decouple their GDP from resource consumption, i.e., domestic material consumption (DMC) growth is lower than GDP growth. In some cases, some nations have moved to being absolutely decoupled, i.e., a negative relationship, where GDP growth is positive and DMC growth is negative. However, much of this resource productivity gain fades when using Material Footprint, a consumption-based metric, since resource use is repatriated to the consuming nation. Thus, similar metrics can tell different stories; proper context in measurement is critical so that circular economy progress is effectively managed. Secondly, the increased distance in time and space of the globalized economy can make the system harder to predict and manage. Looming deficits or surpluses in one part of the system may not be evident to another part until the system no longer functions properly; distance tends to lead to delays, so the ability to fix issues rapidly is diminished. For example, the stock of a critical component at a second-tier supplier may gradually run down; without visibility of this, there is no warning for the focal firm, which only finds out when stock is no longer available and learns that it has a two-month manufacturing lead time. Thirdly, with increased distance and outsourced components all over the world, it makes it harder to build a financially viable reverse logistics chain that returns the components from end-of-use products to be valorized all over the world, particularly when their availability is not often predictable as the duration products are used for by customers varies widely.

At lower levels of hierarchy, the phases of self-organization of networks have been considered at the *meso* level in industrial symbiosis ecosystems (Chertow and Ehrenfeld, 2012; Domenéch, 2010); these are largely in geographically co-located instances—an example of reduction of distance in time and space. At the micro level, CBM innovation is starting to take effect, with specific industries emerging. Urban mining is becoming viable as the value and availability of e-waste increases. Demand for zero-waste grocery stores has grown as societal views on the disbenefits of plastic packaging has moved from a niche to prevailing one. An enabling condition that is likely to be an underpinning of any scaling of the circular economy is the general trend toward ubiquitous connectivity, meaning partner access and customer sourcing costs are drastically reduced. A manifestation of this is via the sharing economy, with Uber and Airbnb using these macro trends to rapidly become dominant in their respective sectors. Community initiatives have also flourished: online platforms that facilitate peer-to-peer exchange of used products; apps to find shops and eateries that fill up water bottles for free; coordinated beach litter clean-ups; and making building materials by sequestering plastic packaging.

4. Leverage points

4.1 Background

Leverage points are places in a system that, from relatively small inputs, can produce large changes in system behavior. Meadows (1999) offers 12 places to intervene in a system with increasing effectiveness as leverage points; Fig. 9 maps these to the system thinking categories presented earlier.

4.1.1 Elements

Constants and parameters are mechanistic "dials" that can be easily modified within a system, e.g., the levels of taxes, incentives, and standards. While the size of buffers and structure of material stocks and flows are potentially strong levers for changing a system, they tend to be large and physical, meaning changing them is difficult practically, time-consuming, and/or costly, and thus seen as less influential.

4.1.2 Relationships

Feedbacks in systems drive the dynamics of system behavior. Delays act in systems to create oscillations, overshooting and undershooting desired set points—like the shower example, or on much bigger scales, like the over and under capacity of whole industries.

INCREASING LEVERAGE ↓

Category	#	Places to intervene in a system
Elements	12	Constants, parameters, numbers (such as subsidies, taxes, standards)
	11	Sizes of buffers and other stabilizing stocks, relative to their flows.
	10	Structure of material stocks and flows (e.g. transport networks, population age structures)
Relationships	9	Lengths of delays, relative to rate of system change
	8	Strength of negative feedback loops, relative to impacts they are trying to correct against
	7	Gain around driving positive feedback loops
Structure	6	Structure of information flows (who does/does not have access to what information)
	5	Rules of system (such as incentives, punishments, constraints)
	4	Power to add, change, evolve, or self- organize system structure
Purpose	3	Goals of system
	2	Mindset or paradigm out of which system arises (goals, structure, rules, delays, parameters)
	1	Power to transcend paradigms

Fig. 9 Places to intervene in a system.

Negative feedback loops act to stabilize systems, therefore, the strength of these in relation to the stock being controlled is important; significantly weakening or removing them results in far less resilient systems. Finally, gains in positive feedback loops cause exponential growth, and often collapse; typically, there is greater leverage in slowing positive loops than strengthening negative feedbacks.

4.1.3 Structure

The structure of information flows can significantly alter behavior. United States law mandating factories publish their hazardous emissions led to a 40% reduction within 4 years (Meadows, 1999) without the need for fines, bans, or limits—corporations simply did not want to be on the list of worst polluters. It is often much easier and quicker to put in place or restore informational feedback than physical structure. More powerful is the ability to set the rules of the system: constraints, punishments, and incentives. Yet stronger is self-organization, the ability to add or change system structure and adapt. While rules for self-organization exist that govern how system change may occur (e.g., DNA), variation and experimentation are integral for this process and key to innovation.

4.1.4 Purpose

The purpose category represents the most influential places to intervene in systems. System goals are interpreted from what a system does rather than what it says it does—like the leisure cyclist. Changing the goal of a system necessarily changes all the prior system areas: from stocks and flows, through to how they are connected, and how the structure may change over time. Going further, mindsets are what generate the goal of a system; mindsets are the underlying assumptions, e.g., Earth's resources should be exploited for human use. Beyond this, the highest leverage is in transcending paradigms, understanding no worldview is correct or right, and in a sense, an ability to choose the one that helps to achieve a purpose.

4.2 Leverage points for increased circularity

4.2.1 Elements

The current circular economy discussion is generally at parameter level: policy interventions that move the dial to support a reduction in waste generation, increasing recycling rates, and steady gains on resource productivity. While pushing in the right direction, there is an inherent underlying incrementalism. Interventions tend not to instigate structural change in the system nor create step change in enabling conditions for actors to innovate. A "drastic" policy change doubling the cost of landfill overnight may be impetus for some actors; yet if waste management as a proportion of total end user costs rises from 1% to 2%, it may not constitute a big enough nudge for behavior change at system level. Innovations that improve resource productivity at GDP growth rates, say ~3%, are certainly welcome, but this must be considered against the context that transformational change requires this for all products, within a year, every year.

The size of buffers in this context is the stock-in-use: with a 2010 stock of 792 gigatons versus inflows of 78 gigatons/year and outflows of 15 gigatons/year (Krausmann et al., 2017), wholesale replenishment of stock-in-use with resources designed for circularity will thus necessarily be a slow, costly, and enormous exercise. Similarly, the structure of this system means that once in place, the function and behavior of stock-in-use is hard to change—maximizing its utility and longevity becomes the best circular leverage. Where there is still a choice of infrastructure development, it should favor low-resource choices, e.g., mobile telecommunications services negating the need for landline infrastructure or decentralized power generation instead of a national network of electrified masts.

4.2.2 Relationships

If new products perfectly conforming to circular economy principles were developed, marketed, and deployed today, legacy products that are unable to be valorized due to their original design and material choice may take decades to exit the stock-in-use. Product lifetimes define the inherent delay in the physical nature of the system, meaning rapid change to a circular economy is difficult without actively identifying, recalling, and substituting legacy products, i.e., artificially reducing the delay.

A key negative feedback loop has been discussed previously, i.e., as the economy grows, it must be met by a stronger response in the opposite direction by resource productivity. This ensures that absolute resource consumption starts to decline and must be driven by increased investment into CBM innovation. Some evidence suggests that material consumption per capita of specific materials, e.g., cement, steel, and copper, do reach saturation for highly industrialized nations (Bleischwitz et al., 2018); so while not a negative relationship, there is an underlying utility level that nations strive for. Understanding the underlying desired performance and reaching it with the least amount of resource would help to accelerate the decoupling process.

A higher leverage approach than strengthening the negative feedback loop is to dial down the positive feedback loop. As linear and circular BM innovation systems compete, curbing the positive feedback cycle of the more dominant linear BM innovation (e.g., no funding or tax breaks for research into linear BMs) rather than strengthening CBM innovation (e.g., subsidies) is a stronger leverage point for resource productivity by allowing CBM innovation to outpace, catch up, and pass linear innovation.

A more fundamental virtuous cycle is economic growth rate. A no-growth scenario would mean any resource productivity innovations would no longer be used as an offset for growth but directly reducing absolute resource use. However, having built a global economic system with the purpose of growth, self-organization has generated structural elements such as large national debts and pension liabilities that require servicing. Thus, while it may be a relatively high leverage place to intervene, it will face understandably strong resistance.

4.2.3 Structure

General understanding of resource consumption is low as information is not readily available. Thus, simply altering the structure of information so that it is accessible may be a strong lever; for example, firms having to publish the amount of nonrenewable resources that their operations consume annually. By virtue of measuring, firms may better manage their resource consumption, particularly against the backdrop of reputational risk associated with appearing on a "most wasteful" league table. CBM innovation may be accelerated as firms scramble to significantly reduce their resource use. The implicit goal then transitions from maximizing manufacturing plant throughput to sell as many widgets as possible every year, to asset management instead, providing customers with performance, using widgets designed to be reused, remanufactured, and recycled in order to maximize profit. Consumers empowered with information may also steer themselves toward those firms doing better within an industry, or change their personal consumption habits to forgo products from particularly wasteful industries.

Yet higher leverage comes from changing the rules of the game, for example: (i) mandating the stewardship of resources through extended producer liability, where firms are liable for the specific material they put into the market, rather than the less strict extended producer responsibility approach that allows a more contributory and arms-reach obligation over recovering fungible resources; (ii) "resource budgets" allowing only a certain amount of resources for each firm; and/or (iii) transitioning the burden of taxation from renewable labor to nonrenewable resources. These all change the constraints for a firm to operate much more in line with the circular economy.

Still more powerful is self-organization that is unpredictable by nature and potentially transformative. As highlighted above, with ubiquitous smartphones and virtually free connectivity, the rules have been rewritten. From nowhere, Airbnb has become the largest hotel firm by number of beds in just a few years; similarly Uber in the taxi sector. Neither has built a hotel or car, respectively. While there are legitimate concerns around rebound effects and social issues, their ability to tap into underutilization of distributed stock-in-use raises a key question: do we already have the stuff we need to provide all the utility we want?

4.2.4 Purpose

The goal of the economy is growth. As highlighted previously, while perhaps not theoretically impossible, the same goal is difficult to maintain for the circular economy. Without growth, then, we must find a level that is considered "enough." Indeed, there is guidance from behavioral economics with a plateauing of wellbeing as incomes rise

(Kahneman and Deaton, 2010; Jebb et al., 2018); if this plateau were the goal, economic output and resources would exhibit goal-seeking behavior rather than the current exponential model. With stability in stock-in-use, a circular economy becomes far more achievable.

This goal is, in fact, underpinned by a change in mental model. The mindset of sufficiency moves from how much money flow can be generated from natural resource inputs, focusing on incremental productivity improvements and resource efficiency, to one based on generating money flows from stock requiring high utilization of long-lived assets. It is the difference in choosing between working for an income, each year seeking a pay rise to buy more things, or living off the rents generated from current assets. One has a utility maximizing "more" mentality versus a utility preserving "sufficient" one; indeed a voluntary simplicity lifestyle is seen to raise wellbeing (Osikominu and Bocken, 2020). This mindset is enhanced by an adjacent mindset, one that has a stronger connection to nature and recognizes the reliance of society on the environment rather than seeing the planet's resources to be exploited for human use. It ties happiness and wellbeing to nature, helping to lower the "target" above for income and thus, resource consumption. The power of mental model change is that, while beliefs can be deeply rooted and hard to change, they tend not to be constrained by the need for time-consuming physical changes; it is possible simply to switch. This reflects the highest leverage of all: understanding that any paradigm is just one of many. This understanding helps us not to hold on to a single paradigm, thus facilitating switching to one that helps us achieve our purpose.

5. Conclusion

Although the systems perspective is often discussed, systems thinking has seldom been applied to the circular economy context. It has been applied here to address this gap, with the aim of providing a clearer picture of the scope of the challenge.

While the focus is generally on flows, systems thinking highlights the importance of the current system behavior of large annual net additions to stock. If the goal is no longer extracting nonrenewable resources, this behavior must stop. To achieve this, innovation must deliver resource productivity gains greater than economic output growth. At current and expected growth rates, this is a daunting challenge requiring the marshalling of extraordinary investment and commitment. The required effort is eased as the positive feedback of economic growth is reduced, allowing innovation rates to outpace it.

A systems perspective also allows us to shift our mental model of CBMs to one of a dynamic system better suited to implementing CBMs, rather than describing and designing them, and thus help to address the design-implementation gap. Furthermore, it helps us understand the CBM innovation process itself—one founded on dynamic capabilities, deliberately maintaining a semistructure that encourages experimentation, allows for failure, and appreciates the haphazard path to scaled implementation.

An assessment of leverage points highlights that we tend to focus on the weakest areas of elements—generally, gentle nudges toward increased circularity. Relationships such as delays in the system, support for CBM innovation, and reducing support for linear innovation are useful but largely predetermined by structure. Structural changes have greater potential; for example, deploying new informational structure may be powerful, "embarrassing" firms to do better, changing the rules of the game toward reducing material use, or fostering the conditions for self-organization. Finally, purpose allows us to question whether the goal of wellbeing instead of growth presents a more desirable and viable pathway to a circular economy. The highest leverage of all is our own mental models; switching to a sufficiency and nature-centric mindset allows us to set the goal that prioritizes, fosters, and establishes a circular economy.

References

Adner, R., 2017. Ecosystem as structure: an actionable construct for strategy. J. Manag. 43 (1), 39–58. https://doi.org/10.1177/0149206316678451.

Bleischwitz, R., Nechifor, V., Winning, M., Huang, B., Geng, Y., 2018. Extrapolation or saturation—revisiting growth patterns, development stages and decoupling. Glob. Environ. Chang. 48 (December 2017), 86–96. https://doi.org/10.1016/j.gloenvcha.2017.11.008.

Brown, S.L., Eisenhardt, K.M., 1997. The art of continuous change: linking complexity theory and time-paced evolution in relentlessly shifting organizations. Adm. Sci. Q. 42 (1), 1–34. https://doi.org/10.2307/2393807.

Chertow, M., Ehrenfeld, J., 2012. Organizing self-organizing systems: toward a theory of industrial symbiosis. J. Ind. Ecol. 16 (1), 13–27. https://doi.org/10.1111/j.1530-9290.2011.00450.x.

Domenéch, T.A., 2010. Social aspects of industrial symbiosis networks. Available from: http://discovery.ucl.ac.uk/762629/.

Farson, R., Keyes, R., 2002. The failure-tolerant leader. Harv. Bus. Rev. 80 (8), 64–71.

Felgendreher, B., Why are global supply chains becoming more fragile? Available from: https://www.enterprisetimes.co.uk/2016/11/07/12110/ (Accessed 6 May 2020).

Geissdoerfer, M., Vladimirova, D., Evans, S., 2018. Sustainable business model innovation: a review. J. Clean. Prod. 198, 401–416. https://doi.org/10.1016/j.jclepro.2018.06.240.

Geng, Y., Sarkis, J., Bleischwitz, R., 2019. How to globalize the circular economy. Nature 565 (7738), 153–155. https://doi.org/10.1038/d41586-019-00017-z.

Haas, W., Krausmann, F., Wiedenhofer, D., Heinz, M., 2015. How circular is the global economy?: an assessment of material flows, waste production, and recycling in the European Union and the world in 2005. J. Ind. Ecol. 19 (5), 765–777. https://doi.org/10.1111/jiec.12244.

Jebb, A.T., Tay, L., Diener, E., Oishi, S., 2018. Happiness, income satiation and turning points around the world. Nat. Hum. Behav. 2 (1), 33–38. https://doi.org/10.1038/s41562-017-0277-0.

Kahneman, D., Deaton, A., 2010. High income improves evaluation of life but not emotional well-being. Proc. Natl. Acad. Sci. U. S. A. 107 (38), 16489–16493. https://doi.org/10.1073/pnas.1011492107.

Kirchherr, J., Reike, D., Hekkert, M., 2017. Conceptualizing the circular economy: an analysis of 114 definitions. Resour. Conserv. Recycl. 127 (April), 221–232. https://doi.org/10.1016/j.resconrec.2017.09.005.

Krausmann, F., et al., 2017. Global socioeconomic material stocks rise 23-fold over the 20th century and require half of annual resource use. Proc. Natl. Acad. Sci. U. S. A. 114 (8), 1880–1885. https://doi.org/10.1073/pnas.1613773114.

Meadows, D.H., 1999. Places to intervene in a by Donella Meadows. World 91 (7), 21.

Meadows, D.H., Meadows, D.L., Randers, J., Behrens, W.W., 1974. The Limits to Growth: A Report for the Club of Rome's Project on the Predicament of Mankind/Donella H. Meadows … [and Others]. Pan, London.
Osikominu, J., Bocken, N., 2020. A voluntary simplicity lifestyle: values, adoption, practices and effects. Sustainability 12 (5), 1903. https://doi.org/10.3390/su12051903.
Osterwalder, A., Pigneur, Y., 2010. Business Model Generation: A Handbook for Visionaries, Game Changers, and Challengers. John Wiley & Sons, Inc., Hoboken, NJ.
Reike, D., Vermeulen, W.J.V., Witjes, S., 2018. The circular economy: new or refurbished as CE 3.0?— Exploring controversies in the conceptualization of the circular economy through a focus on history and resource value retention options. Resour. Conserv. Recycl. 135, 246–264. https://doi.org/10.1016/j.resconrec.2017.08.027.
Teece, D.J., Pisano, G., Shuen, A., 1997. Dynamic capabilities and strategic management. Knowl. Strategy 18 (March), 77–116. https://doi.org/10.1142/9789812796929_0004.
Zott, C., Amit, R.H., 2011. Designing your future business model: an activity system perspective. SSRN Electron. J. 3. https://doi.org/10.2139/ssrn.1356511.

CHAPTER 4

Conceptualizing the circular bioeconomy

Warren E. Mabee
Department of Geography and Planning/School of Policy Studies, Queen's University, Kingston, ON, Canada

1. Introduction

This chapter uses Canada's forest sector as an example to explore the circular bioeconomy. Two primary research questions guide this chapter. Firstly, what would be the impact of expanding circularity within the forest bioeconomy, particularly in terms of potential for carbon sequestration? Secondly, what are the key strategic elements of policy that could be used to support circularity within the forest bioeconomy?

In the sections that follow, the chapter (a) defines the concept of the circular bioeconomy, (b) provides examples of existing and future circular bioeconomy applications, (c) quantifies environmental costs and benefits associated with such a system, (d) identifies barriers to the implementation of the circular bioeconomy, and (e) proposes ways forward for the concept.

1.1 Methodology

The chapter identifies key characteristics that might be used to define a circular bioeconomy by drawing on the existing literature, and applies them using a real-world example to explore potential impacts and benefits.

For any given location, bioeconomies are shaped by both climate (which influences species mix, growth rate, growing season) and by society (which brings industrial activity as well as local policy and regulation). Thus it is important to be specific in developing examples of how a circular bioeconomy might perform. Canada's forest sector, which currently produces a wide range of harvested wood products and already displays some characteristics of circularity, is used in this chapter. A review of the expected half-life of different wood products produced by Canada's forest bioeconomy was undertaken, again drawing values from previously published studies.

Using the half-life approach, the current flow of carbon through Canada's forest sector was assessed, as was a putative future system in which additional circular elements were imposed upon the system. This approach can be used to assess the ability of circular components to maintain biomass—and thus, by extension, carbon—in solid forms before

Circular Economy and Sustainability, Volume 1
https://doi.org/10.1016/B978-0-12-819817-9.00033-8

ultimate disposal of the material, and long-term recycling of biogenic carbon through the atmosphere and back into growing trees and plants.

This approach identifies a number of challenges, which can be addressed through a holistic policy approach. A number of recommendations for future policy development are made.

2. Circular bioeconomies

A circular economy constantly reuses and recycles materials through a variety of applications, minimizing material inputs to the economy and maximizing utility from each of these inputs. The modern definition of circular economies dates to the late 1970s (Stahel and Reday-Mulvey, 1981) and emphasizes ways in which the products and residues of any given process might be designed to act as inputs to a downstream process. Key elements of a circular economy include the need to design products for a long service life, to reuse materials in a series of products, and to reduce the amount of waste at every stage. Because reusing materials requires additional labor inputs, the circular economy can also be seen as a tool to maximize the amount of work that is done for every unit of material input (Unay-Gailhard and Bojnec, 2019). In order to achieve a truly circular economy, conventional concepts of production and consumption need to be reorganized (Bakker and Bridge, 2006); products need to be designed not just for current use, but for future use and ultimate disposal.

The natural cycle of growth, mortality, and regrowth through the interconnected ecosystems of the planet provide myriad examples of circularity, which can provide a conceptual framework for the circular economy. For instance, carbon atoms pass through multiple states over time—from the atmosphere to terrestrial and marine biomass, and in turn into animals, and ultimately back into the atmosphere, or into long-term storage in soils and rocks. The circular economy must emulate these loops—the goods produced must be designed to be reused and recycled, anticipating future products and ultimate environmental fate, and reducing negative and wasteful outputs, including externalities such as packaging or transport (Winkler, 2011). In a "carbon constrained" future, where greenhouse gas emissions are limited through policy or technology, a circular economy approach can minimize the impact of the population on the planet (Geng et al., 2009).

From the literature (Stahel and Reday-Mulvey, 1981; Winkler, 2011; Geng et al., 2009; Zhen et al., 2011, 2017), a circular economy generally incorporates the following features:

1. The use of inputs that can be reused and/or over multiple applications (Sanchez et al., 2019; Wainaina et al., 2020), and ultimately returned to nature;
2. A focus on long-lasting or durable products (Kaddoura et al., 2019);
3. Selection of renewable forms of energy to power systems, including waste-to-energy options (Halkos and Petrou, 2019; Grim et al., 2020); and

4. Organization across multiple sectors to meet the needs of human populations at local or regional scales, rather than around individual sectors (Skrinjaric, 2020).

A parallel concept to the circular economy is the bioeconomy—an economy that relies upon biological inputs (i.e., wood and agricultural biomass, animals, fish, etc.) to support production, rather than fossil resources or other nonrenewable inputs (Duchesne and Wetzel, 2003; Staffas et al., 2013). A key feature of the bioeconomy is the renewable nature of these inputs; with proper management, they can be made available in a sustainable fashion, serving not just present generations but future populations as well. Sustainable resource management is critical to the bioeconomy; if biomass is taken in an unsustainable fashion, bioeconomy development can lead to negative environmental effects including deforestation, and negative social impacts including loss of local livelihoods (Heimann, 2019). Developing the bioeconomy in a sustainable fashion requires strong governance institutions—policy that can safeguard the resource—coupled with effective mechanisms for stakeholder participation (McCormick and Kautto, 2013).

The concept of a circular bioeconomy expands upon the bioeconomy definition by introducing key elements of the circular economy, as listed above. A circular bioeconomy thus can include elements of sequential product use (i.e., fiber "cascades"), which allows material to be reused or recycled over several applications and/or lifetimes (Jarre et al., 2020). Often authors will refer to a progression or cascade from solid/durable products through to chemical and finally energy products. Circular bioeconomy concepts are usually characterized by the inclusion of energy recovery mechanisms instead of other forms of product and/or residue disposal (Mohan et al., 2016). Finally, authors have increasingly focused on the circular bioeconomy as a pathway to rural employment, adding key jobs in rural locations and empowering these populations (Matiuti et al., 2017).

The term "circular bioeconomy" is relatively new and has only appeared in the literature within the last decade, with one of the first uses examining wood-based economies in Germany (Pannicke et al., 2015). The circular bioeconomy term is used most regularly with regards to the forest sector, as wood is typically used for durable applications and can be recycled relatively easily. Some authors have explored the use of circular bioeconomies with the agricultural sector (Viaggi, 2015; Sagues et al., 2020); in these cases, authors tend to look at ways in which residual wastes can be used as a feedstock for value-added chemicals, fertilizers, or energy applications.

Circular bioeconomies are one of many approaches to economic restructuring that are being proposed in the face of both a changing climate (Sadhukhan et al., 2018; Nayha, 2019) and a growing population with increasing need for raw materials (Battista et al., 2020). The circular bioeconomy concept is seen in part as a way to shift back from a highly mechanized society that treats global ecosystems as both storehouse and garbage bin, to a society more in balance with nature (Szekacs, 2017). The circular bioeconomy essentially would operate as a closed loop—where material inputs are generated and harvested sustainably, used in a series of applications, and then returned to nature to support new growth. For this to happen, however, companies would need to restructure supply chains and anticipate the impacts of every production decision on future uses; ultimately, the biomass needs to be returned to nature (Winkler, 2011).

The circular bioeconomy (and broader aspects of the circular economy) are well supported through an industrial ecology approach, which focuses on connections between companies and products, and draws parallels between these linkages and natural ecosystem functions (Ehrenfeld, 2004; Braungart et al., 2007). The circular bioeconomy will also benefit from other areas of study such as biomimicry; new bioproducts within the circular bioeconomy can adapt or copy structures found in nature (Robinson, 2004).

Some of the first regions to adopt circular bioeconomy approaches in modern times are found in China, where economic growth is often constrained by a lack of access to materials. Effective low-carbon systems have been developed in China to accelerate economic growth, particularly in rural regions such as the northeast of the country (Zhen et al., 2011, 2017). One city in this region (Dalian) has recently focused on circular bioeconomy opportunities in order to manage scarce resources and increasing environmental degradation; the experience in this community highlights significant challenges, including lack of incentives, lack of financial support, and a need for increased awareness of and participation in the circular bioeconomy (Geng et al., 2009).

More recently, the circular bioeconomy has been explored in other regions of the world, including Finland (Nayha, 2019), Sweden (Temmes and Peck, 2020), Germany (Jarosch et al., 2020), Brazil (Farzad et al., 2017), Malaysia (Sadhukhan et al., 2018), Canada (Lemire et al., 2019), and the United States (Watanabe et al., 2019). These studies have supported an expanded view of the circular bioeconomy concept, and highlighted a number of challenges, which are explored in the following sections.

2.1 Examples of the circular bioeconomy

Conceptual representations of the circular bioeconomy can be seen in the forest and agricultural sectors. The forest ecosystem is characterized by the cyclical establishment, growth, mortality, and decay of trees and related plants; this cycle extends over decades or (more often) centuries (Christensen, 2014; Christensen et al., 1996; Braun, 2002). Commercial forest activities interrupt this cycle; biomass is removed from the ecosystem and may be delivered to points around the world, which means that the nutrients and materials found in these trees cannot return to the local ecosystem (Kurth et al., 2020). Agricultural systems are similar, although establishment, growth, harvest, and use of crops takes place in a very compressed timeframe; this cycle is measured in years or (more often) months (Saleem et al., 2020). As in forestry, the removal of crops from the fields lowers the availability of nutrients to the next generation of plants; in agriculture more often than in forestry, fertilization regimes are utilized to replenish nutrients and ensure growth in the future (Greer et al., 2020). The use of fertilizer in agriculture is a challenge to the circular bioeconomy model, as the production of these inputs may rely upon fossil energy and nonrenewable sources of nitrogen or potassium (Barquet et al., 2020). A circular bioeconomy approach to agriculture would necessarily include new ways of recovering or producing fertilizers (Barquet et al., 2020; Beckinghausen et al., 2020).

In the sections that follow, the forest sector and its ability to adopt circular bioeconomy elements are examined using data from Canada.

2.2 Circular economies in Canada's forest sector

It has been noted that the application of circular bioeconomy principles could reinforce the ability of forests to sequester carbon over the long term and thus increase their ability to offset greenhouse gas emissions (Yufang and Yanqing, 2011). To achieve this, the sector would need to maximize the use of every fiber that is harvested, essentially extending the total fiber supply and expanding the number of potential biomass applications while maintaining or reducing current harvest levels.

The Canadian forest economy already employs a number of important circular elements in its cascade of products. Wood fiber tends to be used in a series of products; sawmills (which mill logs into lumber) produce chips, which in turn are taken up by a number of secondary industries, including papermaking. Paper products in turn are regularly recovered through municipal "blue-bin" programs and recycled into new products; in 2016, approximately 3.6 million tons of paper and paperboard were diverted from landfills across Canada (ECCC, 2018). In the same year, the country consumed about 5.2 million tons of paper and paper products (FAOStat, 2020), indicating that recycling programs across the country captured almost 70% of total use, although there are indications that in some locations across the country as much as 25% of recycled materials were contaminated and could not actually be recycled (Chung, 2018). Paper and affiliated products, notably inks, can be viewed as a system that has been extensively redesigned in order to allow circular activities to happen; the onset of recycling led to the development of vegetable-based inks (Mirkovic et al., 2019; Zimmerman and Morkbak, 1997) and new deinking technologies (Valls et al., 2019) to facilitate fiber recovery and reuse.

Canada's forest sector has also become a major producer and user of renewable energy. Approximately 62% of energy needs in the pulp and paper industry are met through the combustion of spent pulping liquors and other processing residues (NRCan, 2020). Residues generated by the sawmill industry that are not currently taken up by other value-added processes (including paper and board production) are increasingly used for wood pellet production, which has doubled since 2012 and reached about 3 million tons per year as of 2018 (Aguilar et al., 2019, 2012).

Proponents of circular economies often refer to the ability of this type of industrial organization to meet the needs of local or regional populations. In Canada, the forest industry provides more than 80% of employment in about 160 remote communities, and is a major employer in hundreds of other communities across the country (FPAC, 2018). Significant decline in the forest sector over recent years, however, has taken its toll on the local benefits that the sector provides. In 2016, Canada harvested only 157 million m^3 of timber despite having an allowable cut of 215 M m^3 (FAOStat, 2020; NFD, 2018), largely due to declining demand for both lumber and paper products. Older mills have been closed, and most sector investment has been in larger centers that serve as transportation and innovation hubs across the country. This has impacted the sectors' contribution to both GDP and employment. Currently, the forest sector contributes about 1.2% of Canada's GDP, down from 1.9% in 2004 (StatsCan, 2018a). Since 2004, direct employment in the forest sector has dropped by ~130,000 employees (from 362,000 to 232,000) (CFS, 2016).

While the Canadian forest sector displays a number of characteristics that are representative of circular economies, there are a number of ways in which these industries act in a more linear fashion. Logs that are harvested in the forest are divided to support a series of products streams that run in parallel, rather than in series. The best portions of the wood are used for lumber or veneer; the chips, sawdust, and other residues from primary processing support the production of boards, paper, and pellets. With the exception of paper, most of these products are used only once. Recent employment data also indicates that the sector is consolidating to focus on specific product streams, and that local benefits of the sector are disappearing as jobs are eliminated or moved to larger hubs.

2.3 Forests, forest products, and the carbon cycle

It is helpful at this point to reconsider the role of forests and harvested wood products in carbon sequestration. The amount of time that it takes trees to grow varies dramatically from location to location; even within Canada, the average rotation age for boreal forests varies significantly as one progresses from east to west and north to south, with anywhere between 60 and 120 years between commercial harvests required to allow trees to reach maturity (Thiffault et al., 2008). Over this period, significant amounts of carbon may be sequestered, but annual increases in the amount of carbon taken up by the forest slows down with age, particularly in northern boreal forests (Lafleur et al., 2018). Significant amounts of carbon may also be released through disturbances including insect outbreaks and fire (Liu et al., 2020), which suggests that even managed forests may not be secure long-term carbon storage options. A study modeled hypothetical carbon stocks in the Canadian province of Ontario's managed forests, and in harvested wood products derived from those forests, between 2001 and 2100 (Chen et al., 2010). This study considered the potential impacts of fire and pests on the forest, carbon releases associated with these disturbances, and the offsetting factor that carbon stored in harvested wood products can play. The study found that total forest carbon stocks would increase by about

48 million tons over the 100 years of the model forecast, even when fire and other disturbances were factored in (Chen et al., 2010), largely due to the role of harvested wood products.

Managing carbon within harvested wood products becomes even more important when one considers that the ability of forests to sequester carbon may be diminishing. One study examining the forests of the United States reported that annual sequestration of carbon in forests is likely to decrease over a 50-year period, from 274 million tons per year in 1990 to 161 million tons per year in 2040 (Skog and Nicholson, 1998). This may be a result of management decisions made by the forest sector; over time, the North American industry has worked to compensate for longer growing periods by favoring faster-growing species to manufacture hybrid composite wood products, rather than solid lumber.

Globally, the forest sector is increasingly aware of the potential to improve carbon sequestration across a range of harvested wood products, and initial studies have identified the particular role that durable and long-lasting products, primarily used for construction, can play (Pieratti et al., 2019). A number of papers have examined the longevity of different classes of harvested wood products, both on their own and in specific applications. It is suggested that net sequestration of carbon in U.S. wood and paper products will increase to as much as 74 million tons by 2040, an increase of 13 million tons over 1990 values (Skog et al., 2004). Similar studies in Canada suggest that in Ontario alone, the total amount of carbon stored in forest products could increase by about 4.2 million tons per year between 2001 and 2100 (Chen et al., 2010) if these products are carefully managed and tracked.

In Table 1, a summary of recent studies is provided showing a range of carbon sequestration values in harvested wood products. These values are reported as half-life values (i.e., the point at which 50% of the products will still be in service) (Skog et al., 2004).

Products shown in the table suggest that wood products fall into three categories: durable construction materials, which tend toward centuries of use; durable applications including furniture, pallets, railway ties, etc., which have a significantly shorter half-life; and finally paper and paper products, which have an extremely short half-life. While not included in this table, energy from forest biomass is also an important product category and has a half-life of only weeks or months (depending upon how long the feedstock is stored). Critical recommendations for increasing net carbon sequestration include: increasing product use-life, increasing recycling of products, and increasing generation of energy at end-of-life (Skog and Nicholson, 1998). These three principles each support development of the circular bioeconomy.

This table highlights one very important distinguishing feature of the circular bioeconomy: it will necessarily extend over very long time periods, to account for the growth and harvesting of wood as well as the extended use of forest products in general (Adams, 1998), and these timespans will almost certainly cross generations and will involve

Table 1 Half-lives in years of selected forest products/applications in Canada.

Product	Skog and Nicholson (1998)	Chen et al. (2010, 2008, 2014, 2018)	Dymond (2012)
Single-family homes	100	85	90
Lumber		79	
Multifamily homes	70	50	75
Nonresidential buildings	67		
Building upkeep	50	25	30
Mobile homes	20		30
Furniture	30		
Railroad ties	30		
Pallets	6		
Paper (free sheet)	6	5	2.5
Paper (all others)	1	5	2.5

different decision-makers at different points in the cycle. Planning for a circular bioeconomy thus needs to be carried out in an anticipatory fashion; decisions made today must plan for multiple pathways for future use, to allow the next generation of decision-makers to ultimately choose how to utilize biomass in its next application.

3. Benefits of a circular bioeconomy

Enacting the circular bioeconomy will involve decisions that span centuries, and thus it is important to be able to demonstrate clear benefits of this approach in order to engage stakeholders and achieve success.

The typical arrangement of the forest products sector is shown in Fig. 1. In this diagram, the industry is described from the perspective of carbon, with arrows representing the subsequent stages of atmospheric residence (*diminishing arrow* from 0 to 100 on the scale), uptake and growth (*increasing arrow* from 0 to 100), and product storage (construction stretching to 300, furniture to 160, and paper use and recycling represented by the nested cycle at 100). Theoretically, the product cycle is circular, with carbon ultimately emitted (through combustion or decay) and returned to the atmosphere, although in practice, a good proportion of solid forest product ends up in landfills and may represent much longer-term sequestration. The sizes of the circles in this diagram represent the approximate half-life of each stage by product, according to data from the literature as described in Table 1. The transfer of carbon from the atmosphere to the forest is spread out over 100 years, reflecting average growth in the Canadian boreal forest. In different types of forests, this period could be considerably longer or shorter.

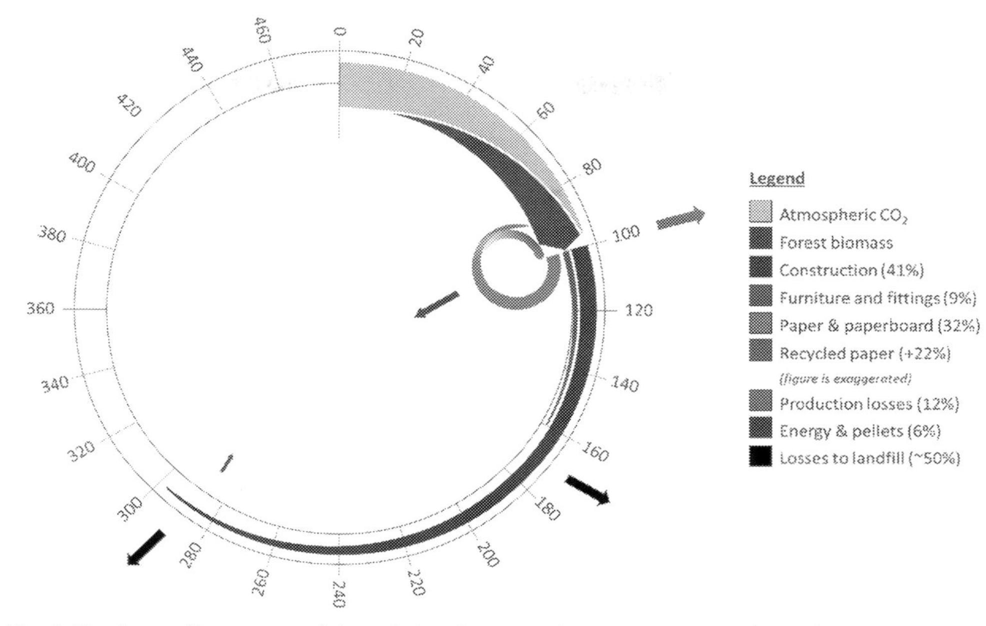

Fig. 1 The "nested" structure of the existing forest products sector. *Image by author.*

Fig. 1 divides the biomass harvested from the forest into long-term durable products (construction), short-term durable products (furniture and fittings), and very short-term products (paper and paperboard). In the legend, the proportion of biomass from forest harvests that are diverted to each of these categories is indicated. These figures are drawn from a review of statistics provided by the Food and Agriculture Organization (FAOStat, 2020), Natural Resources Canada (NRCan, 2018), and Statistics Canada (StatsCan, 2018b).

The figure emphasizes the relatively short sequestration period that the majority of forest products in Canada provides. This is due in large part to the parallel or "nested" structure of the forest industry, where short-lived and longer-lived products are produced from different fractions of the same trees. This ultimately means that a relatively small fraction of biomass is sequestered—less than 20% endures for more than a century, and there is very little recycling of any category outside of pulp and paper. Wood used for energy (*arrows* radiating in at 100, 290) has no sequestration value time associated with product and thus can be seen as an almost instantaneous change from growing trees to atmospheric emission.

The figure also indicates losses to landfill and production losses, which may or may not take the form of atmospheric emissions; recent work has suggested that decomposition rates can be very slow in modern landfill sites (O'Dwyer et al., 2018). In other cases, advanced decay, incineration, or energy recovery may accelerate the release of carbon into the atmosphere.

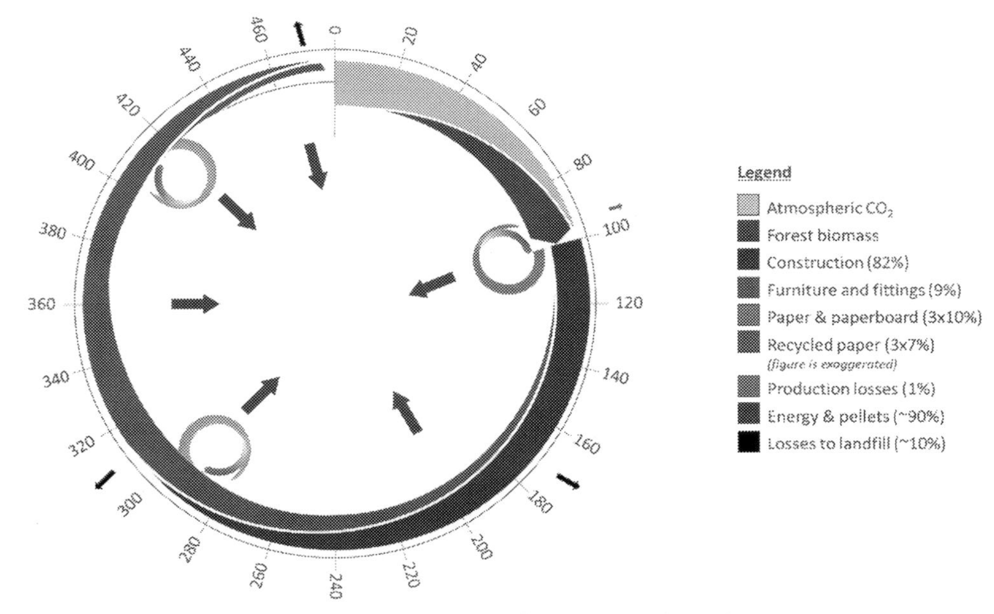

Fig. 2 A "series" approach to forest products manufacture *Image by author.*

Fig. 1 illustrates two important issues that must be addressed in order to advance the concept of the circular bioeconomy. First, the use of clean or "virgin" forest biomass has allowed individual industries within the forest sector to optimize their production based on homogenous inputs. Using these relatively clean and predictable fiber sources has permitted significant mechanization and ultimately, streamlining of the forest industry, particularly in developed countries like Canada.

Because each part of the industry is essentially working in isolation when it comes to product development, it follows that society requires different fibers for each application. Outside of the pulp and paper sector, most fiber that is used in construction or in other manufacturing is very difficult to recycle (Teshnizi, 2019). Because these fibers are not being recycled, more carbon needs to be cycled through the forest ecosystem to meet a variety of different needs, and carbon sequestration cannot be optimized.

Fig. 2 provides a conceptual representation of how the circular bioeconomy might change the forest sector. Existing fiber flows in the Canadian forest sector were placed into a model and these flows were modified to introduce circularity. Instead of a series of parallel production pathways, the model diverts the majority of biomass (80%) taken from the forest into long-lasting construction-oriented products. In this particular model, a constraint that was introduced was that all biomass must be recycled or reused at least once, and the amount of biomass delivered to landfill was further constrained to just 10% over the entire cycle. Thus there are (at least) two sequential cycles of long-term applications, which utilize the majority of the biomass. By implementing these constraints, biomass remains in solid form for an extended period—potentially over 300 years of use.

At the same time, the model recognizes that demand for paper products and packaging will remain. Instead of sourcing all paper production from virgin fiber, two thirds of the fiber is sourced from recycled material that has been used in long-term durable applications. Expanded energy production is also associated with damaged or decayed biomass recovered from these systems; almost all biomass eventually would be utilized for energy production, but in some cases this would not take place until after almost 300 years' worth of service in other applications.

The advantage of this approach over the "nested" forest products approach is threefold. By adding additional processing steps using the same fiber, new labor opportunities are created. The potential to sequester carbon is vastly increased; carbon atoms could spend up to 480 years in a sequestered form (either growing biomass or a series of forest products) across this cycle. The majority of the biomass is ultimately used for energy production after a series of uses. This approach essentially sequesters carbon in harvested wood products for up to 80% of the total cycle, slowing the cycling of carbon through the atmosphere and increasing the total storage of carbon. Finally, the use of each wood fiber in three or more applications extends the supply of biomass to support a circular bioeconomy, and increases the potential contribution that it could make to the overall GDP.

4. Barriers to the circular bioeconomy

The circular economy approach to the forest sector can only be implemented by overcoming a series of challenges. A serious challenge is the timescale involved. It would take literally centuries to achieve all of the benefits of a true circular bioeconomy within the forest sector.

The example of the Canadian forest sector highlights one challenge with circular bioeconomies, and that is the reliance that downstream products have on upstream operations. As an example: when Canada's sawmill industry experiences a downturn, this affects the amount of chips available to secondary manufacturers such as pulp and paper producers. Most often, the disruption that the bioeconomy will experience will be related to changes in consumer demand—product lines become obsolescent and new technologies can dramatically change sectors. This has been most notably observed in the newsprint industry, where the internet and digital media led to a major collapse in demand for mechanical pulps (Latta et al., 2016), leading to a cascading change in the forest products sector. Development of a circular bioeconomy thus requires planning for obsolescence in various product lines and anticipating potential alternative production pathways.

Circular bioeconomies are also challenged by current practices in the way that products are used. In the forest sector, wood used for construction makes heavy use of metal fastenings, petroleum-based resins, and other nonorganic materials. Designing buildings so that components can be reused a century or more into the future requires different design decisions today, such as minimizing the use of petroleum-based resins and paints. This might lead to new designs, such as all wood-based products in a wall panel that could be pulled out of service in one structure and reused in another. The emergence of structural insulated panels, a composite building material used in modularized construction, is an example of the type of design philosophy that might be applied here, with the directive that the panels be sourced solely from wood and natural materials (Asdrubali et al., 2017; Leng and Pan, 2019). To support the circular bioeconomy, new products—particularly for construction—will need to be designed with recyclability in mind.

The circular bioeconomy can only work if the anticipated benefits of long-term carbon sequestration, reduced pressure on ecosystems, and increased employment (particularly at local levels) are realized over time. Measuring these impacts and identifying places where the system is not working is a major challenge. Life cycle assessment (LCA) is a powerful tool that can be used to better understand the impacts of present-day decisions on future environmental, economic, and social indicators (D'Amato et al., 2017, 2020). As a scientific tool, however, LCA is constantly evolving, designed for single applications (and not for comparative work), and is not well suited to monitoring different aspects of the economy. A simplified tool that could track transactions and cumulative benefits, within a well-established framework informed by LCA, would be preferable. A range of tools that might be applicable in this case are currently being developed (Cabernard et al., 2019; Zorpas, 2020). It will be important to evaluate impacts.

A challenge that is regularly raised is the potential loss of circularity that may come with the adoption of industrial practices (Szekacs, 2017; Cattaneo et al., 2018). One of the most challenging issues is designing for future recycling purposes. Engineers design products to perform in an optimum fashion while in use; few companies are willing to compromise on current performance in order to meet goals of circularity. In a recent review of more than 20 Swedish and Finnish biorefinery operations, for example, it was found that the products being generated did not meet the vision of a circular bioeconomy (Temmes and Peck, 2020). The concept of extended producer responsibility, which attempts to integrate real environmental costs into the market price of products (McKerlie et al., 2006), is one approach that could help drive appropriate decision-making for a circular economy. In order to shift production toward a true circular bioeconomy, incentives may be required to help the sector adopt these practices.

5. Ways forward

Development of the circular economy in Canada's forest sector could provide significant benefits, as measured by fiber availability and perhaps in terms of employment. A reorganized forest sector meets the basic principles of the circular economy: its products can be reused and ultimately returned to nature, acting as the basis for a new series of inputs.

This chapter has identified four major recommendations that can better support the concept of the circular bioeconomy, based on the example of the Canadian forest sector:

1. Plan for obsolescence. The circular bioeconomy will rely heavily on sequential operations, and if one component of the circular bioeconomy fails, the entire system may fail. Thus it is essential that the circular bioeconomy anticipate changes in consumer demand and plan for these changes.
2. Design for recyclability. It will be essential that future bioproducts are designed and used in such a way that they can be recycled. Without intelligent design and guidelines for the use of these materials, the cost of recovery and reuse may be too high to facilitate circularity.
3. Evaluate impacts. It will be critical to be able to evaluate the performance of circular bioeconomies in order to ensure that they are performing in a sustainable fashion, and to highlight opportunities to improve the system to increase benefits including carbon sequestration and localized employment.
4. Reward performance. The benefits of the circular bioeconomy do not manifest in year 1, but rather are accrued over centuries of use. Thus, it will be very important that current-day sector players make the right decisions to allow the promise of the circular bioeconomy to be realized. This can be achieved through regulation (such as extended producer responsibility) and through incentives.

The circular bioeconomy will be built on long-lasting products, shifting the focus from rapid production of goods to more effective use of labor. Application of circularity to areas like forestry could provide more diverse jobs across a range of products; a single fiber could provide multiple jobs over decades, as material is used and reused in various applications. The circular bioeconomy will greatly extend biomass availability while increasing carbon sequestration in harvested wood products. Finally, the circular bioeconomy will allow for much expanded production of bioenergy, which ultimately will help reduce demand on fossil fuels while working within the natural biogenic carbon cycle.

Ultimately, adopting the circular economy will require massive change over a long timeframe. Work needs to be done now to establish the appropriate products and designs to allow future generations to take advantage of the circular approach.

References

Adams, B., 1998. Timescapes of Modernity. Routledge, London, UK.

Aguilar, F., Hartkamp, R., Mabee, W.E., Skog, K.E., 2012. Wood energy markets, 2011-2012. In: Fonseca, M., Posio, J. (Eds.), Forest Products Annual Market Review 2011–2012. United Nations, New York and Geneva, pp. 95–106.

Aguilar, F., Abt, K., Glavonjic, B., Mabee, W.E., 2019. Wood Energy Markets, 2018–2019. UNECE/FAO Forest Products Annual Market Review, 2018–19. United Nations, New York and Geneva, pp. 86–97.

Asdrubali, F., Ferracuti, B., Lombardi, L., Guattari, C., Evangelisti, L., Grazieschi, G., 2017. A review of structural, thermo-physical, acoustical, and environmental properties of wooden materials for building applications. Build. Environ. 114, 307–332.

Bakker, K., Bridge, G., 2006. Material worlds? Resource geographies and the 'matter of nature'. Prog. Hum. Geogr. 30 (1), 5–27.

Barquet, K., Jarnberg, L., Rosemarin, A., Macura, B., 2020. Identifying barriers and opportunities for a circular phosphorus economy in the Baltic Sea region. Water Res. 171, 10.

Battista, F., Frison, N., Pavan, P., Cavinato, C., Gottardo, M., Fatone, F., et al., 2020. Food wastes and sewage sludge as feedstock for an urban biorefinery producing biofuels and added-value bioproducts. J. Chem. Technol. Biotechnol. 95 (2), 328–338.

Beckinghausen, A., Odlare, M., Thorin, E., Schwede, S., 2020. From removal to recovery: an evaluation of nitrogen recovery techniques from wastewater. Appl. Energy 263, 15.

Braun, B., 2002. The Intemperate Rainforest: Nature, Culture and Power on Canada's West Coast. University of Minnesota Press, Minneapolis, USA.

Braungart, M., McDonough, W., Bollinger, A., 2007. Cradle-to-cradle design: creating healthy emissions—a strategy for eco-effective product and system design. J. Clean. Prod. 15 (13–14), 1337–1348.

Cabernard, L., Pfister, S., Hellweg, S., 2019. A new method for analyzing sustainability performance of global supply chains and its application to material resources. Sci. Total Environ. 684, 164–177.

Cattaneo, C., Marull, J., Tello, E., 2018. Landscape agroecology. The dysfunctionalities of industrial agriculture and the loss of the circular bioeconomy in the Barcelona Region, 1956–2009. Sustainability 10 (12), 22.

CFS, 2016. Statistical Data—Forest Industry Employment Ottawa, Canada: Canadian Forest Service. Available from https://cfs.nrcan.gc.ca/statsprofile/employment/ca.

Chen, J.X., Colombo, S.J., Ter-Mikaelian, M.T., Heath, L.S., 2008. Future carbon storage in harvested wood products from Ontario's crown forests. Can. J. For. Res. 38 (7), 1947–1958.

Chen, J.X., Colombo, S.J., Ter-Mikaelian, M.T., Heath, L.S., 2010. Carbon budget of Ontario's managed forests and harvested wood products, 2001–2100. For. Ecol. Manag. 259 (8), 1385–1398.

Chen, J.X., Colombo, S.J., Ter-Mikaelian, M.T., Heath, L.S., 2014. Carbon profile of the managed Forest sector in Canada in the 20th century: sink or source? Environ. Sci. Technol. 48 (16), 9859–9866.

Chen, J.X., Ter-Mikaelian, M.T., Yang, H.Q., Colombo, S.J., 2018. Assessing the greenhouse gas effects of harvested wood products manufactured from managed forests in Canada. Forestry 91 (2), 193–205.

Christensen, N.L., 2014. An historical perspective on forest succession and its relevance to ecosystem restoration and conservation practice in North America. For. Ecol. Manag. 330, 312–322.

Christensen, N.L., Bartuska, A.M., Brown, J.H., Carpenter, S., Dantonio, C., Francis, R., et al., 1996. The report of the ecological society of America committee on the scientific basis for ecosystem management. Ecol. Appl. 6 (3), 665–691.

Chung, E., 2018. In: Many Canadians Are Recycling Wrong, and It's Costing Us Millinos. CBC News. April 6.

D'Amato, D., Droste, N., Allen, B., Kettunen, M., Lähtinen, K., Korhonen, J., et al., 2017. Green, circular, bio economy: a comparative analysis of sustainability avenues. J. Clean. Prod. 168, 716–734.

D'Amato, D., Gaio, M., Semenzin, E., 2020. A review of LCA assessments of forest-based bioeconomy products and processes under an ecosystem services perspective. Sci. Total Environ. 706, 11.

Duchesne, L.C., Wetzel, S., 2003. The bioeconomy and the forestry sector: changing markets and new opportunities. For. Chron. 79 (5), 860–864.

Dymond, C.C., 2012. Forest carbon in North America: annual storage and emissions from British Columbia's harvest, 1965–2065. Carbon Balanc. Manag. 7 (8). https://doi.org/10.1186/1750-0680-7-8.
ECCC, 2018. Solid Waste Diversion and Disposal Ottawa, Canada: Environment and Climate Change Canada. Available from: https://www.canada.ca/en/environment-climate-change/services/environmental-indicators/solid-waste-diversion-disposal.html#DSM.
Ehrenfeld, J., 2004. Industrial ecology: a new field or only a metaphor? J. Clean. Prod. 12 (8–10), 825–831.
FAOStat, 2020. FAOStat Forestry Database Rome, Italy: United Nations. Available from http://www.fao.org/faostat/en/.
Farzad, S., Mandegari, M.A., Guo, M., Haigh, K.F., Shah, N., Gorgens, J.F., 2017. Multi-product biorefineries from lignocelluloses: a pathway to revitalisation of the sugar industry? Biotechnol. Biofuels 10, 24.
NRCan, 2020. Forest Bioeconomy, Bioenergy and Bioproducts Ottawa, Canada: Natural Resources Canada. Available from https://www.nrcan.gc.ca/our-natural-resources/forests-forestry/forest-industry-trade/forest-bioeconomy-bioenergy-bioproducts/13315.
FPAC, 2018. Economic Backbone Map Ottawa, Canada: Forest Products Association of Canada. Available from http://www.fpac.ca/canadian-forestry-industry/economy/.
Geng, Y., Zhu, Q.H., Doberstein, B., Fujita, T., 2009. Implementing China's circular economy concept at the regional level: a review of progress in Dalian, China. Waste Manag. 29 (2), 996–1002.
Greer, K., Martins, C., White, M., Pittelkow, C.M., 2020. Assessment of high-input soybean management in the US Midwest: balancing crop production with environmental performance. Agric. Ecosyst. Environ. 292, 10.
Grim, R.G., Huang, Z., Guarnieri, M.T., Ferrell, J.R., Tao, L., Schaidle, J.A., 2020. Transforming the carbon economy: challenges and opportunities in the convergence of low-cost electricity and reductive CO2 utilization. Energy Environ. Sci. 13 (2), 472–494.
Halkos, G., Petrou, K.N., 2019. Analysing the energy efficiency of EU member states: the potential of energy recovery from waste in the circular economy. Energies 12 (19), 32.
Heimann, T., 2019. Bioeconomy and SDGs: does the bioeconomy support the achievement of the SDGs? Earth Future 7 (1), 43–57.
Jarosch, L., Zeug, W., Bezama, A., Finkbeiner, M., Thran, D., 2020. A regional socio-economic life cycle assessment of a bioeconomy value chain. Sustainability 12 (3), 15.
Jarre, M., Petit-Boix, A., Priefer, C., Meyer, R., Leipold, S., 2020. Transforming the bio-based sector towards a circular economy—what can we learn from wood cascading? Forest Policy Econ. 110, 12.
Kaddoura, M., Kambanou, M.L., Tillman, A.M., Sakao, T., 2019. Is prolonging the lifetime of passive durable products a low-hanging fruit of a circular economy? A multiple case study. Sustainability 11 (18), 22.
Kurth, V.J., D'Amato, A.W., Bradford, J.B., Palik, B.J., Looney, C.E., 2020. Assessing the ecological impacts of biomass harvesting along a disturbance severity gradient. Ecol. Appl. 30 (2), 11.
Lafleur, B., Fenton, N.J., Simard, M., Leduc, A., Pare, D., Valeria, O., et al., 2018. Ecosystem management in paludified boreal forests: enhancing wood production, biodiversity, and carbon sequestration at the landscape level. For Ecosyst. 5, 14.
Latta, G.S., Plantinga, A.J., Sloggy, M.R., 2016. The effects of internet use on global demand for paper products. J. For. 114 (4), 433–440.
Lemire, P.O., Delcroix, B., Audy, J.F., Labelle, F., Mangin, P., Barnabe, S., 2019. GIS method to design and assess the transportation performance of a decentralized biorefinery supply system and comparison with a centralized system: case study in southern Quebec, Canada. Biofuels Bioprod. Biorefin. 13 (3), 552–567.
Leng, W.Q., Pan, B., 2019. Thermal insulating and mechanical properties of cellulose nanofibrils modified polyurethane foam composite as structural insulated material. Forests 10 (2), 12.
Liu, J.X., Sleeter, B.M., Zhu, Z.L., Loveland, T.R., Sohl, T., Howard, S.M., et al., 2020. Critical land change information enhances the understanding of carbon balance in the United States. Glob. Chang. Biol. 26, 3920–3929.
Matiuti, M., Hutu, I., Diaconescu, D., Sonea, C., 2017. Rural pole for competitivity: a pilot project for circular bioeconomy. J. Environ. Prot. Ecol. 18 (2), 802–808.
McCormick, K., Kautto, N., 2013. The bioeconomy in Europe: an overview. Sustainability 5 (6), 2589–2608.

McKerlie, K., Knight, N., Thorpe, B., 2006. Advancing extended producer responsibility in Canada. J. Clean. Prod. 14 (6–7), 616–628.

Mirkovic, I.B., Medek, G., Bolanca, Z., 2019. Ecologically sustainable printing: aspects of printing materials. Teh Vjesn 26 (3), 662–667.

Mohan, S.V., Nikhil, G.N., Chiranjeevi, P., Reddy, C.N., Rohit, M.V., Kumar, A.N., et al., 2016. Waste biorefinery models towards sustainable circular bioeconomy: critical review and future perspectives. Bioresour. Technol. 215, 2–12.

Nayha, A., 2019. Transition in the Finnish forest-based sector: company perspectives on the bioeconomy, circular economy and sustainability. J. Clean. Prod. 209, 1294–1306.

NFD, 2018. National Forestry Database Ottawa, Canada: National Forestry Database. Available from http://nfdp.ccfm.org/supply/background_e.php.

NRCan, 2018. Statistical Data Ottawa, Canada: Natural Resources Canada. Available from https://cfs.nrcan.gc.ca/statsprofile.

O'Dwyer, J., Walshe, D., Byrne, K.A., 2018. Wood waste decomposition in landfills: an assessment of current knowledge and implications for emissions reporting. Waste Manag. 73, 181–188.

Pannicke, N., Gawel, E., Hagemann, N., Purkus, A., Strunz, S., 2015. The political economy of fostering a wood-based bioeconomy in Germany. Ger. J. Agric. Econ. 64 (4), 224–243.

Pieratti, E., Paletto, A., De Meo, I., Fagarazzi, C., Migliorini, M.G.R., 2019. Assessing the forest-wood chain at local level: a multi-criteria decision analysis (MCDA) based on the circular bioeconomy principles. Ann. For. Res. 62 (1), 123–138.

Robinson, J., 2004. Squaring the circle? Some thoughts on the idea of sustainable development. Ecol. Econ. 48 (4), 369–384.

Sadhukhan, J., Martinez-Hernandez, E., Murphy, R.J., Ng, D.K.S., Hassim, M.H., Ng, K.S., et al., 2018. Role of bioenergy, biorefinery and bioeconomy in sustainable development: strategic pathways for Malaysia. Renew. Sust. Energ. Rev. 81, 1966–1987.

Sagues, W.J., Assis, C.A., Hah, P., Sanchez, D.L., Johnson, Z., Acharya, M., et al., 2020. Decarbonizing agriculture through the conversion of animal manure to dietary protein and ammonia fertilizer. Bioresour. Technol. 297, 5.

Saleem, S., Levison, J., Parker, B., Martin, R., Persaud, E., 2020. Impacts of climate change and different crop rotation scenarios on groundwater nitrate concentrations in a Sandy aquifer. Sustainability 12 (3), 25.

Sanchez, B., Rausch, C., Haas, C., 2019. Deconstruction programming for adaptive reuse of buildings. Autom. Constr. 107, 13.

Skog, K.E., Nicholson, G.A., 1998. Carbon cycling through wood products: the role of wood and paper products in carbon sequestration. Forest Prod J. 48 (7–8), 75–83.

Skog, K.E., Pingoud, K., Smith, J.E., 2004. A method countries can use to estimate changes in carbon stored in harvested wood products and the uncertainty of such estimates. Environ. Manag. 33, S65–S73.

Skrinjaric, T., 2020. Empirical assessment of the circular economy of selected European countries. J. Clean. Prod. 255, 17.

Staffas, L., Gustavsson, M., McCormick, K., 2013. Strategies and policies for the bioeconomy and bio-based economy: an analysis of official national approaches. Sustainability 5 (6), 2751–2769.

Stahel, W., Reday-Mulvey, G., 1981. The Potential for Substituting Manpower for Energy. Vantage Press, New York, USA, p. 116.

StatsCan, 2018a. Table 36-10-0434-01: Gross Domestic Product (GDP) at Basic Prices, by Industry. Available from https://www150.statcan.gc.ca/t1/tbl1/en/tv.action?pid=3610043401.

StatsCan, 2018b. Human Activity and the Environment 2017: Forests in Canada Ottawa, Canada: Staistics Canada. Available from https://www150.statcan.gc.ca/n1/pub/16-201-x/16-201-x2018001-eng.htm.

Szekacs, A., 2017. Environmental and ecological aspects in the overall assessment of bioeconomy. J. Agric. Environ. Ethics 30 (1), 153–170.

Temmes, A., Peck, P., 2020. Do forest biorefineries fit with working principles of a circular bioeconomy? A case of Finnish and Swedish initiatives. Forest Policy Econ. 110, 12.

Teshnizi, Z., 2019. Vancouver pre-1940 houses: a cache for old-growth forest wood. J. Cult. Herit. Manag. Sustain. Dev. 10 (1), 41–51.

Thiffault, E., Hannam, K.D., Quideau, S.A., Pare, D., Belanger, N., Oh, S.W., et al., 2008. Chemical composition of forest floor and consequences for nutrient availability after wildfire and harvesting in the boreal forest. Plant Soil 308 (1–2), 37–53.

Unay-Gailhard, I., Bojnec, S., 2019. The impact of green economy measures on rural employment: green jobs in farms. J. Clean. Prod. 208, 541–551.

Valls, C., Cusola, O., Vidal, T., Torres, A.L., Roncero, M.B., 2019. A straightforward bioprocess for a cleaner paper decolorization. J. Clean. Prod. 236, 10.

Viaggi, D., 2015. Research and innovation in agriculture: beyond productivity? Bio-based Appl. Econ. 4 (3), 279–300.

Wainaina, S., Awasthi, M.K., Sarsaiya, S., Chen, H.Y., Singh, E., Kumar, A., et al., 2020. Resource recovery and circular economy from organic solid waste using aerobic and anaerobic digestion technologies. Bioresour. Technol. 301, 14.

Watanabe, C., Naveed, N., Neittaanmaki, P., 2019. Digitalized bioeconomy: planned obsolescence-driven circular economy enabled by co-evolutionary coupling. Technol. Soc. 56, 8–30.

Winkler, H., 2011. Closed-loop production systems—a sustainable supply chain approach. CIRP J. Manuf. Sci. Technol. 4 (3), 243–246.

Yufang, W., Yanqing, Z., 2011. Study on the countermeasures of forest resource reserve. Energy Procedia 5, 529–534.

Zhen, G., Qieyi, L., Xiaoxu, W., 2011. On development model based on intra-country cyclic economy under low-carbon economy for Northeast China. Energy Procedia 5, 1553–1557.

Zhen, W., Qin, Q.D., Kuang, Y.Q., Huang, N.S., 2017. Investigating low-carbon crop production in Guangdong Province, China (1993-2013): a decoupling and decomposition analysis. J. Clean. Prod. 146, 63–70.

Zimmerman, W., Morkbak, A.L., 1997. Enzymatic deinking of offset waste paper printed with vegetable oil based ink. Abstr. Pap. Am. Chem. Soc. 213, 64. CELL.

Zorpas, A.A., 2020. Strategy development in the framework of waste management. Sci. Total Environ. 716, 13.

CHAPTER 5

Circular economy and financial performances of European SMEs

Cristina Mocanu[a,b], Eva Militaru[a], Ana Maria Zamfir[a], and Monica Mihaela Maer-Matei[a,c]
[a]National Research Institute for Labour and Social Protection, Bucharest, Romania
[b]University of Bucharest, Bucharest, Romania
[c]Bucharest University of Economic Studies, Bucharest, Romania

1. Introduction

Given the increasing attention on environmental protection, the circular business model has become of great interested for scholars and companies. The circular economy relies on the principle of material balance (Kneese et al., 1970), referring to the fact that circulating matter and energy can reduce the consumption of new inputs (Andersen, 2007). As a circular business model is meant to address both economic growth and sustainability purposes, we are interested in investigating whether there is evidence of positive impact on the performance of firms that have adopted practices related to circular economy. We build on the assumption that the decision to undertake activities that contribute to a circular economy is endogenous to expected performances. Therefore the empirical analysis would suffer from biases and incorrect conclusions if self-selection was not corrected in order to account for unobservable characteristics that could guide firms' decisions to engage in such activities. In order to explore our assumptions, we analyze company level information from the Flash Eurobarometer 441 dataset, focusing on actions relating to optimizing water usage.

Water protection represents a European priority and is targeted by various strategies, pieces of legislation, and programs. Early European water legislation focused on controlling the pollution with dangerous substances of drinking water, fishing waters, bathing waters, and groundwater. In 1991, the policy framework addressed the problems of biological wastewater and pollution by nitrates from agriculture. Only in 1996 was pollution from industrial installations addressed in European policy. The EU Water Framework Directive was adopted in 2000 and includes various actions targeting cleaning polluted waters and keeping the waters clean. Concerns regarding the fact that industrial installations can emit a significant amount of pollution into water led to the adoption of several pieces of EU legislation in this respect that are currently in force. For example, the Industrial Emission Directive establishes principles for the control of industrial installations for

Circular Economy and Sustainability, Volume 1
https://doi.org/10.1016/B978-0-12-819817-9.00024-7

environment protection by taking into account both related costs and benefits. Additionally, a directive adopted in 2009 imposed measures on storage and distribution of petrol. Furthermore, industrial activities represent an important cause of water scarcity, a problem that can also be due to poor water quality caused by industrial pollution.

So, given that economic activities are an important source of water pollution and scarcity, implementing circular economy principles in the field of water consumption could be important as it aims to reduce the overall freshwater consumption and wastewater production by various recycle and reuse possibilities. Changing practices and equipment of companies can lead to water savings through water consumption reduction and increase of internal reuse. Our study aims to explore decisions of companies with respect to the adoption of water-related circular economy practices and the way their financial performances are influenced by such decisions. This chapter starts with a review of the most relevant studies regarding the impacts of circular economy practices on firms' performances and continues with a description of the applied methodology; the next section discusses the obtained results and the chapter ends with our conclusions.

2. Literature review

As the awareness of the impact of economic activities on the environment increased, employing circular economy strategies became desirable for companies, but also studying the benefits of such strategies became a central topic for a developing body of research. The benefits of implementing circular economy strategies were analyzed by taking into account both the impact on the environment and the impact on firms' performances. But putting different endeavors into the same equation to optimize resources usage, as well as the impact of production process on the environment, proved to be quite a difficult mission (Horbach and Rammer, 2019). Some eco-innovations have higher costs and they need longer periods of time to produce effects on different indicators of firms' performances, while policy decisions in order to increase demand for eco-products are necessary in order to create and consolidate markets (Soltmann et al., 2015).

Even if research on this topic is still at the beginning, the literature already made available seems to build on a consensus that employing circular economy strategies sustains a pathway to consolidate economic performances among companies, but also to access new economic niches that are constantly emerging (Endrikat et al., 2014; Ghisetti, 2018; Moric et al., 2020), even if there are papers with completely neutral or negative findings. Analysis on the links between circular economy practices and firms' performances were mainly based on quantitative techniques, and a significant number of papers have emphasized the importance of endogenous factors (perceptions, culture, awareness, etc.) that moderate the effects (Prashar and Sunder, 2020).

Companies interested in adopting circular economy strategies usually employ not only one strategy, but a mix of them, and target different activities in order to reach both the aims of implementing circular economy practices and of benefiting financially from their strategy (Zamfir et al., 2017; Mura et al., 2020). Reducing costs by optimizing the use of resources or reusing waste or water can lead to increased profits, or to consolidating positions in the newly emerging markets (Moric et al., 2020). Improving resource efficiency has proved to be conducive to positive returns and profitability (Ghisetti and Rennings, 2014; Rexhauser and Rammer, 2014), while other strategies are still under debate.

Effects of other eco-innovations or environment regulations are still questioned and provide mixed results. Some papers considering the potential conflict between competitiveness and the environment pointed out to the importance of environmental regulations in improving firms' performances, while others considered regulations as barriers to profitability and competitiveness (Rexhauser and Rammer, 2014). Even if extensive analyses were carried out in relation to the Porter hypothesis that assumes a positive impact of environmental regulation on firms' competitiveness (Porter and Van der Linde, 1995), evidence supporting or rejecting its validity continue to appear, as innovations and regulations are strongly linked with production processes, and thus, to the characteristics of economic sectors (Ghisetti and Rennings, 2014; Rexhauser and Rammer, 2014; Horbach, 2018; Horbach and Rammer, 2019) and, more generally, to the economic development of a specific region or country.

One of the previous studies carried out by the authors made use of the same dataset of Flash Eurobarometer 441 (European Commission, 2016) in order to analyze the links between firms' economic performances and circular economy activities (Zamfir et al., 2017). The results showed that economic performances of companies are influenced by a mix of factors, including country, sector of activity, size, importance of R&D activities, but also the share of turnover invested in the circular economy (Zamfir et al., 2017). If some of the factors are exogenous, the last two are largely endogenous and linked with a managerial vision on the near and future developments of both companies and economies. Moreover, there are studies emphasizing the need for a threshold for investments in circular economy eco-innovations of more than 10% of a company's revenues in order to provide real benefits among SMEs (Demirel and Danisman, 2019).

Papers aiming to analyze the literature in the field and to integrate potential contradictory findings raised a new question—namely, for whom the circular economy sustains financial performance and for whom it does not (Telle, 2006; Ghisetti, 2018)—emphasizing that there are winners and losers in the process, as circular economy innovations have different costs and affect the production processes in different ways. Thus, the focus moves to how to develop adequate policy and financial support in order to sustain the transition to a circular business model, tailoring the policy measures to the features of those that need the most (Rizos et al., 2016).

Studies addressing the links between different types of circular economic strategies adopted by companies and their economic performances are even scarcer and offer even more puzzling results. Among them, studies addressing the influence of optimizing water usage on firms' performances are not very numerous. There are findings pointing out to a positive link between various mixes of strategies and firms' economic performances—using renewable energy and minimizing waste being the most prevalent strategies among European SMEs (Zamfir et al., 2017). These strategies are, to a certain extent, already translated into legal requirements. But for a small cluster of countries—the United Kingdom, Hungary, and Slovakia—optimizing water usage seems to be linked with better financial performances of companies (Zamfir et al., 2017). However, on the contrary, other papers have evidenced a negative correlation between optimizing water usage and firms' turnover development (Horbach, 2018).

The water industry has made important reforming efforts in order to increase both its productivity and efficiency, but the issues of ownership or monopoly in the sector have proved to be an obstacle in providing clear conclusions. Nevertheless, activities aiming to reduce the environmental impact of the sector and wastewater activities could have a significant impact on the sector's efficiency, making them an important subject for further discussion and analysis to produce better-informed policy decisions (Abbott and Cohen, 2009).

Evidences on the progress of achieving environmental sustainability through implementing sustainable management of water are rather limited, even if specific objectives and targets have been established globally and awareness of the fundamental importance of water is high (UNESCO , 2015). The public debate and policy objectives seem to be in favor of adopting many more environmental regulations. So, the present study aims to produce knowledge on how different strategies can influence companies' performances, with focus on strategies for optimizing water usage.

3. Methodology

We build our analysis on microdata from Flash Eurobarometer 441 "European SMEs and the Circular Economy." In 2016, 10,618 companies from the manufacturing, industrial, retail, and service sectors covering the 28 EU member states were interviewed with respect to the adoption of circular economy practices. Our analysis is focused on whether companies have re-planned the way water is used to minimize usage and maximize reuse.

Our methodological approach is built on the assumption that the decision to undertake water optimization strategies is endogenous to expected performance, and therefore, the consequence of not accounting for selection endogeneity is obtaining biased estimates of the effects of strategic choices. In light of this, our proposed methodology is based on four main steps. The first step concerns the estimation of a binary strategy choice equation on the decision to optimize water usage. The second one deals with the modeling of firms' financial performances—taking into account, and if necessary correcting for, selection biases—for firms either adopting water usage optimization strategies or not. The third step implies the estimation of counterfactuals for the financial outcomes if the firm has adopted the opposite strategy, based on the corrected outcome equations resulting from Step 2. And finally, the fourth step concerns the classification of firms based on the relationship between strategic choice and effects on financial performance, judging the effects by comparing with the counterfactual outcomes. In what follows, we describe in detail the theory underlying our methodological approach.

So, first, in order to estimate the water usage strategy choice equation, let us start from the decision—the point at which the manager of a firm/SME has to choose between two possible strategies, in other words he or she has to make a binary choice $\{D_0, D_1\}$:

D_1—adopt water usage optimization strategies,

D_0—do not adopt water usage optimization strategies.

The strategy choice is dependent on a number of observable and unobservable characteristics of firms, and can be estimated through a probit regression.

Second, no matter what the strategy is, there will be economic outcomes at firm level. From all the possible measurable financial outcomes, due to data availability, we use as outcome or performance indicator the turnover evolution during the most recent year, which actually represents 1 year after the decision has been taken. We will denote the outcome variable as π, and for each firm it can have two potential values: π_0 if D_0 is chosen and π_1 if D_1 is chosen.

Third, in order to evaluate the impact of a chosen strategy on the outcome variable, we have to estimate what would had been the outcomes if the other strategy had been adopted; in other words, to estimate the counterfactual and calculate the difference $\pi_1 - \pi_0$ (the strategy or treatment effect), while we observe only one outcome for each firm.

In estimating the effects, we follow the approach proposed by Hamilton and Nickerson (2003), based on Heckman (1974, 1979) and Lee (1978). According to this, for each firm that adopts decision D_1 and for which we observe π_1, we have to estimate $E(\pi_0 | D_1)$, which is the expected counterfactual outcome for firms that adopted D_1 if they had adopted D_0. And similarly, for each firm that adopts decision D_0 and for which we observe π_1, we have to estimate $E(\pi_1 | D_0)$, which is the counterfactual outcome for firms that adopted D_0 if they had adopted D_1.

The estimation technique for the counterfactual outcomes depends on whether the choice is exogenous or not. In the case of an exogenous choice, the average effect of a strategy choice can be estimated by simple univariate regression.

But, if there is endogeneity in a firm's decision, meaning that the firm's choice is based on expected financial outcomes, but not all variables that affect strategy choice and performance are observable, the estimations are not so straightforward. Estimating the effects of strategy choice on performance via simple OLS (ordinary least squares) leads to biased estimations. The estimations proposed by Heckman and Lee deal with this bias and assume that the strategy choice depends on three factors: the expected net benefit of D_1 over D_0, covariates that influence strategy choice but do not influence outcomes, and unobservable factors. Under several assumptions regarding the distribution of errors and the covariance among errors in the models of each strategy choice, they show that the expected values of the error terms in the two equations can be written as:

$$E(\varepsilon_{1i} \mid D_1) = -\sigma_{u1}\varphi[X_i\beta + Z_i\delta]/\Phi[X_i\beta + Z_i\delta] = -\sigma_{u1}\lambda_{1i}, \tag{1}$$

$$E(\varepsilon_{0i} \mid D_0) = \sigma_{u0}\varphi[X_i\beta + Z_i\delta]/(1 - \Phi[X_i\beta + Z_i\delta]) = \sigma_{u0}\lambda_0, \tag{2}$$

where $\varphi[.]$ is the normal density and $\Phi[.]$ is the cumulative normal distribution, X is the set of firms' characteristics that influence strategy choice and also outcome, Z is the set of firms' characteristics that influence strategy choice but not outcome, σ_{u1} and σ_{u0} are the covariance between the error terms of the outcome and strategy choice equations, and λ is known as the inverse Mills ratio.

When the choice is exogenous, $\sigma_{u1} = \sigma_{u0}$ and is equal to zero, and so are the expected values of the error terms shown in relations (1) and (2). On the other hand, when the strategy choice is endogenous, the inverse Mills ratios λ can be estimated by using a probit regression on the strategy choice:

$$D_i^* = X_i\beta + Z_i\delta + u_i, \tag{3}$$

where $u_i = \gamma(\varepsilon_{1i} - \varepsilon_{0i}) + v_i$ and $\beta = \gamma(\beta_1 - \beta_0)$.

Then, the sample selection corrected equations of outcomes are the following:

$$\pi_{1i} = X_i\beta_1 - \sigma_{u1}\varphi\left[X_i\hat{\beta} + Z_i\hat{\delta}\right]/\Phi\left[X_i\hat{\beta} + Z_i\hat{\delta}\right] + e_1, \tag{4}$$

$$\pi_{0i} = X_i\beta_0 + \sigma_{u0}\varphi\left[X_i\hat{\beta} + Z_i\hat{\delta}\right]/\left(1 - \Phi\left[X_i\hat{\beta} + Z_i\hat{\delta}\right]\right) + e_{0i}. \tag{5}$$

Eqs. (4) and (5) can be estimated via OLS and the estimators will not be biased as, by construction, the expected values of the error terms e_{1i} and e_{0i} are both zero due to the inclusion of the inverse Mills ratio terms in Eqs. (4) and (5).

So, finally, based on the proven existence of endogeneity and on the three equations, further on, the observable variables that influence strategy choice and affect or not outcomes, as well as the magnitude and sign of these effects can be discussed. Also, the implications of the estimated covariance terms σ_{u1} and σ_{u0}, coefficients of the inverse Mills ratio terms in Eqs. (4) and (5) can be considered. Thus, when $\sigma_{u1} < 0$, there is a positive selection in strategy D_1, meaning that firms choosing this strategy have above average performance using the chosen strategy, while if firms that chose D_0 had instead chosen D_1, their outcomes would have been worse. On the contrary, when $\sigma_{u1} > 0$, we can speak about a negative selection in strategic decision D_1: firms that chose this strategy have below average performance for the financial outcome used, and if the firms choosing D_0 had chosen D_1, their performance would have exceeded that of the other firms who chose D_1. The same interpretation stands for the situation when $\sigma_{u0} > 0$ or $\sigma_{u0} < 0$, but the other way around, the first case being associated with positive selection and the

Table 1 Typology of strategic choice and performance.

Absolute advantage of firms adopting D_1	Absolute advantage of firms adopting D_0
$\sigma_{u1} < 0$ $\sigma_{u0} < 0$	$\sigma_{u1} > 0$ $\sigma_{u0} > 0$
Comparative advantage	Comparative disadvantage
$\sigma_{u1} < 0$ $\sigma_{u0} > 0$	$\sigma_{u1} > 0$ $\sigma_{u0} < 0$

latter one with negative selection in strategy choice D_0. According to Maddala (1983), based on the signs of these coefficients, a typology of strategy choice and performance can be depicted, as shown in Table 1.

So, we applied the above theoretical considerations, our aim being, as presented in the steps underlying our methodological approach, first of all to test for endogeneity concerning the decision to implement water usage optimization strategies, and then to estimate the sample selection corrected outcomes for firms either adopting water usage optimization strategies or not. The results are discussed in relation to the above typology of strategic choice and performance.

4. Results and discussions

The strategic decisions of firms are based on expected outcomes—this is fundamental in management, while in econometric analysis it is a source of endogeneity. Our empirical research has focused on data collected for more than 10,000 European SMEs through the Flash Eurobarometer 441, our aim being discussion of the impact of water usage optimization strategies upon economic performances of firms. This entails a classical problem of selection-based endogeneity when (1) it is very likely that firms choose to adopt water usage optimization strategies based on their expected financial performances and (2) not all factors that affect both performance and strategy choice are observable. As a consequence, estimating the effect on financial performances of the decision to implement water optimization strategies or not without controlling for self-selection bias leads to biased estimates of treatment effects, thus seriously affecting the conclusions.

Therefore, following Heckman's approach (Heckman, 1974, 1979) to deal with the selection bias, first we have estimated a strategy choice equation on the decision to plan water usage, based on which we have constructed the inverse Mills ratio terms. And second, we estimated the financial performance equations, having as a dependent variable the variation of firms' turnover during the previous year and including the inverse Mills ratio terms among other independent variables.

Before presenting our results, we shall focus first on descriptive statistics on firms' behavior concerning the adoption of water usage optimization strategies. Out of the 10,618 firms covering all EU28 countries, only 12% have adopted strategies to optimize water usage in the last 3 years. Their strategies in this respect refer to the re-planning of the way water is used in order to minimize usage and maximize reuse. We observe that the share of firms that undertake action to optimize water usage increases with firm size; thus, around 15% of the firms with 50–250 employees have implemented water usage optimization strategies. Furthermore, economic activity is to be considered when analyzing the distribution of firms based on implementation of water usage optimization strategies. We note that more than one quarter of firms in the fields of accommodation and food service and mining and quarrying have adopted water usage optimization strategies. On the other hand, less than 10% of the firms in information and communication, professional, scientific and technical activities, or transportation and storage have undertaken water usage optimization strategies. We note that only 17% of the firms the main activity of which is water supply, sewerage, or waste management have implemented water usage optimization strategies in their own company (see Appendix 1, Tables A.1–A.6 for detailed tables with key descriptive statistics). It is also worth noting that the sample does not cover firms whose activity is in the agricultural sector.

On the other hand, financial performances of analyzed firms seem homogenous across economic activities, judging by the signs of turnover variation, and we highlight firms in financial activities as being the most successful, while in manufacturing and trade we find the most significant share of down-trending firms. Moreover, when we compare the financial performance evolutions of firms considering the dichotomous strategy implementation approach, we observe rather small differences between firms who implement and those who do not implement water usage optimization strategies (66.3%–64.1% increased turnover, but also 18.0%–16.1% decreased turnover). Also, we note that firms who implemented water usage optimization strategies tend to be less stable; they either increased or decreased their turnover in the previous year.

Therefore, following this first set of results, we could presume that the decision to undertake water usage optimization strategies could be linked to some unobserved variables, which consequently could affect the expected financial outcomes of firms.

The decision to implement water usage optimization strategies is specific to certain countries and economic activities, as the results of the probit regression shows. It is more likely that firms from France, Belgium, Luxemburg, Ireland, Portugal, and Finland will undertake water usage optimization strategies. The legislative framework in some countries could be stimulative concerning water usage optimization strategies. Firms in mining and quarrying, electricity, gas, steam, and accommodation and food service activities have a higher propensity to implement such strategies, while firms in information and communication activities have the lowest propensity for water usage optimization strategies. As expected, in general, firms where activity is intensively based on the use of water are more concerned about water usage optimization strategies, since reducing the cost of water usage leads to total cost reduction and improved competitiveness. It is worth mentioning that the R&D investment intensity of firms as % of turnover is

Table 2 Water usage optimization strategic decision, explanatory variables and signs, probit regression.

Coefficient sign	Variables
+	R&D intensity (% of turnover) Country dummies: France, Belgium, Luxemburg, Ireland, Portugal, Finland
−	Firms with 1–9 employees, firms with 10–49 employees Country dummies: Greece, Estonia, Lithuania, Slovakia, Slovenia, Bulgaria Economic activity dummies: manufacturing, water supply, sewerage, waste management, construction, wholesale and retail trade, repair of motor vehicles, transportation and storage, information and communication, financial and insurance activities, professional, scientific and technical activities, administrative and support services

Note: $P < 0.10$ for all coefficients.
Source: author's own calculations based on data from the Flash Eurobarometer 441.

a predictor of adoption of water usage optimization strategies, as you can see in Table 2 (see Appendix 2, Tables B.1 and B.2 for a detailed table on coefficients values and signs).

Based on the above probit regression we have estimated the inverse Mills ratio terms for firms who implement and those who do not implement water usage optimization strategies, according to the methodology described in the previous section of this chapter. Then, we have included the inverse Mills ratio terms thus calculated in the specific financial performance equations for each of the two strategies, along with other regressors. The estimations obtained through OLS are unbiased, as the expected values of the error terms are null due to the inclusion of the invers Mills ratio terms in both equations. The estimated turnover evolutions for both strategic groups are corrected for sample selection bias.

The analysis of the outputs of the two equations yields two types of information, as follows. First, we can detect the observed variables, which influence the turnover evolutions for firms who adopt water usage optimization strategies and for firms that do not adopt such strategies. Second, judging by the coefficient of the inverse Mills ratio in the financial performance equations, we can discuss whether firms are positively or negatively selected in the strategy they adopt.

Therefore, the observed variables that seem to have affected the turnover evolutions during the previous year, by water usage optimization strategies, are presented in the table below. The evolution of the turnover is expressed as relative variation during the previous year, measured in percentages. The most important positive effect on the turnover evolution is the value of the turnover in the previous year, which means that stronger firms with higher market shares have a greater potential to increase their businesses regardless of the strategic choice of implementing water usage optimization strategies. Furthermore, as already noted when explaining the results of the probit regression, there

Table 3 Turnover evolution of firms by water usage optimization strategy adopted, explanatory variables and signs, OLS regression.

Firms who ADOPT water usage optimization strategies	Firms who DO NOT ADOPT water usage optimization strategies
+	+
R&D intensity (% of turnover) Turnover value at the end of the previous year (natural logarithm)	Turnover value at the end of the previous year (natural logarithm) Water supply, sewerage, waste management Information and communication Financial and insurance activities Professional, scientific and technical activities Inverse Mills ratio term
–	–
Water supply, sewerage, waste management, construction, information and communication Inverse Mills ratio term	Mining and quarrying Accommodation and food service

Note: $P < 0.10$ for all coefficients.
Source: Author's own calculations based on data from the Flash Eurobarometer 441.

is a link between the R&D intensity and the adoption of water usage optimization strategies, and the propensity to spend on R&D also influences the turnover evolution at firm level for these firms. Regarding the turnover evolution, which can be explained by the economic activity of firms, we remark that carrying out activities such as water supply, sewerage, waste management, or information and communication negatively affect both the selection of firms into the strategy of implementing water usage optimization actions and the financial performances of those firms, while still being predictors of higher financial outcomes for firms who choose not to implement water usage optimization strategies (see Table 3).

The coefficients of the inverse Mills ratio terms are statistically significant in both regressions, meaning that selection into one strategy or another, as well as the financial performances of firms, are influenced alongside the observable variables highlighted above by unobservable characteristics at firm level (these could be managerial quality, organizational culture, other strategies, objectives, etc.).

Moreover, the signs of the coefficients of inverse Milles ratio terms show that, in relation to water usage optimization strategies, there are positive selections in both—adopt or not adopt. Therefore, firms who choose to adopt water usage optimization strategies have above average performance using this strategy, while firms choosing the strategy to not adopt water usage maximization actions have above average performance using their strategy as well. This situation can be defined as a comparative advantage, as firms choose the strategy which provides them a relative advantage. So, if firms that chose not to adopt water usage optimization strategies had chosen to adopt these strategies instead, their financial performance measured by the turnover evolution would have been worse than that of firms actually choosing to adopt water usage optimization strategies, and vice versa.

Our results are consistent with previous findings that indicate that circular economy practices among SMEs across EU countries are very heterogeneous, displaying important variation within and between countries (Bassi and Dias, 2019). In order to achieve success, companies adopting circular economy practices need to adopt innovative business models capable of enabling value creation mechanisms (Lieder and Rashid, 2016; Pádua Pieronia et al., 2018). Many firms still perceive circulatory economy practices as not applicable to them or too costly and risky to be adopted (Cristoni and Tonelli, 2018), and we provide additional evidence that such perceptions have a kernel of truth in them—although various policy instruments support the adoption of circular economy practices among SMEs, several challenges and barriers remain (Rizos et al., 2016; Bressanelli et al., 2019).

5. Conclusions

This chapter has explored the relation between the adoption of circular economy practices and financial performance among European SMEs. We have built our analysis on the assumption that managerial decision in general, and circular economy activities in particular, are endogenous to expected financial performances. Our empirical analysis has been concentrated on firms' behavior concerning water usage optimization strategies. We found evidence on the endogeneity of firms' decisions toward water usage optimization strategies and corrected for this using econometric techniques, as described in the methodological section. But what is more interesting is that whether they choose to implement water usage optimization strategies or not, firms have a relative advantage and their choice has, in the end, a positive influence on their turnover. Additionally, our results show that companies that have a propensity to invest in R&D are more open to adopting water usage optimization strategies. This is consistent with findings of previous studies (Bassi and Dias, 2019). Moreover, we found that, for companies adopting water usage optimization strategies, investing in R&D results in positive effects on their financial performance.

However, we have to mention some limitations of our study. First, the coverage of the Eurobarometer dataset concerning firms' characteristics or other contextual variables is rather low. As a result, due to the limited number of explanatory variables that could be included in the models, the estimated endogeneity could be overrated. Second, the SMEs' financial performances are measured as variation of turnover during the previous year, while some of the company-level benefits generated by circular business models might need longer periods of time to become visible.

A. Appendix 1

Table A.1 Percentage distribution of firms by economic activity and water usage optimization strategies implementation.

	Strategies for water usage optimization		
	Not implemented	Implemented	Total
B. Mining and quarrying	73.3	26.7	100.0
C. Manufacturing	87.5	12.5	100.0
D. Electricity, gas, steam, and air conditioning supply	80.3	19.7	100.0
E. Water supply, sewerage, waste management	83.0	17.0	100.0
F. Construction	88.3	11.7	100.0
G. Wholesale and retail trade, repair of motor vehicles	89.9	10.1	100.0
H. Transportation and storage	90.7	9.3	100.0
I. Accommodation and food service activities	74.8	25.2	100.0
J. Information and communication	93.8	6.2	100.0
K. Financial and insurance activities	87.4	12.6	100.0
M. Professional, scientific, and technical activities	90.9	9.1	100.0
N. Administrative and support service	86.6	13.4	100.0
Total	88.2	11.8	100.0

Source: Author's own calculations based on data from the Flash Eurobarometer 441.

Table A.2 Percentage distribution of firms by size and water usage optimization strategies implementation.

	Strategies for water usage optimization		
	Not implemented	Implemented	Total
1–9 employees	89.3	10.7	100.0
10–49 employees	87.4	12.6	100.0
50–250 employees	84.7	15.3	100.0
Total	88.2	11.8	100.0

Source: Author's own calculations based on data from the Flash Eurobarometer 441.

Table A.3 Percentage distribution of firms by turnover evolution in the previous year and water usage optimization strategies implementation.

	Turnover evolution in the previous year			
Strategies for water usage optimization	Decrease	No change	Increase	Total
Not implemented	16.1	19.9	64.1	100.0
Implemented	18.0	15.7	66.3	100.0
Total	16.3	19.4	64.3	100.0

Source: Author's own calculations based on data from the Flash Eurobarometer 441.

Table A.4 Percentage distribution of firms by economic activity.

Sector of activity (NACE) sections	No.	%
B. Mining and quarrying	30	0.28
C. Manufacturing	1448	13.64
D. Electricity, gas, steam, and air conditioning supply	61	0.57
E. Water supply, sewerage, waste management	100	0.94
F. Construction	1222	11.51
G. Wholesale and retail trade, repair of motor vehicles	3627	34.16
H. Transportation and storage	656	6.18
I. Accommodation and food service activities	757	7.13
J. Information and communication	483	4.55
K. Financial and insurance activities	348	3.28
M. Professional, scientific, and technical activities	1362	12.83
N. Administrative and support services	524	4.94
Total	10,618	100

Source: Author's own calculations based on data from the Flash Eurobarometer 441.

Table A.5 Percentage distribution of firms by size.

Firm size	No.	%
1–9 employees	6687	62.98
10–49 employees	2475	23.31
50–250 employees	1456	13.71
Total	10,618	100

Source: Author's own calculations based on data from the Flash Eurobarometer 441.

Table A.6 Percentage distribution of firms by economic activity and turnover evolution in the previous year.

	Turnover evolution in the previous year			
Sector of activity (NACE) sections	Decrease	No change	Increase	Total
B. Mining and quarrying	13.3	23.3	63.3	100
C. Manufacturing	17.3	20.8	61.9	100
D. Electricity, gas, steam, and air conditioning supply	14.8	24.6	60.7	100
E. Water supply, sewerage, waste management	9.0	25.0	66.0	100
F. Construction	16.7	22.5	60.8	100
G. Wholesale and retail trade, repair of motor vehicles	18.3	17.3	64.4	100
H. Transportation and storage	14.0	22.6	63.4	100
I. Accommodation and food service activities	17.6	15.2	67.2	100
J. Information and communication	13.5	18.0	68.5	100
K. Financial and insurance activities	9.5	17.5	73.0	100
M. Professional, scientific, and technical activities	14.1	22.2	63.7	100
N. Administrative and support services	14.5	18.1	67.4	100
Total	16.3	19.4	64.3	100

Source: Author's own calculations based on data from the Flash Eurobarometer 441.

B. Appendix 2

Table B.1 Probit regression results.

Variables	Coefficient	Variables	Coefficient
R&D intensity (% of turnover)	0.08324	*Economic activity dummies*	
Firm size dummies		Manufacturing	−0.48017
1–9 employees	−0.22681	Water supply, sewerage, waste management	−0.41254
10–49 employees	−0.15172	Construction	−0.50975
Country dummies		Wholesale and retail trade, repair of motor vehicles	−0.59111
France	0.36816	Transportation and storage	−0.59627
Belgium	0.38032	Information and communication	−1.0552
Luxemburg	0.52692	Financial and insurance activities	−0.56444
Ireland	0.64066	Professional, scientific, and technical activities	−0.67307
Greece	−0.21002	Administrative and support services	−0.44699
Portugal	0.58151		
Finland	0.52047	*Wald chi^2*	*515.80*
Estonia	−0.81633	*Prob > chi^2*	*0.0000*
Lithuania	−0.53693	*Pseudo R^2(McFadden)*	*0.0832*
Slovakia	−0.44774		
Slovenia	−0.44843		
Bulgaria	−0.62822		

Note: $P < 0.10$ for all coefficients.
Source: Author's own calculations based on data from the Flash Eurobarometer 441.

Table B.2 OLS regressions results, dependent variable = Turnover evolution for the previous year.

Firms who ADOPT water usage optimization strategies, D1		Firms who DO NOT ADOPT water usage optimization strategies, D0	
Variables	**Coefficient**	**Variables**	**Coefficient**
R&D intensity	6.14562	Turnover value at the end of the previous year (ln)	4.16683
Turnover value at the end of the previous year (ln)	6.56193	Water supply, sewerage, waste management	18.4777
Water supply, sewerage, waste management	−10.72412	Information and communication	7.84213
Construction	−8.99292	Financial and insurance activities	8.88686
Information and communication	−27.94278	Professional, scientific, and technical activities	4.22169
Inverse Mills ratio term	−10.34479	Mining and quarrying	−17.65797
		Accommodation and food services	−5.50280
		Inverse Mills ratio term	22.24483

Note: $P < 0.10$ for all coefficients.
Source: Author's own calculations based on data from the Flash Eurobarometer 441.

References

Abbott, M., Cohen, B., 2009. Productivity and efficiency in the water industry. Util. Policy 17, 233–244.

Andersen, M.S., 2007. An introductory note on the environmental economics of the circular economy. Sustain. Sci. 2, 133–140.

Bassi, F., Dias, J.G., 2019. The use of circular economy practices in SMEs across the EU. Resour. Conserv. Recycl. 146, 523–533.

Bressanelli, G., Perona, M., Saccani, N., 2019. Challenges in supply chain redesign for the circular economy: a literature review and a multiple case study. Int. J. Prod. Res. 57 (23), 7395–7422.

Cristoni, N., Tonelli, M., 2018. Perceptions of firms participating in a circular economy. Eur. J. Sustain. Dev. 7 (4), 105–118.

Demirel, P., Danisman, G.O., 2019. Eco-innovation and firm growth in the circular economy: evidence from European SMEs. Bus. Strateg. Environ. 28 (8), 1608–1618.

Endrikat, J., Guenther, E., Hoppe, H., 2014. Making sense of conflicting empirical findings: A meta-analytic review of the relationship between corporate environmental and financial performance. Eur. Manag. J. 32 (5), 735–751.

European Commission. 2016. Brussels DG communication COMM A1 'strategy, corporate communication actions and Eurobarometer' unit (2016) flash Eurobarometer 441 (European SMEs and the circular economy), ZA6779 data file version 1.0.0; GESIS data archive: Cologne, Germany.

Ghisetti, C., 2018. On the economic returns of eco-innovation: where do we stand? In: Horbach, J., Reif, C. (Eds.), New Developments in Eco-Innovation Research. Springer, Cham, pp. 55–79.

Ghisetti, C., Rennings, K., 2014. Environmental innovations and profitability: how does it pay to be green? An empirical analysis on the German innovation survey. J. Clean. Prod. 75, 106–117.

Hamilton, B.H., Nickerson, J.A., 2003. Correcting for endogeneity in strategic management research. Strateg. Organ. 1 (1), 51–78.

Heckman, J., 1974. Shadow prices, market wages, and labor supply. Econometrica 42, 679–694.

Heckman, J., 1979. Sample selection bias as a specification error. Econometrica 47, 153–161.

Horbach, J., Horbach, J., 2018. The impact of resource efficiency measures on the performance of small and medium-sized enterprises. In: Reif, C. (Ed.), New Developments in Eco-Innovation Research. Springer, Cham, pp. 147–162.

Horbach, J., Rammer, C., 2019. Employment and Performance Effects of Circular Economy Innovations. ZEW discussion papers, no. 19-016, ZEW - Leibniz-Zentrum für Europäische Wirtschaftsforschung, Mannheim, Available at: http://hdl.handle.net/10419/196125.

Kneese, A.V., Ayres, R.V., D'Arge, R.C., 1970. Economics and the Environment: A Material Balance Approach. Resources for the Future, Baltimore, MD.

Lee, L.F., 1978. Unionism and wage rates: A simultaneous equation model with qualitative and limited (censored) dependent variables. Int. Econ. Rev. 19, 415–433.

Lieder, M., Rashid, A., 2016. Towards circular economy implementation: a comprehensive review in context of manufacturing industry. J. Clean. Prod. 115, 36–51.

Maddala, G.S., 1983. Limited-Dependent and Qualitative Variables in Econometrics. Cambridge University Press, Cambridge.

Moric, I., Jovanovic, J.S., Dokovic, R., Pekovic, S., Perovic, D., 2020. The effect of phases of the adoption of the circular economy on firm performance: evidence from 28 EU countries. Sustainability 12, 2557.

Mura, M., Longo, M., Zanni, S., 2020. Circular economy in Italian SMEs: a multi-method study. J. Clean. Prod. 245, 118821.

Pádua Pieronia, M., Blomsmaa, F., McAloonea, T.C., Pigosso, D.C.A., 2018. Enabling circular strategies with different types of product/service-systems. Procedia CIRP 73, 179–184.

Porter, M.E., Van der Linde, C., 1995. Toward a new conception of the environment competitiveness relationship. J. Econ. Perspect. 9 (4), 97–118.

Prashar, A., Sunder, V., 2020. A bibliometric and content analysis of sustainable development in small and medium-sized enterprises. J. Clean. Prod. 245, 118665.

Rexhauser, S., Rammer, C., 2014. Environmental innovations and firm profitability: unmasking the porter hypothesis. Environ. Resour. Econ. 57, 145–167.

Rizos, V., Behrens, A., Van der Gaast, W., Hofman, E., Ioannou, A., Kafyeke, T., Flamos, A., Rinaldi, R., Papadelis, S., Hirschnitz-Garbers, M., Topi, C., 2016. Implementation of circular economy business models by small and medium-sized enterprises (SMEs): barriers and enablers. Sustainability 8, 1212.

Soltmann, C., Stucki, T., Wörter, M., 2015. The impact of environmentally friendly innovations on value added. Environ. Resour. Econ. 62, 457–479.

Telle, K., 2006. "It pays to be green"—a premature conclusion? Environ. Resour. Econ. 35 (3), 195–220.

WWAP (United Nations World Water Assessment Programme), 2015. The United Nations World Water Development Report 2015: Water for a Sustainable World. UNESCO, Paris.

Zamfir, A.-M., Mocanu, C., Grigorescu, A., 2017. Circular economy and decision models among European SMEs. Sustainability 9, 1507.

CHAPTER 6

History and evolution of the circular economy and circular economy business models

Alisha Tuladhar, Konstantinos Iatridis, and Dimo Dimov
School of Management, University of Bath, Bath, United Kingdom

1. Introduction

Climate change is upon us and the clock is ticking. The Intergovernmental Panel on Climate Change (IPCC) Report 2018 warns the public of dire consequences, such as flood hazards, if the Earth overshoots the 1.5 °C temperature increase mark by 2030. Research identifying solutions to combat climate change indicates that while adopting renewable energy, alongside closing fossil fuel plants, tackles 55% of the greenhouse gas (GHG) emissions, the remaining 45% of the "harder-to-abate" emissions can be combatted by a transformation in the "way we design, produce, and use goods" (EMF Ellen MacArthur Foundation, 2019, p. 13). In achieving the latter, the circular economy (CE) plays a significant role.

CE is defined as "an economic system that replaces the 'end-of-life' concept with reducing, alternatively reusing, recycling, and recovering materials in production/distribution and consumption processes" (Schroeder et al. 2019, p.224). The recirculation of materials back into the system allows products to retain value and is therefore a better alternative compared to the prevalent linear economic model of "take-make-dispose." Shifting from a linear to a circular model reduces demand for both raw resource inputs and waste disposal—two highly carbon-intensive activities closely linked with global warming (Pratt and Lenaghan, 2015). Therefore, CE has been recognized as an appropriate mitigation strategy against the problems of climate change, resource scarcity, and waste disposal. Furthermore, CE has not been only acknowledged as an appropriate climate change mitigation strategy, but also as an enabler toward achieving Sustainable Development Goals (SDG) such as Clean Water and Sanitation (SDG 6), Affordable and Clean Energy (SDG 7), Decent Work and Economic Growth (SDG 8), Responsible Consumption and Production (SDG 12), and Life on Land (SDG15) (Schroeder et al., 2019).

Driven by the significant potential of CE, scholars working on the negative externalities of the private sector (Korhonen et al., 2018; Zink and Geyer, 2017; Reike et al.,

Circular Economy and Sustainability, Volume 1
https://doi.org/10.1016/B978-0-12-819817-9.00031-4

2018) have investigated how CE principles can be incorporated in business operations. Thus, a separate stream of literature has emerged investigating circular economy business models (CEBM) or circular business models (CBM). This topic has attracted the interest of practitioners (EMF Ellen MacArthur Foundation, 2013, 2019), policy makers (European Commission, 2019), and researchers (Adam et al., 2018; Bocken et al., 2018; Rizos et al., 2016) (for recent reviews, see Rosa et al., 2019; Lüdeke-Freund et al., 2019; Lieder and Rashid, 2016; Ghisellini et al., 2016). Scholars argue that if the private sector is to lead the transitions toward CE, a much greater emphasis on business models will be needed in future discourses (Kirchherr et al., 2017; Lewandowski, 2016) as it is important to accurately describe how businesses design, create, and deliver value in alignment with CE. As Kirchherr et al. (2017) bluntly state: "A CE understanding lacking business models is one with no driver at the steering wheel in our point of view" (p. 228).

This chapter offers an in-depth analysis of circular business models to allow readers unfamiliar with the topic to make sense of the CE and CBM literature. The chapter is split in four sections. First, we describe the methodology used, how and why we divided the periods into three specific chronological periods. Second, we focus on the history and origins of CE. The third section analyzes circular business models and proposes an integrated *Holistic Circular Business Model Canvas*. The chapter concludes with a future research agenda.

2. Methodology

For simplification, and to allow the reader to make sense of how the narrative has developed over time, we adopted the approach of previous studies (Reike et al., 2018; Blomsma and Brennan, 2017) and split the literature into three periods: (i) pre-1990s; (ii) 1990–2010; and (iii) 2010–present and beyond. We begin with the 1960s, as during this time we observe an acceleration of environmental movements and systematic investigations of the negative environmental impacts of economic activities (Melosi, 2005). Similarly, we opted for 1990 as the starting year of phase (ii) because, post 1990, we witnessed a proliferation of research and policy-making initiatives on the concept of sustainability. Equally, we chose 2010 as the base year for phase (iii) as, following that year, we witnessed a mainstreaming of CE with increased academic scholarly debate and policy-making reports, such as the EMF's 2013 report on the circular economy and the EU's policy endorsements of CE.

3. Circular economy: History, evolution, and definition

Together, these studies (mentioned toward the end of Section 1) provide important insights into how the circular economy as a concept evolved over time. Fig. 1 depicts the evolution of the narrative.

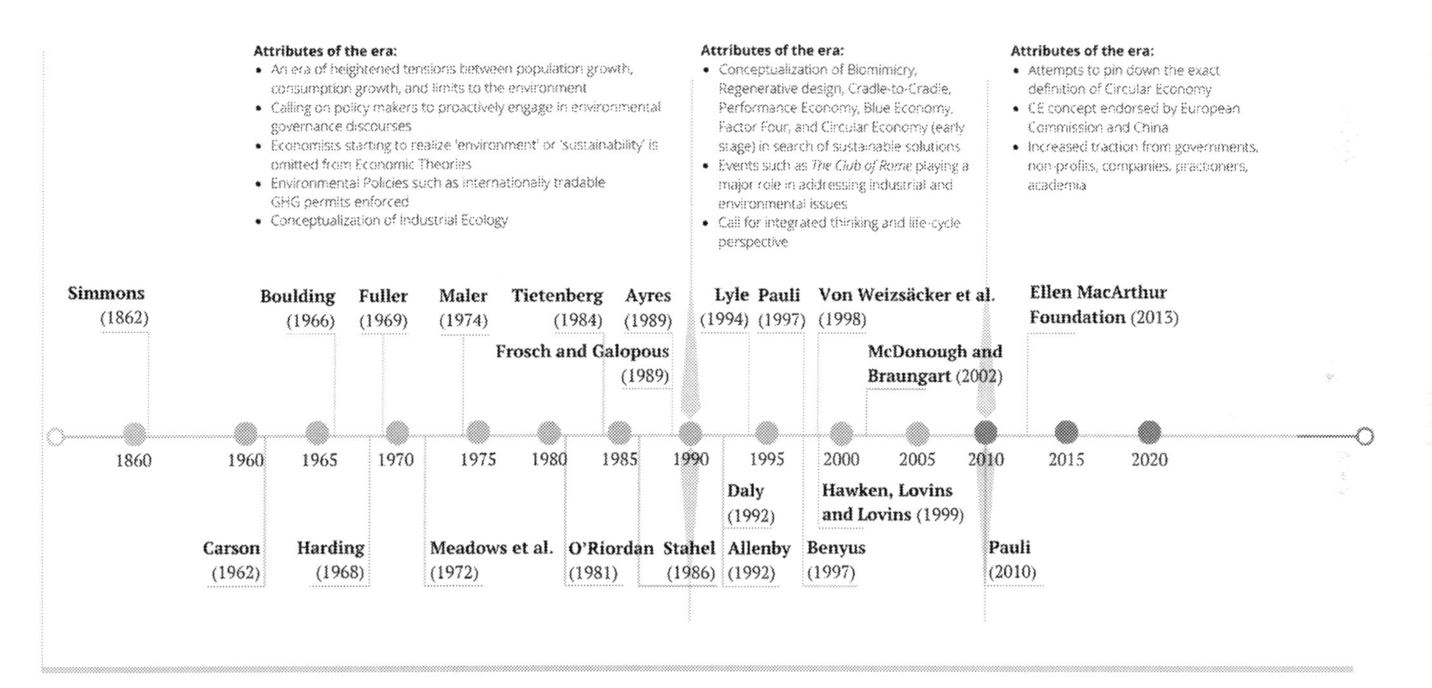

Fig. 1 History of the circular economy concept. *Credit: Author's own.*

Although the term "circular economy" might be relatively new, the concept itself is not entirely novel. It has stemmed from different schools of thought, including the 3Rs, cradle-to-cradle, industrial ecology, biomimicry, and the blue economy. As such, CE acts an umbrella term for myriad concepts, with contributors such as John Lyle, William McDonough, Michael Braungart, and Walter Stahel, to name a few. While the usage of the term has been much popularized by the Ellen MacArthur Foundation (EMF) (2013), especially in the United Kingdom, the antecedents are rooted deep in history and can be traced back to different schools of thought. Although, the seminal works around environmental issues started to proliferate in the 1960s, the early traces of CE date back to as early as the 1860s, when Simmonds (1862) suggested the need for win-win practices to exist, from which both the economy and the environment benefit. The following section illustrates how the narrative of CE developed over the years.

3.1 Pre-1990s

Fig. 1 showcases the evolution of CE concepts. The 1960s was ripe for environmental issues, as industrialization and consumption in Europe and the US were acerbating public opinion. Seminal publications, such as *Silent Spring* (Carson, 1962), *Tragedy of the Commons* (Hardin, 1968), *The Economics of the Coming Spaceship Earth* (Boulding, 1966), and *Operating Manual for Spaceship Earth* (Buckminster, 1969) highlighted the tensions between population growth, consumption growth, and the limited capacity of the environment. Specifically, Boulding presented the notion of an open "cowboy" economy and a closed "spaceman" economy. He described the former as "reckless and exploitative" behavior, with unlimited natural resources leading to high rates of consumption and production, while in the closed "spaceman" economy, he compared the Earth to a single spaceship with limited natural resources, where humanity needs to place itself in a "cyclical ecological system" capable of continuous material reproduction. Boulding asserted that the attitudes toward consumption and production in these two scenarios differ strikingly. While the open system treats infinite production and consumption as a desirable and plausible outcome, the closed system discourages both production and consumption, as there is a limited amount of stock. Instead, it focuses on the maintenance of extant resources. As such, he reasons that the measurement of success is based on the nature, quality, and complexity of the capital stock, and not necessarily on production, consumption, or on some of the classic ways of measuring economic success (Boulding, 1966). This ideology was further echoed in the works of Meadows and colleagues (Meadows et al., 1972) in their seminal report to *The Club of Rome in 1972: The Limits to Growth*. In it, they asserted that humanity's prospects to last another 100 years are impeded since there are limits to growth and the world is finite.

One solution was proposed by Mäler (1974), who, following an analysis of the macroeconomic policies related to externalities and financial incentives for biodiversity preservation, suggested internationally tradable permits to emit greenhouse gases. Similarly, O'Riorden (1981) advocated for a green ideology of environmentalism. He further called upon policy makers to proactively engage in environmental governance discourse, such as environmental planning, pollution control, and resource management. A slow yet concrete call to action was beginning to emerge. Scholars from both economics and environmental sciences started to be more intrigued by and interested in this realm. Tietenberg, an economics professor, focused on economics and environmental issues by challenging the theoretical economic assumptions and definitions of "market efficiency" and "sustainability." He argued that market allocations can be efficient, but not sustainable; sustainable but not efficient; inefficient and unsustainable; and efficient and sustainable. The final option being a win-win situation, providing an opportunity for both current and future generations' welfare increment (Tietenberg, 1984).

Promoting the notion of efficiency and resource longevity, Stahel's work entitled *Product life as a variable: the notion of utilization* (Stahel, 1986) was a fundamental contribution, as it challenged the current linear economic model. He highlighted not only the problems associated with a linear system, such as growing waste volumes and limitations in availability of resources, but also advocated for solutions, such as the self-replenishing system. The latter is a system that "creates spiral-loops that minimizes matter and energy flow, and environmental deterioration without restricting economic growth or social and technical progress" (Stahel 1986, p. 185). Concrete strategies included, but were not limited to, life extension of goods, reuse, repair, and remanufacture. As such, his work on performance economy is generally considered as a solid building block of the CE (Sillanpää and Ncibi, 2019).

While Stahel's work emphasized the life extension of goods, Frosch and Gallopoulos (1989) had a different viewpoint on manufacturing. Following their work with General Motors (GM), they published an article entitled *Strategies for Manufacturing*. In it, the authors highlighted the need to develop and implement an alternative, integrated manufacturing system, one where "wastes from one industrial process can serve as the raw materials for another, thereby reducing the impact of industry on the environment." (Frosch and Gallopoulos 1989, p. 144). A couple of years later, Robert Frosch published the article *Industrial ecology: a philosophical introduction* (Frosch, 1992), and ever since, he has been commonly referred to as the father of industrial ecology (Frosch, 2005). Although industrial ecology and performance economy seem disparate concepts, they fall under the umbrella of CE due to their common ideology or vision of ensuring maximum material utility.

3.2 1990–2010

The 1990s saw the first serious reference to CE with Pearce and Turner's book chapter titled *The Circular Economy* (Pearce and Turner, 1990). This pioneering work challenged the open, linear economy to consider the linkages between the economy and the environment by presenting a complete picture of the economy in a flow diagram illustrating the input-output flows of material, energy, and utility. This was the first graphic illustration of the circular economy, in which they identified three economic functions of the environment: as a resource generator, waste absorber, and direct source of utility. They made a valid point by bluntly stating that "economies exist and the natural environment exists. What we do not know is what needs to occur for them to co-exist in equilibrium" (Pearce and Turner, 1990, p. 42).

Similarly, in the 1990s, other economists began to join this line of thinking by recognizing the flaws in economic theories in addressing environmental issues. In his article, *Allocation, distribution, and scale: toward an economics that is efficient, just, and sustainable*, Daly challenged the extant economic theories, as most omitted the dire reality of environmental issues. As such, he argued that economic theories need to recognize scale issues independently and not just as a mere subset of allocation or distribution. He elaborated by using a boat metaphor—a boat needs both balance and weight; if either is faulty, it will sink. Thus, he concluded that scale along with allocation and distribution should be recognized as fundamental parts of the economic problem (Daly, 1992).

Another concept that gained traction in the same decade is industrial ecology (IE). IE focuses on reducing harmful environmental impacts caused by resource-intensive manufacturing units and energy plants that extract raw materials and natural resources from the Earth to produce products and services for the Earth's population. Allenby, a prolific author in the field of IE, covered several aspects of IE. While, his seminal piece *Achieving sustainable development through industrial ecology* (Allenby, 1992a) addressed industrial ecology in relation to sustainable development, his other works, such as *Industrial ecology: The materials scientist in an environmentally constrained world* (Allenby, 1992b), and a coauthored publication with D. Richards, *The Greening of Industrial Ecosystems* (Allenby and Richards, 1994), investigated IE in relation to materials science. These insights were further developed by other environmental economists, who more recently argued that simply circulating materials is not enough—minimization of inputs, such as scarce metals and resources is also a crucial prerequisite for residual circulation (Andersen, 2007).

In parallel to IE scholarship progression, the notion of regenerative design also immerged as a research area with prominence in the built environment. Lyle expressed the salience of "waste as a resource" and illustrated the myriad options of "strategies for regenerative design" in the realm of landscape architecture (see Lyle, 1994 for more). His proposal of waste as a resource is idealistic, however, and in practice may not be feasible, as the possibility of turning waste to resource is subject to a material's properties. For example, some plastics, such as polyethylene (PET), are recyclable and thus this waste can be turned into resource. However, other plastics, such as polypropylene (PP), hinder closed-loop recycling and therefore may be inapplicable (Eriksen et al., 2019). Hence, the initial material selection, product-design for disassembly, and systems-thinking are crucial in order to ensure that environmental efficiency is maximized before, during, and after usage of the product. Furthermore, Lyle (1994) also addressed the economics of the products and the enabling policy conditions as he wrote "they must cost no more (and preferably less) than competing technologies, and they must be supported by public policy" (p. 307). In today's context, however, in addition to costing less and being supported by public policy, scholars have pointed out that these recycled materials should displace the primary materials' production in a one-to-one ratio or else there would be increased production of unwanted materials, which is counterintuitive (Zink and Geyer, 2016).

One branch of regenerative design is the concept of biomimicry. Biomimicry is, in simple terms, mimicry of biology, which can span to various levels such as mimicry of a specific organism, behavior of the organism, or in relation to the larger context or mimicry of an ecosystem. Benyus compared organizations with the natural ecosystem and provided "ten lessons" for an ecologically focused company, culture, or economy to derive inspirations from nature while seeking solutions to solve problems or simply to reinvigorate products and processes (Benyus, 1997). The commonality between regenerative design and biomimicry is that both cultivate a systems-thinking approach, whereby the entire chain of interactions is considered and is not just narrowed to a single aspect (Meadows and Wright, 2008).

In similar vein to the concepts of IE, regenerative design, and biomimicry, Gunter Pauli developed the concept of the blue economy (Pauli, 2010) in a publication that is generally considered as the main foundation of CE. In his initial piece, *Zero emissions: the ultimate goal of cleaner production*, he presents innovative solutions that lead to zero waste and maintains that the solution lies in forming clusters of industries, where the waste of one is input for another (Pauli, 1997). This idea is further elaborated in the book, *The blue economy: 10 years, 100 innovations, 100 million jobs*, where a plethora of implementable real world solutions are illustrated, the concept of blue economy is developed (Pauli, 2010), and the notion of integrated thinking is emphasized (Bocken et al., 2014).

The importance of integrated thinking is also echoed in *Factor Four*, a seminal publication by Weizacker and colleagues, which was part of the report to The Club of Rome. It illustrated how resource productivity of our economies could grow fourfold, how we could live twice as well, and yet use half as much (von Weizacker et al., 1997). The requirements for factor four implementation were twofold: first, it needed scaling of the proposed technologies, and second, it needed integrated thinking—not reductionism, where the design challenge is disjointed rather than solved holistically. This notion of integrated thinking is considered a vital requirement of CE as it allows an analysis of the entire value chain and the lifecycle.

Similar salience of integrated thinking is also exhibited in the works of McDonough and Braungart, as they developed the concept of cradle-to-cradle, another building block of CE. In their seminal article, *The next industrial revolution*, they invited the readers to envision how the next industrial revolution should look: one where no hazardous materials are generated, one where regulations are not needed as it is fully compliant with nature, and so on (McDonough and Braungart, 1998). Moreover, their 2002 publication *Cradle to Cradle: Remaking the Way We Make Things* solidified the notion of CE. Some of their fundamental principles that carry weight in CE discourse today are: biological nutrients, technological nutrients, waste equals food, use solar energy, and re-design the system so that the ecology, economy, and equity are integrated and not constrained (McDonough and Braungart, 2002).

Natural ecology considerations are also inherent in the concept of natural capitalism. Lovins and Lovins showcased how companies that shifted their business logic and practice to an ecologically sound plan benefited their bottom-line exponentially (Hawken et al., 1999). Natural capitalism can be summarized as comprising three main elements: dramatically increasing the productivity of natural resources, operating in a closed-loop system, and establishing a solutions-based business model (Lovins et al., 1999). They emphasized that with simple tweaks in the design and with a whole-system perspective, the savings and benefits reaped are often exponential.

3.3 2010–present and onwards

The 20th century marks the attempt to pin down the exact definition of CE. One such attempt to produce a definition via analyzing 114 definitions is presented by Kirchherr et al. (2017) who defined CE as follows:

> *an economic system that replaces the "end-of-life" concept with reducing, alternatively reusing, recycling and recovering materials in production/distribution and consumption processes. It operates at the micro level (products, companies, consumers), meso level (eco-industrial parks) and macro level (city, region, nation and beyond), with the aim to accomplish sustainable development, thus simultaneously creating environmental quality, economic prosperity and social equity, to the benefit of current and future generations. It is enabled by novel business models and responsible consumers (pp. 224–225).*

Further to this development, the work of the Ellen MacArthur Foundation is commendable as their initiatives and reports have mobilized the industry toward a CE transition (EMF Ellen MacArthur Foundation, 2019). EMF largely relies on the following five principles (EMF Ellen MacArthur Foundation, 2013):

(1) Think in systems;
(2) Design out waste/design for reuse;
(3) Build resilience through diversity;

(4) Rely on energy from renewable sources; and
(5) Waste is food/think in cascades/share assets (symbiosis).

As such, the current rhetoric of CE differs from the previous century's interpretation in three main ways. First, it encourages a systems thinking approach rather than simply striving for optimization and cleaner production (Mendoza et al., 2017). Second, it moves beyond the idea of eco-industrial parks, where geographical-closeness is a prerequisite, toward establishing reverse logistics and a closed-loop supply chain, enabled via collaboration regardless of geographic proximities (Lüdeke-Freund et al., 2019). Third, the discussion has moved from 3Rs to 10Rs, thus encompassing a larger set of hierarchies (Reike et al., 2018).

The concept has attracted the attention of scholars focusing on large organizations (such as Bocken et al., 2016; Hopkinson et al., 2018; Frishammar and Parida, 2018) and SMEs (Rizos et al., 2016; Zamfir et al., 2017), consultancies (e.g., Accenture, McKinsey), publicly funded organizations (e.g., Switch-Asia), and nongovernmental organizations (e.g., EMF, Tearfund, Chattam House), as well as policy making institutions (e.g., EU Commission), and governments (e.g., China). This wide range of interest has given rise to CE definitions, principles, patterns, archetypes, and business models. Ensuring the transition from theory to practice, circular business models (CBM) are seen to be a growing research area, as they attempt to address the operationalization of CE.

4. Circular business model (CBM)

Following the history and meaning of the concept, the next step is to understand how the circular economy can move from theory to practice. Such a nexus between theory and practice is the business model, which describes the rationale for how an organization creates, delivers, and captures value (Osterwalder and Pigneur, 2010; Teece, 2010). Circular business models (CBMs), in a sense, are a new kind of business logic or business model, where value creation is based on capturing the economic value remaining in products after their use and exploring new market offerings where applicable, which implies that a return flow from consumers to producers is essential (Rosa et al., 2019; Linder and Williander, 2017).

Scholars such as Linder and Williander (2017), Roos (2014), and Den Hollander and Bakker (2016) have provided the main definitions of CBM, upon which Nußholz (2017) built to come up with the following definition:

> *A circular business model is how a company creates, captures, and delivers value with the value creation logic designed to improve resource efficiency through contributing to extending useful life of products and parts (e.g., through long-life design, repair and remanufacturing) and closing material loops (p. 12).*

One advantage of this definition is that it builds on the previous three definitions, and, in so doing, builds on their key concepts and adds the notion of closing material loops. Thus, compared to previous definitions, this definition captures the key aspects of CBMs and allows researchers, policy makers, and practitioners to better understand what CBM is. However, this definition is not free of shortcomings. It does not explicitly refer to performance-economy, sharing economy, or rental/leasing models, which enable resource-life extension of products. These strategies are all inherent in CBM. There is a plethora of ways describing how the economic value of the product and material could be maintained. One of the simplest innovative CE strategies and a major type of CBM is PSS (product-service system) (Tukker and Tischner, 2006), which outlines three main categories (product-, use-, and result-oriented PSSs) and has eight archetypes (Tukker, 2004). In fact, previous literature reviews have identified over 26 CBMs (Lüdeke-Freund et al., 2019) and 9 CBM classification methods (Rosa et al., 2019) (covered in Section 4.3). While covering the entire spectrum of archetypes, typologies, classifications, and types of CBMs available in the literature is beyond the scope of this chapter, we will consolidate and report the seminal works in the field. As such, the discussion of CEBM is divided in three main sections: (i) Integration of Circular Business Model Canvases, (ii) ReSOLVE Framework by EMF, and (iii) other hybrid models (as found in Rosa et al., 2019).

4.1 Holistic circular business model canvas—An integration of circular business model canvases

One of the most widely cited definitions of business model is that of Osterwalder and Pigneur, who state that a business model is the core logic explaining how a company creates, delivers, and captures value (Osterwalder and Pigneur, 2010). With time, the business model canvas has become a popular tool for communicating and envisioning the configuration of essentially nine key components: key partners, key resources, key activities, value proposition, customer relationships, delivery channel, target customer, revenue stream, and cost structure. As such, it is no surprise that the usage of BMC has prevailed in CEBM scholarship in recent years. The two main canvas proposals are: the circular business model canvas proposed by Lewandowski (2016) and the framework for sustainable circular business model (SCBM) innovation by Antikainen and Valkokari (2016). This section will describe these two canvases, highlight their individual contributions, and finally combine the two, proposing an integrated circular business model canvas. In doing so, we combine the micro-level business perspective with a systems-level view of the firm, which includes ecosystem and sustainability factors. A combination of these two models could be a useful and powerful tool that tackles both micro-business level decisions and the larger sustainability impact issues.

The first canvas, Lewandowski's Circular BMC (2016) extended the term Value Proposition in the context of CE by including concepts and models such as "Product Service System (PSS)," "Circular Product," "Virtual service," "Incentives for customers in Take Back System (TBS)." These were drawn from the ReSOLVE framework (EMF, 2015) (also in Section 4.2). This *Circular BMC* (Lewandowski, 2016, p. 21, Fig. 3) in addition has two additional components to the traditional nine-building blocks (cf. Osterwalder and Pigneur, 2010): "Take Back System" and "Adoption Factors." "Take Back System" refers to the establishment of a *reverse logistics channel* or a *closed-loop supply chain* (Osterwalder and Pigneur, 2010). In contrast, "Adoption Factors" are identified as factors that would hinder or enable circular production and consumption, such as internal "Organizational Capabilities" and external "PEST: Political Environment Social and Technological" factors.

The second canvas is the *Framework for Sustainable Circular Business Model Innovation* (Antikainen and Valkokari, 2016, p. 9, Fig. 2) takes a wider perspective and goes beyond the nine-building blocks of Business Model Canvas and integrates multilevels: namely, business ecosystem and sustainability evaluation elements. They added "Business Ecosystem Level," "Sustainability Impact" and iterative cycles of "Sustainability and circularity evaluation of the business model" (signified by bold arrows) around the nine core business level elements to signify system-wide changes. In asserting that a circular business model innovation requires a holistic and systemic thinking approach, they included "Trends and Drivers" under which factors such as new legislation, consumer awareness, resource scarcity, fall. At the "Business Ecosystem Level", the authors also highlight "stakeholder involvement" as being key since this includes external and internal parties such as suppliers, customers, network actors, environment, and society, who have an interest and influential power over the business (Antikainen and Valkokari, 2016).

There are certain advantages and limitations to these models, which are self-acknowledged by the authors. First, Lewandowski's canvas is purely based on a literature review and lacks empirical evidence to support the canvas' findings and, which raises questions about its application. Antikainen and Valkokari's (2016, p. 9, Fig. 2) framework has been applied to a real case, a Finnish social enterprise. They found that due to its simplicity the framework worked well but adding the information required proved to be time consuming.

Since we value both a micro-perspective and a broader ecosystem view, we propose to integrate the two models: Lewandowski's circular BMC and Antikainen and Valkokari's framework for sustainable circular business model innovation. These additions make a case for thinking beyond the core business and reflecting on the drivers and consequences the private sector may have on environment, society, and business. As such, we propose Fig. 2 as a holistic circular business model canvas, followed by a detailed account of the integration.

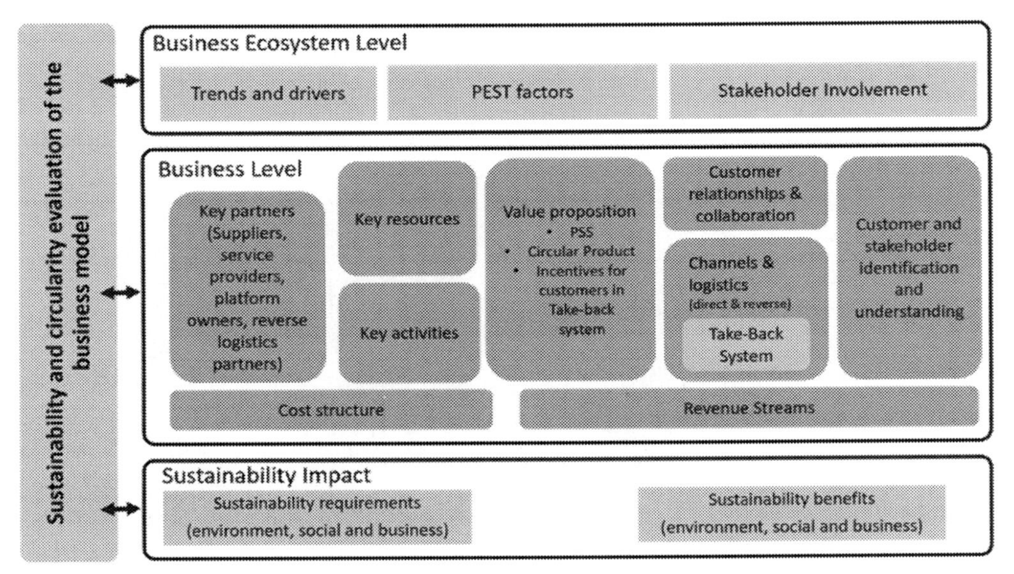

Fig. 2 Holistic circular business model canvas. *Credit: Based on and expanded from Lewandowski, M., 2016. Designing the business models for circular economy—towards the conceptual framework. Sustainability 8:1–28. https://doi.org/10.3390/su8010043; Antikainen, M., Valkokari, K., 2016. A framework for sustainable circular business model innovation. Technol. Innov. Manag. Rev. 6:5–12. https://doi.org/10.22215/timreview1000.*

The Holistic Circular Business Model Canvas (Fig. 2) draws on Lewandowski's Canvas and adds two main subcomponents to Antikainen and Valkokari's (2016, p. 9, Fig. 2) original *Framework for Sustainable Circular Business Model Innovation*. The first addition is the "PEST factors," placed at the *Business Ecosystem Level*. Since Political, Environmental, Social, and Technological factors are elements external to the firm's core activities and beyond the firm's control, it is aptly suited to *Business Ecosystem Level*. The second addition is the *Take-Back System* under *Channels & logistics*. The salience of establishing a *Reverse Logistics* or a *Closed-Loop Supply Chain* is prominent in the context of CE (as previously mentioned), thus this inclusion is deemed important. Finally, a point to emphasize rather than to add, is that of *Value Propositions*. The figure emphasizes few options under *Value Propositions*. Again, it draws from Lewandowski's canvas, *Product Service System (PSS), Circular Product, Incentives for customers in Take-Back System among others*. For additional CE design options and CEBM patterns, refer to Lüdeke-Freund and team's typology of CEBM options (which will be detailed in Section 4.3.

An exclusion from Lewandowski's model, however, is the explicit mention of organizational capabilities. On the basis that organizational capabilities implicitly has the same meaning as key activities and key resources, we did not see the need for it to be a separate block.

Similar to the aforementioned comments on the usability of the previous two canvasses, this canvas, the holistic circular business model canvas, also has its advantages and limitations. The main benefits of using this canvas would be that it allows firms to have an ecosystemic view of the firm in a simple and timely manner. It would also potentially be useful in communicating the business model to various stakeholders, including financers and media. However, we do recognize that the canvas needs to be tested with a real firm and, therefore, recommend future researchers to investigate empirically its practical relevance and usability.

4.2 ReSOLVE framework

The ReSOLVE framework is a list of business actions that could help businesses and government transition and accelerate toward a circular economy (EMF Ellen MacArthur Foundation, 2015). The acronym ReSOLVE stands for regenerate, share, optimize, loop, virtualize, and exchange. Regenerate means to shift from finite to renewable energy; share means to encourage maximum utilization of products so that environmental stress on virgin materials is lessened; optimize is to ensure maximum efficiency with the help of technology and data, alongside waste removal in processes such as production process or supply chains; loop refers to remanufacturing or recycling components; virtualize includes usage of technologies in order to reduce consumption (e.g., skype to avoid travelling, e-books to avoid printing, etc.); and finally, exchange includes replacing old products/services with new efficient technologies/product/services (e.g., multimodal transport).

One question that needs to be asked, however, is whether these actions help or hinder the environment. This argument is also echoed in the works of Zink and Geyer who are critical of the sharing economy being promoted as "green" (Zink and Geyer, 2016). They argue that it depends on various factors: the target market, what the customers would do without the service, and how joining the service changes their behavior. Their example of car-sharing membership service ZipCar is very interesting, as the authors assert that if a company attracts customers who otherwise were reliant on public transport, this *increases* total environmental impact; however, if it attracts customers who shed their cars and drive less, it would *decrease* total environmental impact. Therefore, we recommend a critical approach to be taken while applying the ReSOLVE framework in order to ensure that other undesirable unintended consequences are omitted or minimized. Despite these concerns, the ReSOLVE framework by the EMF has proven to be seminal, as several scholars (Lewandowski, 2016; Hopkinson et al., 2018; Heyes et al., 2018; Manninen et al., 2018; Jabbour et al., 2019; Nerurkar, 2017) have built on it to better understand the notion of CBM in various fields.

4.3 Hybrid forms

The term "hybrid" is used by Rosa and colleagues, as they found that there were other forms of business models that did not fit into the business model canvas (BMC) category nor the ReSOLVE category (Rosa et al., 2019). Thus, under the hybrid classification, the scholarly team classified seven different business models. These models are varied, as they range from having an emphasis on circular design (Bocken et al., 2014; Bocken et al., 2016), sustainability impact (Nerurkar, 2017), public-private partnership (PPP) (Witjes and Lozano, 2016), manufacturer-centric model (Tolio et al., 2017), balanced score card (Janssen and Stel, 2017), sustainable value exchange matrix (Morioka et al., 2018), to morphological matrix (Haanstra et al., 2017). Another systematic literature review (Pieroni et al., 2019) in the field has identified 92 different approaches (including conceptual models, methods, or tools) found in the gray and academic literature. The authors suspect that the upsurge in the number of publications may have been due to earlier calls from scholars expressing the lack of appropriate methods and tools in the field (Ghisellini et al., 2016; Lewandowski, 2016; Bocken et al., 2016; Pieroni et al., 2019).

A subsection of hybrid forms is the CEBM design strategies and options emerging in the literature. The scholarly team of Bocken and colleagues proposed in 2016 a list of CBM strategies. Building on the work of Stahel and Braungart, Bocken et al. (2016) classified the strategies in accordance with two main loops: slowing the loops and closing the loops. The first, slowing the loops, relates to prolonging the current life and the development of product reuse practices, such as the access and performance model and extending product value. The second pertains to using materials considered as waste in a traditional linear model (Bocken et al., 2016), such as industrial symbiosis, where the by-product of one company becomes a resource for another.

This study encouraged designers and business managers to think about capturing both value from a product life-extension perspective and via residual value retainment at the end-of-life. Another advantage of such a classification is that it allows future researchers to identify strategies that are well-researched versus those that are not. For instance, Merli and colleagues cited this work and highlighted that research on innovative business models, especially related to slowing resource loops, is scant (Merli et al., 2018), hence, indicating a potential research area.

Similarly, building on this work, Lüdeke-Freund et al. (2019) published a review and a typology of CEBMs. The review covered closed-loop supply chain (CLSC) and CEBM, the two fundamental research areas of the CE concept. While the field of closed-loop supply chain (CLSC) is established, the literature on CEBMs is emerging (Lüdeke-Freund et al., 2019). The literature review identified 26 CEBM design options, which was further consolidated into six major CEBM patterns: repair and maintenance; reuse and redistribution; refurbishment and remanufacturing; recycling; cascading and repurposing; and organic feedstock business model patterns (Lüdeke-Freund et al., 2019). These categories were then mapped against design strategies, such as: design for long-life products, design for product life extension, design for technical cycles, design for biological cycles, to ensure resource strategy such as slowing or closing to be met, and for the value strategy such as either retaining product value or retaining material value to be achieved.

The authors report on conducting a morphological analysis, mainly grounded on other scholarly works (such as Pauli, 2010; Bocken et al., 2014; Bocken et al., 2016; Tukker, 2004; Clinton and Whisnant., 2014; Beltramello et al., 2013; Bisgaard et al., 2012; Braungart et al., 2007; Planing, 2015). These design options were first coded and mapped consequently against the business model (BM) dimensions with a subdivision into categories for further clarity. For instance, value proposition was divided into products and services, value delivery into target customers and value delivery processes, value creation into partners and stakeholders and value creation processes, and value capture into revenue and costs. Hence, a practitioner could take the morphological box and create a preferred combination. For example, a company selling a long-lasting product could look at these design options and choose an added-value service option of repair and take back, which could target "green" customers and quality-conscious customers. As such, the value-delivery process could include taking back used products in partnership with retailers, thus resulting in a value creation process of reselling the taken-back products if seen fit; hence enabling win back base materials. Thus, these could lead to additional product revenue as the taken-back material is resold and the costs of resource inputs are lower. These business model combinations would work especially well with electronic products where the majority of the internal parts may be reusable. However, it is worth noting the self-acknowledged limitations that these options are not an exhaustive list nor are they a ready-made business model template, rather a compilation of the extant literature to date (Lüdeke-Freund et al., 2019).

Building on the findings of Lüdeke-Freund and colleagues, we argue that the most ideal design strategy from an ecological viewpoint is to design for product-life extension, since it allows retention of both product and material value, or alternatively, to select a combination of both slowing and closing strategies. Other scholars agree that merely slowing resource use means the harm to the environment is being delayed and not rectified; if life extension and reuse is not a possibility, waste hierarchy suggests recycling to be the next best alternative (Nußholz, 2017). Hence, if the economy were to aim toward circularity, then both the product and material value need to be salvaged; thus, both slowing and closing is essential for biological and technical cycles. Ensuring the right design strategy and business model could lead to potential effects of product-value retention or material-value retention, slowing and closing loops, respectively.

5. Future research agenda/conclusion

This study set out to review the history of the circular economy (Section 3), synthesize the narrative of the circular economy over time (Fig. 1), and discuss how the circular economy can move from theory to practice by drawing on insights from the myriad circular economy business models found in the literature (Section 4). In doing so, we propose the integration of previous circular business models into a single one, a holistic circular business model canvas (Fig. 2). The value of this framework lies in capturing the complex ecosystemic view of a firm and synthesizing it in a simple manner, whereby firms encompass both a micro and a macro perspective. As such, this canvas has the potential to aide in communicating a firm's business model to various stakeholders, including financers and media. This fits in well with the recent popularity of ecosystems and recognition of value that resides across firm boundaries.

Following this historical review, we highlight three areas that we deem interesting for further research. First, circular supply chains or reverse supply chains are crucial in ensuring a successful CE transition. How does a transformation from a linear supply chain into a circular supply chain take place? Researchers could build on the work of Tolio et al. (2017) who argue that for a reverse supply chain to be built, manufacturers need to develop a simultaneous design of forward and reverse parts—almost mirroring an image where activities of repair, refurbishment, resell take place. In a globalized world, the reality of such a tight closed loop system is difficult, but not impossible (Tolio et al., 2017).

This brings us to highlighting the importance of empirical research in CE, which can shed light on the complexity of CE transition and thus highlight the various challenges that arise and how these are dealt with. Studying real case-studies and documenting empirical evidence from businesses and governments will allow researchers, practitioners, and policy makers to capture knowledge about how both large organizations and small/ medium firms are transitioning or "born-circular." Exploring the role of partnerships and collaborative efforts amongst firms could guide the CE transition research. Such exploratory studies could be done via action research, case-study, longitudinal study, or ethnography.

Finally, the discussion of impact is significant. Following the common adage in management, "what gets measured, gets done," research on impact measurement is crucial to get CE done. A focus on the three pillars of sustainability: environmental, societal, and economic is an entry point to such research. Prior literature has reported that the societal pillar is often overlooked in CE studies (Merli et al., 2018). While the indirect CE benefits to society are captured—such as better jobs, better health and well-being for people and the environment (EMF Ellen MacArthur Foundation, 2015)—research on how the private sector measures direct impact is, however, scant. Thus, moving forward with a prolific momentum toward CE and CBM, we deem these three areas key in advancing CE research: reverse supply chain, empirical studies supporting CE transition, and impact measurement.

References

Adam, S., Bucker, C., Desguin, S., Vaage, N., Saebi, T., 2018. Taking part in the circular economy: four ways to designing circular business models. SSRN Electron. J. https://doi.org/10.2139/ssrn.2908107.

Allenby, B.R., 1992a. Achieving sustainable development through industrial ecology. Int. Environ. Aff. 4, 56–68.

Allenby, B.R., 1992b. Industrial ecology: the materials scientist in an environmentally constrained world. MRS Bull. 17, 46–51. https://doi.org/10.1557/S0883769400040859.

Allenby, B.R., Richards, D.J., 1994. The Greening of Industrial Ecosystems. National Academy Press, Washington, DC.

Andersen, M.S., 2007. An introductory note on the environmental economics of the circular economy. Sustain. Sci. 2, 133–140. https://doi.org/10.1007/s11625-006-0013-6.

Antikainen, M., Valkokari, K., 2016. A framework for sustainable circular business model innovation. Technol. Innov. Manag. Rev. 6, 5–12. https://doi.org/10.22215/timreview1000.

Beltramello, A., Haie-Fayle, L., Pilat, D., 2013. Why New Business Models Matter for Green Growth. vol. 1 OECD, Paris, https://doi.org/10.1787/5k97gk40v3ln-en.

Benyus, J.M., 1997. Biomimicry: Innovation Inspired by Nature. William Morrow, New York.

Bisgaard, T., Henriksen, K., Bjerre, M., 2012. Green Business Model Innovation: Conceptualisation, Next Practice and Policy. Nordic Innovation, Oslo, Norway.

Blomsma, F., Brennan, G., 2017. The emergence of circular economy: a new framing around prolonging resource productivity. J. Ind. Ecol. 21, 603–614. https://doi.org/10.1111/jiec.12603.

Bocken, N.M.P., Short, S.W., Rana, P., Evans, S., 2014. A literature and practice review to develop sustainable business model archetypes. J. Clean. Prod. 65, 42–56. https://doi.org/10.1016/j.jclepro.2013.11.039.

Bocken, N.M.P., De Pauw, I., Bakker, C., Van Der Grinten, B., 2016. Product design and business model strategies for a circular economy. J. Ind. Prod. Eng. 33, 308–320. https://doi.org/10.1080/21681015.2016.1172124.

Bocken, N.M.P., Schuit, C.S.C., Kraaijenhagen, C., 2018. Experimenting with a circular business model: lessons from eight cases. Environ. Innov. Soc. Trans. 28, 79–95. https://doi.org/10.1016/j.eist.2018.02.001.

Boulding, K.E., 1966. The economics of the coming spaceship earth. In: Jarrett, H. (Ed.), Environmental Quality in a Growing Economy. Essays From Sixth RFF Forum. Johns Hopkins University Press, Baltimore, pp. 3–14.

Braungart, M., McDonough, W., Bollinger, A., 2007. Cradle-to-cradle design: creating healthy emissions—a strategy for eco-effective product and system design. J. Clean. Prod. 15, 1337–1348. https://doi.org/10.1016/j.jclepro.2006.08.003.

Buckminster, F.R., 1969. Operating Manual for Spaceship Earth, first ed. Southern Illinois University Press, Carbondale.

Carson, R., 1962. Silent spring. Penguin, New York.

Clinton, L., Whisnant., R., 2014. Model behavior—20 business model innovations for sustainability. SustainAbility, London.

Daly, H.E., 1992. Allocation distribution and scale: towards an economics that is efficient just and sustainable. Ecol. Econ. 6, 185–193. https://doi.org/10.1016/0921-8009(92)90024-M.

Den Hollander, M., Bakker, C., 2016. Mind the gap exploiter: circular business models for product lifetime extension. In: Proc. Electron. Goes Green, Berlin, Germany, pp. 1–8.

EMF Ellen MacArthur Foundation, 2013. Towards the Circular Economy. Vol. 2: Opportunities for the Consumer Goods Sectorowards a Circular Economy. https://www.ellenmacarthurfoundation.org/publications/towards-the-circular-economy-vol-2-opportunities-for-the-consumer-goods-sector. (Accessed 3 September 2019).

EMF Ellen MacArthur Foundation, 2019. Completing the Picture: How the Circular Economy Tackles Climate Change. http://www.ellenmacarthurfoundation.org/publications/completing-the-picture-climate-change. (Accessed 3 October 2019).

EMF Ellen MacArthur Foundation, SUN, McKinsey Center for Business and Environment, 2015. Growth Within: a Circular Economy Vision for a Competitive Europe. http://www.ellenmacarthurfoundation.

org/publications/growth-within-a-circular-economy-vision-for-a-competitive-europe. (Accessed 19 November 2019).

Eriksen, M.K., Christiansen, J.D., Daugaard, A.E., Astrup, T.F., 2019. Closing the loop for PET, PE and PP waste from households: influence of material properties and product design for plastic recycling. Waste Manag. 96, 75–85. https://doi.org/10.1016/j.wasman.2019.07.005.

European Commission, 2019. Report From the Commission to the European Parliament the Council the European Economic and Social Committee and the Committee of the Regions on the Implementation of the Circular Economy Action Plan; COM/2019/190 Final, Brussels.

Frishammar, J., Parida, V., 2018. Circular business model transformation: a roadmap for incumbent firms. Calif. Manage. Rev., 1–25. https://doi.org/10.1177/0008125618811926.

Frosch, R.A., 1992. Industrial ecology: a philosophical introduction. Proc. Natl. Acad. Sci. U. S. A. https://doi.org/10.1073/pnas.89.3.800.

Frosch, R., 2005. "Father of Industrial Ecology," Receives Prestigious Society Prize From International Society for Industrial Ecology. http://www.belfercenter.org/publication/robert-frosch-father-industrial-ecology-receives-prestigious-society-prize. (Accessed 12 September 2019).

Frosch, R.A., Gallopoulos, N.E., 1989. Strategies for manufacturing. Sci. Am. https://doi.org/10.1038/scientificamerican0989-144.

Ghisellini, P., Cialani, C., Ulgiati, S., 2016. A review on circular economy: the expected transition to a balanced interplay of environmental and economic systems. J. Clean. Prod. 114, 11–32. https://doi.org/10.1016/j.jclepro.2015.09.007.

Haanstra, W., Toxopeus, M.E., Van Gerrevink, M.R., 2017. Product life cycle planning for sustainable manufacturing: translating theory into business opportunities. Procedia CIRP 61, 46–51. https://doi.org/10.1016/j.procir.2016.12.005.

Hardin, G., 1968. The tragedy of the commons. Science (80-) 162, 1243–1248. https://doi.org/10.1126/science.162.3859.1243.

Hawken, P., Lovins, A., Lovins, L.H., 1999. Natural Capitalism: Creating the Next Industrial Revolution. Earthscan, London.

Heyes, G., Sharmina, M., Mendoza, J.M.F., Gallego-Schmid, A., Azapagic, A., 2018. Developing and implementing circular economy business models in service-oriented technology companies. J. Clean. Prod. 177, 621–632. https://doi.org/10.1016/j.jclepro.2017.12.168.

Hopkinson, P., Zils, M., Hawkins, P., Roper, S., 2018. Managing a complex global circular economy business model: opportunities and challenges. Calif. Manage. Rev. 60, 71–94. https://doi.org/10.1177/0008125618764692.

Jabbour, C.J.C., ABL de S, J., Sarkis, J., Filho, M.G., 2019. Unlocking the circular economy through new business models based on large-scale data: an integrative framework and research agenda. Technol. Forecast. Soc. Change 144, 546–552. https://doi.org/10.1016/j.techfore.2017.09.010.

Janssen, K.L., Stel, F., 2017. Orchestrating partnerships in a circular economy—a working method for SMEs. In: XXVIII ISPIM Innov. Conf.—Compos. Innov. Symphosium. The International Society for Professional Innovation Management (ISPIM), Vienna, pp. 1–17.

Kirchherr, J., Reike, D., Hekkert, M., 2017. Conceptualizing the circular economy: an analysis of 114 definitions. Resour. Conserv. Recycl. 127, 221–232. https://doi.org/10.1016/j.resconrec.2017.09.005.

Korhonen, J., Honkasalo, A., Seppälä, J., 2018. Circular economy: the concept and its limitations. Ecol. Econ. 143, 37–46. https://doi.org/10.1016/j.ecolecon.2017.06.041.

Lewandowski, M., 2016. Designing the business models for circular economy—towards the conceptual framework. Sustainability 8, 1–28. https://doi.org/10.3390/su8010043.

Lieder, M., Rashid, A., 2016. Towards circular economy implementation: a comprehensive review in context of manufacturing industry. J. Clean. Prod. 115, 36–51. https://doi.org/10.1016/j.jclepro.2015.12.042.

Linder, M., Williander, M., 2017. Circular business model innovation: inherent uncertainties. Bus. Strateg. Environ. 26, 182–196. https://doi.org/10.1002/bse.1906.

Lovins, A.B., Lovins, L.H., Hawken, P., 1999. A road map for natural capitalism. Harv. Bus. Rev. 77, 145–159.

Lüdeke-Freund, F., Gold, S., Bocken, N.M.P., 2019. A review and typology of circular economy business model patterns. J. Ind. Ecol. 23, 36–61. https://doi.org/10.1111/jiec.12763.

Lyle, J.T., 1994. Strategies for Regenerative Design. John Wiley & Sons, New York.

Mäler, K., 1974. Environmental Economics: A Theoretical Inquiry. Johns Hopkins University Press, Baltimore.
Manninen, K., Koskela, S., Antikainen, R., Bocken, N., Dahlbo, H., Aminoff, A., 2018. Do circular economy business models capture intended environmental value propositions? J. Clean. Prod. https://doi.org/10.1016/j.jclepro.2017.10.003.
McDonough, W., Braungart, M., 1998. The next industrial revolution. Atl. Mon. 282 (4), 82–92.
McDonough, W., Braungart, M., 2002. Remaking the Way We Make Things: Cradle to Cradle. North Point Press, New York.
Meadows, D.H., Wright, D., 2008. Thinking in Systems: A Primer. Chelsea Green Publishing, Vermont.
Meadows, D.H., Meadows, D.L., Randers, J., Behrens, W.W., 1972. The Limits to Growth. Universe Books, New York.
Melosi, M., 2005. Garbage in the Cities: Refuse, Reform, and the Environment, Rev. edn. University of Pittsburgh Press, Pittsburg.
Mendoza, J.M.F., Sharmina, M., Gallego-Schmid, A., Heyes, G., Azapagic, A., 2017. Integrating backcasting and eco-design for the circular economy: the BECE framework. J. Ind. Ecol. 21, 526–544.
Merli, R., Preziosi, M., Acampora, A., 2018. How do scholars approach the circular economy? A systematic literature review. J. Clean. Prod. 178, 703–722. https://doi.org/10.1016/j.jclepro.2017.12.112.
Morioka, S.N., Bolis, I., de Carvalho, M.M., 2018. From an ideal dream towards reality analysis: proposing sustainable value exchange matrix (SVEM) from systematic literature review on sustainable business models and face validation. J. Clean. Prod. 178, 76–88. https://doi.org/10.1016/j.jclepro.2017.12.078.
Nerurkar, O., 2017. A framework of sustainable business models. Indian J. Econ. Dev. 5, 1–6.
Nußholz, J.L.K., 2017. Circular business models: defining a concept and framing an emerging research field. Sustainability 9, 14–17. https://doi.org/10.3390/su9101810.
O'Riorden, T., 1981. Environmentalism. 2nd rev. Pion, London.
Osterwalder, A., Pigneur, Y., 2010. Business Model Generation: A Handbook for Visionaries, Game Changers, and Challengers. Hoboken, NJ, Wiley.
Pauli, G., 1997. Zero emissions: the ultimate goal of cleaner production. J. Clean. Prod. 5, 109–113.
Pauli, G.A., 2010. The Blue Economy: 10 Years, 100 Innovations, 100 Million Jobs. Paradigm Publications, Taos, NM.
Pearce, D.W., Turner, R.K., 1990. The circular economy. In: Economics of Natural Resources and the Environment. Johns Hopkins University Press, Baltimore, MD (chapter 2).
Pieroni, M.P.P., McAloone, T.C., Pigosso, D.C.A., 2019. Business model innovation for circular economy and sustainability: a review of approaches. J. Clean. Prod. 215, 198–216. https://doi.org/10.1016/j.jclepro.2019.01.036.
Planing, P., 2015. Business model innovation in a circular economy reasons for non-acceptance of circular business models. OPEN J. Bus. Model. Innov., 1–11. In press https://www.researchgate.net/publication/273630392; 2015. (Accessed 20 January 2019).
Pratt, K., Lenaghan, M., 2015. The Carbon Impacts of the Circular Economy. Scotland,.
Reike, D., Vermeulen, W.J.V., Witjes, S., 2018. The circular economy: new or refurbished as CE 3.0?—Exploring controversies in the conceptualization of the circular economy through a focus on history and resource value retention options. Resour. Conserv. Recycl. 135, 246–264. https://doi.org/10.1016/j.resconrec.2017.08.027.
Rizos, V., Behrens, A., van der Gaast, W., Hofman, E., Ioannou, A., Kafyeke, T., et al., 2016. Implementation of circular economy business models by small and medium-sized enterprises (SMEs): barriers and enablers. Sustainability 8, 1212. https://doi.org/10.3390/su8111212.
Roos, G., 2014. Business model innovation to create and capture resource value in future circular material chains. Resources 3, 248–274. https://doi.org/10.3390/resources3010248.
Rosa, P., Sassanelli, C., Terzi, S., 2019. Towards circular business models: a systematic literature review on classification frameworks and archetypes. J. Clean. Prod. 236, 117696. https://doi.org/10.1016/j.jclepro.2019.117696.
Schroeder, P., Anggraeni, K., Weber, U., 2019. The relevance of circular economy practices to the sustainable development goals. J. Ind. Ecol. 23, 77–95. https://doi.org/10.1111/jiec.12732.
Sillanpää, M., Ncibi, C., 2019. The circular economy: case studies about the transition from the linear economy. In: Circular Economy. Academic Press, Cambridge, pp. 1–35, https://doi.org/10.1016/b978-0-12-815267-6.00001-3.

Simmonds, P.L., 1862. Waste Products and Undeveloped Substances: Or, Hints for Enterprise in Neglected Fields. R. Hardwicke, London.
Stahel, W.R., 1986. Product life as a variable: the notion of utilization. Sci. Public Policy. https://doi.org/10.1093/spp/13.4.185.
Teece, D.J., 2010. Business models, business strategy and innovation. Long Range Plann. 43, 172–194. https://doi.org/10.1016/j.lrp.2009.07.003.
Tietenberg, T.H., 1984. Environmental and Natural Resource Economics. Scott Foresman, Glenview, IL.
Tolio, T., Bernard, A., Colledani, M., Kara, S., Seliger, G., Duflou, J., et al., 2017. Design, management and control of demanufacturing and remanufacturing systems. CIRP Ann. Manuf. Technol. 66, 585–609. https://doi.org/10.1016/j.cirp.2017.05.001.
Tukker, A., 2004. Eight types of product–service system: eight ways to sustainability? Experiences from SusProNet. Bus. Strateg. Environ. 13, 246–260. https://doi.org/10.1002/bse.414.
Tukker, A., Tischner, U., 2006. Product-services as a research field: past, present and future. Reflections from a decade of research. J. Clean. Prod. 14, 1552–1556. https://doi.org/10.1016/j.jclepro.2006.01.022.
von Weizacker, E., Lovins, A.B., Lovins, L.H., 1997. Factor Four: Doubling Wealth—Halving Resource Use. Earthscan Publications, London.
Witjes, S., Lozano, R., 2016. Towards a more circular economy: proposing a framework linking sustainable public procurement and sustainable business models. Resour. Conserv. Recycl. 112, 37–44. https://doi.org/10.1016/j.resconrec.2016.04.015.
Zamfir, A.-M., Mocanu, C., Grigorescu, A., 2017. Circular economy and decision models among European SMEs. Sustainability 9, 1507. https://doi.org/10.3390/su9091507.
Zink, T., Geyer, R., 2016. There is no such thing as a green product. Stanford Soc. Innov. Rev., 26–31. https://doi.org/10.1016/j.jacr.2015.03.015.
Zink, T., Geyer, R., 2017. Circular economy rebound. J. Ind. Ecol. 21, 593–602. https://doi.org/10.1111/jiec.12545.

CHAPTER 7

A triple-level framework to evaluate the level of involvement of firms in the circular economy (CE)

G. Lanaras-Mamounis[a], A. Kipritsis[a], Thomas A. Tsalis[a], Konstantinos I. Vatalis[b], and Ioannis E. Nikolaou[a]

[a]Business Economics and Environmental Technology Laboratory, Department of Environmental Engineering, Democritus University of Thrace, Greece
[b]Department of Mineral Resources Engineering, University of Western Macedonia, Kozani, Greece

1. Introduction

The concept of CE has lately gained great momentum among scholars, decision makers, and business the community. The academic debate mainly focuses on determining the content of the concept of CE. Scholars have made many efforts to explicate and elaborate the concept of CE by drawing insights and using knowledge from other academic fields (Govindan and Hasanagic, 2018; Veleva and Bodkin, 2018; Daddi et al., 2019). The heart of the idea of CE is based on shifting from conventional linear thinking (e.g., take-produce-distribute and dispose of waste materials) to a new circular economic thinking (e.g., cradle-to-cradle, largely eliminating waste materials) in order to return the end-of-life products and materials back into the production process. The concept of CE seems to be a contemporary response of policy makers (e.g., the European Union) and practitioners (e.g., industrial sectors) to present and future obstacles associated with the obvious "limit to growth" of nature, which could be tackled by both current and future economies through their production processes in order to cover the consumption needs of societies (Leipold and Petit-Boix, 2018).

A widely used framework to present the overall concept of the CE in scientific literature is the three-level economic framework: micro, *meso*, and macro level (Dopfer et al., 2004). The micro level is related to the various practices and projects adopted by business in order to narrow, slow and close the loop of the materials utilized throughout production processes (Bocken et al., 2016). The majority of such practices and projects rest on the principles of reusing, recycling, remanufacturing, and repurposing of materials. The *meso* level is a regional level and mainly examines firms' collaborations with their counterparts in the context of industrial ecology, industrial symbiosis, eco-clusters, and eco-industrial parks. Finally, the macro level includes methodological

Circular Economy and Sustainability, Volume 1
https://doi.org/10.1016/B978-0-12-819817-9.00014-4

frameworks to examine how the principles of CE have been implemented over large geographical regions or at national level.

According to the current literature, many methodological frameworks have been suggested in order to assist market actors in identifying appropriate ways to achieve CE goals at each level (Haupt et al., 2017). The mainstream of such methodological frameworks focuses mainly on the progress of CE on a single level (Franco, 2017; Gómez et al., 2018; Sauvé et al., 2016). Other methodologies evaluate the progress of CE in relation to various measurement units (e.g., financial or nonfinancial), while there are also frameworks that provide information by examining simultaneously one or more CE principles (e.g., reuse, recycle, and redesign). Furthermore, at micro level, it is important to stress that the majority of the suggested methodological frameworks offer indicators to evaluate either the operations of firms or the circularity of products (Linder et al., 2017).

Crucially, although that there are connections between the three levels of CE (i.e., micro, *meso*, and macro), these frameworks avoid examining more than one level. This partial emphasis of the current methodological frameworks highlights a gap in the literature. Actually, there is a need for a holistic approach which will incorporate all levels in order to evaluate how an actor contributes to CE. To this end, this book chapter suggests a triple-level CE methodological framework to evaluate the contribution of firms to CE. In other words, this implies that the suggested framework tries to measure the contribution of a firm to CE at micro level (e.g., operations and production processes), *meso* level (e.g., exchange materials among firms), and macro level (e.g., collaboration with local communities) simultaneously.

The rest of the chapter is organized as follows: in Section 2, the theoretical background regarding the current measurement frameworks of CE is analyzed. Section 3 presents the proposed methodological framework suggested to analyze the contribution of firms to CE. Section 4 presents an application of the suggested framework in a sample of 19 firms operating in the plastics sector, and the final section summarizes the conclusions and highlights some limitations as future research directions.

2. Theoretical background

The concept of CE has lately become very popular among academics and practitioners (Sauvé et al., 2016; Geisendorf and Pietrulla, 2018). The focus for the concept of CE is varied among academics and practitioners. Various scholars examine different techniques and frameworks in order to elucidate the meaning of the concept of CE and the ways in which various market agents (e.g., firms, consumers, households) could achieve the key goals of CE (Anastasiades et al., 2020). Moreover, practitioners adopt CE concepts by investing in related projects (such as recycling, reusing, redesigning) mainly to exploit new profitable opportunities (Makropoulos et al., 2018).

Given this background, a comprehensive analysis of the CE concept is necessary. Many scholars and international institutes have lately launched notable initiatives to outline the boundaries of the content of CE (Geissdoerfer et al., 2017; Korhonen et al., 2018). Thus, a number of definitions have been given regarding the CE concept. Kirchherr et al. (2017) identified 114 different definitions, which place emphasis on various principles of CE such as reuse, recycle, reduce, and recover. They also contend that there is a weak connection between these definitions and the concept of sustainable development. In addition, by reviewing the related literature, Korhonen et al. (2018) grouped CE definitions into two general categories. The first category encompasses definitions that are an evolution of the definition given by the Ellen MacArthur Foundation (EMF) and they are based on key features of the EMF's definition, such as regeneration and restoration of materials. The second category includes various initiatives of scholars to express the content of the concept of CE by focusing on closing economic and ecological loops. One of the contributions of Korhonen et al. (2018) is the new definition for CE that focuses on the combination of the classical components of the economic system, such as production and consumption. Similarly, Bocken et al. (2016) have highlighted some other useful perspectives of CE such as slowing, narrowing, and closing resource loops.

The definition of the CE concept is a crucial step since it offers the basis for selection of indicators, which will guide the evaluation of the contribution of various market actors to CE. Moraga et al. (2019) presented a mapping context to classify current measurement frameworks regarding CE by examining various aspects. A significant weakness of these frameworks is their concentration on a single analysis level (i.e., micro, *meso*, or macro level). Additionally, the literature proposes various indicators and accounting frameworks to evaluate the contribution level of various market actors to CE. Table 1 illustrates CE frameworks based on the level of analysis.

As for the micro level, the literature concentrates on firms' internal operations and production processes. On the one hand, there are methodological frameworks that focus on determining the contribution of firms to CE mainly at organizational and operational levels. In this context, Rossi et al. (2020) suggested a number of indicators by combining the principles of CE with classical business models and various components of sustainable development. De Jesus et al. (2018) propose various cleaner production and eco-design practices to promote alternative organizational techniques that improve work places of firms in the context of promoting the principles of CE. On the other hand, methodological frameworks have been developed that evaluate firms' progress on their product circularity. In this way, Cayzer et al. (2017) designed a framework to measure the performance of products by utilizing the key principles of CE. Through lifecycle thinking, Niero and Olsen (2016) have developed a multifaceted business system integrating CE principles and value creation ideas. In a similar vein, Mestre and Cooper (2017) suggest a framework to assist firms in incorporating CE principles into the lifecycle of their products, developing a four-level framework to achieve through the design of products the

Table 1 CE measurement frameworks.

Analysis level	Emphasis	Indicators- accounting Framework	Relative literature
Micro	Business operations	Eco-innovation, remanufacture, refurbish, recycle, canvas mapping tool	Smol et al. (2017) and Rossi et al. (2020)
	Production	Longevity indicator, eco-efficiency index, lifecycle analysis techniques, material circularity indicator, sustainable manufacturing CE index, CE toolkit	Franklin-Johnson et al. (2016), Figge et al. (2018), Azevedo et al. (2017), and Moktadir et al. (2018)
Meso	Industrial symbiosis, industrial ecology, eco-industrial parks, green supply chain	Eco-efficiency, flow analysis and assessment, substance flow analysis, material flow analysis	Ma et al. (2015), Kazancoglu et al. (2018), Mulrow et al. (2017), and Domenech et al. (2019)
Macro	Local and regional level	CE index, resource consumption intensity, waste emission intensity, waste recycling, and utilization rate	Guo et al. (2017), Ngan et al. (2019), Petit-Boix and Leipold (2018), and Wang et al. (2018)

slowing, closing, bio-inspiring, and bio-basing of the cycle. Furthermore, based on principles such as the initial generation of materials, the period of material restoration, and the period of recycling, scholars have proposed indicators to evaluate the longevity of materials without losing their usefulness and properties during their lifetime (Franklin-Johnson et al., 2016). Figge et al. (2018) have developed a set of indicators to evaluate circularity and durability of materials, as these are considered very important factors for firms and other market actors.

Meso level-related literature includes approaches that focus on appropriate indicators to evaluate the level of cooperation between individual firms for achieving the key principles of industrial ecology and industrial symbiosis. A number of studies examine the cooperation level of firms with other actors of the supply chain regarding CE issues (Govindan and Hasanagic, 2018). Moreover, Kazancoglu et al. (2018) suggested a conceptual framework to facilitate firms to incorporate the principle of CE into the procedures of green supply chain management (GSCM) by developing a hierarchical three-level approach for indicators. Additionally, Wen and Meng (2015) designed a hybrid technique by combining substance flow analysis and resource productivity, aiming to quantify the influence of firms on CE in the eco-industrial parks, which are considered the main way of establishing industrial symbiosis. Similarly, Felicio et al. (2016) have developed an indicator to measure the industrial symbiosis of eco-industrial parks based on various theoretical scenarios.

Finally, at macro level, there are studies that propose indicators to measure CE at a regional level. In more detail, Ngan et al. (2019) developed a CE index combining recycled materials with economic value. Petit-Boix and Leipold (2018) presented a CE index in order to examine various CE principles at municipal level. Avdiushchenko and Zając (2019) have suggested CE indicators to evaluate circular economy strategies within the European Union, while Silvestri et al. (2020) suggested two CE composite indicators to assess the static and dynamic circular economy conditions of European regions.

3. Research methodological framework

This chapter develops a questionnaire that can serve as a methodological tool for investigating the CE strategies implemented by firms using a triple-level approach. Specifically, the methodology consists of two parts, namely, the development of the questionnaire and the assessment of CE strategies.

3.1 Questionnaire development

The development of the questionnaire was a great challenge for the research team. The main goal was that the questionnaire had to include questions that could be regarded as indicators for assessing firms' performance regarding the level of implementation of various CE strategies More specifically, it includes 37 close-ended questions (i.e., indicators) (Appendix 1) divided into eight sections in accordance with the eight main categories of CE principles proposed in the related literature (Marques et al., 2017; Acevedo-García et al., 2020), namely reduce, repair/maintenance, reuse, refurbish/remanufacture, recycle, recover/cascade, extraction of biochemical feed, and renewable and efficient energy (see Table 2). Every single one of the 37 indicators/items corresponds to a particular strategic action (SA) associated with the CE concept and can be implemented by firms.

Table 2 Number of indicators/items (questions) per CE strategy categories.

CE strategy categories	Codification	Number of items (SA)
Reduce	RDCS	5
Repair/maintenance	RPMS	6
Reuse	RSES	6
Refurbish/remanufacture	RFBS	4
Recycle	RCCS	4
Recover/cascade	RCVS	4
Extraction of biochemical feed	EBFS	4
Renewable and efficient energy	REES	4

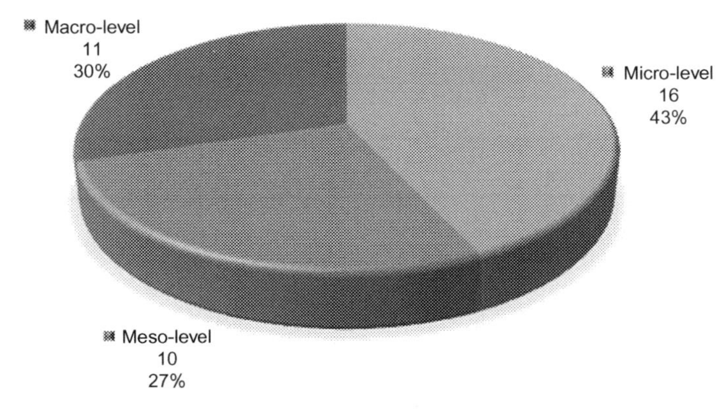

Fig. 1 Number of SA per CE implementation level.

The degree of implementation or development of these actions was examined by using a six-point Likert scale: 0: not implemented, 1: very low level of implementation, 2: low level of implementation, 3: moderate level of implementation, 4: high level of implementation, 5: full implementation. That is to say, all firms were asked to state to what extent they have encompassed these practices in their production process and operations.

According to Table 2, the RDCS category refers to strategies and actions that aim to minimize the consumption of raw materials, energy use, and waste production. The second category, RPMS, is associated with strategies that aim for the technical restoration and maintenance of products and equipment, while the RSES category focuses on reusing materials that have reached the end of their lifecycle for a similar or different purpose. Another category is the RFBS, which is related to the technical and aesthetical restoration of products, in order to bring them back in an almost-as-new state. The RCCS category includes traditional strategies of thermal disintegration of recyclable materials that can be used in the production process, and RCVS refers to the energy recovery from production waste or the reuse of materials for lower valued purposes. The seventh category, EBFS, is associated with strategies ensuring that all materials that return to the biosphere are biodegradable, nontoxic and nonbio-accumulative. Finally, REES refers to strategies that promote the consumption of renewable sources of energy in company's activities (Lieder and Rashid, 2015; Yuan et al., 2006; Ghisellini et al., 2016; Bocken et al., 2016; Lüdeke-Freund et al., 2019; Ellen Macarthur Foundation, 2015; Charter and Gray, 2008).

Fig. 1 depicts the classification of the strategic actions (SA) into the three levels of CE implementation. In particular, the micro level includes 16 SA, which refer to strategic actions implemented within the physical limits of single a firm. 10 SA are associated with the *meso* level and refer to material/energy/waste/equipment exchanges between different firms, while the macro level includes the strategic actions of CE (11 SA) that have impacts on larger geographical areas of operations (i.e., regional or national level) (Geng et al., 2012; Ghisellini et al., 2016; Kirchherr et al., 2017; Park et al., 2010; Lüdeke-Freund et al., 2019).

Fig. 2 depicts the matching of the 37 CE SA with the various processes typically encountered in operational and productive phases of firms. It includes the physical limits of the internal micro level, the main components involved, as well as a number of external interactions of a firm with third-party firms (meso level) and the local environment (macro level). This figure was mainly used as a visual aid to locate and include all related SA of CE with standard industrial entities, according to the triple-level approach.

Additionally, all CE strategies were grouped into material and energy loops—namely closed, slowed, and narrowing loops (Fig. 3). Strategic actions (RDCS, RCCS, and EBFS) of closed-loop CE strategies facilitate the minimization or the elimination of losses of materials and energy from production processes, while the strategic actions of slowed-loop (RPMS and RFBS) and narrowing-loop (RSES, RCVS and REES) extend the lifetime of resources and simply reduce the losses of materials and energy from production processes, respectively (Bocken et al., 2016; Braungart et al., 2007; Lüdeke-Freund et al., 2019; Geissdoerfer et al., 2017).

3.2 Assessment of CE strategies

In this section, the level of implementation of CE strategic activities were ranked based on the answers from the sampled firms. For analysis purposes, the degree of implementation index (DII) (see Eq. 1) was used, which is a slightly modified version of the relative importance index (RII). Actually, RII is a widely used technique for ranking the importance of factors or items according to responses (Wedawatta et al., 2014; Samantha, 2018; Kometa et al., 1994; Joseph et al., 2015; Gündüz et al., 2013; Akintoye et al., 1998).

$$DII = \frac{\sum_{i=1}^{i=N} D}{A^{*}N}, \qquad (1)$$

where D is the degree of implementation of a particular CE strategic action (SA) stated by a firm, A is the highest level of implementation (5 for the Likert scale used) and N is the total number of sampled firms (this index takes values between 0 and 1).

4. Questionnaire survey

4.1 Sample selection

The designed questionnaire was used to investigate the CE practices implemented and actions undertaken by Greek firms that operate in the plastics industry. A questionnaire survey was carried out between May 2019 and October 2019. A fully structured questionnaire was distributed to a sample of 66 firms of which the

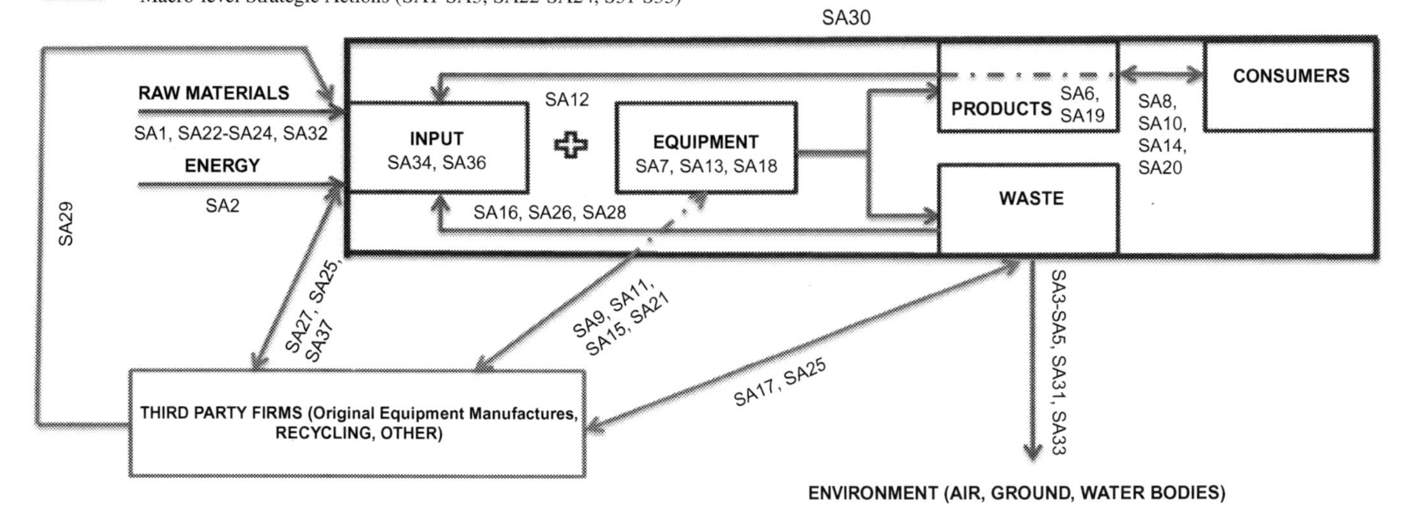

Fig. 2 Distribution of SA in relation to business activities and the implementation levels.

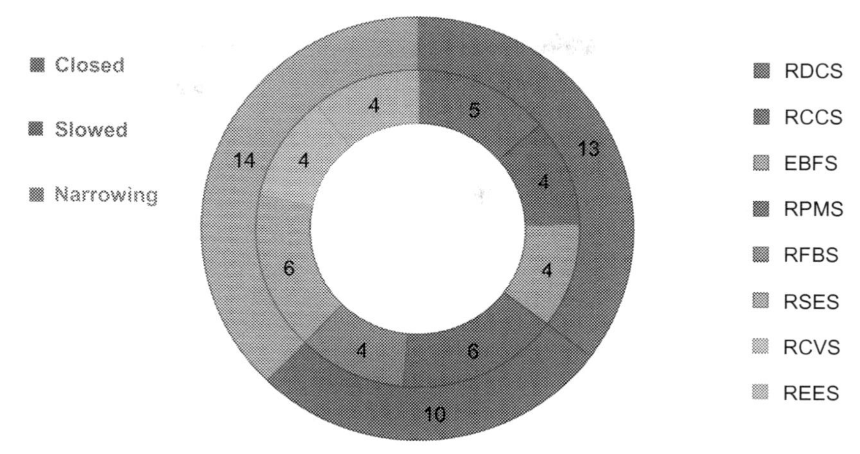

Fig. 3 Classification of CE strategies in material and energy loops.

overwhelming majority are members of the Association of Hellenic Plastic Industries (AHPI) (available at https://www.ahpi.gr/, 10/08/2019). Initially, an email was sent to all sampled firms inviting them to participate in this research. The email included a short description of the objective of the research, a brief introduction to the main concepts of CE, as well as a URL link to the online questionnaire. Then, follow-up calls were made to remind firms to fill out the questionnaire and to ensure the maximum possible response rate. In total, 19 completed questionnaires were received which corresponds to a 28.8% response rate. Although, the sample was small, the results of the questionnaire survey were very useful. First, it provided initial and new evidence for the CE practices adopted by Greek plastics firms, and second, it was a means of testing the structure and the content of the questionnaire.

4.2 Results and analysis

Table 3 presents the total number of firms that stated a particular degree of implementation for each one of the 37 SA. For example, 10 firms declared that they have not implemented (0) SA1, 6 stated a very low degree of implementation (1), and 1 and 2 firms stated a moderate (3) or a high degree (4) of implementation. Table 4 discloses the highest and lowest implemented SA of each CE category by utilizing DII, providing in this way an overview of the firms' attitude towards CE.

According to Table 4, the sampled firms were more familiar with SA associated with the RPMS, RCCS, and EBFS categories. Interestingly, all SA from the EBFS category have high DII scores (above 0.5), which means that firms pay attention to issues associated with the EBFS category. The high awareness of Greek plastics producers regarding the biodegradability and nontoxicity of their products and waste could be attributed to strict European legislation (such as the waste management directive (2008/98/EC)).

Table 3 Total number of firms stating a particular degree of implementation for each SA.

	Degree of implementation							Degree of implementation					
RDCS	**0**	**1**	**2**	**3**	**4**	**5**	**RPMS**	**0**	**1**	**2**	**3**	**4**	**5**
SA1	10	6	0	1	2	0	SA6	3	0	0	2	8	6
SA2	5	10	2	2	0	0	SA7	0	1	1	3	5	9
SA3	7	3	4	0	1	4	SA8	11	4	4	0	0	0
SA4	6	5	1	0	0	7	SA9	3	0	4	6	1	5
SA5	6	4	0	0	1	8	SA10	8	11	0	0	0	0
							SA11	2	2	2	1	8	4
	Degree of implementation							**Degree of implementation**					
RSES	**0**	**1**	**2**	**3**	**4**	**5**	**RFBS**	**0**	**1**	**2**	**3**	**4**	**5**
SA12	5	4	6	1	2	1	SA18	2	2	2	9	2	2
SA13	2	8	5	3	1	0	SA19	9	1	3	3	1	2
SA14	12	2	3	1	1	0	SA20	13	5	1	0	0	0
SA15	13	1	3	2	0	0	SA21	10	5	1	3	0	0
SA16	6	4	2	3	1	3							
SA17	5	5	3	0	2	4							
	Degree of implementation							**Degree of implementation**					
RCCS	**0**	**1**	**2**	**3**	**4**	**5**	**RCVS**	**0**	**1**	**2**	**3**	**4**	**5**
SA22	0	2	2	1	1	13	SA26	16	3	0	0	0	0
SA23	2	5	5	2	2	3	SA27	18	0	0	1	0	0
SA24	1	1	1	4	1	11	SA28	9	6	0	4	0	0
SA25	1	3	2	1	3	9	SA29	8	8	0	2	1	0
	Degree of implementation							**Degree of implementation**					
EBFS	**0**	**1**	**2**	**3**	**4**	**5**	**REES**	**0**	**1**	**2**	**3**	**4**	**5**
SA30	0	0	5	6	1	7	SA34	15	4	0	0	0	0
SA31	3	1	0	0	0	15	SA35	15	3	0	0	1	0
SA32	3	0	2	1	3	10	SA36	15	4	0	0	0	0
SA33	3	1	1	1	3	10	SA37	17	1	1	0	0	0

0: not implemented, 1: very low, 2: low, 3: moderate, 4: high, 5: very high.

Furthermore, SA from RCVS, REES, and RFBS categories are not very popular CE strategic actions for Greek plastics firms. Actually, the implementation of such practices needs investment in technologies, long pay-back periods, and governmental incentives in order to be implemented (Painuly, 2001; Shivakumar et al., 2019), which might be deterrent factors for Greek firms.

Another interesting finding from Table 4 is that the sampled firms mainly choose to design and implement "traditional" CE Strategies, such as RPMS, RCCS, and EBFS,

Table 4 Ranking of SA per CE category.

CE category	DoI	SA	IMPL	DII	CE category	DoI	SA	IMPL	DII
RDCS	High	SA5	Macro	0.51	RCCS	High	SA22	Macro	0.82
		SA4	Macro	0.44			SA24	Macro	0.78
	Low	SA2	Macro	0.21		Low	SA25	Meso	0.71
		SA1	Macro	0.18			SA23	Macro	0.46
RPMS	High	SA7	Micro	0.81	RCVS	High	SA29	Meso	0.19
		SA6	Micro	0.72			SA28	Micro	0.19
	Low	SA8	Micro	0.13		Low	SA27	Meso	0.03
		SA10	Micro	0.12			SA26	Micro	0.03
RSES	High	SA17	Meso	0.41	EBFS	High	SA31	Macro	0.80
		SA16	Micro	0.38			SA32	Macro	0.73
	Low	SA14	Micro	0.16		Low	SA33	Macro	0.72
		SA15	Meso	0.14			SA30	Micro	0.71
RFBS	High	SA18	Micro	0.54	REES	High	SA35	Meso	0.07
		SA19	Micro	0.32			SA34	Micro	0.04
	Low	SA21	Meso	0.17		Low	SA36	Micro	0.04
		SA20	Micro	0.07			SA37	Meso	0.03

D, degree of implementation; *SA*, strategic action; *IMPL*, implementation level.

Table 5 Ranking of SA for each implementation level.

Implementation level	D	SA	CE category	CE loop	DII	Average DII
Micro	High	SA7	RPMS	SL	0.81	0.31
		SA6	RPMS	SL	0.72	
		SA30	EBFS	CL	0.71	
		SA18	RFBS	SL	0.54	
	Low	SA20	RFBS	SL	0.07	
		SA34	REES	NL	0.04	
		SA36	REES	NL	0.04	
		SA26	RCVS	NL	0.03	
Meso	High	SA25	RCCS	CL	0.71	0.30
		SA11	RPMS	SL	0.64	
		SA9	RPMS	SL	0.58	
		SA17	RSES	NL	0.41	
	Low	SA15	RSES	NL	0.14	
		SA35	REES	NL	0.07	
		SA27	RCVS	NL	0.03	
		SA37	REES	NL	0.03	
Macro	High	SA22	RCCS	CL	0.82	0.55
		SA31	EBFS	CL	0.80	
		SA24	RCCS	CL	0.78	
		SA32	EBFS	CL	0.73	
	Low	SA4	RDCS	CL	0.44	
		SA3	RDCS	CL	0.37	
		SA2	RDCS	CL	0.21	
		SA1	RDCS	CL	0.18	

rather than "new" strategies linked with the CE concept (e.g., RDCS). This means that the adaptation of the examined firms in the latest trends towards corporate environmental management and CE is slow.

Furthermore, by utilizing the DII scores, Tables 5 and 6 present the most and least implemented SA by sampled firms in relation to the level of implementation and the CE loops, respectively. These tables are accompanied by the average DII achieved for each level of implementation and each material and energy loop to present the average performances of the sampled firms in each one of the above. With respect to the implementation level, SA that refer to macro level have the highest degree of implementation in comparison to the SA associated with *meso* and macro levels, suggesting a certain level of sensitivity of the sampled companies regarding the implications of their activities in their larger geographical area of operation. The low average DII scores at *meso* and micro levels indicate that there is a lot room for improvement in all strategic CE categories of these levels. According to the average DII, and with respect to the micro level, Greek plastics firms need to perform radical changes in their internal operational and productive

Table 6 Ranking of SA for each CE loop.

CE loop	D	SA	CE category	Implementation level	DII	Average DII
Closed loop	High	SA22	RCCS	Macro	0.82	0.57
		SA31	EBFS	Macro	0.80	
		SA24	RCCS	Macro	0.78	
		SA32	EBFS	Macro	0.73	
	Low	SA4	RDCS	Macro	0.44	
		SA3	RDCS	Macro	0.37	
		SA2	RDCS	Macro	0.21	
		SA1	RDCS	Macro	0.18	
Slowed loop	High	SA7	RPMS	Micro	0.81	0.41
		SA6	RPMS	Micro	0.72	
		SA11	RPMS	Meso	0.64	
		SA9	RPMS	Meso	0.58	
	Low	SA21	RFBS	Meso	0.17	
		SA8	RPMS	Micro	0.13	
		SA10	RPMS	Micro	0.12	
		SA20	RFBS	Micro	0.07	
Narrowing loop	High	SA17	RSES	Micro	0.41	0.17
		SA16	RSES	Micro	0.38	
		SA12	RSES	Micro	0.34	
		SA13	RSES	Micro	0.33	
	Low	SA36	REES	Micro	0.04	
		SA26	RCVS	Micro	0.03	
		SA27	RCVS	Meso	0.03	
		SA37	REES	Meso	0.03	

processes towards CE, in order to achieve better results. Furthermore, a lack of established synergies between firms of the sample and independent firms can be identified, as the low DII of 0.30 indicates an overall low level of cooperation for the exchange of waste/energy/byproducts for achieving the goals of industrial ecology. An improvement is expected in this area, as Greece has taken steps towards the establishment of both local eco-industrial parks (EIPs) and national or even global eco-industrial networks, following the example of the most famous EIP in the world located in Kalundborg, Denmark (Chertow, 2000; Marinos-Kouris and Mourtsiadis, 2013).

Finally, the findings from Table 6 demonstrate that the examined plastics firms have mainly implemented SA related to the closed and slowed loops. Conversely, SA related to the narrowing loops are underdeveloped. This means that the sampled firms have invested in strategies that minimize or even eliminate the loss of valuable materials and energy—and as a result, the entrance of new resources in their processes—and also actions that prolong and extend the lifetime of materials and products.

5. Conclusion and discussion

This chapter proposes a triple-level framework to quantify the contribution of firms to CE. It is a questionnaire-based framework that evaluates the degree of implementation of various strategic corporate aspects and practices in relation to CE, providing an overall picture of corporate CE performance. Taking into consideration the weaknesses of the previous frameworks, the proposed framework encompasses the three implementation levels (i.e., micro, *meso*, and macro levels). Actually, this framework could be regarded as a micro-level approach but it differs from other frameworks due to the fact that it is also closed connected with the *meso* and macro levels of implementation, examining the influence of firms on these two levels. That is to say, it is of critical importance to examine the implication of corporate CE strategies not only at micro level (e.g., reuse and recycle waste materials,), but also at *meso* level (e.g., use of waste materials from other firms) and at macro level (e.g., reduction of waste discharge to local water bodies and landfills).

What is more, the framework contributes to current literature by providing a simple methodology to quantify information about corporate CE strategic actions. The majority of the previous methodological frameworks have focused on developing interesting mathematical indexes, which are very useful but they might not be practical because, for competitive reasons, in many cases firms avoid disclosing the information necessary for the calculations (Linder et al., 2017; Moraga et al., 2019; Vatalis and Kaliampakos, 2006). Thus, the majority of such frameworks are conceptual and normative models, which focus on what should be quantified without providing empirical evidence for the firms' stance on issues related to the concept of CE. On the contrary, the suggested framework develops a new tool for gathering quantified information indirectly (through question-based surveys), which can be widely used for comparison purposes, offering empirical evidence on the firms' CE implemented practices.

Although the suggested framework was not modified in order to include the specific characteristic of the plastics industry, its application offers some interesting findings and useful conclusions about the behavior of the sampled firms toward CE. In particular, the sampled firms seem to place more emphasis on CE strategies at the macro level, whereas micro and meso levels have gained insufficient attention among the sampled firms. This can be explained by the limited interconnection between firms, mainly for competitive reasons, the limited institutional interventions to strengthen the CE behavior of firms, and, finally, the technical weaknesses of firms to reintroduce waste materials in their production process. In addition, examined firms mainly adopt CE strategies related to recycling (RCCS) and extraction of biochemical feed (EBFS) issues; although refurbishing and remanufacturing (RFBS) activities are key to firms' CE strategies (Lyons, 2005; Halabi et al., 2013), the sampled Greek plastics firms do not incorporate such strategies in their operations.

Moreover, the majority of the implemented CE strategies are linked with the closed loop, with firms placing great emphasis on EBFS strategies. In contrast, firms incorporate CE strategic actions that do not have an impact on narrowing the materials circle that they use in their production processes. This could be explained by the fact that firms face difficulties in reusing some of their materials due to technological deficiencies and the limits of thermodynamics laws.

Although the proposed framework offers several contributions to the relevant literature, there are some limitations that could be starting points for future research in order to provide an effective methodological framework for assessing corporate efforts on CE. Specifically, there is a need for new methods for gathering reliable information that will not rely exclusively on the responses and opinions of the firms' employees. Furthermore, techniques need to be developed to assess the accuracy of the information derived from respondents (i.e., firms) to describe the actual corporate performance and management practices in relation to the CE. Another issue that should be investigated is the structure of the questionnaire. All questions/indicators that describe the corporate CE strategic action should be revised in order to enhance their effectiveness in evaluating corporate CE practices. Thus, new questions can be added or other ones can be removed and modified. In addition, sector specific indicators could be developed in order to encompass the peculiarities of various sectors. To facilitate this, more empirical studies need to be carried out in order to improve the suggested framework.

A. Appendix 1

Strategic action (SA)	Description	CE strategy category	Level of implementation	Material/ energy loop
SA1	Reduction in total amount of raw materials used	RDCS	Macro	Closed loop
SA2	Reduction in total amount of energy consumption	RDCS	Macro	Closed loop
SA3	Reduction in total amount of solid waste production	RDCS	Macro	Closed loop
SA4	Reduction in total volume of wastewater discharges	RDCS	Macro	Closed loop
SA5	Reduction in total air emissions	RDCS	Macro	Closed loop
SA6	Design durable, long-lasting products	RPMS	Micro	Slowed loop
SA7	Acquire and maintain durable, long-lasting equipment	RPMS	Micro	Slowed loop
SA8	Provide product warranty and technical support to customers	RPMS	Micro	Slowed loop

Continued

cont'd

Strategic action (SA)	Description	CE strategy category	Level of implementation	Material/ energy loop
SA9	Acquire equipment covered by OEM warranty and technical support	RPMS	Meso	Slowed loop
SA10	Provide incentives to customers to repair/maintain products instead of replacing	RPMS	Micro	Slowed loop
SA11	Incentives provided to company by OEMs to repair/maintain equipment instead of replacing	RPMS	Meso	Slowed loop
SA12	Increase amount of secondary raw materials in total raw material consumption	RSES	Micro	Narrowing loop
SA13	Acquire second-hand equipment	RSES	Micro	Narrowing loop
SA14	Provide incentives to customers for returning unwanted, end-of-life company's products	RSES	Micro	Narrowing loop
SA15	Incentives provided to company by OEMs for returning unwanted, end-of-life equipment	RSES	Meso	Narrowing loop
SA16	Increase amount of reused company's waste in total raw material consumption of production processes	RSES	Micro	Narrowing loop
SA17	Increase amount of "exporting" company's waste for reuse by other companies	RSES	Meso	Narrowing loop
SA18	Acquire upgradeable equipment	RFBS	Micro	Slowed loop
SA19	Increase percentage of aesthetically/functionally refurbished products for sale	RFBS	Micro	Slowed loop
SA20	Provide incentives to customers to return products for refurbishment/remanufacturing	RFBS	Micro	Slowed loop
SA21	Incentives provided to company by OEMs to return equipment for remanufacturing	RFBS	Meso	Slowed loop
SA22	Increase amount of recyclable raw materials	RCCS	Macro	Closed loop
SA23	Increase amount of recycled raw materials	RCCS	Macro	Closed loop

Continued

cont'd

Strategic action (SA)	Description	CE strategy category	Level of implementation	Material/ energy loop
SA24	Increase amount of recyclable packaging	RCCS	Macro	Closed loop
SA25	Increase amount of company's waste ending up in third-party recycling facilities	RCCS	Meso	Closed loop
SA26	Increase amount of energy recovered from company's waste	RCVS	Micro	Narrowing loop
SA27	Increase amount of energy recovered from other companies' waste	RCVS	Meso	Narrowing loop
SA28	Increase amount of raw materials from downcycled materials of company	RCVS	Micro	Narrowing loop
SA29	Increase amount of raw materials from downcycled materials of other companies	RCVS	Meso	Narrowing loop
SA30	Perform lifecycle assessment of products and processes	EBFS	Micro	Closed loop
SA31	Treat solid/water/airborne waste emissions before disposal/ discharge	EBFS	Macro	Closed loop
SA32	Decrease amount of toxic/ nonbiodegradable raw materials in total consumption of raw materials	EBFS	Macro	Closed loop
SA33	Perform waste quality analysis (BOD, COD, CO_2, CH_4, SO_x, NO_x, biodegradable fraction)	EBFS	Macro	Closed loop
SA34	Increase amount of renewable energy in total energy consumption from own facilities	REES	Micro	Narrowing loop
SA35	Increase amount of renewable energy in total energy consumption from external sources	REES	Meso	Narrowing loop
SA36	Increase amount of other nonconventional energy sources in total energy consumption	REES	Micro	Narrowing loop
SA37	Increase amount of other nonconventional energy sources in total energy consumption	REES	Meso	Narrowing loop

References

Acevedo-García, V., Rosales, E., Puga, A., Pazos, M., Sanromán, M.A., 2020. Synthesis and use of efficient adsorbents under the principles of circular economy: Waste valorisation and electroadvanced oxidation process regeneration. Sep. Purif. Technol. 242, 116796.

Akintoye, A., Taylor, C., Fitzgerald, E., 1998. Risk analysis and management of private finance initiative projects. Eng. Constr. Archit. Manag. 5 (1), 9–21.

Anastasiades, K., Blom, J., Buyle, M., Audenaert, A., 2020. Translating the circular economy to bridge construction: lessons learnt from a critical literature review. Renew. Sust. Energ. Rev. 117, 109–522.

Avdiushchenko, A., Zając, P., 2019. Circular economy indicators as a supporting tool for European regional development policies. Sustainability 11 (11), 3025.

Azevedo, S.G., Godina, R., Matias, J.C.D.O., 2017. Proposal of a sustainable circular index for manufacturing companies. Resources 6 (4), 63.

Bocken, N.M., De Pauw, I., Bakker, C., Van Der Grinten, B., 2016. Product design and business model strategies for a circular economy. J. Ind. Prod. Eng. 33 (5), 308–320.

Braungart, M., McDonough, W., Bollinger, A., 2007. Cradle-to-cradle design: creating healthy emissions-a strategy for eco-effective product and system design. J. Clean. Prod. 15, 1337–1348. https://doi.org/10.1016/j.jclepro.2006.08.003.

Cayzer, S., Griffiths, P., Beghetto, V., 2017. Design of indicators for measuring product performance in the circular economy. Int. J. Sustain. Eng. 10 (4–5), 289–298.

Charter, M., Gray, C., 2008. Remanufacturing and product design. Int. J. Prod. Dev. 6, 375–392. https://doi.org/10.1504/IJPD.2008.020406.

Chertow, M., 2000. Industrial symbiosis: literature and taxonomy. Annu Rev Energ Environ 25, 313–337. https://doi.org/10.1146/annurev.energy.25.1.313.

Daddi, T., Ceglia, D., Bianchi, G., de Barcellos, M.D., 2019. Paradoxical tensions and corporate sustainability: A focus on circular economy business cases. Corp. Soc. Responsib. Environ. Manag. 26 (4), 770–780.

De Jesus, A., Antunes, P., Santos, R., Mendonça, S., 2018. Eco-innovation in the transition to a circular economy: an analytical literature review. J. Clean. Prod. 172, 2999–3018.

Domenech, T., Bleischwitz, R., Doranova, A., Panayotopoulos, D., Roman, L., 2019. Mapping industrial symbiosis development in Europe_typologies of networks, characteristics, performance and contribution to the circular economy. Resour. Conserv. Recycl. 141, 76–98.

Dopfer, K., Foster, J., Potts, J., 2004. Micro-meso-macro. J. Evol. Econ. 14, 263–279. https://doi.org/10.1007/s00191-004-0193-0.

Ellen MacArthur Foundation, 2015. Towards a Circular Economy: Business Rationale for an Accelerated Transition. Ellen MacArthur Foundation, Cowes.

Felicio, M., Amaral, D., Esposito, K., Durany, X.G., 2016. Industrial symbiosis indicators to manage eco-industrial parks as dynamic systems. J. Clean. Prod. 118, 54–64.

Figge, F., Thorpe, A.S., Givry, P., Canning, L., Franklin-Johnson, E., 2018. Longevity and circularity as indicators of eco-efficient resource use in the circular economy. Ecol. Econ. 150, 297–306.

Franco, M.A., 2017. Circular economy at the micro level: a dynamic view of incumbents' struggles and challenges in the textile industry. J. Clean. Prod. 168, 833–845.

Franklin-Johnson, E., Figge, F., Canning, L., 2016. Resource duration as a managerial indicator for circular economy performance. J. Clean. Prod. 133, 589–598.

Geisendorf, S., Pietrulla, F., 2018. The circular economy and circular economic concepts—a literature analysis and redefinition. Thunderbird Int. Bus. Rev. 60 (5), 771–782.

Geissdoerfer, M., Savaget, P., Bocken, N.M., Hultink, E.J., 2017. The circular economy—a new sustainability paradigm? J. Clean. Prod. 143, 757–768.

Geng, Y., Fu, J., Sarkis, J., Xue, B., 2012. Towards a national circular economy indicator system in China: an evaluation and critical analysis. J. Clean. Prod. 23, 216–224. https://doi.org/10.1016/j.jclepro.2011.07.005.

Ghisellini, P., Cialani, C., Ulgiati, S., 2016. A review on circular economy: the expected transition to a balanced interplay of environmental and economic systems. J. Clean. Prod. 114, 11–32. https://doi.org/10.1016/j.jclepro.2015.09.007.

Gómez, A.M.M., González, F.A., Bárcena, M.M., 2018. Smart eco-industrial parks: a circular economy implementation based on industrial metabolism. Resour. Conserv. Recycl. 135, 58–69.
Govindan, K., Hasanagic, M., 2018. A systematic review on drivers, barriers, and practices towards circular economy: a supply chain perspective. Int. J. Prod. Res. 56 (1–2), 278–311.
Gündüz, M., Nielsen, Y., Özdemir, M., 2013. Quantification of delay factors using the relative importance index method for construction projects in Turkey. J. Manag. Eng. 29 (2), 133–139.
Guo, B., Geng, Y., Ren, J., Zhu, L., Liu, Y., Sterr, T., 2017. Comparative assessment of circular economy development in China's four megacities: the case of Beijing, Chongqing, Shanghai and Urumqi. J. Clean. Prod. 162, 234–246.
Halabi, A.X., Montoya-Torres, J.R., Pirachicán, D.C., Mejía, D., 2013. A modelling framework of reverse logistics practices in the Colombian plastic sector. Int. J. Ind. Syst. Eng. 13 (3), 364–387.
Haupt, M., Vadenbo, C., Hellweg, S., 2017. Do we have the right performance indicators for the circular economy?: insight into the Swiss waste management system. J. Ind. Ecol. 21 (3), 615–627.
Joseph, R., Proverbs, D., Lamond, J., 2015. Assessing the value of intangible benefits of property level flood risk adaptation (PLFRA) measures. Nat. Hazards 79 (2), 1275–1297.
Kazancoglu, Y., Kazancoglu, I., Sagnak, M., 2018. A new holistic conceptual framework for green supply chain management performance assessment based on circular economy. J. Clean. Prod. 195, 1282–1299.
Kirchherr, J., Reike, D., Hekkert, M.P., 2017. Conceptualizing the circular economy: an analysis of 114 definitions. SSRN Electron. J. 127, 221–232. https://doi.org/10.2139/ssrn.3037579.
Kometa, S.T., Olomolaiye, P.O., Harris, F.C., 1994. Attributes of UK construction clients influencing project consultants' performance. Constr. Manag. Econ. 12 (5), 433–443.
Korhonen, J., Honkasalo, A., Seppälä, J., 2018. Circular economy: the concept and its limitations. Ecol. Econ. 143, 37–46. https://doi.org/10.1016/j.ecolecon.2017.06.041.
Leipold, S., Petit-Boix, A., 2018. The circular economy and the bio-based sector-perspectives of European and German stakeholders. J. Clean. Prod. 201, 1125–1137.
Lieder, M., Rashid, A., 2015. Towards circular economy implementation: a comprehensive review in context of manufacturing industry. J. Clean. Prod. 115, 36–51. https://doi.org/10.1016/j.jclepro.2015.12.042.
Linder, M., Sarasini, S., van Loon, P., 2017. A metric for quantifying product-level circularity. J. Ind. Ecol. 21 (3), 545–558.
Lüdeke-Freund, F., Gold, S., Bocken, N.M.P., 2019. A review and typology of circular economy business model patterns. J. Ind. Ecol. 23, 36–61. https://doi.org/10.1111/jiec.12763.
Lyons, D., 2005. Integrating waste, manufacturing and industrial symbiosis: an analysis of recycling, remanufacturing and waste treatment firms in Texas. Local Environ. 10 (1), 71–86.
Ma, S., Hu, S., Chen, D., Zhu, B., 2015. A case study of a phosphorus chemical firm's application of resource efficiency and eco-efficiency in industrial metabolism under circular economy. J. Clean. Prod. 87, 839–849.
Makropoulos, C., Rozos, E., Tsoukalas, I., Plevri, A., Karakatsanis, G., Karagiannidis, L., Makri, E., Lioumis, C., Noutsopoulos, C., Mamais, D., Rippis, C., Lytras, E., 2018. Sewer-mining: a water reuse option supporting circular economy, public service provision and entrepreneurship. J. Environ. Manag. 216, 285–298.
Marinos-Kouris, D., Mourtsiadis, A., 2013. Industrial symbiosis in Greece: a study of spatial allocation patterns. Fresenius Environ. Bull. 22, 2174–2181.
Marques, A., Guedes, G., Ferreira, F., 2017. Leather wastes in the Portuguese footwear industry: new framework according design principles and circular economy. Procedia Eng. 200, 303–308.
Mestre, A., Cooper, T., 2017. Circular product design. A multiple loops life cycle design approach for the circular economy. Des. J. 20 (Suppl. 1), S1620–S1635.
Moktadir, M.A., Rahman, T., Rahman, M.H., Ali, S.M., Paul, S.K., 2018. Drivers to sustainable manufacturing practices and circular economy: a perspective of leather industries in Bangladesh. J. Clean. Prod. 174, 1366–1380.
Moraga, G., Huysveld, S., Mathieux, F., Blengini, G.A., Alaerts, L., Van Acker, K., de Meester, S., Dewulf, J., 2019. Circular economy indicators: what do they measure? Resour. Conserv. Recycl. 146, 452–461.

Mulrow, J.S., Derrible, S., Ashton, W.S., Chopra, S.S., 2017. Industrial symbiosis at the facility scale. J. Ind. Ecol. 21 (3), 559–571.
Ngan, S.L., How, B.S., Teng, S.Y., Promentilla, M.A.B., Yatim, P., Er, A.C., Lam, H.L., 2019. Prioritization of sustainability indicators for promoting the circular economy: the case of developing countries. Renew. Sust. Energ. Rev. 111, 314–331.
Niero, M., Olsen, S.I., 2016. Circular economy: to be or not to be in a closed product loop? A life cycle assessment of aluminium cans with inclusion of alloying elements. Resour. Conserv. Recycl. 114, 18–31.
Painuly, J.P., 2001. Barriers to renewable energy penetration: a framework for analysis. Renew. Energy 24, 73–89. https://doi.org/10.1016/S0960-1481(00)00186-5.
Park, J., Sarkis, J., Wu, Z., 2010. Creating integrated business and environment value within the context of China's circular economy and ecological modernization. J. Clean. Prod. 18, 1494–1501. https://doi.org/10.1016/j.jclepro.2010.06.001.
Petit-Boix, A., Leipold, S., 2018. Circular economy in cities: reviewing how environmental research aligns with local practices. J. Clean. Prod. 195, 1270–1281.
Rossi, E., Bertassini, A.C., dos Santos Ferreira, C., do Amaral, W.A.N., & Ometto, A.R., 2020. Circular economy indicators for organizations considering sustainability and business models: plastic, textile and electro-electronic cases. J. Clean. Prod. 247, 119137.
Samantha, G., 2018. The impact of natural disasters on micro, small and medium enterprises (MSMEs): a case study on 2016 flood event in Western Sri Lanka. Procedia Eng. 212, 744–751.
Sauvé, S., Bernard, S., Sloan, P., 2016. Environmental sciences, sustainable development and circular economy: alternative concepts for trans-disciplinary research. Environ. Dev. 17, 48–56.
Shivakumar, A., Dobbins, A., Fahl, U., Singh, A., 2019. Drivers of renewable energy deployment in the EU: an analysis of past trends and projections. Energ. Strat. Rev. 26. https://doi.org/10.1016/j.esr.2019.100402, 100402.
Silvestri, F., Spigarelli, F., Tassinari, M., 2020. Regional development of circular economy in the European Union: a multidimensional analysis. J. Clean. Prod. 255, 120218.
Smol, M., Kulczycka, J., Avdiushchenko, A., 2017. Circular economy indicators in relation to eco-innovation in European regions. Clean Technol. Environ. Policy 19 (3), 669–678.
Vatalis, K.I., Kaliampakos, D.C., 2006. An overall index of environmental quality in coal mining areas and energy facilities. Environ. Manag. 38 (6), 1031–1045.
Veleva, V., Bodkin, G., 2018. Corporate-entrepreneur collaborations to advance a circular economy. J. Clean. Prod. 188, 20–37.
Wang, N., Lee, J.C.K., Zhang, J., Chen, H., Li, H., 2018. Evaluation of urban circular economy development: an empirical research of 40 cities in China. J. Clean. Prod. 180, 876–887.
Wedawatta, G., Ingirige, B., Proverbs, D., 2014. Small businesses and flood impacts: the case of the 2009 flood event in Cockermouth. J. Flood Risk Manag. 7 (1), 42–53.
Wen, Z., Meng, X., 2015. Quantitative assessment of industrial symbiosis for the promotion of circular economy: a case study of the printed circuit boards industry in China's Suzhou New District. J. Clean. Prod. 90, 211–219.
Yuan, Z., Bi, J., Moriguichi, Y., 2006. The circular economy: a new development strategy in China. J. Ind. Ecol. 10, 4–8.

CHAPTER 8

Exploring resource-service systems—Beyond product-service systems and toward configurations of circular strategies, business models, and actors

Fenna Blomsma[a], Mike Tennant[b], and Geraldine Brennan[b,c]
[a]Universität Hamburg, Faculty of Business, Economics and Social Sciences, Hamburg, Germany
[b]Imperial College London, Centre for Environmental Policy, Faculty of Natural Sciences, London, United Kingdom
[c]Irish Manufacturing Research Co., Dublin, Ireland

1. Introduction

The concept of (industrial) product-service systems ((i)PSSs, iPS^2) or product-as-a-service (PaaS) and the emergent concept of material-service systems (MSS) have been identified as potential mechanisms for enabling a circular economy (CE) (Aurisicchio et al., 2019; Boons and Bocken, 2018; Stahel, 2006; Zeeuw van der Laan and Aurisicchio, 2020; Yang et al., 2018). PSS include services such as offering spare parts or repair and maintenance services, as well as sharing models, such as tool libraries and car sharing systems (Tukker et al., 2006). MSS includes chemical management services that support the handling of chemicals (Stoughton and Votta, 2003) and chemical leasing or materials-as-a-service offerings.

Collectively, we refer to services with a resource component such as PSS and MSS as resource-service systems (RSS). This framing is aligned with and extends Tukker et al. (2006) to explicitly include materials and components as well as products. RSS thus covers the range of resource services in between pure resource sales (e.g., sales of materials, components, products) on the one hand, and pure service (where no or a negligible resource component is required) on the other. In between these two end-points, there exists a range of RSS where a mix of "tangible *resources*" and "intangible services" designed and combined so that they are jointly capable of fulfilling customer needs "*across the value chain*" (adapted from Tukker and Tischner, 2006 adaptations in italics).

RSSs can be considered circular business models[a]: ways of operating (part of) a business in a manner that is in line with the circular principle of addressing structural

[a] For reasons of simplicity, we limit our discussion here to circular business models, although we are aware that this topic is closely intertwined with sustainable business models. We refer the interested reader to: Schaltegger et al. (2012) and Boons and Lüdeke-Freund (2013).

Circular Economy and Sustainability, Volume 1
https://doi.org/10.1016/B978-0-12-819817-9.00017-X

waste. Addressing structural waste includes tackling obvious as well as unseen and intangible types of waste, while creating improved or new value creation and capture opportunities and reducing value loss and destruction (Blomsma and Brennan, 2017; EMF, 2016). The implementation of RSS requires the involvement of multiple actors within and between organizations. This is because redefining relationships and establishing new ways of working together calls for partners to collaborate and to establish new internal as well as shared processes. After all, "people make flows flow" (Baumann, 2004, 2012). As such, there is a close connection between the development of circular business models and the development of circular value chains.

Although the expectations for RSS to contribute to increased circularity and sustainability are high, their adoption levels are still low (OECD, 2017; Tukker, 2015). Moreover, evidence around the actual impact of the largest of the RSS subdomains, PSS, is mixed (Tukker, 2015), with concerns raised around rebound effects (Zink and Geyer, 2017). This prompts the need to design RSS in a manner that increases adoption levels while simultaneously addressing structural waste.

So far, however, there is a lack of a systematic, detailed, and pragmatic approach to understanding and designing new RSS with the aim to facilitate circularity across different lifecycle stages. Without this, there are limited mechanisms to: (a) understand the role of circular strategies in value creation and capture activities; (b) understand the necessary supporting actor configurations required, including the accompanying business model ecosystem; or (c) understand the barriers that need to be overcome and the enablers that make RSS a success. And thus, RSS-centered circular oriented innovation, or COI (Brown et al., 2019) cannot be effectively enabled.

In this chapter, we explore how to remedy this gap in a manner benefitting both the academic and business communities. First, we provide an analytical framework to guide the generation of deeper scholarly insights into RSS and circularity. Second, for practitioners of COI, we provide a pragmatic approach to understand and explore directions for COI through RSS, including both circular business models and circular value chains.

In the next section, we first explore the nature of existing PSS and MSS approaches and highlight where they are insufficient for understanding how to create and capture circular value through RSS-centered COI. We then put forward two frameworks that, combined with existing RSS thinking, can offer a way forward. Specifically, we discuss the Resource States framework and the Big Five structural wastes framework. Next, we explore the automotive industry through the lens of these frameworks and analyze the illustrative case of Riversimple—a car-as-a-service company. In the discussion section we generalize what we have learned and consolidate our insights by linking it to the broader CE business model and value chain discourses.

2. The resource-service system literature and its limitations

In this section we examine the current resource-service system (RSS) literature, and where and how it engages with CE. We discuss where it falls short in providing insights needed to advance COI.

2.1 RSS concepts and their role in a circular economy

In RSS, suppliers are incentivized to align with resource productivity in two ways. First, where resources are a revenue source, offering through-life support for suppliers means that they can continue to provide revenue for longer (Tukker, 2015). Second, where the resources are a cost, suppliers have an incentive to minimize resource use (Tukker, 2015). For example, in PSS, the provision of spare parts and maintenance services may extend product lifespan, preventing products from reaching their end-of-life prematurely. Through this, the continuity of the customer's activities is supported. Upgrading, refurbishment, and remanufacturing of products in between use-cycles allows the useful life of products to be further extended. Addressing other types of waste are PSS that facilitate pooling and sharing. Such models serve to make better use of un- or underused product capacity, seeking to generate more value from existing assets. This can benefit asset owners, who can generate additional revenue, as well as asset users, who get access to products that may otherwise be too costly for them to own and operate.

Analogous to PSS, MSS' goal is to increase the resource productivity of materials and other chemical resources. This can be achieved through better handling and warehousing, as well as through advice on use: two services that chemical management service (CMS) providers typically provide their customers with to streamline processes and improve their efficiency. CMS or materials-as-a-service concepts can also consist of services that recapture, treat, and redeploy materials. Reusing materials multiple times prevents them from reaching what would otherwise be a premature end-of-life. Such solutions may offer cost reductions, better process outcomes through a better understanding of requirements, secure access to resources that may be experiencing volatile pricing, as well as offering environmental benefits (Buschak and Lay, 2014; OECD, 2017; Stahel, 2006).

Through RSS, resources can be provided with through-life support, their lifetime extended, and/or their capacity for value creation optimized. This addresses structural waste, while offering value creation opportunities.

2.2 Limitation (a): RSS literature does not engage sufficiently with circular strategies, waste, and resources

Although the existing frameworks, tools and approaches are valuable in their respective contexts, they fall short from the perspective of a CE. Current typologies that distinguish between different RSS offerings for PSS (see Tukker, 2004; Tukker et al., 2006; Mont,

2002; Baines et al., 2007; and for MSS see Weigel, 2003, in Mont et al., 2008) typically do not engage with circular strategies, or not with the full range that falls under the CE umbrella (Blomsma and Brennan, 2017; Homrich et al., 2018). Where frameworks do acknowledge circular strategies they are insensitive to the range of potential implementation scenarios that are possible for such strategies (Blomsma and Tennant, 2020). For example, one scenario is when recycling material quality is preserved at a high level, whereas in a downcycling scenario a subsequent use for a material is found for an application that is considered either of lesser quality or of lesser value, which can be regarded as a cascade. Similarly, reuse that involves upgrading, refurbishment, or remanufacturing, is different from reuse that merely entails redistribution of a product to a new user or new use context, since product quality and features are remade in the former, but not the latter. As such, the RSS literature, in its current form, does not make the role of circular strategies explicit in RSS design.

In addition to this, current RSS approaches do not provide insight into the effect on structural waste in a system: they do not invite their users to consider what type or types of waste are present, the role of circular strategies in addressing them, and whether supporting circular strategies are needed to prevent new wastes from emerging. This lack of engagement with structural waste means that there is a risk of applying circular strategies with only limited capacity to solve the root causes of such waste appearing within our economies. For example, an initiative may focus on recycling, whereas reuse and more intensive use of products would be more effective.

Lastly, the RSS literature currently does not comprise a coherent whole. Rather, it consists of PSS and MSS as silos. This is problematic, since systemic COI tends to either create new, adapt or improve existing circular configurations: situations where two or more circular strategies are present and where synergies are sought, or where trade-offs between them have to be managed (Blomsma, 2016; Fischer and Pascucci, 2017). Crucially, circular configurations tend to consist of circular strategies affecting a mix of materials, components, and products. So far, an explicit acknowledgement of the relational nature of combined RSS approaches across these different forms of resources is absent.

There is, however, evidence that the combined application of RSS approaches can be beneficial. For example, Renault and its supplier combined the application of a PSS for cutting machines with an MSS for cutting fluids. This resulted in extending the lifetime of the fluid from a week or a month to a full year, reducing the total cost of ownership for the related processes by 33%, while the supplier was able to improve its margin by 125% (McKinsey and Co, 2016).

Moreover, a clearer distinction between different RSS types may be beneficial. Take Rolls Royce's airplane engine power-by-the-hour scheme (Smith, 2013), or façades-as-a-service models (Azcárate-Aguerre et al., 2018). Although both are typically treated as PSS, it is perhaps more appropriate to think of these as components-as-service (CSS), since they have to fit in and function as part of larger assemblies, respectively airplanes and buildings, which can themselves be operated as PSS. As such, these CSSs and PSSs can operate in tandem, in different parts of value chains.

Since RSS can be linked in these ways, it is necessary to consider the interactions between circular strategies affecting these different forms of resources. Moreover, where such links occur, their success or failure is likely to be driven by value chain dynamics. Therefore, it is crucial that RSS tools allow for exploring and understanding circular value chains, which at present it falls short of enabling.

2.3 Limitation (b): RSS literature does not allow for systematically exploring and understanding the necessary circular actor configurations

A second shortcoming of the RSS literature is that it allows for limited exploration and understanding of different actor perspectives and roles that are relevant for achieving a CE. Although the RSS literature does emphasize the alignment of processes and operating procedures, as well as supporting business models and reward structures, there are a number of dimensions to actor configurations that are particularly salient for circularity that are not currently addressed.

Different world views and mental models play an important role, and are a first step, in how actors prioritize different circular strategies and what success factors they pay attention to. This may obstruct collaboration, depending on whether their perspective is grounded in materials, components, or products (Lapko et al., 2019; Machacek et al., 2017).

Although there is an acknowledgement that creating circular systems can be achieved through linking different value chain actors in a variety of ways (Baumann, 2004, 2012; Brennan and Tennant, 2018), what constitutes an effective mix of links and relationships remains poorly understood. RSS can interface and link with other RSSs (see examples above), as well as with *other forms* of organizing resource flows. For example, RSS may interact with product or material sales models that have no service component. The combined dynamics of these two modes of operating may codetermine whether parts of the system are circular or not. This is of particular importance when CE practices are developing in an environment that still operates in a largely linear way. Similarly, pure services with no or a limited resource component may be needed to support parts of circular systems. The presence or absence of such support services can play an important role in establishing circular systems. Finance and insurance services are a prime example of such support services. Yet, it is remains unclear currently what actor configurations—the sets of actors and their manner of interaction within a value chain—are effective in delivering high levels of circularity as well as creating and capturing value.

In summary, the current RSS literature has shortcomings and blind spots when it comes to RSS-centered COI. At present PSS and MSS literature is limited in its capacity to understand the change that is needed within and between businesses to address structural waste and to capture the opportunities it represents. The main limitations are: (1)

current RSS approaches do not engage in sufficient detail with the role of circular strategies, waste, and resources and (2) current RSS approaches lack the capacity to shed light on the enabling actor configurations to support greater circular value creation and capture through RSS-oriented COI. This is a significant knowledge gap that stands in the way of a greater understanding of the combined resource and actor components of RSS.

3. Research design

In this section, we explore how the knowledge gap identified above can be addressed. First, we introduce the two analytical frameworks used in our exploration. Next, we describe how these two frameworks can be combined as part of an approach to analyzing RSS. Lastly, we explain our choice for a case study in the automotive sector, the results of which are presented in the subsequent section.

3.1 Analytical framework (1) Resource States

Resources "flow" through our economies and take on different forms along the way. As a result, it is not always clear when we say "reuse" whether we mean the reuse of materials, or whether we mean the reuse of products. Likewise, "recycling," in the colloquial sense, can cover both a new life for materials as well as for products (compare, for example, the use of recycling and reuse in (Krystofik et al., 2018 and Lazarevic and Valve, 2017), but also (Corporate Citizenship, 2014). When we describe circular strategies, therefore, it is helpful to have a clear understanding of exactly what it is that cycles.

For this, in line with Blomsma and Tennant (2020), we use "material entropy" (Boulding, 1966) to describe the degree to which materials are distributed and diffuse or are concentrated and organized. This results in distinguishing between three resource states. The first resource state is that of "particles," or elements, substances, molecules, or materials. Through the act of organizing, using technological or logistical means, (bulk) materials are created. Next, "parts" are created, covering the level of components, modules, and (sub)assemblies. This is where materials are given an intermediate level of organization: components are more organized than materials but are not yet sufficiently organized to be useful on their own. Finally, components are assembled and become "products," from which end-users can derive value.

Following the remainder of the industrial life cycle in line with lifecycle thinking, we see that at the end-of-use the former organization processes are partly reversed: products lose their form through a temporary (partial) disassembly to allow for repair, upgrading, or remaking processes. These processes negate a product's limiting state. Finally, when the product reaches its end-of-life it loses its high organization level permanently. At this point, if possible, components are "cannibalized," and the remainder is either landfilled or recycled. In addition to this, circular systems can have outputs to other systems where resources continue to be used. Think of using wastes as inputs, or cascading components

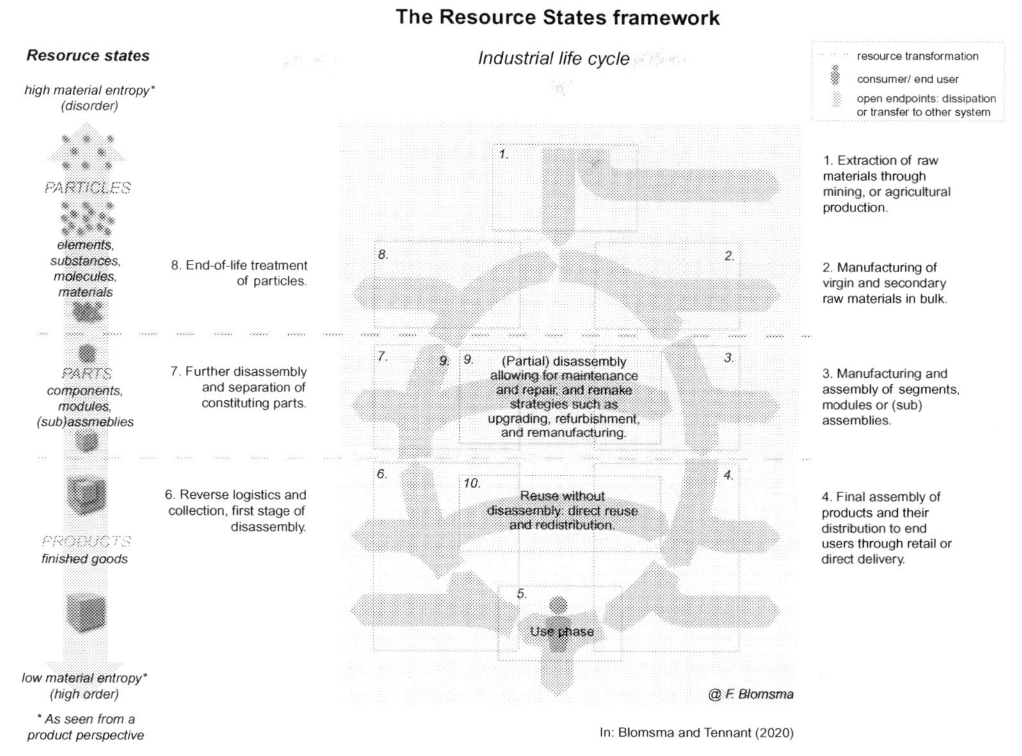

Fig. 1 Resource states framework: depicting the industrial life cycle, the flow of resources through the economy, and the different levels of organization these resources take on during this journey. *Image: Authors.*

and products for alternate use. As such, many different circular strategies can be part of a single system.

The journey of resources through the economy can be visualized as the Resource States framework, depicted in Fig. 1. This framework captures both the different levels of organization of resources, and the dynamism of their transitions as they move through the economy.

3.2 Analytical framework (2) the Big Five structural wastes

In addition to resources, it is necessary to take a systematic approach to waste, as waste exists in various forms. Sometimes it is obvious and clearly visible, such as when resources end up in a bin or in a landfill. Other times, waste is unseen and less straightforward to identify, such as when materials are reused once where multiple cascades could have contributed to increased value extraction, or when products are idle or underused. Collectively, we refer to these different types of waste as "structural waste" (Blomsma and

Tennant, 2020). Fig. 2 contains a typology for structural waste—the "Big Five" structural wastes—which is based on an analysis of a range of circular economy frameworks, such as *Cradle-to-Cradle* by Braungart and McDonough (2002), *The Performance Economy* by Stahel (2006), *The Blue Economy* by Pauli (2010), *Regenerative Design for Sustainable Development* by Lyle (1994), and the industrial symbiosis framework (e.g., Lowe and Evans, 1995) (see, for overview, Blomsma, 2018).

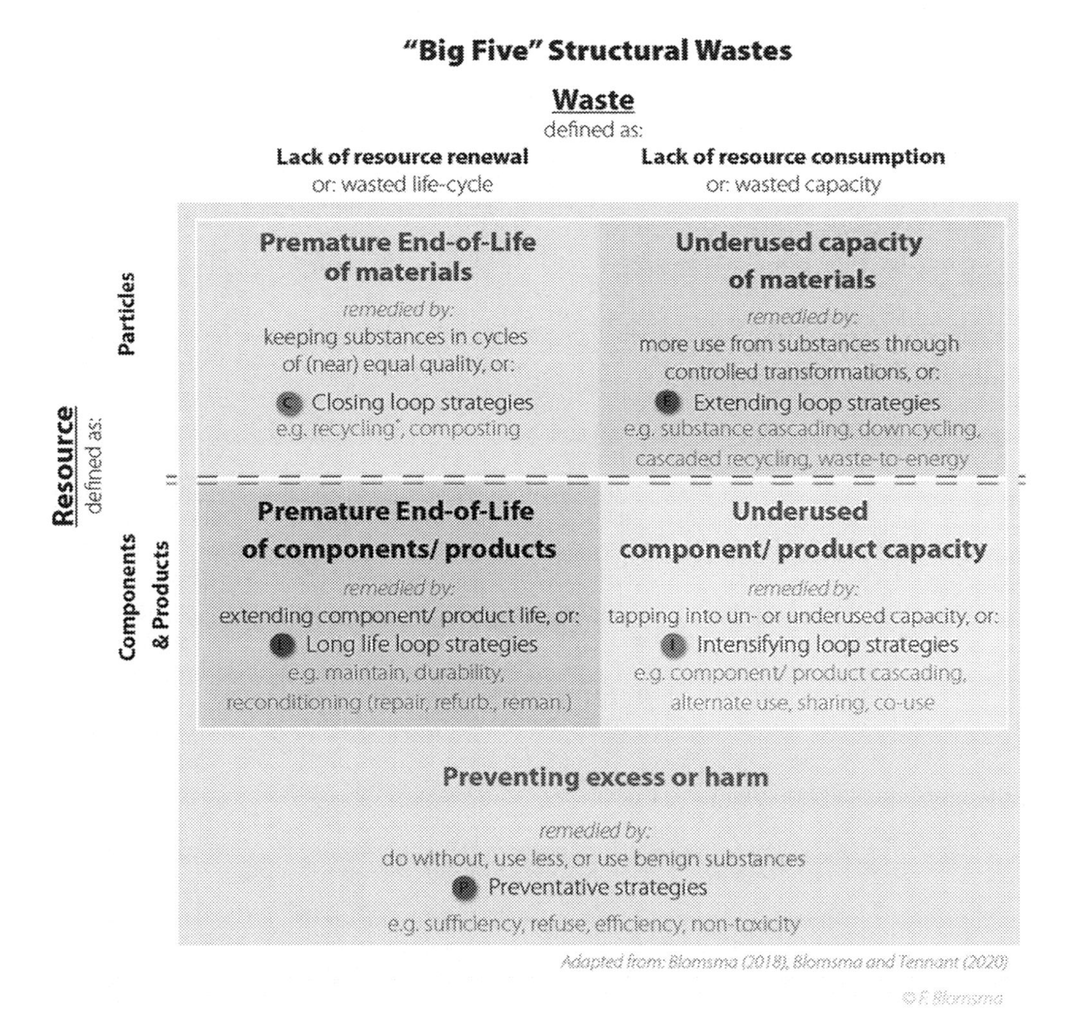

Fig. 2 The "Big Five" structural wastes, depicting in what manner materials, as well as components and products, can be considered a waste or wasted, and which circular strategies address which type of structural waste. *Image: Authors.*

The Big Five structural wastes framework considers materials, as well as components and products, as waste or wasted if they are not renewed or restored. In addition to this, resources can also be regarded as wasted when their capacity for value creation is not fully used. This results in a typology of four structural waste types: premature end-of-life of a material, underused material capacity, premature end-of-life of component or product, and underused component and product capacity. This is supplemented by a fifth type that covers preventative approaches that reduce quantities of resources used—think of "doing without" and "doing more with less,"—or that reduce harm in some other way, such as through nontoxicity. Each type of structural waste is linked to a group of circular strategies. For example, premature end-of-life of materials can be addressed by circular strategies that are aimed at closing loops, such as high-quality recycling and composting. Likewise, underused component and product capacity can be addressed by intensifying component and product use, such as through component cascading and product cascading, or approaches based on sharing and co-use.

3.3 Exploratory case selection: RSS potential in the automotive industry

We have chosen to explore RSS potential using an illustrative case from the automotive sector. Our reason for this is that this sector is, in one sense, mature and therefore already applies many circular strategies. For example: cars are commonly maintained and repaired throughout their life. In addition to this, there is a lively second-hand market for cars, where they are sometimes sold to a third or even fourth owner. Furthermore, within the EU, the ELV (end of life vehicles) Directive dictates that 85% of materials need to be reused or recycled, and 95% of materials to be reused or recovered (including energy recovery) (Directive 2000/53/EC).

Despite these commonly applied circular strategies, significant structural waste is still present. For one thing, cars are not in use 95% of the time,[b] sparking a recent interest in car sharing and mobility-as-a-service offerings. Moreover, potential exists for extending vehicle life, the life of parts and components, as well as to ensure better quality outcomes when it comes to recycling of materials (Allwood and Cullen, 2012; ABN Amro, 2016).

To delve into the potential of the full range of RSS, we unpack the whole system design approach of car-as-a-service provider Riversimple. In this model, Riversimple follows a radical efficiency approach, which entails a complete redesign of both its hydrogen-powered vehicles as well as its upstream value chain. Although Riversimple is currently a prerevenue company in the pilot stage, this case is illustrative of the degree to which circular strategies can be leveraged to address structural waste, while also creating new opportunities for value creation and capture. Data collection was done through a series of interviews and email correspondence with Riversimple employees, and

[b] https://fortune.com/2016/03/13/cars-parked-95-percent-of-time/.

consulting online company presentations as well as other published sources documenting the company's approach.

4. Results—Analysis of Riversimple—A car-as-a-service company

In line with the analytical frameworks used, we present our analysis as visual mappings: see Fig. 3 for the resource configuration, which depicts how resources flow and the role of various circular strategies, and see Fig. 4 for the actor configuration, which depicts touch points of different actors along the system.

4.1 The pillars of the Riversimple model

The core of Riversimple's approach to addressing structural waste can be summarized using three pillars: (a) the cars stay on the same balance sheet from beginning to end-of-life, (b) all operating costs are internalized on that same profit and loss account, and (c) building revenue-generating assets that are operated by the company, instead of products-for-sale. The consequence of this philosophy is that pricing to the customer is not driven by build cost but by lifetime cost or total cost of ownership (TCO). Through this, the sum of all costs in the system as a whole can be minimized. This is as opposed to focusing on minimizing just the build-cost as is the norm in regular for-sale models, where future cost saving as a result of efficiency, longevity, or end-of-life credits are usually discounted, in part because the first buyer does not or only minimally benefits from them. Next, we explain the role of PSS, CSS, and MSS, as well as the implications for actor relationships.

4.2 Product-service system

In the PSS typology of Tukker et al. (2006), the Riversimple model is a use-oriented PSS that can be operated as a product lease or a renting/sharing scheme. This means that use of the cars is intensified by giving more users access to them over a longer period of time, thus reducing idle time. As such, this model addresses the structural waste known as "underused product capacity." The Riversimple model is not unique in this: other car-sharing services, such as peer-to-peer car sharing (e.g., Get Around, Hiya Car), also accomplish this.

However, unlike peer-to-peer car-sharing models, Riversimple operates the cars it has manufactured and can therefore maximize the lifetime of these assets through high-quality maintenance and repair, as well as by providing high-quality spare parts. As such, this model also addresses premature end-of-life of the components it manufactures and the vehicle as a whole. Other car sharing models, such as the third party operated models of Greenwheels and Zipcar, where car-sharing operators buy cars from car

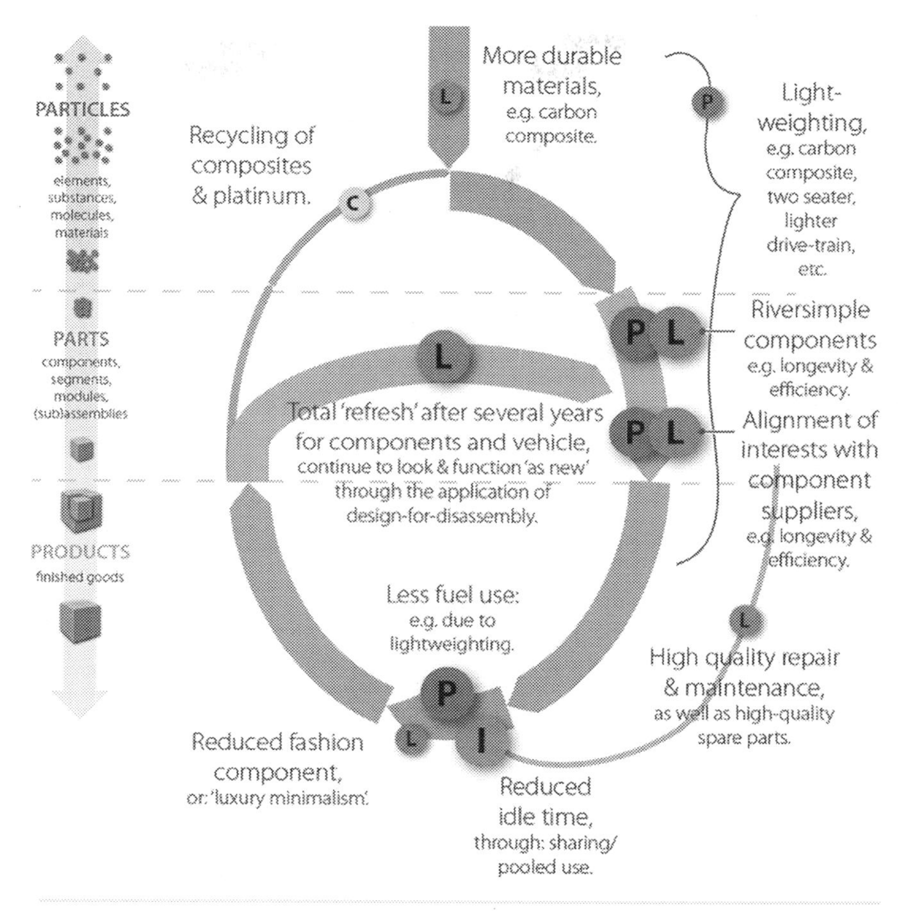

Systems integrator
Riversimple

Exploring different models, among which is renting/sharing. Service will include fuel, parking (in authorized areas), insurance, and maintenance. Depicted here is the set of circular strategies - or the circular configuration, which the company aims to leverage for circular value creation.

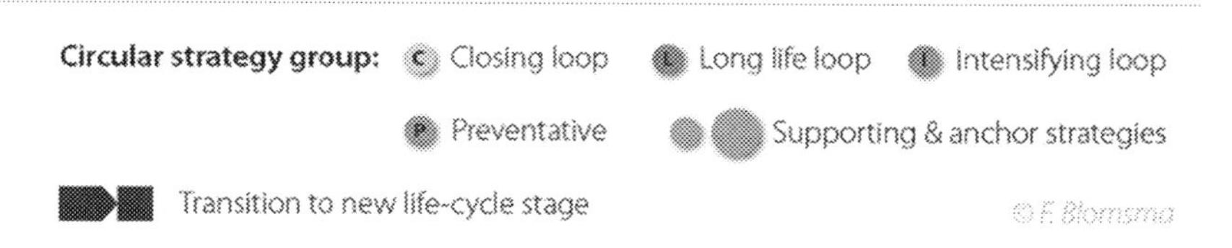

Fig. 3 Resource configuration of the Riversimple case. *Image: Authors.*

manufacturers, can also accomplish this, at least to a degree: once these models operate at scale, they can negotiate better service terms and higher quality spare parts through leveraging their buying power.

However, the Riversimple system is different from most other car-sharing services in that it actively addresses a multitude of structural wastes on top of the above. Specifically, Riversimple's vehicles are among the first to be designed specifically to be used as a service. This means that an emphasis was put on reducing the TCO of the vehicles through applying a whole systems approach, leading to radically improved efficiencies. Through applying this, the company creates a high number of synergistic circular strategies (Blomsma et al., 2018). This is a result of Riversimple's starting assumption that conventional cars are over-specified for conventional use (Bocken and Short, 2016; Wells, 2018).

For a start, mainstream car manufacturers do not have to provide the most efficient cars possible as their profits are not directly correlated to the fuel efficiency of their cars. Typically, only 14%–30% of fuel energy drives vehicles, with the rest lost as a result of power-transmission inefficiencies and powering ancillary items (Wells, 2018). In addition, of the four or five seats, only one or two are frequently used. Finally, the top speed possible is often higher than the actual top speed driven, resulting in larger engines than necessary. In terms of usage patterns, cars are not optimized for the frequent starting, stopping, and accelerating that is characteristic for city road use. Riversimple's hydrogen-powered Rasa prototype is approximately three times lighter than other production fuel cell cars. How they do this? By designing for a maximum of two occupants, using lightweight carbon composite for the body, with lower top speeds, and through decoupling acceleration from cruising demands, removing accessories, and by providing "luxury minimalism." These efforts allow Riversimple to include the fuel as part of the service at a price point that is competitive and reduces the cost of operating their car-as-a-service system. As such, this radical efficiency approach is leveraged by the company as an important contributor to generating healthy profit margins.

In addition to this, several design choices were made by Riversimple aimed at prolonging the productive lifetime of their vehicles: the carbon composite monocoque is not only lightweight, it is also extremely durable; high-speed metal-on-metal components were reduced to make the overall design more robust and long lived; and the vehicles are designed to be "refreshed" every few years, applying variations of refurbishment and remanufacturing enabled by design-for-disassembly, so they can continue to look and function like new. This latter intervention combats both aesthetic and functional premature end-of-life. Where cars normally stay within a renting or leasing system for anywhere between 12 and 18 months before being sold off, Riversimple's cars are meant to stay in operation for a minimum of 20 years.

4.3 Component and material-service systems

As well as these circular strategies described above, Riversimple aims to further extend the resource/service mindset up-stream in their supply chain and stimulate the adoption of use- and result-oriented RSS at their suppliers. The aim of this is to stimulate the development of several more circular strategies in their supply chain in the form of component-service systems and a material-service system. This means making components more efficient, more robust, easier to service, and easier to repair and replace.

An illustrative example in the Riversimple case is the membrane that is found in the fuel cell of hydrogen powered vehicles. Conventionally, producers of such components make more money by selling more parts and by reducing the costs at which those parts are made. This way of operating is fundamentally in conflict with addressing structural waste in two ways: (1) the sales of more parts means that more raw materials are needed to produce them and (2) the part supplier does not have an inherent incentive to provide the most efficient and most robust part possible. The efficiency aspect is particularly problematic, since the overall efficiency of the vehicle, and therefore the total cost of ownership, is significantly influenced by the amount of platinum on the membrane. Since platinum is expensive, the supplier has an inherent incentive to reduce the amount of platinum (platinum loading has gone down in the last 20 years as a result of an industry-wide drive), while it is in Riversimple's interest to increase the amount of platinum in relation to current levels. Especially for smaller companies, it may be difficult to demand that suppliers deliver components with specifications that deviate from current industry norms. As such, the interests between actors in the system are fundamentally not aligned.

Instead, working with suppliers in component-service systems means that Riversimple gets access to the functionality of the components for a fee, while the suppliers retain ownership of them. This incentivizes the development of more robust and long-lived components. In addition to this, Riversimple envisions the creation of a revenue sharing model, where the CSS suppliers are rewarded according to their contribution to the overall efficiency of the system. As a result, the component supplier has an incentive to develop and provide the most efficient components. In the case of platinum, this may mean increasing the platinum loading. Since the supplier will get back the components at the end of their life, the materials will remain in the system.

Riversimple is applying the same logic to its materials and is exploring with material suppliers (e.g., platinum and carbon composite) whether it is possible to get access to these raw materials through a materials-as-a-service mechanism. However, the motivation behind this is subtly different for both materials. As platinum is a high-value and rare metal, it is important to secure access to it to continue to operate. This is not necessarily the case for the carbon composite. At present, this material is difficult to recycle. A materials-as-a-service model would allow a supplier of such a material to plan for its return and develop appropriate recycling technologies in the meantime. In the case of the carbon composite, therefore, MSS serves as a way to develop circular value chains as the company scales.

As such, the Riversimple system design contains resource-service systems in each resource state. That is: it consists of an integrated set of a product-service system, component-service systems, and material-service systems. Here it is also key to differentiate between anchoring circular strategies and supporting circular strategies. By "anchoring" circular strategies we refer to strategies that are central to the model, whereas "supporting" circular strategies are those that enable or support other circular strategies or that prevent new structural wastes from occurring. Fig. 3 gives an overview of the set of circular strategies—or the circular configuration—used in this case, highlighting the different anchoring vs supporting circular strategies.

4.4 Aligning interests across value chains and actor interfaces

Riversimple's aim to integrate several RSSs and to align interests across actors involves a radical redesign of the value chain: all partners involved have to adapt and align their business models, including their approach to value creation and clear procedures for responsibility and risk sharing, so that a new ecosystem of circular business models is created. For this reason, Riversimple has worked closely with suppliers to increase the performance and robustness of their components—taking more effort and time than initially planned, as off-the-shelf solutions were found to be insufficient. In addition to this, new innovative sourcing and contracting mechanisms are being explored. That is: new ways of sharing information to increase transparency (e.g., block chain) are being developed. This system will provide a mechanism for greater visibility and fairness within the value chain, as well as enabling cumulative learning, which can contribute to further efficiency of the whole system. Through Riversimple's novel approach to value chain collaboration, all partners benefit and interests across the value chain become aligned with each other as well as with reduced resource use. Importantly, this means that the relationship of the different actors changes from a transactional relationship, which is the norm in the current linear economy, to having an ongoing relationship throughout the life-time of the resources, as indicated in Fig. 4. That is, the actor interfaces change.

5. Discussion and conclusion—Resource-service systems

Our exploration of the nature of existing PSS and MSS approaches has highlighted a significant knowledge gap related to understanding how to create circular value and capture it through RSS-centered COI, or circularity oriented innovation. We have argued that existing PSS and MSS approaches: (1) do not engage in sufficient detail with resources, waste, and circular strategies and (2) that they lack the capacity to investigate enabling actor configurations from a systemic point of view.

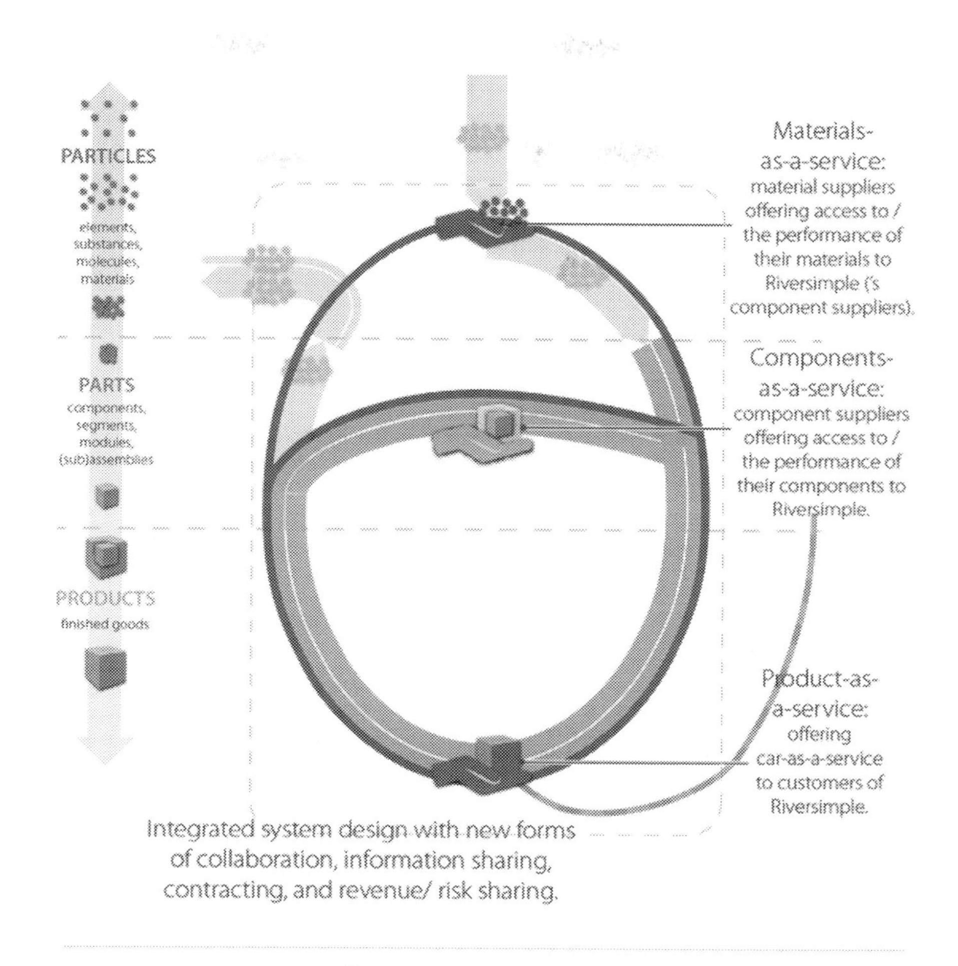

Systems integrator
Riversimple

Exploring different models, among which is renting/sharing. Service will include fuel, parking (in authorized areas), insurance, and maintenance. Depicted here is the set of value chain relationship types, highlighting where resource/service-systems can be found.

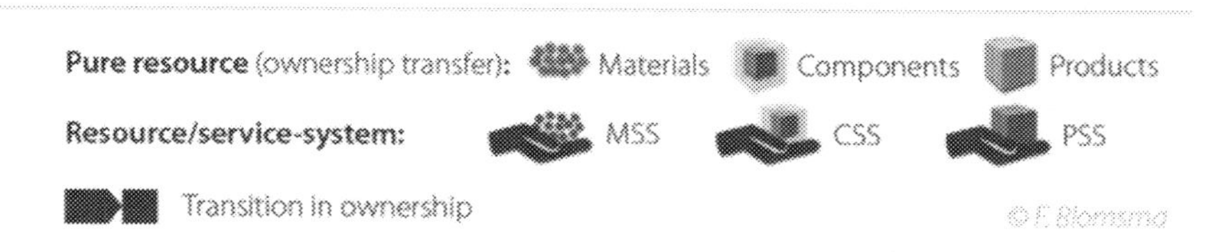

Fig. 4 Actor configuration of the Riversimple case. Note that in particular on the components level, there may in reality be different layers of CSS not depicted here for reasons of simplicity. *Image: Authors.*

Through our exploratory case study of the Riversimple system design—using the Resource States framework and the Big Five structural wastes framework—we have shown how multiple circular strategies can be used synergistically throughout a value chain, and how they can be enabled by a combination of a product-service system, component-service systems, and material-service systems.

Moreover, the case has illustrated how circular strategies can actively be leveraged as sources of shared value creation and capture, for both the actors that are part of the value chain as well as for the environment, through reducing fuel and material consumption. With this, we make a contribution to the circular business model literature, and provide insight into how circular strategies can be leveraged for value creation and capture within businesses (Bocken et al., 2016; Geissdoerfer et al., 2018; Lüdeke-Freund et al., 2019).

In addition to this, the case illustrated how RSS can be used to change actor relationships from linear transaction-based relationships, to on-going collaborative relationships aligned with shared value creation and reduced resource impact. This observation is in line with other recent work on circular economy and stakeholder relationships (see, e.g., Casalegno et al., 2020).

This, however, does not mean that the Riversimple model is the only way of capturing circular value or that capturing further circular advantage is out of reach for other model types that reply on other actor interfaces (e.g., peer-to-peer car sharing, or third party operated models). Rather, the Riversimple case illustrates the range of different actor interfaces that are possible. These and other configurations of actor interfaces, such as buy-back systems or producer responsibility schemes, will have to be explored further and be compared with regards to the degree to which such opportunities offer value creation and capture opportunities. Currently, little knowledge of such topics exists, highlighting the need for further work.

The exploratory case study presented also provides an example of an analytical framework for studying circular value chain phenomena: through mapping where structural waste is present, what circular strategies can be used to address it, and how actor configurations can facilitate it through working in new ways, among which are RSS approaches. This method can also be used by practitioners in COI projects, as it facilitates thinking through the implications of new business model and supply chain designs.

Lastly, we have identified further opportunities to extend and connect current PSS and MSS approaches; in particular, interrogating the distinction between PSS and CSS in the context of circularity. We also contend that to date, PSS and MSS approaches have been studied in isolation, and there is value in studying these as part of larger circular configurations. Comparing and contrasting different configurations with regards to their shared value creation and circularity levels can enable the formulation of recommendations and best-practice, as well as the identification of barriers and how to overcome them. As such, our contribution in this chapter addresses the calls of Lozano et al. (2013) for establishing "collaborative business models," Boons and Bocken (2018) for creating "ecologies of business models," and Hsieh et al. (2017) for a "sustainable business ecosystem." With this, we have made a contribution to the circular value chain literature.

References

ABN Amro, Circle Economy, 2016. On the Road to the Circular Car.

Allwood, J., Cullen, J., 2012. Sustainable Materials With both Eyes Open Magazine of Concrete Research. UIT Cambridge Ltd, Cambridge.

Aurisicchio, M., Zeeuw van der Laan, A., Tennant, M., 2019. Material-service systems for sustainable resource management. In: Proc. EcoDesign 2019 Int. (Symp).

Azcárate-Aguerre, J.F., den Heijer, A., Klein, T., 2018. Integrated façades as a product-service system—business process innovation to accelerate integral product implementation. J. Facade Des. Eng. 6 (1), 41–56.

Baines, T.S., Lightfoot, H.W., Evans, S., Neely, A., Greenough, R., Peppard, J., Roy, R., Shehab, E., Braganza, A., Tiwari, A., Alcock, J.R., Angus, J.P., Bastl, M., Cousens, A., Irving, P., Johnson, M., Kingston, J., Lockett, H., Martinez, V., Michele, P., Tranfield, D., Walton, I.M., Wilson, H., 2007. State-of-the-art in product-service systems. Proc. Inst. Mech. Eng. Part B J. Eng. Manuf. 221, 1543–1552.

Baumann, H., 2004. Environmental assessment of organising: towards a framework for the study of organisational influence on environmental performance. Prog. Ind. Ecol. An. Int. J. 1 (1–3), 292.

Baumann, H., 2012. Using the life cycle approach for structuring organizational studies of product chains. In: 18th Greening of Industry Network Conference, Linköping.

Blomsma, F., 2016. Making Sense of Circular Economy—How Practitioners Interpret and Use the Idea of Resource Life-Extension. Imperial College London, London.

Blomsma, F., 2018. Collective 'action recipes' in a circular economy—on waste and resource management frameworks and their role in collective change. J. Clean. Prod. 199, 969–982.

Blomsma, F., Brennan, G., 2017. The emergence of circular economy: a new framing around prolonging resource productivity. J. Ind. Ecol. 21, 603–614.

Blomsma, F., Tennant, M., 2020. Circular economy: preserving materials or products? Introducing the resource states framework. Resour. Conserv. Recycl. 156, 104698.

Blomsma, F., Kjaer, L., Pigosso, D., McAloone, T., Lloyd, S., 2018. Exploring circular strategy combinations—towards understanding the role of PSS. Procedia CIRP 69, 752–757.

Bocken, N.M.P., Short, S.W., 2016. Towards a sufficiency-driven business model: experiences and opportunities. Environ. Innov. Soc. Trans. 18, 41–61.

Bocken, N.M.P., de Pauw, I., Bakker, C., van der Grinten, B., 2016. Product design and business model strategies for a circular economy. J. Ind. Prod. Eng. 33, 308–320.

Boons, F., Bocken, N., 2018. Towards a sharing economy—innovating ecologies of business models. Technol. Forecast. Soc. Change 137, 40–52.

Boons, F., Lüdeke-Freund, F., 2013. Business models for sustainable innovation: state-of-the-art and steps towards a research agenda. J. Clean. Prod. 45, 9–19. https://doi.org/10.1016/j.jclepro.2012.07.007.

Boulding, K., 1966. The economics of the coming spaceship earth. In: Jarrett, H. (Ed.), Environmental Quality in a Growing Economy—Essays From the Sixth RFF Forum. The Johns Hopkins University Press, Baltimore.

Braungart, M., McDonough, W., 2002. Cradle to Cradle: Remaking the Way We Make Things, first ed. North Point Press, New York.

Brennan, G., Tennant, M., 2018. Sustainable value and trade-offs: exploring situational logics and power relations in a UK brewery's malt supply network business model. Bus. Strateg. Environ. 27, 621–630. https://doi.org/10.1002/bse.2067.

Brown, P., Bocken, N., Balkenende, R., 2019. Why do companies pursue collaborative circular oriented innovation? Sustainability 11, 635.

Buschak, D., Lay, G., 2014. Chemical industry: servitization in niches. In: Servitization in Industry. Springer International Publishing, Cham, pp. 131–150.

Casalegno, C., Civera, C., Mosca, F., Freeman, R.E., 2020. Circular Economy and Relationship-Based View. Symphonya. Emerg. Issues Manag. 149. https://doi.org/10.4468/2020.1.12casalegno.civera.mosca.freeman.

Corporate Citizenship, 2014. Ahead of the Curve: How the Circular Economy Can Unlock Business Value. (Self-published).

EMF, 2016. The New Plastics Economy—Rethinking the Future of Plastics. The Ellen MacArthur Foundation.

Fischer, A., Pascucci, S., 2017. Institutional incentives in circular economy transition: the case of material use in the Dutch textile industry. J. Clean. Prod. 155, 17–32.

Geissdoerfer, M., Morioka, S.N., de Carvalho, M.M., Evans, S., 2018. Business models and supply chains for the circular economy. J. Clean. Prod. 190, 712–721.

Homrich, A.S., Galvão, G., Abadia, L.G., Carvalho, M.M., 2018. The circular economy umbrella: trends and gaps on integrating pathways. J. Clean. Prod. 175, 525–543. https://doi.org/10.1016/j.jclepro.2017.11.064.

Hsieh, Y.-C., Lin, K.-Y., Lu, C., Rong, K., 2017. Governing a sustainable business ecosystem in Taiwan's circular economy: the story of spring Pool glass. Sustainability 9, 1068.

Krystofik, M., Luccitti, A., Parnell, K., Thurston, M., 2018. Adaptive remanufacturing for multiple lifecycles: a case study in office furniture. Resour. Conserv. Recycl. 135, 14–23. https://doi.org/10.1016/j.resconrec.2017.07.028.

Lapko, Y., Trianni, A., Nuur, C., Masi, D., 2019. In pursuit of closed-loop supply chains for critical materials: an exploratory study in the green energy sector. J. Ind. Ecol. 23, 182–196.

Lazarevic, D., Valve, H., 2017. Narrating expectations for the circular economy: towards a common and contested European transition. Energy Res. Soc. Sci. 31, 60–69. https://doi.org/10.1016/j.erss.2017.05.006.

Lowe, E.A., Evans, L.K., 1995. Industrial Ecology and Industrial Ecosystems. J. Clean. Prod. 3, 47–53.

Lozano, R., Carpenter, A., Satric, V., 2013. Fostering green chemistry through a collaborative business model: a chemical leasing case study from Serbia. Resour. Conserv. Recycl. 78, 136–144.

Lüdeke-Freund, F., Gold, S., Bocken, N.M.P., 2019. A review and typology of circular economy business model patterns. J. Ind. Ecol. 23, 36–61.

Lyle, J.T., 1994. Regenerative Design for Sustainable Development. John Wiley & Sons, New York, NY.

Machacek, E., Richter, J., Lane, R., 2017. Governance and risk–value constructions in closing loops of rare earth elements in global value chains. Resources 6, 59.

McKinsey & Co. 2016. No Title.

Mont, O., 2002. Clarifying the concept of product–service system. J. Clean. Prod. 10, 237–245.

Mont, O., Singhal, P., Fadeeva, Z., 2008. Chemical management services in Sweden and Europe: lessons for the future. J. Ind. Ecol. 10, 279–292.

OECD, 2017. Economic features of chemical leasing, series on risk management no. 37, environment, health and safety, environment directorate, OECD.

Pauli, G., 2010. The Blue Economy: 10 Years, 100 Innovations, 100 Million Jobs. Paradigm Publications.

Schaltegger, S., Freund, F.L., Hansen, E.G., 2012. Business cases for sustainability: the role of business model innovation for corporate sustainability. Int. J. Innov. Sustain. Dev. 6, 95. https://doi.org/10.1504/IJISD.2012.046944.

Smith, D.J., 2013. Power-by-the-hour: the role of technology in reshaping business strategy at Rolls-Royce. Technol. Anal. Strateg. Manag. 25, 987–1007.

Stahel, W., 2006. The Performance Economy, second ed. Palgrave MacMillan, London.

Stoughton, M., Votta, T., 2003. Implementing service-based chemical procurement: lessons and results. J. Clean. Prod. 11, 839–849.

Tukker, A., 2004. Eight types of product–service system: eight ways to sustainability? Experiences from SusProNet. Bus. Strateg. Environ. 13, 246–260.

Tukker, A., 2015. Product services for a resource-efficient and circular economy—a review. J. Clean. Prod. 97, 76–91.

Tukker, A., Tischner, U., 2006. Product-services as a research field: past, present and future. Reflections from a decade of research. J. Clean. Prod. 14 (17), 1552–1556.

Tukker, A., van den Berg, C., Tischner, U., 2006. Product-services: a specific value proposition. In: Tukker, A., Tischner, U. (Eds.), New Business for Old Europe. Greenleaf, Sheffield. 2001.

Wells, P., 2018. Degrowth and techno-business model innovation: the case of Riversimple. J. Clean. Prod. 197, 1704–1710.

Yang, M., Smart, P., Kumar, M., Jolly, M., Evans, S., 2018. Product-service systems business models for circular supply chains. Prod. Plan. Control 29, 498–508.

Zeeuw van der Laan, A., Aurisicchio, M., 2020. A framework to use product-service systems as plans to produce closed-loop resource flows. J. Clean. Prod. 252, 119733.

Zink, T., Geyer, R., 2017. Circular economy rebound. J. Ind. Ecol. 21, 593–602.

CHAPTER 9

Complementing circular economy with life cycle assessment: Deeper understanding of economic, social, and environmental sustainability

Mehzabeen Mannan and Sami G. Al-Ghamdi
Division of Sustainable Development, College of Science and Engineering, Hamad Bin Khalifa University, Qatar Foundation, Doha, Qatar

1. Introduction

The fundamental philosophy of circular economy (CE) is the concept of "cradle-to-cradle" in production and consumption systems through the application of reuse, recovery, and recycling of materials and energy. This concept is gaining popularity worldwide as it aims to increase resources efficiency through closed loop technologies. However, the closed loop concept of CE does not always ensure environmental benefits. Hence, comprehensive environmental assessment tools such as life cycle assessment (LCA) are quite beneficial for careful assessment of the available options of CE in product design. The combination of LCA and CE can result in more in-depth analysis and better understanding of economic, social, and environmental sustainability. Hence, this chapter gives an overview of the opportunities and benefits of implementing LCA in CE evaluation. The following sections will introduce the basics of LCA along with the integration of LCA into the CE concept. Finally, five case studies were assessed where LCA has been used as an assessing tool for CE.

2. LCA: A holistic approach

LCA is a comprehensive, robust scientific tool that addresses the environmental aspects and potential environmental impacts of any product or process throughout its life cycle, starting from raw material extraction to final disposal or recycling (ISO, 2006a; Guinee, 2002; Finnveden et al., 2009). This "cradle-to-grave" approach refers to a holistic view of environmental interactions that compiles all input and output data associated with any product system or process and analyzes environmental loadings over its life cycle linked to relevant inputs and outputs (Klöpffer, 2003; Yoonus et al., 2020). These inputs can be

Circular Economy and Sustainability, Volume 1
https://doi.org/10.1016/B978-0-12-819817-9.00032-6

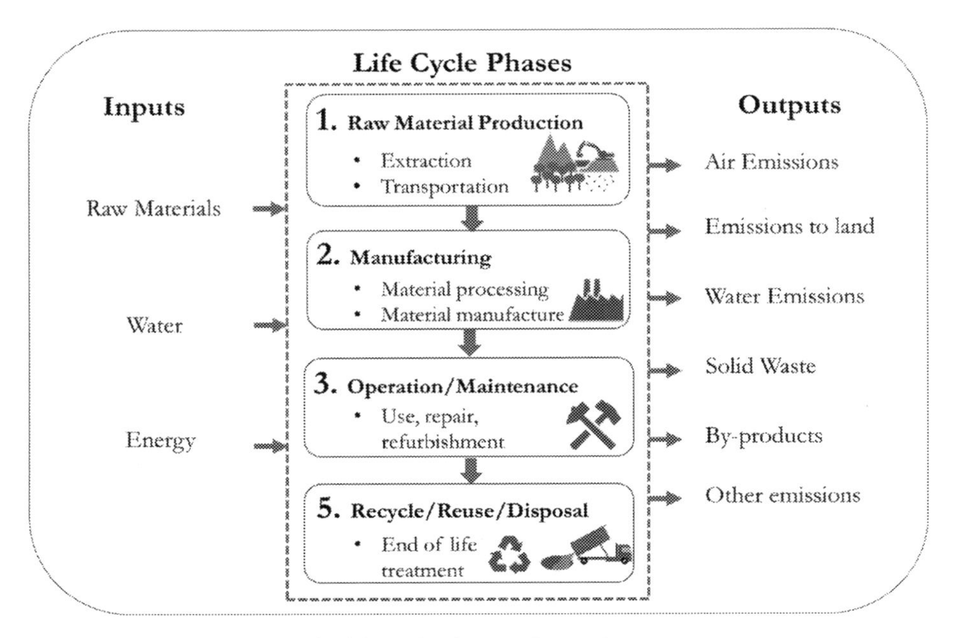

Fig. 1 Input and output streams for life cycle phases of a product.

listed as extracted raw materials from the earth required for any product stream or process, as well as the energy and water used, while outputs include emissions to land, air, and water; solid wastes; by-products; and other emissions (Fig. 1). At the end of the evaluation, LCA summarizes the environmental impact of products or processes in the area of resources used, human health hazards, and ecosystem damage (Baumann and Tillman, 2004).

Evaluation of all life cycle phases of a product enables LCA to estimate the cumulative impact results; this, in turn, allows selection of the product or production route that is most environmentally friendly (Mannan and Al-Ghamdi, 2019, 2020). Therefore, LCA tends to provide comparative conclusions, helping policymakers find better alternatives by comparing all key environmental impacts (Seidel, 2016).

2.1 ISO standards for LCA

The International Organization for Standardization (ISO) has declared international standards for most technological fields. During the 1980s and 1990s, when environmental issues started to be handled with great care, more than 350 international standards were published by the ISO. Regarding LCA, the series of standards from ISO 14040 to ISO 14049 are mainly connected to LCA, with ISO 14040 and ISO 14044 being especially familiar to LCA practitioners as foundational standards (Table 1).

The first standard, ISO 14040:2006, provides basic principles and a framework for LCA, including goal and scope definition, description of life cycle inventory analysis

Table 1 LCA standards by ISO.

Standard numbers	Standard title	References
ISO 14040:2006	Environmental management—Life cycle assessment—Principles and framework	ISO (2006b)
ISO 14044:2006	Environmental management—Life cycle assessment—Requirements and guidelines	ISO (2006c)
ISO 14046:2014	Environmental management—Water footprint—Principles, requirements, and guidelines	ISO (2014)
ISO 14047:2012	Environmental management—Life cycle assessment—Illustrative examples of application of ISO 14044 to impact assessments	ISO (2012a)
ISO 14048:2002	Environmental management—Life cycle assessment—Data documentation format	ISO (2002)
ISO 14049	Environmental Management—Life cycle assessment—Examples of application of ISO 14041 to goal and scope definition and inventory analysis	ISO (2012b)

and impact assessment phases, critical review, limitations, relationships between phases and conditions for use of value choices, and optional elements without detailed LCA techniques or methodologies for individual phases. ISO 14041, 14042, and 14043 have been revised by ISO 14044:2006, which has been designed to parallel the structure of ISO 14040:2006 with more details. The next standard, ISO 14046:2014, describes an LCA framework for assessing the water footprint of any product, process, or organization. It provides clear guidelines on assessing and reporting a water footprint analysis either as a standalone assessment or as part of a comprehensive evaluation. ISO 14047:2012 provides illustrations on how to conduct Life cycle impact assessment (LCIA) by ISO 14044:2006, reflecting key elements of the LCIA phase. Guidelines for data documentation have been provided by ISO 14048:2002, which enables LCA practitioners to document transparent data and exchange LCA and LCI data for consistent data documentation. Examples of how to carry out LCI have been demonstrated through ISO 14049:2012, while nevertheless, reflecting only a portion of LCI.

2.2 Stages in LCA

Nowadays, LCA is known as a sophisticated decision support tool that is both highly transparent and reliable. As discussed above, the stages involved in evaluating impacts with LCA are defined by the ISO. Typical LCA plans for any product or process generally include four main stages. Fig. 2 represents the components involved in LCA.

2.2.1 Goal and scope definition

The objective of an LCA study is generally set up at the beginning of the goal and scope definition stage. Based on the study's aims, the scope of assessment is determined,

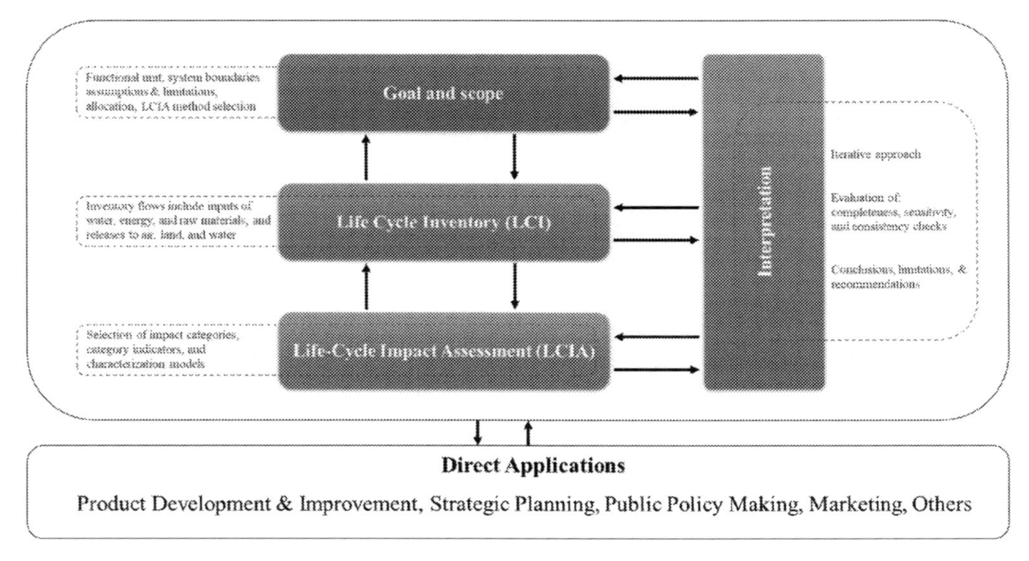

Fig. 2 Graphic representation of the four main stages in LCA.

including modelling options, functional unit of the project, selection of impact assessment method and impact categories, allocation regulations, measuring data quality, and formulation of system boundaries. When the study goal is fixed, it is necessary to define which specific product, process, or technological options will be studied through LCA, along with the functional unit. Selection of the functional unit should be reasonable and allow for comparison of all available alternatives in the study. Choice of impact assessment method and impact categories should be made during this stage, as these choices guide the life cycle inventory stage in data collection. System boundary sets the limits within which the investigation should be conducted since it is often impractical to include all input and output data in LCA studies.

Selection of the precise goal and scope during the initial stage is crucial for any LCA study. Ideally, all requirements for LCA modelling should be made clear in this stage to avoid inaccuracy in subsequent stages. However, in practice, LCA is an iterative process; therefore, decisions are often needed during a later stage for questions that were not evident in the initial stage. Typically, in an ISO-standardized LCA, two parties are involved in the goal and scope definition stage: the commissioner and the practitioner. The commissioner is the person who initiates an LCA study, while the practitioner is the person or group involved in carrying out the study.

2.2.2 Life cycle inventory (LCI)

The second stage of LCA includes the collection of all environmental inputs and outputs that are highly associated with the product or process being analyzed (Fig. 3). This stage

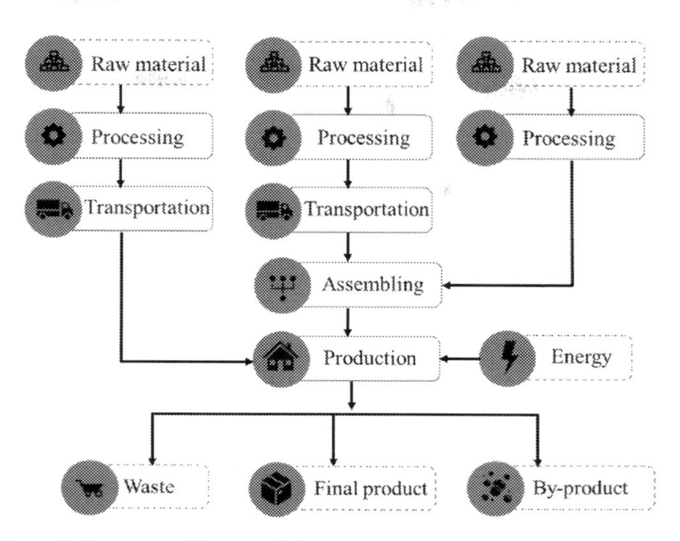

Fig. 3 Sample life cycle inventory flow model.

helps to create a comprehensive picture of the product life cycle and how the environment is affected during each life cycle phase.

The LCI analysis stage is intended for constructing a detailed flowchart of the technical process with consideration of all environmentally relevant process flows. This complex stage deals with multiple unit processes, as well as hundreds of flows going in and out of the system boundary for the study. For all specific unit processes, data quantifying the use of extracted raw materials, energy, and water are listed. Emissions affecting air, land, and water are also identified in this stage. These data are gathered and adjusted based on the functional unit set during the goal and scope definition stage. Flowcharts can be simple when based on an analyzed system, or highly complicated for industrial scale projects, such as chemical industries, that involve numerous recycling loops in their systems. Development of the flowchart is considered an iterative process since data is added at different times during the data collection process in the LCI stage. Collected data can be categorized as primary or secondary data, where primary data is collected directly from the supply chain process, while secondary data is mostly robust data found in the literature. Validation of collected data can be performed by following ISO 14041.

2.2.3 LCIA

The third stage of LCA aims to create a comprehensive evaluation by translating environmental loads found in the LCI stage into several environmental impacts. This stage is beneficial for converting massive LCI data into easily readable results that offer increased convenience for further communication and comparison. Several LCIA methods are available these days to perform LCA, e.g., CML, 2000, 2001, Eco-indicator 99, IPCC, 2007, Impact 2002+, ReCiPe, and USETox (Rejane Rigon et al., 2019).

The LCIA stage begins with aggregation of all LCI data into impact categories, selected during the beginning of the goal and scope definition stage, based on the purpose of the LCA. In traditional LCA, the LCIA stage broadly translates environmental burdens associated with a product or process into potential impacts on three areas of protection or safeguard subjects, including natural resources use, human health, and the natural environment (Dewulf et al., 2015). For clearer observations, these three categories have been subdivided into several impact categories, e.g., global warming, ozone depletion, photochemical oxidant formation, acidification, eco-toxicity, human toxicity, resource consumption, land use, etc. Some typical impact categories in LCA have been discussed below.

2.2.3.1 Global warming

Global warming, commonly known as "climate change," indicates an increasing global temperature effect on the lower atmosphere (Stranddorf et al., 2005). Naturally, radiation from the sun heats up our planet and a portion of this is reflected into the atmosphere. However, carbon dioxide and other greenhouse gases (GHSs) in the atmosphere (e.g., methane and nitrous oxides) absorb and reflect the radiation again toward the Earth's crust, causing global warming. In most LCIA methods listed above, the global warming effect is calculated with global warming potentials (GWP) for all elements responsible for the greenhouse effect present in the study (Andrić and Al-Ghamdi, 2019). The unit for GWP is CO_2-equivalent.

2.2.3.2 Ozone depletion

The concentration of ozone gases is significant in the stratosphere, which acts as a protective shield from toxic ultraviolet B (UVB) radiation entering the Earth's atmosphere (Morales-Méndez and Silva-Rodríguez, 2018). Depletion of the ozone layer causes damage to the stratosphere and, therefore, has severe consequences for life on Earth; numerous chlorine- and bromine-based substances (e.g., chlorofluorocarbons, hydrochlorofluorocarbons, methyl bromide, and halons) are held responsible for this depletion (Renzulli, 1991). In LCA, the impact of ozone depletion is quantified through ozone depletion potentials, developed by the World Meteorological Organization (WMO) (Solomon and Albritton, 1992).

2.2.3.3 Acidification

Acidification, considered as a regional effect, is caused by the release of acidifying chemicals such as sulfur oxides, nitrous oxides, hydrochloric acid, and ammonia. These elements have a significant impact on surface and ground water resources, soils, ecosystems, and organisms (Singh et al., 2018). Generally, the acidification potential in LCA is presented as SO_2-equivalent, reflecting the total acidification caused by a product or process (Al-Thawadi and Al-Ghamdi, 2019).

2.2.3.4 Toxicity

Toxicity includes several impacts, such as toxicity of organic solvents or heavy metals, as well as carcinogenic impacts. This complicated category is often divided into two subcategories: eco-toxicity and human toxicity. Human toxicity is defined by the potential health consequences caused by a chemical release in the environment (Hertwich et al., 2001).

There are several ready-to-use LCIA methods available on the market that allow the practitioner to carry out the LCA without looking in-depth at the different steps, such as characterization or classification. However, each LCIA method carries its own measuring principles. Some available ready-to-use LCIA methods are Ecoindicator'99, EPS, EDIP, CML, ReCiPe, and Traci.

2.2.4 Interpretation

Refinement of the results is critical for quantitative LCA studies to obtain useful, presentable, and validated data aligned with the initially defined goal and scope. This process of refinement for conclusive results is known as "interpretation" in traditional LCA (Yoonus and Al-Ghamdi, 2020). Typically, this stage indicates impact hotspots of the analyzed system. Generally, data quality analysis, sensitivity analysis, and uncertainty analysis are performed during the interpretation stage. Moreover, limitations and future recommendations are made in this stage.

3. LCA in CE

The concept of CE was introduced just a few years ago but has gained huge attention in the fields of economics, sustainable environment, waste management, and so on, both in the public and private sector bodies in this short period of time. The formulation of CE is based on several concepts and approaches, ranging from performance economy to industrial economy (Ghisellini et al., 2016).

3.1 Environmental sustainability and CE

Since its inception, industrial economy has been inclined to the fundamental linear economic model. The existing linear economy is based on a model starting with raw material extraction, progressing to production, and finally, disposal by customers after use, allowing billion of tons of raw materials to enter into the production system. The principle of linear economy is oriented toward pursuing economic and social benefit over environmental consequences, ultimately pushing toward resource depletion. To foster sustainable environmental, social, and economic growth, the European Union adopted the CE package in 2015. According to the Ellen MacArthur Foundation, "The circular economy refers to an industrial economy that is restorative by intention; aims to rely on renewable energy; minimizes, tracks, and eliminates the use of toxic chemicals; and eradicates waste through careful design. The term goes beyond the mechanics of production and consumption of goods and services in the areas that it seeks to redefine (examples include rebuilding capital, including social and natural, and the shift from consumer to user)" (The Ellen MacArthur Foundation, 2012). This economic structure has the potential to redefine economic systems into less vulnerable and more sustainable ones. This concept has been recognized as an umbrella concept (Blomsma and Brennan, 2017), originally created by Hirsch and Levin, defined as a "broad concept or idea used loosely to encompass and account for a set of diverse phenomena" (Hirsch and Levin, 1999).

The most frequently asked question during incorporation of CE strategies is: "Among CE strategies, which one is most beneficial for the environment, as well as for the product?" Although the novel approach of CE is appealing and receiving huge interest these days, the lack of quantitative measures for assessing environmental performance is one reason to hold the CE concept back from being implemented into practice. Currently, mass-based metrics are able to evaluate the circularity of any process or product but are not capable of assessing environmental perspectives. For example, the Material Circularity Indicator (MCI) is an established way to determine the success of CE. However, this strategy is incapable of answering which CE approach (either reuse, recycling, or other recovery options) is the best option for the environment. Along with this, the availability of few indicators capable of including end-of-life treatment is highlighted as a major issue in terms of CE assessment (Moraga et al., 2019). Implementation of a sustainable CE requires consideration of economic, social, and environmental circularity strategies (Niero and Rivera, 2018). In other words, for successful implementation of the CE concept, material circularity and environmental performance should be assessed simultaneously, as circularity is not always equivalent with environmental sustainability (Haupt and Hellweg, 2019). On the verge of this uncertainty about choosing the best solution—one that considers both material circularity and environmental impact—a comprehensive assessment such as LCA can act as a perfect tool for concisely evaluating all options.

3.2 Implementation of LCA

The CE concept aims to create value for society and the economy while reducing environmental impacts, whereas LCA is a robust scientific tool for assessing the environmental load of any process or product (Fig. 4). Thereby, a combination of CE principals and the robustness of LCA can be beneficial and has potential for providing a holistic approach that could strengthen available CE alternatives.

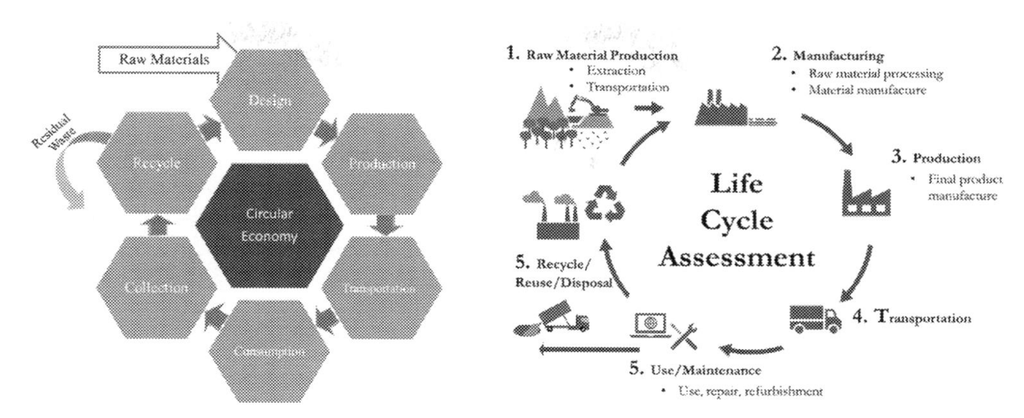

Fig. 4 Simplified representation of circular economy and life cycle assessment.

LCA has been recognized as one of the most complete assessment processes, covering four out of five CE requirements, including a reduction in input and use of natural resources, increase in renewable and recyclable resources, reduction in emissions, and a reduction in valuable material losses (Haupt and Hellweg, 2019). Among these four dimensions, LCA can provide direct evaluation for all except the second.

While implementing LCA in CE evaluation, LCA can be utilized to compliment CE in different stages. First, for any hypothetical product or process (macro, micro, or regional level) the pros and cons of CE can be analyzed through LCA. Second, possibly better alternatives for the entire life cycle can be identified through LCA once the limitations of a circular model are recognized during the first stage. Finally, LCA continuously sets goals and aims for improving circularity in real-life scenarios. Therefore, combining CE and LCA ensures mobilizing business with proven benefits for the environment and society.

According to PRé Sustainability, a leading software company regarding sustainability metrics, integration of the material circularity index (MCI) and LCA can be performed using "SimaPro" software, which ensures the best CE strategy for any specific scenario (Valencia, 2017a). They designed seven steps for integrating CE and LCA (Fig. 5). This approach is an iterative process for selecting the most suitable strategy.

4. CE and LCA: Case studies

The combination of CE and LCA provides an opportunity for product manufacturers to compare CE strategies along with complete environmental assessments, and thus ensures a balanced environmental and ecological system for the new circular or looped product design method. In recent years, the concept of CE has been promoted by the European Union (EU), and diverse national governments are adopting this concept as well,

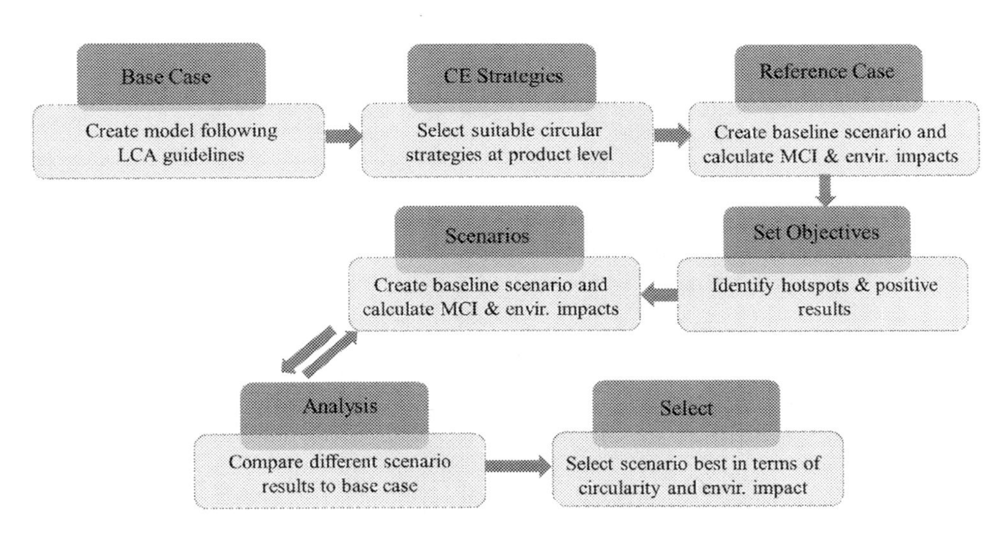

Fig. 5 Step by step integration of LCA and CE. *(From Valencia, E., 2017. Why Circular Economy Business Models Need LCA (Part 2): Seven Steps to Conduct a Circular Economy and LCA Study [Internet]. https://simapro.com/2017/7-steps-to-a-combined-circular-economy-lca-study-in-simapro/.)*

including France, the Netherlands, Finland, China, the United Kingdom, Japan, Canada, and Sweden, along with several private business organizations globally (Korhonen et al., 2018). Nearly all industries responsible for high energy consumption and emissions are nowadays committed to shifting toward more sustainable strategies and much ongoing research is tackling these issues.

4.1 Case study 1: Beverage packing sector

In Europe, aluminum cans are recognized as the second highest volume packaging material for beer after plastics (European Economics, 2016). However, the aluminum industry is highly responsible for GHG emissions. On a global scale, this industry emits around 1.1% of global GHG, and waste generation is considered a huge concern by the International Energy Agency (2009). To investigate the possibility and practicality of recycling aluminum (Al) cans while considering all alloy materials (can-to-can closed loop), an LCA study was performed for 30 loops of Al cans (Niero and Olsen, 2016). A total of five life cycle stages were considered during the modelling of the LCA framework, including production of the lid, production of the can body, manufacturing process of Al can and filling, end-of-life, and finally, material regeneration. After assessing available options through LCA, the study concluded that the option of a closed product loop that considered the use of beverage can scraps resulted in lower impacts on climate change compared to other available recycling options. The study also summarized future recommendations based on the analysis, including the reduction of lid weight and development of a method for separating the lid and body of the can at the collection point. The model presented in this study is based on the UK market; however, it has potential for use in other markets as well.

Further comparative analysis of the closed loop Al can recommended using a life cycle sustainability assessment (LCSA) framework to examine CE strategies over two other frameworks: MCI and C2C (cradle-to-cradle design protocol) (Niero and Hauschild, 2017). The comprehensive nature of the LCSA framework and its wide range of impacts make it more suitable for assessing CE strategies. Moreover, prevention of the burden shift among stakeholders was indicated as another benefit of utilizing a LCSA framework. In the later stage, a simulation study on eight types of beer packaging was performed in two different locations (the UK and India), considering the coupling of material circularity-based and life cycle-based indicators (Niero and Kalbar, 2019). A multicriteria decision analysis (MCDA) was employed to couple the two sets of indicators and allow for measurement of the analyzed system in terms of both CE and environmental context. Analysis of the results through TOPSIS (technique for order by similarity to ideal solution) showed the benefit of integrating two sets of indicators. Therefore, this study proposed that using both material circularity and life cycle indicators of the MCDA system at product level can be useful for future advancement in CE strategies.

4.2 Case study 2: Bulk waste management

For many European countries, management of bulky waste has become a serious issue. Generally, objects such as furniture, mattresses, and the other large discards are known as bulky waste, and pose both environmental and logistical burdens. The atypical composition of various materials used to make these bulky objects makes treatment of these wastes critical and hence, in accordance with CE strategy, material recycling and energy recovery are considered ideal options for this sector.

The URBANREC project in Europe has begun to evaluate recycling options for bulky waste, using LCA to assess the fragmentation process of these wastes (Samson-Bręk et al., 2019). Laminated cutting technology using water jets has been employed during the fragmentation process in this project, and LCA performed for bulky wastes, including polyurethane (PU) foam and mattresses. SimaPro 8.5.2.0 software and the ReCiPe LCIA method have been used for the analysis, in which the functional unit was set as 1 mg of each type of waste. Analysis of recirculation of theses wastes showed a positive contribution in reducing emissions of GHG and consumption of fossil fuels.

4.3 Case study 3: Construction and demolition waste

Globally, construction and demolition waste (CDW) is considered as an ever-increasing commodity that constitutes a key portion of all solid waste generation (Rao et al., 2007).

In Europe, CDW alone represents about 25%–30% of total municipal waste produced and most of it ends up in landfills despite its huge recycling and reutilization potential (Iacoboaea et al., 2019). In fact, proper management of CDW can turn waste into valuable secondary raw materials while protecting against overuse of natural resources (e.g., the use of CDW as an aggregate in concrete production). However, landfilling still dominates, even as EU regulations and CE objectives suggest that recycling CDW should be the primary method for CDW management.

To demonstrate the economic and environmental benefits of recycling and reutilizing CDW to replace partial natural aggregates in concrete production, a study was performed in the municipality of Bologna, Italy (Zanni et al., 2018). Four CDW management scenarios were analyzed using SimaPro software version v.7.3.3, and IMPACT 2002+ was selected as the preferred LCIA method. Among the four CDW management options, scenario 3 (involving a temporary storage facility, sorting, and fraction recovery for recycling) was found to be the most favorable option. LCA of the CDW-based aggregative production suggested that the use of renewable energy in this process can enhance environmental performance of the overall system. A comparative LCA of an eco-designed concrete mix with the standard one confirmed that replacement of 25% of natural aggregates leads to a decrease of about 39% in land occupation indicators. Overall study results for the recycling of CDW promote the circular design of building construction elements.

4.4 Case study 4: Tire end-of-life management

In Brazil, around two third of the used tires end up in the energy recovery process in the cement industry, whereas the rest is used for material recovery (Lonca et al., 2018). Therefore, to analyze the end-of-life management strategies for tire materials, as well as to test the improvement of the circular flow, a case study was performed using both MCI and LCA (Lonca et al., 2018). Assessment of the two different end-of-life strategies confirmed that retreading of tires has the potential to shift the burdens on natural resources along with human health as a result of extra fuel consumption due to rolling resistance, where retreading provides improved result in material circularity with resources preservation, enabling avoidance of environmental burden.

4.5 Case 5: Second-hand use of laptop computers

This case study investigates the environmental burdens of a real-life business project related to second-hand use of laptop computers, with a focused concern on metal resources (André et al., 2019). Investigation results of reusing laptops indicated environmental benefits in all impact categories when compared to manufacturing new laptops, as impacts from all work stages related to enabling reuse of laptops were found to be negligible. In the case of the climate change impact category, the production of printed circuit boards (PCBs) was found mainly responsible. Second-hand laptops resulted in around 54% less net climate change impact compared to the new laptops. Similarly, in the human toxicity impact category, PCB was found to be the major element for emissions and the net impact of second-hand laptops was nearly half that of new.

5. Summary

Worldwide, CE has gained a high level of attention as it offers a vision for mobilizing the economy while reducing the environmental burden. CE is a type of restorative or regenerative strategy designed to create value for both economic and social life by redefining the concept of a product's end-of-life through the mantra of "reduce, reuse, and recycle." The basic principle of CE has the aim of eradicating waste, which is in contrast to the traditional linear economy that is based on the "take, make, and dispose" theory of production. This concept offers great opportunity for remodeling existing economic structures, ultimately increasing resource efficiency. Recently, global policymakers have shown interest in this regeneration and restoration-based economy over the extraction and consumption-based economic style.

However, there is still a lack of clear views about how to adopt CE strategies in real practice and how to ensure that this strategy will be successful enough to reduce environmental impact (Contreras, 2015). The application of LCA is beneficial in this respect since it can assess the environmental impact of CE. Hence, a combination of CE and LCA is the preferred method for realizing development of products and services for companies, product designers, and sustainability teams, as well as consumers, are closed loops are not always the preferable method from an environmental viewpoint (Valencia, 2017b). LCA integration thus can boost the propositions of CE and vice versa. The combination can result in more in-depth analysis and understanding of economic, social, and environmental sustainability (Szita, 2017).

LCA is a science-based environmental assessment tool that has potential for complimenting the strong vision of CE. LCA can help improve CE strategies in the following ways:

- Testing CE models for product or process, as in many cases, closed loops produce negative environmental impacts.
- Discovering the limitations, impact hotspots, and inaccurate assumptions made during CE strategy, and thereby finding alternatives.
- Continuously improving CE strategies to implement them in real business cases.

Thus, LCA continuously provides insights and assists with setting key performance indicators. It enables the defining of a baseline scenario and thus helps to evaluate progress or any drawbacks while promoting circularity in all possible sectors (Fig. 6).

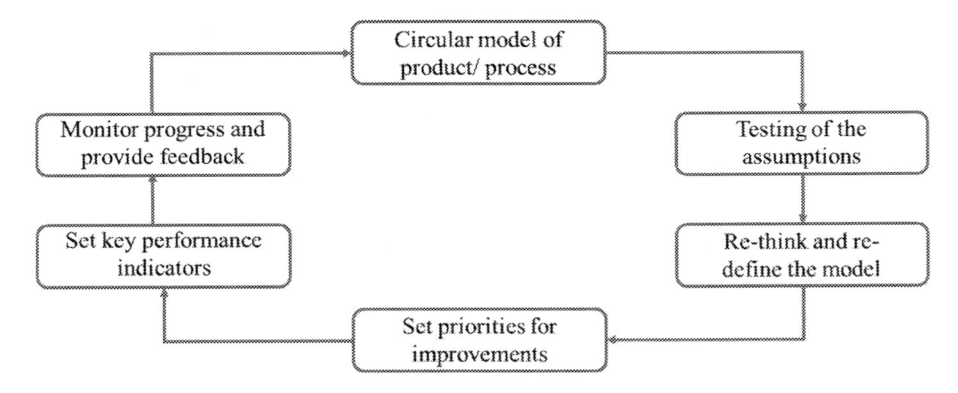

Fig. 6 Steps for continuous development of CE strategies using LCA. *(From Contreras, S., 2015. Complementing the Circular Economy With LCA [Internet]. Pré. https://www.pre-sustainability.com/news/complementing-the-circular-economy-with-lca.)*

References

Al-Thawadi, F.E., Al-Ghamdi, S.G., 2019. Evaluation of sustainable urban mobility using comparative environmental life cycle assessment: a case study of Qatar. Transp. Res. Interdiscip. Perspect. 1, 100003.

André, H., Ljunggren Söderman, M., Nordelöf, A., 2019. Resource and environmental impacts of using second-hand laptop computers: a case study of commercial reuse. Waste Manag. 88, 268–279.

Andrić, I., Al-Ghamdi, S.G., 2019. The impact of climate change on urban environment in GCC countries and related energy systems: mitigation measures and associated challenges. In: Int Conf Sustain Infrastruct 2019 Lead Resilient Communities through 21st Century—Proc Int Conf Sustain Infrastruct 2019, pp. 100–109.

Baumann, H., Tillman, A.M., 2004. The Hitch Hiker's Guide to LCA [Internet]. Studentlitteratur Lund, p. 542. http://www.amazon.ca/exec/obidos/redirect?tag=citeulike09-20&path=ASIN/9144023642%5Cnhttp://www.amazon.co.uk/Hitch-Hikers-Guide-LCA-Applications/dp/9144023642.

Blomsma, F., Brennan, G., 2017. The emergence of circular economy: a new framing around prolonging resource productivity. J. Ind. Ecol. [Internet] 42 (January), 2011. http://onlinelibrary.wiley.com/doi/10.1111/j.1467-9744.2007.00877.x/abstract%0Ahttps://doi.org/10.1111/j.1467-9744.2007.00877.x.

Contreras, S., 2015. Complementing the Circular Economy With LCA. [Internet]. Pré https://www.pre-sustainability.com/news/complementing-the-circular-economy-with-lca.

Dewulf, J., Benini, L., Mancini, L., Sala, S., Blengini, G.A., Ardente, F., et al., 2015. Rethinking the area of protection "natural resources" in life cycle assessment. Environ. Sci. Technol. 49 (9), 5310–5317.

European Economics, 2016. The contribution made by beer to the European economy [Internet]. The Brewers of Europe. https://brewersofeurope.org/uploads/mycms-files/documents/publications/2016/EU_economic_report_2016_web.pdf.

Finnveden, G., Suh, S., Koehler, A., Hauschild, M.Z., Ekvall, T., Guinée, J., et al., 2009. Recent developments in life cycle assessment. J. Environ. Manag. [Internet] 91 (1), 1–21. http://www.ncbi.nlm.nih.gov/pubmed/19716647%0Apapers3://publication/doi/10.1016/j.jenvman.2009.06.018.

Ghisellini, P., Cialani, C., Ulgiati, S., 2016. A review on circular economy: the expected transition to a balanced interplay of environmental and economic systems. J. Clean. Prod. 114, 11–32.

Guinee, J.B., 2002. Handbook on life cycle assessment operational guide to the ISO standards. Int. J. Life Cycle Assess. 7 (5), 311–313.

Haupt, M., Hellweg, S., 2019. Measuring the environmental sustainability of a circular economy. Environ. Sustain. Indic. 1–2, 100005.

Hertwich, E.G., Mateles, S.F., Pease, W.S., McKone, T.E., 2001. Human toxicity potentials for life-cycle assessment and toxics release inventory risk screening. Environ. Toxicol. Chem. 20 (4), 928–939.

Hirsch, P.M., Levin, D.Z., 1999. Umbrella advocates versus validity police: a life-cycle model. Organ. Sci. 10 (2), 199–212.

Iacoboaea, C., Aldea, M., Petrescu, F., 2019. Construction and demolition waste—a challenge for the European Union? Theor. Emperical. Res. Urban. Manag 14 (1), 30–52.

IEA, 2009. Energy Technology Transitions for Industry. Energy Technol Transitions Ind.

ISO, 2002. ISO/TS 14048:2002 Environmental Management—Life Cycle Assessment—Data Documentation Format (Internet) https://www.iso.org/standard/29872.html.

ISO, 2006a. Environmental Management—Life Cycle Assessment—Principles and Framework (Internet) https://www.iso.org/obp/ui/#iso:std:37456:en.

ISO, 2006b. ISO 14040:2006 Environmental Management—Life Cycle Assessment—Principles and Framework [Internet]. https://www.iso.org/standard/37456.html.

ISO, 2006c. ISO 14044:2006 Environmental Management—Life Cycle Assessment—Requirements and Guidelines [Internet]. https://www.iso.org/standard/38498.html.

ISO, 2012a. ISO/TR 14047:2012 Environmental Management—Life Cycle Assessment—Illustrative Examples on How to Apply ISO 14044 to Impact Assessment Situations [Internet]. https://www.iso.org/standard/57109.html.

ISO, 2012b. ISO/TR 14049:2012 Environmental Management—Life Cycle Assessment—Illustrative Examples on How to Apply ISO 14044 to Goal and Scope Definition and Inventory Analysis [Internet]. https://www.iso.org/standard/57110.html.

ISO, 2014. ISO 14046:2014 Environmental Management—Water Footprint—Principles, Requirements and Guidelines [Internet]. https://www.iso.org/standard/43263.html.

Klöpffer, W., 2003. Life-cycle based methods for sustainable product development. Int. J. Life Cycle Assess. 8 (3), 157–159.

Korhonen, J., Honkasalo, A., Seppälä, J., 2018. Circular economy: the concept and its limitations. Ecol. Econ. 143, 37–46.

Lonca, G., Muggéo, R., Tétreault-Imbeault, H., Bernard, S., Margni, M., 2018. A bi-dimensional assessment to measure the performance of circular economy: a case study of tires end-of-life management. In: Benetto, E., Gericke, K., Guiton, M. (Eds.), Designing Sustainable Technologies, Products and Policies. Springer, Cham, https://doi.org/10.1007/978-3-319-66981-6_4.

Mannan, M., Al-Ghamdi, S.G., 2019. Life-cycle assessment of thermal desalination: environmental perspective on a vital option for some countries. In: World Environ Water Resour Congr 2019 Groundwater, Sustain Hydro-Climate/Climate Chang Environ Eng—Sel Pap from World Environ Water Resour Congr 2019, pp. 449–460.

Mannan, M., Al-Ghamdi, S.G., 2020. Environmental impact of water-use in buildings: latest developments from a life-cycle assessment perspective. J. Environ. Manag. 261.

Moraga, G., Huysveld, S., Mathieux, F., Blengini, G.A., Alaerts, L., Van Acker, K., et al., 2019. Circular economy indicators: what do they measure? Resour. Conserv. Recycl., 452–461.

Morales-Méndez, J.D., Silva-Rodríguez, R., 2018. Environmental assessment of ozone layer depletion due to the manufacture of plastic bags. Heliyon 4 (12), e01020.

Niero, M., Hauschild, M.Z., 2017. Closing the loop for packaging: finding a framework to operationalize circular economy strategies. Procedia CIRP 61, 685–690.

Niero, M., Kalbar, P.P., 2019. Coupling material circularity indicators and life cycle based indicators: a proposal to advance the assessment of circular economy strategies at the product level. Resour. Conserv. Recycl. 140, 305–312.

Niero, M., Olsen, S.I., 2016. Circular economy: to be or not to be in a closed product loop? A life cycle assessment of aluminium cans with inclusion of alloying elements. Resour. Conserv. Recycl. [Internet] 114, 18–31. http://www.sciencedirect.com/science/article/pii/S0921344916301604.

Niero, M., Rivera, X.C.S., 2018. The role of life cycle sustainability assessment in the implementation of circular economy principles in organizations. Procedia CIRP 69, 793–798.

Rao, A., Jha, K.N., Misra, S., 2007. Use of aggregates from recycled construction and demolition waste in concrete. Resour. Conserv. Recycl. 50 (1), 71–81.

Rejane Rigon, M., Zortea, R., Alberto Mendes Moraes, C., Célia Espinosa Modolo, R., 2019. Suggestion of life cycle impact assessment methodology: selection criteria for environmental impact categories. In: Petrillo, A., De Felice, F. (Eds.), New Frontiers on Life Cycle Assessment - Theory and Application. IntechOpen, https://doi.org/10.5772/intechopen.83454. Available from: https://www.intechopen.com/books/new-frontiers-on-life-cycle-assessment-theory-and-application/suggestion-of-life-cycle-impact-assessment-methodology-selection-criteria-for-environmental-impact-c.

Renzulli, J.J., 1991. The regulation of ozone-depleting chemicals in the European community. In: Symposium on European Community Environmental Law [Internet]. https://lawdigitalcommons.bc.edu/cgi/viewcontent.cgi?article=1493&context=iclr.

Samson-Bręk, I., Gabryszewska, M., Wrzosek, J., Gworek, B., 2019. Life cycle assessment as a tool to implement sustainable development in the bioeconomy and circular economy. In: Biernat, K. (Ed.), Elements of Bioeconomy. IntechOpen, https://doi.org/10.5772/intechopen.84664. Available from: https://www.intechopen.com/books/elements-of-bioeconomy/life-cycle-assessment-as-a-tool-to-implement-sustainable-development-in-the-bioeconomy-and-circular-.

Seidel, C., 2016. The application of life cycle assessment to public policy development. Int. J. Life Cycle Assess. 21 (3), 337–348.

Singh, V., Dincer, I., Rosen, M.A., 2018. Life cycle assessment of ammonia production methods. Exergetic, Energ. Environ. Dimens., 935–959.

Solomon, S., Albritton, D.L., 1992. Time-dependent ozone depletion potentials for short- and long-term forecasts. Nature 357 (6373), 33–37.

Stranddorf, S.K., Hoffmann, L., Schmidt, A., 2005. FORCE technology. Impact categories, normalisation and weighting in LCA. Environ. News [Internet] 78, 90. http://www2.mst.dk/udgiv/publications/2005/87-7614-574-3/pdf/87-7614-575-1.pdf.

Szita, K., 2017. The application of life cycle assessment in circular economy. Hungarian. Agric. Eng. 31, 5–9.

The Ellen MacArthur Foundation, 2012. Towards a Circular Economy—Economic and Business Rationale for an Accelerated Transition. Greener Management International.

Valencia, E., 2017a. Why Circular Economy Business Models Need LCA (Part 2): Seven Steps to Conduct a Circular Economy and LCA Study (Internet) https://simapro.com/2017/7-steps-to-a-combined-circular-economy-lca-study-in-simapro/.

Valencia, E., 2017b. Why Circular Economy Business Models Need LCA (Part 1) | PRé Sustainability [Internet]. https://www.pre-sustainability.com/news/why-circular-economy-business-models-need-lca.

Yoonus, H., Al-Ghamdi, S.G., 2020. Environmental performance of building integrated grey water reuse systems based on life-cycle assessment: a systematic and bibliographic analysis. Sci. Total Environ. 712, 136535.

Yoonus, H., Mannan, M., Al-Ghamdi, S.G., 2020. Environmental performance of building integrated grey water reuse systems: Life Cycle Assessment perspective. In: World Environ Water Resour Congr 2020 Water, Wastewater, Stormwater Water Desalin Reuse—Sel Pap from Proc World Environ Water Resour Congr 2020, pp. 1–7.

Zanni, S., Simion, I.M., Gavrilescu, M., Bonoli, A., 2018. Life Cycle Assessment applied to circular designed construction materials. Procedia CIRP 69, 154–159.

CHAPTER 10

Life cycle costing as a way to include economic sustainability in the circular economy. New perspectives from resource-intensive industries

M. Sonia Medina-Salgado[a], Anna Maria Ferrari[b], Davide Settembre-Blundo[c], Marco Cucchi[c], and Fernando E. García-Muiña[a]
[a]Department of Business Administration (ADO), Applied Economics II and Fundamentals of Economic Analysis, Rey-Juan-Carlos University, Madrid, Spain
[b]Department of Sciences and Methods for Engineering, University of Modena and Reggio Emilia, Reggio Emilia, Italy
[c]Gruppo Ceramiche Gresmalt, Sassuolo, Italy

1. Introduction

In recent years, the theme of the circular economy (CE) has been the subject of numerous studies. The Ellen MacArthur Foundation defines CE as an economic system aimed at the reuse of materials in subsequent production cycles, minimizing waste (MacArthur, 2013). In contrast, a linear economy model focuses on a linear process of extraction, production, consumption, and waste. The linear economy paradigm emphasizes economic objectives at the expense of the ecological and social dimensions. On the contrary, the CE paradigm consists of a closed-loop regenerative economic model aimed at reducing the impact of production processes (Sauvé et al., 2016). This model assumes primary importance in the current context of scarcity of resources and of increasing attention to environmental issues.

The concept of CE is strongly connected to the concept of sustainability. Some authors identify CE as the optimal solution or, in other cases, a necessary condition for a path of sustainability. Other academic papers, instead, consider CE as one among several approaches to achieve sustainable development goals (Geissdoerfer et al., 2017). In order to connect the two concepts and to define a practice of circularity as sustainable, it is not enough to limit the analysis to a mere environmental assessment, it is essential to consider the three pillars of sustainability: environmental, economic, and social. Indeed, in a business context, some circular practices may have a positive environmental impact but might not be economically sustainable, reducing the competitive advantage of the company and its ability to create value. The environmental sustainability assessment should

Circular Economy and Sustainability, Volume 1
https://doi.org/10.1016/B978-0-12-819817-9.00005-3

therefore always be supported by an equally comprehensive economic sustainability assessment.

In order to address this issue, the life cycle sustainability assessment (LCSA) is one of the most widely used methodologies in sustainability evaluation (Guinée, 2016). The LCSA is an integration framework of different models that measures the performance of a product or process with respect to the triple bottom line. The LCSA combines three tools that meet the three pillars of sustainable development: the life cycle assessment (LCA) for the environmental impact, the life cycle costing (LCC) for the economic impact, and the social life cycle assessment (S-LCA) for the social impact (Kloepffer, 2008). LCC allows evaluation of the economic consequences of a decision on the life cycle of a product, in terms of costs, revenues, and cash flows (Bierer et al., 2015). Despite the potential of this tool, its application in manufacturing contexts is still limited due to the complexity of the analysis and the lack of consensus.

The purpose of this research is to verify, through an operational case, the effectiveness of the LCC in assessing circularity from an economic perspective. To achieve this objective, an aggregate LCC calculation model will be applied to a major tile manufacturer in the Sassuolo ceramic district in Italy. The model will focus on six ceramic body scenarios and it will evaluate a specific practice of circularity: the reintroduction of fired scrap in the production process. Through this practice, the relationship between circularity and economic sustainability will be investigated and the effectiveness of the LCC as an impact assessment tool will be analyzed.

The chapter is organized as follows. Section 2 presents a concise literature review on the themes of CE and LCC as a tool for assessing economic sustainability. Section 3 illustrates the research objectives of the chapter and the methodological framework used to address them. Section 4 introduces the case study of an important ceramic tile manufacturing company in which LCC tools have been operationally applied. Section 5 illustrates the results obtained from the application of the LCC to the manufacturing context proposed in the previous section. Finally, Section 6 offers some concluding remarks and highlights the main limitations of the research.

2. Literature review

2.1 CE and impact assessment

In a social context increasingly sensitive to environmental issues, the CE is receiving more and more consideration from policymakers and scientific researchers.

The CE consists of a regenerative industrial system that replaces the end-of-life concept with sustainable practices aimed at the use of renewable energy, the elimination of toxic chemicals, the reuse of resources, and the elimination of waste (MacArthur, 2013). Geissdoerfer et al. (2017), after reviewing several contributions, define CE as "a regenerative system in which resource input and waste, emission, and energy leakage are minimised by slowing, closing, and narrowing material and energy loops" (Geissdoerfer et al., 2017, p. 6). The notion of CE, therefore, starts from a different approach to that of traditional production methods. In a circular model, what is commonly considered a production waste is transformed back into a resource, triggering a virtuous cycle that regenerates itself (Lieder and Rashid, 2016). The final aim is, therefore, to minimize virgin materials consumption and waste production by closing resource flows into loops (D'Adamo, 2019).

The growing interest in the issues of CE is associated with the increasing responsibility assumed by companies regarding the environmental impact of their activity. In this concern, the implementation of circularity enables firms to undertake a path of sustainable development and to demonstrate their commitment to these issues (Genovese et al., 2017). In the Italian ceramic tile sector, for instance, companies have been implementing similar practices for years, such as the reuse of water deriving from production processes and the cogeneration of electricity. These measures, however, require an assessment of the environmental and economic impact of the production process (Takata et al., 2019). The absence of an impact assessment could lead to circularity practices that increase the company's environmental impact (Haupt and Zschokke, 2017). Alternatively, some practices could improve the environmental performance, but reduce the competitive advantage of the company by increasing its costs. In support of this theme, a large amount of research is being developed on the impact assessment of CE strategies (Elia et al., 2017). Among the methodologies proposed in the evaluation of circularity, the LCSA is one of the most widespread and used tools (Ferrari et al., 2019). As one of the LCSA tools, LCC is proposed as an operational tool for assessing economic sustainability.

Proposition 1. Impact assessment tools help us to understand whether circular economy practices are sustainable for a company or not.

2.2 LCC

LCC was first adopted by the United States Department of Defense in the mid-1960s. From the mid-1980s, US public agencies and private owners started to use a LCA in order to compare different design options in buildings and their relative benefits (Goh and Sun, 2015). Based on these approaches, LCC has been used over the years as an economic impact tool for capital equipment or long-lasting products of great economic value. Despite this, the industrial accounting version of LCC has never been transformed into a general methodology valid for all types of industry. The first international standard adopted for LCC was ISO15686-5 (2017), the aim of which was establishing clear terminology and a common methodology for the economic impact evaluation in the building sector. Following the ISO standard, the LCC tool must consider account costs or cash flows in a life cycle perspective (acquisition-disposal) and it must be performed over a specific period of analysis. Over the years, the LCC approach has been segmented into at least two different categories: the conventional LCC, which concerns a purely industrial accounting analysis, and the environmental LCC, which concerns the internalization of negative externalities (Hunkeler et al,. 2008).

The conventional LCC focuses on internal costs and is a pure economic evaluation taking into consideration the different stages in a life cycle approach. The costs considered are the ones directly covered by the main producer or user and they usually refer to acquisition costs (R&D and investments costs) and ownership costs (operating, maintaining, and disposal costs). At this stage of the analysis, environmental externalities are not considered unless they are directly related to new taxes or subsidies (Hunkeler et al., 2008).

The environmental LCC (*E*-LCC) is a further step compared to conventional LCC. It considers all the internal costs covered by one or more actors in the product life cycle and, in addition, it includes the externalities that will be internalized in future decisions (Hunkeler et al., 2008). *E*-LCC is not considered as a stand-alone assessment, but is usually conducted in connection with the LCA tool for the environmental impact analysis, following ISO 14040 and 14044 (2006, b) (Moreau and Weidema, 2015). From the LCA study, the researcher identifies and quantifies the main environmental damages along the life cycle of a product or a process. Once the LCA study is completed, the E-LCC should quantify, in economic terms, the previous damage and should allocate it to those who have to bear the cost. Each cost item should be quantified in terms of a functional unit, chosen during the LCA analysis and then maintained in the E-LCC.

Although LCC appears to be a comprehensive tool, most papers in the literature deal separately with the economic assessment of industrial costs and environmental externalities. This circumstance arises from certain limitations of the LCC in describing production processes, such as the risk of processing unreliable data and the lack of consensus on this approach, which lead scholars to not combine the two approaches (Neugebauer et al., 2015).

Proposition 2. The LCC quantifies externalities and internal costs of a product during its entire life cycle. Despite its limitations, the LCC is potentially a suitable tool for assessing economic sustainability.

3. Methodological framework

The literature analysis has highlighted the importance for companies to take a path of sustainability. In this regard, CE practices are one of the alternatives for achieving sustainability goals. Despite the great interest in the subject, CE is not a model that necessarily has to be adopted to declare a business as sustainable (Andersen, 2006). CE, applied without an impact assessment strategy, can lead companies to make wrong investments. In this context, an economic sustainability assessment assumes a key role in the decision-making process.

LCC is proposed in this chapter to address this issue. From an academic perspective, LCC is considered as an effective tool in assessing economic sustainability. From a manufacturing perspective, however, there is a lack of LCC tool applications that assess circularity with a detailed analysis of industrial costs and environmental externalities. For the purpose of exploring this topic, two research questions have been formulated:

RQ1: Is LCC adequate to assess the impact of circularity in a manufacturing process?

RQ2: Is LCC a comprehensive tool in describing the economic pillar of sustainability?

In order to answer these questions, an aggregate LCC calculation model applied to the ceramic tile sector will be proposed. In specific, the model will be implemented in a ceramic tile manufacturing company in the ceramic district of Sassuolo, Italy. The model will be based on an assessment of industrial costs (conventional LCC) and an assessment of environmental externalities (environmental LCC). The purpose of the research will be to evaluate the economic sustainability of the ceramic manufacturing process, considering in the analysis the most significant circularity practices implemented by the company.

The main focus of the model will be the analysis of different scenarios based on six combinations of ceramic body, with different percentages of local raw materials. In some of the ceramic body scenarios, a small percentage of fired scrap will be inserted. The reintroduction of these fired scraps is one of the circularity practices that could be implemented by a ceramic manufacturing company. The results of this analysis will allow determination, through an operational case, of the importance of economic sustainability in the evaluation of circularity.

On data collection, the company involved in the research has been growing strongly in turnover in recent years and it has invested significantly in new "Industry 4.0" technologies. These investments have allowed the installation of a set of sensors for data collection in the production plant. The data obtained can be used in the compilation of the LCC model. For the purpose of completing the model, the boundaries of the analysis will be limited to the production process, as the manufacturer does not have data on the other phases of the tile life cycle, such as the laying of the tile and the end of life.

4. The aggregate LCC calculation model

As previously introduced in literature analysis, LCC is an instrument for assessing costs throughout the life cycle of a product. In an industry oriented toward sustainable development, it is crucial to have a tool able to assess different sustainability practices in economic terms.

In the ceramic tile sector, entrepreneurs have several options to achieve their sustainable development goals. CE practices are a means of achieving these goals but, if poorly evaluated, they can result in a negative environmental or economic impact for the company. On the contrary, if the company evaluates these practices from a perspective that

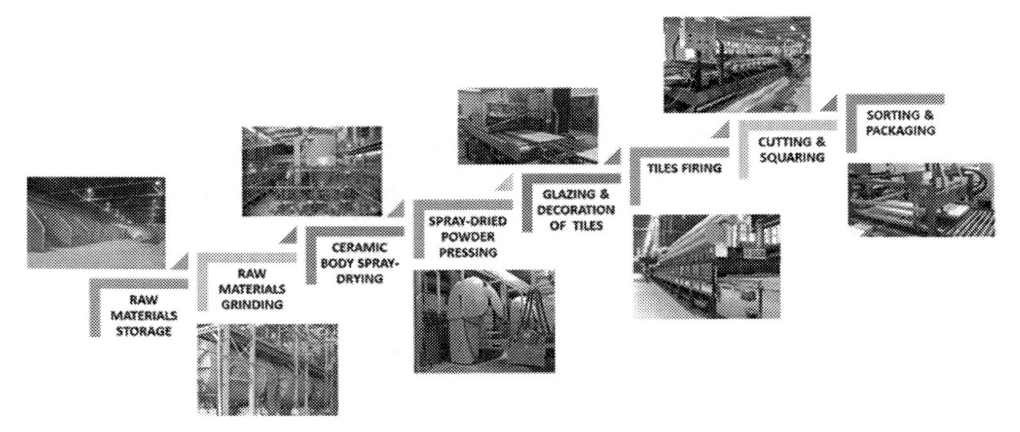

Fig. 1 Ceramic production process layout. *(From Garcia-Muiña, F., González-Sánchez, R., Ferrari, A., Settembre-Blundo, D., 2018. The paradigms of industry 4.0 and circular economy as enabling drivers for the competitiveness of businesses and territories: the case of an Italian ceramic tiles manufacturing company. Soc. Sci. 7(12), 255).*

involves all three pillars of sustainability, this can lead to the creation of corporate value that results in greater competitive advantage.

In order to address this issue, an aggregate LCC calculation model was created from a flow approach. The calculation model was made on an Excel sheet and represented the whole production process of the tile, as described in Fig. 1.

The production process begins with an order of raw materials. Depending on the choice of the ceramic body, the received raw materials are then loaded on conveyor belts and transferred to mills for milling, together with water as a milling vehicle. The milling process reduces raw materials to a compound called slip. Afterwards, the slip is sent to the spray drier and it is sprayed with a flow of hot air. The result of this process is a formation of round agglomerates of fine particles, which is called powder. The powder is then transported to the hydraulic pressing machines and is pressed until a raw compacted product is obtained. In the last part of the process, the tile is then glazed and subsequently decorated using very high-resolution digital inkjet printers. After the decoration, the tile enters the firing phase in long kilns that fire in cycles of 30–50 min at temperatures of about 1200°C. After firing, tiles proceed to the cutting and grinding phase (if necessary) and, subsequently, to the sorting line, in which a size and flatness control is conducted. In conclusion, tiles move to the packaging phase and are palletized in preparation for the following selling stage (Garcia-Muiña et al., 2018).

Based on the production process above, a model of LCC was created including both an analysis of industrial costs (conventional LCC) and an analysis of the negative environmental externalities produced (environmental LCC). The LCC study is based on a

Ceramic body scenarios	P 01	P 03	P 04	P 15	P 17	P 19
EU clay	25%	20%	45%	50%	28%	29%
Local clay	0%	0%	0%	0%	30%	30%
Local Feldspar	5%	19%	10%	24%	10%	11%
Local Feldspar Sand	7%	10%	11%	0%	10%	10%
Fired Scrap 90m	0%	8%	5%	3%	3%	0%
Extra EU clay	25%	25%	10%	5%	0%	0%
Extra EU Feldspar	38%	18%	19%	18%	19%	20%

Fig. 2 Scenarios of ceramic body compositions. *(Source: Own elaboration).*

comparison of six different scenarios characterized by six different ceramic body compositions (Fig. 2).

Fig. 2 illustrates, for each of the six ceramic bodies, the composition of raw materials in percentage terms. The decision to investigate raw material procurement scenarios stems from the fact that this phase is one of the most environmentally impacting (Settembre Blundo et al., 2018). Starting from scenario P01, the following scenarios are characterized by a growing percentage of local raw materials, with the exception of P17 and P19, which are very similar.

Raw materials represented in green are materials that are sourced from domestic suppliers or from European suppliers. The supply of these raw materials is more sustainable as they are transported over short distances or by using more sustainable means of transport, such as rail. The other raw materials are instead provided by suppliers outside Europe, typically located in Turkey and Ukraine. These raw materials are more impactful because they are transported over long distances and by more polluting means of transport. In addition, as previously mentioned, some ceramic bodies (P03, P04, P15, and P17) contain a small percentage of fired scrap, which is generated during the production process. In order to be reintroduced into the production process, the waste undergoes an additional milling process.

Starting from these body compositions, the following paragraphs will be dedicated to the description of the model and the evaluation of the results.

4.1 Conventional LCC

The conventional LCC focuses on internal costs and is a pure economic evaluation taking into consideration the different stages in a life cycle approach. For the conventional LCC analysis of the ceramic tile producer under investigation, based on the structure of the production process described in the previous chapter, it has been decided to divide the calculation model into seven phases:

1. Design, testing, planning
2. Body preparation and spray drying

3. Glaze preparation
4. Pressure, drying, glazing, firing
5. Cutting, squaring, lapping, sorting
6. Packing

ISO15686-5 (2017) represents the standard that is commonly considered in a conventional LCC analysis. Using the ISO standard as a reference, the costs analyzed are those directly related to the production process, such as design, raw material costs, labor costs, energy costs, and plant maintenance. In addition, some specific costs have been considered for each step of the process.

Once the process costs have been identified, data have been extracted from the company's management software. Data collection is normally one of the most critical points in a process cost analysis. Many manufacturing companies do not have data gathering systems that provide for detailed data throughout the production process. In this specific business case, due to the investments in Industry 4.0 technologies, the company's factories are equipped with several tools for automatic data collection throughout the production process. The sensors for data collection installed in the production facilities are of two types: wired sensors that collect production data (e.g., atomized humidity or square meters of tiles produced) and radio frequency sensors that collect data on consumptions (e.g., consumption of electricity and methane). The data gathered by the sensors are stored in an MES (manufacturing execution system) software, a system capable of collecting data and offering tools to visualize them. The main part of the data collected in the MES software is then transferred to the company ERP system.

The extracted data is then used to compile a spreadsheet containing a table for each process step. Fig. 3 shows the data obtained from the body preparation and spray drying phase.

Fig. 3 represents the conventional LCC costs of the body preparation and spray drying phase, considering as an example the scenario of ceramic body P 01. The first column represents the cost items. The cost items referring to sustainable raw materials (local and European) and to the company circularity practices are highlighted in green. The second column is the annual cost of each of the cost items, referring to the year 2018. The total cost of the body preparation and spray drying phase (28,470,000.00€) and the number of square meters produced (10,000,000 m^2) are shown at the end of the table. The ratio between the total costs and the square meters provide the cost per square meter of this phase.

The same procedure has been applied to each phase of the production process for the scenario P01. The conventional LCC costs of each process phase have subsequently been added together to achieve a total cost. The results of this process are shown in Fig. 4.

Then, the calculation model has been applied in each of the six ceramic bodies scenarios illustrated previously. Since the production process is not influenced by the ceramic body used, raw material costs have been the only item to vary in the results of conventional LCC. All other cost items have remained unchanged regardless of the scenario considered.

Cost items	Monetary units
EU clay	4,800,000.00 €
Local clay	- €
Local Feldspar	630,000.00 €
Local Feldspar Sand	880,000.00 €
Fired Scrap 90 m	- €
Extra EU clay	7,800,000.00 €
Extra EU Feldspar	6,550,000.00 €
Electricity	1,050,000.00 €
Electricity from co-generation (Internal or from the Cluster "Smart Grid")	1,200,000.00 €
Mill Depreciation-Maintenance-Cleaning	500,000.00 €
Milling Pebbles	260,000.00 €
Human Resources (Cost of labour)	2,300,000.00 €
Fuel from spray drying	2,500,000.00 €
Subtotal	28,470,000.00 €
m^2 produced in one year	10,000,000 m^2
€/ m^2	2.85 €

Fig. 3 Body preparation and spray drying phase, body scenario P 01. *(Source: Own elaboration.)*

Conventional LCC phases	€/square metre
Design - testing - planning	0.32 €
Body preparation and spray drying	2.85 €
Glaze preparation	0,92 €
Pressure - drying - glazing - firing	2.09 €
Cutting - squaring - lapping	0.15 €
Sorting - packing	0.41 €
Total	**6.74 €**

Fig. 4 Total conventional LCC costs for the ceramic body P 01. *(Source: Own elaboration.)*

4.2 Environmental LCC

The aim of the environmental LCC analysis is to quantify environmental externalities from a life cycle perspective.

For the evaluation of environmental externalities generated by the ceramic production process under research, reference has been made to the environmental LCC data obtained from the paper of Garcia-Muiña et al. (2019). The paper presents the same case study and illustrates some first results obtained with the application of LCC to the same ceramic body compositions that will be considered in this chapter. In order to conduct an *E*-LCC analysis, Garcia-Muiña et al. (2019) initially perform an environmental impact analysis of a ceramic manufacturing process using the LCA methodology. LCA analysis is generally the starting point for an E-LCC assessment and this approach will be followed in this chapter as well, considering one square meter of ceramic tile as a functional unit of reference.

In this chapter, for each of the ceramic body compositions previously illustrated, data have been collected on a number of factors including material flows, waste generation, energy, methane, and emissions. The inventory analysis has been performed using the SimaPro® software by PRé Consultants. Based on the data collected, an LCA predictive analysis has assessed the environmental impact of the company's production process from the cradle to the gate, on the basis of 24 impact categories.

The impact categories of the LCA assessment have then been summarized in six damage categories for the E-LCC evaluation, which are as follows:

1. Ecosystem services
2. Access to water
3. Biodiversity
4. Building technology
5. Human health
6. Abiotic resources

These six different cost items have been selected following the EPS 2015dx calculation method (Steen in Garcia-Muiña et al. 2019). The EPS2015dx was created in 1990 to assess the environmental performance of different design options (Steen, 1999).

Considering the model's impact items, the ecosystem services refer to the production capacity of the ecosystem, for example, regarding the categories of yields of crops, fish and meat, wood, and freshwater. Access to water refers to the production capacity of water for domestic and irrigation use. Biodiversity, on the other hand, corresponds to the number of species becoming extinct each year. Since assessing the exact number of extinct species is challenging, the EPS method provides an estimate based on the probability of extinction of red-listed species. In this regard, for the estimation of this cost item, all practices related to hunting, harvesting, emissions of toxic substances, habitat reductions, and similar, have been taken into consideration. Building technology is an impact category based on housing availability, expressed in square meters. For the impact assessment of human health, the main indicators considered by the EPS method are levels of life expectancy, malnutrition, diarrhea, aggravation of angina pectoris, working capacity, asthma cases, severe COPD, cancer, skin cancer, impaired vision, poisoning, and intellectual disability. Finally, the abiotic resources item represents the depletion of resources such as minerals, natural gas, oil, and coal. The result of this cost item is the damage resulting from the extraction of these materials (Steen, 2015).

The same software used for the LCA analysis has enabled us to summarize the data of the 24 environmental impact categories in an economic value for each of the six environmental LCC impact categories of the method EPS 2015dx. The values have been reported in environmental load units (ELU), where 1 ELU is equal to 1€ (Steen, 2015).

5. Interpretation and discussion of the results

For the case of the ceramic body P 01, Fig. 4 illustrates the results of the exhaustive analysis of the conventional LCC.

The figures in Fig. 4 are the costs per square meter for each step of the production process in ceramic body scenario P 01. Each cost obtained from the calculation model has been added up until a definitive conventional LCC value has been determined.

As expected, Fig. 4 illustrates that some phases of the production process have significantly higher costs than other phases. One of the most cost-intensive stages is the body preparation and spray drying phase, which includes the cost of raw materials, a relevant cost source for a ceramic tile company. In addition to the body preparation and spray drying phase, the pressure, drying, glazing, firing phase shows one of the higher costs, due to the presence of the most energy-intensive phases of the entire process. The results obtained from this analysis are in line with previous papers concerning ceramic sectoral analyses (Ferrari et al., 2019). Similar considerations have also emerged in different contexts from the ceramic one of Sassuolo, as for example that of Ye et al. (2018).

For the same ceramic body P 01, Fig. 5 illustrates the results of the monetary quantification of the environmental damage of environmental LCC.

Fig. 5 represents, for each of the six impact categories of the EPS2015dx method, the environmental damage in monetary terms of the production process by opting for the P01 ceramic body scenario.

From these results, the costs of human health and abiotic resources are the highest. In particular, the abiotic resources category is affected by 69.95% of the total damage, especially due to the construction of the facilities for the extraction of clay. The human

Environmental LCC	€/square metre
Ecosystem services	0.031900 €
Access to water	0.001880 €
Biodiversity	0.000103 €
Building technology	0.000280 €
Human health	1.320000 €
Abiotic resources	3.140000 €
Total	**4.49 €**

Fig. 5 Total environmental LCC costs for ceramic body P 01. *(Source: Own elaboration.)*

Ceramic body scenario	Conventional LCC (€/square metre)	Environmental LCC (€/square metre)
P 01	6.74	4.49
P 03	7.03	4.35
P 04	6.76	4.18
P 15	6.57	4.16
P 17	6.22	4.08
P 19	6.07	4.11

Fig. 6 Summary table of conventional and environmental LCC costs for each ceramic body scenario. *(Source: Own elaboration.)*

health category is affected by 29.28% of the total damage, especially due to the emissions of carbon dioxide from fossil sources, arising mainly during the transport of raw materials by barge.

Fig. 6 shows the total costs of conventional LCC and environmental LCC for each of the six ceramic bodies.

As can be seen, the higher the percentage of local raw materials, the lower the economic damage produced by the environmental impact. The use of local raw materials makes it possible to reduce the impact generated by road transport of raw materials imported from outside Europe. These results, already reported in the paper of Garcia-Muiña et al. (2019), confirm that transportation is one of the most impactful phases from an environmental perspective. Similar results are reported by Quinteiro et al. (2014), who equally highlight the importance of environmental damage in the transportation phase, despite using a different methodology.

Regarding conventional LCC, there is a downward trend in industrial costs as the proportion of local raw materials increases. The change in costs is only imputable to the change in the percentage of raw materials of the ceramic body scenarios because, as previously mentioned, the production process does not vary with ceramic body variations. The results reveal that the selection of local raw materials, in addition to a lower economic impact in environmental terms, enables the company to reduce the industrial cost of its production process.

Some relevant considerations on the practice of circularity, i.e., the reuse of fired scrap in the production process, are possible. To address this, it is necessary to focus on the comparison between the two scenarios, P17 and P19. The two ceramic body compositions are extremely similar in terms of percentage of raw materials. However, the P17 mixture contains a small percentage (3%) of fired scrap (see Fig. 2). The inclusion of fired scrap in the production process reduces the environmental LCC value from 4.11 to 4.08 euros per square meter. In contrast, the industrial costs of the P17 scenario increase, compared to those of the P19 scenario, from 6.07 to 6.22 euros per square meter.

The results show that the reuse of fired scrap is not necessarily positive from the perspective of economic sustainability. All the same, to assess which of the two scenarios is preferable from the company's side is not simple. The costs arising from the analysis of conventional LCC are totally within the competence of the company. However, the environmental LCC costs (from the cradle to the gate) are produced by more than one actor in the production chain. The impossibility of identifying the part of the environmental damage for which the company is competent makes a direct comparison between conventional and environmental LCC unreliable.

6. Conclusions

Overall, this chapter has contributed to research on the topic of economic sustainability, an issue that is not always sufficiently explored in the academic literature. A limited number of papers have already attempted to conduct economic sustainability assessments with different criteria. For instance, Zorn et al. (2018) assess economic sustainability through the use of financial indices. Cetiner and Edis (2014), on the other hand, evaluate economic sustainability of building renovation operations using LCA methods. In this regard, with reference to the ceramics sector, Lv et al. (2019) and Ye et al. (2018) analyze economic sustainability using LCSA tools. Within this research field, this study deals with the economic pillar of sustainability through a highly detailed analysis, supported by the technological level of the company plants under analysis.

In the chapter, an LCC model has provided an assessment of the level of economic sustainability of the production process in a ceramic manufacturing company, focusing on the circularity practices applied in this business context. In response to the first research question (RQ1), the LCC model has evaluated the role of circularity in the ceramic manufacturing context of reference. In the assessment of conventional and environmental LCC, some circularity practices (e.g., the recycling of wastewater and the cogeneration of electricity) already implemented by the company in its production process have been considered. In addition, the comparison between scenario P17 and P19 has allowed evaluation of the introduction of fired scrap into the production process. The results have demonstrated that this practice of circularity, although desirable from an environmental perspective, may not be convenient from an industrial one. This outcome suggests that a practice of circularity does not necessarily result in a higher level of sustainability. In this respect, an economic sustainability analysis is therefore of considerable importance in the evaluation of circularity.

Despite the significant results, the comparison between the P17 and P19 ceramic body scenarios has revealed some of the limits of the LCC model in the evaluation of circularity. The introduction of fired waste into the production process has an impact on both the industrial costs of conventional LCC and the environmental costs of environmental LCC. Nevertheless, the model does not offer comparability between the two cost categories because it is not possible to quantify the percentage of total costs that the company must bear for the environmental LCC. The aggregate LCC model estimates the total monetary damage of the environmental impact of the process but fails to divide the damage between the actors in the supply chain.

In response to the second research question (RQ2), LCC, as developed in this chapter, appears to be incomplete in describing the economic pillar of sustainability. The analysis of conventional LCC performed in the model is a good starting point for the assessment of industrial business costs. In order to be complete, however, the model should consider additional costs including long-term investments in plant and machinery. To include these categories of cost, the short-term perspective of the functional unit of LCC should be abandoned. Regarding environmental LCC, the model has not allowed the quantification of the environmental damage produced solely by the company. The environmental LCC results of the model are limited to a monetary description of the total environmental damage and falls within the environmental pillar of sustainability. In order to describe the economic pillar, future research on the subject should focus on researching a methodology for attributing the economic damage of environmental externalities to the different actors involved in the production process.

Finally, the limitations of this research are underlined. Firstly, the analysis concerned a single case study of a ceramic manufacturing company. Secondly, according to ISO standards, an LCC analysis should evaluate the entire life cycle of a process or product, from the cradle to the grave. However, difficulties in data collection have led us to evaluate only the production process, from the cradle to the gate. To conclude, an economic sustainability analysis should also evaluate the impact of circularity on business revenue and not only on costs. All these aspects will be the subject of future research already planned.

Acknowledgments

The authors would like to acknowledge the editor and an anonymous reviewer for his helpful suggestion to improve the paper.

Funding

This research was co-funded by the European Union under the LIFE Programme (LIFE16 ENV/IT/000307: LIFE Force of the Future-Forture).

References

Andersen, M., 2006. An introductory note on the environmental economics of the circular economy. Sustain. Sci. 2 (1), 133–140.

Bierer, A., Götze, U., Meynerts, L., Sygulla, R., 2015. Integrating life cycle costing and life cycle assessment using extended material flow cost accounting. J. Clean. Prod. 108, 1289–1301.

Cetiner, I., Edis, E., 2014. An environmental and economic sustainability assessment method for the retrofitting of residential buildings. Energy Build. 74, 132–140.
D'Adamo, I., 2019. Adopting a circular economy: current practices and future perspectives. Soc. Sci. 8 (12), 328.
Elia, V., Gnoni, M., Tornese, F., 2017. Measuring circular economy strategies through index methods: a critical analysis. J. Clean. Prod. 142, 2741–2751.
Ferrari, A., Volpi, L., Pini, M., Siligardi, C., García-Muiña, F., Settembre-Blundo, D., 2019. Building a sustainability benchmarking framework of ceramic tiles based on life cycle sustainability assessment (LCSA). Resources 8 (1), 11.
Garcia-Muiña, F., González-Sánchez, R., Ferrari, A., Settembre-Blundo, D., 2018. The paradigms of industry 4.0 and circular economy as enabling drivers for the competitiveness of businesses and territories: the case of an Italian ceramic tiles manufacturing company. Soc. Sci. 7 (12), 255.
Garcia-Muiña, F., González-Sánchez, R., Ferrari, A., Volpi, L., Pini, M., Settembre-Blundo, D., 2019. Identifying the equilibrium point between sustainability goals and circular economy practices in an industry 4.0 manufacturing context using eco-design. Soc. Sci. 8 (8), 241.
Geissdoerfer, M., Savaget, P., Bocken, N., Hultink, E., 2017. The circular economy—a new sustainability paradigm? J. Clean. Prod. 143, 757–768.
Genovese, A., Acquaye, A., Figueroa, A., Koh, S., 2017. Sustainable supply chain management and the transition towards a circular economy: evidence and some applications. Omega 66, 344–357.
Goh, B., Sun, Y., 2015. The development of life-cycle costing for buildings. Build. Res. Inf. 44 (3), 319–333.
Guinée, J., 2016. Life Cycle Sustainability Assessment: What Is it and What Are its Challenges? Taking Stock of Industrial Ecology. Springer International Publishing, pp. 45–68.
Haupt, M., Zschokke, M., 2017. How can LCA support the circular economy?—63rd discussion forum on life cycle assessment, Zurich, Switzerland, November 30, 2016. Int. J. Life Cycle Assess. 22 (5), 832–837.
Hunkeler, D., Lichtenvort, K., Rebitzer, G., Ciroth, A., 2008. Environmental Life Cycle Costing. SETAC, Pensacola, FL.
International Organization for Standardization 14040, 2006. Environmental Management. Life Cycle Assessment, Principles and Frameworks. Geneva, Switzerland.
International Organization for Standardization 14044, 2006. Environmental Management. Life Cycle Assessment, Requirements and Guidelines. Geneva, Switzerland.
International Organization for Standardization 15686-5, 2017. Buildings and Constructed Assets, Service Life Planning—Part 5: Life-Cycle Costing. Geneva, Switzerland.
Kloepffer, W., 2008. Life cycle sustainability assessment of products. Int. J. Life Cycle Assess. 13 (2), 89–95.
Lieder, M., Rashid, A., 2016. Towards circular economy implementation: a comprehensive review in context of manufacturing industry. J. Clean. Prod. 115, 36–51.
Lv, J., Gu, F., Zhang, W., Guo, J., 2019. Life cycle assessment and life cycle costing of sanitary ware manufacturing: a case study in China. J. Clean. Prod. 238, 117938.
MacArthur, E., 2013. Towards the circular economy. J. Ind. Ecol. 2, 23–44.
Moreau, V., Weidema, B., 2015. The computational structure of environmental life cycle costing. Int. J. Life Cycle Assess. 20 (10), 1359–1363.
Neugebauer, S., Martinez-Blanco, J., Scheumann, R., Finkbeiner, M., 2015. Enhancing the practical implementation of life cycle sustainability assessment—proposal of a Tiered approach. J. Clean. Prod. 102, 165–176.
Quinteiro, P., Almeida, M., Dias, A., Araújo, A., Arroja, L., 2014. The Carbon Footprint of Ceramic Products. Assessment of Carbon Footprint in Different Industrial Sectors. Vol. 1 Springer, Singapore, pp. 113–150.
Sauvé, S., Bernard, S., Sloan, P., 2016. Environmental sciences, sustainable development and circular economy: alternative concepts for trans-disciplinary research. Environ Dev 17, 48–56.
Settembre Blundo, D., García Muiña, F., Pini, M., Volpi, L., Siligardi, C., Ferrari, A., 2018. Lifecycle-oriented design of ceramic tiles in sustainable supply chains (SSCs). APJIE 12 (3), 323–337.

Steen, B., 1999. A Systematic Approach to Environmental Priority Strategies in Product Development (EPS): Version2000-Models and Data of the Default Method. Chalmers University of Technology, Environmental Systems Analysis, Göteborg.
Steen, B., 2015. The EPS 2015d Impact Assessment Method: An Overview. Swedish Life Cycle Center, Report number 2015:5, Chalmers University Of Technology, Göteborg.
Takata, S., Suemasu, K., Asai, K., 2019. Life cycle simulation system as an evaluation platform for multitiered circular manufacturing systems. CIRP Ann. 68 (1), 21–24.
Ye, L., Hong, J., Ma, X., Qi, C., Yang, D., 2018. Life cycle environmental and economic assessment of ceramic tile production: a case study in China. J. Clean. Prod. 189, 432–441.
Zorn, A., Esteves, M., Baur, I., Lips, M., 2018. Financial ratios as indicators of economic sustainability: a quantitative analysis for Swiss dairy farms. Sustainability 10 (8), 2942.

CHAPTER 11

Circular economy during project life cycle

Ibtisam Sulaiman Alhosni[a], Omar Amoudi[b], and Nicola Callaghan[c]
[a]Civil Engineering Department, Middle East College, Knowledge Oasis Muscat, Al Rusayl, Oman
[b]College of Engineering, National University of Science and Technology, Muscat, Oman
[c]School of Computing, Engineering and Built Environment, Glasgow Caledonian University, Glasgow, Scotland, United Kingdom

1. Introduction

The construction sector is one of the highest consumers of raw and natural resources around the world, using 3 billion tons per annum (The Ellen MacArthur Foundation, 2013). In addition to this high consumption level, it is also responsible for the majority of the waste in the majority of countries throughout the world. For instance, in the UK, waste from the construction industry covers 50% of the total percentage of waste (PricewaterhouseCoopers Network, 2015). In the USA, the Environmental Protection Agency (2009) reported that the construction industry generates approximately 26% of total waste. Therefore, any model that focuses on consumption rather than restoration of resources will have a negative impact on the value chain and will incur critical losses of resources including losses in the production stage, end of product life, disrobing of ecosystem services, as well as imbalances affecting economic growth (Cheshire, 2016). While linear consumption has numerous effects on areas including economic, social, and environmental aspects, CE plays a critical role in minimizing these losses within the construction sector.

The aim of this chapter is to determine the CE principles over the construction project life cycle at the micro, meso, and macro levels. There will be a critical focus on CE principles relating to buildings and materials. A conceptual framework will then be developed to demonstrate the interactions between CE principles and project life cycle stages.

The rest of the chapter briefly presents the methodology adopted, concept and principles of CE in the built environment at various levels, various techniques, tools, and circular models used to enhance circular economy (CE) in the built environment considering various project life cycle phases, and at the end, a conceptual circular model is developed to demonstrate the interactions and linkages between CE principles and project life cycles.

Circular Economy and Sustainability, Volume 1
https://doi.org/10.1016/B978-0-12-819817-9.00012-0

2. Methodology

This chapter is concerned with understanding CE during project life cycles. The chapter begins with a deep literature discussion on CE principles at three levels—micro, meso, and macro—as well as a detailed discussion at buildings and materials levels. A wide range of studies on CE and its adoption in the built environment were analyzed, taking into account the main principles and techniques used in the construction industry to enhance CE aspects over a project's life cycle. This concludes with several critical approaches to CE principles at buildings and materials levels, including reducing demand for buildings and/or materials, circular design of buildings, and circular business models. At the end of this chapter, a conceptual framework is developed to clarify the interactions between CE principles and project life cycle stages and how these principles can be incorporated into project life cycles. This is further discussed to demonstrate the possibility of implementing CE principles over five stages of project life cycle.

3. CE in construction/built environment

Although the literature is rich in principles of and approaches to CE, there is gap in the current knowledge, which involves the steps needed to move toward CE implementation in the built environment. The majority of studies classify CE principles at three levels, as discussed below.

3.1 Principles of the CE at micro, meso, and macro levels

A detailed review carried out by Ghisellini et al. (2016) categorized the implementation of CE in the built environment into three levels: micro (single consumer, company, or organization), meso (eco-industrial parks), and macro (regions, cities, and provinces). Table 1 summarizes the classification levels by defining the scope/focus of each level and the ways to adopt CE principles.

The following section details the adoption of CE at the buildings and materials level.

3.2 Principles of the CE at buildings and materials levels

To start the discussion on CE principles on buildings, it is essential to initiate materials levels as it is a fundamental part on meso level, building level. Although there is a limited focus of CE at the meso level of individual buildings, some studies suggest different principles. Cheshire (2016) outlined in detail the CE principles in building design. The essential principles introduce three internal circles (retain, refit, and refurbish) relating to retaining the existing structure—which is considered the best option for resource efficiency to achieve environmental and economic benefits—through refit and refurbishment of the existing structure instead of demolition (Cheshire, 2016). Cheshire (2016) describes several structures in London as role models of refurbished and refitted

Table 1 Classification of micro, meso, and macro levels.

	Focus	Adopting CE through
Micro level	Single consumer, companies, or organizations (Winkler, 2011) Product or materials level (Kirchherr et al., 2017)	Green designs, regenerative designs, design for reconciliation (Gómez et al., 2018), eco-designs, and eco-products (Van Berkel et al., 1997)
Meso level	Eco-industrial parks or industrial symbiosis (Ghisellini et al., 2016) Building level "circular buildings" (Pomponi and Moncaster, 2017)	Different production industries, such as energy, water, material, and supply chain, will work together to achieve environmental benefits (Gwehenberger et al., 2003) and economic benefits (Chertow, 2000)
Macro level	Regions, provinces, and cities level (Abreu and Ceglia, 2018)	Eco-city, "zero-emission," legislation supports recycling innovation, waste management, and water management (Geng et al., 2009)

structures, including 55 Baker Street, as well as the Tea Building, which was previously a "white collar factory." Beyond these three circles, the remaining options are reclaim, remanufacture, and recycle.

However, Geldermans (2016) suggests a "stepwise approach," which discusses the building's circularity value at the products and materials level and the building design level. These could be grouped into three actions: reducing the demand for buildings and/or materials, reusing existing buildings and/or materials, and circular design of buildings and/or materials. Fig. 1 has been created to summarize these three elements.

3.2.1 Reduce demand for buildings and/or materials

Several studies have discussed the various techniques that could help to reduce the demand for buildings and/or materials as the first step toward minimizing waste. Examples include:

- *Lean principles*. The lean principle in construction encourages the concept of designing the building to use less materials and resources, therefore reducing waste and negative impacts on the environment over the building's lifetime (Esa et al., 2017). For example, engineering services in buildings, mechanical, electrical, ventilation, etc., should be designed to eliminate any wasted energy or resources (Cheshire, 2016). In 2011, the restoration of Unity House, which was built in 1867 in Wakefield, Yorkshire, is a good example of lean principles that minimized electricity usage and reused the existing service systems, such as ventilation (CIBSE Journal, 2014).
- *Reduce site waste*. According to Lu and Yuan (2012), the demand for materials could be reduced by off-site activities. Moreover, a study done by Garmulewicz et al. (2018) demonstrated that designing the structure using 3D-software could reduce the waste of off-cuttings due to improper sizing or shape. For example, carpet tiles produced

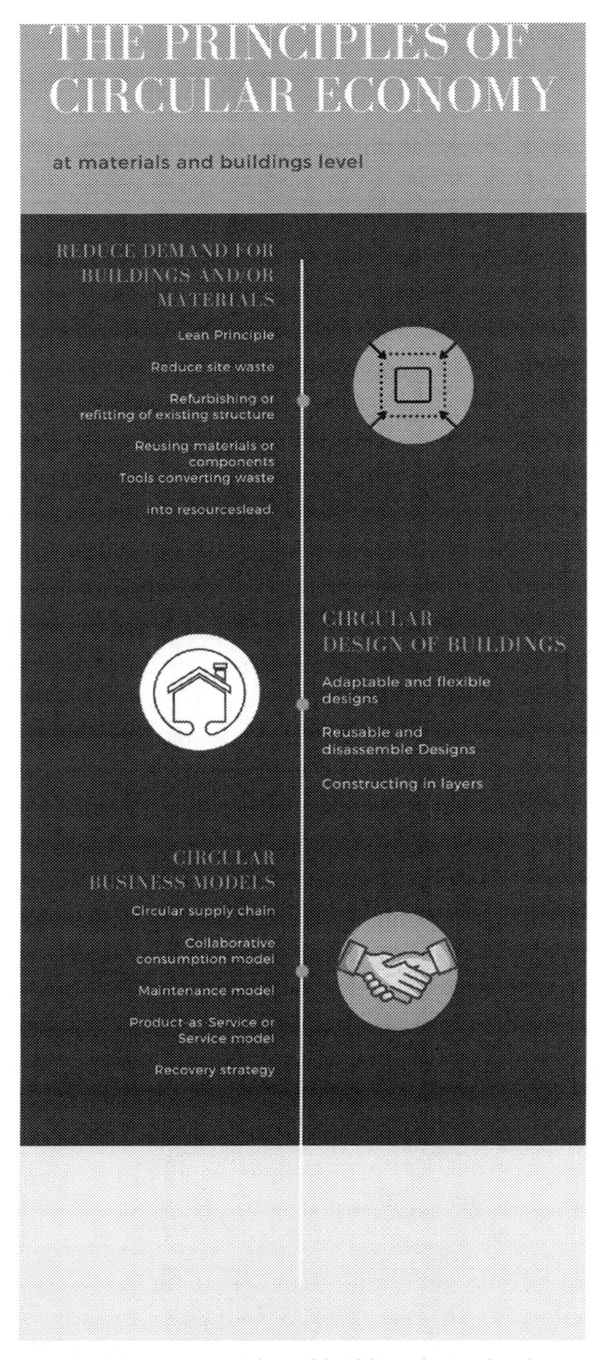

Fig. 1 Circular economy principles at materials and building design levels.

using a "random design" that allows them to be located in any orientation, thereby improving insulation while also reducing cutting (Cheshire, 2016).

- *Refurbishing or refitting of existing structure*. Reducing the demand for new buildings through reusing existing structures will help to eliminate waste through refurbishing or refitting; for example, the regeneration of 55 Baker Street in London. The demolition of the 1950s offices building, of 75,000 m^2 was believed to be essential as it had a low floor to ceiling height, which was unattractive commercially (Expedition Engineering, 2018). However, the structural company, Expedition Engineering, and the designers, Make Architects, suggested a creative idea to redevelop and refurbish the existing buildings (Lomholt, 2019), potentially saving the project time and money. The refurbishment in 2008 created a new urban center, containing retail spaces, leisure areas, residential spaces, and offices (Expedition Engineering, 2018).
- *Reusing materials or components*. Geldermans (2016) states that "reuse thinking" usually starts from the design stage, including the selection of materials according to their lifespan. For example, if the structure is designed for a long period, with internal fixtures that have a short lifespan, this will result in higher waste at the end of its life (Esposito et al., 2018). As a result, these materials should be made of biological components and nontoxic materials that can be returned to the environment safely, or which are designed to be reused or remanufactured (Geldermans, 2016). Elements and components with long lifespans should (Esposito et al., 2018) be designed to be resilient and durable as well as being suitable to upgrade, repair, maintain, remanufacture, or disassemble (Ball and Melton, 2015).
- *Tools converting waste into resources*. Leising et al. (2018) explain the potential of three critical methods to turn waste into resources to reduce the demand for new materials as follows:
  - *Raw materials passport*. Rau and Oberhuber (2016) proposed a "raw materials passport," which is a digital description defining waste as a raw component to facilitate reusing, recycling, reclamation, or maintenance (Goens et al., 2018). A study carried out by Akanbia et al. (2018) on adopting building information modeling (BIM), which can be used to support the idea of the raw materials passport to facilitate CE function, found that BIM offers the ability to create detailed data of each material and share it with all parties through evaluating the performance of each material through the whole lifespan and at the end of life. Similarly, Ness et al. (2015) suggest a digital service that tracks the steel elements in the structure to simplify the reuse process in the new structure.
  - *Banks of materials*. Following the identification of all materials in the previous step, Rau and Oberhuber (2016) explains that the building will be a bank of materials or temporary storage of materials, where all its materials could be easily reused again for different iterations due to their high value (Rosengren, 2017; Dietvorst, 2016; Geldermans, 2016). According to Stephan and Athanassiadis (2017) and

Ortlepp et al. (2016), the reusable material is dependent on the condition of existing materials stock, which is the bank of the raw materials.
- *Salvaged materials market*. After documenting the building materials in the raw materials passport, a marketplace is essential to supply the salvaged materials. The market will offer connections between different sectors that are interested in material circulation (Leising et al., 2018). For instance, a project funded by the EU from 2015 called "Buildings as Material Banks (BAMB)" was established to provide a platform to maximize the use of valuable components (Goens et al., 2018).

Although the previous techniques could help to reduce the demand for new buildings and materials through minimizing waste in the production stage and reusing existing components, they could also have potential if considered at the design stage. The following section discusses the circular design of buildings.

3.2.2 Circular design of buildings

The circular design of buildings and materials are discussed by several authors and there is a focus on considering this principle at the beginning of the design stage. The main points associated with the circular design principle can be summarized as follows.

- *Adaptable and flexible designs*. Adaptable and flexible design means that a building could be designed to be modified to elongate its lifespan to meet the new users' needs (Addis and Schouten, 2004). Designing an adaptable building will help to extend the building function and its economic service, as its utility to meet new needs will increases (Galle, 2017). There are several proposed concepts of design for adaptability. Examples of these are:
  - *Open buildings*. Open building is a common concept proposed by Habraken in the 1960s (Kilpatrick, 2014). Its main objective is to design buildings with the idea of segregation with infill elements between supportive structural elements to allow changes at any level without affecting the other elements, similar to the "constructing in layers" concept, discussed below (Geldermans, 2016).
  - *Industrial, flexible, and demountable (IFD)*. After implementing the open buildings concept, the Dutch government established a program called "industrial, flexible, and demountable (IFD)" (Gassel, 2011). It demonstrated the importance of adaptability, customization, and standardization through several techniques, such as prefabricated panels and demountable partition wall with spaces to allow flexible services (Gassel, 2011).
  - *Multispace concept*. The multispace concept was created by 3DReid which proposed several design parameters for various types of buildings to study the possibility of adaptable designs (Beadle et al., 2008). The main parameters include structural design, floor to ceiling height, depth, cladding layout, core design, and servicing (Beadle et al., 2008).

- *Reusable and disassemble designs*. Studies carried out by Sanchez and Haas (2018), Conejos et al. (2015), and Douglas (2006) demonstrate that preplanning for reuse of the building at the inception stage is the most sustainable strategy to achieve positive environmental, social, and financial performance, rather than just trying to minimize the negative impacts on the environment. Furthermore, Sassi (2008), in his research on closed loop material cycle construction proposed a technical evaluation of components and materials that could be disassembled, and set several proposals that help designers to evaluate and test their components to be disassembled at the end of life (Sassi, 2008).
- *Constructing in layers*. According to Cheshire (2016), each element in the structure has a different lifespan that should be recognized and be independent to facilitate the process of replacing or "peeling-off" without affecting the adjacent layers. Several studies propose different approaches, for example, Duffy (1992) classified building layers into four sections (Shaviv and Pushkar, 2014):
  1. Shell: the structure of the building.
  2. Services: ducts, wires, and pipes insulations.
  3. Scenery: internal outfit.
  4. Set: equipment, fittings, and furniture.

Each layer has a different lifespan, with the shell layer having the longest lifespan and the shortest being the set layer (Shaviv and Pushkar, 2014). Although the shell layer has the longest lifespan, it should be provided with flexible service space to facilitate maintenance or addition of new services (Cheshire, 2016); whereas the set layer should be made from biological components and designed for the short-term, to then be recycled or reused (UK Green Building Council, 2017).

A more expanded and detailed classification was outlined in Brand (1994). It layered the building into six parts according to the lifespan of the layers (Alter, 2015). While the "structure" layer has a long life, the "stuff" layer has a short one, which is sensible and motivates people to rethink the way of handling various layers (Alter, 2015).

Following the detailed criteria of circular designs, circular business models are becoming essential to enhance the circularity level in the construction industry.

3.2.3 Circular business models

> *A circular business model describes how an organization creates, delivers, and captures value in a circular economic system, whereby the business rationale needs to be designed in such a way that it prevents, postpones or reverses obsolescence, minimizes leakage and favors the use of "resources" over the use of "resources in the process of creating, delivering and capturing value".*
>
> ***Hollander and Bakker (2016, p. 2)***

Although there is limited literature that defines the circular business model, it can be understood from the above definition that circular business models can play an essential role in value creation in the supply chain in the construction industry. Arup and Bam (2017) state that CE models will play a critical role as they will allow the securing and

Table 2 Circular business models.

Circular business models	Study was done by	Through	Outputs
Circular supply chain	Lacy and Rutqvist (2015), Esposito et al. (2018)	Using inputs that are totally recyclable, bio-gradable, and renewable	Buildings will be designed with low cost and valuable materials
Collaborative consumption model	Ghisellini et al. (2016), Kjaer et al. (2018)	Shared ownership of products by several users (e.g., renting, trading, gifting, sharing, lending, bartering)	Benefits of the materials and buildings will be maximized as all the construction parties will focusing on sharing profits
Maintenance model	Van Sante (2017), Jonker et al. (2017), Stahel (2012)	Elongate the lifespan of materials and buildings through maintenance, upgrading, refurbishment, and repair	Materials and buildings will be designed to be adaptable for changes over long life span
Product-as-service or service model	Stahel (2012), Minunno et al. (2018), Esposito et al. (2018)	Selling performance or services rather than products. The manufacturer being responsible for the product until the end of its life	It will encourage the manufacturer to design products for durability, disassembly, or reuse
Recovery strategy	Lacy and Rutqvist (2015), Esposito et al. (2018)	Maximize the value of products by turning the waste into new valuable materials	Waste will be reduced and recovery market will serve the industry with low cost and valuable materials

controlling of resources. Lacy and Rutqvist (2015) highlight that applying a circular business model could reduce the consumption of materials by 90%; therefore, the gross profit of a company will increase by 50%.

Table 2 summarizes various studies that suggested several circular business models.

The five models presented in Table 2 are interlinked to each other; starting with "circular supply chain," which focuses on reusing and recycling construction materials which are a result of a "recovery strategy" that serves the market with new valuable materials created from waste. Between them, the new circular thinking "collaborative consumption model," "maintenance model," and "product-as-service" or "service model" focus on maximizing the benefits of materials or buildings through sharing the responsibilities of all the parties until the end of its life. Several attempts have been made to focus on servicing the construction industry with these models; for example, Delta Development enforced a new business model that allows suppliers to offer their service for rent and retain ownership of their materials—such as, LED lighting, tiles, and office furniture—that are used in buildings, giving them the potential to recover the materials at the end of the building's life (Cheshire, 2016).

Although there are various studies and practices regarding circular business models, the next section provides a simplified picture that shows the possibility of implementation of CE principles over a project's life cycle.

3.3 Interaction between CE principles and construction stages

Fig. 2 shows a conceptual framework that shows the potential interactions of implementation of CE principles over five stages (discussed previously in detail as three main principles: reduce demand for buildings and/or materials, circular design of buildings, and circular business models) of a building project's life cycle (i.e., design, material production, logistics, construction, and operation). The design stage starts with selecting circulated materials and reducing use of raw materials, as well as designing for durability, disassembly, etc. At the production stage, CE requires minimizing waste through lean thinking and applying circular supply chains that depend on recyclable, bio-gradable, and renewable inputs. At the stage of logistics, in implementing a circular business model it is essential to share services, for example, selling performance or service rather than products in the operation stage. The construction stage must minimize site waste and ensure the use of programs to track building elements over their lifespan to facilitate repairing, upgrading, reproduction, remanufacturing, and reusing. At the operation stage, new technologies have the potential to save data on reusable materials through, for example, BIM, banks of materials, and raw materials passports, which will feed salvaged materials back to the market, providing circulated products that will increase reusing and recovery strategies: the market for resale of salvaged materials will feed the market with circulated products of high quality.

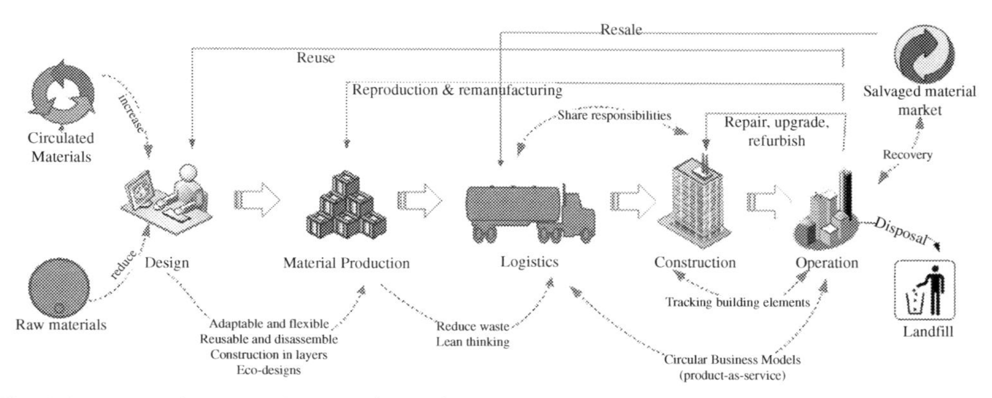

Fig. 2 Interaction between CE principles and construction stages.

4. Conclusions and discussion

Based on previous studies, the three main principles for implementing CE over the five stages of the project life cycle are: reduce demand for buildings and/or materials, circular design of buildings, and circular business models. Therefore, the chapter ends with a clear view of the main results through a framework that clarifies the interactions of implementation of CE principles over different project stages (i.e., design, materials production, logistics, construction, and operation). Starting with the design stage, reduce demand for buildings or materials through selecting circulated materials sources, as well as designing for durability, disassembly, etc. At the materials production stage, working on minimizing waste through lean thinking is essential. Circular business models are fundamental at the logistics stage to ensure responsibilities and services sharing. At the construction stage, rather than only focusing on minimizing construction waste, tracking building elements over their lifespan to facilitate repairing, upgrading, reproduction, remanufacturing, and reuse is also critical. Lastly, the operation stage plays an essential role through saving data of reusable materials and circulated products to feed the salvaged materials market with high-quality products. To upgrade the concept, new technologies, such as BIM and raw materials passports, would facilitate the adoption of CE interactions over a project's life cycle.

In the end, it is highly recommend to adopt CE principles in the built environment and academics and industry players should work together and invest more effort to explore and develop efficient and practical techniques and tools that enhance CE in order to enhance sustainability aspects at large.

References

Abreu, M.K., Ceglia, D., 2018. On the implementation of a circular economy: the role of institutional capacity-building through industrial symbiosis. Resour. Conserv. Recycl. 138, 99–109.

Addis, W., Schouten, J., 2004. Principles of Design for Deconstruction to Facilitate Reuse and Recycling. CIRIA, London.

Akanbia, L., Oyedele, L., Akinade, O., Ajayi, A., Delgado, M.D., Bilal, M., et al., 2018. Salvaging building materials in a circular economy: a BIM-based whole-life performance estimator. Resour. Conserv. Recycl. 129, 175–186.

Alter, L., 2015. There's a lot to learn from Stewart Brand's "How Buildings Learn" (Video) [online]. Treehugger https://www.treehugger.com/green-architecture/watch-stewart-brands-how-buildings-learn.html. (Accessed 18 February 2015).

Arup and Bam, 2017. Circular business models for the built environment [online]. Arup and Bam https://www.bam.com/sites/default/files/domain-606/documents/8436_business_models-606-1490949582322972999.pdf. (Accessed 18 February 2015).

Ball, J.D., Melton, P., 2015. Circular economy at scale: six international case studies. Environ. Build. News 24 (10), 1–19.

Beadle, K., Gibb, A., Austin, S., Madden, P., Fuster, A., 2008. Adaptable futures: setting the agenda. In: Proceedings of the 1st I3CON International Conference, Loughborough University, Loughborough, UK, 14–16 May.

Brand, S., 1994. How Buildings Learn. Viking, New York.

Chertow, M.R., 2000. Industrial symbiosis: literature and taxonomy. Annu. Rev. Energy Environ. 25 (1), 314–334.

Cheshire, D., 2016. Building Revolutions: Applying the Circular Economy to the Built Environment. RIBA Publishing, British Library.

CIBSE Journal, 2014. Trimming the fat. CIBSE J., 34–35.

Conejos, S., Langston, C., Smith, J., 2015. Enhancing Sustainability Through Designing for Adaptive Reuse From the Outset: A Comparison of adaptSTAR and Adaptive Reuse Potential (ARP) Models. vol. 33 Emerald Group Publishing Limited, pp. 531–552. issue 9/10.

Dietvorst, C., 2016. What can the sector learn from circular economy visionary Thomas Rau? . Viawater [online]. 9 June https://www.viawater.nl/news/what-can-the-sector-learn-from-circular-economy-visionary-thomas-rau. (Accessed 2 December 2018).

Douglas, J., 2006. Building Adaptation. second ed Elsevier Ltd, UK.

Duffy, F., 1992. The Changing Workplace. Phaidon Press, London.

Ellen MacArthur Foundation, 2013. Towards the circular economy, economic and business rationale for an accelerated transition [online]. Ellen MacArthur Foundation https://www.ellenmacarthurfoundation.org/assets/downloads/publications/Ellen-MacArthur-Foundation-Towards-the-Circular-Economy-vol.1.pdf. (Accessed 20 December 2018).

Esa, M.R., Halog, A., Rigamonti, L., 2017. Strategies for minimizing construction and demolition wastes in Malaysia. Resour. Conserv. Recycl. 120, 219–229.

Esposito, M., Tse, T., Soufani, K., 2018. Introducing a circular economy: new thinking with new managerial and policy implications. Calif. Manag. 60 (3), 5–19.

Expedition Engineering, 2018. Transforming buildings [online]. Expedition https://expedition.uk.com/projects/transforming-buildings/. (Accessed 20 December 2018).

Galle, W., 2017. Design for Change, Towards a Circular Economy (Presentation Handout). Vrije Universiteit Brussel, Brussels.

Garmulewicz, A., Holweg, M., Veldhuls, H., Yang, A., 2018. Disruptive technology as an enabler of the circular economy: what potential dose 3D printing hold? Calif. Manag. Rev. 60 (3), 112–132.

Gassel, F.V., 2011. Experiences With the Design and Production of an Industrial, Flexible, and Demountable (IFD) Building System. Eindhoven University of Technology, Department of Architecture, Building and Planning, The Netherlands, pp. 209–214.

Geldermans, R.J., 2016. Design for change and circularity—accommodating circular material & product flows in construction. In: SBE16 Tallinn and Helsinki Conference; Build Green and Renovate Deep, Tallinn and Helsinki Delft University of Technology, pp. 301–311.

Geng, Y., Zhu, Q., Doberstein, B., Fujita, T., 2009. Implementing Chain's circular economy concept at regional level: a review of progress in Dalian, China. Waste Manag. 29, 996–1002.

Ghisellini, P., Cialani, C., Ulgiati, S., 2016. A review on circular economy: the expected transition to a balanced interplay of environmental and economic systems. J. Clean. Prod. 114, 11–32.

Gómez, A.M., González, F.A., Bárcena, M.M., 2018. Smart eco-industrial parks: a circular economy implementation based on industrial metabolism. Resour. Conserv. Recycl. 135, 58–69.

Goens, H., Capelle, T., Henrotay, C., Steinlage, M., 2018. D13 prototyping + feedback report [online]. BAMB2020 https://www.bamb2020.eu/wp-content/uploads/2018/10/20180425-BAMB-WP4-D13.pdf. (Accessed 20 December 2018).

Gwehenberger, G., Erler, B., Schnitzer, H., 2003. In multi-strategy approach to zero emissions. In: Article presented at Technology Foresight Summit 2003, Budapest 27–29 March. (United Nations Industrial Development Organization).

Hollander, D.M., Bakker, C., 2016. Mind the gap exploiter: circular business models for product lifetime extension. In: In Proceedings of the Electronics Goes Green, Berlin, pp. 1–8.

Jonker, J., Stegeman, H., Faber, N., 2017. The Circular Economy, Backgrounds, Developments, Concepts, and the Search for Matching Business Models. White Paper, Nijmegen School of Management, Radboud University, The Netherlands.

Kilpatrick, I., 2014. Inter-generational living: open building architecture and the importance of choice & independence [online]. Regeneration http://sieplcoatesstudio.weebly.com/uploads/2/3/3/0/23301256/kilpatrick_book_portion.pdf. (Accessed 3 November 2018).

Kirchherr, J., Reike, D., Hekkert, M., 2017. Conceptualizing the circular economy: an analysis of 114 definitions. Resour. Conserv. Recycl. 127, 221–232.

Kjaer, L.L., Pigosso, D.C.A., Niero, M., Bech, N.M., McAloone, T.C., 2018. Product/service-systems for a circular economy the route to decoupling economic growth from resource consumption? J. Ind. Ecol., 1–14.

Lacy, P., Rutqvist, J., 2015. Waste to Wealth: The Circular Economy Advantage. Palgrave Macmillan, Basingstoke, Hampshire.

Leising, E., Quist, J., Bocken, N., 2018. Circular economy in the building sector: three cases and a collaboration tool. J. Clean. Prod. 176, 976–989.

Lomholt, I., 2019. Baker Street London, Office Building, Former M&S HQ, English Project [online]. E-architect https://www.e-architect.co.uk/london/55-baker-street. (Accessed 20 December 2018).

Lu, W., Yuan, H., 2012. Off-site sorting of construction waste: what can we learn from Hong Kong? Resour. Conserv. Recycl. 69, 100–108.

Minunno, R., O'Grady, T., Morrison, G.M., Gruner, R.L., Colling, M., 2018. Strategies for applying the circular economy to prefabricated buildings. Buildings 8 (9), 125.

Ness, D., Swift, J., Ranasinghe, D.C., Xing, K., Soebarto, V., 2015. Smart steel: new paradigms for the reuse of steel enabled by digital tracking and modelling. J. Clean. Prod. 98, 292–303.

Ortlepp, R., Gruhler, K., Schiller, G., 2016. Material stocks in Germany's non-domestic buildings: a new quantification method. Build. Res. Inf. 44 (8), 840–862.

Pomponi, F., Moncaster, A., 2017. Circular economy for the built environment: a research framework. J. Clean. Prod. 143, 710–718.

PricewaterhouseCoopers Network 'PwC', 2015. Corporate sustainability lessons learned, going circular, towards 100% reuse and recycling [online]. PwC UK Organizations https://www.pwc.co.uk/assets/pdf/pwc-going-circular-2015.pdf. (Accessed 20 December 2018).

Rau, T., Oberhuber, S., 2016. Material Matters. Bertram + de Leeuw Uitgevers, Haarlem.

Rosengren, C., 2017. 3 ways to boost the circular economy in NYC's construction industry [online]. Waste Drive https://www.wastedive.com/news/3-ways-to-boost-the-circular-economy-in-nycs-construction-industry/445010/. (Accessed 20 December 2018).

Sanchez, B., Haas, C., 2018. Capital project planning for a circular economy. Constr. Manag. Econ. 36, 303–312.

Sassi, P., 2008. Defining closed-loop material cycle construction. J. Build. Res. Inf. 36 (5), 509–519.

Shaviv, E., Pushkar, S., 2014. Green building standards—visualization of the building as layers according to lifetime expectancy. Energy Procedia 57, 1696–1705.

Stahel, W.R., 2012. The Business Angle of a Circular Economy—Higher Competitiveness, Higher Resource Scarcity and Material Efficiency. The Product-Life Institute, Geneva, pp. 1–10.

Stephan, A., Athanassiadis, A., 2017. Quantifying and mapping embodied environmental requirements of urban building stocks. Build. Environ. 114, 187–202.

UK Green Building Council, 2017. Practical how-to guide: build circular economy thinking into your projects [online]. UK Green Building Council. https://www.ukgbc.org/sites/default/files/How%20to%20build%20circular%20economy%20thinking%20into%20your%20projects.pdf. (Accessed 20 December 2018).

U.S. Environmental Protection Agency, 2009. Buildings and their impact on the environment: a statistical summary [online]. EPC https://archive.epa.gov/greenbuilding/web/pdf/gbstats.pdf. (Accessed 20 December 2018).

Van Berkel, R., Willems, E., Lafleur, M., 1997. The relationship between cleaner production and industrial ecology. J. Ind. Ecol. 1, 51–65.

Van Sante, V., 2017. Circular construction [online]. ING https://think.ing.com/uploads/reports/ING_EBZ_Circular-construction_Opportunities-for-demolishers-andwholesalers_juni-2017.pdf. (Accessed 20 December 2018).

Winkler, H., 2011. Closed-loop production systems—a sustainable supply chain approach. CIRP J. Manuf. Sci. Technol. 4, 243–246.

CHAPTER 12

The role of ecodesign in the circular economy

Karine Van Doorsselaer
Department Productdevelopment, University of Antwerp, Antwerp, Belgium

1. Introduction

The circular economy is based on closing loops, according to the hierarchical order: product reuse, components reuse, and finally, material reuse. To fulfill these principles the products need to be repairable, upgradeable, and easy to maintain and refurbish, and the materials should be suitable for recycling processes. Whether products fit into the circular economy starts at the very beginning of a product's life, which is the design phase of the product.

The development of each new product or the adjustment of an existing product starts with an idea. This idea is transformed into a detailed product design that fulfills all the specifications: the technical aspects, the economic issues, the social- and human-related needs. The decisions concerning "what," "how," and "for whom" are taken in the early phase of the design process. The products can be toys, packaging, electronic equipment, furniture, and all the other products that are used by the consumer. In addition, products for public services, like shelters at bus stops, or industrial products, like medical equipment, run through the steps of the product development process.

By implementing ecodesign into the product development process, the designer will think about the environmental impact of the product for each phase of the life cycle. According to the Ecodesign Directive of the European Commission, it is estimated that 80% of all product-related environmental impacts of energy-related products are determined during the design phase of a product. The production, distribution, use, and end-of-life management of energy-related products is associated with important impacts on the environment, such as the consequences of energy and other materials/resources consumption, waste generation, and release of hazardous substances. The implementation of the principles of ecodesign will support the achievement of Sustainable Development Goal 12 Sustainable Production and Consumption, in which ecodesign is also to referred as a key to success (Timmermans and Katainen, 2019; United Nations, n.d.).

Not only the environmental impact of products is determined in the design phase; whether the product fits into the circular economy is also determined by the designer. Decisions made in the early phase of the life cycle can make products easier to repair,

Circular Economy and Sustainability, Volume 1
https://doi.org/10.1016/B978-0-12-819817-9.00018-1

upgrade, and remanufacture, and improve the recyclability of the materials. As part of its "circular economy" package, the European Commission presented in December 2015 an action plan for the circular economy. In this action plan, the Commission indicated it would promote the reparability, upgradability, durability, and recyclability of products by extending the scope of ecodesign requirements beyond energy efficiency.

Ecodesign is a key to success in the circular economy and will get more and more attention in the design process, supported by legislation.

In this chapter, we discuss the principles of ecodesign and the different "design for X" guidelines that support the designer to develop products that fit into the circular economy. Regarding the butterfly-model of the Ellen MacArthur Foundation, we discuss the ecodesign guidelines according to the different ways to close loops: use and reuse products for as long as possible, reuse components, and recycle or compost materials. In addition, a few qualitative tools based on guidelines a designer can use to implement ecodesign into the design process are discussed.

2. Ecodesign

The definition of ecodesign can be summarized with one word: lifecycle thinking (Fig. 1).

During the product development process, the designer will pay attention to the environmental impact for each phase of the life cycle of the product: raw materials, production of materials out of the raw materials, the production techniques to transform the

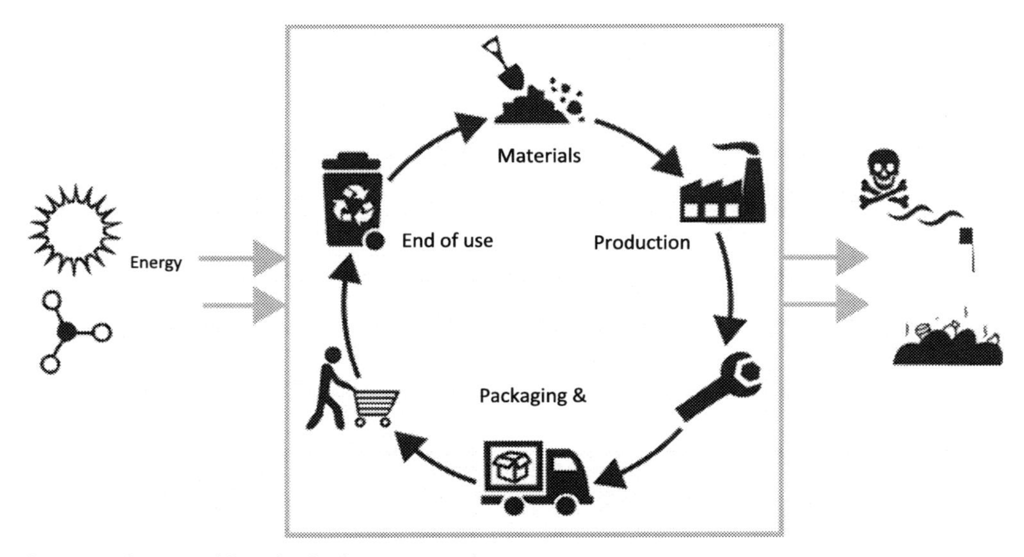

Fig. 1 Ecodesign, a lifecycle thinking approach.

materials into components, assembly of these components into products, the distribution and packaging of the products, the use phase (the impact of consumables and energy consumption), and finally, the end if use phase by the consumer. Regarding the principles of the circular economy, the designer will support the possibilities to extend the lifespan of the product as long as possible by taking into account reuse, repairing, and refurbishing possibilities of the product, and improve the recyclability of the material. For each step in the life cycle the designer will try to reduce the input of energy and materials/auxiliaries and the output of emissions and waste.

Implementation of the ecodesign guidelines in the design process is also key to success in closing the loops in the circular economy.

The designer will search for the best compromise between the other needs and specifications for the product and the principles of ecodesign. The ecodesign guidelines are the copilot in the development process.

To implement ecodesign in the product development process, the designer will take the various "design for X" guidelines into account, where X stands for the different design strategies, such as "design for disassembly," "design for repair," "design for recycling," etc.

In the following sections we discuss these ecodesign guidelines, which support the development of products that fit into the circular economy. We realize that the list of ecodesign guidelines is not complete. For each product group there might be some specific guidelines. The aim of the list is to get started and to invite the designer to extend the list regarding specific products or services.

2.1 Design for assembly/design for disassembly (DfA/DfD)

To extend the lifespan of products, these products should be easy to maintain, repair, and upgrade. To support these possibilities the products should be easily disassembled and assembled. Furthermore, to recycle the materials once the product is no longer usable, one crucial aspect for success is that all the components/materials are easy to disassemble. Consequently, implementing the DfA and DfD guidelines is a necessity for successfully closing the loops.

The designer will take into account who will disassemble the products—whether it will be the consumer him/herself, the technician of a general repair service, or the technician of the production company.

For safety reasons, not all products should be disassembled by the consumer; for example, high voltage electronics. In addition, for protection of intellectual property, sometimes it is not appropriate for a product to be disassembled by a technician who is not connected to the company. The use of very special tools will assure that only appropriate technicians are able to disassemble the product. Regarding the new business model "product-service systems," in which the company retains ownership of the products, it is also important to implement the DfD guidelines, as this will make it cost-efficient for the products to be dismantled quickly and easily. Regarding who will disassemble the product, the general ecodesign guidelines listed below must be considered in relation to the individual context.

The general ecodesign guidelines concerning DfD and DfA are:

- Use standard joining components and tools like standard screws and screwdrivers.
- Use snap fits, which are easy to disassemble.
- Avoid glue.
- Make joints visible and reachable.
- Minimize the amount of parts.
- Design modular structures of the product with subassemblies that are easy to assemble and disassemble.
- Assemble in one direction to avoid unnecessary turning of the product during assembly. And make sure that each component is fixed in the assembly so the component can't move during further assembly.
- Make a manual to assemble and disassemble the product (digital).
- Think about the implementation of active disassembly techniques, e.g., thermally reversible adhesive sprays, which make it possible to disassemble the product just by heating.

2.2 Design for maintainability (DfM)

To extend the lifetime of a product, it's crucial that the product can be easily and cheaply maintained. Not all users take good care of their products, and limit themselves to the most necessary maintenance. Products that are difficult to clean will quickly look unkempt, so that the user will be inclined to discard them.

To improve the maintenance of the product, the designer must think about the disassembly and assembly of the product, and also about:

- easy cleaning of the product, no blind spots;
- locating components that are susceptible to wear in an accessible place in the product;
- providing a manual with practical tips for the user concerning the maintenance of the product;
- provide a maintenance program or service, e.g., digital support; and
- making the product upgradeable, especially for electronic equipment—it's not good enough to sell products that are digitally supported for only 2 years.

2.3 Design for repair

The first requirement for a product that needs to be repaired is that the product can be disassembled so the broken component(s) can be repaired or replaced. In this regard, the designer should be guided by the DfA and DfD guidelines.

In a circular economy in a globalized world, we assume that a product should be locally repaired in an easy way. The designer should take the following ecodesign guidelines into account:

- Think about which component or group of components is most susceptible to wear and tear, and assemble these in a visible and easy to remove place.
- Provide a repair kit and manual with repair guidelines (digital) and tools.
- Guarantee the availability of components over a long period of time.

2.4 Design for remanufacture

This strategy is very useful for the product-service systems business model, where the products remain the property of the manufacturing company. The company has every interest in making products that are quick, easy, and cheap to upgrade, and which can be refurbished. The ecodesign DfD guidelines are even more valuable in this context.

2.5 Design for recycling and design for composting

After the product and the components have been optimally used and reused, finally the material will be recycled. Bio-based materials might end up as feedstock for biomass fermentation or will be composted. One general, very important rule for closing material loops is that the use of chemicals should be limited in the circular economy. The risk of accumulation of chemicals due to closing material loops should be avoided because the environmental and health impact of the cocktail of these chemicals is unpredictable. Even when it is possible to purify the recycled material, chemicals should be kept to a minimum to limit costs and to prevent pollution and health risks at every stage of the life cycle of the product.

2.5.1 Design for recycling

The designer will take two criteria into account for the selection and use of recyclable materials: first, can the material be recycled, ideally without loss of properties; and second, are there existing collection and recycling facilities available, so that the material can be collected and recycled. In spite of some materials being perfectly fit for remelting/recycling, often there is a lack of collecting, sorting, and recycling facilities. The reasons are many:

- Collecting, sorting, and recycling process are too expensive in combination with a low economic value of the recycled material.
- The volume of the waste stream of the specific material is too low to make recycling profitable.
- Lack of a general waste treatment program in various countries worldwide. A very big hurdle is that collecting, sorting, and recycling facilities differ so much from region to

region, from country to country. A general, uniform scheme would be advisable, but that will take a lot of time and is the responsibility of governance worldwide.

It is the task of the designer to keep abreast of the evolution of various recycling opportunities for different materials worldwide, to improve the recyclability of products.

The designer can support the recycling industry by choosing recycled material, but then he or she should be aware of the possibilities and risks of using recycled material. As an example we discuss the recyclability of plastics.

2.5.1.1 Recyclability of plastics

Plastics are divided into three categories: thermoplastics, thermosets, and rubbers. Related to the chemical structure of these kinds of plastics, only the thermoplastics and the thermoplastic rubbers can be remelted and have the potential to be recycled, according to the state of the art today. Vulcanized rubbers and the thermosets are not remeltable, so it is not possible to close the material loops in a proper way. Nowadays, these materials are milled and used as fillers or they are burned, thus recovering the caloric value of the materials through "thermal recycling." A lot of investigation is going into the potential recycling of these forms of plastic, on an ongoing basis.

Although the thermoplastics are theoretical suitable to recycle by remelting, there is a big hurdle to doing it in a proper way: the properties of the recycled plastic are less than the virgin plastic, caused by degradation during the remelting process.

The quality of the recycled plastic is also influenced by the presence of additives and other chemicals, for example, those used for surface treatment. The mixture of the additives and chemicals in postconsumer plastic waste is unknown; there might even been additives present that were previously used, e.g., pigments based on cadmium, which are now on the prohibited list in Europe. On this European list of prohibited substances are additives that are still used on other continents, such as stabilization additives based on lead. Due to the globalization of trade and the inability to control all incoming products into Europe, the material flow of plastics becomes contaminated.

Another aspect that might contaminate plastic waste is the use of the products by the consumer, e.g., a consumer might use a PET-bottle to rinse paint brushes with white spirit, through which traces of paint and solvent enter the material stream. It is also known that plastic products absorb substances during the use phase, e.g., shampoo is included in the PE bottle, or gasoline penetrates the wall of a gas tank.

Plastic waste is commonly contaminated with all kinds of chemical substances, which remain in the recycled plastic. Quality control of the recycled material of postconsumer waste is very difficult, as the input stream changes minute by minute.

This obstacle can be minimized when the sorting process is at the product level, e.g., separate collection of PET bottles. Another solution is to implement a material passport for all the products and components in combination with digital separation techniques so the input stream of the recycling process is very well controlled. Still, the history of the product and what the consumer has done with the product remain a problem to get recycled plastic that contains only known chemicals. Furthermore, the degradation molecules present an obstacle to obtaining good recycled material.

Sometimes an extra component is added to the recycled plastic, a compatibilizer, to improve the properties and to achieve the quality of virgin plastic. It is clear that this approach does not fit in the circular economy, as the plastic is even more contaminated.

Because the properties of the recycled thermoplastics are not that good, the recycling industry is facing the problem that there is very little market for the recycled material. To solve this economic problem, governments promote the use of recycled content in materials. From my point of view, this is a very dangerous strategy, as the recycled material is contaminated with degradation molecules—unknown chemicals that can cause environmental and health problems during the use of the product based on material with recycled content. To counter the risks of migration of molecules from the recycled material, the recycled plastic is used as an intermediate layer with virgin plastic on both sides. This solution might fit in the recycling economy, but is not an option in the circular economy, because the recycled material contaminates the material stream.

To counter all these difficulties, the solution for recycling plastics is not remelting the material (mechanical recycling), but chemical recycling. With chemical recycling of plastics, the large molecules are depolymerized into monomers or other chemicals, which can be used to make new plastics or in other chemical processes. To make the distinction in the use of recycled plastics generated either by mechanical or by chemical recycling process, we can use the term "recycled contend" for remelted plastic and "recycled feedstock" when the plastic is polymerized with recycled monomers.

A lot of investigation is going on about that topic, and there are already a number of successful projects (e.g., Ioniqa in the Netherlands (Ioniqa, 2019)).

2.5.1.2 Design for recycling guidelines

- Select standard materials (the higher the volume of one type of material, the more cost efficient the recycling process will be).
- Use mono-material, avoid mixtures of different kinds of materials, e.g., composites of PE + wood fibers or plastic multilayers used in packaging.
- Avoid additives, fillers, and surface treatments, which can cause contamination of the material.
- Use as few different types of materials as possible.
- Select materials that suit the collecting and recycling programs of the regions the product is sold.
- Use material passports (digital).

2.5.2 Design for composting

In analogy with the definition of recyclable materials, compostable materials also have to fulfill two criteria: the material must be compostable, and suitable collection and composting facilities must be available.

Note: there is a big difference between the definitions "biodegradable" and "compostable." Biodegradable means that the material degrades in nature. Compostable means that the material only degrades in specific conditions (humidity, temperature, etc.), which are found in composting installations.

2.5.2.1 Theoretical compostability of materials

The compostability of materials is tested and certified by the European norm EN 13432. We have some remarks and concerns about this certification process, which are useful for the designer to know.

First, it's important to mention that the certification is product-based and not material-based. EN 13432 requires the compostable plastics to disintegrate after 12 weeks and completely biodegrade after 6 months. The time of composting depends of the wall thickness of the product. That's why, for example, a plastic bag made of compostable material will get the certificate while a toy made from the same material will not be certificated.

Secondly, we have big concerns regarding the toxicity tests on the composed material. In Flanders (Belgium) the toxicity of the composted material is tested by the growth of plants, like cress. The germination and growth of cress in the tested compost is compared with a reference dish filled with pure compost. More thorough testing is advisable. To avoid accumulation of harmful substances, legislation might be much stricter. Furthermore, the digestibility of the degradation fragments of the product, when they are eaten by animals, should be tested, especially for compostable products, which are used in agriculture or are home-composted.

Regarding the transition to the bio-based economy, we should take the precautionary principle seriously into account. We should be careful that, by using compostable materials, especially plastics, within some decades, we don't end up with plastic mud, like we have now with plastic soup.

2.5.2.2 Collecting, sorting, and composting facilities

Although products have the "OK" compost certification label, it is not guaranteed that the products will be collected to be composted. Like the lack of recycling facilities in some parts of the world, composting installations are not available everywhere. Furthermore, in some countries, certificated products are not allowed in the collection of the compostable waste fraction. In Belgium, for instance, certificated compostable plastic products are not allowed. The reason for this regulation is that the compostable plastic bags hinder the sieving process and to avoid the risk of wrong sorting behavior by consumer causing compostable plastic to enter the compostable stream. In addition, by wrong sorting behavior of the consumer it is possible that compostable plastic ends up in the plastic recycling stream, which will cause a lot of problems in the recycling process.

The conclusion is that currently, biodegradable plastics are only useful to use for applications where biodegradability has a function, like plant clips used in agri- and horticulture, but certainly not for packaging.

It is vital that the use of biodegradable material does not encourage consumers to think it is safe to litter. The use and communication of biodegradable/compostable materials might give the perception to the consumer that the product may be thrown into nature. This behavior must, of course, be avoided.

2.5.2.3 Design for composting guidelines

- Get your certification for the product produced from compostable material. The certificate is product-based—the product is not automatically compostable because the material is biodegradable; e.g., a foil made out of compostable material might get the certificate, while a housing of a product that is thicker than a foil will not fulfill the composting criteria.
- Be aware that some vegetable fibers are sometimes mixed with a noncompostable glue to make a construction material, so the bio-based material might be toxic when composted.

2.6 Design for sustainable behavior

The consumer has a very big role in closing the loops of the circular economy. Whether the products will be used as long as possible, whether the product will be repaired when it's broken, whether the product will be discarded in a proper way by the consumer, are largely dependent on the attitude and behavior of the consumer.

The consumer might be stimulated by financial incentives to act in the right way. But the designer can also improve the sustainable behavior of the consumer. "Design for sustainable behavior" is a user- and use-centered design approach that can be applied to design of products and services to steer the consumer, consciously or unconsciously, toward sustainable behavior. In the publication (Selvefors and Renström, 2018) the design for sustainable behavior approach is limited to a few design guidelines with focus on the product:

- Strengthen the relationship between the user and the product so that he/she cherishes the product and does not tend to discard it. This might be possible by making it possible that the consumer add a personal touch to the product. The personalization of the products makes the user realizes he/she has a unique product so that the relationship with the product will be strengthened.

- Make products that are not too much linked to trends. Products are discarded because they don't fit in a certain fashion trend. The challenge for the designer is to find a balance between trend-sensitive and classic design.

In the context of this publication, it's not relevant to dig deeper in the theory of the design for sustainable behavior strategy. We refer the reader to the literature for more information about "nudging" and design for sustainable behavior (Lockton et al., 2010).

2.7 Design for the sharing economy

The circular economy is supported be the increasing interest of consumers/users in the sharing economy. For former generations it was important to buy and have products for themselves, while the younger generation mainly want to pay for the function of the product. Developing products for the sharing economy will challenge the designer to generate other product requirements. The products will be very durable and robust to realize a long lifespan. Products that fit into public services like the service of sharing bicycles need other specifications. The designer will take into account that not all the users will handle the product with respect. The list of specifications that the designer makes up at the start of the design process will be adjusted to the multiuse of the product.

3. Ecodesign tools

The tools the designer uses during the product development process are based on the ecodesign guidelines as discussed in previous chapters. These tools are also used for screening existing products in order to look for opportunities to reduce their environmental impact or to make the product more suitable for the circular economy.

The qualitative tools based on the ecodesign guidelines are preferred compared with qualitative tools like "life cycle assessment" (LCA) studies. Experience has shown that the results of LCA studies are dependent on the parameters taken into account, the many assumptions and the weighting factors that are time- and location-dependent. For example, for the calculation of the environmental impact based on the Eco-Indicator 95 method according to Goedkoop (1995), the depletion of the ozone layer was 40 times worse than climate chance. Twenty-five years later the scientific knowledge had changed, so in the current frequently used ReCiPe method, the calculation of the total environmental impact has been adapted with the new knowledge (Huijbregts, 2016).

To implement ecodesign in the design process we prefer the ecodesign guidelines. These ecodesign guidelines are brought together in the different qualitative tools on a structured way based on the steps in the life cycle of the product. We discuss three of these tools to explain the principle.

3.1 LiDS-wheel

A qualitative tool that has been used since the 1990s is the LiDS-wheel or Eco-wheel. This tool was developed by Prof. Brezet of the technical university of Delft (Brezet and van Hemel, 1997).

LiDS-wheel stands for "life cycle design strategies" wheel. The tool is especially used to screen existing products concerning their environmental impact. Based on the results of the scan, the designer chooses a few "strategies"—opportunities that can be approved by the redesign of the product.

The axes of the wheel correspond with the phases of the life cycle. For each phase, there are general guidelines. The LiDS-wheel consists of four concentric circles. The designer will give a score of how good or bad the product is based on the evaluation of the ecodesign guidelines. The higher the score, the better the product already is. The score given is especially for the visual aspect, so one can see with a glimpse where the product is already quite good and in which phase of the product the greatest opportunities are to improve. In Fig. 2 the results of the scan of the product is marked in *dark orange* (*dark gray* in the print version). After the evaluation of the product, the designer will

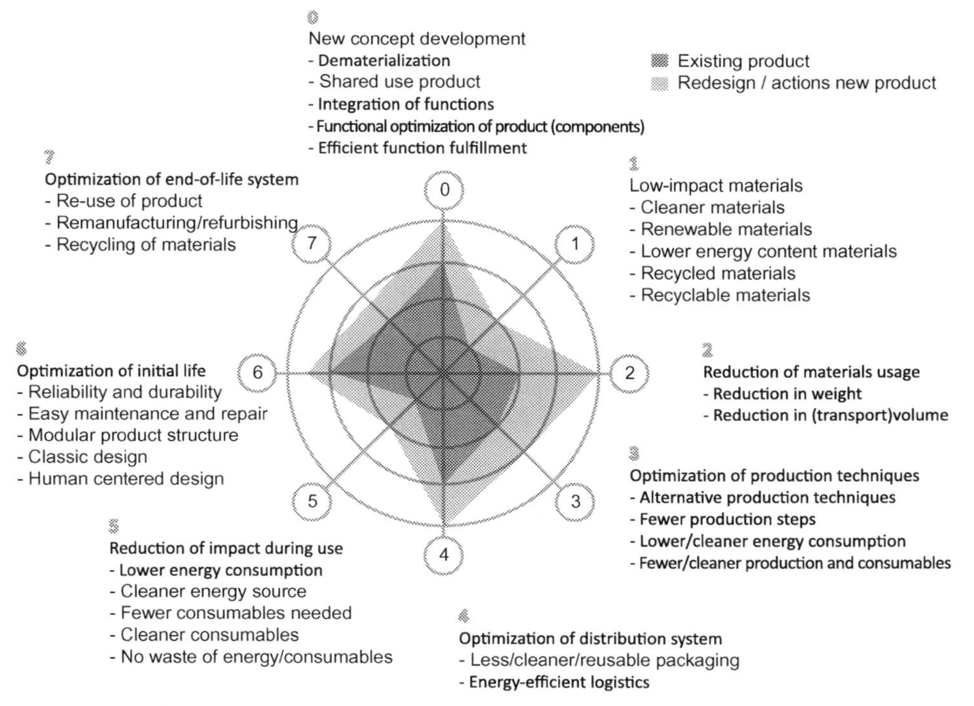

Fig. 2 LiDS-wheel.

choose two or three strategies that will be focused on during the redesign process. The result of the scan of the re-designed product is marked in *light orange* (*light gray* in the print version) on Fig. 2. In using the ecodesign guidelines the designer will think about the environmental impact of each phase of the life cycle and will be invited to look for alternatives to do better.

The guidelines of the LiDS-wheel can, of course, also been used when the designer is developing a totally new product. The guidelines make the designer aware of the environmental impact for the whole life cycle of the product and will help to find the best compromise between the environmental impact and all the other specifications of the product.

3.2 Eco-Star

During the years we used the LiDS wheel and were advising companies about ecodesign and trying to convince them to implement awareness of the environment in the design process, we encountered some resistance. Often the answer was: Implementing and following the ecodesign guidelines costs time and money, so we are not interested.

To get rid of that perception in industry I developed an ecodesign tool that screens the ecological and economic criteria and benefits when ecodesign is implemented in the design process. Economic and ecologic guidelines are brought together to scan the product. These guidelines are often linked which each other to create the awareness that ecology and economy can go hand in hand, e.g., the guideline "use less material" gives economic and ecologic benefits.

The tool is called the Eco-Star (Fig. 3) (Van Doorsselaer and Du Bois, 2018).

The Eco-Star consists of the phases of the life cycle of a product. To make it visually attractive, each phase has its own color. Each colored diamond is split into two triangles. The inner triangle for each phase corresponds with the economic guidelines; the outer triangle corresponds with the ecological guidelines. For each triangle, the economic and the ecological one, there are five guidelines, which will be evaluated regarding the screened product. Step by step all 30 ecodesign guidelines and the 30 economic guidelines are discussed and evaluated. For each positive answer on a guideline, the triangle will be colored as follows: starting from the baseline where the two triangles touch each other, the triangle is colored up to the first dot. For each subsequent positive answer for the other guidelines, the triangle is colored up to the next dot. After the scan, the Eco-Star is colored as, for example, the office chair in Fig. 4.

By coloring the Eco-Star, the economic and ecologic opportunities to improve the product are visualized. Much more important in using the Eco-Star is that the designer will scan the product in a very structured and accurate way to determine opportunities to improve the product for economic and ecological criteria. The Eco-Star can be used to scan existing products to find opportunities to improve or the list with guidelines can be used during the development of new products.

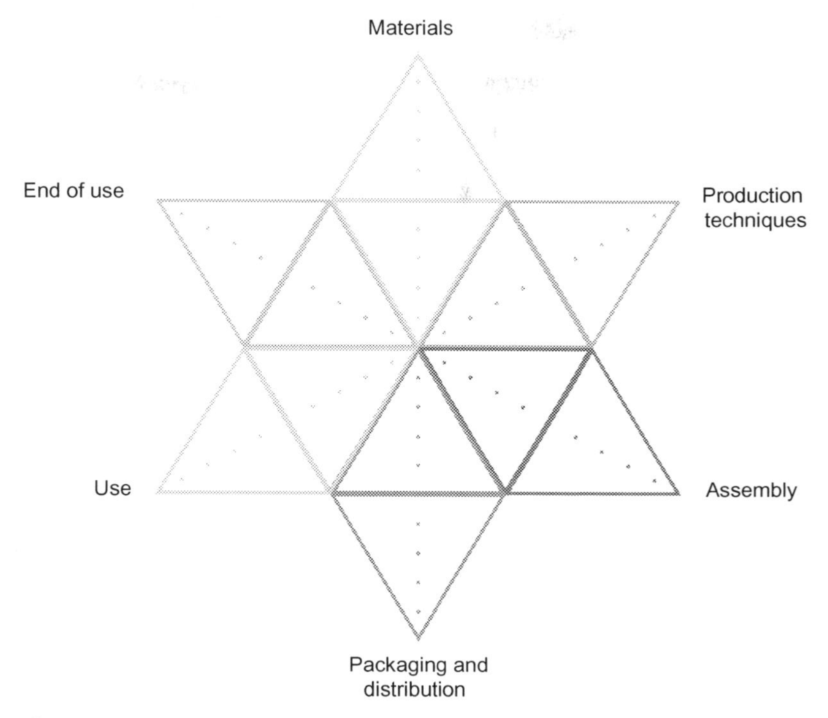

Fig. 3 Eco-Star.

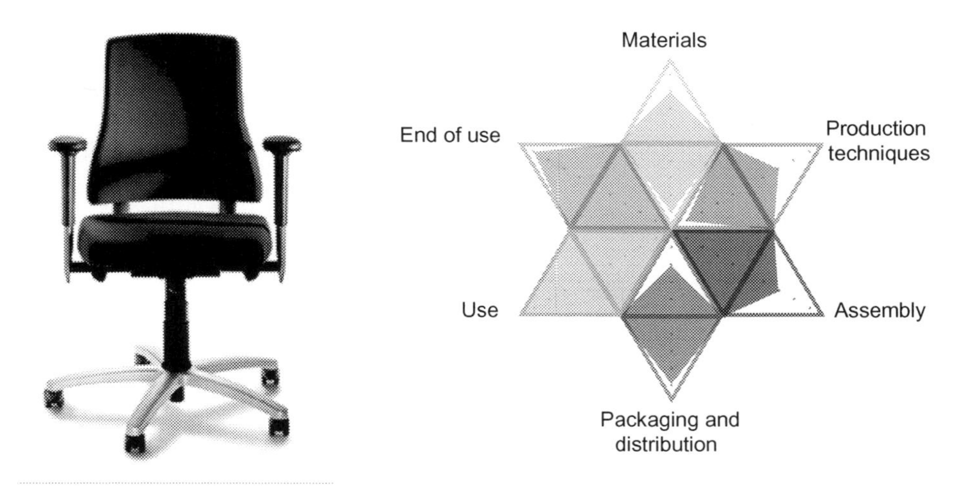

Fig. 4 Example of a scan with the Eco-Star.

3.3 Ecodesign checklist

Working with the LiDS-wheel and the Eco-Star, I noticed that it was sometimes extremely difficult to give an overall score for the complete product regarding the several phases of the life cycle. The solution is to split the product into functional blocks or components and do the ecodesign scan for each component. The ecodesign guidelines stay the same.

The ecodesign checklist looks like a matrix for each phase of the life cycle; the example in Fig. 5 gives the material selection phase of the product.

The steps for working with the ecodesign checklist are:

1. Select the relevant components of the product, e.g., housing, electrical component, cooling device, etc.
2. For each guideline on the checklist, the component must be thoroughly studied with a view to critical aspects and possibilities for improvement.
3. All the critical aspects are written down together with possible improvements. In order to make the results visual, a score is given in the matrix, from "un-ecological" to "very ecological." As with the LiDS-wheel and the Eco-Star, the score given in the ecodesign checklist is especially a visual support to see the bottlenecks and opportunities at a glance.

Checklist Product/components

Very on-eco | 0 | Very eco

Materials

	Comp.1	Comp. 2	Comp. 3	...
	......	......	......	
Renewable raw materials?				
Use of recycled materials?				
Limited number of materials ?				
No use of harmful substances (Surface treatments, glue,...) ?				

Fig. 5 Part of the checklist as an example of the approach (full checklist available from the author).

Working with an ecodesign checklist on a component level is more efficient because the designer will go more into detail. The visual outcome of the evaluation shows where the opportunities for improvement are.

In addition to use the ecodesign checklist during a scan of an existing product, the checklist can, of course, also be used during the product development process of new products. The guidelines stimulate an overall approach in a structured way.

3.4 Conclusions on ecodesign tools

The qualitative ecodesign tools guide the designer through the life cycle of the product and invite him or her to reflect on all the ecodesign guidelines to minimize the environmental impact of the product for the complete life cycle, and will give answer to the question "does the product fit into the circular economy?"

We realize that to evaluate the ecodesign guidelines the designer needs a lot of knowledge and information, which is linked with the whole value chain. Therefore, it is recommended that the designer evaluate the products together with the various departments of the company (purchasing, production, etc.), and that he or she inform him- or herself thoroughly with the other stakeholders. Last but not least, the designer should keep informed about new knowledge and technical developments. As technology, digitalization, and legislation is constantly in evolution, the designer should have the taste for lifelong learning.

4. The role of the designer in the value chain

A key to success to the transition to the circular economy is chain collaboration. There should be transparency in the value chain about the critical aspects of the product. For example, to improve the recyclability of the material, the recycling company should know what the composition of the material is, so there is no risk of contamination the recycled material.

As mentioned before, the designer should inform him- or herself about the stakeholders of the complete value chain, so he or she can evaluate the ecodesign guidelines and make the correct decisions. The designer might be a kind of intermediary conversation partner between all the stakeholders. This position of the designer in the value chain is shown in Fig. 6.

Ecodesign is lifecycle thinking. During the different phases of this life cycle, the different stakeholders in the value chain are involved. So it's obvious that when the designer wants to implement ecodesign guidelines, he or she needs to collaborate with the stakeholders.

Each stakeholder in the value chain will, of course, focus on his or her own benefit. It's an enormous challenge for the designer to convince every stakeholder to help closing the cycles. The issue of different interests and return on investments will be captured with

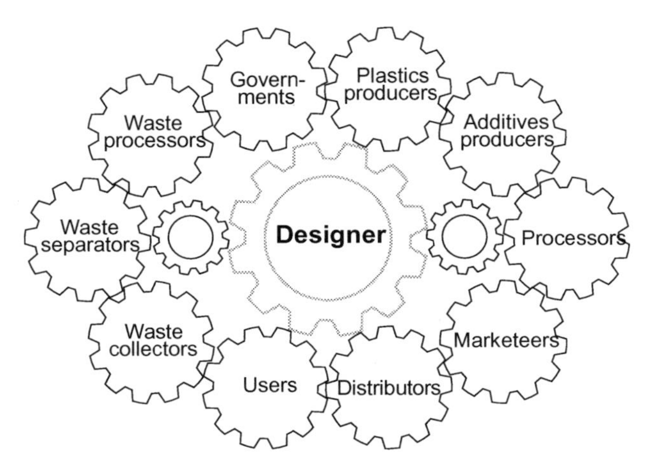

Fig. 6 The position of the designer in the value chain.

the development and integration of new business models, such as product-service systems. Implementing ecodesign in these new business models will greatly contribute to success.

Of course, the transition to the circular economy should be stimulated by governance and can be supported be the digital possibilities, like the block chain technology, which are topics of other chapters in this book.

5. Conclusion

The implementation of the principles of ecodesign supports the design of products so they fit into the circular economy. The lifecycle thinking approach of ecodesign stimulates the designer to minimalize the environmental impact of the product regarding the complete life cycle of the product. The designer can also implement the ecodesign guidelines to optimize the lifespan of a product by making products repairable and easy to maintain. In addition, the recyclability of the materials is stimulated by the decisions of the designer, which are based in the design for recycling guidelines.

The ecodesign guidelines can be used to evaluate existing products to look for opportunities to improve the possibilities to extend the lifespan of products and to minimalize the environmental impact concerning the complete life cycle of the product. These ecodesign guidelines can also be used during the product development process of new products or services that fit in the circular economy. Especially for the development and implementation of new business models like product service systems, the use of the ecodesign guidelines will stimulate economic success.

The ecodesign guidelines are often brought together in ecodesign tools, so the designer can use these guidelines in a structured way. The most common tools are the LiDS-wheel and the Eco-star, which make the link between ecological and economic opportunities, and checklists, which evaluate the product on a component level.

To evaluate the ecodesign guidelines, the designer will familiarize him- or herself thoroughly with all the stakeholders of the value chain. For example, the designer will be informed about legislation, follow up the recycling and composting possibilities (worldwide), keep informed of the new digital possibilities to implement into the product to optimize the closing of the loops, etc.

Ecodesign can be considered as the copilot during the product development process. The designer will always look for the best compromise between the ecodesign guidelines and the other requirements of the product (technical, economic, and human-related). Implementing the ecodesign principles is a key to success for the circular economy.

References

Brezet, H., Hemel, C.v., 1997. EcoDesign: A Promising Approach to Sustainable Production and Consumption. UNEP, France.

Goedkoop, M., 1995. The eco-indicator 95. RIVM Report 9523. RIVM, Bilthoven.

Huijbregts, M.A.J., 2016. ReCiPe 2016 a harmonized life cycle impact assessment method at midpoint and endpoint level report I: characterization RIVM report 2016-0104. RIVM, sl.

Ioniqa, 2019. https://ioniqa.com/. (Accessed 3 May 2021).

Lockton, D., Harrison, D., Stanton, N.A., 2010. Design with Intent Toolkit: 101 Patterns for Influencing Behavior Through Design. Equifine, sl.

Selvefors, A., Renström, S., 2018. Design sustainable behaviour. Hållbarhetsguiden, EcoDesign circle (online) https://sustainabilityguide.eu/methods/design-sustainable-behaviour/.

Timmermans, F., Katainen, J., 2019. Towards a Sustainable Europe by 2030. European Commission, Brussels.

United Nations, n.d. Sustainable Development Goals [Online]. United Nations. https://www.un.org/sustainabledevelopment/sustainable-consumption-production/. (Accessed 3 May 2021).

Van Doorsselaer, K., Du Bois, E., 2018. Ecodesign. Ontwerpen voor een duurzame en circulaire economie. 2nd, Academia Press, sl.

Further reading

Van Doorsselaer, K., Koopmans, J.R., 2020. Ecodesign. A Life Cycle Approach for a Sustainable Future. Hanser. In press.

CHAPTER 13

Sustainable finance and circular economy

Anastasios Sepetis
Department of Business Administration, University of West Attica and Hellenic Open University, Athens, Greece

1. Introduction

By approving the political Treaty of Paris[a] and the 2030 Agenda of the UN for Sustainable Development in 2015, societies and governments have chosen a more sustainable route for our planet and our economy. The 2030 Agenda of the UN is mainly focused on the 17 targets of Sustainable Development. For the following 15 years these goals will lead us, so that we can prepare for a future that will ensure stability, the health of our planet, fair and resilient societies without any exclusions, and economies that flourish. Regarding sustainable finance at a global level, the financial initiative of the UN Environmental Program (UNEP FI[b]) refers to a corporate relationship between UNEP and the national financial sector, in order to mobilize the finance of public and private sector for sustainable development. Moreover, for circular economy, the UNEP circularity platform[c] provides an understanding of the circularity concept, its scope, and how it contributes to promoting sustainable consumption and production patterns. It also offers a wide range of resources and features stories illustrating how various stakeholders have successfully adopted circular approaches. The EU Taxonomy Regulation Platform on sustainable finance[d] and the Circular Economy Finance Support Platform[e] confirm Europe's determination to play a leading role globally in the fight against climate change and in the implementation of the Paris Treaty.

In recent years, financial stakeholders, and especially banks and institutional and pension funds investors, have worked intensively on environmental, social, and governance

[a] http://unfccc.int/paris_agreement/items/9485.php.

[b] https://www.unepfi.org/about/.

[c] https://www.unenvironment.org/circularity.

[d] Article 20 of the Taxonomy Regulation creates a "Platform on sustainable finance." The platform will be an advisory body composed of experts from the private and public sector. This group of experts will advise the Commission on the technical screening criteria for the EU Taxonomy, the further development of the EU1NBSP.taxonomy, and sustainable finance more broadly. In addition, the platform will monitor and report on capital flows toward sustainable investments.

[e] https://circulareconomy.europa.eu/platform/en/financing-circular-economy.

Circular Economy and Sustainability, Volume 1
https://doi.org/10.1016/B978-0-12-819817-9.00002-8

(ESG) risk and circular risk. Within this framework, these particular organizations have developed initiatives and new banking and investment products that encourage an improved ESG and circular performance on behalf of their clients and provide to sustainable businesses an easier and lower cost access to capital. Moreover, the new pressure from institutional investors to evaluate ESG factors and circular business models (CBMs), as well as the new social responsibility investment (SRI) funds and "green" or "social" bonds and pension funds that are connected with these kinds of strategies and can possibly lead to the best pricing of ESG risk and circular risk, lead in this direction. Furthermore, institutional investors are, due to the large number of capital movements, in the position to negotiate directly and affect the decisions of company boards related to ESG factors and CBMs.

Regarding the firms, and the importance that the financial background of a firm plays for its administration and the stakeholders, it becomes obvious that one of the most important keystones to form the ESG factors and the CBMs is the connection between its financial operation with ESG risk and circular risk priorities and purposes. Under conditions of a free and competitive market, this can only happen if the market itself gets "addicted" to giving ESG and circular "signals" to companies and especially to those companies that mainly turn to external funding in order to leverage their business plans.

The global financial and business system recognizes the decisive role the capital market plays in achieving these basic environmental and social goals of sustainable finance and circular economy, as well as the need to mobilize large amounts of private capital in order to succeed in this change. In the following sections we will describe how this change can be achieved according to the Sustainable Development Goals (SDGs) and simultaneously operate under optimum sustainable capital market efficiency.

2. Sustainable finance and circular economy policies

The sustainable finance of the Sustainable Development Goals (SDG 8, 9, 13, 14 and 15) is a very important initiative of the team created by the 20 richest countries of the planet (G20). The Financial Institutions Participation Statement of UNEP FI for Sustainable Development was the initial base of the financing initiative when it was created right after the Rio Summit in 1992. By signing this declaration, the financial institutions openly recognized the role of finance in the services sector, in order to make the economy and our way of living sustainable and commit to the integration of environmental and social problems in all aspects of their activities. Today, UNEP FI cooperates with more than 300 members, banks, insurers, and investors, and over 100 supporting institutions, in order to contribute to the creation of a financial sector that serves people and the planet and should have a positive impact. UNEP FI's target is to inspire, inform, and provide the possibility for financial institutions to improve the human quality of life without endangering the conditions of future generations. Utilizing the role of the UN, the UNEP FI accelerates sustainable financing. UNEP FI supports the principles of the global financing sector. In this framework, it has created or cooperated with financing and institutional investors globally.

In Europe the ministers of finance, regulatory bodies, and the financial sector are actively looking for ways to support UNEP FI. The European Commission established the EU High-Level Group on Sustainable Finance (HLEG) for promoting SDGs to help develop an overarching and comprehensive EU roadmap on sustainable finance.[f] In December 2016, the EU Commission adopted a package of measures implementing several key actions announced in its Sustainable Finance Action Plan (SFAP) report. The Report argues that sustainable finance is about two urgent imperatives: (1) improving the contribution of finance to sustainable and inclusive growth by funding society's long-term needs; and (2) strengthening financial stability by incorporating ESG factors into investment decision-making (EU, 2019a).

An analysis of the history of the development of sustainable finance is proposed by Schoenmaker, who claims that a first step in sustainable finance "Sustainable Finance 1.0," is for financial institutions to avoid investing in companies involved in trades with very negative impacts on health (e.g., tobacco), international relations (e.g., cluster bombs), or the environment and wildlife/natural world (e.g., whale hunting). Some firms have started to include social and environmental considerations in the stakeholder model "Sustainable Finance 2.0." While financial firms have started to avoid unsustainable companies from the risk perspective of Sustainable Finance 1.0 and 2.0, the frontrunners are now increasingly investing in sustainable companies and projects to create value for the wider community—"Sustainable Finance 3.0" (Schoenmaker, 2017).

Seen from the company's point of view, Serafeim by Harvard Business School suggests that companies don't win over investors just by issuing sustainability reports and engaging in other standard ESG practices. They need to integrate ESG efforts into strategy and operations. In that article, Serafeim makes five recommendations: identify the material issues in the industry and develop initiatives that set the firm apart from rivals; create accountability mechanisms to ensure the board's commitment; infuse the whole organization with a sense of purpose and enthusiasm for sustainability and good governance; decentralize ESG activities throughout operations; and communicate regularly and transparently with investors about ESG matters (Serafeim, 2020). According to Sepetis, the move toward sustainable finance policies demands brave steps, the final choice for the implementation of SDGs under sustainable capital market optimum efficiency lies in the coordinated efforts of decision-makers, of businesses, and of capital market stakeholders. Those, instead of engaging themselves in fruitless confrontations, can focus their judgments on the prediction of basic compensatory benefits. Their prediction would be according to the current situation in the capital market, having an additional relative

[f] https://ec.europa.eu/info/business-economy-euro/banking-and-finance/sustainable-finance_en.

awareness of the consequences that sustainable finance policies can cause in the future in the effective operation of the capital market (Sepetis, 2020a).

On the other hand, the EU Circular Economy action plan, which was presented by the Commission in December 2015, has as a goal supporting the transition to a stronger and more circular economy, where resources are used in a sustainable way, bringing benefits for both the environment and the economy (EU, 2020c). The Ellen MacArthur Foundation has described the circular economy in a diagram comprising two cycles: a biological cycle, in which residues are returned to nature after use, and a technical cycle, where products, components, or materials are designed and marketed to minimize wastage. Such a circular system aims at maximizing the use of pure, nontoxic materials and products designed to be easily maintained, reused, repaired, or refurbished to extend their useful life, and later to be easily disassembled and recycled into new products, with minimization of wastage at all stages of the extraction-production-consumption cycle (Ellen MacArthur Foundation, 2013).

Achterberg and coauthors presented the Value Hill: a circular business strategy tool that provides companies with an understanding of how to position their business in a circular context and to develop future strategies for a circular economy. Additionally, the Value Hill provides an overview of the circular partners and collaborations essential to the success of a circular value network. The Value Hill proposes a categorization based on the lifecycle phases of a product: pre, in, and post use. This allows businesses to position themselves on the Value Hill and understand possible circular strategies they can implement as well as identifying missing partners in their circular network (Achterberg et al., 2016). Pieroni and coauthors suggest that in implementing circular economy, manufacturers will need to develop new business models. Available approaches are granular, generic, infrequently focusing on advice or implementation, and lacking practical demonstration. They propose the usage of a tool in order to cover these gaps. Based on design research methodology, 22 conceptual and practical requirements were identified and translated in functions for the development of the Circular Economy Business Model Configurator, a tool that supports manufacturers in strengthening proposals of business models for circular economy (Pieroni et al., 2020).

A review of CBMs by Bocken et al. (2019) concludes with the suggestion of several avenues for future research in order to support the operationalization and mainstreaming of CBMs. They suggest that future research can contribute to the trialing of new CBMs to find the most suitable ones for businesses, as well as supporting the organizational change dynamics of transforming businesses' dominant business models for the CE. This may be best supported by action-oriented research approaches that are underpinned by strong theoretical insight and practice review. For this to work in practice, business practitioners would need to be open to higher levels of research involvement, and different types of interactions would need to be designed into research projects. Moreover, they propose that academics should become more effective at translating theoretical insight into effective CBMI tools, processes and support—from ideation and design to implementation and testing, and evaluating and improving—to guide CE operationalization in a way that it lives up to "sustainability expectations" and avoids negative side effects. This is needed in order not to "dilute" sustainability or circularity objectives in favor of more conventional business cases that lack a clear positive environmental and societal impact (Bocken et al., 2019). The European Circular Economy Stakeholder Platform[g] published research by Fischer and Achterberg that proposed 10 steps that business managers need to take into account when developing a financeable CBM. These steps can support businesses that wish to establish a sound CBM and to overcome the financing barriers that such models face (Fischer and Achterberg, 2016).

The G20 Resource Efficiency Dialogue 2019 aims to raise awareness on the linkages between the circular economy and sustainable finance[h] and proposes key messages on linkages for financial stakeholders, which are as follows.

(a) The finance industry has key roles to play in facilitating a shift to a circular economy. It can, for example, provide resources for circular investments, offer insurance products suitable for circular practices, such as leasing and sharing, and develop rating systems and information disclosure requirements that can help improve transparency around sustainability-related business risks.

(b) Banks and other lenders and investors can raise awareness through dialogue with clients. They can help client companies identify various kinds of risks associated with linear models and make them aware of the opportunities of transitioning to more circular solutions.

(c) Many circular economy business models have risk profiles that are hard to assess and generate low short-term return on investment. In addition, securities are often in the form of contracts, which banks consider riskier than physical assets. New models for risk assessment are called for.

(d) The finance industry has done significant work on how to assess and communicate risks related to climate change. As a result, the industry has developed knowledge and scenario analysis tools, and gained experience in working with natural science data. This will make it easier to address other sustainability issues, including circularity and resource-related risks, in a more holistic manner.

(e) Some, generally large, companies use green bonds to raise funding for sustainability-related investments. However, a large share of the companies in the waste management and recycling sector are SMEs and lack the capacity and size to issue their own bonds. The recycling industry also suffers from an image problem that contributes to weak interest from investors.

[g] https://circulareconomy.europa.eu/platform/en.

[h] The G20 Resource Efficiency Dialogue 2019 and Follow up of the G20 Implementation Framework for Actions on Marine Plastic Litter took place in Tokyo (Japan) on October 8–10, 2019.

Nowadays, international financing institutions, the European Union, and national financial stakeholders shape strategies in order to minimize consequences that endanger financial stability, which derive from the changes of sustainable finance and circular economy for the financial system. The Working Group FINANCE proposes that a flipside of circular risk is businesses' increasing dependency on virgin resources and corresponding exposure to price volatility. Moreover, environmental legislation can be expected to increase in the coming years, which will put the current linear system under stress. This raises another concern in risk analysis, one that so far has been neglected: the opportunities of circular economy to provide a solution to what can be called "linear risk" (Working Group FINANCE, 2016). According to the European Banking Authority (EBA), sustainable finance aims at integrating ESG factors into financial services, and supporting sustainable economic growth. It also aims at increasing financial actors' awareness and transparency about the need to mitigate ESG risks via appropriate management, considering in particular the longer-term nature of such risks and the uncertainty on their valuation and pricing (EBA, 2020). Furthermore, the circular risk in the EU's Circular Economy Finance Guidelines for, for example, the Netherlands banks, are voluntary process guidelines, which recommend transparency and disclosure and promote integrity in the debt and equity market for the circular economy (EU, 2017).

Taking the constantly increasing ESG or circular risk of the market in total for granted, and the appearance of the increasing number of firms that concentrate capital by the systematization of risk, and sustainable innovative products and services, we should ask ourselves, whether environmental, social, or circular management policies are against or harmonize with positive results, as far as the assessment of the shareholder value of firms and financial stakeholders value is concerned.

3. The environmental, social, and governance (ESG) risk and the circular risk

The financial stakeholders that leverage sustainable finance and the circular economy promote and improve the transparency of risks related to ESG factors and the CBMs that may impact the financial system, and the mitigation of such risks through the appropriate taxonomy of financial and corporate factors.

According to the study of Sepetis and coauthors in 2011, as well other studies, it is claimed that firms should be taxonomized and evaluated according to ESG criteria (Gillan et al., 2010; Sepetis et al., 2011; Hoepner, 2013; Yu et al., 2018). Moreover, Sepetis, in another paper, has suggested an holistic financial evaluation model for ESG risk (Sepetis, 2020b). Gunnar and coauthors' paper found that scholars and investors have published more than 2000 empirical studies and a number of critical studies to evaluate this relationship. They reach the conclusion that 90% of their studies find a nonnegative relationship between ESG investment and corporate financial performance (CFP).

The most important conclusion is that a large majority of those studies reported positive findings. They point out that the positive impact of ESG investment in CFP seems to be stable over time (Gunnar et al., 2015).

According to Piney and coauthors, in the last 3 years, great progress has been made toward the incorporation of ESG issues in the capital market (Piney et al., 2019). The key issue of the survey of the CFA Institute and UNEP PRI was to highlight a number of considerations that financial professionals and investors should have in mind when integrating ESG factors into the investment process:

(a) There is no single agreed-upon definition of ESG or best practice for ESG integration.
(b) ESG integration looks at risks and opportunities revealed by the analysis of environmental (E), social (S), and/or governance (G) issues that are material for a company or country.
(c) Investors should focus on ESG analysis, not ESG investing.
(d) ESG integration is consistent with a manager's fiduciary duty to consider all relevant information and material risks in investment analysis and decision-making.
(e) To date, one of the main drivers of ESG integration globally has been client demand, largely from institutional investors.
(f) Asset owners and asset managers should strive to do a better job of educating each other about how and why they integrate ESG data into the investment process (CFA Institute and UNEP PRI, 2019).

On the other the hand, for the circular risk in CBMs, Sonerud suggested that investors who are aware of circular economy business models "all agree conceptually that they can capture some value from this," but the "key barrier so far is risk and lack of data." Investors are wary about getting involved in new business models when it is hard to quantify the risks and returns (Sonerud, 2014).

The European Commission's Expert Group on Circular Economy Financing has analyzed the status of barriers to the transition to a circular economy in the EU. The experts have identified major challenges and, to address these challenges, have given their expert opinions and recommendations to the main stakeholders, including policy makers, financial institutions, and project promoters. The issue of risk and the unfavorable risk/revenues profile of circular economy projects dominated the discussions of the experts, regardless of whether they focused on policy, finance, or business management. It seems to be counterintuitive that the approach that preserves the economic value of materials and products faces the problem of revenue generation and uncertainties, resulting in a high financial and financing risk. It is the experts' conclusion that circular economy projects are not necessarily inherently riskier than linear projects, especially from the long-term perspective. The regulatory system, markets and financial risk assessment are distorted and biased in favor of the financing of linear projects. In order to correct this distortion, a number of incentives should be provided through a series of well-designed and coordinated actions. Furthermore, the Expert Group on Circular Economy Financing has identified the following key problems:

(a) insufficient clarity on the financial/industrial scope of a "circular economy" project (related to a definition and a taxonomy);
(b) lack of or insufficiently developed risk assessment methodologies for circular projects and businesses;
(c) how a linear investment can be made circular, by, for instance, changing an existing company's supply chain and production process to eliminate negative impacts on the environment and reach zero waste or by changing the eligibilities/requirements of new/existing financial instruments; and
(d) how a level playing field can be created between linear circular investment decisions by including the financial and nonfinancial impacts of the project's delineation (externalities) (EU, 2019a,b).

To tackle these problems, the European Investment Bank (EIB) published a series of Circular Economy Guides to support the transition to a circular economy, and aiming to:

(a) Promote a common understanding of the CE concept and related challenges and opportunities among financial and project partners.
(b) Raise awareness about and promote circular solutions among project promoters and other stakeholders.
(c) Facilitate and harmonize due diligence of and reporting on CE projects by financial and project partners.
(d) Communicate the vision of how the EIB can further support the transition to a circular economy (EIB, 2020).

In the last few years, despite the steps forward, the ESG factors and the circular economy factors have been a source of considerable confusion for the financial stakeholders of firms. This is due to the fact that linear ESG risk of firms, as well as the circular risk of CBMs, is interpreted differently in different types of financial institutions. Commercial banks are primarily concerned with ESG and CBM factors of firms, the borrower's creditworthiness, and avoiding additional ESG and CBM costs. Investment banks are turning their attention to this parameter, plus all the ESG and CBM effects of firms that have an impact on the demand or supply of business debt securities. Insurers are only concerned about linear risk, taking into account the probability of accidental environmental or social events of firms and seek to avoid, by internal procedures, the risk weighting and uncertainty that their client's environmental or insurance claims impart by transferring them to the market. The institutional investor or venture and pension fund manager can account for a wide range of environmental and social threats of firms, many of which are qualitative rather than quantitative.

When hearing the term ESG or circular risk, the commercial banker tends to associate it with balance sheet data (downgraded territory, reputation, and customer difficulty in repaying the loan); the insurer, with environmental accidents. Institutional investors (pension funds, mutual funds) and capital investment managers are generally more interested in what environmental or circular and governance regulations entail in business processes and less in the making of profitable investment decisions. Busch and coauthors examine to what extent financial markets strengthen and make sustainable business practices easier. The authors point out that their current role is rather "mediocre." They conclude that in the "old paths" there is a paradoxical situation, where on the one hand the financial market participants incorporate more and more ESG factors in their investing decisions and on the other hand, as long as their organizing abilities are concerned, there does not seem to be a turn toward more sustainable practices. The authors spot two main challenges in the field of sustainable financial services and investments that are related to entering new paths that may help overcome this situation. First of all, reorientation to a long-term model for sustainable financial services and investments is very important. Second, the ESG data must become more trustworthy (Busch et al., 2015).

According to Eccles and Klimenko (2019), business leaders understand that they should play a crucial role in confronting emergency challenges, such as climate change. But many of them believe that pursuing a sustainability program is against the wishes of their shareholders and the impression among business leaders is that ESG just hasn't gone mainstream in the investment community. However, they conclude that a sea change in the way investors evaluate companies is under way. Its exact timing can't be predicted, but it is inevitable. Large corporations, whose shares are owned by the big passive asset managers and pension funds, will feel the change the soonest. But it won't be long before mid-cap companies come under this new scrutiny as well. All companies, though, should seize the opportunity to partner with investors willing to reward them for creating long-term value for society as a whole (Eccles and Klimenko, 2019).

Porter and coauthors suggest the social purpose of investment. A serious problem is that many in the investment community have moved away from fundamental investing and its powerful social purpose, seeing algorithm-driven strategies and trading on market movements as ends in themselves. In the process, the connection between capital investment and improvement of society is lost. Understanding deeper insights into economic-value creation, and deepening conventional investment analysis by adding shared-value thinking, will unlock growth, accelerate innovation, drive productivity, and improve shareholder returns. Moving in this direction would restore investing as a true profession, with a higher purpose—generating greater profit by expanding opportunity for all, instead of extracting short-term profit for the few at society's long-term expense (Porter et al., 2019).

As the next step, the capital market should operate under the optimum evaluation of the ESG and circular risk and should go forward toward an efficient sustainable capital market. According to the Ellen MacArthur Foundation, the circular economy financing market is taking off, with a steep increase in activity over the last 18 months. Increasingly recognized as a crucial part of the solution to climate change and other ESG issues, circular economy also offers significant opportunities for new and better growth. Now is the time for finance to capitalize on this industrial transformation, and help scale the circular economy (Ellen MacArthur Foundation, 2020). The UNEP FI report Financing Circularity: Demystifying Finance for the Circular Economy explores strategies and actions that financial institutions can take to manage related risks and opportunities (UNEP FI, 2020). Moreover, the European Commission has already taken a series of initiatives in this respect for sustainable finance, including integrating the circular economy objective under the EU Taxonomy Regulation, and carrying out preparatory work on EU Ecolabel criteria for financial products (EU, 2020a). Furthermore, the EU's CE Finance Expert Group, in March 2020, published "a generic, sector-agnostic circular economy categorization system that defines distinct categories of activities substantially contributing to a circular economy; a set of minimum criteria to be met by activities under each defined category in order to be considered as substantially contributing to a circular economy; and methodological guidance including an indicative list of typical investments/projects for each circular economy category" (EU, 2020b).

Additionally, sustainable finance and circular economy policies should synchronize with a new sustainable capital market that should operate under optimum efficiency.

4. Sustainable finance and circular economy for a sustainable capital market

The EU Commission defines "sustainable finance" as the process of taking due account of environmental and social considerations when making investment decisions, leading to increased investment in longer-term and sustainable activities. More specifically, environmental considerations refer to climate change mitigation and adaptation, as well as to the environment more broadly and to the related risks (e.g., natural disasters). Social considerations may refer to issues of inequality, inclusiveness, labor relations, and investment in human capital and communities. The governance of public and private institutions, including management structures, employee relations, and executive remuneration, plays a fundamental role in ensuring the inclusion of social and environmental considerations in the decision-making process.

All three components—environmental, social, and governance—are integral parts of sustainable economic development and finance (EU, 2019a,b). In this framework, regarding capital markets, the European Union has already run certain initiatives, such as for example, green bonds and the promotion of long-term investments. Moreover, a financial system has been put in place that ensures a high level of transparency for investors, which is related to the publishing of nonfinancial information regarding a diversity of issues such as the environment, issues that are related with social issues and employers, respect for human rights, the fight against corruption and bribery, and others. Shaping of a coherent international, European, and national financial strategy for sustainable financing will contribute to the definition, prioritization, and organization of ways with which the reform of financing policy of the financial institution can encourage and strengthen the capital flow (public and private) for sustainable investments (ESMA, 2019a,b).

On the other hand, the European Commission, in the Circular Economy Finance Support Platform, will enhance the link between existing instruments, such as the European Fund for Strategic Investments (EFSI) and the InnovFin—EU Finance for Innovators—initiative backed by Horizon 2020, and potentially develop new financial instruments for circular economy projects. The platform will bring together the Commission, the EIB, national promotional banks, institutional investors, and other stakeholders, raising awareness of circular economy investment opportunities and promoting best practices among potential promoters, analyzing projects and their financial needs, and providing advice on structuring and bankability (EU, 2017).

According to a number of academic researchers and financial stakeholders to date, there are strong linkages between climate change mitigation, sustainable finance, and circular economy, which can help to stimulate risk analysis of investments in ESG factors and circularity criteria. They need an in-depth holistic financial analysis of the ESG and circular risk and a generally accepted definition for the sustainable capital market. According to Sepetis, sustainable capital market should be defined "as the capital market that promotes sustainable development with the implementation of mandatory or voluntary sustainable management policies. These policies have as a goal to develop sustainable innovation and sustainable technological progress and at the same time to leverage and promote sustainable firm's management in the same firms and in the same financial stakeholders of the capital market, while at the same time they will contribute to the optimum efficiency of the capital market." This goal will be achieved when the applied sustainable management policies are holistically designed and are identified for having as their main goal to not cause asymmetric information, price failures, and permanent oligopoly conditions in the capital market, where they are applied (Sepetis, 2020a).

Regarding ESG risk and circular risk, the basic financial theory suggests two basic trends in order to shape investment decision-making in the capital market: (a) avoiding uncertain situations that could raise the risk grade of an investment, and (b) spotting new investment opportunities. These trends seem to be used by investors, when they incorporate additional ESG risks or circular risks. This is documented by a series of theoretical and practical researches that have been mainly conducted in the United States and in Europe, but also in other areas (developing economies). These researches suggest two concepts or case studies in order to explain the potential connection between environmental, social, and circular efficiency, which is a result of corresponding business strategies, in order to explain the potential connection between environmental, social and circular efficiency, which is a result of the implementation of equivalent business strategies and the shareholders' value that are under negotiation in the capital markets. These concepts are:

1. The implementation of sustainable finance or circular economy policies raises new risks and uncertainties in the capital market. Financial stakeholders recognize that ESG factors or CBMs can cause uncertain situations and unpredictable costs for a company.
2. The investors simultaneously recognize the investing opportunities created by sustainable finance or circular economy policies in the capital market. The innovative companies that implement ESG factors or CBMs claim that they systematize the ESG risk or circular risk and create sustainable innovative opportunity.

The above two mentioned hypothesis, which appear contradictory to each other, but may be complementary in the end, are the cornerstone for the typical business behavior model of the capital market system. This dynamic correlation of these two counterbalanced behaviors and predictions result in the dynamic of capital market balance to the point that each time it satisfies the optimum conditions.

This is the well-known dynamic and the terms of efficient balance of the capital market, without the interference of exogenous political maneuvers that are related to the implementation of sustainable finance or circular economy policies. When the abovementioned policies appear, the capital market system reacts, at first by creating asymmetric information, price rigidity, and oligopolistic situations, until the right correction moves are made, so that a new balance point is reached, which will have incorporated the "nuisance" caused by the exogenous intervention of sustainable finance or circular economy policies.

In their research in 2019, Elena Escrig-Olmedo and coauthors studied and evaluated the timeless course of financial organizations, which evaluate, manage, and analyze the ESG risks regarding years from 2008 until 2018. In their study, they reach the conclusion that the organizations that evaluate, manage, and analyze ESG risk are evolving, developing, and multiplying. Moreover, in the last decade, they have played an active and dynamic role in the financial market, as they take an important part in the effort in order to achieved sustainable development in the capital market. Nevertheless, a detailed analysis of these criteria shows that the evaluation, management, and financial analysis institutions of these ESG criteria are not fully satisfied by the incorporation of sustainability principles in the evaluation process of ESG factors of businesses (Escrig-Olmedo et al., 2019). Clementino and Perkins (2020) conducted pioneering research on the growing number of firms that are being evaluated on ESG criteria by sustainability rating agencies (SRAs); the data point being that firms may react very differently to being rated. In their analysis they yield a fourfold typology of corporate responses. The typology captures conformity and resistance to ratings across two dimensions of firm behavior. Furthermore, it is claimed that corporate responses depend on managers' beliefs regarding the material benefits of adjusting to and scoring well on ESG ratings and their alignment with corporate strategy (Clementino and Perkins, 2020).

The European Securities and Markets Authority (ESMA) and, more particularly, the regulatory body of securities markets of the European Union, published in the summer of 2019 technical advice on sustainability aspects in the credit rating market (ESMA, 2019a) and the final guidelines (ESMA, 2019b) regarding the notification demands that are valid for the credit rating evaluations. ESMA evaluated the level of examining ESG factors, in particular, actions of credit rating evaluations, as in concrete credit rating evaluation actions, as well as in the credit rating evaluation market in total. ESMA reached the conclusion that while the credit rating evaluation organizations (credit rating agencies (CRAs)) examine the ESG factors in their analysis, nevertheless the evaluation level of ESG criteria may vary depending on the different methodology that each agency/organization follows. However, given the special role that credit ratings have in the regulatory framework of the European Union in order to estimate credit risk, the European Commission claims that it would be currently inappropriate for the credit rating evaluation CRA to be modified and explicitly impose the examination of sustainability characteristics in all its credit rating capacity. Instead, ESMA suggests that the European Commission should evaluate if there are enough tax and regulatory safeguards in the financial products to satisfy the demand for a clear sustainable evaluation (ESMA, 2019a,b).

According to Achterberg and Rens Van Tilburg, in order to accelerate the transition toward a circular economy for financial decision-makers, six guidelines are proposed:

(a) assess different securities;
(b) emphasize relationship-based financing;
(c) value natural capital gains;
(d) become a knowledge partner;
(e) have a long-term vision; and
(f) become a financial chain director.

They concluded that CBMs require multiple forms of capital, by collaborating with other financiers (e.g., through sharing information and co-funding circular projects) risks can be spread, durations can be matched, and specific financial needs of these types of businesses can be met. Additionally, money flows within the supply chain can be regulated by offering supply chain finance services (Achterberg and van Tilburg, 2016).

According to the EU Action Plan for Circular Economy 2020, accelerating the green transition requires careful yet decisive measures to steer financing toward more sustainable production and consumption patterns. In addition, the Commission suggests it will:

(a) Enhance disclosure of environmental data by companies in the upcoming review of the nonfinancial reporting directive;
(b) Support a business led initiative to develop environmental accounting principles that complement financial data with circular economy performance data;

(c) Encourage the integration of sustainability criteria into business strategies by improving the corporate governance framework;
(d) Reflect objectives linked to circular economy as part of the refocusing of the European semester and in the context of the forthcoming revision of the State Aid Guidelines in the field of the environment and energy;
(e) Continue to encourage the broader application of well-designed economic instruments, such as environmental taxation, including landfill and incineration taxes, and enable member states to use value added tax (VAT) rates to promote circular economy activities that target final consumers, notably repair services (EU, 2020c).

Every capital market seeks to create new proper conditions, in order to attract the biggest possible number of "clients," which are firms and investors. In the framework of this effort, competing with any other capital markets, the latter develop business strategies in order to raise the volume of exchanges and participations and at the same time are trying these new strategies to establish that they do not have a negative effect on themselves or on the local market in which they are incorporated.

The success of these two, ESG factors and CBMs, as new business strategies depends on the potential and the insight of the market to promote sustainable innovation and its final sustainable technological progress accomplishments. A number of financial analysts claim that the environmental, social, and circular risk management dynamic provides a main index of the general best business model practices within a firm. The challenge for the financial analysts is to examine the type of business, the ESG factors, and the CBM that is applied, and moreover, to present their conclusions regarding the effect on environmental, social, and circular efficiency, which is directly related to economic effectiveness. On the other hand, the case of recognizing environmental, social, and circular opportunity that is promoted by companies will assist only little if the investing, the banking, and insuring research and the investment reports of those institutional investors fail to recognize, to evaluate, and promote this information sufficiently and effectively in the capital market. As a result, the capital market and the economic stakeholders (investors, analysts, and shareholders) must be in the position to define investing risk and the opportunity, by incorporating environmental, social, and circular data in their financial analysis, in order to support their decisions, according to a number of environmental, social, and circular investing models.

The ESG factors or CBM classification and disclosure of information are very important factors in shaping rational information, and affect the estimation of information costs, the businesses, and the investors. They contribute, in the framework of symmetrical information, to the creation of optimum competition conditions in the capital market. Based on the above, it is suggested that ESG factors or CBM revenue and expenditure that correspond to sustainable finance or circular economy policies must be incorporated in international accounting standards and in national accounting systems, and have their own category of accounts with clear input and rules. In mentioning rules, we suggest ensuring a systematic and full methodology for registering new environmental, social, and circular information, which is compatible with every kind of business in order to ensure the comparability of registrations. These kinds of steps by the indirect stakeholders, which are the outsiders in the capital markets, as well as by the direct economic stakeholders, the investors themselves, would significantly strengthen the reliability of that case and could result in the leverage of businesses and the market toward sustainability.

In recent years, there has been major revolution by international, European, and national bodies in order to classify, evaluate, and distribute the optimum information for ESG factors or CBMs' efficiency for companies in the capital market. Nevertheless, a basic problem today lies in the fact that, with the existing policies that are mainly characterized by ambiguity in the equivalent classification, evaluation and notification criteria, as well as the lack of organization within the controlling mechanisms, predispose companies to voluntarily discover environmental, social, and corporate governance or CBMs. Moreover, the capital market today partly supports investing decision-making on the basis of voluntary reports by the financial consultants and analysts, and relies on their ability to discover the real environmental, social, and circular efficiency of companies.

This will result, despite the big changes that have been made in recent years for the classification, evaluation, and notification of ESG factors or CBMs in capital market efficiency: (a) in the creation of asymmetric information for the ESG factors or CBMs in the capital market; and (b) the financial stakeholders to not pay the equivalent attention to the ESG factors or CBMs when taking their investing decisions.

In conclusion, regarding sustainable finance and circular economy policies in the capital market, many steps have been made in recent years. However, during the first steps of those policies, asymmetric information and temporary (not yet permanent) green, social, or circular oligopoly markets arose in the capital market with ESG factors and circularity criteria. Based on the already given definition of sustainable capital market, asymmetric information and temporary oligopoly markets have created a temporary problem in the optimum efficiency of the market until the right correction moves are made, so that the new balance point will have incorporated the nuisance made by the exogenous intervention of sustainable finance or circular economy policies.

5. Results and discussion

In a previous paper regarding the ESG risk evaluation, Sepetis suggested a holistic sustainable financial model for the sustainable capital market (Sepetis, 2020b). Furthermore, in this paper, for the evaluation of ESG and circular risk, three (key) sectors that can lead to the dissemination of sustainable finance and circular economy policies in the operation of the capital market and bring new optimum efficiency conditions are suggested:

(a) The design of holistic fostering of transparency and taxonomy for the long-term evaluation and notification of the ESG risk and circular risk models by the financial stakeholders. Sufficient regulatory safeguards should exist in every financial product to satisfy the demand for a clearer sustainable evaluation. Presuming that the capital market leverage toward sustainability is mainly based on the definition of sustainable capital market as it was defined above, we propose the incorporation of a percentage of ESG and circular risk in the market discount interest rate, when this will be shaped in the international, European, or national financial action plan for financing sustainable development by the financial institutions, such as the European Central Bank. The banks' discount interest rate is determined by the central financial organization mechanisms (such as the European Central Bank and World Bank) and has the possibility to accelerate the risk range of sustainable (ESG factors and CBMs) adjustment. We suppose that the risk range of the sustainable adjustment grade reflects the percentage in which sustainable innovation and sustainable technology progress reduces sustainable degradation that is produced with the per unit produced product of firms. The ESG and circular risk percentage is calculated by the average risk range sustainable performance of firms, for example, of the member states in the total of the European market, as this is determined by the annual progress reports for SDGs of the European Union and its member states. The sustainable interest rate of each member state may adapt according to ESG and circular risk in the annual progress reports for SDGs or in each firm's sector and will be financed, respectively, by the policies and the European Union goals, according to the grade of adjustment of sustainable development that state wishes to achieve.

(b) The holistic taxonomy, evaluation, and notification of sustainable performance of ESG factors and CBM information in the accounting standards and financial statement of firms and the creation of reliable financial ESG factors and a circular benchmark index during the decision-making process. During the procedure of taking an investment decision, the capital market is not based in voluntary progress reports, such as a voluntary sustainability report, but mainly in mandatory data register of the firms in its official accounting standards, such as the balance sheet statement and the financial annual reports that are submitted to the stock exchange. In order to calculate market's and firm's sustainable performance, all data (economic, social, environmental, and circular) that determine the definition of sustainability should be registered in the official accounting standards and financial statement of firms, and during the evaluation phase, a commonly accepted index of sustainable performance used, which will derive from the firm's official data. In order to determine sustainable performance of firms and of the market in total, we assume, for the needs of the financial model, that inside a market, it is obligatory that companies register and evaluate their economic, sustainable changes, and the financial and nonfinancial data (economic, social, environmental, circular data) according to an originally holistic sustainable accounting standard. Given this assumption, all companies in a market are obliged to register and recognize their sustainable situation (or in the case that we wish to leverage a firm sector or a firm portfolio), in order to be able to characterize them by the market as "sustainable" according to the calculation of relevant sustainable performance evaluation indexes.

(c) The design of pioneer holistic financial models for taxonomy, evaluation, and notification of the ESG factors and CBMs in the firm's shareholder value.

The progress in these three key sectors will accelerate, if the deficiencies of the financial risk models and their involution with sustainable issues are examined in a systematic way, according to the generally accepted principles of international accounting standards—these are determined by international organizations and relevant regulations of each capital market. This includes the development of a theoretical framework and a measurement system for the environmental, social, and circular performance in the co-management of firms that is based on the decisions of impartial financial experts. In their turn, they should present the arguments and should depict the needs of the funding investors in a more consistent and balanced way, by committing independent financial brands of institutional investors, promoting the environmental-, social-, and circular-relevant decisions.

6. Conclusion

The international, European, and national financial, environmental, and social conditions today synchronize, and SDGs are established and are connected in multiple ways with public policies for sustainable finance and circular economy, with modern business strategy and international financial stakeholders. These goals are connected to motives that define the basic ESG factors, and CBM policies, as well as the regulatory legislation and policies, where the basic goal is serving the social or general interest and sustaining natural resources, and not the business or financial expediencies of the economic way of thinking in order to maximize profit.

It is a fact that, today, companies systematically try to achieve better economic results and adapt to the new conditions of supply of natural resources, as well as to the new conditions of consuming and investing demand that are shaped and are connected with ESG factors or CBM parameters. As a result, the ESG risk or circular risk are unavoidably connected with a constantly intensifying pace with business strategy and the financial and investing system. Moreover, they gradually become an organic part of the shaping of businesses value in the capital markets. These developments put a number of compatibility questions on the table that are related to the efficient operation of the capital market—the basic regulator of capital goods—to the ESG factors or CBM policies of a business strategy and of the financial and investment stakeholders, as well as with the priorities of the relevant public policy for sustainable development.

The financial system, the investing community, and the capital market are generally opposed to theoretical arguments that cannot be translated into action in a simple and distinct way. The financial system and the investing community cannot be passive viewers in this "game" of international, European, and national strategies for sustainable finance and circular economy policies. Unless they can be persuaded of the practical importance of this effort that aims at the incorporation of sustainability in capital markets, and consequently respond to the inaccurate "signals" as if they were correct, without any further investigation, then it is very probable that a feedback of failing signals in the capital markets can be created. The result of this failure will be to gradually discourage the procedure of internalization of the sustainable capital market and in the end will work as a brake on the transition toward sustainability.

The capital market does play an important role in the promotion of policies of the SDGs with an international, European, and national meaning and range. It is part of the monetary measures field and financial regulations. Under the proper organization and tactics, its can co-assist toward the right direction and sustainable development, by incorporating its goals in the sense of sustainable finance and circular economy policies efficiency. Furthermore, the capital market is a crucial part of the operation and balance of the financial system related to the ESG risk or circular risk. In this capacity, it is already displayed in the official texts of international organizations and of the European Union, as a risk factor that can provide complementary political solutions to assist in the gradual incorporation of sustainable finance or circular economy policies to the overall financial and investing market system, by contributing to the holistic dissemination of rules and priorities of the sustainable capital markets.

Common sense claims that responsible people in their daily lives have their faces turned to the future, despite the imperative needs of the present. We work, we prepare, we save money, we insure ourselves, and we invest for our future needs and for our retirement years away from the action. The same principles should also apply for SDGs. ESG factors and CBMs, in cooperation with the typical financial analysis methods of decision-making, which use mathematical calculations, statistical methods, and financial models to identify the optimum action, can potentially name the changes that society and the planet today can take, in order to avoid future danger. Relevant practices are the financial analyses that have significantly contributed to the improvement of legislation, of regulation, and of investments. A relative example is our national economic policy. The meanings that analysts have introduced in the decades of 1930 and 1940, as well as the unemployment rate, the current account deficit, and the gross national product are common senses. Based on those analyses, governments today have managed to avoid the drastic exchanges between big economic expansion and consequent recession, which was a common practice in the 19th and beginning of 20th century. Today, the next steps and challenges of sustainable finance and circular economy policies and financial analysis are to evaluate the ESG and the circular risk in the sustainable capital market that should operate under optimum efficiency.

Conflicts of Interest

The author declares no conflicts of interest regarding the publication of this paper.

References

Achterberg, E., van Tilburg, R., 2016. 6 Guidelines to empower financial decision-making in the circular economy. https://circulareconomy.europa.eu/platform/en/knowledge/6-guidelines-empower-financial-decision-making-circular-economy. (Accessed 10 November 2020).

Achterberg, E., Hinfelaar, J., Bocken, N.M.P., 2016. Master circular business with the value hill. In: Circle Economy. Amsterdam, The Netherlands https://circulareconomy.europa.eu/platform/en/knowledge/master-circular-business-value-hill. (Accessed 10 November 2020).

Bocken, N., Strupeit, L., Whalen, K., Nußholz, J., 2019. A review and evaluation of circular business model innovation tools. Sustainability 11, 2–25.

Busch, T., Bauer, R., Orlitzky, M., 2015. Sustainable development and financial markets: old paths and new avenues. Bus. Soc. 55 (3), 303–329.

CFA and UNEP PRI, 2019. ESG integration in Europe, The Middle East, and Africa: markets, practices, and data. CFA Institute.

Clementino, E., Perkins, R., 2020. How do companies respond to environmental, social and governance (ESG) ratings? Evidence from Italy. J. Bus. Ethics. https://doi.org/10.1007/s10551-020-04441-4.

EBA, 2020. Discussion paper on management and supervision of ESG risks for credit institutions and investment firms. https://eba.europa.eu/financial-innovation-and-fintech/sustainable-finance. (Accessed 10 November 2020).

EIB, 2020. Circular economy guide for supporting the circular transition. https://www.eib.org/en/publications/the-eib-in-the-circular-economy-guide. (Accessed 10 November 2020).

EU, 2017. Circular economy: commission delivers on its promises, offers guidance on recovery of energy from waste and works with EIB to boost investment. https://ec.europa.eu/commission/presscorner/detail/en/IP_17_104. (Accessed 10 November 2020).

EU, 2019a. EU technical expert group on sustainable finance (TEG) the EU taxonomy to define sustainable economic activities. https://ec.europa.eu/info/publications/sustainable-finance-technical-expert-group_en. (Accessed 10 November 2020).

EU, 2019b. Accelerating the transition to the circular economy—improving access to finance for circular economy projects. https://ec.europa.eu/info/publications/accelerating-transition-circular-economy_en. (Accessed 10 November 2020).

EU, 2020a. The EU classification system for environmentally sustainable activities.

EU, 2020b. Categorisation system for the circular economy: a sector-agnostic categorisation system for activities substantially contributing to the circular economy. https://op.europa.eu/en/publication-detail/-/publication/ca9846a8-6289-11ea-b735-01aa75ed71a1. (Accessed 10 November 2020).

EU, 2020c. A new circular economy action plan for a cleaner and more competitive Europe. https://ec.europa.eu/environment/circular-economy/. (Accessed 10 November 2020).

Eccles, G.R., Klimenko, S., 2019. The investor revolution. Harv. Bus. Rev. 97, 106–116.

Ellen MacArthur Foundation, 2013. Towards the circular economy. https://www.ellenmacarthurfoundation.org/assets/downloads/publications/Ellen-MacArthur-Foundation-Towards-the-Circular-Economy-vol.1.pdf. (Accessed 10 November 2020).

Ellen MacArthur Foundation, 2020. Financing the circular economy capturing the opportunity. https://www.ellenmacarthurfoundation.org/assets/downloads/Financing-the-circular-economy.pdf. (Accessed 10 November 2020).

Escrig-Olmedo, E., Fernández-Izquierdo, M.Á., Ferrero-Ferrero, I., Rivera-Lirio, J.M., Muñoz-Torres, M.J., 2019. Rating the raters: evaluating how ESG rating agencies integrate sustainability principles. Sustainability 11, 915–935.

ESMA, 2019a. Final report. Guidelines on disclosure requirements applicable to credit ratings. https://www.esma.europa.eu/sites/default/files/library/esma33-9320_final_report_guidelines_on_disclosure_requirements_applicable_to_credit_rating_agencies.pdf. (Accessed 10 November 2020).

ESMA, 2019b. Technical advice. ESMA technical advice to the European commission on sustainability considerations in the credit rating market, https://www.esma.europa.eu/sites/default/files/library/esma33-9-321_technical_advice_on_sustainability_considerations_in_the_credit_rating_market.pdf (Accessed 10 November 2020).

Fischer, A., Achterberg, E., 2016. Create a financeable circular business in 10 steps. https://circulareconomy.europa.eu/platform/en/knowledge/create-financeable-circular-business-10-steps. (Accessed 10 November 2020).

Gillan, S., Hartzell, J., Koch, A., Starks, L., 2010. Firms' environmental, social and governance (ESG) choices, performance, and managerial motivation. In: Unpublished Working Paper. Texas Tech University and University of Texas at Austin. http://www.pitt.edu/~awkoch/ESG%20Nov%202010.pdf. (Accessed 10 November 2020).

Gunnar, F., Busch, T., Bassen, A., 2015. ESG and financial performance: aggregated evidence from more than 2000 empirical studies. J. Sustain. Finan. Invest. 5 (4), 210–233.

Hoepner, A.G.F., 2013. Environmental, social, and governance (ESG) data: can it enhance returns and reduce risks? Deutsche AWM Global Financial Institute. https://eur-lex.europa.eu/legal-content/en/HIS/?uri=CELEX%3A52018PC0353. (Accessed 10 November 2020).

Pieroni, M.P., McAloone, T.C., Pigosso, D.C.A., 2020. Business model innovation for circular economy: integrating literature and practice into a process model. In: In: Proceedings of the Design Society: DESIGN Conference. vol. 1. Cambridge University Press, pp. 2119–2128.

Piney, C., McCorkle, C., Lawrence, S., Lau, S., 2019. Sustainability in Capital Market: A Survey From Current Progress and Practices [ebook]. High Meadows Institute, pp. 3–40.

Porter, M.E., Serafeim, G., Kramer, M., 2019. Where ESG fails. In: Institutional Investor. https://www.institutionalinvestor.com/article/b1hm5ghqtxj9s7/Where-ESG-Fails. (Accessed 10 November 2020).

Schoenmaker, D., 2017. From risk to opportunity: a framework for sustainable finance. In: RSM Positive Change Series. http://hdl.handle.net/1765/101671. (Accessed 10 November 2020).

Sepetis, A., 2020a. Sustainable Finance: The Contribution of Financial Methods of Sustainable Development to the New Welfare State. Papazisis.

Sepetis, A., 2020b. A holistic sustainable finance model for the sustainable capital market. J. Finan. Risk Manag. 9 (2), 99–125.

Sepetis, A., Katsikis, I., Nikolaou, I., 2011. Environmental, social and corporate governance: a framework of evaluation for financial stakeholders. J. Reg. Soc. Econ. Bus. 1, 5–23.

Serafeim, G., 2020. Social-impact efforts that create real value: they must be woven into your strategy and differentiate your company. Harv. Bus. Rev.

Sonerud, B., 2014. Meeting the financing needs of circular businesses. In: Ellen MacArthur Foundation's Schmidt-MacArthur Fellowship Programme. http://www.ellenmacarthurfoundation.org/higher_education/global_campus/schmidt. (Accessed 10 November 2020).

UNEP FI, 2020. Financing circularity: demystifying finance for the circular economy. https://www.unepfi.org/publications/general-publications/financing-circularity/. (Accessed 10 November 2020).

Working Group FINANCE, 2016. Money makes the world go round (and will it help to make the economy circular as well?). https://circulareconomy.europa.eu/platform/en/knowledge/money-makes-world-go-round. (Accessed 10 November 2020).

Yu, E.P., Guo, C.Q., Luu, B.V., 2018. Environmental, social and governance transparency and firm value. Bus. Strategy Environ. 27 (7), 987–1004.

CHAPTER 14

How to advance sustainable and circular economy-oriented public procurement—A review of the operational environment and a case study from the Kymenlaakso region in Finland

R. Husgafvel[a], L. Linkosalmi[b], D. Sakaguchi[c], and M. Hughes[a]
[a]Wood Material Technology Research Group, Department of Bioproducts and Biosystems, School of Chemical Technology, Aalto University, Espoo, Finland
[b]Clean Technologies Research Group, Department of Bioproducts and Biosystems, School of Chemical Technology, Aalto University, Espoo, Finland
[c]Faculty of Health Science, Department of Human Care Engineering, Nihon Fukushi University, Aichi, Japan

1. Introduction

1.1 Research approach and previous research

Circular economy (CE) is an important focus area within both the EU and national policy frameworks in Finland. CE is expected to play a major role as an accelerator for sustainable recovery, and enabling CE requires focus on both policy and finance (WCEFonline, 2020). This study focuses on the advancement of sustainable and CE-oriented public procurement through a case from the Kymenlaakso region in Finland covering both procurement organization and market development aspects. In general, this study focuses on the key features, elements, drivers, and barriers of the international, EU level, and national operational environment and city/municipality perspectives in the Kymenlaakso region in Finland on advancing sustainable and CE-oriented public procurement. The structure of the study encompasses: (1) description of the research approach and a review section about previous study of sustainable and circular economy-oriented public procurement and development in Finland, (2) aims of the study, (3) material and methods, including description of the survey/questionnaire study, (4) results covering all the research themes, (5) discussion, and (6) conclusions, including recommendations for further action and future research.

Circular Economy and Sustainability, Volume 1
https://doi.org/10.1016/B978-0-12-819817-9.00015-6

Specifically, the following main themes were the focus of the research approach:

1. Consideration of sustainability and circular economy aspects in procurement planning and implementation by procurement organizations.
2. Use of guidebooks in procurement.
3. Application of environmental labels or management systems, standards, reporting systems, or certificates in procurement.
4. Local measures to promote sustainable and circular economy-oriented public procurement.
5. Incentives and taxation related means to promote sustainable and circular economy-oriented public procurement.
6. Procurement organization measures to promote sustainable and circular economy-oriented public procurement.
7. Issues that should be taken into account in procurement planning and subject definition.
8. Issues that should be taken into account in the procurement phase and in the benchmarking of bids.
9. Issues that should be taken into account in the contract terms and follow-up and in reporting.
10. Indicators for the assessment of economic sustainability as a part of the procurement process and its different phases.
11. Indicators for the assessment of environmental sustainability as a part of the procurement process and its different phases.
12. Indicators for the assessment of social sustainability as a part of the procurement process and its different phases.
13. Indicators for the assessment of circular economy as a part of the procurement process and its different phases.
14. Areas/fields that offer best opportunities to promote sustainable and circular economy-oriented procurement in the future.
15. Measures related to combating climate change that could be an effective part of sustainable and circular economy oriented public procurement in the future.
16. The most important next steps for the development of sustainable and circular economy oriented public procurement.

Pollice (2018) noted that circular procurement is a relatively new concept (there is no commonly agreed definition), that public procurement is significant for the advancement of CE, and that addressing CE challenges requires more than focusing on value chains, reverse logistics, or sustainable sources. Sustainable public procurement is encouraged internationally and it is also expected to play a role in the advancement of sustainability in the private sector (Brammer and Walker, 2011). CE is a sustainable development strategy driven by principles of reducing, reusing, and recycling materials, energy, and waste (Heshmati, 2015). The implementation of the CE concept requires support from all stakeholders (Lieder and Rashid, 2016). Potočnik (2017) noted the importance of supportive market and public policy efforts driven by political will and that the implementation of Sustainable Development Goals (SDGs) (cf. UN, 2019) requires leadership, system change, and improved global governance.

Steiner (2017) highlighted that it is increasingly important to use resources wisely to reduce environment impact and to use less resources to support economic growth. Noronha (2017) concluded that the achievement of SDGs often requires sustainable resource management and that it can be promoted through circularity in production and consumption supported by new multistakeholder partnerships such as much broader and large-scale public-private partnerships. Most public sector organizations are using some sustainability criteria in their procurement, and environmental aspects are typically well addressed. However, there is variation in how well, e.g., safe practices in the supply chain, support of human rights, and purchasing from diverse suppliers are taken into account (Brammer and Walker, 2011).

Hartley et al. (2020) presented policy recommendations based on a life cycle perspective and noted that this perspective is necessary for a transition toward a circular economy. They concluded that for practitioners, the life cycle perspective is about preparation of products for reuse, recycling, or remanufacturing, and highlighted the requirement for an integrated strategy starting from product design. Kirchherr et al. (2017) noted that CE must be understood as a fundamental systemic change and that there are explicit linkages between CE definitions and sustainable development, excluding future generations and social equity, which are almost always neglected. CE implementation requires addressing the role of market barriers and associated failed governmental interventions and focus on new ideas, such as policies that promote circular products and associated reduced value-added taxation (Kirchherra et al., 2018).

Ormazabal et al. (2018) noted that public institutions currently form a major barrier to CE development and that policy instruments, financial stimulation, and technological modernization are essential in this context. Ghiselli et al. (2016) noted the need for the involvement of and collaboration between all actors in society and Mylan et al. (2016) noted that CE is about moving toward a more sustainable society including acknowledgment of the role of domestic consumption and associated practices. Guohui and Yunfeng (2012) stated that CE should create economic, environmental, and social benefits including focus on local activities and impacts. Environmental advantages can be gained through the integration of CE principles into supply chain management and there are benefits from enhanced emphasis on sustainable supply chain management (Genovese et al., 2017).

Snider et al. (2013) noted that firms that are involved in public procurement have a more legally oriented approach to corporate social responsibility and a less voluntary orientation. Bratt et al. (2013) noticed that public procurement has the potential to promote corporate sustainability and to steer the decisions of both procurers and producers in a more sustainable direction (expert bodies that are responsible for the development of procurement criteria play a major role). It is important to focus on entrepreneurial innovations in addressing market imperfections and to address the possibilities to support more sustainable markets (Cohen and Winn, 2007).

A previous study by Pöyhönen (2017) on CE oriented around public procurement in Kouvola city in the Kymenlaakso region suggests that there is a significant lack of CE knowledge and expertise in the public procurement sector. Associated challenges also include availability of products and services, problems with interaction and communication, city planning and land use issues (e.g., recycling of materials), functionality of and lack of skills about new techniques, tender competence of small companies (e.g., about environmental criteria), change of thinking and resources in education, and lack of cooperation, division of responsibilities, and too tight sectoral boundaries in procurement organizations. Currently, green public procurement (GPP) criteria (e.g., EU, 2019; Motiva, 2019b) are not typically used. There is a need for training of procurers and collaboration among actors supported by dissemination of good practice and capacity building through pilot experiments. In addition, it is essential to promote CE principles in the public procurement sector (Pöyhönen, 2017).

Previous research on company perspectives in Finland indicates that, for example, sustainability and long lifespan of products, components, and materials and holistic system-level thinking are important for the future development of CE and that sustainable, long-lasting, and fixable products, wastes as raw material, and new services (including substitution of products) and products are among the important drivers and opportunities. The main barriers to CE development include, for example, lack of information (for example, product/supply chain) and profitability and CE can improve economic, social, and environmental sustainability through, for example, supply chain management, staff skills and training, and recycling and reuse (Husgafvel et al., 2021). In addition, companies have positive perspectives on future CE development and they are willing to implement measures to advance CE such as the improvement of their material and energy efficiency and the development of sectoral guidelines and best practice. Companies perceive that challenging areas of CE development comprise, e.g., development of international guidelines and best practice and sectoral cooperation and interaction, whereas the most promising innovation focus areas include, e.g., energy efficiency and fuels, clever products and services, intelligent production and processes, and new recycling/reuse innovations (Husgafvel et al., 2018b).

Common public and circular procurement challenges encompass product design including disassembly, new business models, cycling of raw materials and products, minimal value loss, use of renewable resources, harmful substances and chemicals, and service intensity of goods and services (Pollice, 2018). More research is needed on how public procurement affects corporate social responsibility in companies dealing with state and local governments, including informing policy-makers about how their decisions influence firms (Snider et al., 2013). There is still limited research on the main aspects of CE, such as bridging production and consumption activities (Witjesa and Lozano, 2016).

Ormazabal et al. (2018) noted that future research is needed, for example, on best practices of small and medium-sized enterprises (SMEs) and their new business opportunities provided by CE. Preuss (2009) noted that further research is needed on public sector sustainable supply chain management and that local authorities have applied sustainability risk assessments and dissemination of sustainability information (Preuss, 2009). Future research should focus on alignment of sustainable procurement and e-procurement policies, development of e-procurement tools and e-business to support sustainable procurement and trade-offs between different aspects of sustainable procurement and their reconciliation (Walker and Brammer, 2012).

Future research should address CE barriers in business models and in various sectors (Kirchherra et al., 2018) and focus on the neglected dimensions such as consumers and business models as enablers of CE (Kirchherr et al., 2017). According to Lozano (2012) further research is needed on embedding sustainability into company systems, including all dimensions of sustainability. Moreno et al. (2016) stated that more focus is needed on design for new circular business models. Further research is needed on sustainability considerations and on conditions in which public policy positively influences sustainable entrepreneurship (Hall et al., 2010). Procurement is among the least addressed parts of company systems in terms of sustainability (Lozano, 2012). More focus is needed on clearer definition of sustainability objectives and broader impact perspective in the context of public procurement (Bratt et al., 2013). Further research should focus on sustainable entrepreneurship and the relationship between entrepreneurial opportunities and market imperfections (Cohen and Winn, 2007). The design and management of sustainable business models has received insufficient research focus and there is a need to consider the development of a research agenda to address the creation of sustainable value through business models for sustainable innovation (Boons and Lüdeke-Freund, 2013).

1.2 Sustainable and circular economy-oriented public procurement

CE principles encompass designing out waste, building resilience through diversity, relying on renewable energy sources, waste as food, and thinking in systems. Strengthening of the education of future generations of procurement officers is important and governments can use their own procurement and material handling to promote circular approaches (Ellen MacArthur Foundation, 2013a). Strong focus is needed on bio-based products and full redesign of supply chains. There are circular opportunities all along the value chain and both municipal authorities and corporations need to develop a new set of circular capabilities along their supply chains. Developments are needed in resource markets, technology, information systems, and consumer preferences. New technologies and business models are both needed (Ellen MacArthur Foundation, 2013b). CE aims at keeping products, materials, and components in circulation and at their highest utility level at all times including both biological and technical materials and cycles (Ellen MacArthur Foundation, 2015a). Cities are facing the challenges of the linear economy and the global economic system that is based on the "take-make-dispose" model encompassing structural waste and negative environmental impacts such as greenhouse gas emissions and low air quality. There is often a lack of holistic urban management (Ellen MacArthur Foundation, 2017a). A vision for CE in cities focuses on planning, design, making, accessing, operating, and maintaining aspects (Ellen MacArthur Foundation, 2019). The elements of a circular city are likely to include a built environment, energy systems, an urban mobility system, an urban bioeconomy, and production systems (Ellen MacArthur Foundation, 2017a). CE in cities can be promoted through digital technologies such as asset tagging, geospatial information, big data management, and connectivity (Ellen MacArthur Foundation, 2017b).

The EU-level CE development encompasses the development of innovative production and consumption approaches and a sustainable, low carbon, and resource-efficient economy. There are links to the global sustainable development agenda (e.g., achievement of sustainable competitive advantages) in European-level CE development (EC, 2015). It is essential to measure progress toward the SDGs and the planning, and evaluation of sustainable development at the national level encompasses natural capital, human capital, and produced capital (UNU-IHDP and UNEP, 2014). Sustainability assessment and reporting can be supported by standards and metrics encompassing economic, environmental, and social performance and impacts (GRI, 2019; Husgafvel et al., 2017). For example, sustainability reporting on procurement can focus on management approach to procurement practices and proportion of spending with local suppliers (GRI, 2016). A growing number of businesses are starting CE initiatives and they need indicators to measure their impact, including both internal (monitoring performance) and external (stakeholder relations) use of indicators. Current indicators are mainly focused on waste information, and the development of new indicators should focus on, e.g., identification of materials priorities covering the life cycle of products and services, the selection of a set of indicators that address relevant CE aspects, indicators that are tailored to the business strategy and continuously updated, engagement of employees and stakeholders, and incorporation of indicators into operational processes (EpE, 2019).

The Directive on Public Procurement (EU, 2016a) clarifies how the contracting authorities can contribute to the protection of the environment and the promotion of sustainable development (and to ensure best value for money for their contracts) including the achievement of sustainability objectives, application of life cycle costing, and use of performance criteria linked to the sustainability and life cycle of the production processes of supplies, services, and works. Public procurement plays a major role in the advancement of CE (EC, 2015). The promotion of green public procurement, including inclusion of circularity requirements in procurement and the application of GPP criteria, is among the key objectives of EU-level actions to advance CE (EC, 2017a).

Environmental labels can provide environmental criteria to support public procurement including life cycle thinking and costing (UNOPS, 2009).

The main CE stakeholders (e.g., policy makers, financial institutions, and project promoters) should focus on characterization of CE projects through metrics and taxonomy (e.g., development of metrics and indicators for national, regional, and corporate levels by nonfinancial policy makers), promotion and clarification of the enabling role of public authorities (e.g., public procurement, subsidies, taxation, and funding), and capacity building (EC, 2019a). The monitoring framework for CE aims at measuring progress toward CE, covering its multiple dimensions at all stages of the lifecycle of resources, products, and services (EC, 2018). The 10 developed indicators are categorized into 4 stages and aspects of CE: (1) production and consumption, (2) waste management, (3) secondary raw materials, and (4) competitiveness and innovation. Green public procurement is among the indicators (under production and consumption) and is marked as an indicator that requires further development. Public procurement amounts to a large share of consumption and it can be a driver of CE including links to EU Public Procurement Strategy and voluntary criteria for green public procurement (EC, 2018). Sustainable and innovative public procurement can be advanced through precommercial procurement that focuses on purchasing research and development services to promote, e.g., the introduction of new solutions and technologies (EC, 2007). The EU Ecolabel framework strongly supports CE and associated life cycle thinking and assessment (EU Ecolabel, 2019). The EU (2019) GPP criteria are also based on a life cycle approach and, for example, the EU legal framework for the ecodesign of energy-related products emphasizes the design phase, taking into account environmental considerations and performance and the full life cycle perspective (EC, 2009). The promotion of CE requires focus on incentives for a level playing field, collaboration along the value chain, the creation of long-term value, market participation, the integration of the public good, the build-up of finance knowledge, and first movers with new business models (EC, 2019a).

Public procurement can play a major role in the advancement of CE, including the integration of CE principles into all phases of procurement to advance a more holistic approach to sustainability (EC, 2017b). Circular procurement can be implemented through system- (contractual methods to ensure circularity, such as product service systems), supplier- (circularity in supplier systems and processes to ensure that their products and services meet circular procurement criteria), and product-level (focus on supply chains and product technical specifications) models. Procurement organizations can develop a circular procurement policy or incorporate CE principles into existing SPP or GPP policy. Sustainability improvements can be made through a procurement hierarchy that is based on the European Waste Hierarchy (EC, 2008b) including the following parts: reduce (e.g., procure less, reduce packaging, consider needs and smarter contracts), reuse (products designed for reuse and end-of-life considerations), recycle (design for recycling and use of recycled materials), and recover (tender specifications and criteria on design for recovery and procurement of recovered products). Integration of circular procurement into existing procurement systems and practices requires strategic thinking (EC, 2017b).

The EU promotes socially responsible public procurement and defines it as being about influencing the market-place and setting an example, including, e.g., encouragement of socially responsible management among companies and enhancement of compliance with community values and needs (EC, 2010). Essential social considerations encompass employment opportunities, compliance with labor and social rights, decent work, equal opportunities, social inclusion, and accessibility design for all. Sustainability criteria (including ethical trade issues and voluntary compliance with corporate social responsibility) should be taken into account and they (including social criteria) can be included in the technical specifications of public tenders (these criteria must be linked to the subject of the contract) and in the contract performance conditions (they must be linked to contract performance). Social and green considerations can jointly form an integrated approach to sustainable public procurement (EC, 2010).

In the Nordic countries, circular procurement supports the goals of sustainable procurement and promotes value retention and closed material loops, including multiple product cycles and materials remanufacturing and reuse (Alhola et al., 2017). Currently, circular procurement has not been implemented systemically (only in pilot cases) and the principles of CE are supported and promoted by public procurement more than the broader CE ideal. The CE principles and closed loops can be promoted in public procurement through, for example, the following approaches to circular procurement (Alhola et al., 2017):

1. Use of GPP-based circular criteria in procurement (better quality products);
2. Procurement of new circular products and materials (new and innovative products);
3. Procurement of new business concepts and services (new business concepts); and
4. Procurement promoting circular ecosystems (circular ecosystems).

New policy measures to support CE include proactive public procurement, support of new business models, resource efficiency goals, and measuring the environmental impacts of resource use. Taxation system should be moved toward the taxation of the consumption of nonrenewable resources (Wijkman and Skånberg, 2015).

The Ellen MacArthur Foundation (2015b) defined a circular public procurement approach as being achieved when public organizations meet their goods and services needs in a way that achieves value for money throughout the life cycle for both wider society and the organization while minimizing environmental impacts and material losses. The promotion of CE requires cross-functional involvement (including procurement departments) and involvement of the entire cross-supply cycle (Ellen McArthur Foundation, 2014). At the European, national, and city level, public procurement could role-model the transition toward CE and support the scaling of circular products and services. Governments can support local healthy food supply chains with less waste though public procurement or demand stimulation (Ellen MacArthur Foundation, 2015c). In general, CE is about a new economic model that seeks to decouple economic growth from the consumption of finite resources based on principles that highlight better management of natural capital, optimization of resource yields through circularity, and system-level effectiveness (Ellen MacArthur Foundation, 2015a).

At the European level, green public procurement (GPP) is seen as an important tool for the achievement of environmental policy goals such as climate change, sustainable consumption, and production and resource use. The EU has developed GPP criteria (EC, 2019b) for 21 product and service groups to support contracting authorities in their tendering (EU, 2016b). Government bodies and the broader public and private sectors can influence markets to produce more sustainable goods and services, contributing to better resource efficiency and a more sustainable economy (Department of Environment, Food, and Rural Affairs, 2006). Green Public Procurement (GPP) has been defined (EC, 2008a) as "a process whereby public authorities seek to procure goods, services and works with a reduced environmental impact throughout their life cycle when compared to goods, services and works with the same primary function that would otherwise be procured." Sustainable Public Procurement (SPP) has been defined (EC, 2019c) as "a process by which public authorities seek to achieve the appropriate balance between the three pillars of sustainable development, economic, social and environmental, when procuring goods, services or works at all stages of the project." The UK Department of Environment, Food and Rural Affairs (2006) defined Sustainable Procurement[a] as "a process whereby organisations meet their needs for goods, services, works and utilities in a way that achieves value for money on a whole life basis in terms of generating benefits not only to the organisation, but also to society and the economy, whilst minimising damage to the environment." The UNEP (2011) defined sustainable procurement as being based on the pillars of sustainable development: economic factors (e.g., costs of products and services over their entire life cycle), social and labor factors (e.g., core labor standards, fair working conditions, and developing local communities), and environmental factors (e.g., climate change, biodiversity, and natural resource use over the whole product life cycle). Potential key performance indicators for the assessment of sustainable procurement encompass, e.g., use of nonrenewable resources and energy sources, supply chain management, and company/organization-level sustainability performance assessment (UNEP, 2011).

Circular public procurement (EC, 2017b) can be defined as "a process by which public authorities purchase works, goods or services that seek to contribute to the closed

[a] Sustainable Procurement should consider the environmental, social, and economic consequences of: design; nonrenewable material use; manufacture and production methods; logistics; service delivery; use; operation; maintenance; reuse; recycling options; disposal; and suppliers' capabilities to address these consequences throughout the supply chain.

energy and material loops within supply chains, whilst minimising, and in the best case avoiding, negative environmental impacts and waste creation across the whole life-cycle." In the Nordic context, circular procurement has been defined as (Alhola et al., 2017): "The procurement of competitively priced products, services or systems that lead to extended lifespan, value retention and/or remarkably improved and non-risky cycling of biological or technical materials, compared to other solutions for a similar purpose on the market." In addition, circular procurement is part of green and/or sustainable procurement (circular procurement elements/aims should not compromise the overall goal of sustainable procurement) and it aims at value creation, social wellbeing, and environmental improvements through closed and safe material loops (Alhola et al., 2017).

Steurer et al. (2007) noted that SPP allows governments to implement their commitment to sustainable development, which is mainly about integrating economic, environmental, and social issues. Therefore, the scope of SPP and associated initiatives should reflect these three dimensions of sustainable development. The successful implementation of SPP initiatives requires high-level political commitment. National SPP initiatives are influenced by both EU-level policies and regional/municipal approaches (both bottom-up and top-down). In addition, they noted that the top-down political commitment to SPP must be complemented by bottom-up ownership and a commitment to learning among the public procurement practitioners, including, for example, involvement of procurers and other stakeholders, including businesses, in the development of SPP initiatives. SPP policies and its most comprehensive instruments should include supportive measures such as securing high-level political commitment, facilitating stakeholder involvement, addressing possible obstacles, and bottom-up ownership (Steurer et al., 2007).

Municipalities can create markets for circular products and services through public procurement, including the use of circular criteria and life cycle costing, and also use other instruments such as zoning and economic support (Circle Economy, 2019a). The Ellen MacArthur Foundation (2015b) recognized that public procurement and infrastructure is one policy intervention type to promote CE and possible additional interventions could include, for example, integration of guidelines on the circularity of products and materials into public procurement policy. Policymakers can address regulatory and market failures to promote enabling conditions for CE, including actively steering and stimulating market activity through setting targets and implementing circular and total cost of ownership-oriented public procurement. CE standards can be incorporated into procurement law or guidelines and lists of preferred materials or suppliers can be developed (Ellen MacArthur Foundation, 2015b). In the Nordic context, CE promotes the development of sustainable cities and regions and public procurement is one way to implement the strategic aims of CE. CE and the implementation of circular procurement require strategic thinking involving cooperation between different departments/functions of municipalities and procurement units. There is a need to engage procurers in innovative circular procurement processes, including focus on life cycle perspective and use of circular criteria/requirements in procurement. Circular aspects can be integrated into procurement through extended product lifespan, efficient/intensive use, biological/technical material cycles, clean and nonrisky cycles, and application of tools such as GPP/Ecolabel criteria and life cycle costing (Alhola et al., 2017).

Municipalities can employ a series of regulatory, economic, and soft instruments to promote the societal benefits of CE and sustainable development of cities encompassing circular policies, strategy and targets, loans and subsidies, and networks and information sharing practices to advance the social capital of the circular business community (Circle Economy, 2019a). Procuring departments can build capabilities and skills in measures of material circularity and in concepts such as total cost of ownership (Ellen MacArthur Foundation, 2015b). For example, the city of Amsterdam aims at being a circular city, and policy instruments for the circular city encompass regulation, legislation, fiscal frameworks, direct financial support (including procurement), economic frameworks, knowledge, advice and information, collaboration platforms, and infrastructure and governance. Levers for the circular city include digitalization, true and fair pricing, innovation networks, systems thinking, experimentation, logistics, and jobs and skills (Circle Economy, 2019b). The strength of institutions is important for successful public procurement actions and those actions could encompass, for example, pilots demonstrating the existing benefits of circular materials and construction techniques, stimulating new materials and techniques, and developing guidance and procedures to incorporate new elements. New actions could also include the adoption of performance models in procurement, in which the customer pays for the use/performance of a product (and not for the product itself), supported by adjusted procurement rules/procedures with defined circularity criteria or key performance indicators (Ellen MacArthur Foundation, 2015b). In addition, procurement of products or services offered through new innovative business models such as performance based business models could decrease public expenditure and relieve pressures on municipal services and budgets (Ellen MacArthur Foundation, 2017a).

1.3 Development in Finland

CE is receiving a lot of attention in Finland, and public procurement is one of the key focus areas. Finland's Road Map to the CE 2.0 (Sitra, 2019a) states that the implementation of CE requires achievement of the following strategic cross-sectoral goals: (1) renewal of the foundations of competitiveness and vitality, (2) transfer to low-carbon energy, (3) natural resources are regarded as scarcities, and (4) everyday decisions working as a driving force for change. All sectors of government need to work together across administrative silos and CE needs to be integrated into structures (Sitra, 2019b). Innovations and private companies play a major role in the advancement of CE (Sitra, 2019c). CE is based on sustainable use of resources and renewable energy. The public sector should shift its purchasing practices toward supporting the CE. Public sector purchases usually have a major impact on private sector service providers' operational development. Product/process design, cooperation across sectors, and new business models are very important for CE development (Sitra, 2015). Sitra (2016) defined that a CE aims at maximizing the circulation of products, components, and materials and the value bound to them as much as possible in the economy. CE offers environmental, economic, and social benefits and moving toward CE means systemic change. Public procurement is among the Finnish CE focus areas. For example, the selection of wood-based products and other products made from renewable raw materials should be encouraged in public procurement when life cycle analysis indicates that they are more sustainable choices, and associated actions encompass changing attitudes and increasing awareness in the units of national and local organization that are responsible for procurement. Green public procurement manuals need to be compiled that support national targets on a procurement sector basis. In addition, the use of secondary raw materials should be promoted in public procurement and public infrastructure construction. The role of public procurement is also important for the promotion of the use of sustainable food and for promoting the development of more sustainable transport (Sitra, 2016).

The contracting authorities are legally obliged to treat participants in procurement procedures in an equal, nondiscriminatory, transparent, and proportionate manner (Act on Public Procurement and Concession Contracts, 2016; EU, 2016a). Development of platforms for experimenting with and testing sustainable and innovative public procurement is among the priorities of the Finnish government action plan (Ministry of the Environment, 2021). The Act on Public Procurement and Concession Contracts (2016) aims at promoting innovative and sustainable procurement, taking into account environmental and social aspects (e.g., labels can be used to prove environmental, social, or other characteristics, management systems/standards/certificates may be used to report on environmental impact management, life cycle costs can be used as a criterion for evaluating the costs, and innovative, environmental, and social aspects can be included in the procurement agreements as special terms and conditions) and environmental, social, and labor laws. Cities and towns are expected to become important enablers of CE (Sitra, 2019b). Cities and municipalities need to implement public procurement according to CE principles and encourage companies to participate in the CE. In the future, the share of procurement including sustainable development and/or circular economy criteria will be significantly increased in municipal procurement (Sitra, 2019d).

In the future, companies are expected to offer solutions instead of products and they will apply CE principles and business models (including renewability, sharing platforms, product as a service, product-life extension, and resource efficiency and recycling) characterized by customer-specific solutions, deeper customer relations, and more precise use of materials (Sitra, 2019e). CE requires approaches that cross the boundaries of administrative sectors and levels. Both legislative and economic steering methods are needed to promote the change toward CE, and measures such as sustainable development taxes could have a major impact. The focus of taxation should change from labor and entrepreneurship to negative environmental impacts (Sitra, 2019b). Procurement decisions have an impact on how much material and energy is used, what greenhouse gas emissions are produced, and how much waste is generated during the life cycles of products and services (Motiva, 2019a). Sustainable procurement is also about influencing markets (in a more environmentally friendly and socially sustainable direction) and acting as an example. Development of more sustainable products and services can be advanced through active market dialogue with product and service suppliers (Helsingin Kaupunki, 2015). Product as a service is considered to be an especially important CE business model because it is expected to encourage more durable and higher quality products, and CE business models require new kinds of cooperation between companies and between the public sector and companies (Sitra, 2019e).

Best practices for the promotion of CE encompass public procurement guidelines and requirements (e.g., criteria and specification/contract performance clauses on recycling, repair, and maintenance), extended producer responsibility, subsidies, fiscal incentives, fees, charges, and payment schemes. It is possible to design policy measures to simultaneously achieve multiple goals such as stimulation of the economy, job creation, and reduction of final energy consumption and harmful emissions. The current regulatory system is a major obstacle for the transition toward a CE-based society and the promotion of CE requires a regulatory system that encourages CE business models such as recycling, sharing, service-based business, and sustainable consumption (Green Budget Europe, The Ex'tax Project, Institute for European Environmental Policy, Cambridge Econometrics, 2018). Sustainable development tax reform shifts the focus of taxation toward emissions and natural resources, and steering of CE solutions is a major part of this development (Tamminen et al., 2019). Taxation is a strong societal steering instrument, and fiscal instruments (e.g., shifting taxation from labor to resource use and pollution) can be used to promote carbon-neutral CE goals, including increased environmental taxes and reduction of environmentally harmful subsidies (Green Budget Europe, The Ex'tax Project, Institute for European Environmental Policy, Cambridge Econometrics, 2018).

Municipal and city public procurement are among the main CE promotion and low carbon solutions focus areas in Finland, including advancement of sustainable and innovative procurement and actions such as joint procurement (e.g., solar power and electric/gas vehicles) and focus on smart construction and infrastructure, sustainable transport and logistics, and sustainable food systems (Hiilineutraalisuomi, 2021; Valtioneuvosto, 2019). From the Kymenlaakso region, Hamina (Hiilineutraalisuomi, 2019b), Kouvola (Hiilineutraalisuomi, 2019a), and Kotka are involved in a network that promotes carbon neutral municipalities including greenhouse gas emission reduction targets. There are many national initiatives and support networks for the promotion of sustainable and CE-oriented public procurement (Circwaste, 2019; KEINO, 2019; Kestävä kaupunki, 2019; Materiaali- ja energialoikka, 2019; Materiaalitori, 2019). In the city of Helsinki, sustainable public procurement is defined as a procedure in which the procured goods, services, or construction projects have smaller environmental, social, and ethical impacts during their life cycle than the conventional options (Helsingin Kaupunki, 2015). Sustainability can be taken into account in each phase of public procurement (e.g., minimal requirements for applicability and the procured subject, basis for comparison, and terms of contract), and sustainable procurement aims at: (1) reducing energy and material use and harmful environmental impacts during the whole life cycle of products, services, or buildings; (2) creating incentives for the development and implementation of new cleantech solutions; and (3) ensuring that social and ethical principles are realized in procurement (Helsingin Kaupunki, 2015).

Public procurement plays a major role in the promotion of energy efficiency and sustainable choices and can advance the reduction of both costs and emissions. Energy efficiency can be included in the assessment of needs and purpose, procurement planning, market analysis, minimum requirements, and comparison of tenders (TEM, 2016). Public green procurement is a major demand factor that can create markets for new resource-efficient, low carbon, and green growth-oriented products and services, but currently there is no common and established definition for that concept (Seppälä et al., 2016). Socially responsible public procurement refers to taking into account local, national, or global societal impacts in procurement (procurement organizations can include them in all phases of procurement and they can also be linked to innovations) including positive and negative impacts of the procured item and its production (TEM, 2017).

Public procurement can promote more energy- and material-efficient products with less harmful environmental impacts, encouraging the innovativeness of companies and the achievement of environmental goals such as the reduction of carbon dioxide emissions; and social sustainability (e.g., human rights) requirements, to address the working conditions in the production of products and services and to influence market development. Full lifecycle perspective can be gained through the application of life cycle costing covering the whole life cycles of procured products, services, and buildings/construction works (Motiva, 2019a). Motiva (2019b) has a procurement environmental criteria databank that can be used in both the drafting of minimum requirements and comparison of tenders. The approach to social responsibility is similar to environmental criteria, i.e., focus on the social aspects and impact in the product lifecycle and development of social criteria (Motiva, 2019c). Procurement organizations can require specific environmental, social, or other labels for procured products and services, and environmental labels can be used in the minimum requirements as a part of the description of the procured item or in quality comparison for giving extra points (Joutsenmerkki, 2017).

Central government can encourage change toward CE and create demand for CE solutions (Sitra, 2019b). State procurements must be sustainable and compliant with the principles of the CE from a full life cycle perspective. Moreover, all public procurement must encompass sustainable development and CE-related goals and criteria (Sitra, 2019b). Local authorities can, for example, increase the procurement of services and user rights (not physical products), including such solutions as property developers as owners of property to encourage more durable and energy and resource efficient construction (Sitra, 2019d). It is challenging to monitor progress toward CE because it is a systemic change affecting the whole economy and all products and services (Sitra, 2019f). It is not easy to develop a good CE barometer and indicators, and lack of knowledge and of a measurement system form a challenge. In addition, many CE phenomena take place in the company-sector interface. A good barometer would produce new information about CE including, e.g., systemic changes, resource cycles, and sharing among consumers (Luoma et al., 2015). The measurement of CE requires new initial data and statistical methods in addition to available statistics. It is noteworthy that many CE events take place, for example, within companies and outside the old systems and that no data has been gathered on all subareas of CE. The indicators (based on the key indicators for green growth (cf. Seppälä et al., 2016)) that could provide information on national CE development encompass the share of national value added related to CE business, monitoring of patent applications related to CE, monitoring the resource productivity or the value added obtained through the expenditure of unit resource, total raw material consumption by material categories and the share of renewable raw materials of the total consumption, the volumes and reuse of industrial, construction, and municipal waste, the share of renewable and low-carbon energy of final use, and monitoring the carbon footprint of the average citizen (Sitra, 2019f).

Informed decision-making at the national level (with supportive links to regional-, sectoral-, and company-level development) about green growth and material and resource efficiency can be supported by a set of key indicators (low carbon and resource efficiency, ecosystem services, economic opportunities and steering, and societal aspects) that are continuously developed based on improved data and changing societal priorities (Seppälä et al., 2016). A CE barometer is based on and indicates the development of different CE perspectives covering, e.g., use of natural resources, resource productivity, material cycles, CE business and innovations, consumption patterns, changes in the economy, and CE drivers and enablers. The potential indicators for further development encompass resource productivity, material loss, new consumption patterns, values and attitudes, CE business, and sustainability of material cycles (Luoma et al., 2015).

Public procurement should focus on purchasing solutions and products that support the CE and the aim of sustainable procurement is to produce zero or minimal waste and to support use of secondary materials and recycling (Sitra, 2016). For example, sustainable services can be purchased and leasing and rental business models be used instead of buying products made of nonrenewable raw materials. The public procurement context is related to both the EU-level new procurement directive development and national Act on Public Procurements. Organizations that are responsible for public procurement need to integrate principles and targets that promote the inclusion of CE solutions into existing procurement processes, and towns need to include principles and goals to promote CE solutions in their local procurement or service strategies. Supportive measures include the development of CE criteria and guidelines (for example for the estimation of life cycle costs) to include CE and material efficiency in public procurement (Sitra, 2016).

2. Aims of the study

This study is aimed at reviewing the key features, elements, drivers, and barriers of the international, EU-level, and national operational environment for CE, and assessing city/municipality perspectives in the Kymenlaakso region in Finland on advancing sustainable and CE-oriented public procurement. The idea was to support both informed decision-making and development of best practices in the public sector. In addition, the intention was to provide valuable insights into public procurement aspects for companies and to support the market development in the public-private interface toward more sustainable and CE-oriented activities and interaction (e.g., bidding competition and tendering). The aims and set-up of this study are linked to the overall CE development at the EU and national level, previous research including our focus on company perspectives, and the development of sustainability management and assessment.

3. Material and methods

The applied research approach encompassed both review and survey elements. The review of the operational environment consisted of a review of multiple sources and previous studies that are covered by both the introduction and discussion sections of this study. The method of the regional study was questionnaire survey (Patten, 2011; Hirsjärvi et al., 2007) and it aimed at addressing the main development factors and challenges. The questionnaire also included open questions. The cities included in the study were Hamina, Kotka, and Kouvola, whereas the municipalities were Iitti, Miehikkälä, Pyhtää, and Virolahti. The benefits of questionnaires include short implementation time, easy analysis, and a lot of information from a lot of people (Gillham, 2007).

Hirsjärvi et al. (2015) noted that the qualitative research approach makes it possible to select appropriate research subjects and objects in accordance with research goals. Oppenheim (1997) defined research design as referring to the strategy, plan, and logic of the research (forming a basis for the research specification). In this study, research design was based on the characteristics of sustainable and CE-oriented public procurement with links to associated frameworks such as CE and sustainability assessment.

Online survey was applied as an information-collection method to describe, explain, and compare, e.g., knowledge and preferences (Fink, 2009). The survey was implemented via email and the key concepts were defined at the beginning of the survey. Overall, the questionnaire was mostly formal and structured with multiple choice closed questions, as defined by Hirsjärvi et al. (2007) and Gillham (2007). There was one open question at the end.

The design was as professional as possible and the questions were clear, as recommend by Sudman and Bradbrun (1982). Survey data analysis uses both qualitative and statistical methods to describe, interpret, and compare preferences, attitudes, values, and behavior of respondents (Fink, 2009). Face validity (peer review) was used to check the quality of the questionnaire, as recommended by Saris and Gallhofer (2014). The construction of effective questions often requires both creativity and expertise supported by review of both opportunities and problems (Saris and Gallhofer, 2014). Peterson (2000) defined closed-end questions as including all possible responses, as specified by the researchers, and Oppenheim (1997) noted that closed questions (1) are easy to compare and process, (2) do not require much time, and (3) are good for testing specific hypotheses.

The elements of the survey encompassed and were influenced by, for example, international, EU-level, and national CE development (e.g., EC, 2015, 2017a,b, 2018; Sitra, 2015, 2016, 2019a,b,c,d), international-, European-, and national-level procurement criteria and performance indicators associated with CE and sustainability, sustainable and green public procurement, previous research on regional sustainable CE development (e.g., Husgafvel et al., 2018a, b, 2021), previous research on sustainability and CE metrics (e.g., Husgafvel et al., 2017), GRI sustainability reporting standards (GRI, 2019), the EU Ecolabel framework, practical measures to address climate change challenges (Hawken, 2017), EU and national procurement policy and law, local procurement guidance, and previous research on CE development and sustainability (e.g., Lakatos et al., 2016; Lieder and Rashid, 2016; Moreno et al., 2016; Sauvé et al., 2016; Witjesa and Lozano, 2016).

Overall, the theoretical frameworks of sustainable engineering (Abraham, 2006; Allen and Shonnard, 2012; Allenby, 2012; Graedel and Allenby, 2010; Rosen, 2012), sustainability assessment (Hak et al., 2007; Bell and Morse, 2008; Sheate, 2010; Dalal-Clayton and Sadler, 2011), and sustainability science (Kates et al., 2001; Komiyama and Takeuchi, 2006, 2011; De Vries, 2012) were taken into account in this study with focus on system-level aspects and economic, social, and environmental dimensions of sustainability. Loorbach (2019) recognized that researchers in the sustainability area need to rethink how they do research.

Questionnaires were sent both to administrative persons and to persons responsible for procurement in the target cities and municipalities in the Kymenlaakso region. The response rate was 19% and the characteristics of the respondents are presented in Figs. 1 and 2. The questionnaire was structured in a way that it included questions

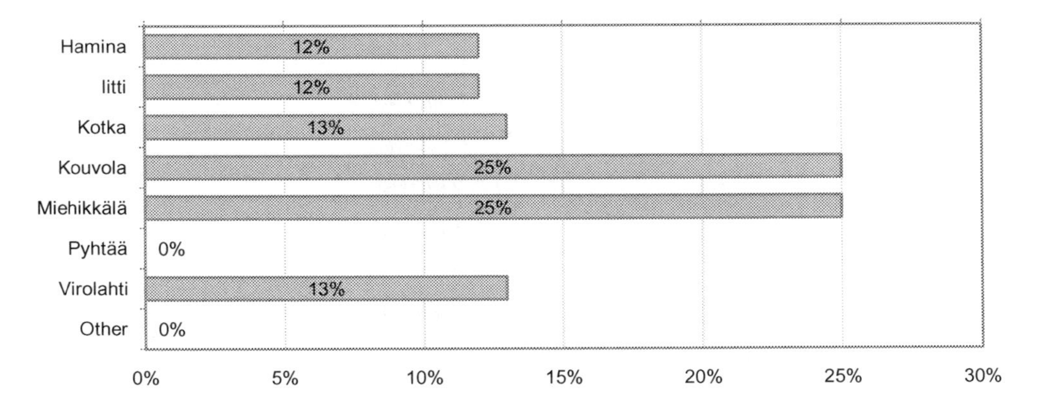

Fig. 1 Responses by the municipalities and cities in the Kymenlaakso region.

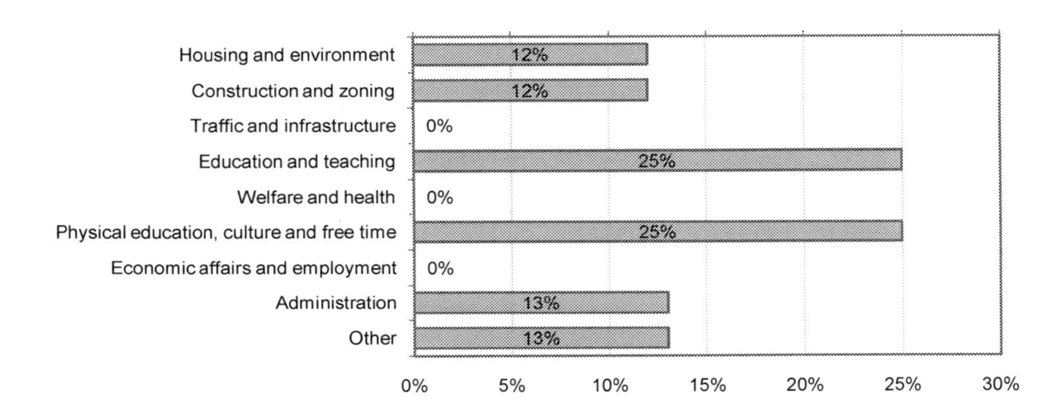

Fig. 2 Fields of administration of the respondents.

and a selection of answering options (choose openly or limited to four options) or options and attached importance scaling. The survey questionnaires links were sent directly to the personal addresses of the respondents by email. The results are presented as figures based on the chosen key themes and associated answers to the survey questions with the idea to both identify and indicate the most preferred options in each theme.

4. Results

The results indicate (Fig. 3) that procurement organization take sustainability and circular economy considerations into account in procurement planning and implementation mainly through environmental criteria, management commitment, strategy, goals, sufficient resources, responsibilities and monitoring, consideration of sustainability and

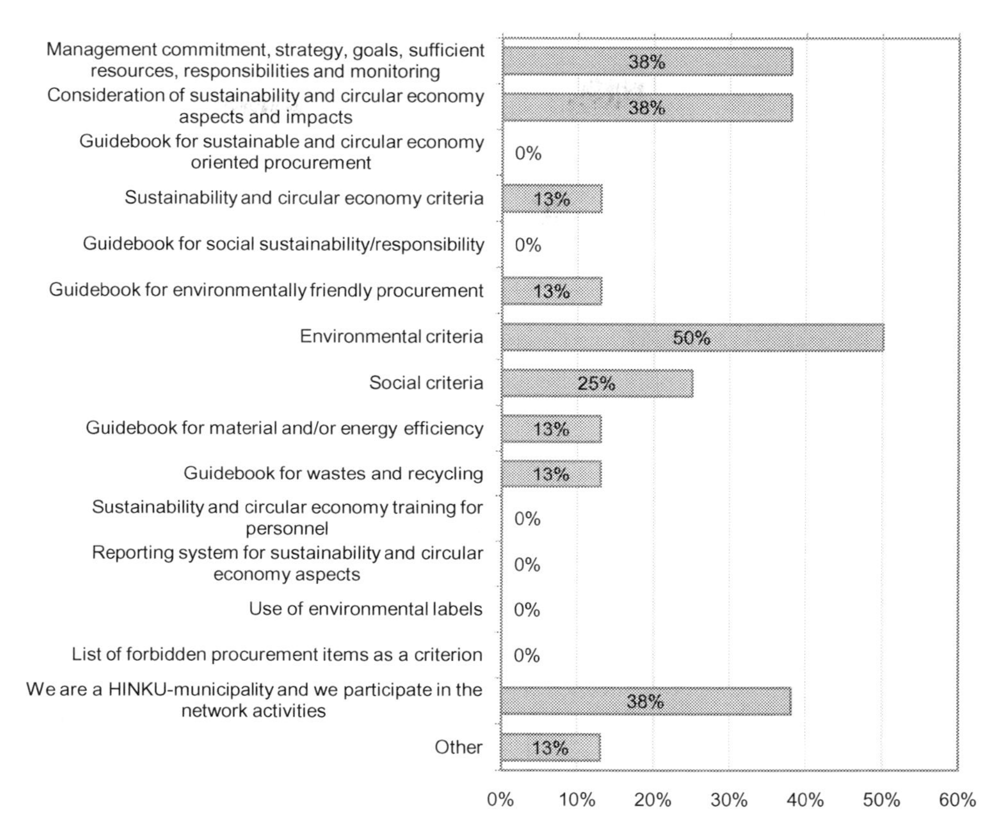

Fig. 3 Does your procurement organization take sustainability and circular economy considerations into account in procurement planning and implementation in any of the following ways? ($n = 20$).

circular economy aspects and impacts, role as a HINKU-municipality, and associated participation in the network activities and social criteria. In addition, recycled furniture, proper maintenance, and consideration in buying and recycling in education materials were among the applied measures. The applied guidebooks (Fig. 4) in procurement have included focus on energy efficiency in public procurement, guidance based on the participation in the carbon neutral municipality (HINKU) project and forum (including procurement initiatives), the Motiva guidebook on and database for sustainable public procurement and guidance from the competence center for sustainable and innovative public procurement (KEINO). The Nordic environmental label (the Nordic Swan) and the EU environmental label have been applied in procurement (Fig. 5).

National and EU legislation and development of national-level standards, criteria, labels, and certificates were considered to be the most efficient measures for the promotion of sustainable and circular economy-oriented public procurement locally (Fig. 6).

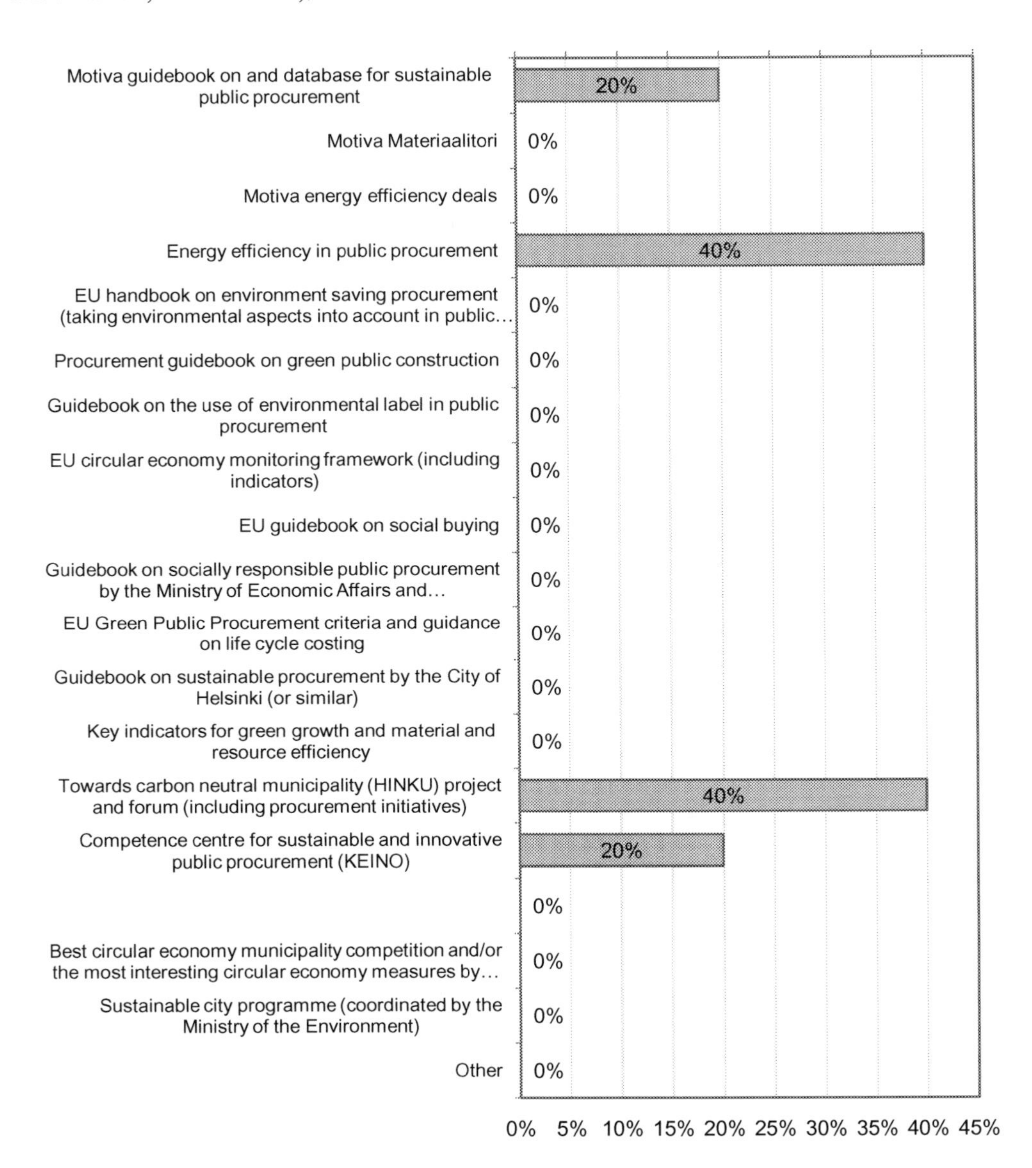

Fig. 4 Have you used any of the following guidebooks in your procurement? ($n = 6$).

Potentially beneficial promotion measures included EU-level strategy, goals, guidance, and tools, development of EU-level standards, criteria, labels, and certificates, development of EU-level assessment and reporting system (including all levels), development of national-level assessment and reporting system (including all levels), and new EU- and national-level incentives and removal of disincentives. Renewable energy subsidies,

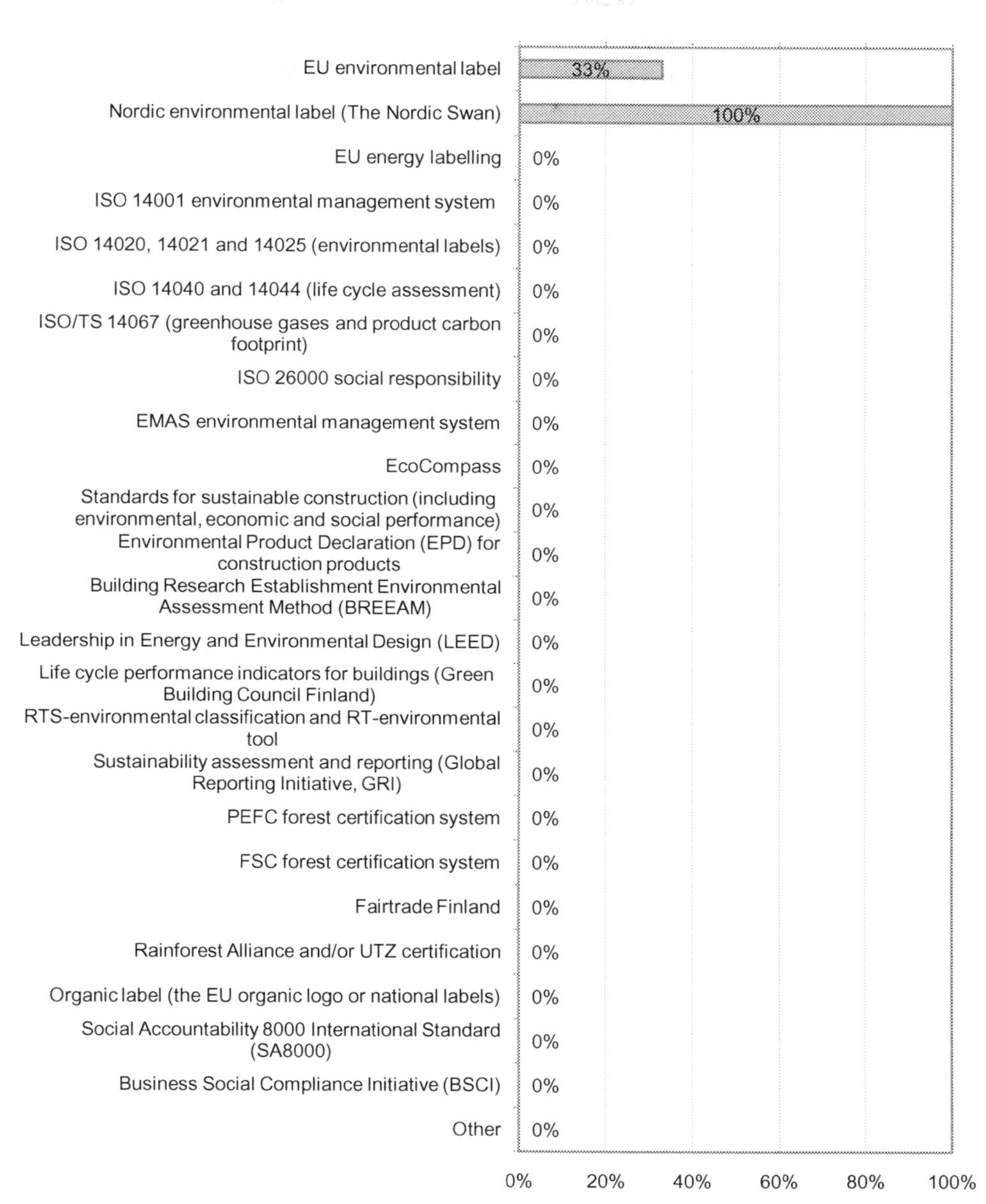

Fig. 5 Do you apply the following environmental labels or management systems, standards, reporting systems, or certificates in your procurement? ($n = 4$).

taxation of waste incineration and waste payments (encouragement of sorting and recycling), pricing, and taxation of greenhouse gas emissions were the main incentives and taxation-related means that could promote sustainable and circular economy-oriented

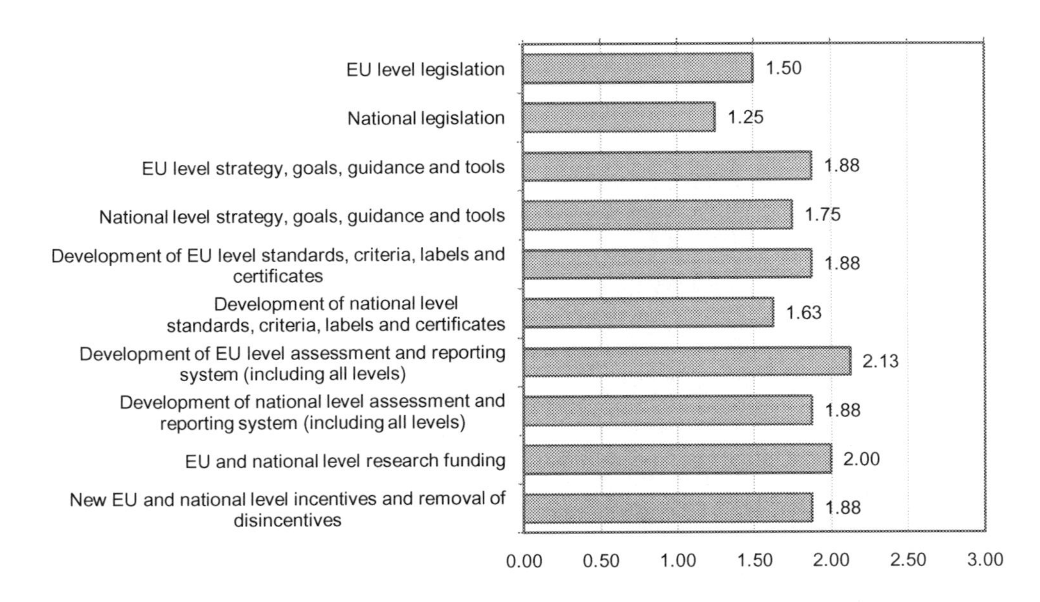

Fig. 6 Which of the following measures are most suitable for the promotion of sustainable and circular economy-oriented public procurement locally?

public procurement (Fig. 7). In addition, fossil fuel taxation changes and removal of subsidies, renewable raw material subsidies, and identification and removal of harmful subsidies were also considered to be important. It was also recognized that the viewpoint of sustainable and circular economy-oriented public procurement should be a crosscutting theme in municipal procurement guidebooks and training.

The most suitable procurement organization measures (Fig. 8) for the promotion of sustainable and circular economy-oriented public procurement encompassed management commitment, sufficient resources, strategy, goals, staff training, responsibilities, internal and sectoral cooperation, interaction, communication and monitoring, market cooperation, engagement of companies, preprocurement consultation, innovation partnerships, new business models and development of local networks and ecosystems (including the whole supply chain), life cycle assessment and life cycle costing training, and consulting and development of best practices and pilot projects. Identification of and goals for sustainability and circular economy issues (including procurement notices and design competitions), sustainability and circular economy criteria, environmental impacts during the whole life cycle (life cycle assessment) and their emphasis, energy and material efficiency, waste minimization, and transformability and flexibility were the main issues (Fig. 9) that should be taken into account in procurement planning and subject definition (minimum requirements).

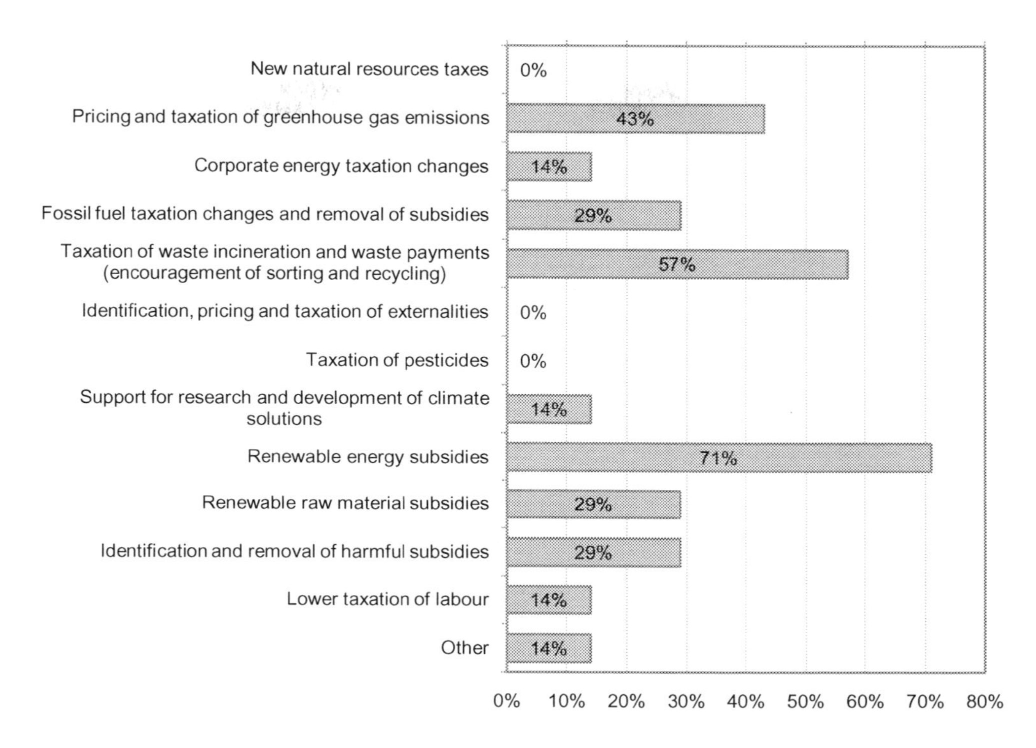

Fig. 7 Which of the following incentives and taxation related means would promote sustainable and circular economy oriented public procurement? (Choose the most important ones, max 4.) ($n = 22$).

The most important issue that should always be taken into account in the eligibility and feasibility criteria for bidders (Fig. 10) was ensuring that the bidder has previously taken care of sustainability and circular economy obligations (for example recycling, wastes, emissions, violations, taxes, social security payments, and pension insurance payments). In addition, sustainability and circular economy competence and references (including management and reporting), sustainability and circular economy standards, labels and certificates (if available), and acting as a part of circular economy-supporting value/supply chain network/ecosystem were among the main issues that should always be considered. The main issues to be taken into account based on the characteristics of specific situations were environmental management system (for example ISO 14001), environmental standards, labels and certificates, action in national circular economy networks, and social standards, certificates, and labels.

The main issues that should always be taken into account in the procurement phase and in the benchmarking of bids (Fig. 11) encompassed life cycle costs, technical reports and/or testing reports focusing on sustainability and circular economy, and sustainability and circular economy issues in procurement notices, invitations for tenders, and market

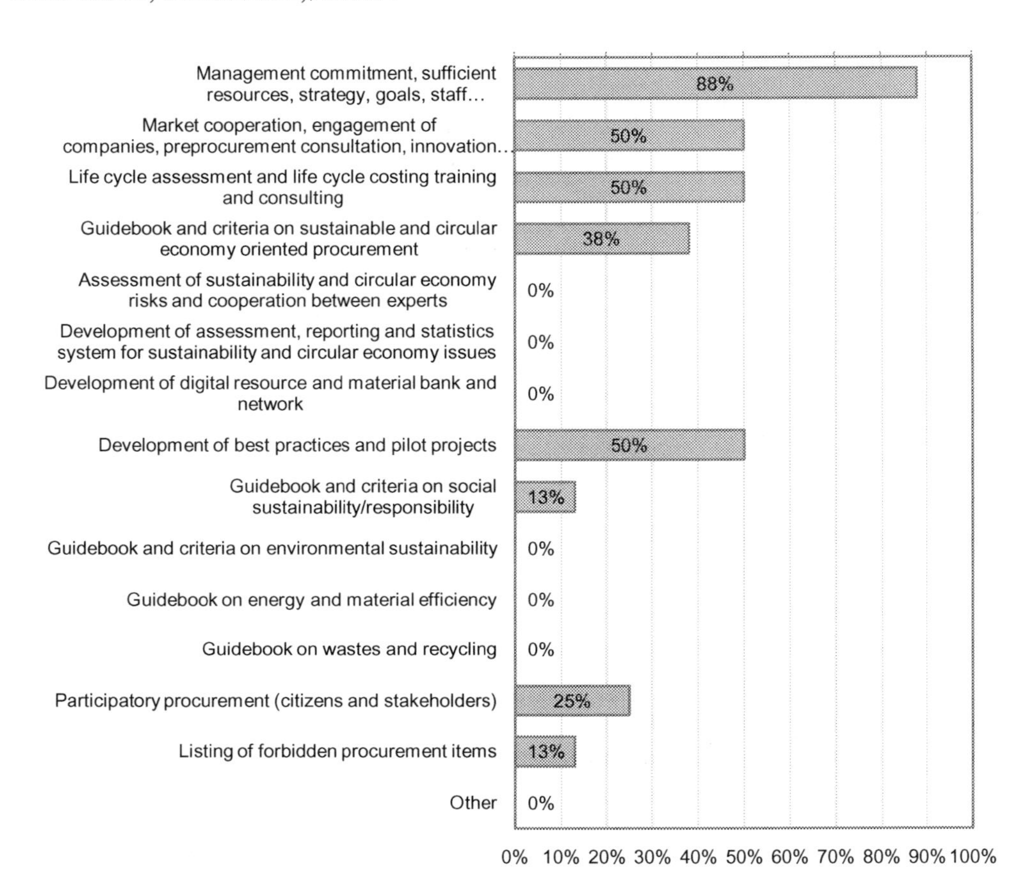

Fig. 8 Which of the following procurement organization measures are most suitable for the promotion of sustainable and circular economy-oriented public procurement? (Choose the most important ones, max 4.) ($n = 26$).

interaction. The main issues to be taken into account based on the characteristics of specific situations were environmental label and/or certificate (extra points), sustainability label and/or certificate (if available, extra points), circular economy label and/or certificate (if available, extra points), and social label and/or certificate (extra points). Sustainability and circular economy aspects, goals, and impacts (economic, environmental, and social aspects), responsibilities for taking care of products and materials at the end of the contract/their use life, and continuous improvement, bonuses, and sanctions related to sustainability and circular economy were the main issues that should always be taken into account in the contract terms and follow-up and in reporting (Fig. 12). The main issues to be taken into account based on the characteristics of specific situations were separate

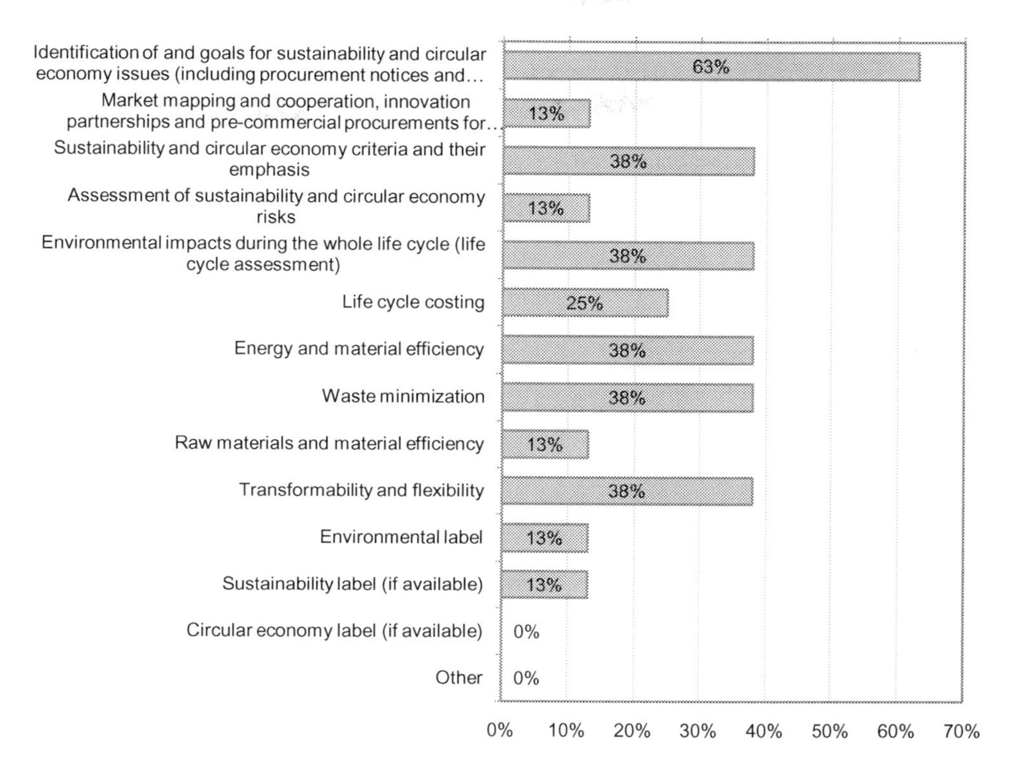

Fig. 9 Which of the following issues should be taken into account in procurement planning and subject definition? (Choose the most important ones, max 4.) ($n = 27$).

sustainability and circular economy report, circular economy label (if available), and development of assessment and reporting system for sustainability and circular economy issues.

The most suitable indicators for the assessment of economic sustainability as a part of the procurement process and its different phases (Fig. 13) included life cycle costs, management and impacts of the supply chain, suppliers, and subsuppliers, and management of the operational environment and risks. The most suitable indicators for the assessment of environmental sustainability as a part of the procurement process and its different phases (Fig. 14) comprised waste quantity, reduction, recycling, and utilization (municipal, industrial, and construction waste), leadership, strategy, management, reporting and code of conduct, energy consumption and the share of renewable energy (whole life cycle), use of raw materials and the share of renewable raw materials (whole life cycle), and transport and traffic emissions and impacts. The most suitable indicators for the assessment of social sustainability as a part of the procurement process and its different phases (Fig. 15) encompassed leadership, strategy, management, reporting and code of conduct, promotion of

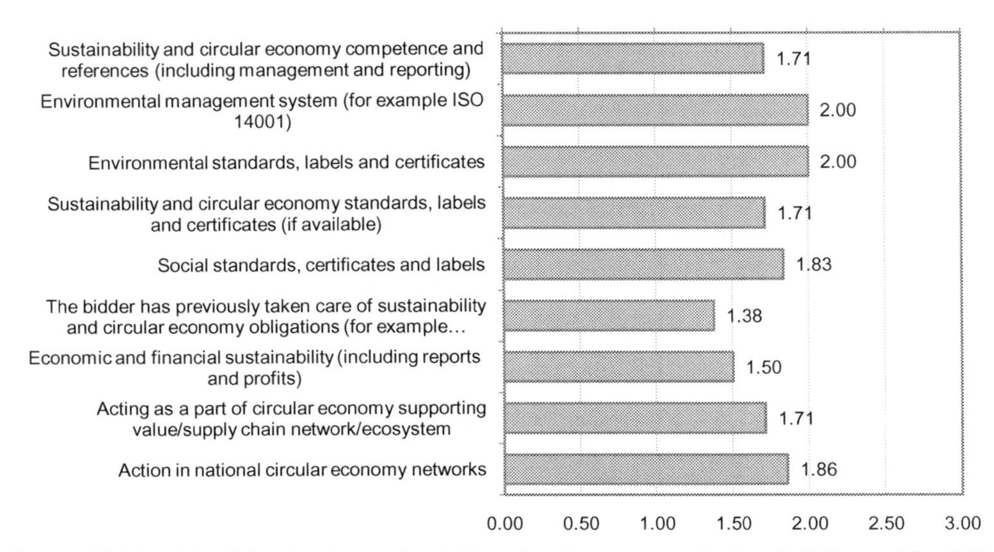

Fig. 10 Which of the following issues should be taken into account in the eligibility and feasibility criteria for bidders?

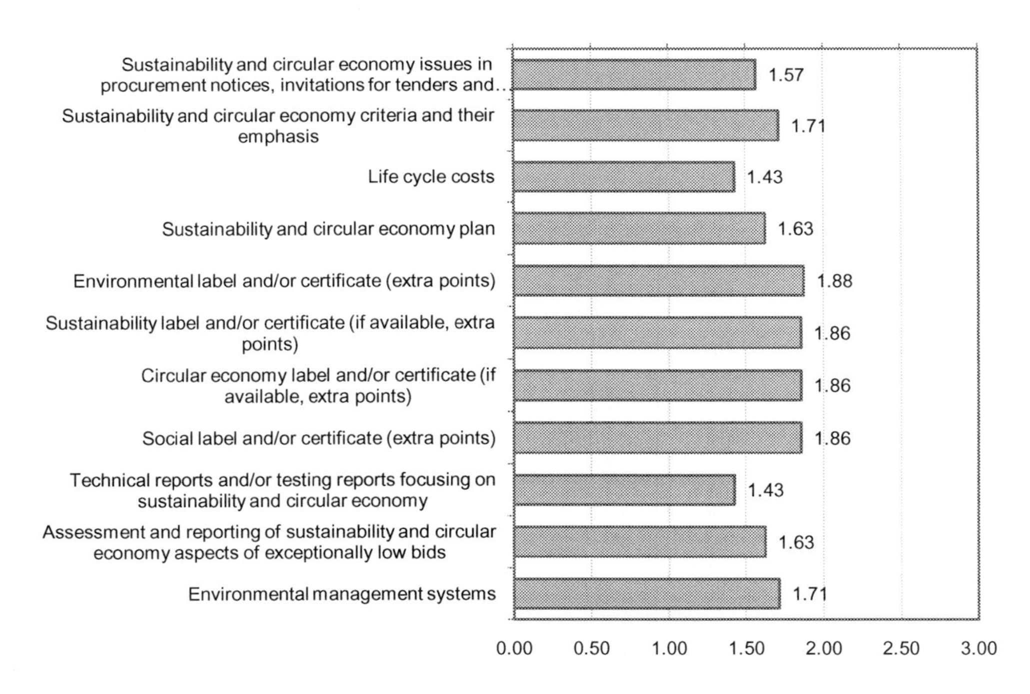

Fig. 11 Which of the following issues should be taken into account in the procurement phase and in the benchmarking of bids?

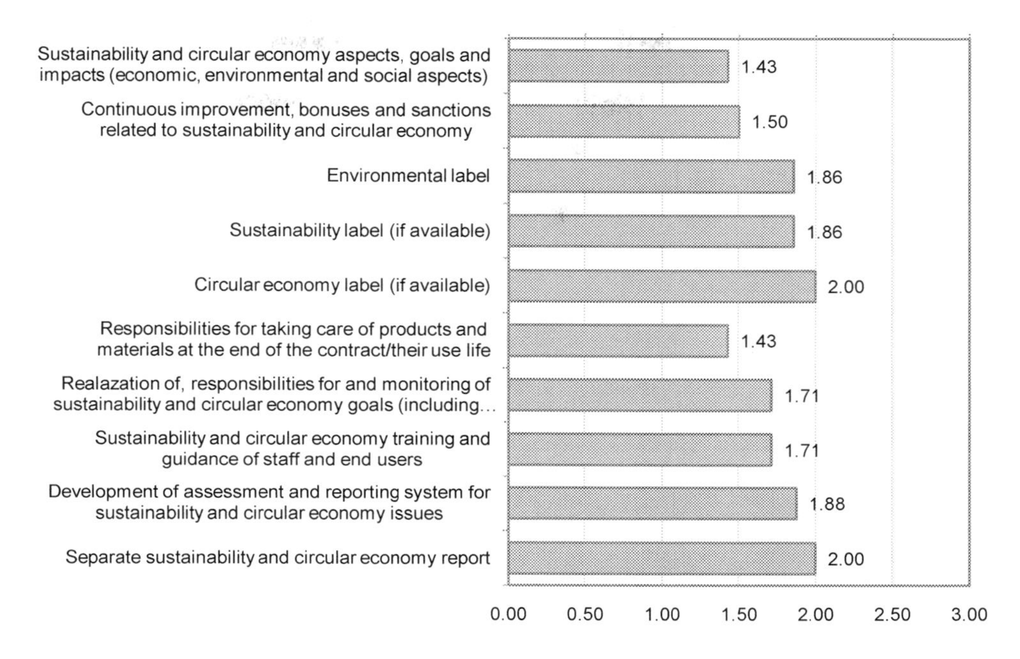

Fig. 12 Which of the following issues should be taken into account in the contract terms and follow-up and in reporting?

employment opportunities (including young and unemployed people), labor practices, conditions, and social and work rights (for example standards, salary, vacations, rights, supervision, collective bargaining, freedom of assembly, and work time), and collaboration with and engagement of the local community.

The most suitable indicators for the assessment of circular economy as a part of the procurement process and its different phases (Fig. 16) included life cycle thinking, use/working life and durability/sustainability, waste quantity, reduction, recycling, and utilization (municipal, industrial, and construction waste), raw materials, material efficiency and material wastage/loss (including use of renewable raw materials and reused materials), recyclability, repairability, refurbishment, remanufacturing, and reuse, packaging material and practices, and recycling/reuse of packages. The areas/fields that offer the best opportunities to promote sustainable and circular economy oriented procurement in the future encompassed construction and buildings, waste management and recycling, and energy (Fig. 17). The measures to combat climate change that were considered to be effective parts of sustainable and circular economy-oriented public procurement in the future (Fig. 18) comprised reduction of food waste, wind and solar power, public transport, LED lightning, updating of old buildings and spaces, and remote access.

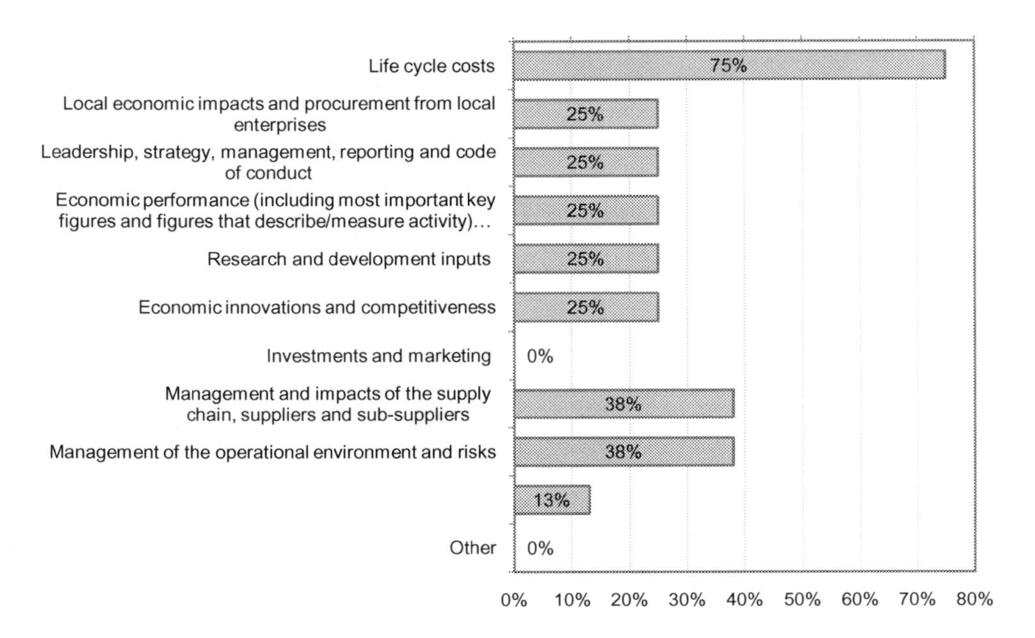

Fig. 13 Which of the following indicators would be most suitable for the assessment of economic sustainability as a part of the procurement process and its different phases? (Choose the most important ones, max 4.) ($n = 23$).

The most important next steps for the development of sustainable and circular economy-oriented public procurement (Fig. 19) included focus on recycling, reuse, refurbishment, and remanufacturing of products, components, and materials, incorporation of sustainability and circular economy issues into local policy, strategy, and goals, focus on the life cycle, use/working life, and sustainability of products, components, and materials, and national recyclability requirement for packages and plastics (reuse as raw materials or products). The potential important next steps encompassed the development of sustainability and circular economy guidebooks, standards, and labels (including indicators), the development of digitalization, digital technology (for example tagging, data management, and connectivity), and information systems, and the assessment of sustainability and circular economy risks.

5. Discussion

Previous studies have addressed the implications of the EU operational environment for CE development. Hartley et al. (2020) studied EU-level transition toward a CE and framed their policy recommendations based on life cycle stages. The recommendation for the production/product design stage was the adoption of circular design standards

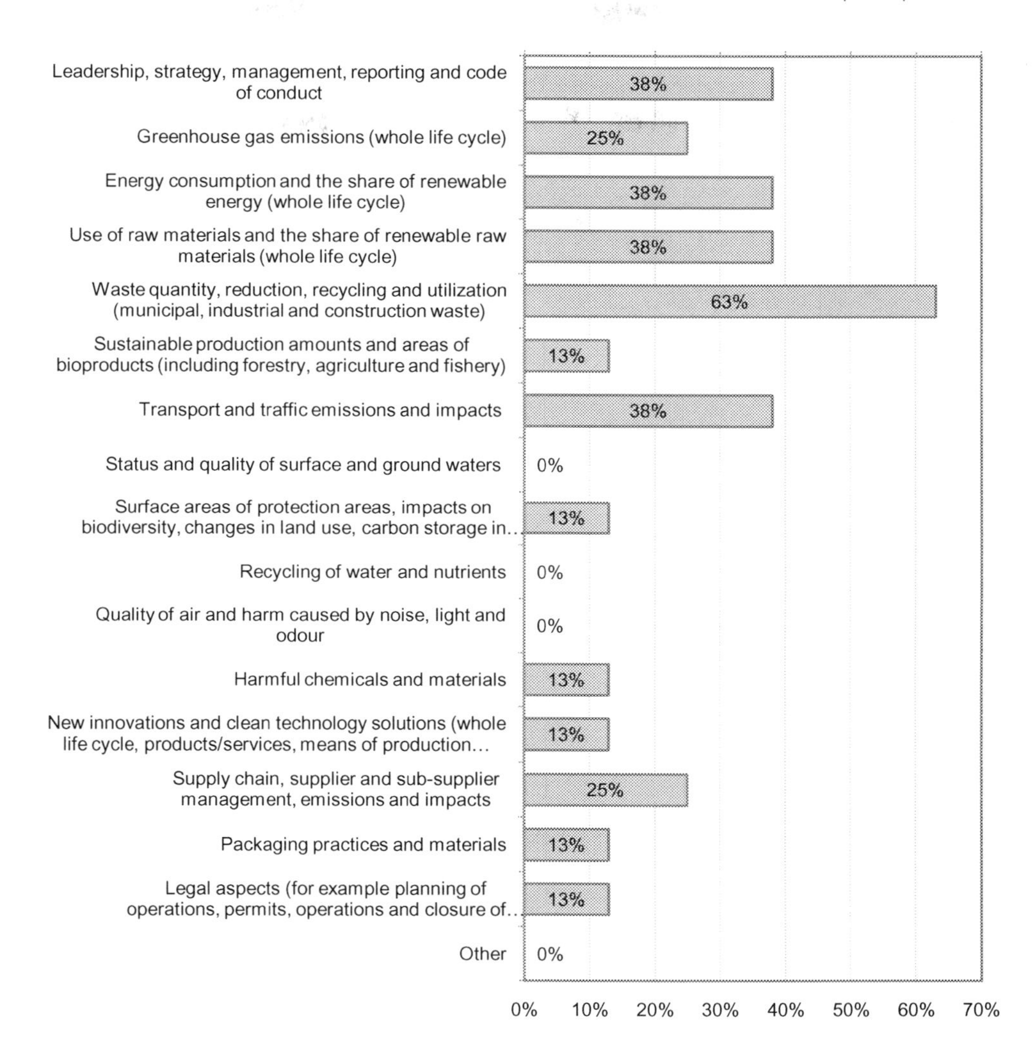

Fig. 14 Which of the following indicators are most suitable for the assessment of environmental sustainability as a part of the procurement process and its different phases? (Choose the most important ones, max 4.) ($n = 27$).

and norms, whereas the recommendation for the use phase/consumption stage was circular procurement. The recommendations for the end-of-life/waste stage encompassed reduced VAT for reused products and products with recycled content, liberalization of waste trading, stimulation of the development of circular trading platforms, and creation of eco-industrial parks. The recommendations for the resource-circulation stage were a circular economy marketing and promotion campaign and material flow accounting (MFA) database (Hartley et al., 2020).

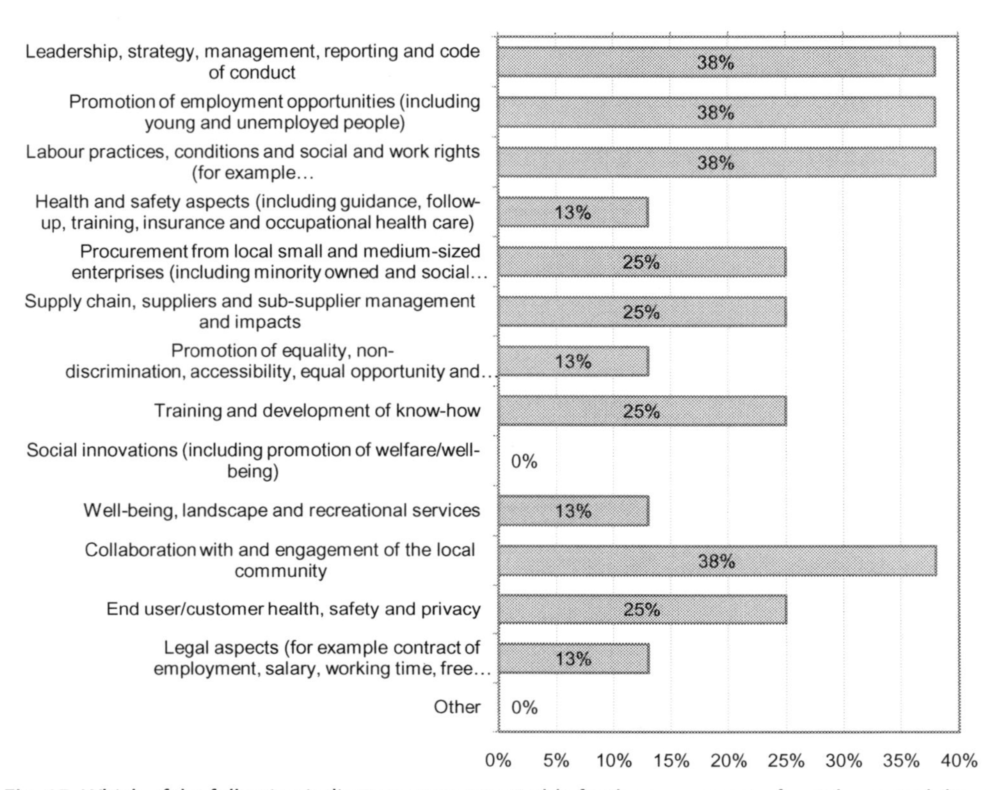

Fig. 15 Which of the following indicators are most suitable for the assessment of social sustainability as a part of the procurement process and its different phases? (Choose the most important ones, max 4.) ($n = 24$).

Leadership is a significant factor in public procurement and senior managers need to be engaged to realize implementation by the purchasing team. Financial concerns are the main barrier to sustainable procurement, including resistance to pay more to buy sustainably (Brammer and Walker, 2011). Van Buren et al. (2016) noted that the promotion of CE requires the involvement of multiple public and private actors, whereas Sauvé et al. (2016) noticed the importance of sustainability and life cycle considerations for addressing practical CE challenges. The promotion of CE development through new consumption patterns depends on incentives and benefits (Lakatos et al., 2016). Khosla (2017) noted that SMEs play an important role in promoting CE and new sustainable solutions to local needs. Public sector organizations are more likely to implement sustainable procurement if there is a supportive governmental policy and legislation environment, and shared learning about sustainability practices across regions is likely to beneficial for all actors (Brammer and Walker, 2011).

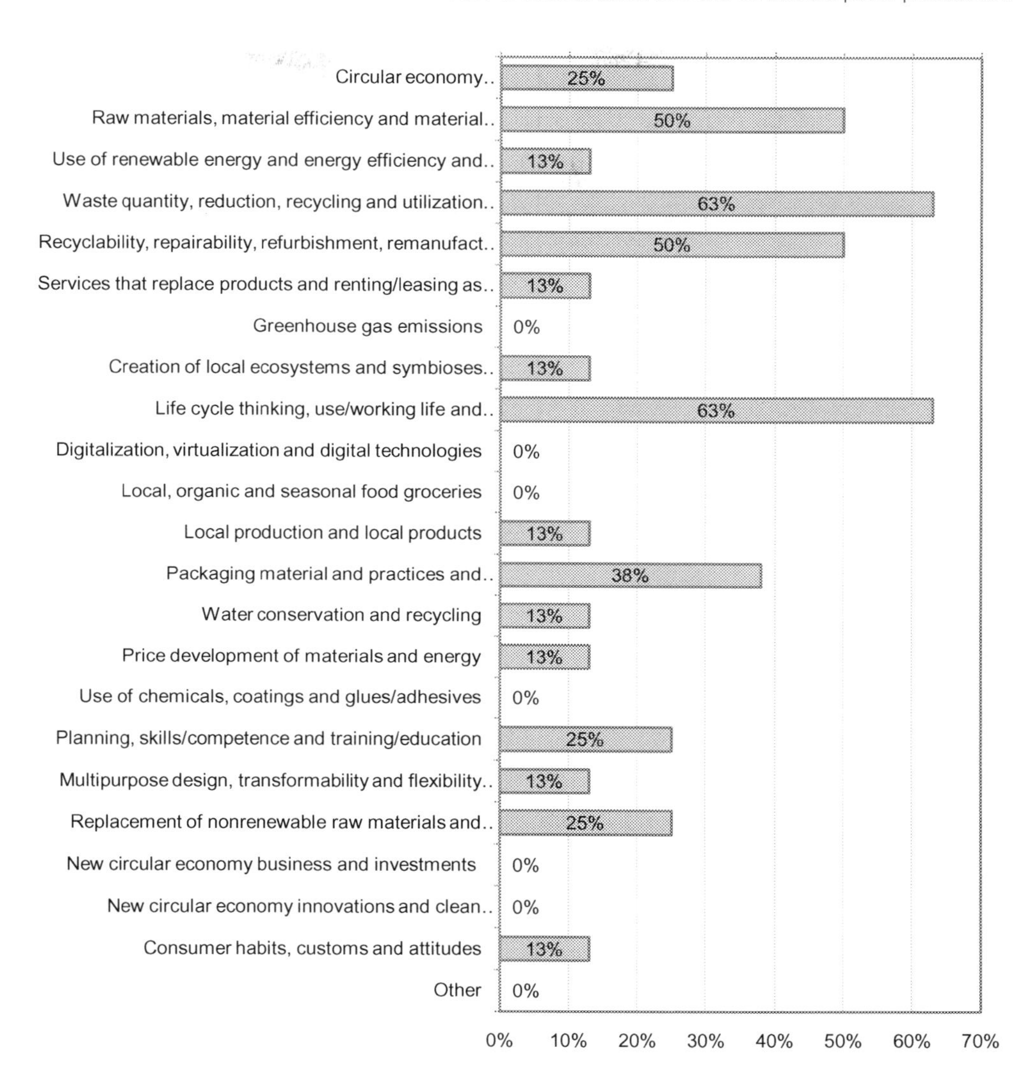

Fig. 16 Which of the following indicators are most suitable for the assessment of circular economy as a part of the procurement process and its different phases? (Choose the most important ones, max 5.) ($n = 35$).

In addition, Maignan et al. (2002) noted that a proactive approach to socially responsible buying can create business benefits, be an expression of corporate values and help to avoid stakeholder pressures and negative publicity. Identification of barriers to change toward more sustainable activities, products, and services can help to incorporate and institutionalize corporate sustainability (Lozano, 2013). Liu and Bai (2014) concluded that policy makers should use regulations and incentives to support the building of

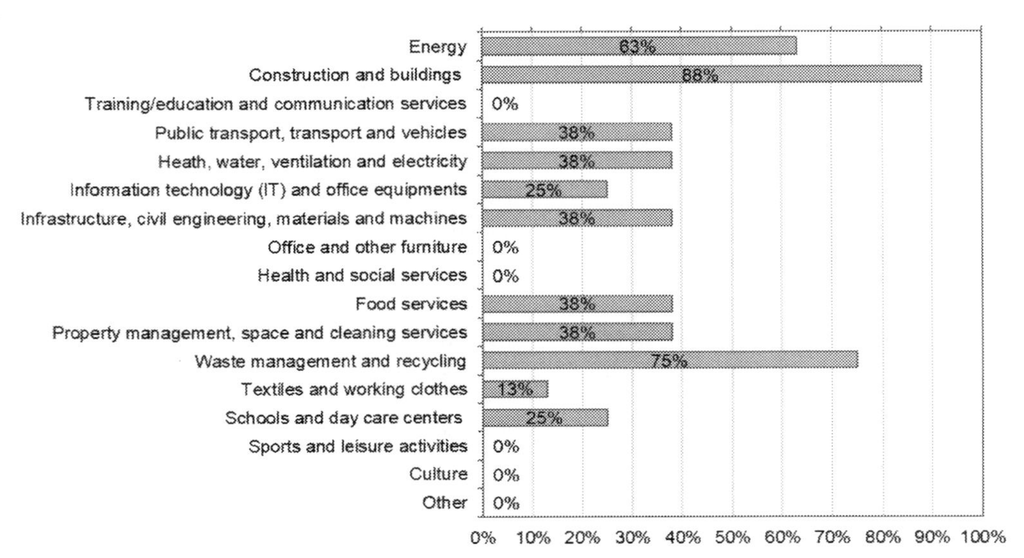

Fig. 17 Which of the following areas/fields offer best opportunities to promote sustainable and circular economy-oriented procurement in the future? ($n = 38$).

modern corporate governance system (to promote CE development among companies) and noted that knowledge, awareness, and environmental market and customer pressures alone are not sufficient preconditions.

Previous studies have indicated the importance of social and environmental criteria in the context of procurement processes and supply management to address sustainable development goals (Preuss, 2009; Srivastava, 2007). Maignan et al. (2002) noted companies can incorporate social responsibility criteria into their purchasing decisions encompassing, e.g., proactive socially responsible buying strategy, implementation and marketing of responsible purchasing practices, and purchasing policies and principles based on organizational values. The adoption of new consumption patterns specific to CE business models would require direct or indirect incentives and benefits in addition to awareness-raising and educational campaigns (Lakatos et al., 2016). Circular SMEs can be supported through, e.g., the development of voluntary circular metrics to measure and communication progress and appropriate policy frameworks that support innovation and competitiveness (Gerholdt, 2017).

Pöyhönen (2017) noted that CE can be advanced in public procurement (case Kouvola city) through new kinds of thinking and attitudes, commitment of leaders and sufficient resources, adoption of CE-oriented procurement criteria, enhanced cooperation inside procurement organizations and across sectors, enabling operational environment (e.g., deregulation and more research), material selection that support wood construction (renewable materials), cooperation with companies, technical development, use of renewable energy choices (e.g., in school transport), innovative use of empty buildings

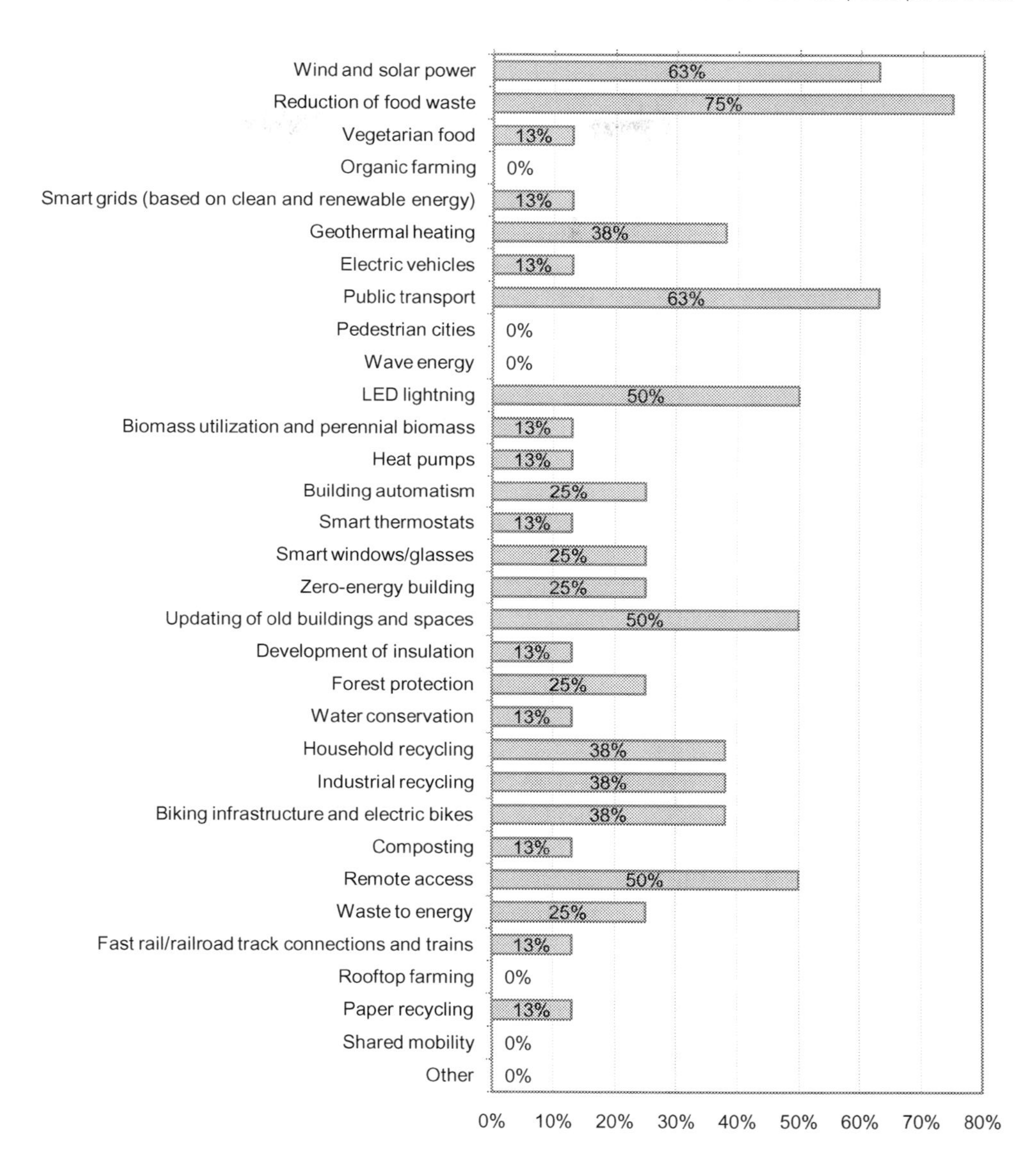

Fig. 18 Which of the following measures related to combating climate change could be an effective part of sustainable and circular economy-oriented public procurement in the future? ($n = 61$).

and spaces, new construction for multiple purposes, creation of an enabling environment for locally produced food, digitalization in education, use of leasing finance (e.g., real estate and lobby services), more efficient use of existing resources (e.g., better asset management) and extension of their lifecycles, and establishment of a digital resource databank (Pöyhönen, 2017).

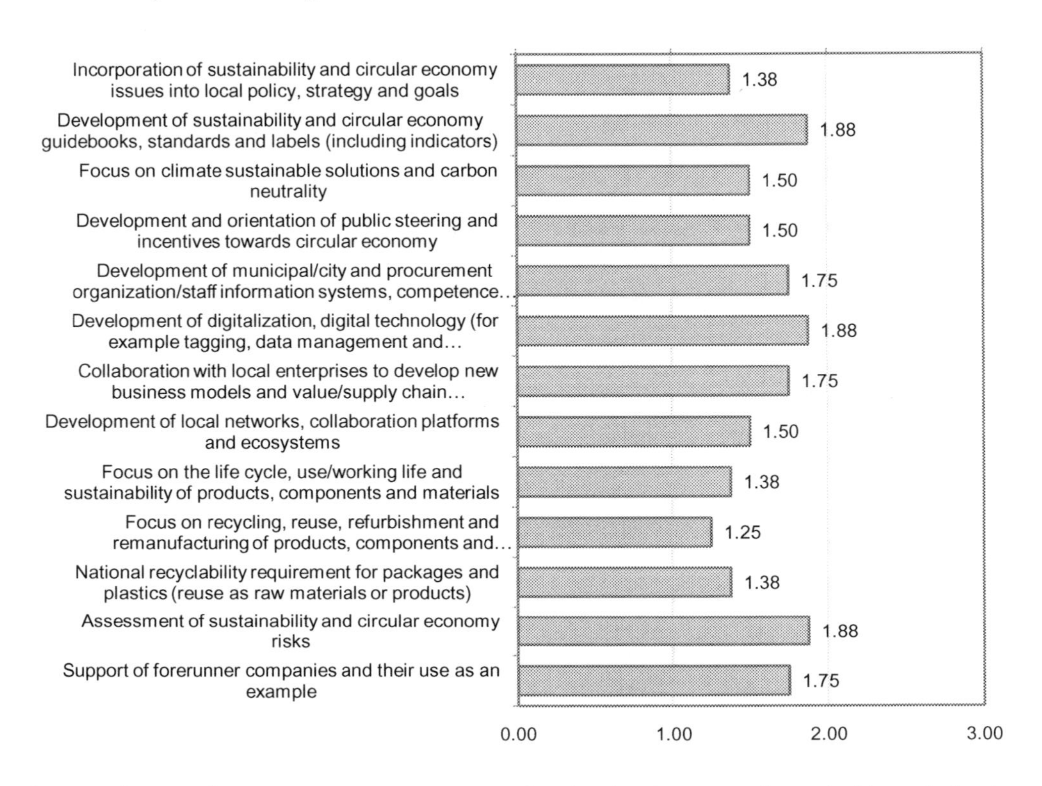

Fig. 19 What are the most important next steps for the development of sustainable and circular economy oriented public procurement?

For example, vehicle, energy, heath, and food service procurements have major environmental impacts. Life cycle costs can be applied, e.g., in the context of vehicles, building and insulation materials, outdoor and indoor lightning, information technology, and energy-using devices (Motiva, 2019a). Important future focus areas encompass cooperation and market dialogue between supply chain actors, overall chain of manufacturers, suppliers, logistics, reprocessing, end markets and consumers, and education of procurers to promote the adoption of new practices (Alhola et al., 2017).

Circular procurement can be promoted through performance-based procurement, the use of life cycle costing, and criteria concerning the reuse and recycling of materials, and it can be advanced in many sectors encompassing construction and renovation of buildings (recycling of material and demolition waste according to contract criteria and conditions), utilization of secondary materials (from site or nearby) in road infrastructure construction, energy, and transportation (creation of local circular systems based on bioenergy), and waste water treatment (more efficient recycling of nutrients including development of related technology and business concepts), and in furniture, appliance, and textile product groups (Alhola et al., 2017).

Enhanced collaboration in procurement and supply practices and more service-oriented systems could advance both CE and sustainability goals (Witjesa and Lozano, 2016). Making decisions about sustainable production and consumption requires scientifically based information on sustainability, and decision-makers should take part in the development of desired indicator sets (Dong and Hauschild, 2017). Sustainability assessment can direct decision-making toward sustainability and it emphasizes bringing about positive net sustainability gains now and into the future. Moreover, sustainability assessment: (1) needs to integrate learning into the process; (2) can take many forms; (3) needs to include the views of all affected and interested parties in the framing of the assessment; and (4) can be directed to any type of decision-making including different levels, contexts, and approaches (Bond et al., 2012).

Walker and Brammer (2012) concluded that public sector communication with suppliers and e-procurement supports some aspects of sustainable procurement (environmental, labor, and health and safety), that buying from small local firms (not e-enabled) may be hindered by e-procurement, and that communication has positive links with buying from local suppliers and SMEs. Pollice (2018) concluded that a new circular procurement mindset for procurement and supply management teams, complete integration of circular procurement concepts and practices in the procurement organizations and structures, new performance measurement systems, new skills and competencies (e.g., thinking in systems and development of new business models), and development of strategic suppliers (to provide and manage material chains over time) are needed.

The key enabling factors of circular procurement encompass buyer control over the whole portfolio (to develop and implement a new sourcing strategy), collaboration and long-term relationships with nontraditional suppliers, circular product redesign involving research and development, circular performance metrics (measurement of the success of circular sourcing strategies) and performance indicators, and an overall ecosystem integration approach, which is about total integration covering a vast network of companies allowing truly circular elements from suppliers to end consumers with close loops in every stage of the production and consumption chain (Pollice, 2018).

Petit-Boix and Leipold (2018) noted that waste management is a key strategy that cities promote and that it is a focus area of academic research. In addition, they stated that urban planning activities need to include environmental accounting. There is a need to define the environmental impacts of the circularity strategies of cities and collaborative research and practical efforts could be focused on the prioritization of the CE strategies of each urban area. Integrated assessments could help to identify the sustainability of CE strategies accompanied by the identification of opportunities and barriers by the key stakeholders (Petit-Boix and Leipold, 2018).

Wang et al. (2018) noted that urban CE and residential circularity grew significantly in many cities in China. Guo et al. (2017) noted that many megacities of China have improved their CE development. They also noted that there are differences among them and proposed targeted supportive measures such as financial and technical assistance, promotion of the recycling of waste waters, the application of integrated waste management, and encouragement of the participation of local residents.

New et al. (2002) recognized that the responses of both public and private sectors to environmental challenges of the supply chain are strongly influenced by organizational structure, patterns of decision-making, and flow of information. In addition, they noted that the success of green supply initiatives seems to be strongly dependent on the ability of the organization to align activity with dominant corporate objectives. They stated that green supply practices need to be implemented with regard to organizational structure and strategy. They also observed a very wide range of skills, enthusiasm, and intellectual understanding in the studied purchasing and selling communities. For example, they found that there was considerable diversity of opinion among public purchasers about how much their scope for actions were constrained by regulation (particularly European Union public procurement rules).

Coggburn (2004) noted that green procurement can accomplish managerial values and that it can be an efficient (for example use of ecolabels) and economical (for example life cycle assessments or joint purchasing) approach. Green procurement can also improve environmental conditions (for example, reduce greenhouse gas emissions) in an efficient and economical way (for example, self-regulation). Successful implementation of green procurement requires that each department is committed to the improvement of the quality of life of residents and citizens including the promotion of a clean and sustainable environment. In addition, broad stakeholder involvement is necessary for the establishment of the legitimacy of green procurement programs (Coggburn, 2004). Swanson et al. (2005) presented a priority-setting tool to promote greener state purchasing and associated development of priorities including consideration of purchase volume, environmental impacts, potential for improvement, and institutional factors such as existing state policies and upcoming contract renewals.

Preuss (2007) concluded that local authorities in England were active in a wide range of activities aimed at addressing the challenges of sustainable development covering environmental, social, and economic development aspects. Local authorities could cooperate with other public sector organizations and with the private sector to learn from sustainability experience elsewhere. Local government has addressed many environmental sustainability issues, such as phasing out hazardous materials in procured products and services, by requiring a minimum recycled content for some goods, and insisting on lower energy or fuel consumption for buildings and vehicles. They could also support local companies and particularly small and medium-sized enterprises. Developments in the field of social sustainability include new types of service providers such as social enterprises and voluntary organizations. Addressing the sustainability aspects of procurement requires focus on the strategic approach to sustainability and the costs of sustainability measures. Local government could design a sustainability strategy and include it in the procurement policy. In addition, it could promote know-how sharing with other buying organizations in both public and private sectors (Preuss, 2007).

SPP initiatives should proactively use the value for money argument for their purposes and governments should take into account the full range of economic, social, and environmental costs and benefits of public procurement. The key elements of a comprehensive SPP policy are laws and action plans and public procurers need to be educated and trained to implement them. It is important to build capacities and capabilities for SPP, and informal SPP instruments such as guidelines and trainings need to be compatible with existing routines of public procurers, up-to-date, speak the similar language, and take into account the legal uncertainties and economic concerns of public procurers (Steurer et al., 2007).

Thomson and Jackson (2007) studied sustainable procurement in the United Kingdom and noted that, green procurement has been encouraged through legislation and provision of information and dismantling of barriers. However, the horizons of green procurement could be expanded through new information to link up green procurement with organizational goals. Implementation of new instruments such as a specific action plan for sustainable procurement can ensure that green procurement becomes embedded within government procurement. Motivated individuals in the procurement decision-making process are a very important driving force for green procurement at the operational level (Thomson and Jackson, 2007).

Walker and Brammer (2009) studied sustainable procurement in the United Kingdom public sector and concluded that the implementation of sustainable procurement practices at the organizational level is influenced by familiarity with policies, perceived inefficiencies/cost of policy, supplier availability/resistance, and organizational incentives/pressures. They noted that financial pressures and costs are the most important barriers to the implementation of sustainable procurement and that local authorities typically buy from local and small suppliers. Many aspects of sustainable procurement and particularly those associated with purchasing from small businesses and local suppliers are well embedded in the United Kingdom. However, there is only modest focus on environmentally oriented aspects of sustainable procurement within public sector organizations (Walker and Brammer, 2009).

Walker and Preuss (2008) addressed the promotion of sustainable development through public sector sourcing from small and medium-sized enterprises and noticed that it can contribute to local economic development through provision of innovative green products and services (particularly in the food sector) and that it can help public sector organizations to better align their operations with their communities. They noted that sourcing from small businesses can benefit from a partnership approach and that it has implications for the governance of sourcing processes. Sourcing from local small businesses can make a contribution to sustainable development at a regional or local level.

Small businesses can be supported through innovative tools (for example partnership approach) applied by public procurers, training of procurement staff (for example on legal issues and EU procurement rules) and more transparent public sector contracting opportunities (Walker and Preuss, 2008).

Meehan and Bryde (2011) recognized that procurement has a key role in sustainability and that policies and practices need to extend beyond the boundaries of an organization, including the incorporation of their whole supply chains. They noted that decision making about procurement that takes into account environmental, economic, and social elements can be promoted through guidelines on sustainability. The results of their study in the United Kingdom suggest that there was a failure to overcome inertia in relation to sustainable procurement and that only environmental elements were considered in the few examples of established sustainable procurement practices. Widespread establishment of sustainable procurement had not taken place even though the studied organizations had sustainability-related issues in their missions, and there were both external and internal pressures to embed sustainability. Suggested measures to promote sustainable procurement encompass development of a small number of sustainable development indicators for procurement, taking experience from other areas, which place emphasis on interorganizational relationships (such as innovation management) and focus on measures to overcome inertia and to promote changes in procurement staff behavior (Meehan and Bryde, 2011).

Preuss and Walker (2011) addressed psychological barriers to implementing sustainable development in procurement process by local government and health care authorities in the United Kingdom and concluded that these barriers include both individual and organizational factors, adaptation processes within both small groups and organizations, and external adaptation processes between organizations. They recognized that there is a need to consolidate information on sustainable procurement (for example, a single website that has links to multiple websites and includes guidance and tools for practitioners), for procurement training that emphasizes the importance of sustainable procurement for practitioners, for organizational guidance on how to meet both financial and sustainability targets, for senior management commitment and training for senior managers, to set mandatory sustainable development targets and for organizations to share best practices, and to consider how sustainable procurement criteria can be incorporated in outsourcing contracts. In addition, procurement practitioners require more clear policy guidance on sustainable procurement in the context of EU rules, including the way in which sustainable procurement can support local economic development issues. There is also a need for increased collaboration between public sector organizations, including the involvement of key suppliers to the public sector, facilitated by regional sharing of best practice on sustainable procurement (Preuss and Walker, 2011).

Krause et al. (2009) addressed sustainable supply chain management and the relationship between sustainable development and purchasing. They highlighted the role of purchasing in the sustainability efforts of companies and particularly noted that a company is no more sustainable than its supply chain. In addition, they concluded that many sustainability efforts contribute to long-term cost reductions, especially in cases that take into account life cycle considerations. Companies that emphasize sustainability as a competitive priority benefit from communicating and documenting their supply chain management efforts to both their customers and the markets. Supply chain management practitioners may need to be proactive in their communication about the measures to advance more sustainable supply chains because it may not be easy to identify the components of sustainability (Krause et al., 2009).

Saidani et al. (2017) studied the assessment of product circularity performance and noted that it is important to address tools and methods to evaluate and eventually enhance product performance in the light of circular economy. They noted that product circularity performance indicators could contribute to the enhancement of product design and development phases, circularity of products, and associated value chains, including the identification of hotspots and areas for improvement and development of circular targets for products.

Previous research on regional and sectoral company perspective in Finland suggests that public procurement can advance CE through acknowledgement of the whole product chain and life cycle, legislation, guidance, obligatory recycling of products, components and materials, and replacement of plastics with renewable raw materials in all possible procurement (Husgafvel et al., 2021). Geng et al. (2012) noted that environmental, economic, and social benefits can be gained through the implementation of CE indicators and that they can provide objective and credible information on the status of CE implementation. They also noticed that CE indicators are important metrics for both policy and decision-making and that they can support the achievement of CE goals. In addition, they noted the importance of a comprehensive set of sustainability indicators.

Public procurement actions can influence market development, including company approaches. It is noteworthy that previous research shows that companies are interested in investing in CE to a gain better grasp of sustainability (economic, social and environmental) and life cycle thinking, significant improvement of resource efficiency, better management of the whole supply/product chain, and to use renewable energy and raw materials (Husgafvel et al., 2021). Interestingly, previous research on process industry suggests that the issues that need further sustainability management attention encompass CE training, specific sustainability management and reporting approaches, need for better performance in supply chain sustainability, monitoring of greenhouse gas emissions, and life cycle thinking (Husgafvel et al., 2017).

In the city of Helsinki, the viewpoints of sustainable public procurement encompass environmental (e.g., emissions, energy and material efficiency, waste sorting and reduction, and chemicals), social and ethical (fair and working life practices, governance and consumer aspects, human rights, participation in local community activities, social involvement, equal opportunities, design for all and promotion of employment opportunities, humane work, and compliance with social and labor laws), and innovativeness aspects and procurement decisions should always include focus on the environmental impacts and costs of the procured products, services, or systems during their whole life cycles (Helsingin Kaupunki, 2015).

CE-supportive governance measures include, e.g., taxation and guidance, business measures include, e.g., new business models, profitable products and services (innovated by forerunners), and use of renewable raw materials, and technological measures include, e.g., the capability to handle all materials in recycling and product design (Husgafvel et al., 2021). Companies consider that the advancement of CE requires support from public steering measures including the roles of energy and fuel policy, financing, investments and innovation, the local operational environment, and communication and education. Moreover, focus is needed, e.g., on development of both international and sectoral guidelines and best practices, life cycle thinking and assessment, recycling/reuse, whole supply chains, material and energy efficiency, waste minimization, and utilization of by-products and side flows (Husgafvel et al., 2018b).

Informed decision-making and procurement procedures can be supported by tools such as sustainability indexes and metrics (Husgafvel et al., 2017). Waas et al. (2014) noted that sustainable development must be considered as a decision-making strategy by all stakeholders (from local to global level) and sustainability assessment and indicators can be very important decision-supporting tools to guide development toward a more sustainable society. Previous research suggests that cross-border CE can be advanced, for example, through company cooperation, agreements between neighboring countries, and reassessment of the whole supply chain (Husgafvel et al., 2021). Hák et al. (2018) concluded that sustainability indicators (to provide clear information and to assess and communicate key sustainability trends) and social norms (reflect sustainability principles and goals) are among the key missing enabling conditions to promote sustainability in the context of decision-making. Dahl (2012) noted that the transition toward sustainability needs to be supported by a new set of values-based indicators and that indicators can be important tools to identify and enable the management of environment- and society-related dimensions. Life cycle thinking in companies can be advanced through, for example, the assessment of full product/service life cycles and whole supply chain assessments jointly with other actors. The development of cooperation between product manufacturers and service producers is also important (Husgafvel et al., 2018b). A set of reliable CE indicators plays a major role in the measurement of progress toward CE (EC, 2015). It is important to assess the environmental and social impacts of new products, services, or business models (Moreno et al., 2016).

For example, a recycling system of components of automobiles has been developed in Japan (ERCC, 2015; Fujii, 2017), in addition to systems for plastics and electronic devices, and Onoda (2019) noted that the current main barriers are the cost efficiency as well as processing of various materials used in each component. Regarding plastics, Saito et al. (2018) noted that the main goal is to minimize the environmental reduction and social cost, thus contributing to a sustainable material life cycle. Akiba (2019) noticed that large-scale renewable energy systems can lead to the establishment of added-value systems with circular supply chain management. Utilization of renewable energy could significantly contribute to the reduction of CO_2 as well as environmental load (Shimizu et al., 2018). The establishment of platforms (e.g., Eco Value Interchange, EVI) and mobility sharing can promote linking industry, enterprise, and users through environmental approaches and platforms can contribute to user awareness about CE and environmental protection (Kato, 2019).

The EU encourages all stakeholders to take part in measures to promote the transition toward a CE including associated sharing of best practices, systemic change aspects, public-private partnerships, specific measures such as extended producer responsibility, best practices on the cascading use of biomass, recycling of construction and demolition waste (including, e.g., the development of indicators to assess environmental performance of buildings, taking into account full lifecycle considerations), and voluntary business approaches taking into account the important role of SMEs (EC, 2015).

In the preparation stage of procurement, it is important to assess the actual needs taking into account, e.g., repair/renting as an option for owning, and comparison of energy use, waste generation, and recyclability and negative impacts of materials. The issues that can be considered in addition to price encompass, e.g., quality, technical merits and support, environmental, social and ethical aspects, life cycle costs, and maintenance services (Helsingin Kaupunki, 2015).

6. Conclusions

The findings of this study suggest that procurement organizations take sustainability and circular economy considerations into account in procurement planning and implementation already in many ways, but there is lack of focus on, for example, guidance and training. The application of both international/EU and national guidebooks could be enhanced and the use of environmental labels, management systems, standards, reporting systems, or certificates could be significantly increased in procurement. The local role of both EU and national law was highlighted and multiple EU-level approaches were considered to be potentially beneficial for local development. Renewable energy subsidies, taxation of waste incineration, and waste payments, including encouragement of sorting and recycling, and pricing and taxation of greenhouse gas emissions were considered to be important. It was also recognized that the viewpoint of sustainable and circular economy-oriented public procurement should be a crosscutting theme in municipal procurement guidebooks and training.

The findings suggest that management commitment, sufficient resources, strategy, goals, staff training, responsibilities, internal and sectoral cooperation, interaction and communication, and monitoring, as well as market cooperation, engagement of companies, preprocurement consultation, innovation partnerships, new business models, and development of local networks and ecosystems including the whole supply chain are important focus areas. Procurement planning and subject definition should particularly address the identification of, and goals for, sustainability and circular economy issues including procurement notices and design competitions and, for example, sustainability and circular economy criteria. The eligibility and feasibility criteria for bidders should particularly ensure that the bidder has previously taken care of sustainability and circular economy obligations covering, for example, recycling, wastes, emissions, violations, taxes, social security payments, and pension insurance payments.

The procurement phase and the benchmarking of bids should include consideration of life cycle costs and technical reports and/or testing reports focusing on sustainability and circular economy. Sustainability and circular economy aspects, goals and impacts (economic, environmental and social aspects), and responsibilities for taking care of products and materials at the end of the contract/their use life should be taken into account in the contract terms and follow-up and in reporting. Life cycle costs, management and impacts of the supply chain, suppliers and subsuppliers, and management of the operational environment and risks were among the most suitable indicators for the assessment of economic sustainability as a part of the procurement process and its different phases. The main indicators for the assessment of environmental sustainability include waste quantity, reduction, recycling and utilization, including municipal, industrial, and construction waste, leadership, strategy, management, reporting and code of conduct, energy consumption, and the share of renewable energy covering the whole life cycle, use of raw materials, and the share of renewable raw materials covering the whole life cycle, and transport and traffic emissions and impacts.

The main indicators for the assessment of social sustainability encompass leadership, strategy, management, reporting and code of conduct, promotion of employment opportunities including young and unemployed people, labor practices, conditions and social, and work rights such as, for example, standards, salary, vacations, rights, supervision, collective bargaining, freedom of assembly, and work time, and collaboration with and engagement of the local community. The most suitable indicators for the assessment of circular economy include life cycle thinking, use/working life, and durability/sustainability, waste quantity, reduction, recycling, and utilization, including municipal, industrial, and construction waste, raw materials, material efficiency, and material wastage/loss, including use of renewable raw materials and reused materials, recyclability, repairability, refurbishment, remanufacturing, and reuse, packaging material and practices, and recycling/reuse of packages.

Construction and buildings, waste management, and recycling and energy were the main areas/fields that offer the best opportunities to promote sustainable and circular economy-oriented procurement in the future. Interestingly, training/education and communication services were not considered important in this context. The reduction of food waste, wind and solar power, public transport, LED lightning, updating of old buildings and spaces, and remote access were the main measures to combat climate change that could be future elements of sustainable and circular economy-oriented public procurement. The most important next steps for the development of sustainable and circular economy-oriented public procurement include focus on recycling, reuse, refurbishment, and remanufacturing of products, components, and materials, incorporation of sustainability and circular economy issues into local policy, strategy, and goals, focus on the life cycle, use/working life and sustainability of products, components, and materials, and national recyclability requirement for packages and plastics to promote their reuse as raw materials or products.

In sum, the results suggest that sustainable and circular economy-oriented public procurement could be promoted at many levels and through multiple individual and joint measures with active involvement of multiple partners and stakeholders. Both public and private sector efforts are important, including focus on the application of governance, business, and technology oriented solutions. Informed decision-making and management can be supported by the development and application of appropriate sustainability and circular economy indicators. Similarly, sustainability and circular economy criteria, standards, guidance, and labels can be developed to support sustainable and circular economy-oriented public procurement. For example, the viewpoint of sustainable and circular economy-oriented public procurement could be a crosscutting theme in municipal procurement guidebooks and training. Future research could focus on, for example, local policy on and management of sustainable and circular economy-oriented public procurement and on associated management and procurement guidance, training, and tools. In addition, interesting focus areas encompass the broader international, EU, and national frameworks, and associated instruments and tools, as well as local networks/ecosystems. In general, research should cover governance, business, and technological aspects and the combination of public and private sector measures with involvement of all partners and stakeholders.

Acknowledgments

This study was supported by the Kymenlaakso Regional Fund of the Finnish Cultural Foundation.

References

Abraham, G. (Ed.), 2006. Sustainability Science and Engineering: Defining Principles. Elsevier, London.

Act on Public Procurement and Concession Contracts, 2016. 1397/2016. https://www.finlex.fi/fi/laki/kaannokset/2016/en20161397.pdf (Accessed 11 January 2020).

Akiba, T., 2019. Large-scale hydrogen energy system using renewable energy (in Japanese). J. Jpn. Weld. Soc. 88 (1), 50–55. https://doi.org/10.2207/jjws.88.50.

Alhola, K., Salmenperä, H., Ryding, S.-O., Busch, N.J., 2017. Circular public procurement in the Nordic countries. Nordic Council of Ministers. TemaNord 2017, 512. http://norden.diva-portal.org/smash/get/diva2:1092366/FULLTEXT01.pdf. (Accessed 11 January 2020).

Allen, D.T., Shonnard, D.R., 2012. Sustainable engineering. In: Concepts, Design, and Cases Studies. Upper Saddle River, NJ, Prentice Hall, p. 223.

Allenby, B.R., 2012. The Theory and Practice of Sustainable Engineering. Pearson Education Limited, England.

Bell, S., Morse, S., 2008. Sustainability Indicators: Measuring the Immeasurable? Earthscan, Routledge.

Bond, A., Morrison-Saunders, A., Pope, J., 2012. Sustainability assessment: the state of the art. Impact Assess. Project Appraisal 30 (1), 53–62. https://doi.org/10.1080/14615517.2012.661974.

Boons, F., Lüdeke-Freund, F., 2013. Business models for sustainable innovation: state-of-the-art and steps towards a research agenda. J. Clean. Prod. 45, 9–19. https://doi.org/10.1016/j.jclepro.2012.07.007.

Brammer, S., Walker, H., 2011. Sustainable procurement in the public sector: an international comparative study. Int. J. Oper. Prod. Manag. 31 (4), 452–476. https://doi.org/10.1108/01443571111119551.

Bratt, C., Hallstedt, S., Robèrt, K.H., Broman, G., Oldmark, J., 2013. Assessment of criteria development for public procurement from a strategic sustainability perspective. J. Clean. Prod. 52, 309–316. https://doi.org/10.1016/j.jclepro.2013.02.007.

Circle Economy, 2019a. The role of municipal policy in the circular economy. Investment, jobs and social capital in circular cities. https://assets.website-files.com/5d26d80e8836af7216ed124d/5d26d80e8836af7603ed12af_Circle%20Economy%20-%20The%20role%20of%20municipal%20policy%20in%20the%20circular%20economy.pdf. (Accessed 19 January 2020).

Circle Economy, 2019b. Building block for the new strategy Amsterdam Circular 2020–2025. Directions for a thriving city within the planetary boundaries. https://assets.website-files.com/5d26d80e8836af2d12ed1269/5de954d913854755653be926_Building-blocks-Amsterdam-Circular-2019.pdf. (Accessed 19 January 2020).

Circwaste, 2019. Kiertotalouden palvelukeskus—työkaluja ja tukea. http://www.materiaalitkiertoon.fi/fi-FI/Circwaste/Kiertotalouden_palvelukeskus/Kiertotalouden_palvelukeskus__tyokaluja_(48133. (Accessed 19 January 2020).

Coggburn, J.D., 2004. Achieving managerial values through green procurement? Public Perform. Manag. Rev. 28 (2), 236–258. https://doi.org/10.1080/15309576.2004.11051834.

Cohen, B., Winn, M.I., 2007. Market imperfections, opportunity and sustainable entrepreneurship. J. Bus. Ventur. 22, 29–49. https://doi.org/10.1016/j.jbusvent.2004.12.001.

Dahl, A.L., 2012. Achievements and gaps in indicators for sustainability. Ecol. Indic. 17, 14–19. https://doi.org/10.1016/j.ecolind.2011.04.032.

Dalal-Clayton, B., Sadler, B., 2011. Sustainability Appraisal: A Sourcebook and Reference Guide to International Experience. Earthscan, London.

De Vries, B.J.M., 2012. Sustainability Science. Cambridge University Press, Cambridge.

Department of Environment, Food and Rural Affairs, 2006. Procuring the future sustainable procurement national action plan: recommendations from the Sustainable procurement task force. United Kingdom. https://assets.publishing.service.gov.uk/government/uploads/system/uploads/attachment_data/file/69417/pb11710-procuring-the-future-060607.pdf (Accessed 19 January 2020).

Dong, Y., Hauschild, M.Z., 2017. Indicators for environmental sustainability. The 24th CIRP conference on life cycle engineering. Proc. CIRP 61, 697–702. https://doi.org/10.1016/j.procir.2016.11.173.
EC, 2007. Communication from the Commission to the European Parliament, the Council, the European Economic and Social Committee and the Committee of the Regions. Pre-commercial procurement: driving innovation to ensure sustainable high quality public services in Europe. COM(2007) final. https://eur-lex.europa.eu/LexUriServ/LexUriServ.do?uri=COM:2007:0799:FIN:EN:PDF. (Accessed 14 January 2020).
EC, 2008a. Communication from the Commission to the European Parliament, the Council, the European Economic and Social Committee and the Committee of the Regions Public procurement for a better environment {SEC(2008) 2124} {SEC(2008) 2125} {SEC(2008) 2126} /* COM/2008/0400 final. https://eur-lex.europa.eu/legal-content/EN/TXT/HTML/?uri=CELEX:52008DC0400&from=EN.
EC, 2008b. Directive 2008/98/EC on waste (Waste Framework Directive). https://eur-lex.europa.eu/legal-content/EN/TXT/?uri=celex%3A32008L0098. (Accessed 14 January 2020).
EC, 2009. Directive 2009/125/EC of the European Parliament and of the Council of 21 October 2009. Establishing a framework for the setting of ecodesign requirements for energy-related products. https://eur-lex.europa.eu/legal-content/EN/TXT/HTML/?uri=CELEX:32009L0125&from=FI. (Accessed 14 January 2020).
EC, 2010. Buying social. A guide to taking into account social considerations in public procurement. https://publications.europa.eu/en/publication-detail/-/publication/cb70c481-0e29-4040-9be2-c408cddf081f/language-en. (Accessed 12 January 2020).
EC, 2015. Communication from the Commission to the European Parliament, the Council, the European Economic and Social Committee and the Committee of the Regions. Closing the loop—An EU action plan for the CE. COM(2015)614 final. http://eur-lex.europa.eu/legal-content/EN/TXT/?uri=CELEX:52015DC0614. (Accessed 12 January 2020).
EC, 2017a. Report from the Commission to the European Parliament, the Council, the European Economic and Social Committee and the Committee of the Regions on the implementation of the CE Action Plan. Brussels, 26.1.2017 COM(2017) 33 final. https://op.europa.eu/en/publication-detail/-/publication/391fd22b-e3ae-11e6-ad7c-01aa75ed71a1/language-en/format-PDF. (Accessed 12 January 2020).
EC, 2017b. Public procurement for a circular economy. Good practice and guidance. European Commission. Environment. http://ec.europa.eu/environment/gpp/pdf/CP_European_Commission_Brochure_webversion_small.pdf. (Accessed 12 January 2020).
EC, 2018. Communication from the Commission to the European Parliament, the Council, The European Economic and Social Committee and the Committee of the Regions on a monitoring framework for the circular economy. COM(2018) final. https://eur-lex.europa.eu/legal-content/EN/TXT/PDF/?uri=CELEX:52018DC0029&from=EN (Accessed 12 January 2020).
EC, 2019a. Accelerating the transition to the circular economy. Improving access to finance for circular economy projects. https://ec.europa.eu/info/sites/info/files/research_and_innovation/knowledge_publications_tools_and_data/documents/accelerating_circular_economy_032019.pdf. (Accessed 10 January 2020).
EC, 2019b. The EU GPP criteria. http://ec.europa.eu/environment/gpp/eu_gpp_criteria_en.htm. (Accessed 12 January 2020).
EC, 2019c. Green and sustainable public procurement. http://ec.europa.eu/environment/gpp/versus_en.htm. (Accessed 12 January 2020).
Ellen MacArthur Foundation, 2013a. Towards circular economy. Economic and business rational for an accelerated transition. vol. 1. https://www.ellenmacarthurfoundation.org/assets/downloads/publications/Ellen-MacArthur-Foundation-Towards-the-Circular-Economy-vol.1.pdf. (Accessed 17 January 2020).
Ellen MacArthur Foundation, 2013b. Towards the circular economy. Opportunities for the consumer goods sector. Vol. 2. https://www.ellenmacarthurfoundation.org/assets/downloads/publications/TCE_Report-2013.pdf. (Accessed 17 January 2020).
Ellen MacArthur Foundation, 2014. Towards the circular economy. Accelerating the scale-up across global supply chains. vol. 3. https://www.ellenmacarthurfoundation.org/assets/downloads/publications/Towards-the-circular-economy-volume-3.pdf. (Accessed 17 January 2020).

Ellen MacArthur Foundation, 2015a. Towards a circular economy: business rational for an accelerated transition. https://www.ellenmacarthurfoundation.org/assets/downloads/TCE_Ellen-MacArthur-Foundation_9-Dec-2015.pdf. (Accessed 17 January 2020).

Ellen MacArthur Foundation, 2015b. Delivering the circular economy—a toolkit for policymakers. https://www.ellenmacarthurfoundation.org/assets/downloads/publications/EllenMacArthurFoundation_PolicymakerToolkit.pdf. (Accessed 17 January 2020).

Ellen MacArthur Foundation, 2015c. Growth within: a circular economy vision for a competitive Europe. https://www.ellenmacarthurfoundation.org/assets/downloads/publications/EllenMacArthurFoundation_Growth-Within_July15.pdf. (Accessed 17 January 2020).

Ellen MacArthur Foundation, 2017a. Cities in the circular economy: an initial exploration. https://www.ellenmacarthurfoundation.org/assets/downloads/publications/Cities-in-the-CE_An-Initial-Exploration.pdf. (Accessed 17 January 2020).

Ellen MacArthur Foundation, 2017b. Cities in the circular economy: the role of digital technology. Google. Ashima, Sukhdev, Julia Vol, Kate Brandt and Robin Yeoman. https://www.ellenmacarthurfoundation.org/assets/downloads/Cities-in-the-Circular-Economy-The-Role-of-Digital-Tech.pdf (Accessed 17 January 2020).

Ellen MacArthur Foundation, 2019. Circular economy in cities. In: Project guide. https://www.ellenmacarthurfoundation.org/assets/downloads/CE-in-Cities-Project-Guide_Mar19.pdf. (Accessed 20 January 2020).

EpE, 2019. Circular economy indicators for businesses. Entreprises pour l'Environnement (EpE). http://www.epe-asso.org/en/circular-economy-indicators-for-businesses-february-2019/. (Accessed 16 January 2020).

ERCC, 2015. Report of evaluation and review of the enforcement status of automobile recycling system (in Japanese), Government Environmental Special Report, 50(5), 57–92, https://www.env.go.jp/council/03recycle/y033-43/mat03_2.pdf (Accessed 16 January 2020).

EU Ecolabel, 2019. http://ec.europa.eu/environment/ecolabel/the-ecolabel-scheme.html. (Accessed 16 January 2020).

EU, 2016a. Directive 2014/24/EU of the European Parliament and of the Council of 26 February 2014 on public procurement and repealing Directive 2004/18/EC. https://eur-lex.europa.eu/legal-content/EN/TXT/HTML/?uri=CELEX:32014L0024&from=FI. (Accessed 20 January 2020).

EU, 2016b. Buying Green! A Handbook on Green Public Procurement, third ed. http://ec.europa.eu/environment/gpp/pdf/Buying-Green-Handbook-3rd-Edition.pdf.

EU, 2019. EU GPP criteria. http://ec.europa.eu/environment/gpp/eu_gpp_criteria_en.htm. (Accessed 16 January 2020).

Fink, A., 2009. How to Conduct Surveys: A Step-by-Step Guide, fourth ed. Sage, Los Angeles, p. 125.

Fujii, K., 2017. Report of current status of automobile recycling (in Japanese). In: ARC Report (RS-1021), pp. 1–20. https://www.asahi-kasei.co.jp/arc/service/pdf/1021.pdf. (Accessed 21 January 2020).

Geng, Y., Fu, J., Sarkis, J., Xue, B., 2012. Towards a national circular economy indicator system in China: an evaluation and critical analysis. Government initiatives. J. Clean. Prod. 23, 216–224. https://doi.org/10.1016/j.jclepro.2011.07.005.

Genovese, A., Acquaye, A.A., Figueroa, A., koh, L., 2017. Sustainable supply chain management and the transition towards a CE. Omega 66, 244–257. https://doi.org/10.1016/j.omega.2015.05.015.

Gerholdt, J. 2017. Supporting growth of circular SMEs. Senior Director, CE and Sustainability Programs, U.S. Chamber of Commerce Foundation. World CE Forum, June 5th, Helsinki, Finland. https://www.slideshare.net/WorldCircularEconomyForum/jennifer-gerholdt (Accessed 28 January 2020).

Ghiselli, P., Cialani, C., Ulgiati, S., 2016. A review on CE: the expected transition to a balanced interplay of environmental and economic systems. J. Clean. Prod. 114, 11–32.

Gillham, B., 2007. Developing a Questionnaire, second ed. Continuum International Publishing Group, London, p. 112.

Graedel, T.E., Allenby, B.R., 2010. Industrial Ecology and Sustainable Engineering, International Edition. Pearson Education Inc., Upper Saddle River, Prentice Hall.

Green Budget Europe, The Ex'tax Project, Institute for European Environmental Policy, Cambridge Econometrics, 2018. Aligning fiscal policy with the circular economy roadmap in Finland. https://ex-tax.com/wp-content/uploads/2019/09/Aligning_Fiscal_Policy_with_the_Circular_Economy_Roadmap_in_Finland_Extax_Sitra_GBE_IEEP_Final_report_final-08-01-19.pdf. (Accessed 20 January 2020).

GRI, 2016. GRI 204: procurement practices. https://www.globalreporting.org/standards/gri-standards-download-center/?g=55ce6876-3838-4cc1-8400-980bba82ea6d. (Accessed 27 January 2020).

GRI, 2019. The GRI standards download center. https://www.globalreporting.org/standards/gri-standards-download-center/?g=55ce6876-3838-4cc1-8400-980bba82ea6d. (Accessed 27 January 2020).

Guo, B., Geng, Y., Ren, J., Zhu, L., Liu, Y., Sterr, T., 2017. Comparative assessment of circular economy development in China's four megacities: the case of Beijing, Chongqing, Shanghai and Urumqi. J. Clean. Prod. 162, 234–246. https://doi.org/10.1016/j.jclepro.2017.06.061.

Guohui, S., Yunfeng, L., 2012. The effect of reinforcing the concept of CE in West China environmental protection and economic development. Procedia Environ. Sci. 12, 785–792. https://doi.org/10.1016/j.proenv.2012.01.349.

Hak, T., Moldan, B., Dahl, A.L. (Eds.), 2007. Sustainability Indicators: A Scientific Assessment (Scientific Committee on Problems of the Environment (SCOPE) Series). Island Press, Washington, DC.

Hák, T., Janouškováa, S., Moldana, B., Dahl, A.L., 2018. Closing the sustainability gap 30 years after "Our Common Future", society lacks meaningful stories and relevant indicators to make the right decisions and build public support. Ecol. Indic. 87, 193–195. https://doi.org/10.1016/j.ecolind.2017.12.017.

Hall, J.K., Daneke, G.A., Lenox, M.J., 2010. Sustainable development and entrepreneurship: past contributions and future directions. J. Bus. Ventur. 25 (2010), 439–448. https://doi.org/10.1016/j.jbusvent.2010.01.002.

Hartley, K., van Santen, R., Kirchherr, J., 2020. Policies for transitioning towards a circular economy: expectations from the European Union (EU). Resour. Conserv. Recycl. 155, 104634. https://doi.org/10.1016/j.resconrec.2019.104634.

Hawken, P. (Ed.), 2017. Drawdown: The Most Comprehensive Plan Ever Proposed to Reverse Global Warming. Penguin Books, New York, p. 240.

Helsingin Kaupunki, 2015. Helsingin kaupungin kestävien hankintojen opas. http://www.hel.fi/wps/wcm/connect/48db87b4-f27c-46bc-982a-2afe995f8875/Helsingin_kaupungin_kestavien_hankintojen_opas%E2%80%932015.pdf?MOD=AJPERES&useDefaultText=0&useDefaultDesc=0 (Accessed 13 January 2020).

Heshmati, A., 2015. A review of the CE and its implementation. In: Discussion Paper No. 9611. Institute for the Study of Labor.

Hiilineutraalisuomi, 2019a. Kouvolasta Kymenlaakson toinen Hinku-kunta. https://www.hiilineutraalisuomi.fi/fi-FI/Ajankohtaista/Kouvolasta_Kymenlaakson_toinen_Hinkukunt(49686). (Accessed 23 January 2020).

Hiilineutraalisuomi, 2019b. Voisiko kuntani lähteä mukaan Hinkuun – case Hamina. https://www.hiilineutraalisuomi.fi/fi-FI/Hinku/Mukaan_Hinkuun/Voisiko_kuntani_lahtea_mukaan_Hinkuun__C(49477). (Accessed 23 January 2020).

Hiilineutraalisuomi, 2021. Hinku network - Towards Carbon Neutral Municipalities. https://www.hiilineutraalisuomi.fi/en-US/Hinku. (Accessed 23 January 2020).

Hirsjärvi, H., Remes, P., Sajavaara, P., 2007. Tutki ja kirjoita. 13. painos. Tammi, Helsinki, p. 448.

Hirsjärvi, S., Remes, P., Sajavaara, P., 2015. Tutki ja kirjoita. 20. Painos. Tammi, Helsinki.

Husgafvel, R., Poikela, K., Honkatukia, J., Dahl, O., 2017. Development and piloting of sustainability assessment metrics for arctic process industry in Finland—the biorefinery investment and slag processing service cases. Sustainability 9 (10), 1693. https://doi.org/10.3390/su9101693.

Husgafvel, R., Linkosalmi, L., Hughes, M., Kanerva, J., Dahl, O., 2018a. Forest sector circular economy development in Finland: a regional study on sustainability driven competitive advantage and an assessment of the potential for cascading recovered solid wood. J. Clean. Prod 181, 483–497. https://doi.org/10.1016/j.jclepro.2017.12.176.

Husgafvel, R., Linkosalmi, L., Dahl, O., 2018b. Company perspectives on the development of the circular economy in the seafaring sector and the Kainuu region in Finland. J. Clean. Prod. 186, 673–681. https://doi.org/10.1016/j.jclepro.2018.03.138.
Husgafvel, R., Linkosalmi, L., Hughes, M., Dahl, O., 2021. Company perspectives on sustainable circular economy development in the South Karelia and Kymenlaakso regions and the publishing sector in Finland (submitted).
Joutsenmerkki, 2017. Opas ympäristömerkin käyttämiseen julkisissa hankinnoissa. https://joutsenmerkki.fi/wp-content/uploads/2017/02/Julkiset_hankinnat__opas_huhtikuu_2017.pdf. (Accessed 14 January 2020).
Kates, R.W., Clark, W.C., Corell, R., Hall, M.J., Jaeger, C.C., Lowe, I., McCarthy, J.J., et al., 2001. "Environment and development: sustainability science." Policy Forum. Science 292 (5517), 641–642. https://doi.org/10.1126/science.1059386.
Kato, H., 2019. Contribution of eco policy CE and decarbonated society (in Japanese). Environ. Meet. 51, 50–55.
KEINO, 2019. Competence Centre for Sustainable and Innovative Public Procurement. https://www.hankintakeino.fi/en. (Accessed 23 January 2020).
Kestävä kaupunki, 2019. http://www.kestavakaupunki.fi/fi-FI. (Accessed 23 January 2020).
Khosla, A. 2017. Meeting the SDGs needs a circular and inclusive economy. SMEs are the key. World CE Forum, June 5th, Helsinki, Finland. Https://www.slideshare.net/worldcirculareconomyforum/ashok-khosla-world-circular-economy-forum-2017-helsinki-finland (Accessed 28 January 2020).
Kirchherr, J., Reike, D., Hekkert, M., 2017. Conceptualizing the CE: an analysis of 114 definitions. Resour. Conserv. Recycl. 127, 221–232. https://doi.org/10.1016/j.resconrec.2017.09.005.
Kirchherra, J., Piscicellia, L., Boura, R., Kostense-Smitb, E., Mullerb, J., Huibrechtse-Truijensb, A., Hekkerta, M., 2018. Barriers to the CE: evidence from the European Union (EU). Ecol. Econ. 150 (2018), 264–272.
Komiyama, H., Takeuchi, K., 2006. "Sustainability science: building a new discipline." Editorial. Sustain. Sci. 1, 1–6. https://doi.org/10.1007/s11625-006-0007-4.
Komiyama, H., Takeuchi, K., 2011. Sustainability science: building a new academic discipline. In: Komiyama, H., Takeuchi, K., Shiroyama, H., Mino, T. (Eds.), Sustainability Science: A Multidisciplinary Approach. United Nations University Press, Tokyo, pp. 2–20.
Krause, D., Vachon, S., Klassen, R., 2009. Special topic forum on sustainable supply chain management: introduction and reflections on the role of purchasing management. J. Supply Chain Manag. 45 (4), 18–25.
Lakatos, E.S., Dan, V., Cioca, L.I., Bacali, L., Ciobanu, A.M., 2016. How supportive are Romanian consumers of the CE concept: a survey. Sustainability 2016 (8), 789. https://doi.org/10.3390/su8080789.
Lieder, M., Rashid, A., 2016. Towards CE implementation: a comprehensive review in context of manufacturing industry. J. Clean. Prod. 115, 36–51.
Liu, Y., Bai, Y., 2014. An exploration of firms' awareness and behavior of developing CE: an empirical research in China. Resour. Conserv. Recycl. 87, 145–152. https://doi.org/10.1016/j.resconrec.2014.04.002.
Loorbach, D., 2019. Sustainability transitions research. Science in transition. In: Presentation in the Sustainability Science Days 9–10 May, Aalto University. Drift for Transition. Dutch Research Institute for Transitions. https://aalto.cloud.panopto.eu/Panopto/Pages/Viewer.aspx?id=f15d76a3-bf26-41f8-b50d-aa48009d1699. (Accessed 20 January 2020).
Lozano, R., 2012. Towards better embedding sustainability into companies' systems: an analysis of voluntary corporate initiatives. J. Clean. Prod. 25, 14–26. https://doi.org/10.1016/j.jclepro.2011.11.060.
Lozano, R., 2013. Are companies planning their organisational changes for corporate sustainability? An analysis of three case studies on resistance to change and their strategies to overcome it. Corp. Soc. Responsib. Environ. Manag. 20, 275–295. https://doi.org/10.1002/csr.1290.
Luoma, P., Larvus, L., Hjelt, M., Päällysaho, M., Aho, M., 2015. Miten kiertotalouden kehitystä mitataan? Esiselvitys kiertotalouden kansallisen barometrin kehittämisestä. Gaia Consulting Oy. https://media.sitra.fi/2017/02/27174938/Miten_kiertotalouden_kehitysta_mitataan-2.pdf. (Accessed 15 January 2020).

Maignan, I., Hillebrand, B., McAlister, D., 2002. Managing socially-responsible buying: how to integrate non-economic criteria into the purchasing process. Eur. Manag. J. 20 (6), 641–648. https://doi.org/10.1016/S0263-2373(02)00115-9.

Materiaali- ja energialoikka, 2019. http://www.energialoikka.fi/. (Accessed 15 January 2020).

Materiaalitori, 2019. https://www.materiaalitori.fi/. (Accessed 15 January 2020).

Meehan, J., Bryde, D., 2011. Sustainable procurement practice. Bus. Strategy Environ. 20 (2), 94–106. https://doi.org/10.1002/bse.678.

Ministry of the Environment, 2021. Government Resolution on the Strategic Programme for Circular Economy. https://ym.fi/documents/1410903/42733297/Government+resolution+on+the+Strategic+Programme+for+Circular+Economy+8.4.2021.pdf/309aa929-a36f-d565-99f8-fa565050e22e/Government+resolution+on+the+Strategic+Programme+for+Circular+Economy+8.4.2021.pdf?t=1619432219261 (Accessed 3 May 2021).

Moreno, M., De los Rios, C., Rowe, Z., Charnley, F., 2016. A conceptual framework for circular design. Sustainability 2016 (8), 937. https://doi.org/10.3390/su8090937.

Motiva, 2019a. Kestävät julkiset hankinnat. Hyvän hankinnan abc. https://www.motiva.fi/julkinen_sektori/kestavat_julkiset_hankinnat/hyvan_hankinnan_abc. (Accessed 15 January 2020).

Motiva, 2019b. Kestävät julkiset hankinnat. Tietopankki. https://www.motiva.fi/julkinen_sektori/kestavat_julkiset_hankinnat/tietopankki. (Accessed 18 January 2020).

Motiva, 2019c. Kestävät julkiset hankinnat. Sosiaalinen vastuu julkisissa hankinnoissa. https://www.motiva.fi/julkinen_sektori/kestavat_julkiset_hankinnat/tietopankki/sosiaalinen_vastuu_julkisissa_hankinnoissa. (Accessed 18 January 2020).

Mylan, J., Holmes, H., Paddock, J., 2016. Re-introducing consumption to the 'CE': a sociotechnical analysis of domestic food provisioning. Sustainability 2016 (8), 794. https://doi.org/10.3390/su8080794.

New, S., Green, B., Morton, B., 2002. An analysis of private versus public sector responses to environmental challenges of the supply chain. J. Public Procurement 2 (1), 93–105. https://doi.org/10.1108/JOPP-02-01-2002-B004.

Noronha, L., 2017. Momentum for change—perspectives from the UN environment. Director, UN Environment Program. World CE Forum, June 5th, Helsinki, Finland. https://media.sitra.fi/2017/02/08113108/momentum_of_change_UNEP_LN.pdf (Accessed 19 January 2020).

Onoda, H., 2019. Current status and issues of efforts to improve the quality of automobile recycling system (in Japanese). Enermix 98 (1), 78–82. https://doi.org/10.20550/jieenermix.98.2_9802anno_1.

Oppenheim, A.N., 1997. Questionnaire Design, Interviewing and Attitude Measurement. Pinter, London and Washington, DC, p. 303.

Ormazabal, M., Prieto-Sandoval, V., Puga-Leal, R., Jaca, C., 2018. Circular economy in Spanish SMEs: challenges and opportunities. J. Clean. Prod. 185, 157–167. https://doi.org/10.1016/j.jclepro.2018.03.031.

Patten, M.L., 2011. Questionnaire Research: A Practical Guide. Pyrczak Pub, Glendale, CA.

Peterson, R.A., 2000. Constructing Effective Questionnaires. SAGE Publications, Inc. https://www-doi-org.libproxy.aalto.fi/10.4135/9781483349022.

Petit-Boix, A., Leipold, S., 2018. Circular economy in cities: reviewing how environmental research aligns with local practices. J. Clean. Prod. 195, 1270–1281. https://doi.org/10.1016/j.jclepro.2018.05.281.

Pollice, F., 2018. The new role of procurement in a circular economy system. In: 22nd Cambridge International Manufacturing Symposium., https://doi.org/10.17863/CAM.31713.

Potočnik, J., 2017. Global use of natural resources—in crisis or not? Co-chair, UNEP International Resources Panel. World CE Forum, June 5th, Helsinki, Finland. https://www.slideshare.net/WorldCircularEconomyForum/janez-potonik-world-circular-economy-forum-2017-helsinki-finland (Accessed 18 January 2020).

Pöyhönen, T., 2017. Kiertotaloutta edistävät julkiset hankinnat—case: Kouvolan kaupunki. Ammattikorkeakoulun opinnäytetyö. Kestävä kehitys. Hämeen ammattikorkeakoulu (HAMK). 88 pp.

Preuss, L., 2007. Buying into our future: the range of sustainability initiatives in local government procurement. Bus. Strategy Environ. 16 (5), 354–365. https://doi.org/10.1002/bse.578.

Preuss, L., 2009. Addressing sustainable development through public procurement: the case of local government. Supply Chain Manag. 14 (3), 213–223. https://doi.org/10.1108/13598540910954557.

Preuss, L., Walker, H., 2011. Psychological barriers in the road to sustainable development: evidence from public sector procurement. Public Adm. 89 (2), 493–521. https://doi.org/10.1111/j.1467-9299.2010.01893.x.

Rosen, M.A., 2012. Engineering sustainability: a technical approach to sustainability. Sustainability 4 (9), 2270–2292.

Saidani, M., Yannou, B., Leroy, Y., Cluzel, F., 2017. How to assess product performance in the circular economy? Proposed requirements for the design of a circularity measurement framework. Recycling 2 (1), 6. MDPI https://doi.org/10.3390/recycling2010006.

Saito, Y., Kumagai, S., Kameda, T., Yoshioka, T., 2018. Current issues and future prospects in plastic recycling (in Japanese). Mater. Cycl. Waste Manag. Res. 29 (2), 152–162. https://doi.org/10.3985/mcwmr.29.152.

Saris, W.E., Gallhofer, E.N., 2014. Design, Evaluation and Analysis of Questionnaires for Survey Research, second ed. John Wiley & Sons.

Sauvé, S., Bernard, S., Sloan, P., 2016. Environmental sciences, sustainable development and CE: alternative concepts for trans-disciplinary research. Environ. Dev. 17, 48–56. https://doi.org/10.1016/j.envdev.2015.09.002.

Seppälä, J., Kurppa, S., Savolainen, H., Antikainen, R., Lyytimäki, J., Koskela, S., Hokkanen, J., Känkänen, R., Kolttola, L., Hippinen, I., 2016. Vihreän kasvun sekä materiaali- ja resurssitehokkuuden avainindikaattorit. Valtioneuvoston selvitys- ja tutkimustoiminnan julkaisusarja 23/2016. https://tietokayttoon.fi/documents/10616/2009122/23_Avainindikaattorit.pdf/9cb50a04-7e40-4405-b065-4886692ba6b2?version=1.0 (Accessed 17 January 2020).

Sheate, W.R. (Ed.), 2010. Tools, Techniques and Approaches for Sustainability: Collected Writings in Environmental Assessment Policy and Management. World Scientific Publishing Co. Ptv. Ltd., Singapore.

Shimizu, K., Hondo, H., Moriizumi, Y., 2018. CO2 reduction potential of solar water heating systems on a municipal basis (in Japanese). J. Jpn. Inst. Energy 97 (6), 147–159. https://doi.org/10.3775/jie.97.147.

Sitra, 2015. The opportunities of a circular economy for Finland. Sitra Stud. 100, 72. https://media.sitra.fi/2017/02/28142449/Selvityksia100.pdf. (Accessed 17 January 2020).

Sitra, 2016. Leading the cycle. Finnish road map to a CE 2016–2025. Sitra Stud. 121, 56. http://www.sitra.fi/sites/default/files/sitra_leading_the_cycle_report.pdf. (Accessed 15 January 2020).

Sitra, 2019a. The critical move. In: Finland's Road Map to the Circular Economy 2.0. https://www.sitra.fi/en/projects/critical-move-finnish-road-map-circular-economy-2-0/#challenge. (Accessed 2 May 2019).

Sitra, 2019b. Central government works across silos in the circular economy. https://www.sitra.fi/en/articles/central-government-works-across-silos-in-circular-economy/. (Accessed 2 May 2019).

Sitra, 2019c. WCEF2019 in brief. https://media.sitra.fi/2019/07/05150211/wcef2019summaryfinalfixed.pdf. (Accessed 18 January 2020).

Sitra, 2019d. Municipalities enable the important moves in the circular economy. https://www.sitra.fi/en/articles/municipalities-enable-important-moves-circular-economy/. (Accessed 22 January 2020).

Sitra, 2019e. New business models play a key role in enterprises's strategies. https://www.sitra.fi/en/articles/new-business-models-play-key-role-enterprises-strategies/. (Accessed 22 January 2020).

Sitra, 2019f. How are we progressing? https://www.sitra.fi/en/articles/how-are-we-progressing/. (Accessed 22 January 2020).

Snider, K.F., Halpern, B.H., Rendona, R.G., Kidalova, M.V., 2013. Corporate social responsibility and public procurement: how supplying government affects managerial orientations. J. Purch. Supply Manag. 19 (2), 63–72. https://doi.org/10.1016/j.pursup.2013.01.001.

Srivastava, S.K., 2007. Green supply-chain management: a state-of-the-art literature review. Int. J. Manag. Rev. 9 (1), 53–80. https://doi.org/10.1111/j.1468-2370.2007.00202.x.

Steiner, A. 2017. CE as means to achieve the UN Sustainable Development Goals and reduce poverty. Director, Oxford Martin School. World CE Forum, June 5th, Helsinki, Finland. https://www.slideshare.net/WorldCircularEconomyForum/achim-steiner (Accessed 19 January 2020).

Steurer, R., Berger, G., Konrad, A., Martinuzzi, A., 2007. Sustainable public procurement in EU member states: overview of government initiatives and selected cases. In: Final Report to the EU High-Level Group on CSR. European Commission, Brussels.

Sudman, S., Bradbrun, N.M., 1982. Asking Questions: A Practical Guide to Questionnaire Design. The Jossey-Bass Series in Social and Behavioural Sciences. Jossey-Bass Publishers. 397 p.

Swanson, M., Weissman, A., Davis, G., Socolof, M.L., Davis, K., 2005. Developing priorities for greener state government purchasing: a California case study. J. Clean. Prod. 13 (7), 669–677. https://doi.org/10.1016/j.jclepro.2003.12.011.

Tamminen, S., Honkatukia, J., Leinonen, T. and Haanperä, O., 2019. Kestävän kehityksen verouudistus. Kohti päästötöntä Suomea. Sitra muistio. https://media.sitra.fi/2019/04/11153727/kestavan-kehityksen-verouudistus.pdf (Accessed 20.May 2019).

TEM, 2016. Energiatehokkuus julkisissa hankinnoissa. Työ- ja elinkeinoministeriön ohjeet. Työ- ja elinkeinoministeriö. https://tem.fi/documents/1410877/2795834/Energiatehokkuus+julkisissa+hankinnoissa/1f3d1ad9-f7a9-4169-95a5-6a96414e9a29 (Accessed 15 January 2020).

TEM, 2017. Opas sosiaalisesti vastuullisiin julkisiin hankintoihin. TEM oppaat ja muut julkaiset 3/2017. http://julkaisut.valtioneuvosto.fi/bitstream/handle/10024/80010/3_2017_Opas_Sosiaalisesti_vastuulliset_hankinnat_31052017_WEB.pdf (Accessed 23 January 2020).

Thomson, J., Jackson, T., 2007. Sustainable procurement in practice: lessons from local government. J. Environ. Plan. Manag. 50 (3), 421–444. https://doi.org/10.1080/09640560701261695.

UN, 2019. Sustainable Development Goals. The United Nations. https://sustainabledevelopment.un.org/?menu=1300. (Accessed 18 January 2020).

UNEP, 2011. Buying for a better world. A guide on sustainable procurement for the UN system. In: The United Nations Environment Programme. https://www.ungm.org/Areas/Public/Downloads/BFABW_Final_web.pdf.

UNOPS, 2009. A Guide to Environmental Labels—for Procurement Practitioners of the United Nations System. The United Nations Office for Project Services. https://www.ungm.org/Areas/Public/Downloads/Env_Labels_Guide.pdf. (Accessed 16 January 2020).

UNU-IHDP and UNEP, 2014. Inclusive wealth report 2014. Measuring progress toward sustainability. Summary for decision-makers. UNU-IHDP, Delhi. http://www.ihdp.unu.edu/docs/Publications/Secretariat/Reports/SDMs/IWR_SDM_2014.pdf. (Accessed 23 January 2020).

Valtioneuvosto, 2019. Kokeilut ja julkiset hankinnat kiertotalouden toimenpideohjelman kärkenä. Ympäristöministeriön, Työ-jaa elinkeinoministeriö, Maa- ja metsätalousministeriön ja Sitran tiedote. https://valtioneuvosto.fi/artikkeli/-/asset_publisher/1410837/kokeilut-ja-julkiset-hankinnat-kiertotalouden-toimenpideohjelman-karkena (Accessed 23 January 2020).

Van Buren, N., Demmers, M., Van der Heijden, R., Witlox, F., 2016. Towards a CE: the role of Dutch logistics industries and governments. Sustainability 8, 647. https://doi.org/10.3390/su8070647.

Waas, T., Hugé, J., Block, T., Wright, T., Benitez-Capistros, F., Verbruggen, A., 2014. Sustainability assessment and indicators: tools in a decision-making strategy for sustainable development. Sustainability 2014 (6), 5512–5534. https://doi.org/10.3390/su6095512.

Walker, H., Brammer, S., 2009. Sustainable procurement in the United Kingdom public sector. Supply Chain Manag. 14 (2), 128–137. https://doi.org/10.1108/13598540910941993.

Walker, H., Brammer, S., 2012. The relationship between sustainable procurement and e-procurement in public sector. Int. J. Prod. Econ. 140 (1), 256–268. https://doi.org/10.1016/j.ijpe.2012.01.008.

Walker, H., Preuss, L., 2008. Fostering sustainability through sourcing from small businesses: public sector perspectives. J. Clean. Prod. 16 (15), 1600–1609. https://doi.org/10.1016/j.jclepro.2008.04.014.

Wang, N., Lee, J.C.K., Zhang, J., Chen, H. and Li, H. 2018. Evaluation of urban circular economy development: an empirical research of 40 cities in China. J. Clean. Prod., 180, 876–887. doi: https://doi.org/10.1016/j.jclepro.2018.01.089.

WCEFonline, 2020. The world circular economy forum online. https://www.youtube.com/watch?v=VE4H1_zOj1M&list=PL5XRiYm1aQFNmA-mlWhZaWagxK0QRG-KD.

Wijkman, A., Skånberg, K., 2015. The circular economy and benefits for society: jobs and climate clear winners in an economy based on renewable energy and resource efficiency. A study pertaining to Finland, France, the Netherlands, Spain and Sweden. A study report at the request of the Club of Rome with support from the MAVA Foundation. The Club of Rome. https://www.clubofrome.org/wp-content/uploads/2016/03/The-Circular-Economy-and-Benefits-for-Society.pdf. (Accessed 17 January 2020).

Witjesa, S., Lozano, R., 2016. Towards a more CE: proposing a framework linking sustainable public procurement and sustainable business models. Resour. Conserv. Recycl. 112, 37e44. https://doi.org/10.1016/j.resconrec.2016.04.015.

CHAPTER 15

A framework to integrate circular economy principles into public procurement

Ioannis E. Nikolaou[a], Thomas A. Tsalis[a], and Konstantinos I. Vatalis[b]
[a]Business Economics and Environmental Technology Laboratory, Department of Environmental Engineering, Democritus University of Thrace, Greece
[b]Department of Environmental Engineering and Antipollution Control, Technological Educational Institute of Western Macedonia (TEI), Kozani, Greece

1. Introduction

Today, public procurement is considered a good tool for providing incentives to the business community for the adoption of environmental practices. It is estimated that public procurement is responsible for roughly 18% of the gross domestic product of European countries (Nikolaou and Loizou, 2015), which implies that the incorporation of green criteria into public procurement will have a considerable impact on potential suppliers (i.e., firms). The adoption of practices in order to satisfy CE criteria will be a prerequisite for the participation in public procurement procedures. Thus, a business will have to incorporate certain environmental practices (e.g., ISO 14001, EMAS) in order to improve its environmental performance in order to be in a better position against their competitors, in the context of public procurgyement procedures, which have a poor environmental and economic performance or do not implement any environmental strategy (Liu et al., 2019a,b).

The concept of CE has lately gained great momentum among scholars and decision makers, who have recommended a number of tools to promote the principles of CE, which also contributes to the protection of the natural environment (Marrucci et al., 2019; Roos Lindgreen et al., 2020). To this end, some scholars have suggested techniques to incorporate CE criteria into public procurement in order to encourage businesses (i.e., suppliers) to change their production and operational behavior (Marrucci et al., 2019). Some scholars emphasize key benefits for the public sector from implementing practices related to the sharing economy, which is a critical aspect of CE, such as decrease in waste production, efficient utilization of assets, and improvement in the local employment rate (Ganapati and Reddick, 2018; Hofmann et al., 2019). Other scholars have examined various institutional aspects to design economic incentives and public procurement motivations to promote eco-design (Kalmykova et al., 2018). Furthermore, Sönnichsen and

Circular Economy and Sustainability, Volume 1
https://doi.org/10.1016/B978-0-12-819817-9.00020-X

Clement (2020) have identified several impediments that public authorities face in their efforts to incorporate CE criteria into public procurement, such as lack of knowledge about these criteria and the higher costs of CE products and equipment. Braulio-Gonzalo and Bovea (2020) suggested a four-stage model, which offers a systematic way to assist public authorities in determining necessary CE criteria in order to evaluate tenders.

However, some key aspects need to be taken into consideration in order to prepare a framework suitable to select suppliers in relation to their contribution to CE. The criteria for public procurement need to achieve some key principles, such as clarity of definition, simplicity, independent certification, and wide applicability, which implies an evaluation method that encompasses suitable and applicable criteria to ensure a "fair, open and transparent evaluation" (E.C., 2017). Another important issue which should be examined so as to develop an effective framework is to analyze how previous methodological frameworks deal with the principles of CE. Actually, what should be examined is the "circularity" of the current criteria of public procurement.

Given this background, this chapter proposes a methodological framework to assist in incorporating criteria regarding CE principles for awarding contracts and selecting tenderers (mainly in technical services). It is based on CE principles, eco-labels, and public procurement literature, in order to design various CE criteria. Also, a scoring-rating system is developed to evaluate an overall score, which assesses suppliers' CE performance. The proposed methodological framework was tested by performing some computational examples, providing valuable feedback on the applicability and effectiveness of the suggested framework, highlighting its positive and negative points.

The rest of the chapter consists of four sections. Section 2 analyzes the current theoretical background of the public procurement and CE. Section 3 develops the basic steps of the suggested methodological framework. Section 4 analyzes some computational examples to show its applicability, and Section 5 discusses the main conclusions and implications of the chapter.

2. Theoretical underpinnings

The current literature considers public procurement as an important mechanism that supports economic development, with many findings showing that public procurement is often able to restart an economy and move away from a stagnant phase (van Winden and Carvalho, 2019; Thiankolu, 2019). Furthermore, many institutional organizations have suggested public procurement as a useful tool, not only to support economic growth, but also to protect the natural environment (Palmujoki et al., 2010; Lundberg and Marklund, 2013), and lately, to promote CE (Witjes and Lozano, 2016; Lozano et al., 2016); these aspects are defined as green public procurement (GPP) and circular public procurement (CPP), respectively. In particular, the GPP approach implies that public authorities use environmental criteria to select suppliers who have good environmental performance and satisfy these criteria, in order to supply public authorities with environmentally friendly products (Palmujoki et al., 2010). As for the CPP, public organizations adopt techniques for incorporating suitable criteria in order to evaluate candidates (i.e., suppliers) regarding their CE practices.

A major point of analysis concerning CPP is the accurate definition of the term CE and how it is linked with the public procurement procedures. In this respect, Sönnichsen and Clement (2020) maintain that the concept of CPP contributes to environmental, social, and ethical behavior of suppliers and consumers. Similarly, Crafoord et al. (2018) have stated that some CPP practices can help businesses to embrace the concept of CE and incorporate principles into their production practices and consumption behavior by developing processes for product remanufacturing. It is worth noting that the European Commission (EC, 2017) have stated that CPP is an effective mechanism for greening public procurement in order to play a critical role in the efforts of modern economies to shift from a linear to a CE phase. Public organizations should invest in goods and services that focus on closing the cycle of their supply chains and minimize environmental impacts and waste in all stages of product life cycle. Alhola et al. (2017) outline the concept of circular procurement as an approach to determine "competitively priced products, services or systems" that gives priority to the lifespan of materials, to close the biological or technical loops of materials. Finally, Jones et al. (2018) define CPP as an approach that assures the products of selected suppliers are produced according to the principles of the CE, such as durability, reparability, and recycling.

It is of critical importance to refer an important part of the literature, which analyzes the institutional context that stipulates public procurement, as it is known that any change in public procurement procedures should be based on an appropriate institutional regime, sufficient regulatory realm, social culture, and balance of cost-effectiveness approach (Dale-Clough, 2015). By analyzing European Union policy tools, there are three general regulatory principles on which public procurement should be based, namely, transparency throughout tender procedures, fair treatment to all tenderers, and contestability for noncontrol of market prices (Engelbrekt, 2011). Public procurement is not only the process of using public funds to provide quality public services for a short-term period, but should have a positive impact on the public over the long-term.

In this debate, European legislation provides, inter alia, a number of technical specifications to help public authorities to integrate green/environmental and CE criteria into their public procurement procedures, with the intention to facilitate businesses to adopt environmental and circular-friendly strategies and practices that can contribute to a more sustainable economy (Dubey et al., 2019). As regards the GPP, there are a number of European directives that offer opportunities for public authorities to integrate environmental criteria regarding products and services in order to define businesses that are eligible to participate in public procurement procedures. The ongoing debate on greening public procurement has also focused on the concept of CE, where a number of approaches have begun to be proposed. The European Commission (E.C., 2017) suggests four priorities that can facilitate public organizations to incorporate the principles of circularity into the processes of public procurement: reducing the amount of products and services that are purchased, reusing equipment, purchasing reused materials and products, and recycling and recovering products and equipment. Based on the key principles of institutional theory, Zeng et al. (2017) have identified positive relationships among institutional pressures, sustainable supply chains, and circular economy management by drawing information and data from the operation of eco-industrial parks in China.

Another important issue examined in the relative literature focuses on how CPP affects the activities of suppliers. More specifically, Witjes and Lozano (2016) have suggested a normative model to align the procedures of public procurement with sustainable business models. They suggest a strong collaboration among actors across supply chains with the intention to reduce the use of raw materials and the amount of waste as well as to facilitate the firms to adopt more sustainable business models. Similarly, Alhola et al. (2019) have identified that the adoption of CE principles in public procurement plays a critical role in changing business models, since the viability of a number of sectors has strong connections with the public authorities and their strategies (e.g., construction, transportation, food, and furniture sectors) Thus, such sectors have to adopt strategies to modify their operations so as to be able to participate in public procurement procedures. To develop the relationship between CE and public procurement, Sönnichsen and Clement (2020) highlighted three areas of impacts of public procurement on suppliers: organizational, behavioral reaction, and operational levels.

Finally, literature suggests mechanisms for incorporation of CE criteria into public procurement, shifting suppliers from the linear economy to CE. By examining good practices from various countries and public authorities, Alhola et al. (2019) have pointed out four types of public procurement in relation to the criteria of circularity. The first category of CPP focuses mainly on promoting products or services that meet principles such as recyclability, reusability, reduction in energy consumption, and use of renewable energy. The second category includes types of CPP that promote new ways of cooperation between public organizations and suppliers through product-service systems and leasing strategies. The third category refers to new products, technologies, and services that adopt CE practices, such as the utilization of recycled materials and longevity of products. The final category includes collaboration between actors from public procurement who operate as ecosystem promoting closed loops and creating networks of exchanging waste among various actors.

3. Methodology

This section analyzes the suggested methodological framework through five consecutives steps, as depicted in Fig. 1. In particular, the first step presents the aim of this methodological framework, which is a crucial step since it describes the reason for the construction of this framework. The second step aims at determining the principles of CE that are necessary in order to evaluate the CE performance of suppliers and their products. The third step refers to the measurement (scoring-rating) system that calculates the total performance score of each supplier, which can be used in awarding a contract. The fourth step presents computational examples to examine the applicability of the proposed methodological framework, while the last step describes the outcomes from the computational examples and the functionality of the overall methodological framework.

3.1 The objective of the methodological framework

Numerous methodological frameworks have lately been suggested to define CE criteria for selecting suppliers with a good CE performance (Sönnichsen and Clement, 2020; Ferasso et al., 2020; Hartley et al., 2020). Migliore et al. (2020) have suggested a number of criteria to verify the proportion of recycled materials used in production. Moreover, CE literature includes various methodological frameworks for evaluating products with respect to various circularity criteria (Linder et al., 2017). In particular, Linder et al. (2017) have suggested a ratio of recirculated economic value as a part of a product to the total product value. Similarly, Mestre and Cooper (2017) have suggested a conceptual

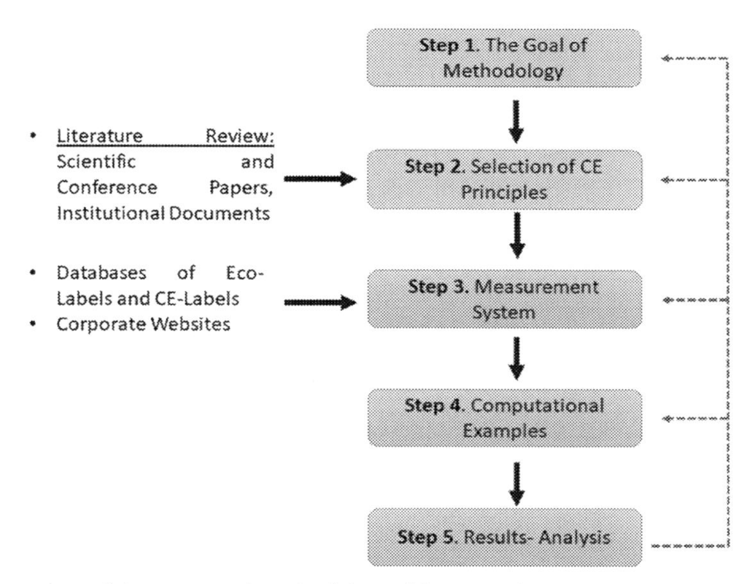

Fig. 1 The structure of the proposed methodological framework.

framework to provide a set of criteria for designing a circular product. In general, the majority of such frameworks propose indicators (Vatalis et al., 2012) that are in line with specific principles of CE such as the 3R model (reduce, recycle, and reuse).

Even though previous methodological frameworks are based on scientific-based criteria, some additional regulatory requirements may have to be taken into consideration so that these frameworks can be easily applicable for the processes of public procurement. Therefore, all selected criteria should be clearly defined and easily understandable to all interested parties (e.g., suppliers), which also can be certified or evaluated by external bodies avoiding unfair competition. In addition, such criteria should not focus solely on a specific principle of CE, but should incorporate as many CE criteria as possible, which ensures a greater degree of equity and offers a reward to those candidates who achieve a superior performance in relation to CE. In the light of these requirements, the proposed methodological framework aims to develop an award criteria-based technique in order to evaluate suppliers' contribution to CE by utilizing scientific-based, regulatory-oriented, and market focused criteria.

3.2 Selection of CE principles

This step concerns the definition of the content and the scope of CE. Different definitions of the concept of CE have been given, which focus on various aspects of CE such as recycle, remanufacture, reuse, repurpose, redesign, and refurbish. Kirchherr et al. (2017) have identified 114 definitions of CE through key principles such as 3R and 9R models (recover, recycle, repurpose, remanufacture, refurbish, repair, reuse, reduce, and refuse). Particularly, in the period before 2012, they identified that the main interest was in materials and products recycling, while in the period from 2012 to 2014 the interest has shifted from the principle of recycle to reduce. In the period after 2014, emphasis has been given to the principle of reuse. Korhonen et al. (2018) classify the current CE definitions according to whether or not they are based on Ellen MacArthur Foundation (EMF) recommendations. Specifically, definitions which rest on the EMF focus on definitions of regeneration and restoration of materials, while the other definition focus on terms from CE models (e.g., 3R and 9R models).

Additionally, there are several other authors who have made efforts to clarify the content of CE, focusing on specific principles and proposing models for the application of the CE concept (see Table 1). In general, the CE models have gradually evolved from 3R to 9R. More specifically, the letter "R" refers to the principle used to evaluate the CE contribution of suppliers. Taking into consideration the needs of the proposed methodological framework, the definition of CE should be operational, offering the basis for quantifying the CE contribution. Thus, the suggested framework is in line with the models that adopt specific CE principles.

Table 1 CE models.

Model	Principles of CE	Literature
3R	Reduce, Reuse, and Recycle	Hongyan (2010), Liu et al. (2017), and Huang et al. (2018)
4R	Reduce, Reuse, Recycle, and Recover	Hu et al. (2011)
5R	Reduce, Reuse, Recycle, Remanufacturing, and Refurbish	Mishra et al. (2019)
6R	Reduce, Reuse, Recycle, Remanufacturing, Redesign, and Recover	Jawahir and Bradley (2016) and Bradley et al. (2018)
7R	Reduce, Repair, Reuse, Rethink, Refurbish, Recycle, Recover	Kazancoglu et al. (2020)
8R	Reduce, Repair, Reuse, Rethink, Refurbish, Recycle, Recover, Re-mine	Kirchherr et al. (2017) and Poponi et al. (2020)
9R	Reduce, Reuse, Recycle, Refuse, Rethink, Repair, Refurbish, Remanufacture, Recover	Kirchherr et al. (2017)

3.3 Measurement system

The core of the proposed methodology is the measurement system. Obviously, for a transparent evaluation of all potential suppliers, accurate and reliable information is required. Therefore, environmental certifications and eco-labels could be used as indicators for the assessment of the principles of CE adopted by potential suppliers. Even though, the determination of new and explicit criteria could be more methodically accurate to introduce effectively the principles of the CE into public procurement processes, nevertheless it may create conditions for unfair competition between suppliers (e.g., between large and small firms). Also, problems are associated with defining whether a supplier satisfies more than one principle of the CE. This problem arises from the variety of definitions and models of CE that appear in the literature.

It is noteworthy that evaluation frameworks that merely focus on a single principle, such as remanufacture or recycle, are unable to provide accurate assessment of suppliers' CE performance who adopt in their operations more than one principle of CE. Such suppliers are likely to offer products and services with a greater contribution to the CE than suppliers that focus exclusively on recycled products. As for the use of eco-labels, the main concerns associated with the fact that some labels have a direct focus on the CE (such as Cradle to Cradle Certified (CM) Products Program), while other labels emphasize the protection of the natural environment and cover certain principles of the CE (e.g., EU Eco-flower, ISO 14001).

Considering the characteristics of labels and their links with the CE principles, the proposed measurement system classifies each label or standard into the nine principles of CE (see Table 2) and then it utilizes a three-point scoring scale to rate the CE contribution of firms based on the labels and implemented standards. (see Table 2).

Table 2 Measurement system.

Levels	Points	Description	Example
A	1	For each firms' certification	A firm is certified with EU flower label
B	2	When a certification is linked with more than one a principle of CE	ISO 14001 covers Reduce, Reuse, and Recycle principles
C	3	When a certification has a closed association with a particular principle of CE	The Green Shape is a label for products that implies they are made of at least 90% organic cotton or recycled materials
D	4	When a certification has a reference to the CE in its title	ISO/TC 323 and BS 8001

4. Computational examples and results

In this section, some computational examples are presented in order to demonstrate the applicability of the proposed methodological framework. Table 3 shows the evaluation results from four hypothetical suppliers (S_A, S_B, S_C, and S_D). According to the hypothetical scenario, supplier S_A is certified with two labels that pay attention to environmental issues (e.g., ISO 14001, Nordic Swan Ecolabel), the s S_B supplier has three labels, each of which has an emphasis on a particular principle of CE (e.g., Electronic Product Environmental Assessment Tool, Global Recycle Standard, and Green Flag Program). Similarly, S_C and S_D suppliers have three labels, where two of them focus on environmental aspects and the other one focuses on CE, such as Cradle to Cradle Certification Program and GreenCircle and BS 8001, and two labels that focus on CE, respectively.

Based on the proposed methodology, Table 3 shows total scores achieved by the four hypothetical suppliers. The S_c Supplier achieves the highest score in relation to the other three suppliers. In addition, it is obvious that the proposed methodology pays attention to the number of labels, the particular focus of the labels on the CE concept, and the number of CE principles that labels cover. Such evaluation results can provide useful input for the overall evaluation of each candidate in order to select the supplier who has the great contribution to CE and meets financial and technical criteria.

5. Conclusion and discussion

The suggested methodological framework attempts to make more operational the concept of CE in the context of public procurement. Moreover, it aims to assist in overcoming limitations of previous normative models of CPP that focus on "what should be made" approaches (Lozano et al., 2016). Specifically, it suggests an approach that

Table 3 Results from the hypothetical suppliers.

		CE principles (Level B)												
Suppliers	**Certification**	**R1**	**R2**	**R3**	**R4**	**R5**	**R6**	**R7**	**R8**	**R9**	**LA**	**LC**	**LD**	**TS**
S_A	C1	2	2	2		2		2			1			11
	C2	2	2	2		2					1			9
						Total score								20
S_B	C1	2		2		2					1	2		9
	C2			2							1	2		5
	C3	2	2	2							1	2		9
						Total score								23
S_C	C1		2								1	3		6
	C2	2	2	2							1	3		10
	C3	2				2	2				1		4	11
						Total score								27
S_D	C1		2								1	3		6
	C2	2	2	2							1	3		10
						Total score								16

LA: Level A, LC: Level C, LD: Level D. TS: Total score.

facilitates public organizations to deal with the CPP. It is based on scoring-rating systems that are regarded as practical tools for evaluating the CE performance of firms.

The contribution of the proposed methodological framework to the relative literature is threefold. First, it provides an up-to-date tool, which offers an objective evaluation process for supplier selection using specific criteria that assess suppliers' performance in the CE and consequently, their impacts on the natural environment.

The second contribution of the proposed assessment framework is associated with the utilization of certification labels, which ensures that the supplier selection is based on accurate and transparent information. Actually, the majority of the previous methodological frameworks rest on a specific principle regarding CE, such as recycle, reuse, or reduce. Indeed, many of such methodological frameworks are designed for evaluating products produced from recycled or reused materials (Gerner et al., 2005). The suggested methodological framework aims at rewarding suppliers who satisfy more than one principle of CE through third party certification schemes in order to ensure that communities will gain the maximum benefit from the public procurement.

Although, in the context of public procurement procedures, the sole focus on one principle is very useful and an easy way to obtain products and select suppliers, there are many other aspects of suppliers' performance that should be taken into account. Without assessing all aspects of candidates' CE performance, the procurement procedure is unable to contribute to CE and protect natural resources, causing negative impacts on the overall economy and society.

The last contribution of the suggested methodological framework is that it promotes sustainable consumption. Public organizations can choose materials and equipment by adopting CE principles in their consumption behavior. Even though public procurement is a policy tool, it is a consumption process that has to overcome some difficulties such as the asymmetric information between consumers and sellers. It should be noted that the business eco-labels have been created to assist in eliminating the phenomenon of asymmetric information (Nikolaou and Kazantzidis, 2016). Consequently, the proposed methodological framework, which adopts the environmental/ecological quality labels and certifications, helps to solve the problem of asymmetric information.

The findings from computational examples have shown that decision makers could have a clear picture about the CE performance of suppliers. The measurement system assists in selecting a supplier with a better performance regarding CE, which satisfies many principles of CE, and it is certified with various labels and standards.

Obviously the suggested framework has some limitations. The first limitation is the absent of a clear definition regarding the CE concept. It is necessary to carefully and systemically analyze each principle of CE and provide a complete and precise definition of

CE in order to be functional and measurable. This could also be a good starting point for future research. Another limitation is related to the application of computational examples. Although these examples are useful to present the structure and the logic of the suggested methodological framework, future empirical researches are necessary to test the applicability level of the proposed methodology and highlight possible drawbacks.

References

Alhola, K., Salmenperä, H., Ryding, S.-O., Busch, J.N., 2017. Circular Public Procurement in the Nordic Countries. Available at: https://norden.diva-portal.org/smash/get/diva2:1092366/FULLTEXT01.pdf (access: 12/09/2020).

Alhola, K., Ryding, S.O., Salmenperä, H., Busch, N.J., 2019. Exploiting the potential of public procurement: opportunities for circular economy. J. Ind. Ecol. 23 (1), 96–109.

Bradley, R., Jawahir, I.S., Badurdeen, F., Rouch, K., 2018. A total life cycle cost model (TLCCM) for the circular economy and its application to post-recovery resource allocation. Resour. Conserv. Recycl. 135, 141–149.

Braulio-Gonzalo, M., Bovea, M.D., 2020. Criteria analysis of green public procurement in the Spanish furniture sector. J. Clean. Prod. 258, 120704.

Crafoord, K., Dalhammar, C., Milios, L., 2018. The use of public procurement to incentivize longer lifetime and remanufacturing of computers. Procedia CIRP 73 (1), 137–141.

Dale-Clough, L., 2015. Public procurement of innovation and local authority procurement: procurement modes and framework conditions in three European cities. Innov. Eur. J. Soc. Sci. Res. 28 (3), 220–242.

Dubey, R., Gunasekaran, A., Childe, S.J., Papadopoulos, T., Helo, P., 2019. Supplier relationship management for circular economy: influence of external pressures and top management commitment. Manag. Decis. 57 (4), 767–790.

E.C, 2017. Public procurement for a circular economy: good practice and guidance. European Commission. Available at https://ec.europa.eu/environment/gpp/pdf/CP_European_Commission_Brochure_webversion_small.pdf (access: 12/09/2020).

Engelbrekt, K., 2011. EU enlargement and the emboldening of institutional integrity in central and Eastern Europe: the 'tough test' of public procurement. Eur. Law J. 17 (2), 230–251.

Ferasso, M., Beliaeva, T., Kraus, S., Clauss, T., Ribeiro-Soriano, D., 2020. Circular economy business models: the state of research and avenues ahead. Bus. Strategy Environ. 29 (8), 3006–3024.

Ganapati, S., Reddick, C.G., 2018. Prospects and challenges of sharing economy for the public sector. Gov. Inf. Q. 35 (1), 77–87.

Gerner, S., Kobeissi, A., David, B., Binder, Z., Descotes-Genon, B., 2005. Integrated approach for disassembly processes generation and recycling evaluation of an end-of-life product. Int. J. Prod. Res. 43 (1), 195–222.

Hartley, K., van Santen, R., Kirchherr, J., 2020. Policies for transitioning towards a circular economy: expectations from the European Union (EU). Resour. Conserv. Recycl. 155, 104–634.

Hofmann, S., Sæbø, Ø., Braccini, A.M., Za, S., 2019. The public sector's roles in the sharing economy and the implications for public values. Gov. Inf. Q. 36 (4), 101399.

Hongyan, Z.H.A.N.G., 2010. Application of 3R principles of cyclic e-conomy on the waste area of suburb. Ecol. Econ. 2, 405–407.

Hu, J., Xaio, Z., Deng, W., Wang, M., Ma, S., 2011. Ecological utilization of leather tannery waste with circular economy model. J. Clean. Prod. 19, 221–228.

Huang, B., Wang, X., Kua, H., Geng, Y., Bleischwitz, R., Ren, J., 2018. Construction and demolition waste management in China through the 3R principle. Resour. Conserv. Recycl. 129, 36–44.

Jawahir, I.S., Bradley, R., 2016. Technological elements of circular economy and the principles of 6R-based closed-loop material flow in sustainable manufacturing. Procedia CIRP 40 (1), 103–108.

Jones, M., Sohn, I.K., Bendsen, L.A.-M., 2018. Circular Procurement Best Practice Report, Regional Networks for Sustainable Development. Available at: file: ///C:/Users/user/Desktop/Book%20Chapters/Circular%20Public%20Procurement/CIRCULAR%20ECONOMY%20POLICY/Circular%20Procurement%20Best%20Practice%20Report%20(3).pdf (access: 12/09/2020).

Kalmykova, Y., Sadagopan, M., Rosado, L., 2018. Circular economy—from review of theories and practices to development of implementation tools. Resour. Conserv. Recycl. 135, 190–201.

Kazancoglu, I., Kazancoglu, Y., Yarimoglu, E., Kahraman, A., 2020. A conceptual framework for barriers of circular supply chains for sustainability in the textile industry. Sustain. Dev. 28 (5), 1477–1492.

Kirchherr, J., Reike, D., Hekkert, M., 2017. Conceptualizing the circular economy: an analysis of 114 definitions. Resour. Conserv. Recycl. 127, 221–232.

Korhonen, J., Nuur, C., Feldmann, A., Birkie, S.E., 2018. Circular economy as an essentially contested concept. J. Clean. Prod. 175, 544–552.

Linder, M., Sarasini, S., van Loon, P., 2017. A metric for quantifying product-level circularity. J. Ind. Ecol. 21 (3), 545–558.

Liu, L., Liang, Y., Song, Q., Li, J., 2017. A review of waste prevention through 3R under the concept of circular economy in China. J. Mater. Cycles Waste Manage. 19 (4), 1314–1323.

Liu, J., Shi, B., Xue, J., Wang, Q., 2019a. Improving the green public procurement performance of Chinese local governments: from the perspective of officials' knowledge. J. Purch. Supply Manag. 25 (3), 100–501.

Liu, J., Xue, J., Yang, L., Shi, B., 2019b. Enhancing green public procurement practices in local governments: Chinese evidence based on a new research framework. J. Clean. Prod. 211, 842–854.

Lozano, R., Witjes, I.S., van Geet, C., Willems, M., 2016. Collaboration for Circular Economy: Linking Sustainable Public Procurement and Business Models. Collaboration for Circular Economy: Linking sustainable public procurement and business models. Utrecht University, Copernicus Institute of Sustainable Development.

Lundberg, S., Marklund, P.O., 2013. Green public procurement as an environmental policy instrument: cost effectiveness. Environ. Econ. 4 (4), 75–83.

Marrucci, L., Daddi, T., Iraldo, F., 2019. The integration of circular economy with sustainable consumption and production tools: systematic review and future research agenda. J. Clean. Prod. 240, 118–268.

Mestre, A., Cooper, T., 2017. Circular product design. A multiple loops life cycle design approach for the circular economy. Design J. 20 (Suppl 1), S1620–S1635.

Migliore, M., Talamo, C., Paganin, G., 2020. Circular economy and sustainable procurement: the role of the attestation of conformity. In: Strategies for Circular Economy and Cross-sectoral Exchanges for Sustainable Building Products. Springer, Cham, pp. 159–173.

Mishra, S., Singh, S.P., Johansen, J., Cheng, Y., Farooq, S., 2019. Evaluating indicators for international manufacturing network under circular economy. Manag. Decis. 57 (4), 811–839.

Nikolaou, I.E., Kazantzidis, L., 2016. A sustainable consumption index/label to reduce information asymmetry among consumers and producers. Sustain. Prod. Consumpt. 6, 51–61.

Nikolaou, I.E., Loizou, C., 2015. The green public procurement in the midst of the economic crisis: is it a suitable policy tool? J. Integr. Environ. Sci. 12 (1), 49–66.

Palmujoki, A., Parikka-Alhola, K., Ekroos, A., 2010. Green public procurement: analysis on the use of environmental criteria in contracts. Rev. Eur. Commun. Int. Environ. Law 19 (2), 250–262.

Poponi, S., Arcese, G., Mosconi, E.M., Arezzo di Trifiletti, M., 2020. Entrepreneurial drivers for the development of the circular business model: the role of academic spin-off. Sustainability 12 (1), 423.

Roos Lindgreen, E., Salomone, R., Reyes, T., 2020. A critical review of academic approaches, methods and tools to assess circular economy at the micro level. Sustainability 12 (12), 4973.

Sönnichsen, S.D., Clement, J., 2020. Review of green and sustainable public procurement: towards circular public procurement. J. Clean. Prod. 245, 118–901.

Thiankolu, M.K., 2019. Using public procurement as a tool of economic and social development policy in Kenya: lessons from the United States and South Africa. Finan. Dev. 1 (1), 1–150.

van Winden, W., Carvalho, L., 2019. Intermediation in public procurement of innovation: how Amsterdam's startup-in-residence programme connects startups to urban challenges. Res. Policy 48 (9), 103–789.

Vatalis, K.I., Manoliadis, O.G., Mavridis, D.G., 2012. Project performance indicators as an innovative tool for identifying sustainability perspectives in green public procurement. Procedia Econ. Finan. 1, 276–285.

Witjes, S., Lozano, R., 2016. Towards a more circular economy: proposing a framework linking sustainable public procurement and sustainable business models. Resour. Conserv. Recycl. 112, 37–44.

Zeng, H., Chen, X., Xiao, X., Zhou, Z., 2017. Institutional pressures, sustainable supply chain management, and circular economy capability: empirical evidence from Chinese eco-industrial park firms. J. Clean. Prod. 155, 54–65.

CHAPTER 16

The role of public policy in the promotion of sustainability by means of corporate social responsibility: The case of the chemicals sector worldwide

Joana Costa[a], Manuela Castro Silva[b], and Tânia Freitas[b]
[a]DEGEIT, Universidade de Aveiro, GOVCOPP and INESCTEC, Aveiro, Portugal
[b]FEP—Universidade do Porto, Porto, Portugal

1. Introduction

In recent years, firms' strategic management has been transforming—stakeholder demands do not comprise merely profit maximization (Fortanier et al., 2011). Such events prompt corporate social responsibility (CSR) into the agenda. According to the European Commission (2018a), CSR corresponds to organizations' concern for their effect on society. It can be conducted by obeying the law or taking into consideration not only environmental questions, but also social, ethical, and human rights issues related to their daily business. This matter's growing relevance has led many firms listed on the Fortune 500 list employing full-time staff to undertake CSR activities, impacting business, annually, in millions of dollars (Robinson and Wood, 2018). Moreover, country differences are found, pointing toward the importance of the entrepreneurial ecosystem in the promotion of these practices (Maignan and Ralston, 2002).

Despite appearing to be very novel, the roots of CSR go back to the 1960s, and several definitions were posed. At first, the concept comprised a three-layered framework (economic, legal, and ethical), and, more recently the concept was revisited, including additional features and connecting the social aspect through philanthropic responsibilities (Carroll, 2016).

Corporate financial performance is linked to the economic dimension of CSR, whose measurement is performed via examination of financial statements (Knight and Bertoneche, 2001), and firm- and country-level aspects play a significant role on financial performance (Salah, 2018).

Apart from previous requirements, stakeholders also require accountability and transparency on this matter (Paun, 2018). Such prospects are bounded by the geographic

Circular Economy and Sustainability, Volume 1
https://doi.org/10.1016/B978-0-12-819817-9.00029-6

environment (Einwiller et al., 2016). CSR reporting fulfils these requirements and already has several global standards, such as the Global Reporting Initiative (GRI) and the Global Compact by the United Nations (UN), among others (Fortanier et al., 2011). Additionally, the latter provides a context for CSR analysis, identifying three main areas—environment, social, and governance—and subsequent issues (United Nations Global Compact, 2018a, b, c).

Chemicals are crucial elements in our daily lives; however, misusing the resources may cause health and environmental problems (European Commission, 2018b). The chemicals industry is estimated to create jobs for around 20 million people (International Labor Organization, 2018). Besides this, the global turnover of the sector was around 3475 billion euros in 2017 (European Chemical Industry Council, 2018). These numbers reinforce the strategic relevance of this industry for economies worldwide (International Labor Organization, 2018). In order to prevent potential problems caused by this industry, improving security and helping the expansion of the communication channels with stakeholders, the initiative Responsible Care was born (European Chemical Industry Council, 2018).

Weber (2008) identified five major areas of business benefits arising from CSR practices: positive effects on company image and reputation; employee motivation, retention, and recruitment; costs reduction; increased revenue from higher sales and market share; and a decrease in CSR-related risk. Two of these benefits are related to corporate financial performance (CFP), proving the strong connection between CSR and CFP. Therefore, it is important to address whether CSR commitment of firms does affect their financial performance. Given the lack of research in this field, in particular, for the chemicals industry, this research aims to fill in that gap; trying to understand to what extent CSR affects the financial performance of these companies. In order to address the research question, and connect CSR practices with financial performance, a quantitative analysis was performed, using a sample which includes the top 50 chemicals companies in the world plus 14 chemicals firms present on the Russell 1000 index over the time frame of 8 years (2009–2017).

The rest of the chapter is structured as follows: Section 2 revises the existing literature in the field; Section 3 provides data collection and methodology; Section 4 presents the econometric estimations and the empirical results; and Section 5 concludes and provides policy recommendations to enhance CSR practices.

2. Literature review

The aim of this section is to summarize the extant of research on CSR and CSP, thus, giving an overview on the matter and providing the background for the research. In this vein, three subsections are included: corporate social responsibility; the relationship between CSR and CFP; and the country of origin effect on CSR. The first one aims to summarize the literature existent on CSR, along with a firm commitment to the topic. Then we provide a synopsis of the three main perspectives present in literature, regarding the correlation between CSR and CFP: positive, negative, and null.

2.1 Corporate social responsibility

From a macroeconomic perspective, corporations have grown their wealth—of the world's 100 biggest economic entities, 69 are companies (Global Justice Now, 2016)—becoming even more powerful actors. Therefore, there is an increased attention on how firms act toward society—if their voice is used in a socially responsible form.

The idea flourished in major business minds and presented itself in numerous forms in American economic circles around 1920 (Frederick, 2006). This time frame, the period prior to the 1950s, was described by Carroll (2008) as a philanthropic one—firms' social contributions were mainly through donations. The CSR concept has gained several definitions over the years and Bhaduri and Selarka (2016) identified six phases to its development: the first and second ones cover the 1950s and 60s, when the idea was brought into academic spotlight; in the 1970s there was a rapid evolution; during the 1980s, stakeholder theory and business ethics were linked to CSR; in the 1990s firms involved CSR activities in their businesses; lastly, from 2000s onwards, scholars concentrated on the effect of CSR in business strategy.

During the 1950s, three major aspects were appraised: the social trust of managers due to their visibility and reputation, the engagement with social causes through philantropy, and the accompany of the competitors in terms of good practices (Frederick, 2006). Carroll (2008) described this decade as more theoretical than practical and pointed out Bowen's major contributions to the academic field. Bowen (1953) defined CSR as an obligation of businessmen to meet social standards.

In the following decade, McGuire (1963) extended this point of view to firms' not having only legal and economic duties, but also other responsibilities toward society. According to Frederick (1960), society's (human and economic) resources should not be used for self or firms' interest, but for major social causes. Davis (1960) set the definition in a managerial context—when business executives' behavior goes outside the corporate interest (economic or technical), they are acting in a socially responsible way. Walton (1967) argued that the intimate strings between the firm and society should be taken into consideration by top management while they are pursuing their objectives. The authors disagreed on the return these actions might have—Davis (1960) suggested they might pay off in the long run; on the other hand, Walton (1967) stated there might be no measurable economic return from them. During this time, CSR was nourished by extrinsic and socially conscious drivers (Carroll and Shabana, 2010).

The CSR pyramid, firstly proposed by Carroll (1979), was restyled to encompass four dimensions: economic, legal, ethical, and philanthropic, to be followed by all companies.

The updated framework combines economic responsibility targets with the business requirement to provide profit. Legal responsibility is linked to companies obeying laws and regulations. Ethical responsibility is to act beyond legal requests. Philanthropic responsibility consists of engaging voluntarily in social activities, such as charitable donations. The Committee for Economic Development (CED) (1971) revealed that the public was increasingly concerned over environmental problems (air and water pollution) and that enterprises were not paying enough attention to societal problems. In addition, society had widened expectations over business' responsibilities, which might be represented graphically in three concentric circles: in this framework, the inner circle represents the effective accomplishment of the economic purpose (employment, products, and the economy's positive development); the intermediate circle refers to the fulfillment of the economic function, keeping in mind changeable social values and priorities, such as environmental preservation, fair treatment of both customers and employees, health care for employees, and rigorous disclosure of information for costumers; lastly, the outer circle encompasses emergent public concerns related to poverty and urban disfigurement, for example, and the belief that business has access to the resources to will help solving these problems.

One of the seminal works presented in the 1970s gathered the main arguments in favor of and against CSR, and is still accepted in general terms (Davis, 1973). Relating to the positive aspects of CSR, the first lies in a company's long-run self-interest—both social goods and programs will boost firm profitability. Apart from this, cost reduction in terms of production relies on them in the future. The second regards public image, intimately connected with the previous idea, which may become favorable, due to the support of social goods, leading to an increased number of costumers and an improved workforce. Thirdly, business viability refers to the industry. The institution only prevails that provides worthy services to society. If the firm desires to become both a powerful and a social actor, it needs to fulfil society's expectations. Regarding avoidance of government regulation, when the businessman accomplishes both private and public goods, there is no need for new laws, thus, he retains power and flexible decision making.

Sociocultural norms play a relevant role on society and firms' conduct. Therefore, management decisions will incline toward a socially accepted behavior, which expresses a form of social responsibility. Shareholder interest is to endeavor toward socially responsible actions, because whenever a firm fails to act this way, it will restrain the amount of benefit received by the stockholder. Several organizations were ineffective while attempting to solve societal problems; hence, it was decided to give business a try. Related to this subject, resource endowment (talent management, innovative mindset, and increasingly productive capital) are expected to provide solutions to social problems and may also yield a profitable outcome—problems can become profits. For instance, chemical firms' waste could be turned into profit in some specific scenarios. Lastly, from an economic point of view, this type of question should be handled prior to any negative consequence—prevention is better than cure.

In contrast, there are also arguments against CSR. Firstly, it is stated that economic criteria are the only valid ones to quantify a firm' success, as its sole goal is profit maximization. Business is forced into social activities, which presents costs and, subsequently, ousts minor firms from the industry. This is reported to have occurred in the chemicals industry. Additionally, the literature highlights as hindering factors to successful CSR strategies: businessman lacking social skills, dilution of business primary purpose (profit maximization), weakened balance of payments, market power, lack of accountability, and lack of broad support (CED, 1971). Nowadays, reputational risk also represents a threat, when there is an abuse from companies, even if it is in a remote place (Smith, 2003). Moreover, CSR can be seen to present a risk to shareholders, when there is special attention given to the environment, leading to an increase in production costs, causing firms to be at a disadvantage regarding their competitors (Nguyen and Nguyen, 2015).

CSR was appraised from a different angle that firms' only social responsibility is to use its resources to obtain more profit (known as the stockholder theory) by Friedman (1970). Jones (1980) argued that companies had the duty toward other groups in society, apart from stockholders (company owners). Such groups are identified as stakeholders—costumers, suppliers, employees, media, government, global competitors, among others (Freeman, 1984). Tuzzolino and Armandi (1981) built a theoretical framework to evaluate organizational social responsibility. When a firm fulfils physiological and safety desires, it is addressing its stockholders; while accomplishing the affiliative demand, it is responding to its peers; whenever the enterprise achieves self-actualization, it satisfies all the claimants. They argued this type of analysis might assist in organizational governance.

Falck and Heblich (2007) affirmed that CSR was a path for both societies and firms to thrive. In another perspective, Smith (2003) claims CSR benefits corporations through (in)direct economic efficiencies. Additionally, CSR might help in enhancing corporative reputation, when it pursues a differentiation strategy (McWilliams and Siegel, 2011).

The ethical dimension was reinforced by Bhaduri and Selarka (2016) as a pillar of CSR, as formerly presented by Epstein (1989), which defined business ethics as the rules that delimit the correct way to act for firms' executives, being influenced by societal values, institutional policies, and moral significance. He considered CSR to offer guidelines for managers' actions and attempts to assess business' social performance bringing the concept of corporate social responsiveness into the spotlight, which shed a light on handling strategically societal values as business is structuring policymaking processes.

Carroll (2008) remarks that there were no major contributions to the construction of CSR as a concept; instead, there was a growth in the literature field of complementary themes, such as stakeholder theory, business ethics, sustainability, corporate citizenship, and corporate social performance (CSP). Wood (1991) conceptualized CSP as a firms' composition of social responsibility prepositions, social responsiveness techniques and societal-correlated policies/programs, and its respective observable results. Wood (1991) also articulated three principles for social responsibilities—institutional, organizational, and individual levels—considering CSP as a wider notion than CSR.

Nowadays, CSR is vital concept for corporations, impacting costumers in their buying choice, and other stakeholders (Moravcikova et al., 2015). CSR activities promote values and opportunities to society and organizations (Verboven, 2011). Therefore, it is important to communicate them properly. CSR reporting helps in this matter.

As Verboven (2011) declared, the public is aware of the less positive situations the chemicals industry may create, but not of the benefits. Hence, it is essential to the industry to build a sustainable corporate image and reputation. This is verified on the growth of CSR reporting for this industry—according to KPMG (2017) it was around 8% between 2015 and 2017. Nevertheless, one of the main difficulties currently existent is how to measure the level of companies' CSR commitment. Several external elements influence how multinational enterprises engage with CSR: industry sensitivity, local institutional constrains, and/or the gap between home and host country (Orudzheva and Gaffney, 2018). The same authors argued that CSR initiatives must take into consideration the global hierarchies between developed and developing world, which influences the effectiveness of multinational enterprises (MNEs) gaining competitive advantage from the developing group, reducing the existent inequality between the developed and developing world. Whilst analyzing CSR policies of 37 MNEs, Bondy and Starkey (2014) discovered that global strategies, as well as integrated internationalization strategies, do not separate local and global CSR topics, and that the issues identified by headquarters are prioritized and local ones are disregarded.

2.2 The relationship between CSR and CFP

Currently, socially responsible behavior is seen as a way to increase competitive advantage (López et al., 2007). Thus, it is essential to verify whether this also translates into a better financial performance. Existent literature has found different results regarding the correlation between CSR and CFP. Furthermore, Griffin and Mahon (1997) suggest that research should be done in one industry-specific context to boost outcomes' internal corroboration.

CFP embodies business performance when economic/financial goals are achieved and is operationalized by accounting and market measures (Venkatraman and Ramanujam, 1986; Orlitzky, 2008). Orlitzky (2008) gathered the casual (benefic) mechanics that associate CSR and CFP in the literature: enhancing organizational reputation; improving internal resources and skills; increasing rivals' costs; attracting a more productive workforce; boosting sales revenues; and reducing business risk; and also concluded that organizational size does not influence their relationship.

2.2.1 Positive association between CSR and CFP

First, it will be determined which scholars have a found a positive relation between these two concepts. Preston and O'Bannon (1997) supported this idea in their study about social and financial performance on major United States enterprises. 10 years before these findings, Bruyn (1987) pointed out that taking into consideration the social factor on investment decision making raises the chances for economic return. Waddock and Graves (1997) reached the same conclusion, analyzing 469 companies from the Standard and Poor's 500 list, and connected this with slack resources and good management theory, using the Kinder Lydenberg Domini (KLD) rating system for CSR. Orlitzky et al. (2003) led a 52 studies metaanalysis, which concluded how CSR is positively associated with CFP—bidirectional and coexistent, CSR is more interrelated to CFP's accounting-based measures than market based-measures, reputation is a relevant intermediary for this interaction, and that a firm might establish an equally beneficial relation with stakeholders. Bird et al. (2007) explored how the market reacts to CSR activities and revealed there was a valuation of proactive employee-related endeavors when environment and diversity requisites were achieved at a minimum level.

2.2.2 Negative association between CSR and CFP

Mittal et al. (2008) found a negative relationship between CSR and CFP in their study in the Indian context. Crisóstomo et al. (2011) discovered a negative linkage between CSR and firm value for the Brazilian scenario. Looking at the United States framework, Fisher-Vanden and Thorburn (2011) discovered a negative response from the market, when firms adhere as members to the Environmental Protection Agency Climate Leaders voluntary program aiming at greenhouse gas reduction.

2.2.3 Null association between CSR and CFP

The research conducted by Auperlee et al. (1985) created its own measure for CSR, based on the CSR pyramid presented by Carroll (1979) (legal, economic, ethical, and philanthropic dimensions) and it found no correlation between profitability and social responsibility. Ullmann (1985) claimed no trend could be discovered while studying the association between these concepts; he explained this as being due to a deficiency in theory, inadequate definition of the keywords, and lack of empirical data available.

2.3 Country of origin effect on CSR

The culture concept is commonly used for nations and organizations and it has six dimensions: power distance; uncertainty avoidance; individualism versus collectivism; masculinity (assertive) versus femininity (caring); long term versus short term orientation; and lastly, indulgence versus restrain. Countries' classification under these criteria are relative, hence they are obtained in comparison to others (Hofstede, 2011). These dimensions might also be applied to firms' stakeholders—consumers, shareholders, and employees—as a way to explain their behavior. Apart from this, consumers' propensity to "punish" firms when they do act responsibly can be related to the Hofstede framework (Williams and Zinkin, 2008).

Ding et al. (2019) argue that CSR's broad scenario may be justified by the enterprise headquarters' home culture. Matten and Moon (2008) highlight that historical and well-established institutions are the reason for differences in CSR in several countries (European and United States); given that, the authors compiled the motives for CSR system discrepancies—political/financial/education/labor and cultural systems, nature of the firm, organization of market procedures, and coordination and control arrangements. Cultural background and business ethics culture possess strategic relevance—management with forward-looking values obtains increased relevance in a global economy, as it can prevent cultural misunderstandings and breach of mutual values (Palazzo, 2002). The author enhanced the importance of acknowledging intercultural dissimilarities and respective consequences on the journey for trustworthy economic partnership, while simultaneously considering ones' own cultural background and other cultures' values. Bondy and Starkey's (2014) findings showed that despite MNEs recognizing the relevance of local host country cultures, none used this whilst architecting their CSR policy—the firms adapted the general one to the local culture.

Existent research focuses mainly on developed economies, as stated by Orudzheva and Gaffney (2018), providing less consideration to what firms from developing countries perform outside their country. There are important differences between German and United States multinational enterprises (MNEs), and those who follow global reporting standards have higher similarities in their reports. Moreover, international institutions that create such standards promote events, provoking a growth in interaction among adherent enterprises (Einwiller et al., 2016). For firms aiming to enhance their CSR commitment, adopting global guidelines will provide them easier access to knowledge on CSR implementation, management, and reporting. Additionally, it assists in discovering ways to handle countries (home/host) and international demands (Fortanier et al., 2011). There is also some evidence of country of origin effects—German MNEs' reports focus more on environmental matters; US reports emphasize the role of society. Conversely, from local firms, MNEs are subject to pressure both in home and host countries. Apart from this, their image is pictured as socially responsible agents promoting sustainable development with the aim of achieve licensing to operate in foreign markets. Lastly, high levels of corporate social performance will lead MNEs to raise their reputation/legitimacy in their operational areas and thus, increase their revenues and financial performance level (Aguilera-Caracuel et al., 2015). When comparing MNEs whose headquarters is located in developed and developing countries, the latter face greater challenges while using CSR as a manner to increment their competitive advantage whenever measures exceed legal compliance, particularly when the host country forms part of a distinct (higher/lower) hierarchical group from their own country of origin. Furthermore, from a strategic point of view, MNEs rooted in developing countries would struggle to institute CSR initiatives that strengthen positioning, rather than just copycat industry's best practices (Orudzheva and Gaffney, 2018).

The assessment of CSR should include the degree to which companies sustain (or oppose) public policies that contribute to sustainability (Lyon et al., 2018). Vogel (2005) states that there were few firms that took any position on political initiatives, because they were apprehensive that this may increase government regulatory requirements. Apart from this, he recommends enterprises should reevaluate their relationship to government if they want to commit truly to acting responsibly. In order to help this endeavor, Lyon et al. (2018) suggest three paths: transparence on corporate political activity, align the latter with public statements and CSR efforts, and become an advocate for public policies that will empower the private sector to better chase sustainability efforts and commitments (without acquiring competitive disadvantages). The effects from country of origin related to CSR and its respective dimensions also play a role on CFP as well, particularly, the contextual factors. National labor market and financial systems provide the framework that will bound firms' resources and capabilities (Bobillo et al., 2010).

3. Methodology

The main goal for this research is to determine how corporate social responsibility, along with country of origin, influences corporate financial performance in the chemicals' industry. Several studies have documented the relationship between corporate social responsibility and corporate financial performance. The empirical part of this research will include a predominantly quantitative approach to study the association between CSR and CFP in the chemicals industry.

Galant and Cadez (2017) have compiled the previously used measurements for both CSR and CFP amid the existent literature. Regarding CSR, the frequent methodologies are: reputation indices (SC KLD 400 social index, Fortune magazine reputation index, Dow Jones Sustainability Index and Vigeo Index) developed by rating agencies based on several criteria—economic, environmental, social, governance, and ethics, among others; content analyses on the firms' communication; questionnaire-based surveys; and one-dimensional measures (philanthropy, for instance). Concerning CFP, research found it is assessed via accounting-based, return on assets (ROA), return on equity (ROE), return on capital employed, return on sales, net (operating) income, and/or market-value indicators (stock returns and company market value, for example).

Timbate and Park (2018) tested this relationship, using the Standard & Poor's 500 firms, retrieving price, audit, and accounting data from Thomson Reuters Eikon and CSR information published by Corporate Responsibility Magazine, which ranks the 100 top corporate citizens with data collected from company websites, and sustainability and annual reports. They intended to examine the quality reporting information, respective market measure, and accrual quality and CSR effects; the control variables used were firm, industry, and year.

On the other hand, Kalish (2015) investigated this relationship across German and United States firms listed in the Reputation Institute's Global RepTrak 100 for a 3 year spectrum (2011, 2012, and 2013) and compared the data results obtained for each country. The linkage between CSR and CFP for the chemicals industry will be appraised in this line, analyzing several regions and concluding on possible differences/similarities among them.

An econometric analysis will be performed, encompassing the top 50 chemical enterprises according to Chemicals and Engineering News (2018), covering a 9 year (2009–2017) time frame. Due to missing data for variables described below, and given the need to balance the panel, the list was shortened to 29 firms. In order to analyze the influence of country of origin effect, the sample was divided into regions, related to their home continent: Africa, America, Asia, or Europe.

In the panel estimation, CSR, the independent variable, was obtained via Thomson Reuters Environmental (resources use, emissions, and innovation), Social (workforce, human rights, community, and product responsibility and governance) Management, Shareholders, and CSR Strategy (ESG) score, which has available data since 2002 for some enterprises—the score results from a weighted evaluation of all 178 indicators—34% for environmental pillars, 35.5% for social, and 30.5% for governance. The information is obtained considering publically reported figures by the firms themselves (Reuters, 2019).

In relation to the dependent variable, CFP, the chosen measurement unit was ROA, which is a profitability indicator (Corporate Finance Institution, 2019a) and this choice was based upon Kalish' methodology (2015). For the control variables, firm size (total assets) and debt to equity ratio—which is calculated via dividing total debt over shareholders equity (Corporate Finance Institution, 2019b) and provides information regarding risk—were chosen, drawing upon Maqbool and Zameer (2018).

4. Econometric estimations

In sum, the model will encompass a CSR index (ESG score; 0–100 score); firms' ROA (net income/total assets); debt to equity ratio (total debt/shareholders' equity); and size (total assets). All variables were collected from the Thomson Reuters Datastream. The data extracted corresponds to the end of each fiscal year and will be applied to all distinct geographic areas, as well as to the total sample, with a twofold objective: what is the impact of CSR in financial performance, and does this impact change geographically?

Concerning geography, the firms included in the sample are distributed: 10 in Europe, 6 in America, 12 in Asia, and 1 in Africa, which will be our benchmark (Table 1).

Additionally, the VIF tests provided information on the absence of multicollinearity problems (Table 2). The panel data estimation will respect the following equation:

$$ROA_{i,t} = \beta_0 + \beta_1 CSR_{i,t} + \beta_2 TotalAssets_{i,t} + \beta_3 DtE_{i,t} + \varepsilon_{i,t}$$

The estimation of both panels shed light on important areas as, contrary to what was expected, variables such as CSR and total assets do not influence the economic performance of the firm. Moreover, when the geographic dimension is included, only debt to equity remains significant.

Evidence proves that CSR is independent from firm financial performance. Therefore, profits are not directly linked to the adoption of these good practices, further reinforcing the need for precise policy actions to positively discriminate those who implement them. The sample mostly comprises European firms, and, these results evidence the need for reevaluating the policy instruments in use to tackle social responsibility strategies among firms. Moreover, the fact that the country of origin effect does appear to be significant is also very interesting, showing that there is scant incentive to promote desirable CSR practices. Most of the large firms in the chemicals sector operate directly or indirectly worldwide, and to some extent they collude in the absence of CSR practices. Given the inexistence of competition pressure due to the natural barriers to entry, these firms are perhaps not afraid of being censored by reduced demand because of their practices, due to lack of alternatives.

These finding reinforce the need for a deeper analysis of the proxies in terms of performance and model building to provide policy makers the correct insights about the actions to be taken to reinforce CSR practices in such a sensitive sector.

Table 1 Descriptive statistics.

Variable	Observations	Mean	Std. Dev.	Min	Max
Year	261	2013	2587	2009	2017
ID	261	15	8383	1	29
ROA	261	0.553	0.038	−0.0834	0.1896
DtE	261	0.7785	0.6007	0.01	3.39
Total assets	261	33,4213	54,6213	17,088	314,1947
CSR	261	6,351,313	1,644,797	1,284,683	90,302
America	261	0.2069	0.4059	0	1
Europe	261	0.3448	0.4762	0	1
Asia	261	0.4138	0.4935	0	1

Table 2 Econometric estimations.

Variables	ROA	ROA_continent
CSR	−0.00031	−0.00025
	(0.00019)	(0.00019)
Total assets	−0.00002	−0.00005
	(0.00007)	(0.00006)
DTE	−0.03163***	−0.03380***
	(0.00493)	(0.00487)
America		0.00145
		(0.02112)
Europe		−0.03045
		(0.01987)
Asia		−0.02020
		(0.02002)
Constant	0.10002***	0.11746***
	(0.01330)	(0.02315)
Observations	261	261
Number of id	29	29
Standard errors in parentheses		

***$P < 0.01$, **$P < 0.05$, *$P < 0.1$.

5. Conclusions and policy recommendations

Extant literature found mixed results regarding the relationship between corporate social responsibility and corporate financial performance. Currently, there is still a lack of research on the chemicals industry regarding corporate social responsibility, corporate financial performance, and how their linkage changes geographically. Hence, the present research used panel data estimations with the top 50 chemicals companies in the world as a sample for a period between 2009 and 2017. The sample was reduced to 29 companies, due to the need for panel balancing in the time frame, which represents an unavoidable limitation; a broader analysis will provide more robust estimations.

According to the existing evidence, mixed results may be justified due to the following reasons: measurement difficulties, weak theoretical groundwork for CSP conceptualization, neglecting relevant variables on model construction, and the unclear causality direction (Galant and Cadez, 2017). Chen et al. (2015) conducted a study comparing different manufacturing industries, which revealed no major differences among CSR practices.

As further steps, there should be a deeper analysis for a larger sample computing more factors for the CFP, such as market value indicators (stock returns and company market value, for example) in order to complement the accounting-based analysis used. ROE may be applied in further analysis, as an alternative measure of performance.

Since the available information was based on data disclosed by the firms, there is always some biasedness and some make-up over imperfections to meet the politically correct standard. In order to address this matter, Galant and Cadez (2017) suggest standardization in CSR reporting and mandatory data disclosure. In order to ease the process for data extraction and analysis, listed firms should publish their key results on relative terms (percentage results); including CSR information, not only mentioning the indicatives in their reports.

Both corporate social responsibility and country of origin have been presented in the literature as influencing factors for firms' financial performance; however, there have been few that explored this for the chemicals industry. Research needs to be considered as a first step in such an unexploited field, as what has been done so far still brought up new insights on the matter and provides a basis for future research.

References

Aguilera-Caracuel, J., Guerrero-Villegas, J., Vidal-Salazar, M.D., Delgado-Márquez, B.L., 2015. International cultural diversification and corporate social performance in multinational enterprises: the role of slack financial resources. Manage. Int. Rev. 55 (3), 323–353.

Auperlee, K., Carroll, A.B., Hatfiled, J.D., 1985. An empirical examination of the relationship between corporate social responsibility and profitability. Acad. Manag. J. 28 (2), 446–463.

Bhaduri, S.N., Selarka, E., 2016. Corporate Social Responsibility Around the World—An Over-View of Theoretical Framework, and Evolution. Springer, Singapore.

Bird, R., Hall, A.D., Momentè, F., Reggiani, F., 2007. What corporate social responsibility activities are valued by the market? J. Bus. Ethics 76 (2), 189–206.

Bobillo, A.M., López-Iturriaga, F., Tejerina-Gaite, F., 2010. Firm performance and international diversification: the internal and external competitive advantages. Int. Bus. Rev. 19 (6), 607–618.

Bondy, K., Starkey, K., 2014. The dilemmas of internationalization: corporate social responsibility in the multinational corporation. Br. J. Manag. 25 (1), 4–22.

Bowen, H.R., 1953. Social Responsibilities of the Businessman. Harper, New York.

Bruyn, S.T., 1987. The Field of Social Investment. Cambridge University Press, Cambridge.

Carroll, A.B., 1979. A three-dimensional conceptual model of corporate social performance. Acad. Manag. Rev. 4 (4), 497–505.

Carroll, A.B., 2008. In: Moon, J., Siegel, D.S. (Eds.), A History of Corporate Social Responsibility in the Oxford Handbook of Corporate Social Responsibility, Andrew Crane, Abigail McWilliams, Dirk Matten. Oxford University Press Inc.

Carroll, A.B., 2016. Carroll's pyramid of CSR: taking another look. Int. J. Corp. Soc. Responsib. 1 (3), 1–8.

Carroll, A.B., Shabana, K.M., 2010. The business case for corporate social responsibility: a review of concepts, research and practice. Int. J. Manag. Rev. 12 (1), 85–105.

CED (Committee for Economic Development), 1971. Social Responsibilities of Business Corporations. CED, New York.

Chemicals and Engineering News, 2018. C&EN's Global Top 50 Chemical Companies. Available at https://cen.acs.org/business/finance/CENs-Global-Top-50-chemical/96/i31. Accessed 4th January 2019.

Chen, L., Feldmann, A., Tang, O., 2015. The relationship between disclosures of corporate social performance and financial performance: Evidences from GRI reports in manufacturing industry. Int. J. Prod. Econ. 170 (B), 445–456.

Corporate Finance Institution, 2019a. ROA Formula/Return on Assets Calculation. Available at https://corporatefinanceinstitute.com/resources/knowledge/finance/return-on-assets-roa-formula/. Accessed 18th July 2019.

Corporate Finance Institution, 2019b. What is the Debt to Equity Ratio? Retrieved from https://corporatefinanceinstitute.com/resources/knowledge/finance/debt-to-equity-ratio-formula/. Accessed 25th July 2019.

Crisóstomo, V.L., Freire, F.S., Vasconcellos, F.C., 2011. Corporate social responsibility, firm value and financial performance in Brazil. Social Responsibility Journal 7 (2), 295–309.

Davis, K., 1960. Can business afford to ignore social responsibilities? Calif. Manag. Rev. 2 (3), 70–76.

Davis, K., 1973. The case for and against business assumption of social responsibilities. Acad. Manag. J. 16 (2), 312–322.

Ding, D.K., Ferreira, C., Wongchoti, U., 2019. The geography of CSR. Int. Rev. Econ. Financ. 59, 265–288.

Einwiller, S., Ruppel, C., Schnauber, A., 2016. Harmonization and differences in CSR reporting of US and German companies: analyzing the role of global reporting standards and country-of-origin. Corp. Commun. Int. J. 21 (2), 230–245.

Epstein, E.M., 1989. Business ethics, corporate good citizenship and the corporate social policy process: a view from the United States. J. Bus. Ethics 8 (8), 583–595.

European Chemical Industry Council, 2018. Facts & Figures of the European Chemical Industry 2018. European Chemical Industry Council, Belgium.

European Commission, 2018a. Corporate Social Responsibility. Retrieved from https://ec.europa.eu/growth/industry/corporate-social-responsibility_en. Accessed 1st November 2018.

European Commission, 2018b. Chemicals Are Everywhere. Retrieved from http://ec.europa.eu/environment/chemicals/index_en.htm. Accessed 18th December 2018.

Falck, O., Heblich, S., 2007. Corporate social responsibility: doing well by doing good. Business Horizons 50 (3), 247–254.

Fisher-Vanden, K., Thorburn, K.S., 2011. Voluntary corporate environmental initiatives and shareholder wealth. Journal of Environment Economics and Management 62 (3), 430–445.

Fortanier, F., Kolk, A., Pinske, J., 2011. Harmonization in CSR reporting. Manag. Int. Rev. 51, 665–696.

Frederick, W.C., 1960. The growing concern over business responsibility. Calif. Manag. Rev. 2 (4), 54–61.

Frederick, W.C., 2006. Corporation, Be Good!: The Story of Corporate Social Responsibility. Dog Ear Publishing, Indianapolis.

Freeman, R.E., 1984. Strategic Management: A Stakeholder Approach. Pitman, Boston.

Friedman, M., 1970. The Social Responsibility of Business is to Increase its Profits. Retrieved from http://umich.edu/~thecore/doc/Friedman.pdf. Available at 18th December 2018.

Galant, A., Cadez, S., 2017. Corporate social responsibility and financial performance relationship: a review of measurement approaches. Economic Research-Ekonomska Istraživanja 30 (1), 676–693.

Global Justice Now, 2016. 10 Biggest Corporations Make More Money Than Most Countries in the World Combined. Retrieved from https://www.globaljustice.org.uk//news/2016/sep/12/10-biggest-corporations-make-more-money-most-countries-world-combined. Available at 1st December 2018.

Griffin, J.J., Mahon, J.F., 1997. The corporate social performance and corporate financial performance debate: twenty-five years of incomparable research. Bus. Soc. 36 (5), 5–31.

Hofstede, G., 2011. Dimensionalizing cultures: the Hofstede model in context. Online Readings in Psychology and Culture 2 (1).

International Labor Organization, 2018. Chemical industries. Retrieved from https://www.ilo.org/global/industries-and-sectors/chemical-industries/lang- -en/index.htm. Accessed 18th December 2018.

Jones, T.M., 1980. Corporate social responsibility revisited, redefined. Calif. Manag. Rev. 22 (3), 59–67.

Kalish, H., 2015, July. Does CSR pay?—The impact of CSR on financial performance. A comparison between Germany and the US. In: Paper presented at 5th IBA Bachelor Thesis Conference. University of Twente, Enschede, The Netherlands.

Knight, R., Bertoneche, M., 2001. Financial Performance. Butterworth-Heinemann, Oxford.

KPMG, 2017. The KPMG Survey of Corporate Responsibility Reporting 2017. KPMG, Switzerland.

López, V., Garcia, A., Rodriguez, L., 2007. Sustainable development and corporate performance: a study based on the Dow Jones sustainability index. J. Bus. Ethics 75, 285–300.

Lyon, T.P., Delmas, M.A., Maxwell, J.W., Bansal, P., Chiroleu-Assouline, M., Crifo, P., Durand, R., Gond, J.P., King, A., Lenox, M., Toffel, M., Vogel, D., Wijen, F., 2018. CSR needs CPR: corporate sustainability and politics. Calif. Manag. Rev. 60 (4), 5–24.

Maignan, I., Ralston, D., 2002. Corporate social responsibility in Europe and the U.S.: insights from businesses' self-presentations. J. Int. Bus. Stud. 33, 497–514.

Maqbool, S., Zameer, M.N., 2018. Corporate social responsibility and financial performance: an empirical analysis of Indian banks. Future Business Journal 4 (1), 84–93.

Matten, D., Moon, J., 2008. "Implicit" and "explicit" CSR: a conceptual framework for a comparative understanding of corporate social responsibility. Acad. Manag. Rev. 33 (2), 404–424.

McGuire, J.W., 1963. Business and Society. McGraw-Hill, New York.

McWilliams, A., Siegel, D.S., 2011. Creating and capturing value: strategic corporate social responsibility, resource-based theory, and sustainable competitive advantage. J. Manag. 37 (5), 1480–1495.

Mittal, R.K., Sinha, N., Singh, A., 2008. An analysis of linkage between economic value added and corporate social responsibility. Manag. Decis. 46 (9), 1437–1443.

Moravcikova, K., Stefanikova, L., Rypakova, M., 2015. CSR reporting as an important tool of CSR communication. Procedia Economics and Finance 26, 332–338.

Nguyen, P., Nguyen, A., 2015. The effect of corporate social responsibility on firm risk. Social Responsibility Journal 11 (2), 324–339.

Orlitzky, M., 2008. Corporate social performance and financial performance. In: Crane, A., McWilliams, A., Matten, D., Moon, J., Siegel, D.S. (Eds.), The Oxford Handbook of Corporate Social Responsibility. Oxford University Press Inc.

Orlitzky, M., Schmidt, F.L., Rynes, S.L., 2003. Corporate social and financial Performance: a meta-analysis. Organ. Stud. 24 (3), 403–441.

Orudzheva, L., Gaffney, N., 2018. Country-of-origin and CSR initiatives: a social dominance perspective. Social Responsibility Journal 14 (3), 501–515.

Palazzo, B., 2002. U.S.-American and German business ethics: an intercultural comparison. J. Bus. Ethics 41 (3), 195–216.

Paun, D., 2018. Corporate sustainability reporting: an innovative tool for the greater good of all. Business Horizons 61 (6), 925–935.

Preston, L.E., O'Bannon, D.P., 1997. The corporate social-financial performance relationship: a typology & analysis. Bus. Soc. 36 (4), 419–429.

Reuters, T., 2019. Thomson Reuters ESG Scores. Thomson Reuters, Canada.

Robinson, S., Wood, S., 2018. A "good" new brand—what happens when new brands try to stand out through corporate social responsibility. J. Bus. Res. 92, 231–241.

Salah, W., 2018. The impact of country-level and firm-level on financial performance: a multilevel approach. International Journal of Accounting and Taxation 6 (2), 41–53.

Smith, N.C., 2003. Corporate social responsibility: whether or how? Calif. Manag. Rev. 45 (4), 52–76.

Timbate, L., Park, C.K., 2018. CSR performance, financial reporting, and investors' perception on financial reporting. Sustainability 10 (2), 522–538.

Tuzzolino, F., Armandi, B.R., 1981. A need-hierarchy framework for assessing corporate social responsibility. Acad. Manag. Rev. 6 (1), 21–28.

Ullmann, A.A., 1985. Data in search of a theory: a critical examination of the relation-ships among social performance, social disclosure, and economic performance of U.S. firms. Acad. Manag. Rev. 10 (3), 540–557.

United Nations Global Compact, 2018a. Environment. Available at https://www.unglobalcompact.org/what-is-gc/our-work/environment. Accessed 16th December 2018.

United Nations Global Compact, 2018b. Social Sustainability. Retrieved from https://www.unglobalcompact.org/what-is-gc/our-work/social. Accessed 16th December 2018.

United Nations Global Compact, 2018c. Governance. Retrieved from https://www.unglobalcompact.org/what-is-gc/our-work/governance. Accessed 16th December 2018.

Venkatraman, N., Ramanujam, V., 1986. Measurement of business performance in strategy research: a comparison of approaches. Acad. Manag. Rev. 11 (4), 801–814.
Verboven, H., 2011. Communicating CSR and business identity in the chemical industry through mission slogans. Bus. Commun. Q. 74 (4), 415–431.
Vogel, D., 2005. The Market for Virtue: The Potential and Limits of Corporate Social Responsibility. The Brookings Institution, Washington.
Waddock, S.A., Graves, S.B., 1997. The corporate social performance-financial performance link. Strateg. Manag. J. 18 (4), 303–319.
Walton, C.C., 1967. Corporate Social Responsibilities. Wadsworth Publishing Co, Inc., Belmont, CA.
Weber, M., 2008. The business case for corporate social responsibility: a company-level measurement approach for CSR. Eur. Manag. J. 26 (4), 247–261.
Williams, G., Zinkin, J., 2008. The effect of culture on consumers' willingness to punish irresponsible corporate behaviour: applying Hofstede's typology to the punishment aspect of corporate social responsibility. Business Ethics: A European review 17 (2), 210–226.
Wood, D.J., 1991. Corporate social performance revisited. Acad. Manag. Rev. 16 (4), 691–718.

CHAPTER 17

Awareness-led social lab on circular economy in Switzerland: Exploring serendipity

Darya Gerasimenko[a,b,c] **and Erica Mazerolle-Castillo**[d,e]
[a]Professor of Sustainable Development at Samara National Research University (SSAU), Russia
[b]Lecturer in Economics at St Gallen University (HSG), Switzerland
[c]Co-facilitator, co-creator, *Science and Mindfulness Lead* (as a Senior Researcher at École polytechnique fédérale de Lausanne (EPFL)) of "*Beyond Waste: Circular Resources Lab 2018*" in canton Vaud, Switzerland
[d]Social Innovation & Communities of Practice Lead at Impact Hub Lausanne/Geneva (IHL/IHG), Switzerland
[e]Co-facilitator, co-creator of "Beyond Waste: Circular Resources Lab 2018", Switzerland

1. Introduction

The complexity of UN Sustainable Development Goals (SDGs) as well as the limited time to reach them signifies a completely new meaning, quality, and depth of collective intention, partnerships, and co-creation. At the time of writing this chapter, we are a decade away from the deadline to achieve the global SDGs, 2030 (UN, Sustainable Development Goals (SDGs), 2015). An opinion piece in *Nature* coauthored by the former executive secretary of the UNFCCC and a member of the IPCC, among others, indicates that avoiding catastrophic and irreversible damage to human prosperity requires radical collaboration across unusual partners by 2020 (Figueres et al., n.d.). Sustainable Development Goal 17 on partnerships has become key for successful work on all other SDGs (UN, Partnerships for Sustainable Development Goals, 2019). The topic of how to create meaningful and effective partnerships and what is needed for them is indeed gaining momentum around the world (Stibbe et al., n.d.).

Our action research within the awareness-led social lab "Beyond Waste: Circular Resources Lab" presented in this chapter focuses on the human capabilities necessary for effective co-creation valuing unity in diversity. What does it take to co-create a space of trust for bridging people, siloed organizations, and institutions to achieve meaningful dialogues and a deep understanding of each other's worldview?

The circular economy approach relates to improvements in all Sustainable Development Goals, and particularly, but not exclusively, SGD12 on consumption and production. Several literature reviews have demonstrated the variety and complexity of the meaning of circular economy (Reike et al., 2018; Korhonen et al., 2018 and others). The circular economy concept belongs to the so called "essentially contested" concepts,

Circular Economy and Sustainability, Volume 1
https://doi.org/10.1016/B978-0-12-819817-9.00025-9

implying that there is agreement on the goals of this concept but disagreement on how to define it (Korhonen et al., 2018).

One of the recent definitions presents circular economy through the necessity to "develop systems approaches to the cooperation of producers, consumers and other societal actors in sustainable development work" (Korhonen et al., 2018). This approach explicitly emphasizes the role of cooperation and partnerships in circular economy transformation. However, that precise aspect of partnerships for circularity has not been well studied yet, and more research is needed along those lines (Schroeder et al., 2018). The weakness of social and institutional dimensions in the narrative around circular economy was well emphasized in Moreau et al. (2017).

Circular economy requires societal innovation as much as it requires technological innovation. Current experiences on circular economy transformation demonstrates that such a degree of complexity and interdependency requires high levels of trust as well as appreciation of the value of co-creation among stakeholders. A new quality and even forms of governance for such transitions, e.g., social labs and/or living labs, are gaining momentum in different parts of the world. Multistakeholder approaches are central to social innovation spaces, where, following the quadruple helix model of Open Innovation 2.0, representatives from all parts of society co-create solutions together: academia, business, government, and civil society (Salmelin and Curly, n.d.). Those labs are becoming safe spaces for experimentation and learning for sustainable solutions across all types of diversities: age, sector, gender, nationality, education, etc. Given the challenge of achieving co-creation in such diverse groups, the following question arises: how to orchestrate social fields for the co-creation of true circular economy solutions? With this question in mind, we started our social lab journey in the Canton of Vaud in Switzerland.

New governance values are entering the stage of sustainable transformation: agile governance (WEF, Agile Governance, n.d.-a), collective leadership by Kuenkel (2016), Sociocracy 3.0 by Bockelbrink et al. (2019), and many others. They allow us to experience new patterns and depths of human interactions for a better world. Muff (2018) recently offered a great overview of the participatory methodologies for reaching the SDGs. Moreover, the theory U process developed by Scharmer (2016, 2018) as well as Art of Hosting Community (n.d.-b), Dragon Dreaming (n.d.-c), and other schools and techniques have opened new opportunities for ecosystem work, offering frameworks and methodologies to practically foster meaningful connections and partnerships for sustainable transformation.

The evolution of the social/living lab concepts over time has recently entered a new stage: awareness-led social labs, where awareness-based practices such as Social Presencing Theatre (SPT), mindfulness, meditation, unity consciousness calibration work, and others have been applied alongside the traditional multistakeholder tools such as design thinking, frame innovation, etc. Recent research from the Max Planck Institute in Berlin led by Prof. Singer within the ReSource Project in the field of social neuroscience demonstrates the value of mindfulness and meditation for the co-creation of a "caring" economy and cooperation in general (Singer and Engert, 2019; Singer, 2018; Trautweina et al., n.d.). Such depth in the quality of social lab spaces opens up new opportunities to explore and understand how to use partnerships (SDG 17) as a means to reach other SDGs, as well as what potential awareness-based approaches contains for sustainable transformation in general.

Awareness-led social labs in the context of the urgent need for change are growing steadily around the world. However, the literature on these experiences is rather scarce (Oxford Lab, 2018). Our lab is one of the first of such explorations and more work needs to be done on how and why those spaces work, and how to use them effectively.

The aim of this chapter is to share the experience and learning from our awareness-led social lab, Beyond Waste: Circular Resources Lab 2018. We share our insights on the values set and self-work needed to experience healing interconnectedness in the context of increasing complexity and diversity through the magic of serendipity. This chapter contributes to the growing literature on this subject and aims to improve the quality of such spaces by discussing their challenges and opportunities. It does not follow a particular "methodology," but rather opens up space to study possibilities of new formats of human interaction at this stage of our action research. This social lab is itself a prototype; however, it is a step forward as it offers new insights and opens a scientific discussion for further exploration.

The chapter is organized as follows: the first section provides a brief description of our social lab experience in Switzerland, its process, and main milestones. The second section presents the serendipitous learnings from our lab hosting team. The third section offers the cohort participants' views on this first prototype. The final section concludes with a discussion and open questions for further research.

2. Beyond Waste: Circular Resources Lab 2018[a]

Beyond Waste: Circular Resources Lab 2018 has been a co-creation process of the following actors: Chair of Green Economy and Resources Governance at the Swiss Federal Institute of Technology in Lausanne (EPFL), Impact Hub Lausanne (IHL), Impact Hub Geneva (IHG), collaboratio helvetica (cohe), and the UN Sustainable Development Solutions Network (SDSN) in Switzerland, with the financial support of Migros Pioneer Fund.

Twenty-five participants were selected through a rigorous online application process calling for the co-creation of circular economy solutions for the Canton of Vaud in Switzerland, utilizing the Theory U process and other awareness-based tools to allow

[a] More information is available at www.circularresourceslab.ch.

serendipity[b] to emerge for our collective learning and healing. Four prototype solutions for circular economy were formed within this journey from September to December 2018, in addition to the social lab prototype itself.

Both authors of this paper have been the primary responsible co-creators for this lab, with the Swiss Federal Institute of Technology in Lausanne (EPFL) represented by Dr. Gerasimenko (Science and Mindfulness Lead) and Impact Hub Lausanne/Geneva by Mazerolle-Castillo (Catalyst).

This social lab is, to our knowledge, the first prototype of an awareness-led social space within the circular economy domain, marking the beginning of such exploration. It is envisaged that the learnings and open questions from this first round will contribute to the growing knowledge on social innovation for sustainable transformation (e.g., circular economy).

This experiment took shape in a unique setting. First, both the steering committee and the co-design and co-facilitation team came together in a cross-sector collaboration without any structure or hierarchy, a challenge in itself given their different organizational cultures and norms. Therefore, cross-sector learning on the orchestration of such a trust-based space already represents a significant value added on its own, even before the start of the social lab itself. Second, we were free to work with any topic within the CE domain (SDG 12) and any methodologies and ideas as a pure experiment to explore the potential of the social lab format. Third, we were not bound by any particular "expected results," hence the value added has been a social lab prototype in itself. Most important was the exploration of such spaces and their potential for SDG solutions in general, using the topic of circular economy as a focus.

2.1 Our social lab journey

Phase 0: Learning about synchronicities around social lab initiatives on circular economy in the Canton of Vaud (September 2017–March 2018)

Synchronicities are a real sign of "readiness in the air" to act on certain issues in a certain way. Parallel to each other, two groups of people started working on very new narratives within the Swiss Romandie context: social labs, awareness-led innovation (i.e., Theory U and its U-process), and circular economy transformation in one form or another. Both groups had been working under already existing living/social lab frameworks inside their institutions, both in Lausanne in the Canton of Vaud. They were the Chair on Green Economy and Resource Governance (GERG) led by Prof. Dr. Bruno Oberle at the Swiss Federal Institute of Technology in Lausanne (EPFL) and Impact Hub Lausanne (IHL). Other institutions interested in this experimentation were collaboratio helvetica (cohe) and SDSN Switzerland.

[b] Serendipity in this context refers to favorable outcomes/events that emerge without having been anticipated or expected and as a result (in our case) of collective intelligence activation.

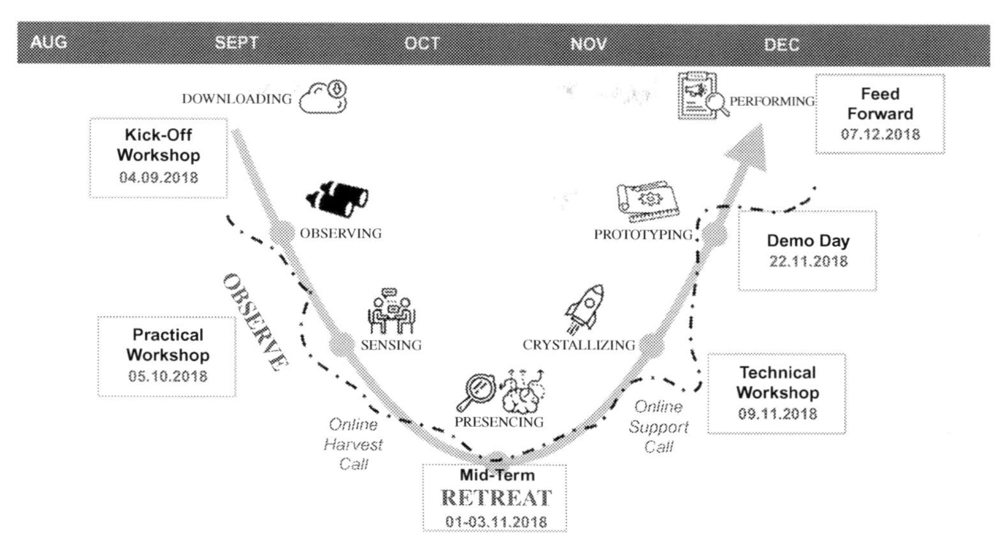

Fig. 1 Beyond Waste: Circular Resources Lab journey. *Source: Authors' own depiction.*

Since it was a new initiative both for Switzerland and the world, there was no best practice to refer to in terms of how to deliver such a complex and innovative project. The nature of such a project is a safe co-creation space with common values and readiness to go beyond the comfort zone and grow in all domains: intellectually, socially, spiritually, etc. At the joint meeting in April 2018, we decided to start with ourselves and create the first partnership of this journey between IHL/IHG and GERG-EPFL. This is how the titles of two separate projects became one: "Beyond Waste: Circular Resources Lab."

Phase 1: Hosting team co-creation and preparation (April–August 2018)
A website was set up to provide visibility about the process, the methodologies, and the overall journey. More than 50 applications were received for 25 places available. Upon selection, co-creators (as we called participants) were asked to share statements answering the following questions: What experience of yours do you bring to this lab [mind]? What excites you about this lab [heart]? What do you expect to happen as a result of this lab [will]? This method of sharing motivation and intention virtually was a conscious choice to spark the co-creators' co-initiation process. Co-initiation, the first out of five of the U-Process' stages, consists of bringing together a group of stakeholders around a shared purpose and creating a safe container in which collaboration can flourish, leading to the co-creation of solutions. The co-initiation phase began with the information session, continued virtually with the application and selection process, and culminated in the Kick-Off Workshop in September 2018 (Fig. 1).

Phase 2: The social lab process (September–December 2018)
The U-Process for leading personal, organizational, and societal change comprises five main stages: co-initiating, co-sensing, presencing, co-creating (incl. crystallizing and

prototyping) and co-evolving (Scharmer, 2016). Within the social lab process, one full-day workshop was offered for each of these five stages, including a 2.5-day workshop with overnight stay for the presencing phase at the bottom of the U. The frequency of workshops was every 3 weeks at the beginning, and picked up speed in the second half of the process, taking place every 1–2 weeks. The time periods in between workshops were also considered as important parts of each stage, as co-creators were expected to actively engage with each other and work on their challenges using the methods and practices learned in person. We set up a Slack channel (an instant messaging platform) to enable ongoing communication and teamwork among the co-creators, including the hosting team.

2.2 Kick-off workshop

The Kick-off workshop was the culmination point of the co-initiating phase, the process through which a diverse group of people gather around a shared intention and begin to form as a social body. It started with welcoming words describing the context and purpose of the lab, followed by a mindfulness session focused on the feeling of "gratitude." It was then structured in two main parts. The morning icebreaker used storytelling to create a group dynamic and map the ecosystem, nourishing the relational aspect, and bringing collective awareness to the social field. The afternoon was focused on content, specifically aimed at ascertaining the status quo surrounding waste and circular economy in the local region (Canton of Vaud) and pertaining to the three preselected fields: plastics, water, and electronic waste. Experts in the fields of plastics and eco-design, water management, and electronic waste were invited to provide an overview of the state of affairs in their respective fields. One session also provided input on the circular economy concept, as well as on the Theory U process and the social lab approach. The last session of that day used an indigenous methodology (Dragon Dreaming circle) to weave individual intentions into a collective vision, co-creating the lab's manifesto (collective intention map).

2.3 Practical workshop

The Practical Workshop marked the beginning of the co-sensing phase, which is dedicated to exploring the system from different angles with the intention of understanding it through a wide range of perspectives, particularly those of voices at the margins. The morning mindfulness session focused on sensing the feeling of "deep listening." Overall, the workshop's aims were twofold: first, to equip participants with tools and methods that awaken perceptive abilities beyond observing with the analytical mind, such as deep listening, sensing with the heart (empathy), and experiencing generative states of conversation (co-creation). Second, for co-creators to identify which specific topic to address within the priorities we identified for our region and form teams around them.

The tools and methods shared and practiced included the Iceberg Model, Levels of Listening, Dialogue Walks, and Case Clinics. Those methodologies prepared participants for "sensing"—the ability to integrate information in a more purposeful way, to inform our next course of action through intuition rather than only analysis. This was extremely new and unusual, pushing many co-creators out of their comfort zone. Yet it contributed to developing genuine relationships, a sense of community, and the discovery of a novel way of addressing complexity. As one co-creator commented at the end of the workshop:

> *I think it's very positive to implement activities focused on human development in order to address very real problems. In an academic or scientific context, we value rational-individual thought above emotional and collective aspects. The Beyond Waste exercise is very audacious and at the same time organic. You are introducing humanity into a problem-solving context. Many of us engineers are completely new to this and you are helping us to learn how to walk in this new environment. It's like a complete rewiring of our brains.*

In the 3-week period between the Practical Workshop and the Midterm Retreat, co-creators were tasked with going on learning journeys and conducting stakeholder interviews, a cornerstone sensing method, to collect insights from actors in their field of enquiry (Picture 1).

2.4 Midterm retreat

The Midterm Retreat corresponded to the presencing phase of the U-Process, where one connects to the source and accesses one's most authentic self, letting go of expectations and old convictions to understand one's purpose in the broader context. We chose a retreat center outside the city. The workshop spread over 2.5 days to allow for spacious reflection and inspired action.

The first morning's mindfulness session focused on "reconnection to self." The following sessions aimed to explore a new mindset and its potential: unity consciousness rather than separation consciousness. Through a sensory walk in the forest, collective art sessions, and guided journaling, we experienced bridging the three major divides that are at the root of the problems facing humanity today: the divides between self and nature, self and others, and self and self (Hubbard, n.d.; Scharmer, 2016).

Up until this stage, the group had not reached full clarity on which teams would be formed and what specific topics would be handled. Through a sharing circle, we set our intentions for shifting from reflection to action mode; new topics and teams were born. We then proceeded with 3D modelling—a technique to map the system in order to better apprehend it, and identify a direction of change to start prototyping. To conclude, we all shared what we were ready to let go of, and what we were committing to or letting come.

2.5 Technical workshop

The Technical Workshop marked the beginning of the co-creation phase, where we used prototyping to translate an idea or a concept into experimentation and action. It

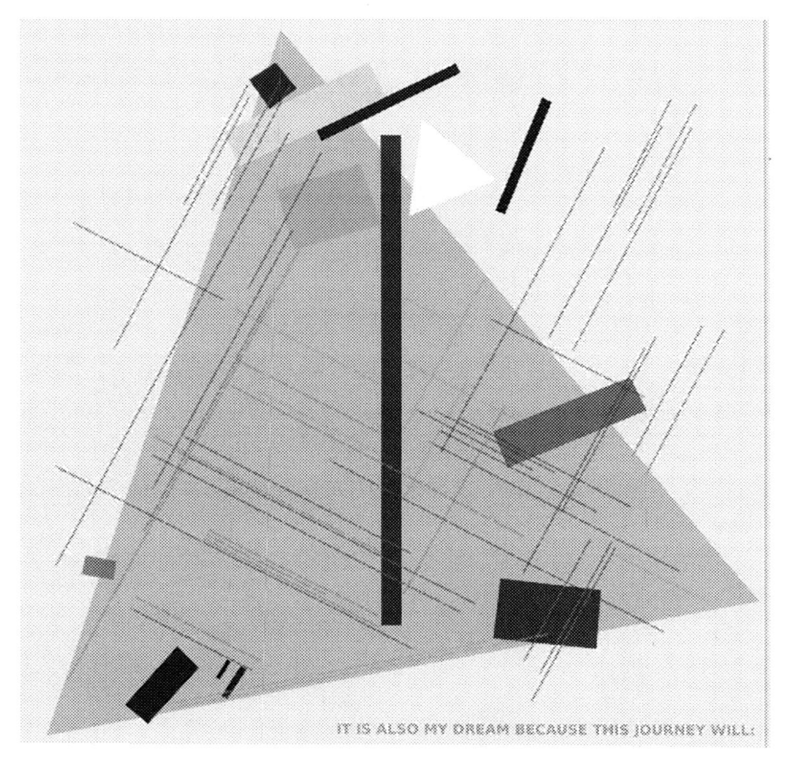

Picture 1 Collective manifesto of our shared dream for this journey, co-created during the Kick-off workshop (those are the individual statements that contribute to the big shared vision). *Source: The art piece of our Collective Manifesto created by our social lab participant, artist Katalin Hausel.*

is about being inclusive and responsive, operating with heightened energy levels, and tuning to future possibilities, functioning like a vehicle for the new that wants to emerge. We began with a mindfulness session focusing on "interconnectedness," followed by an expert's input on prototype design. Working through the steps of the Design Thinking process, we learned to embrace the prototyping mindset of iterative design. After an intensive design session using Empathy Maps and User Journeys, among other tools, we had a whole-group exchange on the progresses of prototype solutions, bringing to the surface the synergies and collaboration opportunities present within the cohort.

2.6 Demo day

The Demo Day aimed to present the prototype solutions for the first time to a public audience, in view of receiving feedback to continue iterating. Throughout the process, co-creators had reframed the problem they were tackling, defined who they were designing a solution for, and deconstructed the solution into manageable pieces for more focused action and development.

The day started with a mindfulness session focusing on "true human empowerment." Then, a pitching training was offered by one of our experienced co-creators, and teams reworked their presentations before the evening event. It was important to constantly remind each other that we were still in the co-creation phase of the U, where generating feedback is the main objective (rather than being judged). As expressed by members of the cohort:

I enjoy this open and nonjudgemental space we created here.

Important is not the objective but the whole journey that we went through together.

Something next to prototypes emerged, that is difficult to explain, but it is here.

2.7 Feed forward

The Feed Forward Workshop corresponded to the sixth and the last phase of the U-Process, co-evolving. While the active part of the journey ended, in reality, it signified the beginning of a new reality where actors actively cooperate with each other to develop solutions to common challenges. We thus invited various stakeholders from the Swiss public administration (federal and cantonal) as well as politicians to include the government dimension in our social lab prototype, as this had been missing up to that point. Before our 20 guests arrived for the public dialogue, we started the day with a meditation on "absolute love frequency."

The second part of the day was dedicated to a conversation with external guests at the intersection of public space and personal responsibility. Overall, the dialogue sought a collective reflection on awareness-led innovation to achieve the Sustainable Development Goals (SDGs). Three distinctive threads emerged from the discourse:

First, the gap between how the current system works and what we actually want to achieve.

Second, a consensus about the need for transformative change, and spaces that support it, as a prerequisite for achieving the SDGs. There seemed to be "an experience gap" between the acknowledgement of what is needed, and the actual knowledge of how to achieve it. Overcoming the fear of failure and changing the way success is measured were emphasized as drivers of greater experimentation.

Third, a realization that systemic transformation implies a paradigm change, which itself departs from individual self-responsibility; achieving systems change requires openness, willingness, and courage to interact at a deeper level to tackle the root causes of existing problems.

The day ended with a reflection among the cohort members about ecosystem awareness as an alternative philosophical framework to political economy.

Phase 3: Reflection and Research (January–September 2019)
How to study this type of social innovation space is itself an open question. What has become clear is that the classical evaluation frameworks based on quantitative outputs, predefined expectations, and parameters are no longer appropriate. They do not show the entire spectrum of possibilities and serendipity elements that are present in such spaces. Those impacts are nonlinear and vary along the time spectrum. Moreover, such social spaces of this quality of presence should not be judged until we learn more about what they truly are, what they are capable of brining, as well as how to navigate their benefits.

The quality of such spaces depends on many factors: facilitation quality and experience, level of engagement of all actors, their background and "readiness" to experience a "change," as well as the willingness to actually follow that change potential. From our experience, great quality social spaces work as an opportunity for change to happen, which depends on each participant's personal decision on how and to what extent they will allow it to unfold: at what pace, when, in which area, with whom, and how. What also became evident is that high levels of trust, compassion, and nonjudgement are capable of supporting everyone in the direction of their own needs and aspirations. Hence, the benefits of the experience are very individual and can vary greatly from person to person depending on their own departure point, intention, and the current life question or struggle they are facing. This became clear through our usage of various harvesting methodologies during the process.

We collected information mainly through entry/exit questionnaires, post-its and handwritten notes, personal blog entries published by some co-creators on our website shortly after the lab, as well as interviews conducted half a year later that relied on the Most Significant Change (MSC) methodology (Dart and Davies, 2007; Limato et al., 2018). From our experience, even the MSC methodology does not allow us to access the full value and range of the experience and its benefits. A special interview protocol should be designed to study those spaces (some suggestions that will come in the next publications and the continuation of this work). As we have just started working with such spaces in various contexts (topics, countries, etc.), we are convinced that it will show more of its facets through more practice and reflections. The next section shares the main insights and learnings collected thus far.

3. The serendipitous learnings from the experience of our cohosting team

An important learning in this entire process has been how to allow serendipity to emerge by removing expectations, silos, and limiting beliefs. First, we have created a space where the transformative journey of the cohort starts with ourselves, our deep trust, and unconditional support among us as a cohosting and organizing team.

There were two circles originally involved into the orchestration of this social lab process: a cohosting team of facilitators consisting of Dr. Darya Gerasimenko (EPFL), Erica Mazerolle-Castillo (IHL), Cynthia Kracmer (IHG), and Isabelle Ruckli (cohe), as well as the steering committee represented by Michael Bergöö (SDSN Switzerland), Dr. Darya Gerasimenko (EPFL), Prof. Bruno Oberle (EPFL), Felix Stähli (IHG/IHL), and Nora Wilhelm (cohe). Both circles changed during the process. With a blank canvas to start, observing the emergence of this orchestration and learning from it throughout was very revealing. We committed to make this unique experiment work and trusted it addressing and healing our own egos along this process.

Academic literature on the Theory U process within the lifelong education domain focuses on what participants learn through their U journey (Pomeroy and Oliver, n.d.). We would like to add here that the facilitation team and steering committee also went through a transformative learning U process and more, and further scientific exploration of the governance/orchestration aspect is essential to learn its potential. In what follows, we offer learnings pertaining to three domains where serendipity unfolded from the perspective of the cohosting team: governance of the space; mindfulness and its role in the process; as well as emergence of prototype solutions.

3.1 Governance (Orchestration) based on respect and trust

Before the partnership between Impact Hub (business and civil society) and GERG-EPFL (academia) took shape, each institution was envisaging realizing their social labs independently. The collaboration between Impact Hub, GERG-EPFL and then cohe and SDSN Switzerland, contributed to an interesting positioning, which attracted a high diversity and quality of participants. More specifically, social innovation facilitation and entrepreneurship expertise (provided by Impact Hub and cohe), and rigorous science and mindfulness inputs (provided by GERG-EPFL), as well as the global network of SDSN enabled the unique design and delivery of the innovative social lab format.

Originally, the cohosting and facilitation (core) team consisted of four people: the catalyst and community of practice lead, the social innovation design lead, the documentation and storytelling lead, and the science and mindfulness lead. However, no dedicated role was fully or officially responsible for the governance of the lab, i.e., managing partnership agreements and relationships between partners and stakeholders, as well as fundraising. A few times during the process, extra meetings were held to address related issues, draining energies away from the main focus of delivering a high-quality social lab process, as we still operated in an old framework of "management needs." We thought that governance/management was needed to ensure smooth operations, and not having a dedicated person for that role posed a challenge. However, the pure experimentation and exploration nature of the lab allowed us to prototype a new quality of cross-sector interaction where no structural governance was established or even needed per se. It opened new horizons to explore further how to work across sectors from a place of trust, unconditional support and love, openness, and without any protocols and "obligations." Work is distributed and done according to "I can handle it well," not because "I have to do it" or someone told me to deal with it. This opened a whole new perspective of what unity field navigation or orchestration actually is, as opposed to management.

Our partner and one of our sponsors—SDSN Switzerland—through its powerful international network offered us an opportunity to apply, and as a result of the selection process, to present at the United Nations SDSN Global Solutions Forum 2019 at Columbia University, where we shared our insights on awareness-led innovation and received valuable feedback (Gerasimenko and Mazerolle, 2019).

Another serendipitous element was a new discovery within our core facilitation team. Krista Kaufmann (IHL) was supposed to help us with logistics during our workshops, and to our great surprise, she became our graphic facilitator when from the very first workshop she voluntarily started drawing the social field and its evolution. This is how by the Midterm Retreat we had unexpectedly become a group of five facilitators. This awareness-led social lab space allowed Krista to discover and develop her amazing graphic recording talent. The graphic recording book of our lab is also available online (Picture 2).

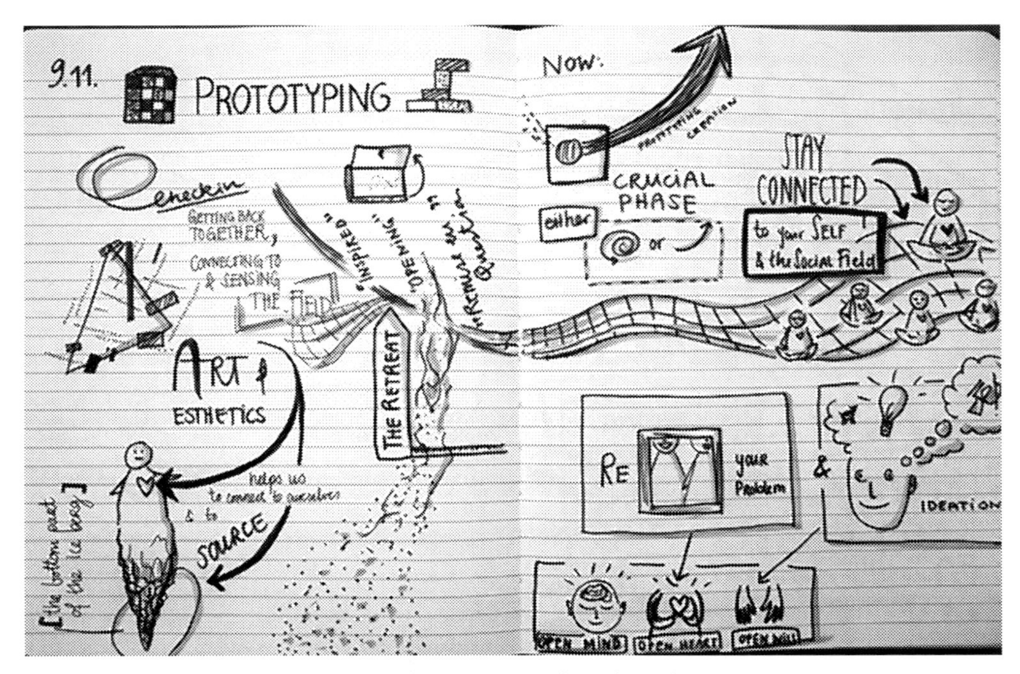

Picture 2 Generative Scribing piece during our Technical Workshop. *Source: By our Generative Scribe—Krista Kaufmann.*

3.2 Mindfulness and meditation

The topic of mindfulness in relation to innovation is gaining momentum in scientific research. In September 2018, the Copenhagen Business School held a "Mind, Meditation, and Innovation Conference," exploring mindfulness as a facilitation tool for innovation activity around the world (Copenhagen Business School, Mind, Meditation and Innovation Conference, 2018). The special issue on Mindfulness in the academic journal *Current Opinion in Psychology* in August 2019 edited by Bernstein et al. (n.d.) spawned a wide discussion in academia. In total, 60 academic papers were presented to explore mindfulness in the broad spectrum of its applications: from health benefits to education, business innovation, and even politics. In November 2019, neuroscientists of the Brain Institute of the Russian Academy of Sciences together with the Moscow State University opened the "Meditation and Altered Consciousness Research Centre" in India, working together with several Buddhist monasteries and in cooperation with the Dalai Lama (News, 2019). Just to name a few more trends adding to the impressive work of Max Plank Institute and Prof. Singer described in the introduction (Singer and Engert, 2019; Singer, 2018; Trautweina et al., n.d.).

While each full-day workshop along the U process was unique in its design, common elements recurred. They always began with a mindfulness session led by Dr. Gerasimenko (Science and Mindfulness Lead) to enhance self-awareness and clarity, and to tune into the unity consciousness field through particular themes along the U process: Kick-Off Workshop—gratitude, Practical Workshop—deep listening, Midterm Retreat—reconnection to self, Technical Workshop—sense of interconnectedness, Demo Day—true human empowerment, Feed Forward—absolute love. While only 1/3 of participants had had prior experience with meditation, an overwhelming majority of the participants[c] found it useful and valuable, however, in their own unique way given their background and prior mindset.

> *Lara (48, entrepreneur): "The mindfulness sessions were very useful, it's something that we learn for life ... So, these little "pockets of breathing", these moments of meditation are very good."*

> *Annabel (27, governance and social innovation officer): "I found also quite interesting the mix between rational and emotional that it was really expressed without any shame. Now it is getting less surprising, but these mindfulness sessions ... people might think kind of "what is going on here? Are we here to work?" However, in the age of too much information in the 21st century, artificial intelligence, IT, etc. it was important to reconnect to what makes us human and unique, me or you, everyone ... Reconnecting to what is fragile and unique in humans. For decades it was*

[c] For anonymity, the names of the interviewees in this section have been changed; however, the age, sex, and background are real.

science, math, to find solutions ... but what if actually the key is our emotions, our empathy, is our ability to connect and relate to each other?"

Richard (40, chemical engineer): "I think what was really important for me was the type of methodology that was used—deep listening exercises, the meditation portion that was really helping me to concentrate on what would come next. I have a tendency in my daily life to really jump from one topic to the other and kind of treat things superficially because you need to move on, to do task one, task two, task three ... and then you move to task four, you don't have time to come back to it. And in this sense, having those breaks of meditation, those moments when we were sharing quite freely and openly, it was really a very appreciative change in the way you can interact with other individuals."

Mike (52, engineer): "I knew before this experience that I was not going to be someone for meditation, I tried a lot of techniques of meditations with experts such as mindfulness, hypnosis, etc. and I have never really succeeded in finding sense, or the feeling that I was succeeding in those exercises. So, what was interesting in the change in me, was more in discovering different ways of meeting people, talking to people and creating relationships within a group of people from different backgrounds and age pretty well. And feeling comfortable sharing things that you would not share in a normal work relationship."

As for the cohosting team (five people), we actively utilized mindfulness and meditation practices for our own healing. Our facilitation team's unity field calibration allowed us to co-create a healthy and attractive space for us first, and therefore, for the cohort participants. An important element of our overarching U process was a 1-day mini U process for the facilitation team, which we did on October 1, 2018, after our Kick-Off Workshop.

3.3 Serendipity in the evolution of prototypes

An initial exploration of the circular economy field in the Canton Vaud was conducted through a dozen stakeholder interviews to gather inputs from government, business, academia, and civil society. Based on those interviews, we selected three focal topics for this social lab journey: water, plastics, and electronic waste. However, the clustering exercise in the Practical Workshop resulted in the emergence of new priority topics better reflecting the current needs: consumer behavior, circular business models, and electronics.

During the Midterm Retreat, however, by tuning in to the field on what is truly needed in order to foster circular economy transformation in Switzerland, topics naturally evolved away from new goods or technologies towards services and education, placing focus on human development and evolution through knowledge and networks.

Brief descriptions of the prototypes that were finally born during the third workshop along the U and onwards are as follows (more information is available on the webpage):

3.4 Circular human incubator

Integrating practical trainings on circular economy into the social welfare and unemployment insurance system in order to harness underutilized human potential and build human and intellectual capacity within an often-marginalized segment of the population.

3.5 Atelier des Futurs

A dedicated community space nestled within a commercial area, which promotes non-transactional consumption and postcapitalist ways of living, thinking, and interacting.

3.6 Circular academy for construction (CA4C)

An educational program on circular economy opportunities in the construction sector, targeting SMEs.

3.7 Magic mushrooms: Exploring the potential of the spent mushroom substrate

Valorization of (usually wasted) mushroom substrate for soil regeneration and a service that develops customized organic fertilizing solutions for farmers.

As one participant recalls:

> *We are not doing the thing that we were discussing in the Lab, the proposal initially discussed doesn't work from a business point of view, but the good thing is that the interaction with the people in BWL gave me a different perspective, that I'm evaluating further. In simple words, what happened is that instead of offering a product, I decided to move towards services. This is another thing I've learned around what CE is, that businesses could move more towards this area of services, creating higher value by finding different ways to do things without the need to produce more stuff. This is something that I learned … So, I switched the way to do things and I hope we continue in the right path with that.*

Those were the first prototypes that were born out of this open space by a group of strangers coming together on the social lab journey (as a prototype) with two of them having a "case giver" role among our participants. However, for further iterations, additional support mechanisms (financial, people, network, etc.) should be included in the entire design of such long-term spaces for the future. The information about those prototypes is online for all interested people who could be interested to develop them further.

4. Selection of the serendipitous insights from the lab cohort members

Here we give space to our co-creators' voices without any major analysis from our side, allowing readers to draw their own conclusions.

Excerpts from the blog entries written within 1 month of the end of the awareness-led social lab (names in this section are real)

Katalin (49, artist, designer): "The journey felt like the organic process of forming a community. The structure proved to be flexible to allow for adjustments, but rigorous enough to hold us in place and achieve the goals that were set at the beginning. It allowed us to see how we could work together, and how we can support each other in the future. In the end, I felt a strong conviction that we started something important, and that we all will be part of the unavoidable change, working together; and also felt extremely privileged to have met such a group of exceptional people."

Alexis (26, program manager): "As it turns out, letting yourself go and trusting others to lead you through a new experience can allow you to connect to the issue—and yourself—in ways you had not expected … In fact, looking back, it was this emotional connection to nature and to each other that made this Lab so successful: and is the key ingredient that is missing from so many other incubators looking to broach the same issues. It is this very disconnect from the heart, the mind, and community, that has led us into the linear economic model—towards the ego and away from the eco … if I could go back and tell myself one thing before I started, or to any others who are looking to get involved in Social Labs with the Impact Hub or others: the unconventional techniques may be exactly what you need to get out of the headspace of stale and tried ideas and into a different sphere of creativity and connection. You just might surprise yourself!"

Sascha (57, engineer, entrepreneur, educator): "Before we started, it was far from obvious that 25 so different people, co-creators and organisers (this distinction became less and less important throughout the process), with different backgrounds, lives, experiences and extremely divergent ways of thinking would find anything in common at all, let alone build such a community and produce something coherent and potentially very useful. A bit of patience, a general interest in the topic of circular economy (CE), a good deal of openness and curiosity and broadly aligned values was the seeds that grew into a real community based on deep respect and friendship, as well as beginnings of promising projects . . . . In my personal opinion, openness and value alignment were the two key ingredients."

Jérôme (41, chemical engineer): "On the first workshop days, instead of starting with the classical 'We have a problem, let's get a solution', I am asked to practice meditation, share emotions and feelings as they come, co-create a manifesto, practice deep listening, presencing theatre … My first thoughts were: 'Where are the solutions in all that?', 'That's not an engineer approach to finding solutions … When are we starting the real work'? Needless to say—way out of my zone of comfort, but I decided to stay on board … And that was only to get even further away from my usual way of functioning: putting an engineer in a centre of mindfulness for two days was in my head like throwing a cat in cold water. But the group was so open, the intentions so genuinely positive and the overall goal (make circular economy progress) so appealing that I felt I should be part of this adventure. As interactions continued, the answers to my initial questions started to appear: the 'real work' started as soon as we met and got to know each other, to build a real circle of trust such I had never thought you could do with 'strangers' in such a short time. It all went by small steps, exercises, exchanges, discussions, writings, listening … But through all that and thanks again to the openness of each and every participant, things aligned more and more clearly in my head and in my heart."

Testimonies from the interviews[d] conducted in July 2019 (half a year after the lab) For the readers, we provisionally grouped those testimonies into four major categories: personal development and new skills; community building and ecosystem network; deeper understanding of CE and the awareness-based practices; and taking action on CE and climate.

4.1.1 Personal development and new skills

Samuel (37, microbiologist, entrepreneur): "The main change is to realise that you are not the carrier of some truth, everything is relative and yeah, you can learn much more putting down your own position and listening. Listening is an active process that I'm applying now … You can learn a lot just by observing and listening in an active way. And you catalyse more things doing this effort, yeah. And in practice, well for my project, my start-up project, actually it was positively influenced because I had the opportunity to test several ideas with people I would not have met in other conditions, so it was a good place to do that, and yeah, you can catalyse your hypothesis and test it faster. Super cool for me."

Paul (28, microengineer): "The most significant change for me was to discover and learn tools to co-construct a different world. I was able to experience co-creation and feel the power of collective intelligence. I rediscovered the importance of human relationships and experienced activities that enable rich and sincere interactions. It seems as though the solutions to create the world we hope for are not futuristic and technical, but are simply found within us and revealed in the magic of being together."

Martina (34, doctoral researcher): "What struck me the most is that by doing things differently and by collaborating differently and having that space of the lab, we had different results. It was really different. First, it was a bit frustrating, I had the feeling we felt a frustration not to have the expected outputs, also from people outside of the lab, there was nothing quantitative and yeah I had the feeling it was not good enough or not what they wanted or what I thought would be a good output. But actually, given the process and the problems at stake, I had the feeling the output was the right one, I realised actually that it is by doing things differently that you create different solutions."

Richard (41, chemical engineer): "Things I have changed as a result of this lab … the way I am interacting with people, trying to listen much more deeply. Seeing what they feel and not only what they say. I used to respond quite quickly to what people say or even tried to anticipate what people say or even to complete their sentence. I felt if you refrain from that if you wait a few seconds after they are finished, people say things differently and they speak it more truly or at least you understand them in a different way. This is one. And I also changed some things connected to circular economy, how people were functioning, what am I buying, where I am going, why I am making those choices, when I am planning some trips for holidays, when I am planning to eat. Just connection to self. I was really working in the way that my day is always filled with things, and now I tend to take a bit of time to just sit, think, maybe I realised that I don't need so many activities, that many things, and I go with what I have. I changed the basic way I am reasoning in general …"

[d] For anonymity, the names of the interviewees in this section have been changed; however, the age, sex, and background are real.

4.1.2 Community building and network ecosystem

Elena (49, artist, designer): "I think for me the biggest change was finding a community, or connecting with different people and some of these relationships really proved long-lasting, for example with Zita, with whom we are still a little bit working together and became really good friends … I think also to a certain extent I can imagine that my mandate with cohe is partially thanks to the fact that I was working with Tanya there so we began to know each other, so when she was hired, she wanted to work with me. So really on a personal relationship level this was the biggest impact … created my own little sub-ecosystem … because you know I'm not Swiss, it kind of opened my eyes to all the things that are happening around."

Caroline (38, environmental engineer): "My perception of groups really changed by knowing that we can breathe life into a group, influence each other and walk as one single being. In autumn, we see birds migrating to warm climates, they move together through the sky. Doing exercises all together was a powerful experience for me, it opened something, gave me hope. Injecting this kind of energy into a group cannot be done alone, there needs to be many people, especially as we did not know each other. Knowing that this can be done was significant for me because I didn't know it was possible and it gives me hope."

Michelle (32, fiduciary accountant): "Also the fact that I met people I would have never met in my everyday life. Because when I go to seminars or when I do activities, it's always the same kind of people. Whereas there it was people who are not at all in my universe … Really at the personal level it brought me a lot, even if the project with my team will not continue, but for me at the personal level it brought me a lot …"

Christophe (33, industrial engineer): "Having some time to think and discuss with others without having any expectation or any framework it was positive, it was just that I was expecting something that I shouldn't have expected before … I'm thankful for what you did … we felt very unique, for me I was taking it seriously …"

4.1.3 Deeper understanding of circular economy and the awareness-based practices

Annabel (27, governance and social innovation officer): "This experience brought me new understanding how we can build a circular economy. It is not only about industries; it is going deeper to trigger profound deep understanding of civilisation and of society and societal change. Going beyond green growth rhetoric and even how high tech can save us. Try to find more coherence in everything and go beyond fancy trendy words. So, yes, I met a lot of people in the lab that were also aware how the words can close down the thinking and how it is beneficial growing in our thoughts … I really became much more vegan than before; I am only buying now second hand. I am doing now my morning yoga (10–30 min). I was doing it before but not in this consistency. I am happy for this positive lifestyle. I really needed it to deal with all new things in my life. The Lab helped me to understand it also meeting more people in my life. It really thought me how important it is to meet people."

Alexandre (57, sustainability professor): "When we have a very complex societal problem, it's important to work on it together. Not just to interview stakeholders, identify and interview stakeholders, understand their position and then try to design some solution that will take them as much as possible into account, but actually involving people in the process of devising a solution … That's the awareness that arose in me as a result of the process … I work in teaching, in research but especially trying to figure out how to best help the societal transition or transformation to sustainability. The whole process we went through showed me that we shouldn't try to design a solution but we should design a process that will replicate this collective solution process, this collective understanding and solution seeking process as wide as possible. In other words, some variation of what we did in the lab is exactly what we need to solve the problem of the societal condition of sustainability … the most significant thing that I'm aware of is less linked to the content that we discussed than it is to the process we went through."

Tanya (33, community manager): "The most significant change for me was a deeper understanding of the process. In the sense that I really appreciate the Theory U, the process that they describe, and it gave me a much deeper understanding from where to act and from where to not act. Because usually we jump from the same patterns and the same underlying assumptions … And I feel that when I'm going through my own processes, it gave me more of, a better sense of where I am … It basically gave me a map of orientation and I feel I'm much more patient, I feel that I'm at the moment at the bottom of the U, I don't have this impatience like "I should do something now, I need to act" It's not so strong there anymore, I feel that I have more trust in the process, like, 'aaah, now I am here and now it's time to stick with it and wait, not to do anything with it.'"

Yannick (31, student, entrepreneur): "I realised throughout the process that I needed to be much more radical in my view of the change that the society needs. That means that CE is just one element inside of a bigger picture where there needs to be a larger shift. I was thinking also that it was very interesting to hear some of the professionals that you had invited … That was a shock for me during the lab that most of the people who are professionals of this field, including university professors, will not be capable of making the change that we need. Actually, many people from civil society have more of the tools to do that … As we know that we need more emotional intelligence and spiritual awakening for the future, I am not saying the social lab provides all of this, but I think it's an easy entry into all of these questions … So, I can only recommend and encourage you to continue doing that great work."

4.1.4 Taking action on circular economy and environment

Jessica (31, senior business grower): "The lab helped us to understand each other at the personal and human level in the context of a lack of trust in the project, which was at an experimental phase in my company. Some people in the team did not want to do a proof of concept, some information was being withheld, etc. … In the end, this lab contributed to establish circular economy as a topic in one of the biggest banks of Switzerland and to build trust between a Dutch start-up and various Swiss circular economy actors."

Yannick (31, student, entrepreneur): "It was pretty timely how things happened when our Social Lab ended, that is when the climate strikes started in Switzerland, and I actually became very involved in that movement, since the beginning of 2019. That was a really good follow-up … Citizens need to take things into their own hands because politicians are not really representing their interests anymore, or at least they're not representing as fast as needed, the shift in consciousness that is happening. And so yeah, I got very involved in that because it fitted perfectly what I thought was needed to make a change happen."

Elena (49, artist, designer): "my daughter completely became a vegetarian and a climate change fighter, and her boyfriend too, I interviewed her twice, so somehow those interviews work, just by asking these questions it clicked in her head". [Comment from the authors: Elena shares here her experience on her Learning Journey's Stakeholder Interviews within the social lab process.]

4.2 Suggestions for the further iterations of social labs/spaces

Numerous reflection rounds during and after the lab, both within the team and with the cohort, yielded some suggestions for further social labs. It was a first prototype of orchestrating such complexity on many levels: a cross-sector partnership among the organizing team without a structured governance; the vertical and horizontal diversity within the cohort team; the complexity of the circular economy topic in general; the novelty of the mindfulness approach; etc. We have never claimed it was perfect—it was just the first prototype!

Four people left the lab process at different times for different reasons, however, only 1 person out of 25 fully lost contact with us (and he is still alive). We find it a very good rate for such an innovative experimental approach over several months long. Since it is indeed a transformational process of deep learning, every person takes it at his/her own pace and time—that can include initial push-back (and we accept it how it is with gratitude).

From various feedback rounds we suggest here several directions to improve the long-term quality of such learning social spaces if resources allow.

First, a case or challenge owner should offer the problem to deal with and be responsible for advising on prototype solutions that are being developed, as is the case in already existing formats such as hackathons, innovation camps, etc. In this case, such awareness-based methodologies and practices discussed above would support a deeper dive into the reconnection to self, others, and nature. However, allowing an open space for the emergence of specific topics can bring interesting outcomes, as they produce new insights on what the problems actually are, and what needs to be addressed and healed within a certain field (e.g., circular city, education, etc.). This path should also continue its evolution to learn more about collective intelligence, unity consciousness calibration, collective healing, and others.

Second, the timing of a long-term lab's active phase could be extended to 6–8 months, with longer time periods in between modules to enable extensive learning journeys and reflections (if we talk about a full-scale social lab process). In our case, the descending side of the U process was lengthier and the ascending side shorter (due to some circumstances), which was challenging for us all.

Third, to include more peer mutual learning and idea generation practices as well as knowledge sharing among participants. A greater focus should be placed on the importance of learning journeys and the time they require to be practiced to a certain quality.

Fourth, more support needs to be provided in the area of coaching, networking, and space-holding in-between workshops if resources allow.

Fifth, as one participant has shared about mindfulness, being able "to accompany the different types of personalities and people along this journey in a way that makes it, for those who are not used to it, not seem too weird" is important. Therefore, exploring different communication styles regarding awareness-led innovation, taking into account diverse backgrounds and mindsets is key. More experience and research are needed to address this.

Shorter and more condensed experiences of such social spaces are also important to explore, given the limited time and resource availability in many countries to tackle problems. Thus, 2- or 3-day social spaces are also an interesting format for unity consciousness calibration work. More exploration and experimentation need to be done on these shorter formats.

Prof. Dr. Gerasimenko has been exploring several such prototypes at Samara University in Russia during 2016–19: awareness-led social space for 1 week as a master course on social innovation for circular economy (Markelova et al., 2019), one-and-a-half day social space during the cross-sector conference on circular economy in Samara (Grekov, n.d.), as well as a 3-day executive course using an awareness-led social space (NiaSam, News Agency, n.d.-d) to explore the mindfulness approach to unfold the potential of the innovation ecosystem in the Samara region with diverse cross-sector representatives, sponsored by the regional Ministry of Economic Development (n.d.-e).

It has become evident that such spaces are sources of never-ending learning for every person that is part of them. They always generate new depths and dimensions to explore. Through collective learning as a process (and as a form of partnership for sustainability), everyone gets from this space what she/he needs in order to progress in the personal domain and therefore support sustainable transformation in the workplace, city, etc. The benefits of such spaces manifest in a nonlinear way at different times and on many different levels and areas of life. We are still engaged in the exploration of the spectrum of possibilities that those spaces can bring us to advance on SDGs.

5. Discussion

One of our cohort members said the following during our process: "We are not here to change the world but more to start the process of internal change." This process of internal change has many angles and manifestations in the outside world. Eventually, what we observed was that every person got different takeaways from this process in terms of how his/her "change" has manifested: from new experiences of how co-creation works to new jobs, new projects and even new habits, values, and mindsets.

We cannot judge our capacity to explore space (the cosmos) and its value for us before we know how to build proper tools, equipment, space shuttles, and rockets to support such exploration. The same is true for these social spaces. We should not judge the efficiency of such "quality of presence" spaces before we know and understand how to co-create proper conditions necessary for this work. Our first steps show a great potential for further practice and research.

The challenges we are facing globally as well as the current SDGs to achieve within 10 years force us to go out of our comfort zone, exploring very new ways of interaction, co-creation as well as the depth and quality of partnerships to co-create true solutions based on trust, deep listening, compassion, nonjudgement, and unconditional love. Instead of a conclusion for this chapter, we are suggesting more exploration and open various discussion domains for further reflections and research.

5.1 "Rethinking" success

New patterns, new neuro-connections in the brain, changed behavior, as well as upgraded mindsets are needed to achieve the SDGs. All that takes time to learn and truly master; it requires patience and a new definition of what the success of such spaces means, moving away from judgement and predetermined outcomes towards an open mind, where serendipity can emerge to show what the real problems are, as well as what is truly needed to solve them.

5.2 New evaluation frameworks

Given the nonlinear effects of such an awareness-led social space, it becomes clear that we need a new evaluation framework for these experiences, opening a whole new research trend for exploration of the potential of such spaces for SDG work. We therefore recommend having a workshop with funding agencies to explore further possibilities of how to evaluate such spaces, given their unusual multidimensional nonlinear impact. What would be an appropriate approach given the explorative stage of practice and sensing within such spaces on one side, and necessity for evaluation on the other?

5.3 Ecosystem facilitation

The role of ecosystem facilitation in holding such spaces is an extremely important research topic. Theory U and other methodologies are a great setting or framework to offer a supportive transformative process; however, what really makes it a transformative space is the quality of presence. Therefore, the ecosystem facilitation (of several people) becomes key to a "quality experience" for the entire cohort. The healing process inside the cohosting team is fundamental for the success of the entire process through trust and congruence. It is also important to add that although the experience of facilitators as space holders is very important; in the end, this space is co-created together with everyone involved. This ecosystem space-holding capacity needs to be further explored.

5.4 Ecosystem self-orchestration

Another interesting dimension is the orchestration of such complex cross-sector, awareness-led social spaces for innovation. Of course, we could use a business-as-usual approach trying to "manage" this process. However, we could take the opportunity for self-orchestration to emerge on its own given the common intention of the organizers, the will to make it work in trust, respect, and nonjudgement (that is not easy for everyone, but possible). This would enable us to see what such spaces could bring us in learning to truly master ourselves, first of all. In this case, we would focus on unity consciousness calibration, common values, commitments, and trust rather than on management and expectations. As this space has demonstrated, such an approach has an amazing potential for further exploration.

5.5 Awareness-led social space formats

We are convinced that longer awareness-led social labs are not the only forms of awareness-led innovation experience. Thus, we have already experimented with shorter term labs of 2 or 3 days, as well as 1 week, to explore the full range of awareness-led formats. The results and the participants' feedback so far are very promising to continue exploration of the mindfulness dimension for awareness-led innovation activities in various settings for a variety of topics.

We would like to express our gratitude to everyone who has supported us on this journey, as well as to the ones who did not, as we have learned a lot from that on what nonjudgement is, how to improve ourselves, and how to build bridges. It has been a long and large journey in many ways. We are especially grateful to our cohort participants, cohosting team members, steering committee members, to all our partners and sponsors, as well as to our ambassadors and volunteers. We are looking forward to continuing this exciting exploration. We are inviting everyone interested to co-create together and to better understand those healing "unity in diversity" spaces.

Acknowledgments (Financial support)

We are grateful to the funds for this project provided by SDSN Switzerland, by Migros Pioneer Fund (for the work of Impact Hub Lausanne-Geneva), as well as by the Chair of Green Economy and Resource Governance Prof. Dr. Bruno Oberle (at the Swiss Federal Institute of Technology in Lausanne (EPFL)).

References

Art of Hosting Community. n.d. Available from: https://www.artofhosting.org/.
Bernstein, A., Vago, D.R., Barnhofer, T., 2019. Special issue on mindfulness. Curr. Opin. Phycol. 28 (August), 1–326.
Bockelbrink, B., Priest, J., David, L., 2019. Sociocracy 3.0—A Practical Guide. Available from: https://sociocracy30.org/guide/.

Anon., 2018. Copenhagen Business School, Mind, Meditation and Innovation Conference. Available from: https://www.tilmeld.dk/mindmeditation/overview.html.

Dart, J., Davies, R., 2007. The 'Most Significant Change' (MSC) Technique: A Guide to Its Use. Available from: http://www.managingforimpact.org/resource/most-significant-change-msc-technique-guide-its-use.

Dragon Dreaming n.d. Project management is inspired by the ancient aboriginal wisdom. http://www.dragondreaming.org.

Figueres, C., Schellnhuber, H., Whiteman, G., et al., 2017. Three years to safeguard our climate. Nature 546, 593–595. https://doi.org/10.1038/546593a.

Gerasimenko, D., Mazerolle, E., 2019. Awareness led social lab: exploring serendipity at Global Solutions Forum 2019 (Video). Available from: https://www.youtube.com/watch?v=BmydV8Z8y10&fbclid=IwAR2KoIYzvoxZDHPDInOoCXM_j6ut5nb2fGY2QPqGJPd9XwFPWYztvqletdo.

D. Grekov, Volga News, Media Portal, Discussion on Circular Economy at Samara University (in Russian) (25.04.2019). Available from: https://volga.news/article/503599.html?fbclid=IwAR0SMd3q73RBXsgyFWRUzb19bMC0kulPNZ73mVC5LHDUhuy2kaQZf_bTYoE.

Hubbard, B., 2001. Emergence. The Shift from Ego to Essence. Hampton Roads Publishing. ISBN 10:1571742042.

Korhonen, J., Nuur, C., Feldmann, A., Eshetu Birkie, S., 20 February 2018. Circular economy as an essentially contested concept. J. Clean. Product. 175, 544–552. Available from: https://doi.org/10.1016/j.jclepro.2017.12.111.

Kuenkel, P., 2016. The Art of Leading Collectively—Co-Creating a Sustainable. Socially Just Future, Chelsea Green, USA.

Limato, R., Ahmed, R., Magdalena, A., Nasir, S., Kotvojs, F., 2018. Use of most significant change (MSC) technique to evaluate health promotion training of maternal community health workers in Cianjur district, Indonesia. Eval. Prog. Plan. 66, 102–110. https://doi.org/10.1016/j.evalprogplan.2017.10.011.

Markelova, E., Gerasimenko, D., Esipova, O., November 2019. Multidimensional (complex) education for the specialists of sustainable development at Samara University. In: Proceedings of the Russian Federal Conference of Environmental Education at The Federal Ministry of Environment and Natural Resources. Moscow (in Russian). Available from: http://new.xn- -b1aqm3d.xn- -p1ai/2019/10/24/mnogomernoe-kompleksnoe-obrazovanie-dlya-speczialistov-ustojchivogo-razvitiya-samarskogo-universiteta/?fbclid=IwAR08O6ykmNFA_1SmKkYhw6AAWaifKAnAmnFnidSru7egNB9CUHf72Dts6fo.

Moreau, V., Sahakian, M., van Griethuysen, P., Vuille, F., June 2017. Coming full circle: why social and institutional dimensions matter for the circular economy. J. Indus. Ecol. 21 (3, Special Issue: Exploring the Circular Economy), 497–506. Available from: https://doi.org/10.1111/jiec.12598.

Muff, K., 2018. Five Superpowers for Co-Creators: How Change Makers and Business Can Achieve the Sustainable Development Goals. Routledge.

NiaSam, News Agency, 2019. Samara University conducts a new form of an executive education programme to explore innovation potential of the Samara region (in Russian) (05.11.2019). Available from: https://www.niasam.ru/Obrazovanie/- -Samarskij-universitet-realizuet-programmu-povysheniya-kvalifikatsii-po-issledovaniyam-innovatsionnogo-potentsiala-oblasti- -140068.html?fbclid=IwAR1CWDYjZhQM_Sxh2NvVE9ysiEfwNuIUhU_KZS7wsrUZWGChc-QD3XU81uY.

Oxford Early Years lab, Report, 2018. Available from: https://www.oxfordearlyyears.org/.

Pomeroy, E., Oliver, K., 2018. Pushing the boundaries of self-directed learning: research findings from a study of u.lab participants in Scotland. Int. J. Lifelong Educ. 37 (6), 719–733. https://doi.org/10.1080/02601370.2018.1550447.

Reike, D., Vermeulen, W.J.V., Witjes, S., August 2018. The circular economy: new or refurbished as CE 3.0?—Exploring controversies in the conceptualization of the circular economy through a focus on history and resource value retention options. Resour. Conserv. Recycl. 135, 246–264. Available from: https://doi.org/10.1016/j.resconrec.2017.08.027.

RIA News, 2019. Russian Scientists opened "Meditation and Altered Consciousness Research Centre" in India. (in Russian). Available from: https://ria.ru/20191114/1560904611.html.

Salmelin, B., Curly, M., 2018. Open Innovation 2.0—The New Mode of Digital Innovation for Prospertiy and Sustainability. Springer.

Samara University News, 2019. 'Space of Trust' as new education programme for ecosystem enhancement in the Samara region (in Russian) (23.10.2019). Available from: https://ssau.ru/news/17281-prostranstvo-doveriya-v-obrazovatelnoy-programme-po-razvitiyu-ekosistemy-samarskoy-oblasti?fbclid=IwAR21du1l6ukC4diOtFG77I0kR5rkWfJ3LpMGhKSm8napPwhh2ZJIefsJ9RA.

Scharmer, O., 2016. Theory U: Leading From the Future as It Emerges, second ed. Berrett-Koehler Publishers, Inc.

Scharmer, O., 2018. Essentials of the Theory. Core Principles and Applications. Berrett-Koehler Publishers, Inc.

Schroeder, P., Anggraeni, K., Weber, U., 2018. The relevance of circular economy practices to the sustainable development goals. Res. Anal. 23 (1), 77–95. https://doi.org/10.1111/jiec.12732.

T. Singer, Perspective from contemplative neuroscience on power and care: how to train care and compassion. Power and Care: Towards Balance for our Common Future-Science, Society, and Spirituality by Tania Singer, Matthieu Richard with Kate Karius, MIT Press (2018) 63–75.

Singer, T., Engert, V., August 2019. It matters what you practice: differential training effects on subjective experience, behavior, brain and body in the ReSource Project. Curr. Opin. Psychol. 28, 151–158. Available from: https://doi.org/10.1016/j.copsyc.2018.12.005.

Stibbe, D., Reid, S., Gilbert, J., 2019. Maximising the Impact of Partnerships for the SDGs. A Practical Guide to Partnership Value Creation. Available from: https://sustainabledevelopment.un.org/content/documents/2564Partnerships_for_the_SDGs_Maximising_Value_Guidebook_Final.pdf.

Trautweina, F.-M., Kanskeb, P., Böckler, A., Singer, T., January 2020. Differential benefits of mental training types for attention, compassion, and theory of mind. Cognition 194, 104039. Available from: https://doi.org/10.1016/j.cognition.2019.104039.

UN, Partnerships for Sustainable Development Goals, 2019. Available from: https://sustainabledevelopment.un.org/partnerships/.

UN, Sustainable Development Goals (SDGs), 2015. Available from: https://sustainabledevelopment.un.org/?menu=1300.

WEF, Agile Governance, 2019. Available from: https://www.weforum.org/agenda/archive/future-of-government/.

CHAPTER 18

How circular design at signify brings economic, environmental, and social value

Anton Brummelhuis and Thomas Marinelli
Head of Sustainable Innovation & Sustainable Design, Signify, Eindhoven, The Netherlands

1. Introduction

In 2016, Philips Lighting spun off from Philips and became a separate company. In May 2018, the new company name Signify was launched; the choice of the new name originated from the way light has become an intelligent language, which connects and conveys meaning. The name is a clear expression of Signify's strategic vision and purpose to unlock the extraordinary potential of light for brighter lives and a better world. "Brighter Lives, Better World" is a sustainability program encompassing product as well as operational commitments to improve lives and to have a positive impact on the planet. Over the past 125 years, Signify has pioneered many key breakthroughs in sustainable lighting, being a driving force behind several leading technological innovations, including LED.

In this chapter, we will introduce the sustainability journey of Signify with its endeavors in sustainable design, firstly focusing on environmental engineering, and later, including social dimensions. We will elaborate on Signify's efforts to innovate for a circular economy, adopting and optimizing four enablers: design, business models, reversed logistics, and collaboration; and four return loops: recycling, parts harvesting, refurbishment, and the service loop. Finally, we will touch on breakthrough innovations beyond lighting in circularity like 3D printing and asset tracking, in horticulture, and in data transmission.

It was a challenging journey to arrive where we are today. After the United Nations published the report *Our Common Future* (the so-called Brundtland report) at the end of the 80s, Philips started to explore the unknown territory of ecodesign to improve environmental properties of products from 1993 onwards. In the beginning, the activities had a narrow focus, whereby the lowering of the environmental load over the life cycle of products was prioritized. A practical and implementation-oriented approach began to develop, identifying improvement options in six so-called "green focal areas," including energy efficiency, material application, packaging and transport, chemical content, lifetime, and recyclability. The options generated were prioritized according to their

Circular Economy and Sustainability, Volume 1
https://doi.org/10.1016/B978-0-12-819817-9.00010-7

environmental benefits but also to the added value for companies, consumers, and society in general. It was also discovered that ecodesign could enable substantial cost reduction in many projects; ecodesign was more and more seen as a tangible contributor to competitive advantage (Stevels, 1993-2007). Consequently, the company started to manage ecodesign consciously, standardized procedures were developed, and the underlying principles and requirements became an integral part of the product development process.

Over the last 10 years, lighting has gone through a significant technological transformation, opening new possibilities in using light. The lighting business extended its design principles to include net positive contributions to energy generation through smart systems in combination with Solar LED, design rules for a circular economy (Philips Lighting, 2017a, b), as well as additional benefits beyond lighting, including social benefits.

2. Sustainable design for brighter lives and a better world

In 2016, we decided to perform a major overhaul of the focal areas to include the potential light has to create brighter lives and a better world. As a result, eight sustainable focal areas were defined to cover environmental as well as social areas. With the launch of the second strategic Brighter Lives, Better World program in September 2020, we updated the focal areas to energy and solar, circularity, weight and materials, packaging, substances, food availability, safety and security, and health and wellbeing. In the sustainable innovation program, we look for opportunities to develop products, systems, and services that are beneficial for the environment and society. Light can contribute to solutions for major global challenges like population growth and urbanization, resource challenges, and digitization connectivity. Furthermore, issues like the coronavirus pandemic show that health and wellbeing have the highest priority, and lighting can offer reliable disinfection through several UV-C applications.

Energy efficiency will continue to be a determining factor in the entire environmental footprint and is a vital component of effective management of climate change. Therefore, Signify will continue to drive the efficacy of light engines, lamps, and luminaires, and increase the energy-saving potential of systems. New innovative circular designs of our luminaires enable serviceability and upgradeability; new generations of light engines can be easily exchanged in installed luminaires.

To innovate for a circular economy, Signify has adopted the circular economy transformation model of the Ellen McArthur Foundation with four enablers: design, business models, reversed logistics, and collaboration; and four return loops: recycling, parts harvesting, refurbishment, and the service loop (Ellen MacArthur Foundation, 2013). The company has developed dedicated design rules to prepare new product developments for a circular economy with an emphasis on serviceability and upgradability of luminaires

(Signify, 2021) Breakthrough innovations have been realized in circularity, such as 3D printing (Signify, 2020b) and asset tracking.

One crucial element of the transition from a linear to a circular economy is the establishment of services. More services will trigger significant social benefits like creating local labor demand in the areas of services themselves, refurbishment, parts harvesting, and recycling. Therefore, Signify has introduced new circular business models such as circular lighting (Signify, 2020c), in which the customer gains the benefits of a customized lighting performance without the need to invest in the lighting installation. Since 2005, Signify have played a pivotal role in the foundation of collection and recycling service organizations (CRSOs) (Signify, 2015) in nearly every EU member state. Today Signify is establishing partnerships to get maximum value out of returned products. No single country or business can establish a full circular solution by themselves, so an intensified collaboration program is in progress. Besides the four enablers and four return loops, significant social benefits, like more local employment, are a consequence of a more circular economy. Other sustainable growth areas that are embedded in our design rules are lighting solutions to enhance food availability, health and wellbeing, as well as safety and security.

For a long time, sustainable innovations focused on energy-efficient buildings, followed by net zero buildings, and as the ultimate environmental goal, energy-generating buildings. Today, buildings are becoming more user-centric, which is driving a trend in livable buildings. This is clearly observed in the steep uptake of the WELL standard, in which lighting is one of the seven WELL v1 categories (WELL, n.d.). Our design guidelines have been extended to cater for wellbeing, detailing how people can see better, feel better, and function better (Signify, 2016).

3. Sustainable innovation at the front-end

The concept of good business is to achieve growth and add value through sustainability. By optimizing environmental, social, and economic benefits in our innovations, we bring lasting value to our customers, employees, investors, and society at large. Sustainable innovation encompasses the research and development of new generations of sustainable technologies, products, systems, and services. This way we can make lives brighter, the world better, stimulate productivity, and attract talent.

In 2016, Signify launched its first Brighter Lives, Better World program, setting ambitious targets on sustainable revenues and sustainable operations leading up to 2020 (Signify, 2020d). In this program, the environmental and societal performance of our products and systems is evaluated over their total life cycle. In product development, we focus on reducing energy consumption, increasing resource efficiency, and reducing or eliminating hazardous substances, with a goal of improving overall environmental performance. In parallel we work on societal solutions that enhance food availability, health and wellbeing, and safety and security.

In our innovation process we apply sustainable design principles in product and system design, based on the eight Sustainable Focal Areas (SFAs) mentioned earlier. These are five environmental areas: energy and solar, circularity, weight and materials, packaging, substances, and three social areas: food availability, safety and security, and health and wellbeing. A successful integration of our sustainability objectives in innovation is ensured via well-defined innovation processes, consisting of our 4-year innovation strategy and portfolio management and actual project development executions.

At a company level, the sustainability goals are defined in endpoints, which provide guidance for the mid- to long-term innovation direction. Each year a cross-functional expert group defines and updates these sustainability endpoints and integrates them in the Signify strategic innovation roadmaps. As part of the portfolio planning on a business level, the sustainability requirements are determined with and for the individual businesses or product categories and are incorporated in the portfolio evaluation checklist. Per innovation project, the more detailed sustainability requirements are embedded in the user requirements list per project and are applied in actual product and system designs via our sustainable design process.

As mentioned in our introduction, Signify has adopted the four circular economy enablers design, business models, reversed logistics, and collaboration. By design, we aim to use natural resources more effectively and extend the lifetime of our lighting to making it serviceable. In collaboration with multiple stakeholders, we are driving the transition from a linear to circular business models (Ellen MacArthur Foundation, 2015c; Accenture, 2014) such as circular lighting. For reversed logistics, we drive collection and recycling efforts and improve the use of recycled materials to close the materials loop.

4. Circular lighting solutions to address global challenges

Our world is confronted with resource challenges and the circular economy is a fundamental means to achieving sustainability and carbon neutrality. The linear model cannot be considered sustainable while the population is growing, global middle-class demand for more material consumption is expanding, and the availability of the resources on planet Earth is limited (Ellen MacArthur Foundation, 2015; Lieder and Rashid, 2016). The European commission and national governments are beginning to act, establishing regulations, limits, and targets, in order to preserve natural resources, reduce waste generation, and minimize the impact on the environment (Lieder and Rashid, 2016; European Commission, 2014). The circular economy (European Commission, 2018; Pieroni et al., 2019b) aims to decouple economic growth from the use of resources by using them more effectively. Products are designed and built as part of a value network where they will be

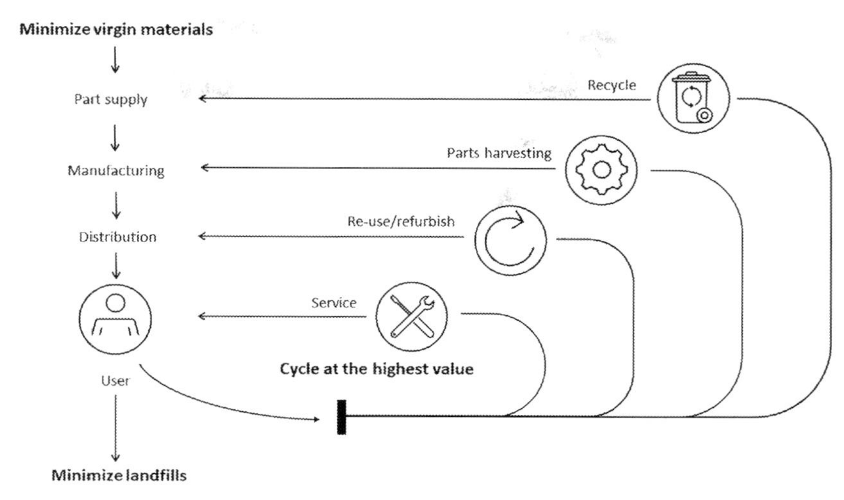

Fig. 1 "Butterfly" model of the circular economy (for amorph materials).

used, serviced, and upgraded to meet customer's needs (Geissdoerfer et al., 2017). Their lifetimes can be prolonged and where possible they can be reused, refurbished, repurposed, or recycled, avoiding waste to landfill. In a circular economy, the more effective use of products, components, and materials is expected to lead to better value for the users (Bakker et al., 2014). Design of circular solutions is the key enabler to further the transition to a circular economy. The circular economy is a vast and broad topic with a variety of definitions, perspectives, and interpretations (Kirchherr et al., 2017). Signify has defined the circular economy as an economic system that maximizes reusability of products, parts, and raw materials, and minimizes value destruction with the aims to preserve value and avoid waste to landfill.

The circular economy is typically described in terms of four concentric loops (World Business Council for Sustainable Development (WBCSD), 2018), as seen in Fig. 1 (Lighting Europe Association, 2017).

Inner circles (like services) describe high value activities and processes and, moving towards the outer circles, residual value is extracted until a product is recycled.

The four loops presented here are:

1. Services—activities aiming at extending the technical and economic lifetime of products.
2. Refurbish—remanufacture products at end of life to sell them again.
3. Parts harvesting—recover valuable parts in products to sell/use them in other products.

4. Recycling—recycle materials to bring them back into the economy.

Signify recognizes three main lifecycle phases for its products and systems: production, use, and return. For the production phase, Signify "walks the talk" by having eliminated all of its production waste. As a matter of policy, Signify steers towards more and high-quality renewable materials for the bill of materials. Via the regulated substances list requirements, hazardous substances are kept out of the bill of materials and, e.g., our packaging requirements ensure the use of at least 80% recycled paper content and phasing out the use of plastic packaging materials for consumer products by 2021. For the use phase, Signify has developed four lighting categories that enable the transition to a more circular economy. These categories are:

1. Replaceable components (lifetime prolonging modules).
2. Serviceable luminaires (luminaire with replaceable driver, controls, and LED board to facilitate lifetime prolongation).
3. Lighting asset management systems (lifetime monitoring and preventive maintenance scheduling).
4. Circular services (services to prolong luminaire lifetimes and managed services with end-of-contract return options).

In cooperation with Lighting Europe (the European lighting association), Signify promotes the use of serviceable luminaires versus the use of sealed-for-life luminaires (Ellen MacArthur Foundation, 2013).

4.1 Circular lighting examples

4.1.1 Circular components upgraded with new D4i standard

Smart city and building projects require faster and wider adoptions of IoT (Internet-of-Things) connectivity in lighting. To facilitate this emerging need, Signify has innovated for a "sensor-ready LED driver portfolio." What helps for a broad adoption in the market is that the connectivity feature has been based upon the new D4i standard interface.

These circular connectivity components enable cooperability and easy exchange between the LED driver and different IoT nodes or sensors on fixtures. This makes it very attractive and easy for customers to switch to connected lighting at a more convenient time. By installing connect-ready luminaires with D4i certified drivers now and upgrading to connectivity later, luminaires stay relevant for longer. The availability of standard compatible modules will ensure easy replacement, prolonging the functional lifetime of fixtures and minimizing waste (Fig. 2).

4.1.2 Signify takes circularity to the streets

The latest Luma generation fulfills the requirements of a circular economy by being energy efficient in application (160 lumens per Watt), serviceable, connected, having

Fig. 2 Circular sensor ready LED-driver upgraded with new D4i standard interface.

a long and extendable lifetime (100 kh), and by being recyclable. The gear flex helps installers to replace and reprogram drivers in 30 s, without the need to touch any wiring. Via the service tag, spare parts and upgradeable components can be easily applied to significantly prolong the product's lifetime (Fig. 3).

4.1.3 Circular lighting solution in the Kortrijk Public Library, Belgium

As a meeting place for people and ideas, the Kortrijk Public Library needed a lighting installation complying with the Innovative Projects for Sustainable Materials Management program run by the city's agency for waste management, highlighting the need

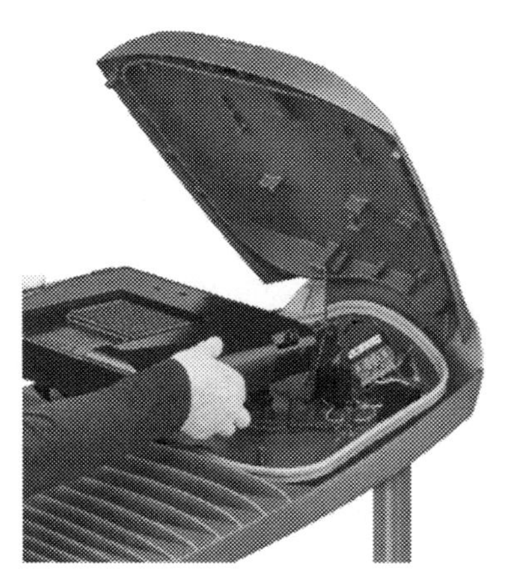

Fig. 3 The new Luma generation is energy efficient, easy-to-service (within 30 s) and connected.

for sustainability. Since financial resources were limited, they required a cost-effective lighting solution with no upfront investment. Signify installed smart LED lighting through its managed services, meaning no upfront cost. The system offers programmable lighting schedules alongside daylight and presence detection, leading to greatly increased energy savings that are in turn used to fund the project. Materials can be reused or recycled after the contract finishes, following the library's climate-friendly ambitions (Fig. 4).

4.2 Lighting innovations beyond illumination

As stated earlier, the purpose of Signify is to unlock the extraordinary potential of light for brighter lives and a better world, offering lighting innovations beyond illumination. To end this chapter we present two global challenges where the characteristics of light enable interesting applications:

- The world is confronted by a growing population, mainly living in cities—predictions are 9.5–10 billion people and 70% urbanization by 2050. Over that time, global food demand will rise by 70%, implying an enormous stretch in sustainable food production, food safety, food accessibility, available natural resources, and GHG emissions. If the current way of producing and consuming food continues, a 70% increase in agricultural yield would be required, while 80% of potential arable land globally is already farmed. Many vegetables are not produced locally and need to be transported over long distances, and approximately one-third of the world's available food is either spoilt or thrown away before it ever reaches a plate. Already, food production and

Fig. 4 Circular lighting solution in the Kortrijk Public Library.

distribution contribute 30% of all GHG emission. Thus, significant changes to our food systems are needed to make sure there is enough safe, healthy, and affordable food in the (near) future. Horticulture LED technologies can provide an answer by realizing extremely efficient farming with high yields, efficient use of space, and enabling local production, at least 90% reduction in water use, and avoiding pollution by pesticides. This is possible because the right light spectrum and growth recipe gives better control over climate and crop and helps to achieve high quality and yield at the right time (Signify, 2019).

- By 2020, there were more than 50 billion devices able to connect to the internet, most of them wirelessly. This is placing wireless communication under increasing pressure; the radio spectrum is becoming congested. In addition, there are areas where radio frequency wireless communication is not permitted, or is not the best fit. Trulifi by Signify is the perfect solution. It is a range of LiFi in addition to this, Trulifi is immune to any electromagnetic interference from, for example, industrial processes. Furthermore, Trulifi has the best latency in the industry. This means that the time between sending data from a device and collecting it at the receiving end is minimal and vice versa.

5. Conclusions

The integration of sustainable design in the product, system, and service development has been a journey of more than a quarter century. Since the mid-1990s, Signify (formerly Philips Lighting) has developed its sustainable design principles and requirements starting from an environmental angle, adding societal focal areas along the road. With the right focus, lighting can offer benefits beyond illumination, in energy efficiency, circular design and packaging, food availability, safety and security, and health and wellbeing. Global challenges like climate change and resource scarcity, as well as demographic change and urbanization, require society to act. A framework of clear regulations, rules, and standards are needed to stimulate sustainable choices and to stop products, systems, and services with a negative impact on planet and society being put on the market. There is still a long way to go and time is ticking. A paradigm shift is needed to turn around the worrying trends; the circular economy is a fundamental means to achieve sustainability and carbon neutrality. The success depends on product design, a new way of collaboration between stakeholders in the value chain, and supportive business models.

The extraordinary potential of light for brighter lives and a better world is further demonstrated by innovations supporting food availability, which requires sustainable food production for a growing global population, and safety and security in a rapidly digitizing society. The possibilities are extensive and promising.

References

Accenture, 2014. Circular Advantage. Innovative Business Models and Technologies to Create Value in a World without Limits to Growth.

Bakker, C., Den Hollander, M., Van Hinte, E., Zijlstra, Y., 2014. Products That Last—Product Design for Circular Business Models. TU Delft Library, The Netherlands.

Ellen MacArthur Foundation, 2013. Towards the Circular Economy Vol. 1: An Economic and Business Rationale for an Accelerated Transition. Retrieved from https://www.ellenmacarthurfoundation.org/assets/downloads/publications/Ellen-MacArthurFoundation-Towards-the-Circular-Economy-vol.1.pdf.

Ellen MacArthur Foundation, 2015. Methodology—an approach to measuring circularity the material circularity indicator. Report. Retrieved from https://www.ellenmacarthurfoundation.org/assets/downloads/insight/CircularityIndicators_Methodology_May2015.pdf.

Ellen MacArthur Foundation, 2015, Deember. Towards a Circular Economy: Business Rationale for an Accelerated Transition. Retrieved from: https://www.ellenmacarthurfoundation.org/publications/towards-a-circular-economy-businessrationale-for-an-accelerated-transition.

European Commission, 2014. Report on Critical Raw Materials for the EU, Report of the Ad Hoc Working Group on Defining Critical Raw Materials.

European Commission, 2018. Monitoring Framework for the Circular Economy. vol. 29 final, 2018. Retrieved from https://ec.europa.eu/environment/circular-economy/pdf/monitoring-framework.pdf.

Geissdoerfer, M., Savaget, P., Bocken, N.M.P., Hultink, E.J., 2017. The circular economy—a new sustainability paradigm? J. Clean. Prod. 143, 757–768.

Kirchherr, J., Reike, D., Hekkert, M., 2017. Conceptualizing the circular economy: an analysis of 114 definitions. SSRN Electr. J.

Lieder, M., Rashid, A., 2016. Towards circular economy implementation: a comprehensive review in context of manufacturing industry. J. Clean. Prod. 115, 36–51.

Lighting Europe Association, 2017. Serviceable Luminaires in a Circular Economy—White Paper. Brussels, Belgium.

Philips Lighting, 2017a. New Business Models in the Circular Economy. White paper.

Philips Lighting, 2017b. Minimize Your Environmental Footprint and Create Instant Savings. Philips Circular Lighting. Retrieved from http://images.philips.com/is/content/PhilipsConsumer/PDFDownloads/Global/ODLI20171010_001-UPD-en_AA-7036_Circular_Lighting_Digi_WTO_02.pdf.

Pieroni, M.P., Mcaloone, T.C., Pigosso, D.C., 2019a. Business model innovation for circular economy and sustainability: a review of approaches. J. Clean. Prod. 215, 198–216.

Pieroni, M.P., Mcaloone, T.C., Pigosso, D.C., 2019b. Business model innovation for circular economy and sustainability: a review of approaches. J. Clean. Prod. 215, 198–216.

Signify, 2015. Collection and Recycling Brochure. Retrieved from: https://www.signify.com/global/sustainability/product-compliance/collection-and-recycling.

Signify, 2016. Human Centric Lighting: The Human Factor in Lighting Designs. Retrieved from https://www.signify.com/global/sustainability/product-compliance/collection-and-recycling.

Signify, 2019. Agricultural Lighting Booklet. Eindhoven, The Netherlands.

Signify, 2020a. Signify Annual Report 2019. Retrieved from: https://www.signify.com/static/2019/signify-annual-report-2019.pdf.

Signify, 2020b. Benefits of 3D Printing. Retrieved from: https://www.signify.com/global/innovation/3d-printing.

Signify, 2020c. Circular Lighting. Retrieved from: https://www.signify.com/global/sustainability/circular-lighting.

Signify, 2020d. Signify Annual Reports 2016–2020. Retrieved from: https://www.signify.com/global/our-company/investors/financial-reports/annual-report.

Signify, 2021. Signify Annual Report 2020. Retrieved from: https://www.signify.com/global/our-company/investors/financial-reports/annual-report.

Stevels, A.L.N., 1993–2007. Adventures in EcoDesign of Electronic Products, first ed. Delft University of Technology: Design for Sustainability Program publication nr. 17.
WELL Standard. https://www.wellcertified.com/certification/v1/.
World Business Council for Sustainable Development (WBCSD), 2018. Circular Metrics—Landscape Analysis. Retrieved from https://www.wbcsd.org/Programs/Circular-Economy/Factor-10/MetricsMeasurement/Resources/Landscape-analysis.

CHAPTER 19

Circular economy and urbanism: A sustainable approach to the growth of cities

Elena Turrado Domínguez, Rafael Hernández López, and M.A. Fernández López
Universidad Camilo José Cela (UCJC), School of Technology and Science, Madrid, Spain

1. The city as a 21st century sustainability challenge

Cities consume 80% of resources (UNEP, 2015) despite occupying less than 3% of the land surface (UNDP, 2020) and, by 2050, they are expected to be home to 67% of the world's population (UN-HABITAT, 2012). They generate approximately 80% of the global CO_2 emissions (IEA, 2008) and are sources of pollution, unsustainable growth, and social inequality (ONU-HABITAT, 2016), which make them a study object in areas such as urban economy, regional science, social physics, transportation science, and complexity science (Batty, 2012).

Thus, it can be said that the concentration of population and the ready access to resources have created an unsustainable urban environment where the demand for products and the consumption of natural resources have shaken the three spheres of sustainability: social, environmental, and economic (Clark et al., 2005; Kates et al., 2005; Ellen MacArthur Foundation, 2012). In this way, urban environments, dominated by the dynamics of an economic system that responds to linear patterns of production and consumption, are one of the causes of the ecological deficit with which humanity now exceeds four of the nine planetary boundaries (Steffen et al., 2015; Rockström et al., 2009).

In this context, the influence of economic progress on urban growth is under debate again, and there is current discussion on how to overcome the indefinite and indeterminate urban expansion that started with the industrial revolution (Zaera-Polo, 2016). Thus, to create a city that is disconnected from market dynamics and attends to the environmental priorities of today's society, this chapter presents circular economy (CE) as a useful system for urban sustainability. We analyze the influence of CE as an emerging economic model that promotes ecological matters and relates environmental sustainability to the dissociation of resource consumption from economic growth (D'Amato et al., 2017; Sauvé et al., 2016), taking into account the urbanistic perspective that will allow rediscovering and claiming the social role of the city. In this way, it will possible to

Circular Economy and Sustainability, Volume 1
https://doi.org/10.1016/B978-0-12-819817-9.00022-3

redefine the city around a strategy that allows understanding it beyond an "abstract space" studied in social sciences, and we will leave behind the "thoughtless fabrication" described in *Teorías e historia de la ciudad contemporánea* (García Vázquez, 2016).

Thus, the aim is to respond to what is proposed in the 2016 Urban Agenda regarding the Sustainable Development Goals (SDG), specifically SDG-11 (sustainable cities and communities): the need for a new paradigm that "redresses the way we plan, finance, develop, govern and manage cities and human settlements, recognizing sustainable urban and territorial development as essential to the achievement of sustainable development and prosperity for all" (pp. 3–4). In this context, the way in which we make sense of urban areas is at a critical point, and our understanding of the cities and the mosaic of urban areas, both in theory and in practice, is at a turning point (Rosales Carreón and Worrell, 2018). Therefore, knowing that cities can play an important role in the development of a sustainable society (Vergragt et al., 2014; Loorbach and Shiroyama, 2016), managing their growth becomes a key factor that will require the coordination of the different disciplines that are involved in urban dynamics (Taylor Buck and While, 2015) and which show the city as a "subject matter" that is too complex to be studied in terms of simple causal chains (Choay, 1970).

2. Structure and methodology

This chapter is framed within an urban context dominated by the overexploitation of resources and urban growth, where the fundamental principles of urban thinking have been forgotten and economic forces have taken control. Therefore, with the aim of obtaining answers that help us to redefine the city of the 21st century, we structured the chapter in the following manner:

- After the introduction presented in Section 1, and Section 2, which describes the structure and methodology followed for the elaboration of the present work, Section 3 presents an approach on urban growth beyond the commodification of architecture. It describes the need to retrain urban thinking in order to attain sustainable urban growth, and it defines ideas for the rethinking of the sustainable city of the 21st century. Thus, far from the urban environment created under an economic regime in which the architect and the urbanist were the least influential agents in the city, the latter is claimed as a social environment.
- In Section 4, we study the influence of CE as an emerging economic model. Analyzing CE from current studies that show its strengths, weaknesses, opportunities, and threats (SWOT analysis) facilitates evaluating CE and defining the requirements for it to become a useful element for sustainable urban growth.

- In Section 5, a descriptive mapping is used to define a new paradigm for sustainable urbanism. Thus, from a multidisciplinary approach, useful concepts for the pursued objective are presented.

  To this end, the research process is based on the design of diagrams, which represent such processes and constitute a dynamic and project framework that allows organizing a theoretical and bibliographic foundation. Their visual capacity, along with their capacity to integrate and connect heterogeneous data, enable the interaction, abstraction, and synthesis needed to map information from several disciplines that, eventually, are integrated in a single study (Montaner, 2014).

  The diagram allows capturing the coherence of thinking through design. In the book *Del diagrama a las experiencias, hacia una arquitectura de la acción* (Montaner, 2014), Montaner asserts that "diagrams are adequate to openly project the future and to respond to the new social, cultural, energy and environmental challenge. Times of change need open and versatile project instruments. . . ." (p. 12).
- Section 6 shows the results of the investigation with a diagram that gathers a conceptual and useful structure for urban growth. It shows CE as a useful category for the processes of urban planning and as a change of paradigm for sustainable urbanism. In this way, addressing a discussion about the meaning of CE in the context of sustainable urbanism, the current research approaches of CE and their application at the urban scale, an intervention framework is shaped with the aim of confirming CE as a theory capable of redefining the city.
- Lastly, Section 7 presents the conclusions of the chapter.

Next, the methodology used is briefly described:

- Phase 1: Review of the state of the art of CE. We analyzed the term with respect to the social issue, its own definition and meaning, sustainability, and its association with the urban environment (macro level of application of CE). To this end, as was previously commented, we identified the strengths, weaknesses, opportunities, and threats (SWOT) of CE to become a useful element for sustainable urban growth.
- Phase 2: Bibliographic review. We conducted a descriptive mapping, developed in two stages and represented in two different diagrams, which gathers references of different scopes, which are connected to redefine, around CE, the fundamentals of a new paradigm for sustainable urbanism.
- Phase 3: Presentation of results and creation of an intervention framework for CE. Its representation through a diagram shows the connections among the fundamentals found in the previous phase and the three dimensions into which the framework is structured.

3. The sustainable urban growth approach

The starting point to address this topic is the problem derived from the existing difference between the growth imposed by economic interests and the growth derived from urban thinking.

The former determines that the term urban is limited to what is built. Therefore, such growth draws cities as study areas that can be isolated from their environment and develops policies that externalize the impact of the city on the biophysical world. In this way, it develops a partial and simplified urban theory around the definition of the city through normative and spatial limits that disregard the space outside of the delimitations created by economic policies. The city, founded under economic requirements, becomes a centralized market for production and consumption (Webber, 2004). In it, the development of the production and movement of goods reduces the complexity of the city to quantifiable and monetary variables (Naredo, 1987), as well as to studies of accessibility, optimal size, and density of the settlement with respect to the spatial distribution of the activities. In this context, the dissemination of settlement units due to the increase of mobility and travel patterns creates a planning that disregards the consequences of infrastructures on the physical space that they occupy, and thus, eventually, the element of space is limited to a mathematical abstraction in which a cost assessment (usually related to distances) allows incorporating them into econometric models.

On the other hand, in contrast to this type of growth that has dominated the development of the 20th-century city, the urbanist growth approach develops the city as an experimental field that combines the theoretical-conceptual world (sometimes utopian) with the practical world. Urban thinking understands that the growth of a city is part of a creative process and an exercise in speculation about the future. In this sense, it is important to highlight that Geddes, in the framework of the urban organicist tradition, from "his abstract syntheses and diagrams, defended an urbanism projected from the study of reality, experience and life" (Montaner, 2014, p. 30). He studied the possibilities of internal renovation and lines of growth of cities from a social perspective. The model defined by Geddes and by his particular view not only does not reject growth, but assumes it as inherent of the new urban condition (Geddes, 1915). Geddes stated that the city can lose its limit and expand in the territory, and, using the "man-reef" metaphor, its internal structures can be constantly adjusted to the environment. Urban structure and morphology are no longer conditioned by geometry, but are understood through biology and governed by networks and flows, thus generating a great conceptual advancement, a system, and a behavioral model that are still current. It is important to highlight that his Valley Section (1905) diagram links, for the first time, human settlements and the means of production to their territorial evolution and location in the landscape, as well as how, years later, such a section was reinterpreted in the diagram "Scale (of complexity) of association" (Smithson and Smithson, 1967), which states that "in order to understand the patterns of human association, we must consider every community in its particular environment" (Smithson and Smithson, 1967, p. 19). Thus, after the functional city as the essence of scientific rationality, the importance of the urban community is introduced again in the urban plot to understand its development and find an urban structure that expresses its needs.

Therefore, it is currently necessary to face the indefinite and indeterminate growth derived from the economic tendency of the 20th century and retrain an urban thinking that can approach the complexity of the city and urban growth from another perspective. Beyond economic agglomeration and the grouping of activities that cause the acceleration of urban obsolescence, other factors must take part in the setting of the spatial development patterns of the city (López de Lucio, 1993). Thus, in this context, the question is: How should the model of the 21st-century sustainable city be?

To respond to this question, we must recover the city as a "meeting place" (Jacobs, 1961; Mumford, 1961) and the ecological-environmental aspect as a fundamental element within the urbanization processes of the 21st-century city. In this way, its logic of "intelligent growth" (Harvey, 2004) leads us to dissolve limits and understand the landscape (urban or not) as a result of specific spatial production processes of the political-economic system (Harvey, 1996). In this context, it is necessary to define an urban and social fabric that is sustainable and habitable (Brenner, 2014) and to conceive the process of urbanization beyond the visible, specific, and physical limits of the city (Brenner and Schmid, 2014; Prieto-Sandoval et al., 2018).

Therefore, it is possible to state that the urban world and everything within it is part of the problem, but also part of the solution. The city is the fundamental environment for the development of people and their sustainable futures (Pomponi and Moncaster, 2017). As was pointed out in Aalborg Charter (1994):

We are convinced that the city or town is both the largest unit capable of initially addressing the many urban architectural, social, economic, political, natural resource and environmental imbalances damaging our modern world and the smallest scale at which problems can be meaningfully resolved in an integrated, holistic and sustainable fashion (p. 3).

Next, we gather two ideas to retrain urban thinking and address the sustainability of the current city. They show the importance of rediscovering the sociocultural role of the city and claiming its social duty.

- A transformation process can only be based on the full identification between the community and the physical and cultural structures of the territory (Geddes, 1915).
- It is necessary to recover the idea of the human being as a biological and psychological entity that proposes the dissolution of the city in the field, as well as the idea of the human being as a social and cultural entity that cannot live outside his/her community (Choay, 1970).

4. Circular economy as an influential concept and useful system

In this section, we analyze the current developmental state of the term CE according to four different aspects. We classified these aspects as strengths, weaknesses, opportunities, and threats (SWOT) in order to evaluate CE as a useful element capable of addressing sustainable urban development and coexisting with the ideas developed in the previous section.

Firstly, we framed the social aspect as a weakness of CE. The coverage of social well-being is very low (EEA, 2015) and it is not known to what extent this could affect subjective wellbeing (Frey and Stutzer, 2002). Its care with respect to the social and local dynamics is very poor (D'Amato et al., 2017). On the other hand, employment creation has become the main reference to social achievements, disregarding the study of the rest of social impacts, which remain unknown (Sauvé et al., 2016; Murray et al., 2017). Therefore, it is necessary to conduct studies on the social dimension, social equality, and equity (Sauvé et al., 2016; Murray et al., 2017; Moreau et al., 2017).

Secondly, the threats presented by CE are determined by the context of this concept. Due to the acceleration of the world economy, based on the overexploitation of resources, and the importance given to economic goals over ecological and social concerns (Sauvé et al., 2016), the main definitions of CE and its objective revolve around two concepts: how to disconnect economic growth from resource consumption (Gregson et al., 2015) and economic prosperity (Kirchherr et al., 2017). However, these simplified definitions and objectives coexist with more open definitions that include interdependencies among systems and a holistic view that can help to change our lifestyle, although it contributes to the ambiguity and confusion of the term. It is worth highlighting the reflection of Korhonen et al. (2018a, b) about the meaning of CE as an economic system that represents a change of paradigm in the way in which human society is interconnected with nature and has the objective of avoiding the depletion of resources, close connections of energies and materials, and facilitates sustainable development. Thus, CE must be analyzed from a holistic perspective that enables the sustainable development of a society that can understand the systemic approach as a tool that involves producers, consumers, and other social agents and makes them collaborate.

Moreover, due to the complexity of the term, partly acquired from its multidisciplinary nature (Prieto-Sandoval et al., 2018), it has been criticized for the unclear definition of its bases and objectives (Gregson et al., 2015; Pomponi and Moncaster, 2017). However, in spite of coexisting with other emerging terms such as bioeconomy (Bugge et al., 2016), green economy (UNEP, 2011), and green growth (OECD, 2011), it can be asserted that CE has a more homogeneous literature and is currently considered as the strongest term (Ellen MacArthur Foundation, 2013b).

Thirdly, the strength of CE lies in its relationship with sustainability. While some authors believe that it could be fundamental for sustainable development (Geissdoerfer et al., 2017), others state that the term CE is essential for it (Prieto-Sandoval et al., 2018; Kirchherr et al., 2017). However, it is necessary to work on the conceptual development (Geissdoerfer et al., 2017) and theoretical framework of CE (Prieto-Sandoval et al., 2018) in order to reach a consensus that guides the practical development and implementation of this model as a useful solution to achieve sustainable development (Xue et al., 2010; Geissdoerfer et al., 2017; Kirchherr et al., 2017). Much more work must be done before CE can be consolidated as a new paradigm of sustainable development for global society (Korhonen et al., 2018a, b). Nevertheless, there is an increasing number of studies, mainly conducted in China, that associate the adoption of measures of CE with sustainable development, whose main objective is to respond to the great environmental, social, and health problems derived from the pattern of economic growth of that country (very rapid and continuous) (Ghisellini et al., 2016; Prendeville et al., 2018).

Lastly, we identified the macro level as an opportunity of CE. Its development in this context is scant, thus the literature on this element is very poor (Ness and Xing, 2017), and the main frameworks are conceptual and lack a practical application to the context of the city (Prendeville et al., 2018; Prieto-Sandoval et al., 2018). However, efforts have been made to specify that the application of CE to cities, regions, or provinces implies the integration and redesign of four systems: industrial, infrastructural, cultural, and social (Ghisellini et al., 2016).

After this analysis and considering the ideas designed in Section 3 (sustainable urban growth approach), we highlight the following aspects for the incorporation of CE as a useful element for sustainable urban growth:

- A conceptual, cohesive and holistic framework is required (Sala et al., 2013, 2015; Murray et al., 2017) to define the strategies that make its application at the urban scale more effective.
- It is necessary to acknowledge the importance of the regeneration of natural systems (fundamental principle of CE) (Ellen MacArthur Foundation, 2012) as a mechanism to respond to the need described by Ghisellini et al. (2016) at the urban scale.

The first of these aspects shows the so-needed search for models and paradigms of sustainability, as it raises awareness of the ecological disturbances caused by our current lifestyle (Sala et al., 2015). It recognizes the need to develop the sensitivity required to remove the perception of the environment as an external element, an inert variable, a source of endless resources, and an instrument of expansion unconnected to the city, of which the latter makes use. It presents the ecological-environmental element as a fundamental condition in the 21st-century city (Harvey, 2004).

The second aspect refers to the search for the "character" and "social personality" of the territory and of the city as essential elements for any hypothesis of positive transformation (Geddes, 1915). It places the ecological element as a benefit for society, as an urban factor of social wellbeing that becomes indispensable to understand the city itself as part of a whole and not as an isolated system. Therefore, it refers to ecology beyond the ecological theoretical framework with which political geography, landscape urbanism, and political ecology remove the barriers between the urban world and the natural world.

5. Defining a new paradigm for sustainable urban planning

Considering the above mentioned, we can assert that the ecological aspect becomes an essential element to project a city that can manage present growth processes, but it also becomes a tool that can be used to solve the negative impacts on wellbeing caused by the growing disconnection between city dwellers and nature (Gullone, 2000).

Therefore, knowing the need for a systemic and multidisciplinary view (Prendeville et al., 2018) capable of evidencing CE as an environmental innovation (Prieto-Sandoval et al., 2018), we conducted a descriptive mapping that can be useful for the necessary change of thinking in the process of urbanization and which can detect the keys to sustainable urban growth.

5.1 Descriptive mapping

The mapping process consisted of creating a first diagram that integrated, in the same theoretical framework, those theories (urban and economic) that recognize ecology as a fundamental element and a connection for the coordination of different disciplines. Thus, we identified the disciplines that led to the creation of a second diagram, and a second image was set as a theoretical framework that gathered the main references and synthesized the keys to urban sustainability.

To shape the first diagram, four events were located and represented on a foundation that refers to the underlying rules of nature. The reading of the diagram was set from two return points (the concept "metabolism" in 1858 and the term "urbanism" in 1867), which served to introduce ecological thinking in the scopes of economy and urbanism (Geddes, 1884, 1915). From these, the dynamics of the diagram recognizes two disciplines (ecological economy and ecological urbanism), which take ecology into account and can be organized to shape a theoretical framework that reflects the keys of intervention, with the aim of achieving a sustainable urban growth created under the integration of CE in the processes of urban planning (Fig. 1).

Next, we present the ideas contained in each of the four events that were previously mentioned:

- The concept "metabolism" was presented as a key concept for the understanding of the interaction between nature and society in 1858 (Marx and Nicolaus, 1993). This concept asserts that humans are surrounded by the natural world, which we cannot be detached from, not even through technological or economic revolutions.
- The term "urbanism" that can be found in the book by Cerdá (1867), *Teoría General de la urbanización*, emerged in the reflecting framework of the spatial impact during the

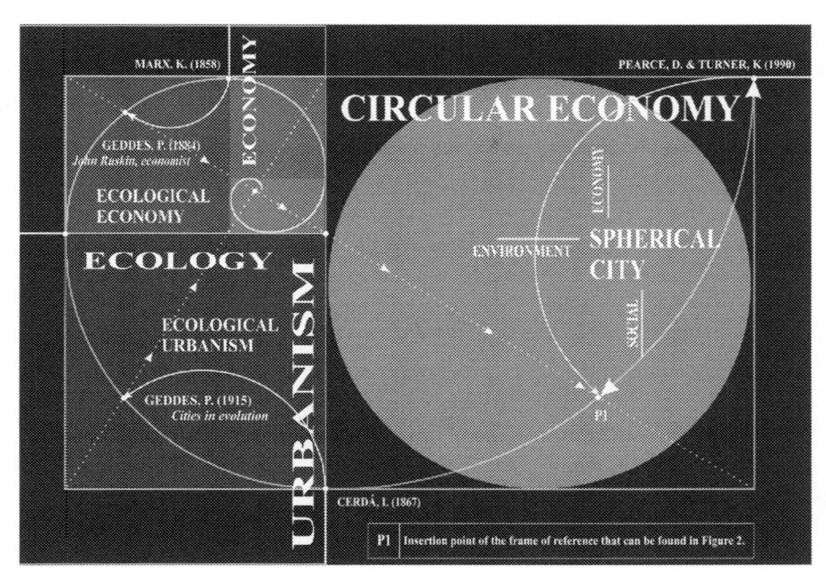

Fig. 1 From events to a "phylogenetic tree." *Compiled by the author.*

industrial revolution, when the city suffered a spontaneous and uncontrollable transformation.

- The article *John Ruskin, economist* (Geddes, 1884), exposed the need of a new approach beyond market forces that control the economy. It also showed the need to focus on the concept of quality of life as something that should take into consideration the biological and aesthetic needs of humans.
- The term "conurbation" was presented as the symbiosis between nature and the built environment (Geddes, 1915). Its meaning was then dispelled from negative and unsustainable aspects that are nowadays related to it. It understood a necessary regional and urban planning based on the integration of environmental resources.

From this point, the second diagram was elaborated around the idea of creating a useful theoretical framework for the necessary change of thinking in the process of urbanization. In turn, with the aim of detecting the keys to sustainable urban growth, we contextualized the ecological tradition of urban thinking, the ecology that studies the environmental impacts and the economy that questions the growth associated with the consumption of resources.

We can assert that the diagram (Fig. 2) acts as an interface for reasoning. Thus, it was created as a flexible framework capable of responding to three different readings: a horizontal reading that showed the chronological evolution of each scope, a vertical reading that allowed a joint view of each decade, and a reading that, by connecting islands,

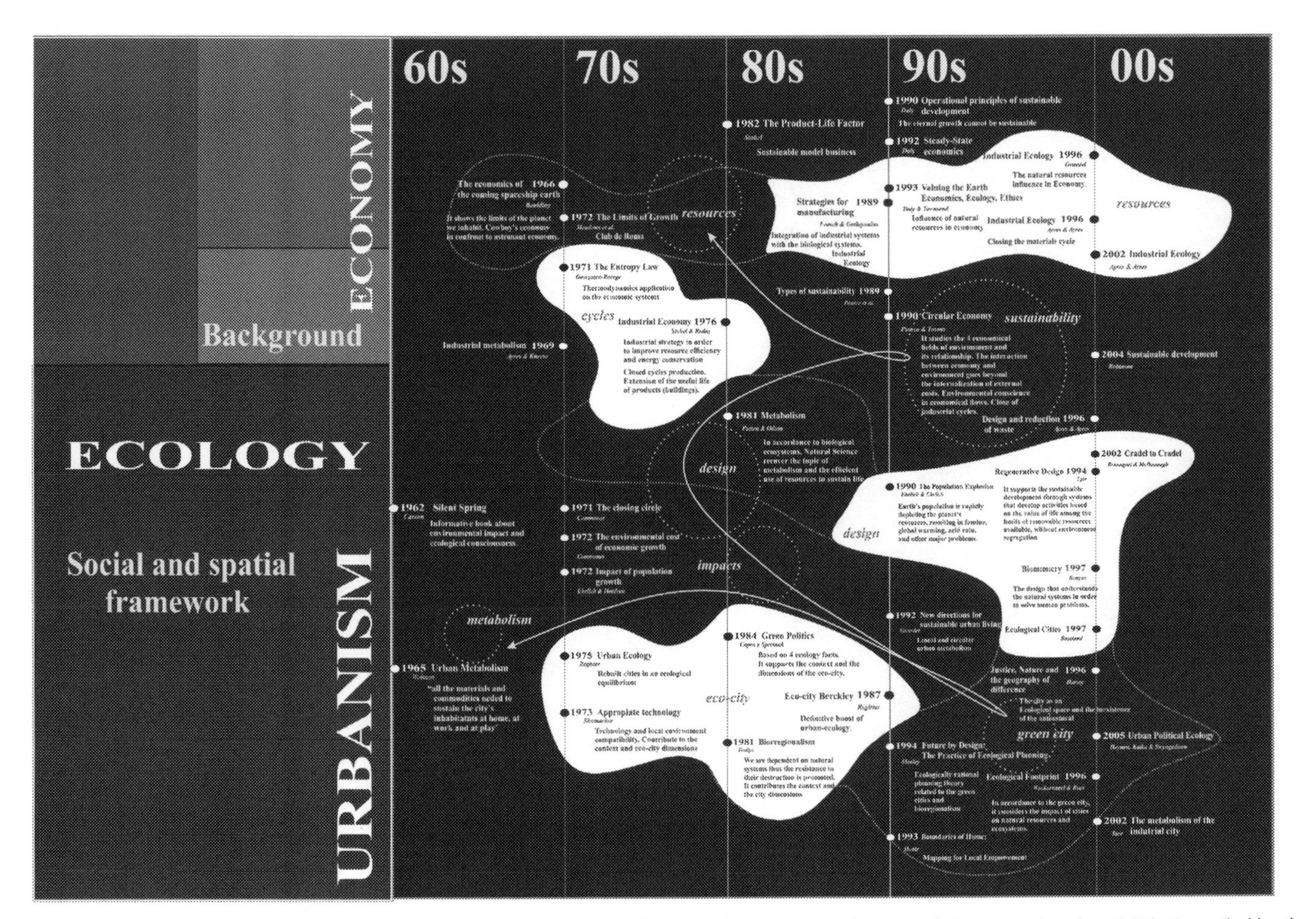

Fig. 2 A "phylogenetic tree." Diagram Inspired by "Evolutionary Tree of Twentieth-Century Architecture" diagram (Jencks, 2000). *Compiled by the author.*

allowed navigating between the synchronous and the asynchronous to carry out disciplinary studies on the idea of paradigm change presented by Kuhn (1962).

Next, following a vertical reading of Fig. 2, we present the facts that allowed us to identify the core of each decade and understand the context of the studies that were conducted in each of them. In this way, the second half of the 20th century was understood as a time in which both the facts and the resulting concerns pushed us to rethink the sustainable growth of the current urban environment.

- During the 1960s, there was an environmental deterioration of industrialized countries. Therefore, the economy showed a rising interest toward environmental issues. Works such as *The Economics of the Coming Spaceship Earth* (Boulding, 1966) are an example of it. As a consequence of this ecological concern, publications such as *Silent Spring* (Carson, 1962) and studies related to industry and cities appeared, exposing ideas such as industrial metabolism (Ayres and Kneese, 1969), and urban metabolism, defined by Wolman, 1965 as: "all the materials and commodities needed to sustain the city's inhabitants at home, at work and at play" (p. 2).
- During the 1970s, the increasing awareness of the depletion of resources led to the publication of *The Club of Rome* (1968), and *The Limits of Growth* (Meadows et al., 1972).

  The application of thermodynamics on economic systems was also studied, causing the emergence of the concept of ecological economy (Georgescu-Roegen, 1971; Ayres, 1999). The work by Stahel and Reday (1976) conceptualized, through industrial economy, economic cycles that describe industrial strategies in order to improve resource efficiency, energy conservation, regional employment, and the prevention of waste. The rules of ecology (Commoner, 1971) were established, the equation "I=P x A x T" was created in order to study environmental impact (I) (Commoner, 1972; Ehrlich and Holdren, 1971; Ehrlich and Ehrlich, 1990), the cities were studied as ecological spaces (Williams, 1973), and the concept of urban ecology emerged (Register in 1975).
- During the 1980s, there was a rise in awareness about the ecological consequences of human activities, together with sociopolitical human rights topics, and also the problematic situation of the lack of resources.

  The context of sustainable development and sustainable society were created and the Brundtland Report (World Commission on Environment and Development, 1987) was written, providing a key definition related to these concepts: "sustainable development is development that meets the needs of the present generations without compromising the ability of future generations to meet their own needs" (p.13).

  About property, Stahel (1982) stated the importance of commerce as a sustainable business model, allowing industries to profit without outsourcing costs and risks related to waste. Ecological industry (EI) appeared as the integration of industrial ecosystems together with biological ecosystems (Frosch and Gallopoulos, 1989; Graedel, 1996; Ayres and Ayres, 2002) and, in the perspective of the cycle of life of products, a closed system of resource flows was represented. At the same time, in order to redesign cities in accordance with ecological concepts (Capra et al., 1984; Roseland, 1997), the concept of urban ecology was implemented, as well as the ecological cities movement (Register, 1987). Therefore, the popular concept of self-sufficient cities emerged among environmental activists from the 1980s (Morris, 1982) and studies about metabolism and the efficient use of resources (Patten and Odum, 1981) were retaken. Moreover, the research on the urban environment was completed by different authors, who wrote about and described city dynamics (Patten and Odum, 1981; Tarr, 2002; Heynen et al., 2005).

- During the 1990s, the United Nations Conference on Environment and Development (1992) took place. Some years later, in 1997, the Kyoto Protocol was signed in order to fight against global warming, and some reactions about the misconception of economic growth models also emerged.

The term "circular economy" was coined by Pearce and Turner (1990), and a context that shows the consequences of the economic flows in the environment was created. Consequently, the concept of closing the industrial cycles was developed and the influence of natural resources in the economy was also studied, taking into account that we need these natural resources to produce goods and the natural systems are treated simply as landfills (Daly and Townsend, 1993). Other concepts were presented, such as "steady-state economy" (Daly, 1992) and the idea that eternal growth is unsustainable (Daly, 1990; Robinson, 2004; Rubin, 2012).

Therefore, due to the anthropocentric and eco-centric focus, the environmental problems became an economic opportunity. Design and waste reduction became main topics (Ayres and Ayres, 1996) and the architectonic regenerative design was created as a precedent concept of "cradle to cradle" (Braungart and McDonough, 2002). Benyus (1997) developed the concept of biomimicry (imitation of natural systems) and some strategies related to regenerative designs were presented, supporting sustainable development (Lyle, 1994).

Furthermore, the linear and circular urban metabolism related to sustainable urban development was studied (Girardet, 1992). Self-sufficient cities were developed, which "learn from the circular metabolism of nature" (Girardet, 1992). The concepts of "green city" and "bioregionalism" (Dodge, 1981) were created around the idea that people depend on natural systems, thus reinforcing the meaning of the "local world" (Aberley, 1993), the history and theory of ecologically rational planning (Aberley, 1994), and the analysis of the concept of an "ecological footprint" presented by Wackernagel and Rees (1996), which considers the impact of cities on natural resources and ecosystems.

Thus, after the study of the previous framework, we extracted the intervention keys (resources, cycles, metabolism, eco-city, sustainability, design, impacts, and green city) that contain fundamental concepts that we must consider to rethink urban growth.

6. Spherical city: A framework for urban circular economy

Next, we respond to the aspects highlighted in Section 4 for the incorporation of CE as a useful element for sustainable urban growth. First, we created a framework with the aim of defining the strategies for the urban application of CE. Second, we took the regeneration of natural systems as a mechanism to specify the application of CE at the urban scale.

Thus, considering the necessary presence of the keys of sustainability found in the previous descriptive mapping, this section presents CE through a structure that can redefine the city and, therefore, as a theory that can posit a change of paradigm within the scope of sustainable urbanism.

6.1 The spherical city concept

This concept was developed to act on the planning processes of the complex urban context. To this end, we established three dimensions of intervention that, in addition, consider the ideas proposed in Section 3 about how to recover the urban thinking and rediscover the sociocultural role of the city.

- *Design*. Is defined as the first dimension to take into account within the urban scale. Its social implications are shown through three concepts: sustainable design (Ellen MacArthur Foundation, 2013a; Prieto-Sandoval et al., 2018), regenerative design (Lyle, 1994), and biomimicry (Benyus, 2003; Helms et al., 2009; Spiegelhalter and Arch, 2010; McGregor, 2013; Taylor Buck, 2017), which allows regenerating ecosystems and connecting citizens with nature (Pedersen Zari, 2015; Zari, 2012). For the design dimension, natural systems become the mechanism that can go from the theory-concept to practice and avoid urban fragmentation. Thus, thanks to design, we can recover urban dynamics that serve human wellbeing, the natural environment, and the quality of urban life.
- *Urban patterns*. Its definition under the influence of the regeneration of natural systems implies acting on the physical characteristics of the urban context and, therefore, in favor of the integration and redesign of four urban systems: industrial, infrastructural, cultural, and social (Ghisellini et al., 2016).

  However, the implementations of the keys of urban sustainability are currently focused on the industrial system through industrial ecology (Lifset and Graedel, 2001; Ayres and Ayres, 2002), the industrial cycles (Xue et al., 2010), the theories that are based on the identification of "R" (Brennan et al., 2015; van Buren et al., 2016; Potting et al., 2017), and the development of industrial parks (Prendeville et al., 2018).

Therefore, in order to achieve the balance of urban factors that can provide a good quality of life for its citizens, we need to develop the other urban systems around a CE that can learn from ecologically and socially responsible planning (McHarg and Steiner, 1998), where the regeneration of natural systems is recognized as a factor of positive social impact.

In this context, in order to define a process of reurbanization from the construction of urban patterns based on CE, we must create a social and sustainable fabric around the necessary local resources/technologies to attain a circular metabolism (Schumacher, 1973; Spiegelhalter and Arch, 2010), focused on the efficient use of resources (Gregson et al., 2015) and the generation of socioeconomic profit (Ness and Xing, 2017). In this way, the urban patterns based on CE will differ from the imbalances (metabolic rift) described by sociologist John Bellamy Foster (2000).

- *Scale*. The perspective of natural ecosystems and their inherent conditions of flexibility, adaptation, and efficiency (Sauvé et al., 2016) will help to understand the city as a plane that includes the biophysical landscape as an instrumental system of support for urban economies that extends, beyond the normative limits, the conventional meaning of infrastructure (Bélanger, 2009). However, research related to the interdisciplinary study of urban metabolism (Broto et al., 2012; Golubiewski, 2012; Kennedy et al., 2007, 2011; Rapoport, 2011) demands a framework that can connect processes and impacts. Thus, in this context, we consider that the flows that take place in cities (Dijst et al., 2018; Lyons et al., 2018), their spatial location, and their periodicity are factors that account for the variation of the characteristics of the places and, therefore, the associated spatial processes that, linked to the studied ecology and to the environmental impacts, can become a tool for the modelling of environmental and socioeconomic systems.

This conceptual structure allows us to establish CE as a useful theory through which a sustainable urban growth can be achieved. It enables us to provide an answer to environmental imperatives (Mostafavi and Doherty, 2016) and value the transformative exchanges between nature and society (Swyngedouw and Kaika, 2003).

Fig. 3 shows how the three studied dimensions are set and presents the meaning of CE in the context of sustainable growth.

As we have explained, this intervention framework defines the strategies for the urban application of CE. Thus, it comprises current research approaches to CE, as well as new research lines that, related to the detected keys of urban sustainability, highlight a necessary interdisciplinary perspective (Murray et al., 2017) for the development of this field of study.

Lastly, through the concept of "spherical city," we presented the urban application of CE as a research scope that can be used in studies on sustainable development at the regional scale (Bulkeley, 2010). Thus, cities based on the connection between the global and local worlds (Geddes, 1915) respond to the change made in the relationship between

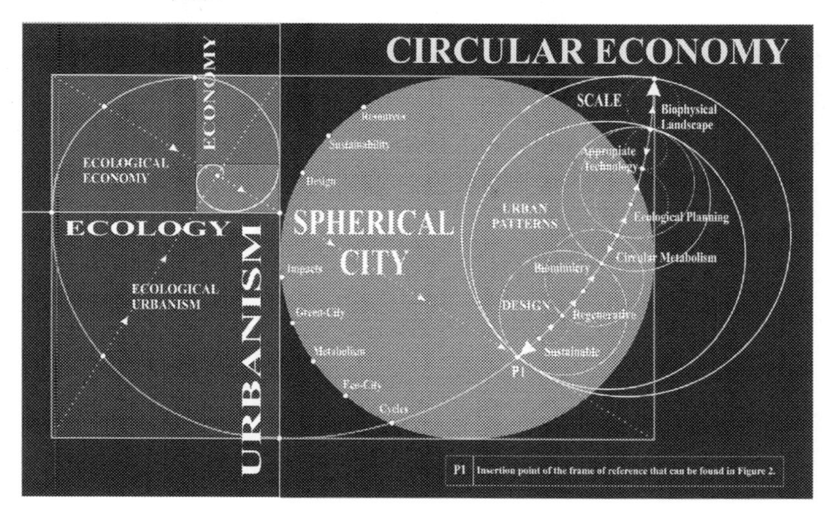

Fig. 3 Mapping the spherical city. *Compiled by the author.*

individuals and the material world (Rosales Carreón and Worrell, 2018). Therefore, the cities that develop considering spatial processes set up a plot that becomes geographical and reflects the complex interactions between the environment, society, and the economy.

7. Conclusions

The ideas and arguments presented in this chapter allow us to draw the following conclusions:

- Regarding the creation of a city that is disconnected from market dynamics.

We understand the importance of recovering urban thinking to move away from the unsustainable patterns of urban growth. To this end, it is fundamental to claim and rediscover the city as a social context.

Thus, we can assert that, in order to respond to the environmental concerns of the society that inhabits it, we must understand the city as a continuous space that allows us to evaluate the impacts as spatial processes, and, in turn, understand the city as an environment that enables the recognition of the proximity to nature of the place that helps us in the definition of identity and belonging.

- Regarding CE and urban sustainability.

CE is understood as a theory that acknowledges the ecological aspects and associates environmental sustainability with the dissociation of resource consumption from economic growth. Therefore, its usefulness is valued in terms of its capacity to redefine urban growth.

In this context, the conducted analysis detected that, while there are few studies focused on the social and urban role of CE, the scope of sustainability is presented as a field of important development for CE. Therefore, taking into account the point tackled above, this chapter presents a study that shows the keys of sustainability to develop both the social aspect of CE and the application of the term at the macro scale.

- Regarding the elaboration of a strategy to redefine the city and create a change of paradigm for sustainable urbanism.

Considering the previous points, CE is consolidated as a useful theory for the model of sustainable city of the 21st century. In it, CE serves, firstly, urban planning, to reformulate tools, and, secondly, urban design, to conceptualize physical qualities that pursue the redefinition of growth from the social and spatial aspects of the city.

A strategy is defined which, based on the regeneration of natural systems and structured around three interconnected dimensions, describes CE as a possible change of paradigm for urban sustainability.

We want to highlight the usefulness of the diagram as a tool of graphical representation. The reasoning implicit in the design of the images that are included in this chapter serves to consolidate the final strategy and guide the reader both in the analysis and in the generation of results. Thus, the development of the research considering its representation can detect deficiencies, gather information from several scopes and reflect on the interdependencies of the city as a system of systems.

References

Aberley, E.D., 1993. Boundaries of Home: Mapping for Local Empowerment. New Society Publishers.

Aberley, D., 1994. Futures by Design: The Practice of Ecological Planning. New Society Pub.

Ayres, R.U., 1999. The second law, the fourth law, recycling and limits to growth. Ecol. Econ. 29 (3), 473–483. https://doi.org/10.1016/S0921-8009(98)00098-6.

Ayres, R.U., Ayres, L.W., 1996. Industrial Ecology: Closing the Materials Cycle. Edward Elgar, Aldershott, UK.

Ayres, R., Ayres, L., 2002. A Handbook of Industrial Ecology. Edward Elgar Publishing. https://doi.org/10.4337/9781843765479.

Ayres, R.U., Kneese, A.V., 1969. Production, consumption, and externalities. Am. Econ. Rev. 59 (3), 282–297.

Batty, M., 2012. Building a science of cities. Cities 29, S9–S16. https://doi.org/10.1016/j.cities.2011.11.008.

Bélanger, P., 2009. Landscape as infrastructure. Landsc. J. 28 (1), 79–95. https://doi.org/10.3368/lj.28.1.79.

Benyus, J.M., 1997. Biomimicry—Innovation Inspired by Nature. William Morrow and Company, New York.

Benyus, J., 2003. Biomimicry. Innovation Inspired by Nature. HarperCollins Publishers.

Boulding, K.E., 1966. The economics of the coming spaceship earth. In: Jarrett, H. (Ed.), Environmental Quality Issues in a Growing Economy. Johns Hopkins University Press, Baltimore, MD.

Braungart, M., McDonough, W., 2002. Cradle to Cradle. Remaking the Way We Make Things. North Point Press, New York.

Brennan, G., Tennant, M., Blomsma, F., 2015. Business and production solutions: closing loops and the circular economy. In: Kopnina, H., Blewitt, J. (Eds.), Sustainability: Key Issues. Routledge, London, United Kingdom.

Brenner, N., 2014. Implosions/Explosions: Towards a Study of Planetary Urbanization. Verlag GmbH, Berlin.

Brenner, N., Schmid, C., 2014. The 'urban age' in question: the 'urban age' in question. Int. J. Urban Reg. Res. 38 (3), 731–755. https://doi.org/10.1111/1468-2427.12115.

Broto, V.C., Allen, A., Rapoport, E., 2012. Interdisciplinary perspectives on urban metabolism. J. Ind. Ecol. 16 (6), 851–861. https://doi.org/10.1111/j.1530-9290.2012.00556.x.

Bugge, M.M., Hansen, T., Klitkou, A., 2016. What is the bioeconomy? A review of the literature. Sustainability 8 (7), 691. https://doi.org/10.3390/su8070691.

Bulkeley, H., 2010. Cities and the governing of climate change. Annu. Rev. Env. Resour. 35 (1), 229–253. https://doi.org/10.1146/annurev-environ-072809-101747.

Capra, F., Spretnak, C., Lutz, R., 1984. Green Politics. Hutchinson.

Carson, R., 1962. Silent Spring. Houghton Mifflin, Boston.

Cerdá, I., 1867. Teoría general de la urbanización y aplicación de sus principios y doctrinas a la reforma y ensanche de Barcelona. Imprenta Española, Madrid.

Choay, F., 1970. El urbanismo. Utopías y realidades, Lumen.

Clark, W.C., Crutzen, P.J., Schellnhuber, H.J., 2005. Science for global sustainability: toward a new paradigm. SSRN Electronic J. https://doi.org/10.2139/ssrn.702501.

Commoner, B., 1971. The Closing Circle: Nature, Man, and Technology. Random House, New York.

Commoner, B., 1972. The environmental cost of economic growth. In: Ridker, R.G. (Ed.), Population, Resources and the Environment. U.S. Government Printing Office, Washington, DC, pp. 339–363.

Daly, H.E., 1990. Toward some operational principles of sustainable development. Ecol. Econ. 2 (1), 1–6. https://doi.org/10.1016/0921-8009(90)90010-R.

Daly, H.E., 1992. Steady-state economics: concepts, questions, policies. GAIA Ecol. Perspect. Sci. Soc. 1 (6), 333–338. https://doi.org/10.14512/gaia.1.6.5.

Daly, H.E., Townsend, K.N., 1993. Valuing the Earth Economics, Ecology, Ethics. MIT Press.

Dijst, M., Worrell, E., Böcker, L., Brunner, P., Davoudi, S., Geertman, S., Harmsen, R., Helbich, M., Holtslag, A.A.M., Kwan, M.-P., Lenz, B., Lyons, G., Mokhtarian, P.L., Newman, P., Perrels, A., Ribeiro, A.P., Rosales Carreón, J., Thomson, G., Urge-Vorsatz, D., Zeyringer, M., 2018. Exploring urban metabolism—towards an interdisciplinary perspective. Resour. Conserv. Recycl. 132, 190–203. https://doi.org/10.1016/j.resconrec.2017.09.014.

Dodge, J., 1981. Living by life: some bioregional theory and practice. Coevolut. Quart. 32, 6–12.

D'Amato, D., Droste, N., Allen, B., Kettunen, M., Lähtinen, K., Korhonen, J., Leskinen, P., Matthies, B.D., Toppinen, A., 2017. Green, circular, bio economy: a comparative analysis of sustainability avenues. J. Clean. Prod. 168, 716–734. https://doi.org/10.1016/j.jclepro.2017.09.053.

EEA (European Environment Agency), 2015. Urban Sustainability Issues—What Is a Resource-Efficient City? https://www.eea.europa.eu/publications/resource-efficient-cities.

Ehrlich, P.R., Ehrlich, A.H., 1990. The Population Explosion. Pocket Books.

Ehrlich, P.R., Holdren, J.P., 1971. Impact of population growth. Science 171 (3977). https://doi.org/10.1126/science.171.3977.1212.

EMF (Ellen MacArthur Foundation) (2012). Circular Economy Report—Towards the Circular Economy. Vol. 1. https://www.ellenmacarthurfoundation.org/publications/towards-the-circular-economy-vol-1-an-economic-and-business-rationale-for-an-accelerated-transition.

EMF (Ellen MacArthur Foundation) (2013a). Circular Economy Overview. Available at https://www.ellenmacarthurfoundation.org/circular-economy/overview/concept.

EMF (Ellen MacArthur Foundation), 2013b. Towards the circular economy. In: Vol. 2: Opportunities for the Consumer Goods Sector. Ellen MacArthur Foundation.

Foster, J.B., 2000. Marx's Ecology: Materialism and Nature. Monthly Review Press, New York.

Frey, B.S., Stutzer, A., 2002. Happinesss & Economics: How the Economy and Institutions Affect Human Well-Being. Princeton University Press, Princeton, NJ.

Frosch, R.A., Gallopoulos, N.E., 1989. Strategies for manufacturing. Sci. Am. 261 (3), 144–152. https://doi.org/10.1038/scientificamerican0989-144.
García Vázquez, C., 2016. Teorías e historia de la ciudad contemporánea. Gustavo Gili.
Geddes, P., 1884. John Ruskin, Economist. W. Brown, Edinburgh.
Geddes, P., 1915. Cities in Evolution: An Introduction to the Town Planning Movement and to the Study of Civics. Williams, London. http://archive.org/details/citiesinevolutio00gedduoft.
Geissdoerfer, M., Savaget, P., Bocken, N.M.P., Hultink, E.J., 2017. The circular economy—a new sustainability paradigm? J. Clean. Prod. 143, 757–768. https://doi.org/10.1016/j.jclepro.2016.12.048.
Georgescu-Roegen, N., 1971. La ley de entropía y el proceso económico. Harvard University Press, Cambridge. https://doi.org/10.4159/harvard.9780674281653.
Ghisellini, P., Cialani, C., Ulgiati, S., 2016. A review on circular economy: the expected transition to a balanced interplay of environmental and economic systems. J. Clean. Prod. 114, 11–32. https://doi.org/10.1016/j.jclepro.2015.09.007.
Girardet, H., 1992. The Gaia Atlas of Cities: New Directions for Sustainable Urban Living. Gaia, London.
Golubiewski, N., 2012. Is there a metabolism of an urban ecosystem? An ecological critique. Ambio 41 (7), 751–764. https://doi.org/10.1007/s13280-011-0232-7.
Graedel, T.E., 1996. On the concept of industrial ecology. Annu. Rev. Energy Environ. 21 (1), 69–98. https://doi.org/10.1146/annurev.energy.21.1.69.
Gregson, N., Crang, M., Fuller, S., Holmes, H., 2015. Interrogating the circular economy: the moral economy of resource recovery in the EU. Econ. Soc. 44 (2), 218–243. https://doi.org/10.1080/03085147.2015.1013353.
Gullone, E., 2000. The biophilia hypothesis and life in the 21st century: increasing mental health or increasing pathology? J. Happiness Stud. 1 (3), 293–322. https://doi.org/10.1023/A:1010043827986.
Harvey, D., 1996. Justice, Nature and the Geography of Difference. Blackwell Publishers, Cambridge, MA.
Harvey, D., 2004. Mundos urbanos posibles. In: Martín, R.A. (Ed.), Lo urbano en 20 autores contemporáneos. Universidad Politécnica de Cataluña, Barcelona, pp. 177–198.
Helms, M., Vattam, S.S., Goel, A.K., 2009. Biologically inspired design: process and products. Des. Stud. 30 (5), 606–622. https://doi.org/10.1016/j.destud.2009.04.003.
Heynen, N.C., Kaika, M., Swygedouw, E., 2005. In the Nature of Cities—Urban Political Ecology and the Politics of Urban Metabolism. ResearchGate. https://www.researchgate.net/publication/275035207_In_the_Nature_of_Cities_-_Urban_Political_Ecology_and_The_Politics_of_Urban_Metabolism.
IEA (International Energy Agency), 2008. World Energy Outlook 2008. OECD/IEA, Paris.
Jacobs, J., 1961. The Death and Life of Great American Cities. Random House, New York.
Jencks, C., 2000. Jenck's theory of evolution: an overview of 20th century architecture. Architect. Rev. 208 (1241), 76–79.
Kates, R.W., Parris, T., Leiserowitz, A., 2005. What is sustainable development? Environ. 47, 9–21.
Kennedy, C., Cuddihy, J., Engel-Yan, J., 2007. The changing metabolism of cities. J. Ind. Ecol. 11 (2), 43–59. https://doi.org/10.1162/jie.2007.1107.
Kennedy, C., Pincetl, S., Bunje, P., 2011. The study of urban metabolism and its applications to urban planning and design. Environ. Pollut. 159 (8), 1965–1973. https://doi.org/10.1016/j.envpol.2010.10.022.
Kirchherr, J., Reike, D., Hekkert, M., 2017. Conceptualizing the circular economy: an analysis of 114 definitions. Resour. Conserv. Recycl. 127, 221–232. https://doi.org/10.1016/j.resconrec.2017.09.005.
Korhonen, J., Honkasalo, A., Seppälä, J., 2018a. Circular economy: the concept and its limitations. Ecol. Econ. 143, 37–46. https://doi.org/10.1016/j.ecolecon.2017.06.041.
Korhonen, J., Nuur, C., Feldmann, A., Birkie, S.E., 2018b. Circular economy as an essentially contested concept. J. Clean. Prod. 175, 544–552. https://doi.org/10.1016/j.jclepro.2017.12.111.
Kuhn, T., 1962. The Structure of Scientific Revolutions. University of Chicago Press.
Lifset, R., Graedel, T.E., 2001. Industrial ecology: Goals and definitions. In: Ayres, R.U., Ayres, L. (Eds.), Handbook for Industrial Ecology. Edward Elgar, Brookfield.
Loorbach, D., Shiroyama, H., 2016. The Challenge of Sustainable Urban Development and Transforming Cities. pp. 3–12. https://doi.org/10.1007/978-4-431-55426-4_1.
López de Lucio, R., 1993. Ciudad y Urbanismo a finales del s.XX. Servei de Publicacions, Universitat de Valencia.

Lyle, J.T., 1994. Regenerative design for sustainable development. In: The Wiley Series in Sustainable Design (USA). http://agris.fao.org/agris-search/search.do?recordID=US9519383.
Lyons, G., Mokhtarian, P., Dijst, M., Böcker, L., 2018. The dynamics of urban metabolism in the face of digitalization and changing lifestyles: understanding and influencing our cities. Resour. Conserv. Recycl. 132, 246–257. https://doi.org/10.1016/j.resconrec.2017.07.032.
Marx, K., Nicolaus, M., 1993. Grundrisse: Foundations of the Critique of Political Economy (Edición: New Ed). Penguin Classics.
McGregor, S.L.T., 2013. Transdisciplinarity and biomimicry. Transdiscip. J. Eng. Sci. 4. https://doi.org/10.22545/2013/00042.
McHarg, I.L., Steiner, F.R., 1998. To Heal the Earth: Selected Writings of Ian L. McHarg. Island Press.
Meadows, D.H., Meadows, D.L., Randers, J., Behrens, W.W., 1972. The Limits to Growth. A report for the club of Rome's projecto on the predicament of mankind, Universe Books, New York.
Montaner, J. M. (2014). Del diagrama a las experiencias, hacia una arquitectura de la acción. http://0-site.ebrary.com.fama.us.es/lib/unisev/Doc?id=10995330.
Moreau, V., Sahakian, M., van Griethuysen, P., Vuille, F., 2017. Coming full circle: why social and institutional dimensions matter for the circular economy: why social and institutional dimensions matter. J. Ind. Ecol. 21 (3), 497–506. https://doi.org/10.1111/jiec.12598.
Morris, D., 1982. Self-Reliant Cities: Energy and the Transformation of Urban America. Sierra Club Books, San Francisco.
Mostafavi, M., Doherty, G., 2016. Ecological Urbanism. Lars Muller Publisher, Baden, Switzerland.
Mumford, L., 1961. The City in History: Its Origins, Its Transformations, and Its Prospects. Harcourt, Brace & World.
Murray, A., Skene, K., Haynes, K., 2017. The circular economy: an interdisciplinary exploration of the concept and application in a global context. J. Bus. Ethics 140 (3), 369–380. https://doi.org/10.1007/s10551-015-2693-2.
Naredo, J.M., 1987. La economía en evolución. Historia y perspectivas de las categorías básicas del pensamiento económico. Siglo XXI, Madrid.
Ness, D.A., Xing, K., 2017. Toward a resource-efficient built environment: a literature review and conceptual model. J. Ind. Ecol. 21 (3), 572–592. https://doi.org/10.1111/jiec.12586.
OECD (Organization for Economic Co-operation and Development), 2011. Toward Green Growth. https://www.oecd.org/greengrowth/49709364.pdf.
ONU-HABITAT, 2016. Conferencia de las Naciones Unidas sobre la Vivienda y el Desarrollo Urbano Sostenible (Hábitat III). Nueva Agenda Urbana http://www.habitat3.org/.
Patten, B.C., Odum, E.P., 1981. The cybernetic nature of ecosystems. Am. Nat. 118 (6), 886–895. https://doi.org/10.1086/283881.
Pearce, D., Turner, R.K., 1990. Economía de los recursos naturales y el medioambiente. John Hopkins University Press, Baltimore.
Pedersen Zari, M., 2015. Ecosystem services analysis: mimicking ecosystem services for regenerative urban design. Int. J. Sustain. Built Environ. 4 (1), 145–157. https://doi.org/10.1016/j.ijsbe.2015.02.004.
Pomponi, F., Moncaster, A., 2017. Circular economy for the built environment: a research framework. J. Clean. Prod. 143, 710–718. https://doi.org/10.1016/j.jclepro.2016.12.055.
Potting, J., Hekkert, M., Worrell, E., Hanemaaijer, A., 2017. Circular Economy: Measuring Innovation in the Product Chain. PBL Publishers, Policy Report.
Prendeville, S., Cherim, E., Bocken, N., 2018. Circular cities: mapping six cities in transition. Environ. Innov. Soc. Trans. 26, 171–194. https://doi.org/10.1016/j.eist.2017.03.002.
Prieto-Sandoval, V., Jaca, C., Ormazabal, M., 2018. Towards a consensus on the circular economy. J. Clean. Prod. 179, 605–615. https://doi.org/10.1016/j.jclepro.2017.12.224.
Rapoport, E., 2011. Interdisciplinary Perspectives on Urban Metabolism: a Review of the Literature. UCL Development Planning Unit (2011). Environmental Institute Working Paper, London.
Register, R., 1987. Ecocity Berkeley. North Atlantic Books, USA.
Robinson, J., 2004. Squaring the circle? Some thoughts on the idea of sustainable development. Ecol. Econ. 48 (4), 369–384. https://doi.org/10.1016/j.ecolecon.2003.10.017.

Rockström, J., Steffen, W., Noone, K., Persson, Å., Chapin, F.S.I., Lambin, E., Lenton, T., Scheffer, M., Folke, C., Schellnhuber, H.J., Nykvist, B., de Wit, C., Hughes, T., van der Leeuw, S., Rodhe, H., Sörlin, S., Snyder, P., Costanza, R., Svedin, U., Foley, J., 2009. Planetary boundaries: exploring the safe operating space for humanity. Ecol. Soc. 14 (2). https://doi.org/10.5751/ES-03180-140232.

Rosales Carreón, J., Worrell, E., 2018. Urban energy systems within the transition to sustainable development. A research agenda for urban metabolism. Resources. Conserv. Recycl. 132, 258–266. https://doi.org/10.1016/j.resconrec.2017.08.004.

Roseland, M., 1997. Dimensions of the eco-city. Cities 14 (4), 197–202. https://doi.org/10.1016/S0264-2751(97)00003-6.

Rubin, J., 2012. The End of Growth. Random House, Toronto, Canada.

Sala, S., Farioli, F., Zamagni, A., 2013. Progress in sustainability science: lessons learnt from current methodologies for sustainability assessment: part 1. Int. J. Life Cycle Assess. 18 (9), 1653–1672. https://doi.org/10.1007/s11367-012-0508-6.

Sala, S., Ciuffo, B., Nijkamp, P., 2015. A systemic framework for sustainability assessment. Ecol. Econ. 119, 314–325. https://doi.org/10.1016/j.ecolecon.2015.09.015.

Sauvé, S., Bernard, S., Sloan, P., 2016. Environmental sciences, sustainable development and circular economy: alternative concepts for trans-disciplinary research. Environ. Dev. 17, 48–56. https://doi.org/10.1016/j.envdev.2015.09.002.

Schumacher, E.F., 1973. Small Is Beautiful: A Study of Economics as If People Mattered. Blond & Briggs, London.

Smithson, A., Smithson, P., 1967. Urban Structuring: Studies of Alison & Peter Smithson. Studio Vista, London.

Spiegelhalter, T., Arch, R.A., 2010. Biomimicry and Circular Metabolism for the Cities of the Future. pp. 215–226. https://doi.org/10.2495/SC100191.

Stahel, W.R., 1982. The Product-Life Factor. An Inquiry into the Nature of Sustainable Societies: The Role of the Private Sector. Houston Area Research Center.

Stahel, W., Reday, G., 1976. The potential for substituting manpower for energy. Report to the Commission of the European Communities.

Steffen, W., Richardson, K., Rockström, J., Cornell, S.E., Fetzer, I., Bennett, E.M., Biggs, R., Carpenter, S.R., de Vries, W., de Wit, C.A., Folke, C., Gerten, D., Heinke, J., Mace, G.M., Persson, L.-M., Ramanathan, V., Reyers, B., Sörlin, S., 2015. Planetary boundaries: guiding human development on a changing planet. Science 347 (6223). https://doi.org/10.1126/science.1259855.

Swyngedouw, E., & Kaika, M. (2003). The environment of the city… or the urbanization of nature. En G. Bridge & S. Watson (Eds.), A Companion to the City. (Pp. 567–580) Blackwell Publishing Ltd. https://doi.org/10.1002/9780470693414.ch47.

Tarr, J.A., 2002. The metabolism of the industrial city: the case of Pittsburgh. J. Urban Hist. 26 (5), 511–545.

Taylor Buck, N., 2017. El arte de imitar la vida: La contribución potencial de la biomimetismo en la configuración del futuro de nuestras ciudades. Environ. Plan. B 44 (1), 120–140. https://doi.org/10.1177/0265813515611417.

Taylor Buck, N., While, A., 2015. Competitive urbanism and limits to smart city innovation: the UK future cities initiative. Urban Stud. 54 (2), 501–519. https://doi.org/10.1177/0042098015597162.

UN-HABITAT, 2012. State of the World's Cities Report. Prosperity of Cities. https://sustainabledevelopment.un.org/content/documents/745habitat.pdf, 2012/2013.

UNDP (United Nations Development Programme), 2020. Sustainable Development Goals. Goal 11. Sustainable Cities and Communities. https://www.undp.org/content/undp/en/home/sustainable-development-goals/goal-11-sustainable-cities-and-communities.html.

UNEP, 2011. Hacia una economía verde: Guía para el desarrollo sostenible y la erradicación de la pobreza—Síntesis para los encargados de la formulación de políticas. United Nations Environment Programme. www.unep.org/greeneconomy.

UNEP, 2015. Global Initiative for Resource Efficient Cities. UNEP Division of Technology, Industry and Economics.

van Buren, N., Demmers, M., van der Heijden, R., Witlox, F., 2016. Towards a circular economy: the role of Dutch logistics industries and governments. Sustainability 8 (7), 647. https://doi.org/10.3390/su8070647.

Vergragt, P., Akenji, L., Dewick, P., 2014. Sustainable production, consumption, and livelihoods: global and regional research perspectives. J. Clean. Prod. 63, 1–12. https://doi.org/10.1016/j.jclepro.2013.09.028.
Wackernagel, M., Rees, W.E., 1996. Our Ecological Footprint: Reducing Human Impact on the Earth. New Society Publisher, Gabriola Island, British Columbia, Canada.
Webber, M., 2004. La era postciudad. In: Martín, R.A. (Ed.), Lo urbano en 20 autores contemporáneos. Universidad Politécnica de Cataluña, Barcelona, pp. 13–23.
Williams, R., 1973. The Country and the City. Oxford University Press, New York, NY.
Wolman, A., 1965. The metabolism of cities. Sci. Am. 213, 179–190.
World Commission on Environment and Development, 1987. Report of the World Commission on Environment and Development: Our Common Future. http://www.un-documents.net/our-common-future.pdf.
Xue, B., Chen, X., Geng, Y., Guo, X., Lu, C., Zhang, Z., Lu, C., 2010. Survey of officials' awareness on circular economy development in China: based on municipal and county level. Resour. Conserv. Recycl. 54 (12), 1296–1302. https://doi.org/10.1016/j.resconrec.2010.05.010.
Zaera-Polo, A., 2016. Ya bien entrado el siglo XXI ¿las arquitecturas del post-capitalismo? El croquis 187, 20.
Zari, M.P., 2012. Ecosystem services analysis for the design of regenerative built environments. Build. Res. Inf. 40 (1), 54–64. https://doi.org/10.1080/09613218.2011.628547.

CHAPTER 20

Overview: The smart sustainable city initiatives and the circular economy

Azadeh Dindarian
Co-founder, NA, Women AI Academy & Consulting GmbH, Berlin, Germany

1. Introduction

In recent years, population growth worldwide has resulted in an increase in urbanization. The UN has projected that 68% of the world's population will live in urban areas by 2050 (United Nation, 2018). Already, two-thirds of the European Union's population lives in urban areas (European Commission, 2011). This current rate of urbanization has put pressure on cities generally in terms of a staggering amount of environmental pollution, waste generation, mobility issues, energy use, and a lack of resources. For example, the waste generated by the population within cities worldwide is expected to increase to 2.2 billion tons per year by 2025 by The World Bank (2014). It is estimated that by 2050, the energy consumption of cities will have reached 730 EJ if the appropriate actions for the reduction in energy consumption are not seriously addressed by stakeholders (Thorpe, 2015). Given these factors, rapid advances in technology and broader use of information and communication technology (ICT) in various aspects of city planning has resulted in the introduction of the terms "smart city," "digital city," "intelligent city," and, more recently, "smart sustainable city," terms that are now widely used by scholars, politicians, and practitioners (Ojo et al., 2016; Akande et al., 2019). In this regard, the use of emerging technologies such as information and communication technology (ICT), the internet-of-things (IoT), cloud computing, big-data analytics, blockchain, and artificial intelligence (AI), has become the center of attention for stakeholders in promoting sustainable and smart cities, so as to create better social and economic experiences for populations (Caragliu et al., 2011; Zhang et al., 2019). It is estimated that, by 2025, the smart city market will have reached $237.6 billion (Grand View Research, 2019). Smart public transport, smart waste management, smart buildings, smart traffic monitoring, smart parking facilities, and e-voting are just a few contributors to the smart city concept.

It is widely believed that the use of ICT-enabled technologies, in particular, IoT devices, within the smart city concept would facilitate access to real-time data and thus increase the transparency of entire operations (Zanella et al., 2014). These real-time data are crucial in gaining a better understanding of the trends and shortcomings within the smart city concept and would be an essential part of the future development of cities and

Circular Economy and Sustainability, Volume 1
https://doi.org/10.1016/B978-0-12-819817-9.00004-1

the goal of sustainable cities (Zanella et al., 2014; Desdemoustier et al., 2019; Batty et al., 2012). Once the data have been collected, it becomes necessary to use artificial intelligence and machine-learning technologies for analyzing and sorting data accordingly (Allam, 2018; Allam and Dhunny, 2019). One challenge—among others—in the implementation of the smart city initiative is the security and privacy of the citizen and the collection and use of data for specific purposes (Van Zoonen, 2016). The use of blockchain technology, therefore, within smart city initiatives, has received significant attention from stakeholders because of the nature of such technology, which can offer security, trustworthiness, tractability, immutability, and transparency, among other benefits (Deloitte, 2018).

It is important to examine the success of the smart city implementation concept from the "technological, human and creative, sustainable, institutional and holistic" angle, as suggested by Desdemoustier et al. (2019). However, given that the fundamental success of the smart city concept depends on the use of emerging ICT, little is known or understood about the impact of current technological advances on the implementation and achievement of a smart sustainable city's objectives, particularly with circular economy concepts in mind. It is important to note that the available literature is mainly in the form of reports by consultancy firms such as Deloitte (2020) and Accenture (2020), and by technology providers, and in various webpages, blogs, and news organs. For this reason, peer-reviewed articles tend to be scarce on available use cases and solutions provided by established companies and start-ups (Petit-Boix and Leipold, 2018). In this chapter, the author has summarized and analyzed available literatures, company reports, and webpages up until April 2020, to give an overview of the concept of the smart sustainable city and its relation to circular economy.

The main objective of this chapter, therefore, is to present an overview of the use cases of emerging technologies in the smart sustainable city concept. In particular, examples are presented of deployments of IoT, AI, and blockchain-enabled technologies and services by smart cities around the world. In this chapter, we will concentrate on the importance of emerging technologies—especially IoT, AI, and blockchain—in the successful implementation of the smart city concept, which facilitates circular economy initiatives. Furthermore, examples are offered of projects and start-ups active within the domain of smart waste management, smart water infrastructure management, and smart energy management.

The remainder of this chapter is organized as follows. Section 2 describes the concept of the smart city and smart sustainable city, using the available definitions within the current literature and reports. Section 3 provides an overview of the circular economy and its elements, along with its impact on the achievement of smart sustainable city objectives. Section 4, introduces use cases in the application of emerging technologies within smart city initiatives and their effectiveness in achieving key CE elements, particularly in relation to projects and solutions on smart water management, smart waste management, and smart energy management. Lastly, Section 5 provides concluding remarks to this chapter.

2. Smart city and smart sustainable city

Over the last 2 decades, the concept of the smart city has emerged among the scientific community and within practitioner circles (Kitchin, 2015; Bibri, 2019). However, it has been widely known that the meaning of the terms "smart city" and "smart sustainable city", and, indeed, the whole concept, remains fuzzy and unclear. In this regard, different definitions are used to explain the terms by scholars, policymakers, and practitioners, while the smart city and smart sustainable city concepts combine numerous characteristics, components, and dimensions (Caragliu et al., 2011; Ojo et al., 2016; Desdemoustier et al., 2019). The definition of smart city has been categorized and investigated based on "technological, human and creative, sustainable, institutional and holistic" criteria by Desdemoustier et al. (2019), who argue that focusing on the technological implementation of ICT is not, on its own, a guarantee of successful smart cities. Thus, the importance of the role of human capital and education in the implementation of the smart city concept cannot be underestimated, and can even be considered crucial (Caragliu et al., 2011; Desdemoustier et al., 2019; Appio et al., 2019).

A distinction is drawn between *soft* smart city strategy and *hard* smart city strategy by Angelidou (2014), who outlines soft smart city strategy as "developing human and social capital through education, culture, social inclusion, social innovation strategy" and hard smart city strategy as "smart buildings, smart energy grids, smart water management, smart mobility" (Angelidou, 2014). In addition, the success of smart city implementation depends on the availability of finance for each project and on the level of participation by citizens (Dameri et al., 2019). Because of this, the needs of citizens within a smart city should not be ignored and ought to be aligned with sustainable goals (Simonofski et al., 2019).

It is important to highlight the aim of smart city initiatives in this space, which is "to harness physical infrastructures, Information Communication Technologies (ICT), knowledge resources, and social infrastructure for economic regeneration, social cohesion, better city administration, and infrastructure management" (Ojo et al., 2014). Furthermore, the success and implementation of a smart city concept can vary from one city to another and from country to country (Praharaj and Han, 2019). For example, the current infrastructure of a city, together with the availability of financial resources, is a critical factor to achieve the goals and implementation of the projects.

Caragliu et al. (2011) believe that a city can become smart only "when investments in human and social capital and traditional (transport) and modern (ICT) communication infrastructure fuel sustainable economic growth and a high quality of life, with a wise management of natural resources, through participatory governance." In this context, participatory governance by stakeholders, with specific collaboration between them, and keeping sustainable resource management within the city is important for the achievement of smart city initiatives. In addition, it is worth pointing out that smart city initiatives should be seen as "local urban strategy," which in particular depends "on both global, standard drivers and local contingencies" (Dameri et al., 2019). Furthermore, the drivers for smart city initiatives can vary from one city to another; the willingness of citizens to participate in the implementation of smart city initiatives could, for example, be seen as one driver for successful implementation (Simonofski et al., 2019).

One of the comprehensive definitions of a smart sustainable city has been developed by the International Telecommunication Union (ITU) and the United Nations Economic Commission for Europe (UNECE): "A Smart Sustainable City is an innovative city that uses information and communication technologies (ICTs) and other means to improve quality of life, efficiency of urban operation and services, and competitiveness, while ensuring that it meets the needs of present and future generations with respect to economic, social, environmental as well as cultural aspects" (ITU, 2015). This definition covers "economic, social, environmental, governance and cultural" as the five main aspects of a smart sustainability city (ITU, 2015). It is, therefore, safe to say that in order for any smart sustainable city initiative to justify itself as such, all five aspects within the definition must be covered.

According to Bibri (2018), it is worth mentioning that the "environmental sustainability" dimension of a Smart City has not been included in most definitions up to now. Furthermore, Yigitcanlar et al. (2019) conclude that a city cannot be a smart city unless it is sustainable in the first place. In addition, Bibri and Krogstie (2017) emphasize the improvement of environmental quality and the social wellbeing of citizens within cities by adapting the correct sustainable urban development strategies. For instance, incorporating circular economy concepts (Ellen MacArthur Fundation, 2019) into sustainable urban development strategies is an important aspect of promoting sustainability within a city.

In order to better understand the current status of smart city initiatives, it is worth highlighting the work of Ahvenniemi et al. (2017), who have analyzed eight smart cities and eight urban sustainability assessment frameworks and concluded that the focus of the frameworks on "modern technologies and smartness in the Smart City frameworks" has been stronger than on "urban sustainability frameworks." In addition, they suggest that "smart sustainable cities" is a more accurate term than "smart cities" and should be used in order to cover both the sustainability and smartness of a city (Ahvenniemi et al., 2017; Kramers et al., 2014).

Furthermore, according to Yigitcanlar et al. (2019), "current Smart City practice fails to incorporate an overarching sustainability goal that is progressive and genuine". In addition, in order to fulfill all aspects of a smart sustainable city, Ahvenniemi et al.

(2017) emphasize the use of lifecycle impact assessment for all smart ICT to be designed for and used in the planning of a smart sustainable city, in order to reduce the unnecessary waste of resources.

Yigitcanlar et al. (2019) point out that the terms "smart" and "sustainable," in the context of smart cities, are "often used as window-dressing or reduced to ancillary aspects". It is, therefore, necessary to look at the smart city concept from a different angle and to shift the focus away from being solely on the application of the latest technologies and their solutions, toward including instead other dimensions such as environmental sustainability (Monfaredzadeh and Berardi, 2015; Taamallah et al., 2017). In particular, the United Nations' Sustainable Development Goals (SDG) (United Nations, 2019) specifically address 17 sustainability issues that societies are facing, which are translated into roadmaps with recommendations to be archived by 2030 by all countries, in order "to promote prosperity while protecting the planet." For instance, SDG 6 (clean water and sanitation), SDG 7 (affordable and clean energy), and SDG 11 (sustainable cities and communities), all directly promote, among others, smart sustainable city initiatives. Specifically, one of SDG 11's targets is a focus on urban air quality and waste management. The aim is to "by 2030, reduce the adverse per capita environmental impact of cities, including by paying special attention to air quality and municipal and other waste management" (United Nations, 2019). Smart sustainable cities are seen by many stakeholders as a way to achieve the SDGs by 2030. For example, a report by McKinsey Global Institute (MGI) suggests that the implementation of smart city initiatives can cut greenhouse gas emissions by 10%–15% and, furthermore, it is estimated that moderate or significant progress toward 70% of the SDGs can be achieved by technologies deployed within smart city initiatives (McKinsey Global Institute (MGI), 2018).

3. Circular economy

Over the last couple of years, the concept of the circular economy (CE) has attracted the attention of scholars and practitioners. The European Commission, in particular, is the biggest promoter of CE within the EU member states (for further reading, see EU Commission, 2014, 2015a, b). In 2015, the EU Commission published guidelines on CE in order to "stimulate Europe's transition towards a circular economy, which will boost global competitiveness, foster sustainable economic growth and generate new jobs" (EU Commission, 2014). In order to achieve the CE and EU SDGs by 2030, the EU commission emphasizes the promotion of resources efficiency and the implementation of new business models, the involvement and support of all stakeholders including private and public entities, and the continued monitoring of progress through developing a CE-specific monitoring framework in partnership with other organizations, such as the European Environment Agency (EEA), with the involvement of EU member states (EU Commission, 2015b). In order to embrace the current situation, it is important to point out that approximately 177 billion tons of resources are expected to be extracted per year by 2050, with cities using 75% of given global resources (de Wit et al., 2018). However, the report by CIRCULARITY GAP (de Wit et al., 2018) shows that only 9.1% of the world's economies are circular, so the need for strategies, new services, and business models for promoting CE is inevitable in order to mitigate climate change.

However, the increased popularity of the CE concept among stakeholders has resulted in many publications on the concept and meaning of CE (Kirchherr et al., 2017; Ghisellini et al., 2016; Korhonen et al., 2018). For example, the Ellen McArthur Foundation has defined the meaning of the CE concept thus: "Looking beyond the current take-make-waste extractive industrial model, a circular economy aims to redefine growth, focusing on positive society-wide benefits. It entails gradually decoupling economic activity from the consumption of finite resources and designing waste out of the system. Underpinned by a transition to renewable energy sources, the circular model builds economic, natural, and social capital. It is based on three principles: Design out waste and pollution, Keep products and materials in use, Regenerate natural systems" (Ellen Macarthur Foundation, 2013). This CE concept, as we can see, incorporates CE business models such as reuse and the promotion of the sharing economy, as well as designing the waste out of the lifecycle of products and services. The core elements of CE are, in short, reuse, remanufacture, and recycle, or the 3Rs (Korhonen et al., 2018; Ellen Macarthur Foundation, 2013; Stahel, 2014). In addition, the circular economy is seen as a way of promoting sustainability practices within organizations and businesses that can benefit from the implementation of the concept, so as to increase resource efficiency and decrease waste in general (Kirchherr et al., 2017; Korhonen et al., 2018).

It is worth noting that Ghisellini et al. (2016) believe that the concept of CE should be applied to processes that are deemed viable, with environmental and economic benefits, and not just "as an approach to more appropriate waste management." Furthermore, they conclude that in order to accomplish CE objectives, it is important to focus on innovative approaches by stakeholders to the processes within their domain (Ghisellini et al., 2016). Consequently, the requirement of innovative business practices, ICT solutions, and government policies will become inevitable in CE initiatives.

Korhonen et al. (2018) offer a scientific-based definition of CE, driven by the three dimensions of sustainable development, namely economic, environmental and social:

> *An economy constructed from societal production-consumption systems that maximizes the service produced from the linear nature-society-nature material and energy throughput flow. This is done by using cyclical materials flows, renewable energy sources and cascading 11-type energy flows. Successful circular economy contributes to all the three dimensions of sustainable development. Circular economy limits the throughput flow to a level that nature tolerates and utilizes ecosystem cycles in economic cycles by respecting their natural reproduction rates.*

Circle Economy (a social enterprise, organized as a cooperative) has defined 7 key elements that have emerged from many CE definitions and terms, used by over 20 organizations, for the unified understanding and application of CE principles by stakeholders (de Wit et al., 2018):

1. Prioritize regenerative resources
2. Preserve and extend what's already made
3. Use waste as a resource
4. Rethink the business model
5. Design for the future
6. Incorporate digital technology
7. Collaborate in order to create joint value

The involvement of all stakeholders and the planned economic returns of CE activities are, therefore, seen as major success factors in adapting and transitioning toward CE within any organization (Ghisellini et al., 2016). On the other hand, cities are seen as facilitators in successfully adapting CE initiatives (Fratini et al., 2019). For example, employing emerging technology in order to increase the lifespan of products, or to provide efficient and effective supply-chain management within the waste industry, can be beneficial in the successful implementation of smart sustainable cities with circular economy elements. Likewise, the use of technology can strengthen collaboration between stakeholders and access to supply-chain data in order to increase transparency and trust among supply-chain stakeholders.

For these reasons, the implementation of CE elements within smart city projects by policymakers and practitioners, including innovative approaches by start-ups and citizens, is worth investigating. In the context of the smart sustainable city, there is a great deal of attention being paid to environmental issues, such as air pollution, resource efficiency (e.g., energy and waste management), urban transport and traffic issues, and sustainable infrastructure and buildings. In the next section, we present some use cases of the application of emerging technology within smart sustainable city initiatives and their impact on achieving CE key elements.

4. Use case of technology in a smart sustainable city

In this section, the application of emerging technologies in industry domains within smart sustainable city projects that contribute to the circular economy concept will be explored. Petit-Boix and Leipold (2018) have identified around 300 CE initiatives in 83 cities around the world, and conclude that most of those cities include urban structure, social consumption, industries and businesses, waste management, energy supply, and green technology in their main strategy agenda. Furthermore, McKinsey Global Institute (MGI) has identified eight industry domains in any Smart City concept, namely mobility, security, healthcare, energy, water, waste, economical housing and development, and engagement and community (McKinsey Global Institute (MGI), 2018). In addition, the interconnectivity between CE elements and current smart city concepts can help in achieving smart sustainable city initiatives.

In the following sections, the author will elaborate in greater depth on the impact of smart water and wastewater management, smart waste management, and smart energy management, together with the use cases of emerging technologies, on the circular economy and on the overall success of a smart sustainable city.

4.1 Smart water and waste-water management

As a direct consequence of urbanization and an increase in population, urban water infrastructures and management have been the focus of considerable attention for many years. As outlined in the previous section, CE initiatives are in place to promote reduce, reuse and recycle strategies for the available resources. Therefore, circular water management within a smart sustainable city is one of the key solutions for efficient and effective water and waste-water management. The benefits of circular water management have been recognized by Renzo Piano and Christoph Kohlbecker, the two architects of Berlin's PotsdamerPlatz development, and implemented within their design (PotsdamerPlatz.de, 2020). In this development, the rainwater collected from roof surfaces is recycled so as to save around 20 million liters of drinking water per year. This rainwater is then used to flush toilets and water the plants and lawns around the development. This set up helps the city of Berlin to move toward a sustainable city.

In this case, for example, employing ICT-enabled equipment, such as IoT devices and sensors, will help in the future to collect real-time data on the quantity and quality of rainwater and to predict demand and supply for this rainwater, based on real-time weather forecasting and usage. The accuracy of the data will help in managing the rainwater infrastructure and to detect leakage in an effective and efficient process; in some cases, the excess rainwater could be used by neighboring developments. Likewise, the data from smart water meters could be used to inform consumers of their consumption behavior, with details of usage cost that would, in return, encourage consumers to use freshwater responsibly and reduce the consumption rate accordingly. The city of Toronto's MyWaterToronto project utilized smart water readers to help households and business to get easy access to their consumption data, and help consumers to increase their usage awareness, to reduce their water usage and detect leakage with the data (MyWaterToronto, 2020).

Recently, it was reported that the introduction of a sensor-based underground drainage (UGD) monitoring system by Greater Visakhapatnam Municipal Corporation (GVMC) in India is expected to further help city stakeholders and government with access to real-time data on flow parameters of sewage so that they can plan and act in accordance with real-time data in an efficient and cost-effective way (India Smart

City Council, 2019b). These real-time data are vital for flood prediction and prevention while increasing resource efficiency and emergency response accordingly.

An example of circular water management is the EcoVolt technology invented and introduced by Cambrian Innovation for the treatment of wastewater, which provides clean-energy and clean-water recovery solutions (Silver et al., 2017). These solutions can help industrial users to reduce their use of freshwater by 40% while producing clean energy onsite (Silver et al., 2017). We can see, therefore, that those technologies designed to enable services such as wastewater treatment to become decentralized and more efficient, would help in achieving the CE element as well as the smart sustainable city concept.

4.2 Smart waste management

Given the current rate of waste generation, it is necessary to consider the use of emerging technologies in reducing the amount of waste generation and the use of resources. In this section, use cases of emerging technologies within the waste management sector are presented. The partnership between the Seoul Metropolitan Government and the Ecube Labs solution in smart waste management has resulted in the installation of 85 clean cubes (smart bins) across the city of Seoul, in South Korea. This solution has "eliminated overflowing waste bins with 66% reduction in collection frequency," resulting in "significantly less litter and cleaner public areas," and has "reduced waste collection costs by 83% [and] increased [the] recycling diversion rate to 46%" (India Smart City Council, 2019a).

Automating waste and recycling logistics in a smart sustainable city can serve to reduce congestion and CO_2 emissions, increase efficiency, and facilitate on-demand waste collection solutions. The Finnish start-up, Enovo (2019), and Slovakian start-up, Sensoneo (2019), both deploy sensors within bins. These sensors are connected to the company platform and to users' dashboards in order to facilitate real-time data on fill levels and to predict pick-up times, while simultaneously aggregating many parameters such as traffic information and the geolocation of pick-up trucks and smart bins. Furthermore, their technology helps to optimize collection routes and resources management accordingly. On the other hand, the availability of trusted data on the quantity and quality of waste would help stakeholders such as recyclers in predicting waste-flow type and facilitate data-based decision-making, so that operations could be optimized accordingly.

On their website, Sensoneo claim that by using their solution, overall waste-collection costs will be reduced by at least 30% and cities will enjoy the benefit of a 60% reduction in related carbon emissions (Sensoneo, 2019). It is worth noting that the quantification of waste at source would help to reduce such waste, and the data collected by recycling companies could be shared among consumers, thus helping them to monitor their behavior toward waste generation. This would, in the process, help consumers to change their consumption habits and hopefully reduce the resultant generation of waste at source.

The deployment of a combination of IoT, sensors, and AI-enabled technologies is part of innovation within the waste management industry. For instance, Recercula is a Spanish-based start-up that is behind RecySmart technology (Recircula Solutions, 2019), which has designed a device that connects to waste containers and interacts with users via a mobile application. This uses sensors and AI technologies to identify waste disposers and the type of waste to be disposed of and provides feedback on the recycling habits of the users. Such technology can help city officials in implementing reward-based or pay-per-use services and, in the long term, it should help waste reduction through changes in consumers' purchase and disposal habits. It would also help producers and packaging industries to move toward more sustainable practices.

The Ministry of Environment of South Korea has set a target for reducing nationwide food waste by 40%, and they have successfully implemented a pay-as-you-throw food policy through the introduction of a range of technologies such as sensors and IoT-enabled bins. A reduction of over 20% in food waste has already been reported in Seoul (Asia Today, 2013).

Winnow Solutions (2019) is a London-based start-up with a mission to reduce food waste within the hospitality industry via AI-based food waste tracking technology. Using Winnow's reporting tools, the hospitality industry can cut food waste significantly and help to reduce carbon emissions, while making businesses more profitable. In this case, big-data analytics and cloud computing would help in achieving smart sustainable city goals.

The concept of the sharing economy can contribute to the creation of a CE within a smart sustainable city and to the reduction of waste generation, along with an increase in the lifespan of products (Esmaeilian et al., 2018). Indeed, the concept can be applied to food, clothing, DIY appliances, and mobility, among others.

One example of this is the UK-based start-up, Too Good To Go (2019), which through its operations in 14 European countries (Go) is on a mission to reduce food waste by connecting consumers via their app to food businesses that buy and collect ready-made food at a reduced price—food that would otherwise have ended up in the bin. It is reported that, to date, users have managed to prevent approximately 6700 tonnes CO_2 emissions.

Bundlee (2019) is a UK-based baby-clothing subscription company on a mission to bring the circular economy to the children's clothing sector by reducing the number of items that end up in landfills or get incinerated. Their business model, based on a monthly fee, involves items being ordered via their website and received by post. Bundlee claims that by renting a bundle of clothes, each renter would help in reducing waste and extend the lifecycle of each piece by 400%, while saving 40,500 L of water and 105 kg of carbon dioxide.

Grover (2019) is a German-based start-up that is offering the latest electronic equipment based on a monthly subscription fee. Their business model helps to reduce electronic waste and promotes the circular economy in this domain. In addition, Grover helps to increase the lifecycle of electronic equipment by having control over ownership of the products. The users of Grover's services can help circularity within cities while reducing the burdens on the waste management industry, since the products would stay in the use cycle for an extended period of time.

4.3 Smart energy management

Higher urban energy consumption is the reason why smart energy within a smart sustainable city concept is at the center of attention. So far, progress in this domain has been tremendous compared to other industries within smart sustainable city projects. One reason could be that regulations and policies around the world are in favor of reducing the planet's dependency on fossil fuels and promoting local, reliable, renewable, and clean energy.

Smart energy uses ICTs, for example, in the form of peer-to-peer (P2P) energy trading, smart grids, smart meters, and smart billings, among others. By 2020, at least 80% of electricity meters will be replaced with a smart meter, which will result in an average energy saving of 3% in the EU (European Commission, 2014). In addition, capturing energy-related data, including demand-and-supply data, data on the availability of assets, and details on the energy production and consumption supply chain is crucial for collating real-time information to be used in decision-making and in implementing smart energy within the smart city concept (O'Dwyer et al., 2019).

City (2019) is a newly-built smart city project in the United Arab Emirates, based on environmental sustainability principles such as a reduction in the energy consumption in buildings, a decrease in carbon emissions, and the use of recycled material and locally-sourced and verified materials in the construction of the city. The 10 MW solar photovoltaic (PV) array and sensor-equipped, 45-m wind towers provide the renewable energy needed to power the city. In this case, the use of renewable energy, with the help of emerging technology, will help the country to become independent from the use of fossil fuel and reduce its carbon footprint in general.

Brooklyn Microgrid (n.d.) is a network of residents and business owners who own solar panels (prosumers) that produce solar energy, with the excess energy being sold on a blockchain-enabled platform and marketplace to consumers within the network. The advantages and benefits of BMG are reported to include support for the local economy, a reduction in greenhouse gas emissions and air pollution, and control over the sources of renewable energy. This practice can be implemented within smart sustainable city concepts in order to increase the production and use of clean technology while, at the same time, promoting the decentralization of energy production and microgrids.

Verv Energy (2019) is a London-based start-up, provider of a cutting-edge smart-energy hub, and green electricity peer-to-peer sharing platform. It uses blockchain and AI technology that could reduce household electricity bills by giving transparent and real-time data on the usage and cost of electricity based on each appliance, and a further reduction of over 20% of carbon emissions is expected by using Verv technology.

The City of Glasgow in Scotland, UK, has a strategy of becoming a circular city (Circle Economy, 2016). It is estimated that its food sector consumes 51% of the city's energy, of which 20% is used up by the boilers in the bakery sector. In a pilot project, the use of heat exchangers allowed the heat generated by ovens to be reused by boilers at the bakeries, thus saving between 15% and 30% of the energy consumed in the baking process, and eliminating 700 tons of CO_2 emissions per year.

Decentralized waste management would help cities and industries to manage their waste locally while reducing their CO_2 emissions and producing energy and secondary resources. For example, the London-based start-up SEaB Energy (SEaB Energy, 2019) uses patented technologies for converting waste to energy and fertilizers onsite, using anaerobic digestion (AD) systems in shipping containers. It is easy and cost-effective to install this decentralized solution in order to achieve circularity within any industry or setting where waste is generated and energy used at the same time.

5. Concluding remarks

In this chapter, the definitions of smart sustainable city concepts are given and the differences discussed. The importance of the application of circular economy elements to the infrastructure of the smart sustainable city has been stressed. A further objective of this chapter was to examine the use cases of the application of the emerging technologies within smart water management, smart waste management, and smart energy management, with examples of the solutions in each domain being given. These examples provide evidence that ICTs, IoT-enabled devices, blockchain technology, big data, AI, and so on, could all help tremendously in achieving environmental sustainability within smart city concepts. Examples of such environmental sustainability include lowering CO_2 emissions, an increase in resource efficiency, a reduction in waste, and overall cost saving. Even though these technologies are, in most cases, still in their infancy or in the pilot phase, the reported results are nevertheless promising that the employment of such technologies is going in the right direction.

The smart sustainable city concepts have attracted interest from every corner of the world, and interested stakeholders include citizens, technology providers, researchers, national and international governments and bodies, and consultancy companies. However, as is widely reported in the literature and discussed in previous sections, each stakeholder has their own interest and agenda in pursuing the smart sustainable city concept, so the barriers to implementation differ widely from one situation to another, and from one region to another (Praharaj and Han, 2019; Voda and Radu, 2019). In this regard, it is important to emphasize that close collaboration is necessary for creating a shared vision toward environmental sustainability of the city between stakeholders in the early stages of development of a smart sustainable city concept, in order to achieve a balanced and implementable specific smart sustainable city. A clear strategy, in tandem with policies from local and national policymakers would, therefore, help to achieve common goals. However, more in-depth research would help our understanding of divergent interests and of how stakeholders can collaborate in this area in order to achieve their common goals, especially given the complexity of implementation.

Future research in this domain should be focused on the short- and long-term impact of technology-based solutions within the entire smart sustainable city concept and its elements. Such research might also look, from a holistic point of view, at the broader impact on the whole system, in an effort to better understand the needs of citizens in the long term. The goal of the smart sustainable city concept, after all, is to connect all the various services with users and to optimize the use of data in overall decision-making.

References

Chapter in Edited Book

European Commission, 2011. Cities of Tomorrow: Challenges, Visions, Ways Forward. Directorate-General for Regional Policy, Publications Office of the European Union.

Ojo, A., Curry, E., Janowski, T., 2014. Designing Next Generation Smart City Initiatives—Harnessing Findings and Lessons From a Study of Ten Smart City Programs.

Ojo, A., Dzhusupova, Z., Curry, E., 2016. Exploring the Nature of the Smart Cities Research Landscape. Springer, Smarter as the New Urban Agenda, pp. 23–47.

Voda, A.I., Radu, L.-D., 2019. Chapter 12 - How can artificial intelligence respond to smart cities challenges? In: Visvizi, A., Lytras, M.D. (Eds.), Smart Cities: Issues and Challenges. Elsevier, pp. 199–216.

Journal Publication

Ahvenniemi, H., Huovila, A., Pinto-Seppä, I., Airaksinen, M., 2017. What are the differences between sustainable and smart cities? Cities 60, 234–245.

Akande, A., Cabral, P., Gomes, P., Casteleyn, S., 2019. The Lisbon ranking for smart sustainable cities in Europe. Sustain. Cities Soc. 44, 475–487.

Allam, Z., 2018. Contextualising the smart city for sustainability and inclusivity. New Design Ideas 2, 124–127.

Allam, Z., Dhunny, Z.A., 2019. On big data, artificial intelligence and smart cities. Cities 89, 80–91.

Angelidou, M., 2014. Smart city policies: a spatial approach. Cities 41, S3–S11.

Appio, F.P., Lima, M., Paroutis, S., 2019. Understanding smart cities: innovation ecosystems, technological advancements, and societal challenges. Technol. Forecast. Soc. Chang. 142, 1–14.

Batty, M., Axhausen, K.W., Giannotti, F., Pozdnoukhov, A., Bazzani, A., Wachowicz, M., Ouzounis, G., Portugali, Y., 2012. Smart cities of the future. Eur. Phys. J. Spl. Topics 214 (1), 481–518.

Bibri, S.E., 2018. The IoT for smart sustainable cities of the future: an analytical framework for sensor-based big data applications for environmental sustainability. Sustain. Cities Soc. 38, 230–253.

Bibri, S.E., 2019. On the sustainability of smart and smarter cities in the era of big data: an interdisciplinary and transdisciplinary literature review. J. Big Data 6 (1), 25.

Bibri, S.E., Krogstie, J., 2017. Smart sustainable cities of the future: an extensive interdisciplinary literature review. Sustain. Cities Soc. 31, 183–212.

Caragliu, A., Del Bo, C., Nijkamp, P., 2011. Smart cities in Europe. J. Urban Technol. 18 (2), 65–82.
Dameri, R.P., Benevolo, C., Veglianti, E., Li, Y., 2019. Understanding smart cities as a glocal strategy: a comparison between Italy and China. Technol. Forecast. Soc. Chang. 142, 26–41.
Desdemoustier, J., Crutzen, N., Giffinger, R., 2019. Municipalities' understanding of the Smart City concept: an exploratory analysis in Belgium. Technol. Forecast. Soc. Chang. 142, 129–141.
Esmaeilian, B., Wang, B., Lewis, K., Duarte, F., Ratti, C., Behdad, S., 2018. The future of waste management in smart and sustainable cities: a review and concept paper. Waste Manag. 81, 177–195.
EU Commission, 2014. Towards a circular economy: a zero waste programme for Europe. COM 2014, 398.
EU Commission, 2015b. Communication from the Commission to the European Parliament, the Council, the European Economic and Social Committee and the Committee of the Regions-Closing the loop-An EU action plan for the Circular Economy. Status of Data 2, 2015.
Fratini, C.F., Georg, S., Jørgensen, M.S., 2019. Exploring circular economy imaginaries in European cities: a research agenda for the governance of urban sustainability transitions. J. Clean. Prod. 228, 974–989.
Kirchherr, J., Reike, D., Hekkert, M., 2017. Conceptualizing the circular economy: an analysis of 114 definitions. Resour. Conserv. Recycl. 127, 221–232.
Kitchin, R., 2015. Making sense of smart cities: addressing present shortcomings. Camb. J. Reg. Econ. Soc. 8 (1), 131–136.
Korhonen, J., Honkasalo, A., Seppälä, J., 2018. Circular economy: the concept and its limitations. Ecol. Econ. 143, 37–46.
Kramers, A., Höjer, M., Lövehagen, N., Wangel, J., 2014. Smart sustainable cities—exploring ICT solutions for reduced energy use in cities. Environ. Model Softw. 56, 52–62.
Monfaredzadeh, T., Berardi, U., 2015. Beneath the smart city: dichotomy between sustainability and competitiveness. Int. J. Sustain. Build. Technol. Urban Dev. 6 (3), 140–156.
O'Dwyer, E., Pan, I., Acha, S., Shah, N., 2019. Smart energy systems for sustainable smart cities: current developments, trends and future directions. Appl. Energy 237, 581–597.
Patrizia, G., Cialani, C., Ulgiati, S., 2016. A review on circular economy: the expected transition to a balanced interplay of environmental and economic systems. J. Clean. Prod. 114, 11–32.
Petit-Boix, A., Leipold, S., 2018. Circular economy in cities: reviewing how environmental research aligns with local practices. J. Clean. Prod. 195, 1270–1281.
Praharaj, S., Han, H., 2019. Cutting through the clutter of smart city definitions: a reading into the smart city perceptions in India. City Cult. Soc. 18, 100289.
Silver, M., Barosky, M., Aviles, C., Rixey, E., 2017. Water Audit Expands Reuse Opportunities at Brewery.
Simonofski, A., Vallé, T., Serral, E., Wautelet, Y., 2019. Investigating context factors in citizen participation strategies: a comparative analysis of Swedish and Belgian smart cities. International Journal of Information Management.
Taamallah, A., Khemaja, M., Faiz, S., 2017. Strategy ontology construction and learning: insights from smart city strategies. Int. J. Knowl. Based Dev. 8 (3), 206–228.
Van Zoonen, L., 2016. Privacy concerns in smart cities. Gov. Inf. Q. 33 (3), 472–480.
Yigitcanlar, T., Kamruzzaman, M., Foth, M., Sabatini-Marques, J., da Costa, E., Ioppolo, G., 2019. Can cities become smart without being sustainable? A systematic review of the literature. Sustain. Cities Soc. 45, 348–365.
Zanella, A., Bui, N., Castellani, A., Vangelista, L., Zorzi, M., 2014. Internet of things for smart cities. IEEE Internet Things J. 1 (1), 22–32.
Zhang, A., Venkatesh, V.G., Liu, Y., Wan, M., Qu, T., Huisingh, D., 2019. Barriers to smart waste management for a circular economy in China. J. Clean. Prod. 240, 118198.

Reference to a Website

Accenture, 2020. Insight Smart Cities. Retrieved 02.01.2020, from https://www.accenture.com/us-en/insight-smart-cities.
Asia Today, 2013. South Korea's Food Waste Solution: You Waste, You Pay. Retrieved 09.12.2019, 2019, from http://www.asiatoday.com/pressrelease/south-koreas-food-waste-solution-you-waste-you-pay.
Brooklyn Microgrid. n.d. From https://www.brooklyn.energy/.
Bundlee, 2019. Bundlee. Retrieved 15.12.2019, from https://bundlee.co.uk/.

City, M., 2019. Masdar City. Retrieved 16.12.2019, from https://masdar.ae/en/masdar-city/the-city.
Deloitte, 2020. Deloitte. Retrieved 03.01.2020, from https://www2.deloitte.com/us/en/pages/consulting/solutions/smart-cities-of-the-future.html.
Ellen Macarthur Foundation, 2013. The Circular Model – Brief History and School of Thought. Retrieved 08.08.2019, 2019, from http://www.ellenmacarthurfoundation.org/circular-economy/circular-economy/the-circular-model-brief-history-and-schools-of-thought.
Ellen MacArthur Fundation, 2019. What is the circular economy? Retrieved 22.09.2019, from https://www.ellenmacarthurfoundation.org/circular-economy/what-is-the-circular-economy.
Enovo, 2019. Enovo. Retrieved 15.09.2019, from https://www.enevo.com/.
EU Commission, 2015a. Closing the Loop: Commission Adopts Ambitious New Circular Economy Package to Boost Competitiveness, Create Jobs and Generate Sustainable Growth. Retrieved 09.08.2019, 2019, from https://europa.eu/rapid/press-release_IP-15-6203_en.htm.
European Commission, 2014. Smart grids and meters. Retrieved 13.12.2019, 2019, from https://ec.europa.eu/energy/en/topics/markets-and-consumers/smart-grids-and-meters/overview.
Too Good To Go, 2019. Too Good To Go. Retrieved 12.12.2019, from https://toogoodtogo.co.uk/en-gb/about-us.
Grand View Research, 2019. Smart Cities Market Size Worth $237.6 Billion By 2025 | CAGR: 18.9%. Retrieved 05.09.2019, 2019, from https://www.grandviewresearch.com/press-release/global-smart-cities-market.
Grover, 2019. Grover. Retrieved 09.12.2019, 2019, from https://getgrover.frontify.com/d/O3V8uPizISFD.
India Smart City Council, 2019a. Case Study City of Seoul. Retrieved 21.10.2019, from https://smartcitiescouncil.com/system/tdf/main/public_resources/Ecube%20city%20of%20seoul.pdf?file=1&type=node&id=2845&force=.
India Smart City Council, 2019b. Now, sensor-based monitoring system to prevent sewage leaking onto roads. Retrieved 12.12.2019, 2019, from https://india.smartcitiescouncil.com/article/now-sensor-based-monitoring-system-prevent-sewage-leaking-roads.
ITU, 2015. Key performance indicators project for Smart Sustainable Cities. Retrieved 06/08/2019, 2019, from https://www.itu.int/en/ITU-T/ssc/united/Documents/SmartSustainableCities-KPI-ConceptNote-U4SSC-website.pdf.
MyWaterToronto, 2020. MyWaterToronto. Retrieved 12.03.2020, 2020, from https://www.toronto.ca/services-payments/water-environment/how-to-use-less-water/mywatertoronto/.
PotsdamerPlatz.de, 2020. A quarter as pioneer in outstanding sustainability. Retrieved 09.01.2020, from https://potsdamerplatz.de/en/sustainability/.
Recircula Solutions, 2019. Recircula Solutions. Retrieved 09.12.2019, from https://recirculasolutions.com/.
SEaB Energy, 2019. SEaB Energy. Retrieved 17.12.2019, from https://seabenergy.com/about-seab/.
Sensoneo, 2019. Sensoneo. Retrieved 12.09.2019, from https://sensoneo.com/waste-management-solution/.
Stahel, W., 2014. Reuse is the key to the circular economy. European Commission. Available at http://ec.europa.eu/environment/ecoap/about-eco-innovation/experts-interviews/reuse-is-the-key-to-the-circular-economy_en.
Thorpe, D., 2015. How Can Cities Cut Energy Use by 50%. Retrieved 07.01.2020, 2020, from https://www.weforum.org/agenda/2015/02/how-can-cities-cut-energy-use-by-50/.
United Nation, 2018. 68% of the world population projected to live in urban areas by 2050, says UN. From https://www.un.org/development/desa/en/news/population/2018-revision-of-world-urbanization-prospects.html.
United Nations, 2019. 17 Goals to Transform Our World. Retrieved 08.08.2019, from https://www.un.org/sustainabledevelopment/.
Verv Energy, 2019. Verv Energy. Retrieved 12.12.2019, 2019, from https://verv.energy/.
Winnow Solutions, 2019. Winnow Solutions. Retrieved 21.11.2019, 2019, from https://www.winnowsolutions.com/company.

Report

Circle Economy, 2016. Circular Glasgow A Vision and Action Plan for the City of Glasgow.
Deloitte, 2018. Super Smart City Happier Society With Higher Quality.
de Wit, M., Hoogzaad, J., Ramkumar, S., Friedl, H., Douma, A., 2018. The Circularity GAP Report. Circle Economy.
McKinsey Global Institute (MGI), 2018. Smart Cities: Digital Solutions for a More Livable Future. McKinsey & Company.
The World Bank, 2014. Results-Based Financing FoR Municipal Solid Waste ExEcutivE Summary. The World Bank.

CHAPTER 21

Transitioning into circular food consumption practices: An analytical framework

Borrello Massimiliano and Cembalo Luigi
University of Naples Federico II, Department of Agricultural Sciences, AgEcon and Policy Group, Italy

1. Introduction

In this chapter, we propose a framework of circular food consumption practice (CFCP). Circular economy (CE) research has devoted so far wide attention to theoretical aspects (Geissdoerfer et al., 2017; Kirchherr et al., 2017; Korhonen et al., 2018), industrial networks (Baldassarre et al., 2019; Mishra et al., 2019; Nogalski et al., 2019), business models (Bocken et al., 2016; Urbinati et al., 2017; Geissdoerfer et al., 2018), and technologies (Blank et al., 2020; Holanda et al., 2020; Kiani et al., 2020). However, the role of individuals for transitioning into a closed-loop-based economic system has been disregarded. Particularly, food is a domain in which CE investigation is struggling to emerge.

CE posits, among its principles, "rethinking consumption" (Moreau et al., 2017, p. 497), but defining meaning and possible implementations of this idea to food consumption is not an easy task. Lack of food circular supply chain implementations have limited the analysis of consumer behavior to researches designed around hypothetical business case studies (e.g., Borrello et al., 2017, 2020). Other scholars seem to be even more disoriented. To illustrate, education on the environmental impacts of food chains, reduction of meat in the diet, and control over labelling have been suggested as CE solutions (Jurgilevich et al., 2016). However, these are general indications for sustainable food consumption, related but not tailored to the CE principle of reconciling economic cycles with natural cycles (Tate et al., 2019) by means of closed-loop material flows. Given that how CE and sustainability concepts relate is still undefined (Geissdoerfer et al., 2017), making them coincide with each other seems to be stretching a point.

In order to reconnect studies on consumer behavior with circular economy principles, current study follows the ecological perspective of consumption supported by Røpke (2009, p. 2495). This author sees human society as "a metabolic organism appropriating resources from the environment, transforming them for purposes useful for humans, and finally discarding them as waste." This perspective is consistent with CE, which evokes the idea that supply chains are "metabolisms" in which waste is

Circular Economy and Sustainability, Volume 1
https://doi.org/10.1016/B978-0-12-819817-9.00016-8

continuously recycled, imitating natural processes such as the carbon biogeochemical cycle (EMF, 2012). Particularly, we approach food consumption as one of the stages of food biological metabolisms—biological metabolisms correspond to supply chains in which materials can return to ecosystems without damaging them (EMF, 2012). Moreover, we consider food packaging as a technical accessory of food products, which is generally not biodegradable, and needs to be included to undertake a comprehensive CE analysis of food consumption.

While in the linear economy, consumers' role in reducing environmental impacts of food supply chains is limited to responsible shopping (Borrello et al., 2016, 2017), in this study, consumers are assumed to actively participate in the creation of circular food systems, contributing to building virtuous material flows; we define consumer behavior in the CE as "circular food consumption" (CFC). Sustainable food consumption is generally analyzed by reducing it to single moments of decision making occurring during food purchase (based on neoclassical or behavioral economics), in which individuals may choose to buy environmentally sustainable products (Mylan, 2015; Mylan et al., 2016). However, while purchasing preferences are fundamental to foster sustainability of food systems, this aspect restricts the role of individuals only to their shopping choices at the end of linearly designed supply chains. In this study, instead, we extend in time and space the analysis of individual behavior by considering food consumption as a practice occurring over a time interval that goes beyond the act of shopping (Warde, 2005). We conceptualize the bundle of activities that characterize CFC as "circular food consumption practice" (CTCP). More specifically, besides food shopping, the CFCP comprises activities undertaken in the domestic sphere in order to prevent the generation of food waste, to foster recycling, and to support circular firms and supply chains. The chance to anchor a CFCP to a desired future novel circular industrial system is conditioned by internal (depending on the individual) and external (depending on the system that provides food and food-related products) (Spaargaren and Van Vliet, 2000) determinants. This means that the feasibility of the emergence of a CFCP depends on factors influenced by consumers' behavior and households' routines and organization, as well as on factors related to the external environment, such as the characteristics of circular business models or the availability of circular solutions (Borrello et al., 2020). To not neglect the relationship between internal and external determinants for the emergence of a CFCP, this study is also informed by literature on sociotechnical transitions (Geels, 2002, 2019). In this way we seek to provide an exploratory assessment of current readiness of the CFCP constitutive elements along an ideal transition pathway.

This chapter unfolds as follows. After having provided the fundamental literature background, the paragraphs of the second section will define the criteria we used to build the CFCP framework. First, concepts coming from food waste management (Schanes et al., 2018) and social practice theory (Reckwitz, 2002) will inform the classification of CFC subpractices. Such classification will follow two distinct criteria: the succession of different stages along the CFCP (vertical practice dimension); and the effort required to consumers to realize the CFCP, in terms of both tangible and intangible elements (horizontal practice dimension). Second, concepts coming from the literature on CE and circular business models (CBM) (EMF, 2012; Bocken et al., 2016; Geissdoerfer et al., 2018) will inform the classification of different strategies by which particular CFC subpractices contribute to create virtuous material flows. Third, literature on sociotechnical transitions (Geels, 2002, 2019) will be used to analyze the CFC subpractices in terms of ease of transition. In the third section, the CFCP framework will be presented and critically described. The fourth section concludes the chapter and provides some suggestions for future research.

1.1 Circular food consumption practice, subpractices and dimensions

CE postulates the need to redesign consumption and the role of individuals in industrial systems. CE considers consumers as active actors of supply chains, participating in the network of interactions needed to close material loops (EMF, 2012). In this study we focus in particular on the CE assumption that consumers restructure their household organization and daily routines by enacting alternative ways of meeting their needs (i.e., in this study, the need of food provisioning—Mylan et al., 2016).

The influence of redesigning consumption upon households' organization and routines is much evident in the case of technical metabolisms. In the CE narrative, technical metabolisms are supply chains in which materials and products unsuitable for returning to ecosystems (e.g., metals, most plastics) are "designed to circulate at high quality without entering the biosphere" (EMF, 2012, p. 22). In these metabolisms, consumers are conceptualized as users, goods as services, and ownership as the result of functionalities. For example, in sustainable product service systems, users return obsolete products to their providers to generate secondary production and distribution cycles (e.g., second-hand and remanufactured products) (Tukker, 2015). As for biological metabolisms, identifying circular ways to redesign consumption of food products is not an easy task. Unlike technical metabolisms—where waste involves, for example, broken, unused, or obsolete items—food waste consists of decaying biological matter. To the best of our knowledge, few studies so far have approached food consumption in the domain of CE, seeking to identify challenges (Borrello et al., 2016), to assess consumers' willingness to participate in circular food supply chains (Borrello et al., 2017, 2020), and highlighting how circularity entails restructuring households' routines concerning food provisioning and food-waste recycling (Mylan et al., 2016; Borrello et al., 2020).

Seeking to enrich this literature, the current chapter proposes a conceptualization of circular food consumption. To this aim, it takes stock of insights generated by other inquiry fields (e.g., sustainable food consumption, food waste management) to identify potential CFC behaviors. Then these behaviors are interpreted according to the school of thought that sees consumption as a practice (Reckwitz, 2002; Schatzki, 2002; Warde, 2005; Shove et al., 2012). The resulting circular food consumption practice framework deduces its conceptual foundations from: claims calling CE to reconnect consumers' analysis to household practices (Corsini et al., 2019; Schulz et al., 2019); accounts seeing food consumption as composed of a set of routinized behaviors occurring beyond the act of purchase (Mylan et al., 2016; Borrello et al., 2020); the CE aim to reduce municipal food waste (Paul et al., 2018; Peng and Pivato, 2019); and studies highlighting that food waste is not composed only of leftovers generated after meals, but is also generated by different households' behaviors, routines, and organization elements (e.g., Roodhuyzen et al., 2017; Schanes et al., 2018).

More specifically, the current study takes its cue from Borrello et al. (2020), who conceptualize food consumption as a practice entailing food waste recycling, reorganization of households' routines, and food provisioning strategies. Along this line, it designs a CFCP, namely: a practice able to minimize food waste and food packaging waste associated to households. Then, following the suggestion of Røpke (2009), the study splits the CFCP into several subpractices, taking place over a certain period, and adopts them as units of analysis (Fig. 1). A subpractice is any behavior that can contribute to limit food

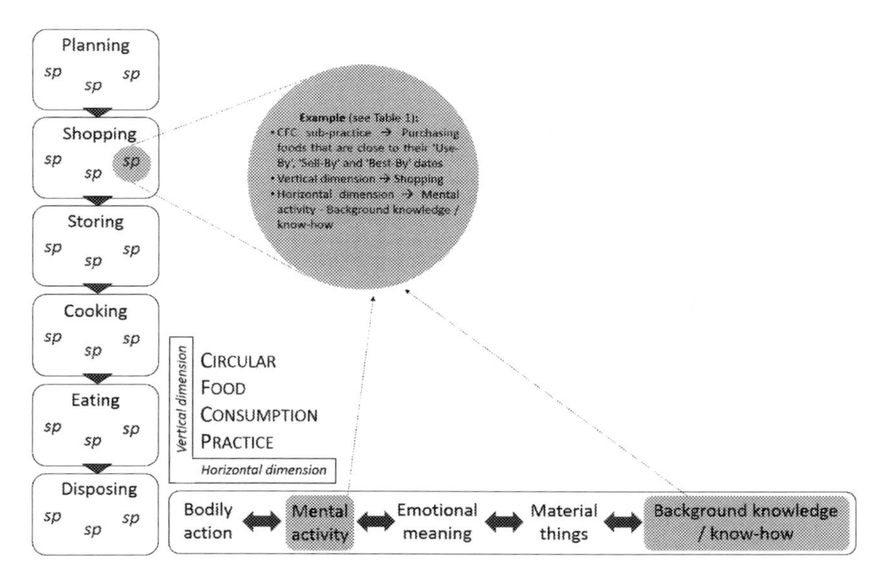

Fig. 1 Vertical and horizontal dimensions of the CFCP (sp = subpractice). *Based on Reckwitz, A., 2002. Toward a theory of social practices: a development in culturalist theorizing. Eur. J. Soc. Theory 5(2), 243–263; Roodhuyzen, D.M.A., Luning, P.A., Fogliano, V., Steenbekkers, L.P.A., 2017. Putting together the puzzle of consumer food waste: towards an integral perspective. Trends Food Sci. Technol. 68, 37–50; Schanes, K., Dobernig, K., Gözet, B., 2018. Food waste matters—a systematic review of household food waste practices and their policy implications. J. Clean. Prod. 182, 978–991.*

and food packaging waste over routinized household activities (e.g., preparing recipes with leftovers, separate collection of waste). Each subpractice is observed in this chapter as standalone, namely its categorization within the CFC framework is independent of other subpractices. Each subpractice is related to two dimensions of the CFCP, namely the vertical and horizontal dimensions. One element of the vertical dimension and one or more elements of the horizontal dimension correspond to each subpractice (Fig. 1).

More specifically, the subpractices are grouped following different phases, namely planning, shopping, storing, cooking, eating, and disposing (Roodhuyzen et al., 2017; Schanes et al., 2018) (CFCP vertical dimension in Fig. 1). These phases are the elements of the household food chain (Roodhuyzen et al., 2017) and extend food consumption in space and time before and after the act of purchasing food products. In each of these phases, individuals' actions play a role in the generation of food and food packaging waste. Therefore, for each phase, we identified a number of beneficial behaviors (CFC subpractices) that can contribute to fostering household circularity.

Besides of its vertical dimension, the CFCP framework is built according to the definition of "practice" proposed by Reckwitz (2002), as summarized by Paddock (2017, p. 124): a practice is "a routinized behavior made up of several interconnecting elements: bodily actions, mental activity, emotional meaning, materials 'things' and background knowledge or 'know-how'." According to this definition, practices are composed of a set of interconnected elements. Body is involved by performing routinized movements (e.g., handling objects, going to a shop); the mind component of practices concerns mental activities (e.g., classifying objects, remembering to do actions), emotional meaning (e.g., value-driven interpretations and aims) and knowledge (e.g., understanding implications of actions, knowing how to perform actions); eventually, practices cannot be performed without material things external to the individual (e.g., objects, tools, technologies). In the proposed framework, the elements included in this definition constitute the CFCP horizontal dimension. We argue that different CFC subpractices are associated, to different extents, to each element of these definitions (CFCP horizontal dimension in Fig. 1). Consistently, each CFC subpractice is associated with one, or more than one, of these elements.

1.2 CE strategies to operate on material loops

Besides the idea to close loops of resources, literature has been also driven by a more general objective, namely the creation of virtuous material flows. This objective concerns generating input savings and maximizing the benefits of energy and natural resource usage (EMF, 2012). Following this perspective, industrial systems design should increase the efficiency and effectiveness of material flows, thus reducing in different ways the need for new resources and the production of waste. For example, the design of better circular processes may operate at different levels by considering desirable properties. These properties were defined as *powers* by the Ellen Mac Arthur Foundation (EMF, 2012, p. 7). The "power of the inner circle" refers to the fact that the shorter the circle, the higher the value retained inside a product (e.g., in terms of energy and person and machine hours). The "power of circling longer" stresses the goal of maximizing the number of consecutive recycling and reuse processes. The "power of cascaded use" refers to diversifying reuse across the value chain by producing different goods and materials through consecutive steps, e.g., involving different enterprises or a bio-refinery to manage material flows. The "power of pure circles" highlights the relevance of producing uncontaminated and quality materials able to circulate longer and without damaging people and the environment.

Along these lines, literature on product and business model design has focused on identifying and classifying different CE strategies to create virtuous material flows. Den Hollander et al. (2017) adopt a classification based on resisting (design for long use), postponing (design for extended use), or reversing (design for recovery) product obsolescence. Bocken et al. (2016) recognize the existence of two fundamental strategies toward the cycling of resources: slowing and closing resource loops. Slowing resource loops concerns the design of products whose life span is extended (e.g., products easy to repair) or intensified (e.g., more durable products). Closing resource loops concerns the recycling of products and materials at the end of their life.

In the proposed CFCP framework, CFC subpractices are classified following and adapting the set of strategies for circular business models suggested by Geissdoerfer et al. (2018): closing loops, slowing loops, intensifying loops, narrowing loops, and dematerializing loops (Fig. 2). In comparison to Bocken et al. (2016), these authors consider intensifying loops as a separate category to "emphasize the importance of a more intense use phase" (p. 713); they include in their framework narrowing loops (use fewer resources per product), disregarded by Bocken et al. (2016) because it is a strategy for resource efficiency applicable also in linear design; they add dematerializing loops, considering the relevance of service solutions and virtual software to reduce resource inputs (EMF, 2016). In our conceptualization we adapt this framework to food consumption as follows: a CFC subpractice closes loops if it contributes to material recycling; it slows loops if it delays the moment at which materials are wasted; it intensifies loops if it contributes to maximize the usage of materials; it narrows loops if it implies the domestic

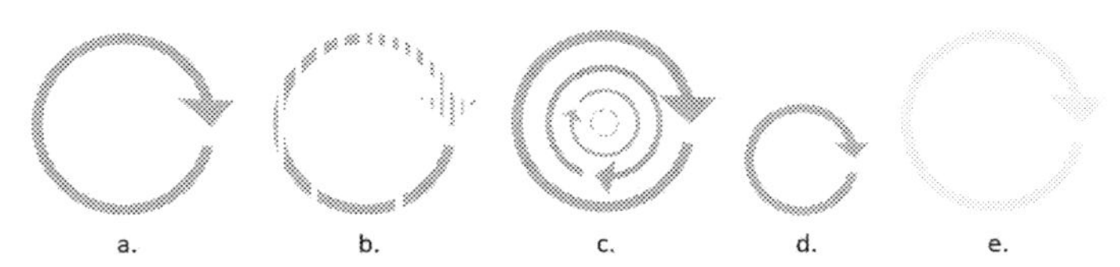

Fig. 2 CE strategies: (A) closing loops, (B) slowing loops, (C) intensifying loops, (D) narrowing loops, and (E) dematerializing loops.

recycling of materials otherwise recycled outside the household; it dematerializes loops if it reduces the need for new material inputs.

1.3 Ease of transition

The emergence of a CFCP is a process of change through which individuals integrate a set of new behaviors into their daily routines. The feasibility of this process is conditioned by the possibility to realize and hold together the elements of the CFCP horizontal dimension conceptualized in this chapter (bodily actions, mental activity, emotional meaning, materials things, and background knowledge or know-how) in the management of the domestic organization of the household. A new practice is more likely to emerge if it is not in conflict with other, already established practices (Spaargaren and Van Vliet, 2000). Therefore, each subpractice that compose the CFCP along its vertical dimension (planning, shopping, storing, cooking, eating, and disposing) involves the elements of the horizontal dimension, thus calling individuals to make an effort to perform certain behaviors inside and outside their household.

Processes of change towards more sustainable configurations are often approached at macro level through the narrative of sociotechnical transitions (Smith et al., 2005; Smith and Stirling, 2010; Sorrell, 2015). Sociotechnical transitions are the processes through which technological innovations replace established modalities of fulfilling human needs. The social component of sociotechnical transitions relates to different levels at which individuals and technology intertwine, such as practices, culture, politics, and institutions. Niche innovations can lead trajectories of change that challenge established sociotechnical configurations. These trajectories are often described through the multilevel perspective (MLP) (Geels, 2002, 2019) (Fig. 3). The MLP is a conceptual tool useful

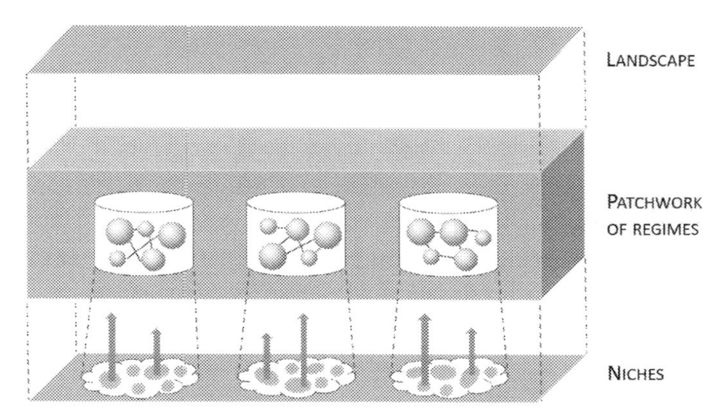

Fig. 3 Multilevel perspective of sociotechnical transitions. *Adapted from Geels, F.W., 2002. Technological transitions as evolutionary reconfiguration processes: a multi-level perspective and a case-study. Res. Policy 31(8–9), 1257–1274.*

to describe the emergence of sustainable patchworks of sociotechnical regimes (i.e., sets of rules gravitating around technologies) from the evolving interaction among niche (radical novelties promoted by innovators), system (technical, political, social, and cultural, business models, or infrastructural configurations), and landscape (e.g., demographics, cultural repertoires, societal concerns, geopolitics, macroeconomic trends, ecological dynamics, wars, financial crises, and oil prices shocks) levels. According to Geels (2019, p. 190), "(a) niche-innovations gradually build internal momentum, (b) niche innovations and landscape changes create pressure on the system and regime, and (c) destabilization of the regime creates windows of opportunity for niche-innovations, which then diffuse and disrupt the existing system."

Notwithstanding the wide spread of transition research, most studies in this field focus on dynamics at the macro level. For example, scholars have used a transition approach to analyze social phenomena (Geels, 2007), cities (Hodson and Marvin, 2010), and industrial sectors, such as energy (Sovacool, 2016), urban water (Bos and Brown, 2012), transportation (Geels, 2012), and food (Jurgilevich et al., 2016). Macro-level analysis is indeed required to address industrial system transitions to configurations based on CE. However, transition research seems to overlook the micro level, the analysis of individuals' behavior in which is crucial to foster transition in systems. As argued by Shove and Walker (2010, p. 476), "discussions of sociotechnical transitions and their governance routinely obscure the central role that practitioners themselves play in generating, sustaining and overthrowing everyday practices." In the words of Watson (2012, p. 2):

> *...practices (and therefore what people do) are partly constituted by the socio-technical systems of which they are a part; and those socio-technical systems are constituted and sustained by the continued performance of the practices which comprise them. Consequently, changes in socio-technical systems only happen if the practices which embed those systems in the routines and rhythms of life change; and if those practices change, then so will the socio-technical system.*

Based on these considerations, the analysis of the emergence of the CFCP proposed in this study is informed by a transition approach. O'Neill et al. (2019) emphasize that the study of transitions and practices complement one another in understanding shifts to sustainable food systems. While the former focuses on the spread of innovations at the regime level, the latter are centered on the integration of novelties in everyday life. To consider this perspective in our framework, we assign a score to each CFC subpractice; this score indicates a certain level of ease of transition depending on current observable integration in the daily routines of households. For each subpractice, the score is conditioned to different extents by the elements of the CFCP horizontal dimension (bodily actions, mental activity, emotional meaning, materials things, and background knowledge or know-how). To illustrate, ease of transition depends on internal factors, such as "competencies and meanings that circulate within the practice" (Watson, 2012, p. 4) and by external factors, such as "the stuff required to accomplish a practice" (Watson, 2012, p. 4). The latter is particularly true for factors related to current spread of CE, such as the existence of technologies, products, and companies required to perform a specific CFC subpractice. Eventually, changes in practices are conditioned by "the way in which one practice bundles together with others" (Watson, 2012, p. 4). Ease of transition is thus assessed, also taking into account how much subpractices conflict with households' extant daily routines and time-space organization.

2. The CFCP framework

In this section we provide a comprehensive description of the CFCP framework. Tables 1 and 2 show, respectively, how the CFCP unfolds in the elements of its vertical dimension (planning, shopping, storing, cooking, eating, and disposing), for the biological (food) and technical (packaging) metabolisms. Each vertical dimension element, in turn, involves a number of subpractices. For example: planning involves planning food shopping and organizing spaces to keep reusable containers and shoppers; shopping involves purchasing imperfect fruit and vegetables and food sold in sustainable types of packaging; storing involves checking food expiry dates and replacing single-use materials with reusable plastic containers to conserve leftovers; cooking involves preparing recipes with leftovers and using only reusable baking trays; eating involves paying attention to plate sizes to limit leftovers and avoiding the use of single-use plastic dishes and glasses; disposing involves domestic composting and separate collection. Each subpractice is able to contribute to virtuous material flows through different CE strategies (closing, slowing, intensifying, narrowing, and dematerializing loops). The subpractices also entail the involvement of one or more element of the CFCP horizontal dimension (bodily actions, mental activity, emotional meaning, materials things, and background knowledge or know-how) (Table 3); consistently with the definition of practice by Reckwitz (2002) (routinized behavior made up of several interconnecting elements), we classify subpractices also as "single elements" or "interconnection," depending on whether the subpractice is composed either of one or more than one element of the horizontal dimension.

Eventually, an ease of transition score is assigned to each subpractice, based on a three-level classification: low, intermediate, high (Fig. 4). The proposed framework of CFCP is meant to provide a conceptual device to analyze the emergence of a bundle of individual behaviors able to promote the creation of biological and technical circular metabolisms at the consumption stage of food supply chains. This framework is useful because it suggests modalities (CE strategy), efforts required for individuals (CFCP horizontal dimension), and implementation potential (ease of transition) of the subpractices, to consider when approaching food consumption in the CE. However, it is worth pointing out that the way in which Tables 1 and 2 were filled in is not expected to provide a comprehensive and reliable picture of CFC—an objective that lies beyond the purposes of this chapter. It represents instead a first attempt to test the CFCP framework through the contextual

Table 1 CFCP framework: food.

Vertical dimension	Subpractice	CE strategy	Horizontal dimension	Single/ Interconnected	Ease of transition	
Planning	Planning food shopping	Dematerializing	Mental activity	Single element	Intermediate	
	Planning meals with foods that are close to their "Use-By", "Sell-By", and "Best-By" dates	Slowing	Mental activity	Single element	Low	
	Controlling food location at home—Smart fridges	Slowing	Bodily action—Mental activity—Material things	Interconnection	Intermediate	
Shopping	Purchasing food coming from circular providers	Closing	Bodily action—Mental activity—Emotional meaning—Background knowledge/know-how	Interconnection	Low	
	Purchasing imperfect fruits and vegetables	Intensifying	Background knowledge/know-how	Single element	Low	
	Purchasing foods that are close to their "Use-By", "Sell-By", and "Best-By" dates	Slowing	Mental activity—Background knowledge/know-how	Single element	Low	12M
Storing	Knowing the difference between "Use-By", "Sell-By", and "Best-By" dates	Intensifying	Background knowledge/know-how	Single element	Intermediate	

	Periodically checking the "Use-By", "Sell-By", and "Best-By" dates	Slowing	Bodily action—Mental activity—Background knowledge/know-how	Interconnection	Intermediate	
	Respecting foods storage guidelines	Slowing	Bodily action—Mental activity—Background knowledge/know-how	Interconnection	Intermediate	
Cooking	Do not over-prepare food (portion control)	Dematerializing	Bodily action—Mental activity	Interconnection	Intermediate	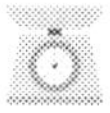
	Preparing recipes with leftovers	Intensifying	Bodily action—Mental activity—Background knowledge/know-how	Interconnection	Intermediate	
	Being flexible on the repertoire of recipes and menu	Slowing—Intensifying	Mental activity	Single element	Intermediate	
Eating	Do not perceive eating leftovers as a sacrifice, thrift	Intensifying	Emotional meaning—Background knowledge/know-how	Single element	Intermediate	
	Proper plate sizes	Dematerializing	Mental activity	Single element	High	
	Managing leftovers	Intensifying	Bodily action—Mental activity	Interconnection	Intermediate	
Disposing	Separate collection	Closing	Bodily action—Mental activity—Material things—Background knowledge/know-how	Interconnection	Intermediate	

Continued

Table 1 CFCP framework: food—cont'd

Vertical dimension	Subpractice	CE strategy	Horizontal dimension	Single/ Interconnected	Ease of transition	
	Solidarity initiatives	Intensifying	Bodily action—Emotional meaning—Background knowledge/know-how	Interconnection	Low	
	Food sharing	Intensifying	Bodily action—Emotional meaning—Material things—Background knowledge/know-how	Interconnection	Low	
	Domestic composting	Narrowing	Bodily action—Mental activity—Emotional meaning—Material things—Background knowledge/know-how	Interconnection	Low	

Note: Icons made by Smashicons, Freepik, iconixar, Vectors Market, Ultimatearm, Becris, photo3idea_studio, Flat Icons, Dimitry Miroliubov, mynamepong from https://www.flaticon.com/.

Table 2 CFCP framework: food packaging.

Vertical dimension	Subpractice	CE strategy	Horizontal dimension	Single/Interconnected	Ease of transition	
Planning	Organize spaces to keep reusable containers/shoppers	Dematerializing	Bodily action—Mental activity—Material things	Single element	Intermediate	
Shopping	Purchasing food filling reusable containers—Scoop and gravity bins	Dematerializing	Material things	Single element	Low	
	Using recyclable or reusable shoppers	Dematerializing	Material things	Single element	Intermediate	
	Purchasing food sold in biodegradable/edible/smart/modified atmosphere packaging	Closing—Slowing	Material things—Background knowledge/know-how	Single element	Low	
Storing	Replacing the use of tinfoil and cling film with reusable containers and biodegradable plastic wraps	Dematerializing	Material things	Single element	Low	
Cooking	Using only reusable baking trays	Dematerializing	Material things	Single element	Intermediate	
Eating	Avoiding the use of single-use plastic dishes and glasses (washing dishes and glasses)	Dematerializing	Bodily action—Material things	Interconnection	Intermediate	

Continued

Table 2 CFCP framework: food packaging—cont'd

Vertical dimension	Subpractice	CE strategy	Horizontal dimension	Single/ Interconnected	Ease of transition	
Disposing	Separate collection	Closing	Bodily action—Mental activity—Material things—Background knowledge/know-how	Single part	Intermediate	
	Upcycling—Trash art	Narrowing	Bodily action—Mental activity—Emotional meaning—Material things—Background knowledge/know-how	Interconnection	Low	
	Deposit return schemes—Reverse vending machines	Closing	Bodily action—Mental activity—Material things—Background knowledge/know-how	Interconnection	Low	

Note: Icons made by Smashicons, Freepik, iconixar, Vectors Market, Ultimatearm, Becris, photo3idea_studio, Flat Icons, Dimitry Miroliubov, mynamepong from https://www.flaticon.com/.

Table 3 Vertical and horizontal dimensions of CFC sub-practices.

Vertical practice dimension	Horizontal practice dimension: Bodily action	Mental activity	Emotional meaning	Material thing	Back. Know. / Know-how
Plan.					
Shop.					
Stor.					
Cook.					
Eat.					
Disp.					

Note: Icons made by Smashicons, Freepik, iconixar, Vectors Market, Ultimatearm, Becris, photo3idea_studio, Flat Icons, Dimitry Miroliubov, mynamepong from https://www.flaticon.com/.

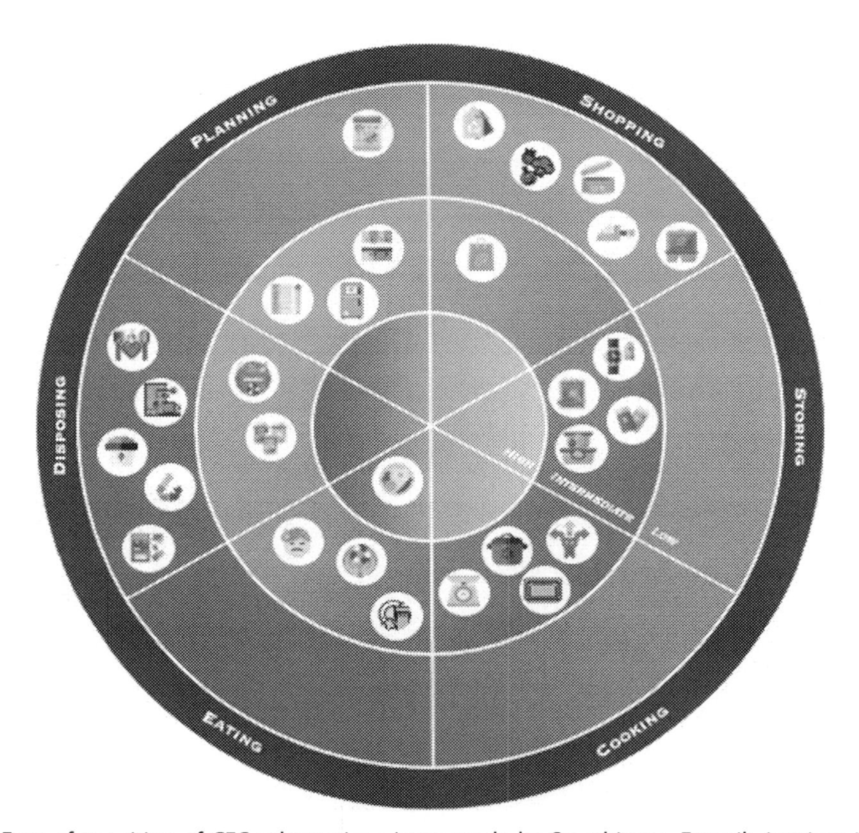

Fig. 4 Ease of transition of CFC subpractices. Icons made by Smashicons, Freepik, iconixar, Vectors Market, Ultimatearm, Becris, photo3idea_studio, Flat Icons, Dimitry Miroliubov, mynamepong from https://www.flaticon.com/.

opportunity of this chapter, thus resulting in an exploratory classification of the CFC subpractices. Therefore, the subpractices included are not to be considered an exhaustive list of consumers' behaviors fostering the creation of virtuous material loops in food systems. The procedure used to classify each subpractice, as for the CE strategy adopted, for the elements of the horizontal dimensions involved, as well as for the ease of transition score, is not rigorous and rests on a high level of subjectivity. To illustrate, to any CFC subpractice, one could have assigned a certain degree of involvement of each element of the horizontal dimension; whereas, we attributed to subpractices one or more of these elements based on their seeming prevalence to condition the enactment of the sub-practice.

As shown in Fig. 5, after having selected a specific CFC subpractice, the CFCP framework allows us to assign it to one of the elements of the vertical dimension and to identify the CE strategies and the elements of the horizontal dimension involved. Based on the analysis of the horizontal dimension and on the potential influence of other household

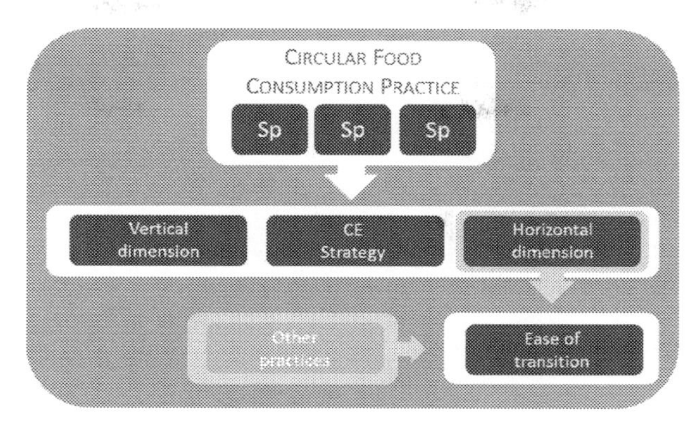

Fig. 5 CFCP framework; determining ease of transition.

practices upon the implementation of the CFC subpractice at hand, an ease of transition score can be assigned (Fig. 4). For the sake of illustration, we report three examples of adoption of the CFCP framework, one for each ease of transition level.

Domestic composting is a do-it-yourself (DIY) activity that individuals can perform to produce a fertilizer domestically by using food leftovers. Domestic composting is a CE strategy able to narrow material loops because food waste is recycled at home, thus reducing the amount of biological materials that end up in landfills or are managed by recycling facilities. However, domestic composting calls practitioners to put into action all the CFCP horizontal dimension elements: it requires physical energy (bodily action), it requires planning and attention (mental activity), it is more likely to attract only individuals finding a sense of achievement in DIY activities (emotional meaning), it needs containers and tools (material things), and its proper realization is conditioned by knowing how foster the mineralization processes of biological matter (background knowledge/ know-how). Furthermore, domestic composting is a space- and time-consuming activity that is likely to conflict with the internal organization of the household. For these reasons and based on our knowledge of its spread among individuals, it is reasonable to assign this activity a low ease of transition score.

Controlling food location at home is a beneficial routine to embed in the planning activities performed in the kitchen that can be also facilitated by new technologies (smart fridges) helping to check stored food. This behavior contributes to slowing material loops, because it helps to avoid spoiled food thanks to good organization of food storage spaces. Transforming this behavior into a routine requires planning and physical performance (bodily action and mental activity), as well as the potential support of smart refrigerators (material things). Based on the seemingly acceptable commitment required to perform it and on the scarce diffusion of smart fridges, we have assigned to this subpractice an intermediate ease of transition score.

Paying attention to appropriate plate sizes foresees the production of leftovers. It contributes to dematerializing material loops, since less resource is consumed at the eating stage. It is a routine that requires a single moments of attention (mental activity) before meals and does not conflict with any other practice. Therefore, a high ease of transition score seems to be appropriate.

3. Discussion

In the current chapter, we have proposed a framework to analyze circular food consumption. Based on recent studies calling on CE to reconnect food consumption to household practices (Mylan et al., 2016; Borrello et al., 2020), the CFCP framework considers circular food consumption as a stage of CE food biological metabolisms, taking place through several activities carried out in different places and times. More specifically, the CFCP is composed of a set of subpractices unfolding, inside and outside the household, over the time that goes from planning purchases to disposing of food waste. The subpractices contribute to CE by creating virtuous material flows, namely by closing, slowing, intensifying, narrowing, and dematerializing resource loops. The subpractices contribute to circularity either through behaviors performed inside the households or through recycling processes connecting consumers to other supply chain actors (Fig. 6). The proposed framework allows us to assess the likelihood (ease of transition) that consumers will undertake circular activities related to food. This potential implementation is the result of the effort required

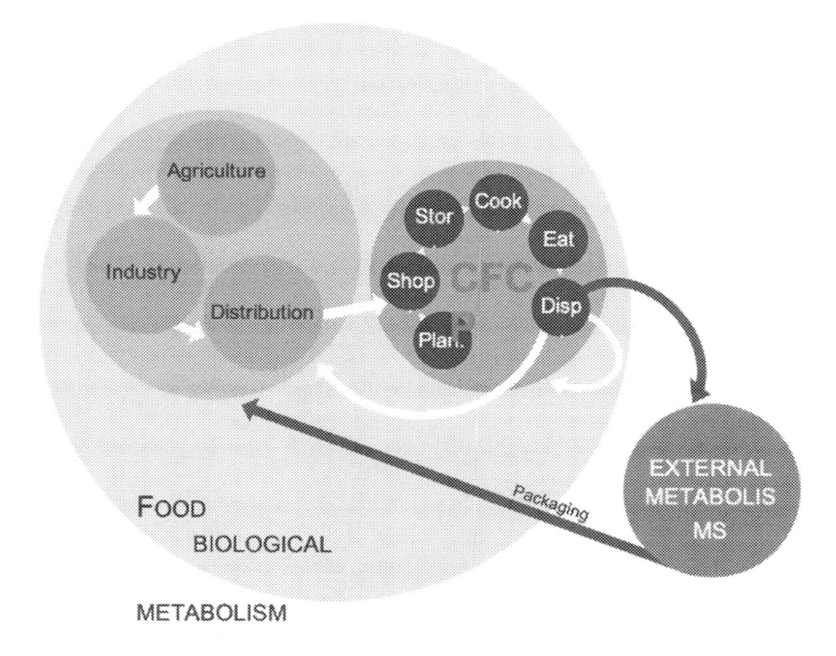

Fig. 6 CFCP within CE metabolisms.

for individuals to carry out each subpractice and of how each subpractice competes with other household routinized behaviors (Fig. 5).

The CFCP framework constitutes a conceptual tool to address food consumption in CE studies. Informed by an ecological perspective of consumption (Røpke, 2009) and a practice approach to consumer behavior, it provides some keynotes regarding the implications of food consumption for CE. On the one hand, under the CFCP framework, individual behavior is reconnected to its environmental impacts by an overarching conceptualization that includes among sustainable consumption activities also those occurring before and after shopping. On the other hand, a practice perspective on consumer behavior sheds light on the constraints on making consumption practices transition into circularity.

In a different light, the current chapter also highlights that seeing CE as a way to create virtuous resource flows allows us not to disregard the role of activities beyond recycling. CE inherently postulates consumers' commitment in new relationships with companies by means of circularly designed business models. For this reason, among CFC subpractices we included activities fostering materials cycles outside the household, such as separate collection, participation in deposit return schemes, using reverse vending machines, as well as purchasing food coming from circular manufacturer and distribution channels. However, other behaviors contributing to reducing resources required and waste generated by supply chains cannot be overlooked. In particular, in the domain of circular food biological metabolisms, rethinking consumption calls for a conceptualization of circular activities encompassing not only recycling food (and food packaging) waste, but also other sustainable behaviors. As stressed by Camacho-Otero et al. (2018), consumption in a CE inherently embeds symbolic meanings (e.g., environmentalism, frugality, community, sharing), contrasting with the materialistic mainstream linear economy. Therefore, we also included among the CFC subpractices participation in nonmarket exchanges (food sharing, solidarity initiatives), household virtuous recycling activities (preparing recipes with leftovers, domestic composting, upcycling, and trash art), as well as several routines falling into the domain of individual frugal and responsible behaviors (e.g., checking expiry dates, using recyclable or reusable shoppers).

Circular biological metabolisms are crucial to reduce the environmental impacts of industrial systems. International institutions currently allocate dedicated resources to promote their spread, such as the future European Circular Bioeconomy Fund.[a] In the current chapter, food consumption was considered as one of the stages of food biological metabolisms, in which individuals can make a significant contribution to creating sustainable food systems. However, so far literature has devoted little attention to consumption in these metabolisms, for instance, by focusing on product design and the end of life of

[a] https://circulareconomy.europa.eu/platform/en/news-and-events/all-news/european-fund-support-circular-bioeconomy

products (Stegmann et al., 2020) or by addressing consumers' engagement in food circular business models (Borrello et al., 2020). We recommend that future research investigates these aspects thoroughly, taking into account the CFCP framework as reference. Transitioning into CE in food systems depends also on consumer behavior through the emergence of a CFCP. This will be conditioned by the spread of circular firms and supply chains, which we have indeed considered in the CFCP framework. However, we highlight that the ease of transition into CFC depends often on how much effort is required to undertake circular household activities, as well as on how these activities conflict with domestic organization and with established daily routines. We then recommend the adoption of the CFCP framework to develop indicators of consumers' readiness to contribute to circularity. Indicators should take into account not only the availability of circular solutions such as circular products or materials, but also aspects related to individuals. Among these, we suggest consideration of a set of variables conditioning sustainable household behavior: environmental awareness, attitudes, and education (Sidique et al., 2010); and sociodemographic characteristics, such as income, type of employment, being an individual household, or having children (Mylan et al., 2016).

4. Conclusions

Food consumption is an under-researched topic in the circular economy literature. The current study has provided a framework to analyze circular food consumption. The CFCP framework represents a methodological tool to classify sustainable household activities complying with CE principles and provide insights on the likelihood that individuals enact these activities. Furthermore, the framework highlights that circular food practices ask consumers for different levels of engagement, depending on how complex performing a practice is, particularly in combination with other practices and established household routines. The main limitation of the current study consists in the method used to assign CE strategies, horizontal dimensions, and ease of transition scores, being highly subjective. However, this limitation is inherent in the realization of a qualitative analytical tool such as the CFCP framework, which relies on users' interpretation when it is practically implemented. The framework is expected to be validated by future research on sustainable consumer behavior, thus helping to foster the adoption of circular practices. Particularly, the framework should inform future studies for the identification of elements influencing the enactment of a practice (i.e., those belonging to its horizontal dimension), thus suggesting the domains in which interventions are required (e.g., providing products, instruments, business models, and knowledge).

References

Baldassarre, B., Schepers, M., Bocken, N., Cuppen, E., Korevaar, G., Calabretta, G., 2019. Industrial symbiosis: towards a design process for eco-industrial clusters by integrating circular economy and industrial ecology perspectives. J. Clean. Prod. 216, 446–460.

Blank, L.M., Narancic, T., Mampel, J., Tiso, T., O'Connor, K., 2020. Biotechnological upcycling of plastic waste and other non-conventional feedstocks in a circular economy. Curr. Opin. Biotechnol. 62, 212–219.

Bocken, N.M., De Pauw, I., Bakker, C., van der Grinten, B., 2016. Product design and business model strategies for a circular economy. J. Ind. Prod. Eng. 33 (5), 308–320.

Borrello, M., Lombardi, A., Pascucci, S., Cembalo, L., 2016. The seven challenges for transitioning into a bio-based circular economy in the agri-food sector. Recent Pat. Food Nutr. Agric. 8 (1), 39–47.

Borrello, M., Caracciolo, F., Lombardi, A., Pascucci, S., Cembalo, L., 2017. Consumers' perspective on circular economy strategy for reducing food waste. Sustainability 9 (1), 141.

Borrello, M., Pascucci, S., Caracciolo, F., Lombardi, A., Cembalo, L., 2020. Consumers are willing to participate in circular business models: a practice theory perspective to food provisioning. J. Clean. Prod. (in press).

Bos, J.J., Brown, R.R., 2012. Governance experimentation and factors of success in socio-technical transitions in the urban water sector. Technol. Forecast. Soc. Change 79 (7), 1340–1353.

Camacho-Otero, J., Boks, C., Pettersen, I., 2018. Consumption in the circular economy: a literature review. Sustainability 10 (8), 2758.

Corsini, F., Laurenti, R., Meinherz, F., Appio, F.P., Mora, L., 2019. The advent of practice theories in research on sustainable consumption: past, current and future directions of the field. Sustainability 11 (2), 341.

Den Hollander, M.C., Bakker, C.A., Hultink, E.J., 2017. Product design in a circular economy: development of a typology of key concepts and terms. J. Indus. Ecol. 21 (3), 517–525.

EMF (Ellen MacArthur Foundation), 2012. Towards the Circular Economy. Vol. 1: An Economic and Business Rationale for an Accelerated Transition. Isle of Wight.

EMF (Ellen MacArthur Foundation), 2016. Growth Within: A Circular Economy Vision for a Competitive Europe. Isle of Wight.

Geels, F.W., 2002. Technological transitions as evolutionary reconfiguration processes: a multi-level perspective and a case-study. Res. Policy 31 (8–9), 1257–1274.

Geels, F.W., 2007. Analysing the breakthrough of rock 'n'roll (1930–1970) Multi-regime interaction and reconfiguration in the multi-level perspective. Technol. Forecast. Soc. Change 74 (8), 1411–1431.

Geels, F.W., 2012. A socio-technical analysis of low-carbon transitions: introducing the multi-level perspective into transport studies. J. Trans. Geogr. 24, 471–482.

Geels, F.W., 2019. Socio-technical transitions to sustainability: a review of criticisms and elaborations of the Multi-Level Perspective. Curr. Opin. Environ. Sustain. 39, 187–201.

Geissdoerfer, M., Savaget, P., Bocken, N.M., Hultink, E.J., 2017. The circular economy–a new sustainability paradigm? J. Clean. Prod. 143, 757–768.

Geissdoerfer, M., Morioka, S.N., de Carvalho, M.M., Evans, S., 2018. Business models and supply chains for the circular economy. J. Clean. Prod. 190, 712–721.

Hodson, M., Marvin, S., 2010. Can cities shape socio-technical transitions and how would we know if they were? Res. Policy 39 (4), 477–485.

Holanda, R.B., Silva, P.M.P., do Carmo, A.L.V., Cardoso, A.F., da Costa, R.V., de Melo, C.C.A., Montini, M., 2020. Bayer process towards the circular economy—soil conditioners from bauxite residue. In: Light Metals 2020. Springer, Cham, pp. 107–114.

Jurgilevich, A., Birge, T., Kentala-Lehtonen, J., Korhonen-Kurki, K., Pietikäinen, J., Saikku, L., Schösler, H., 2016. Transition towards circular economy in the food system. Sustainability 8 (1), 69.

Kiani, S., Kujala, K., Pulkkinen, J., Aalto, S.L., Suurnäkki, S., Kiuru, T., Ronkanen, A.K., 2020. Enhanced nitrogen removal of low carbon wastewater in denitrification bioreactors by utilizing industrial waste toward circular economy. J. Clean. Prod., 119973.

Kirchherr, J., Reike, D., Hekkert, M., 2017. Conceptualizing the circular economy: an analysis of 114 definitions. Resour. Conserv. Recycl. 127, 221–232.
Korhonen, J., Nuur, C., Feldmann, A., Birkie, S.E., 2018. Circular economy as an essentially contested concept. J. Clean. Prod. 175, 544–552.
Mishra, S., Singh, S.P., Johansen, J., Cheng, Y., Farooq, S., 2019. Evaluating indicators for international manufacturing network under circular economy. Manag. Decis.
Moreau, V., Sahakian, M., Van Griethuysen, P., Vuille, F., 2017. Coming full circle: why social and institutional dimensions matter for the circular economy. J. Indus. Ecol. 21 (3), 497–506.
Mylan, J., 2015. Understanding the diffusion of sustainable product-service systems: insights from the sociology of consumption and practice theory. J. Clean. Prod. 97, 13–20.
Mylan, J., Holmes, H., Paddock, J., 2016. Re-introducing consumption to the 'circular economy': a socio-technical analysis of domestic food provisioning. Sustainability 8 (8), 794.
Nogalski, B., Szpitter, A.A., Jablonski, A., Jablonski, M., 2019. Networked business models in the circular economy. In: Networked business models in the circular economy, pp. 1–297.
O'Neill, K.J., Clear, A.K., Friday, A., Hazas, M., 2019. 'Fractures' in food practices: exploring transitions towards sustainable food. Agricult. Human Values 36 (2), 225–239.
Paddock, J., 2017. Household consumption and environmental change: rethinking the policy problem through narratives of food practice. J. Cons. Cult. 17 (1), 122–139.
Paul, S., Dutta, A., Defersha, F., Dubey, B., 2018. Municipal food waste to biomethane and biofertilizer: a circular economy concept. Waste Biomass Valorization 9 (4), 601–611.
Peng, W., Pivato, A., 2019. Sustainable management of digestate from the organic fraction of municipal solid waste and food waste under the concepts of back to earth alternatives and circular economy. Waste Biomass Valorization 10 (2), 465–481.
Reckwitz, A., 2002. Toward a theory of social practices: a development in culturalist theorizing. Eur. J. Soc. Theory 5 (2), 243–263.
Roodhuyzen, D.M.A., Luning, P.A., Fogliano, V., Steenbekkers, L.P.A., 2017. Putting together the puzzle of consumer food waste: towards an integral perspective. Trends Food Sci. Technol. 68, 37–50.
Røpke, I., 2009. Theories of practice—new inspiration for ecological economic studies on consumption. Ecol. Econ. 68 (10), 2490–2497.
Schanes, K., Dobernig, K., Gözet, B., 2018. Food waste matters—a systematic review of household food waste practices and their policy implications. J. Clean. Prod. 182, 978–991.
Schatzki, T.R., 2002. The Site of the Social: A Philosophical Account of the Constitution of Social Life and Change. Penn State Press.
Schulz, C., Hjaltadóttir, R.E., Hild, P., 2019. Practising circles: studying institutional change and circular economy practices. J. Clean. Prod. 237, 117749.
Shove, E., Walker, G., 2010. Governing transitions in the sustainability of everyday life. Res. Policy 39, 471–476. https://doi.org/10.1016/j.respol.2010.01.019.
Shove, E., Pantzar, M., Watson, M., 2012. The Dynamics of Social Practice: Everyday Life and How It Changes. Sage.
Sidique, S.F., Lupi, F., Joshi, S.V., 2010. The effects of behavior and attitudes on drop-off recycling activities. Resour. Conserv. Recycl. 54 (3), 163–170.
Smith, A., Stirling, A., 2010. The politics of social-ecological resilience and sustainable socio-technical transitions. Ecol. Soc. 15 (1).
Smith, A., Stirling, A., Berkhout, F., 2005. The governance of sustainable socio-technical transitions. Res. Policy 34 (10), 1491–1510.
Sorrell, S., 2015. Reducing energy demand: a review of issues, challenges and approaches. Renew. Sustain. Energy Rev. 47, 74–82.
Sovacool, B.K., 2016. How long will it take? Conceptualizing the temporal dynamics of energy transitions. Energy Res. Soc. Sci. 13, 202–215.
Spaargaren, G., Van Vliet, B., 2000. Lifestyles, consumption and the environment: the ecological modernization of domestic consumption. Environ. Polit. 9 (1), 50–76.
Stegmann, P., Londo, M., Junginger, M., 2020. The circular bioeconomy: its elements and role in European bioeconomy clusters. Resour. Conserv. Recycl. X, 100029.

Tate, W.L., Bals, L., Bals, C., Foerstl, K., 2019. Seeing the forest and not the trees: learning from nature's circular economy. Resour. Conserv. Recycl. 149, 115–129.
Tukker, A., 2015. Product services for a resource-efficient and circular economy—a review. J. Clean. Prod. 97, 76–91.
Urbinati, A., Chiaroni, D., Chiesa, V., 2017. Towards a new taxonomy of circular economy business models. J. Clean. Prod. 168, 487–498.
Warde, A., 2005. Consumption and theories of practice. J. Cons. Cult. 5 (2), 131–153.
Watson, M., 2012. How theories of practice can inform transition to a decarbonised transport system. J. Trans. Geogr. 24, 488–496.

CHAPTER 22

From linear economy legacies to circular economy resources: Maximising the multifaceted values of legacy mineral wastes

William M. Mayes[a], Susan L. Hull[b], and Helena I. Gomes[c]
[a]Department of Geography, Geology and Environment, University of Hull, Hull, United Kingdom
[b]Department of Biological and Marine Sciences, University of Hull, Hull, United Kingdom
[c]Food, Water, Waste Research Group, Faculty of Engineering, University of Nottingham, Nottingham, United Kingdom

1. Introduction

1.1 Mining and the Anthropocene

Our modern day, technological societies are built on cycles of production and consumption, in which mining and associated processing of metals and minerals are the basis. The global mining and mineral processing sector is both the biggest energy consumer and waste producer of any sector of society (Oberle et al., 2019). The environmental impacts of the sector stretch across a range of components of the earth system. The mining sector is responsible for over 35% of global carbon emissions according to recent reports, while associated processing sectors such as steel, cement, and alumina production contribute ~17% of global CO_2 emissions (Allwood et al., 2010). Furthermore, atmospheric particulate emissions from metal and nonmetallic extraction industries account for 19% of global particulate matter (PM) health impacts (Oberle et al., 2019). The impacts of mining on the water environment have been well-documented, both in terms of the generation of metal-rich and sometimes acidic waters/sediments that can have intergenerational polluting legacies (Younger et al., 2002), and of being a critical consumer of water and driver of resource and social conflicts in some jurisdictions (Oberle et al., 2019; Salem et al., 2018). These issues have associated impacts on aquatic biodiversity and human health, while metal-rich wastes can negatively affect soil quality (Arvay et al., 2019; Tepanosyan et al., 2018). Mining itself can serve as a conduit for population movement and impact on the biodiversity of sensitive habitats, as is being increasingly documented in parts of Amazonia where unregulated mining brings with it associated logging activities and subsequent deforestation (Caballero Esppejo et al., 2018; Sonter et al., 2017).

Circular Economy and Sustainability, Volume 1
https://doi.org/10.1016/B978-0-12-819817-9.00009-0

Extractive industries have not only disrupted the global carbon cycle but also dramatically altered flows of many metals around the planet (Nriagu and Pacyna, 1988). Recent assessments of global metal flux suggest that anthropogenic metal emissions overwhelm natural cycles of weathering, transport, and deposition (Viers et al., 2009). Some observers have cautioned that even in this era of open (and big) data, there is considerable uncertainty about the scale of metal emissions associated with primary extraction and processing sectors (Hudson-Edwards, 2016). Where national-scale inventories of legacy mining sources have been undertaken, the contribution of long-abandoned mines to pollution fluxes can be found to exceed that from all highly regulated contemporary industry and domestic sources (Mayes et al., 2010).

Such anthropogenic changes to the global circulation of metals around the planet have led to some observers suggesting that early mining activity could be a marker for the Anthropocene Epoch (or stage); delineating in the geological records the indelible marker of humankind in the sedimentary environment (Foulds et al., 2013). Indeed, studies from the Rio Tinto-Odiel basins, draining the massive sulfide deposit of the Iberian Pyrite Belt in Spain, show the signal of early metal mining some 4500 years before present in downstream sediments (Leblanc et al., 2000).

The broader impact of humankind on the earth system has been increasingly quantified and understood in recent decades with various international agencies highlighting worrying trajectories in atmospheric emissions and associated climate breakdown (IPCC, 2018) alongside systemic biodiversity decline through habitat loss and fragmentation (Diaz et al., 2019). These environmental impacts are intrinsically linked with associated impacts on human wellbeing and access to fundamental human rights and needs (e.g., access to clean water, air, and soil; Raworth, 2017). Despite efforts to embed sustainable resource stewardship in national and international governance for several decades (e.g., Brundtland, 1987), the global trajectories on resource use show continued acceleration in recent years, with global resource consumption tripling since the 1970s, at a time when human population has roughly doubled (Oberle et al., 2019). Waste production in the mining sector in particular is set to increase further in the future, given declining ore-grades for many materials (Hudson-Edwards, 2017).

Given this crisis of resource use and environmental impacts driven by traditional linear economic models (take–make–dispose), considerable efforts are now looking at alternative economic models that aim to put human wellbeing and environmental protection on an equal footing with GDP for improving quality of life (e.g., Raworth, 2017; sustainable consumption and production are enshrined in United Kingdom National Ecosystem Assessment, 2005). One major development in conceptualising and reframing current resource use patterns has been the development of discussions around circular economy (CE) that have gained significant traction with researchers and policymakers in the last decade (European Commission, 2015). The CE has many definitions (see Millar et al., 2019) but the most widely used definitions follow that of the Ellen MacArthur Foundation (2013):

> *Looking beyond the current "take, make and dispose" extractive industrial model, the circular economy is restorative and regenerative by design. Relying on system-wide innovation, it aims to redefine products and services to design waste out, while minimising negative impacts.*

Despite the importance of resource extraction on all facets of the earth system, along with the fundamental role it plays in modern-day economies, discourse on CE has largely overlooked these sectors and focused primarily on postconsumer wastes (Sapsford et al., 2019; Velenturf et al., 2019).

1.2 Lessons from history?

Documenting the environmental impacts of mining and mineral processing industries is not a recent development. The environmental impacts of riverine lead pollution from mining were a concern to the UK government in the late 19th century (River Pollution Commission, 1874; Palumbo-Roe et al., 2009), while the earlier accounts of German Philosopher Friedrich Engels in the middle of the 19th century starkly highlight the life-shortening and chronic occupational exposure of mine workers across coal, ironstone, and lead mines of the United Kingdom (Engels, 2005). In the early 21st century, many nations that were among the first to industrialise are now in postindustrial phases. As such, they are in a position to carefully evaluate the long-term legacies of mining and mineral processing that took place under a lax regulatory environment. It is clear that many of these externalities have migrated with the extractive sector to other parts of the planet; a pattern facilitated by international organisations, according to some observers (Marglin, 2008). However, the technologies that have developed in recent years for managing these legacies—pollution prevention measures, sustainable remediation of pollution legacies, and restoration of legacy sites to a range of productive uses—show tentative positivity in improving the environment and potentially contributing to societal wellbeing around extractive and processing industry sites (Younger et al., 2002). Such environmental engineering applications can be seen as a key contributor to the CE by minimising environmental impacts of resource use, potentially driving novel approaches to recovery of value from secondary resources and easing pressure on virgin reserves (Gomes et al., 2019). Quantifying the benefits of such approaches, however, needs to embrace systems that go beyond simple economic analyses and also consider the full environmental and social implications of new resource recovery ventures (Purnell et al., 2019).

The ecosystem service (ES) approach has been widely used in a range of settings to classify and quantify the various benefits humankind reaps from the natural world. Despite the well-justified critiques of the approach (see, for example, Schröter et al., 2014), it can provide a useful model to consider the relative balance of environmental impacts and benefits associated with mining and mineral processing wastes alongside abiotic material provision. There are four broad classes of ES:

- Provisioning services: products directly obtained from the natural world (e.g., food, fiber, timber);
- Regulating services: benefits obtained by regulation of ecosystem processes (e.g., water regulation, climate regulation);
- Cultural services: nonmaterial benefits humankind gains from the natural world (e.g., ecotourism, recreation, cultural diversity, spiritual value);
- Underpinning or supporting services: indirect services necessary for the production of the other three service types (e.g., pollination, soil formation, photosynthesis).

This chapter aims to review recent developments in valorisation of legacy wastes from the mining and mineral processing sector with the CE by considering the varied abiotic material benefits and ESs that can emerge at legacy waste sites. Using case study evidence, some of the potential circular benefits of legacy wastes are considered alongside the potential management conflicts that may be present at legacy sites.

2. Methods

The review is structured according to the different resource types documented to be exploited at mining and mineral processing waste disposal sites. These are defined as abiotic resources (e.g., geological resources such as metals, minerals, and aggregate) and those which fall under the broad classes of ES (provisioning, regulating, cultural, and underpinning) as defined by the United Kingdom National Ecosystem Assessment (2005). Examples are drawn from a range of jurisdictions with a particular focus on those early-to-industrialise nations where environmental problems associated with legacy wastes have been well-documented.

Data for hydropower potential at abandoned metal mine sites in England and Wales is based on that published in Mayes et al. (2010).

3. Review and discussion

3.1 Abiotic material benefits: Minerals, metals, and materials

Keeping materials and by-products in productive use as long as feasible is a critical basis of the CE. Reuse of wastes from the mining and mineral processing sector has a long history and is primarily driven by changes in market value of materials within wastes and the development of recovery technologies. For example, many historical metal mine spoil heaps that date back to the 18th and 19th centuries in northern England associated with early Pb and Zn production were reworked in the 1970s for recovery of fluorspar (CaF_2) and barytes ($BaSO_4$) that were hitherto of little value. Steel and chemical industry expansion drove the demand for fluorspar, while barytes were used in both chemical industries and drilling operations associated with North Sea Oil expansion (Dunham, 1990). Spoil reworking was also commonplace as extraction technologies improved to allow low-grade Pb ores to become economically viable (Palumbo-Roe, 2010).

Bulk reuse of wastes has long been practised in various mining and processing sectors, such as the use of waste mine spoil (sometimes mixed with fly ash to cement materials) to backfill mine voids and reduce the risk of subsidence (Bian et al., 2012), or in broader landscaping to restore original land profiles. Mine waste has also been used in low-grade construction applications (e.g., road bases) and is still deployed in some jurisdictions. Reworking former steel by-product tips to recover both aggregate (blast furnace slag to be used as a cement substitute), weathered basic oxygen furnace slag (for low grade aggregate/fill use), and iron (for steel-making) has long been practised (Gomes et al., 2016a,b, 2019) in what would now be efforts deemed to be contributing to the CE by keeping resources in use and avoiding the need for extracting virgin materials.

More recently, there has been a drive, particularly across the European Union, to improve inventories of legacy waste [e.g., the Mining Waste Directive (MWD)] and invest in technologies for value recovery. This is in part driven by environmental concerns (the MWD was prompted by the mine tailings spills in Baia Mare, Romania and Aznalcóllar, Spain in the late 1990s), but also by both CE initiatives and concerns over security of supply of critical raw materials (EU, 2018). The EU has identified a range of minerals and materials for which supply is deemed either crucial for strategic industries and/or at risk of disruption (Mathieux et al., 2017). These initiatives have seen the development of a range of endeavors looking at landfill mining and prospecting a range of mineral-rich wastes and by-products that were hitherto considered of little economic value. Some examples include bauxite processing residue, which is rich in high-value rare earth elements, and other elements critical for green technologies such as gallium, scandium, and vanadium (Gomes et al., 2016b).

Table 1 summarises some of the resources and minerals that have gained research interest from a recovery perspective in recent years. In the case of steel slag and alumina wastes, Gomes et al. (2019) highlight that despite the considerable efforts to develop recovery technologies, most are at fairly early stages of technology readiness (typically below TRL5, which is technology validation in a relevant environment, usually the laboratory testing of an integrated system). While bulk reuse of materials like steel slag can be as high as 80% in some jurisdictions (Euroslag, 2012), such afteruses are not enough to meet all by-products arising. Indeed in the case of alumina by-products (bauxite processing residue), less than 3% of the annual production currently has a viable afteruse, despite 3000 patents on reuse technologies (Evans, 2016). As such, while the landfill mining discipline is one that is rapidly growing, challenges of translating viable reuse and recovery technologies to full-scale operation are a priority to address for the environmental engineering community.

Table 1 Summary of resources and minerals, the recovery of which have gained interest in the last 5 years.

Waste	Target metal	Recovery technology	Maximal recovery	Ref.
Steel slag	V	Bioleaching	92%	Mirazimi et al. (2015)
	V	Ion exchange	>99% (batch tests), 72% (column tests)	Gomes et al. (2017)
	Al and Si	Alkaline leaching	40%	Nikolić et al. (2016)
	V	Extraction with primary amine	90%	Ning et al. (2016)
	Cr	Alkaline pressure leaching	46%	Kim et al. (2016)
	Al, Cr, V	Bioleaching	100%	Gomes et al. 2018a,b)
	–	Bulk reuse as construction materials	–	Galán-Arboledas et al. (2017), Fakhri and Ahmadi (2017)
Flotation tailings of polymetallic ores	Cu, Zn, Au	Three-step process (acid leaching + ferric leaching + biooxidation)	96%	Muravyov and Fomchenko (2018)
	Cu, Zn, Au	Biooxidation	85%	Muravyov (2019)
	Cu, Zn, Pb, Ag	Bioleaching and brine leaching	96%	Romero-García et al. (2019)
Mine water	Sb	Electrocoagulation	82%	Arnold et al. (2019)
Solid metalliferous wastes (sludges, dusts, residues, slags, red mud, and tailing wastes)	Zn, Cu, Fe, Pb	Biohydrometallurgy	90%	Sethurajan et al. (2018)
Copper slag	Fe, Cu and Mo	Carbothermal reduction	100%	Sarfo et al. (2017)
Iron ore tailings	–	Reuse in pigments for sustainable paints	–	Galvão et al. (2018)
Waste rocks	–	Reuse as aggregates in fly-ash-based polymers	–	Capasso et al. (2019)

Modified from Gomes, H.I., Rogerson, M., Courtney, R., Mayes, W.M., 2019. Integrating remediation and resource recovery of industrial alkaline wastes: case studies of steel and alumina industry residues. *In: Resource Recovery From Wastes,* pp. 168–191.

Beyond recovery of bulk material or component minerals or elements, there are also a range of environmental engineering applications reusing waste material in niche applications. These typically make use of the specific chemical and physical properties of mineral-rich wastes in water remediation or land contamination applications. For example, alkalinity-generating media such as steel slag and bauxite residue have been tested at various scales for neutralising acidic waters and as sorbents for nutrient-rich water (Bowden et al., 2009; Zijlstra et al., 2010). Meanwhile, ferric oxyhydroxides (ochre) from coal mine water treatment systems have been widely tested for metal and phosphorus removal given their very high specific surface area and sorption capacity (e.g., Dobbie et al., 2009; Mayes et al., 2009b).

3.2 Provisioning services: Energy

Energy recovery from mining and mineral wastes has traditionally focused on fossil fuel recovery, such as coal slime from coal tailings. Such fuel sources have additional environmental burdens in emissions and may, depending on full life cycle assessment, represent negative reuse of mining wastes (Bian et al., 2012). The potential for thermal resources from abandoned mines is not limited to solid wastes, however, and mine water itself has been a focus of research interest in recent years. Given mining gives rise to large technogenic (i.e., human-made) aquifers with a relatively stable groundwater thermal regime, opportunities for heat recovery have been explored over a range of scales (Banks et al., 2019). Groundwater temperature is typically above soil temperature in deep mines and rises 1–3 °C with every 100 m depth (Banks, 2012). Various configurations of heat exchanger have been tested at abandoned mine sites (see Banks et al., 2019, for examples). With each technology and configuration, there are various trade-offs between thermal efficiency and issues such as energy requirement for pumping and issues of clogging with iron oxyhydroxides when waters are oxygenated in open systems, adding significant maintenance burdens (Banks et al., 2019). Some large-scale trials have demonstrated the efficacy of using mine waters for thermal resource recovery, most notably at Heerlen in the Netherlands (Verhoeven et al., 2014). Heat recovery can also extend to environmental treatment systems, for example, Bailey et al. (2016) suggest a potential 47.5 MW of thermal energy in treated coal mine water discharges, which compares favorably with the 2.3 MW of electricity used in powering some of the treatment systems (e.g., pumps and dosing units), suggesting opportunities for circularity in energy use at remedial sites.

A range of engineered hydraulic structures are often still present at abandoned mine sites. In upland settings, these typically encompass drainage adits (shallow-gradient tunnels beneath the target ore that drain water from the area to be mined) that can be several kilometers in length and underdrain large areas (e.g., Banks et al., 1997). Some of these have high flow rates and generally good potable water quality (modest zinc concentrations are not problematic for drinking water quality compliance, but are an issue for instream environmental quality standards due to the sensitivity of salmonids to Zn; Mayes et al., 2009b). As such, use of drainage adits for water supply (provisioning and regulating service) has been practised in many areas, such as the Pennine Orefields of England (Banks et al., 1997). Given the predictable flow rates and hydraulic heads associated with drainage adits, there is also potential for using the water for micro-hydro power generation, in some cases to support remedial efforts (Choi and Song, 2017). Fig. 1 shows the range of flow and hydraulic head values for noncoal mine discharges on the national databases in England and Wales (after Mayes et al., 2010). This shows a relatively large number of sites plotting within the operating ranges of common micro-hydro generation technologies such as Archimedes screw systems, propeller (e.g., Kaplan), and turbine-based (Cross-Flow) systems. Where local (and often remote) communities are located in close-proximity to mine discharges, there may be advantages of looking at mine drainage as opposed to in-channel drainage, given consistent flow rates and reduced disturbance to instream habitat connectivity. Energy generation from legacy waste sites has also considered their use for photovoltaic power generation or wind power, as well as growth of biofuel crops (e.g., Song et al., 2015; Milbrandt et al., 2014). Milbrandt et al. (2014) highlight that across the >11,000 km^2 of abandoned mined lands in the continental USA, there is potential for development of 44 GW of solar power, 4.8 GW of wind power, and 3.4 GW of biomass power.

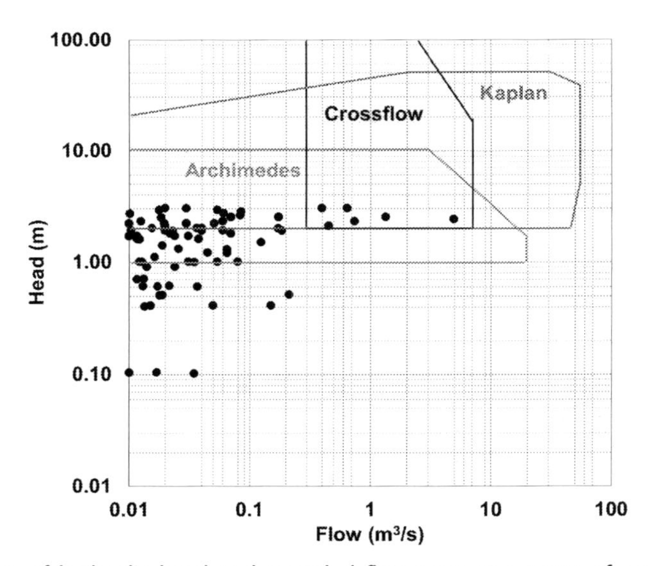

Fig. 1 Comparison of hydraulic head and recorded flow rates at a range of noncoal mine water discharges. Modified from *Mayes, W.M., Potter, H.A.B., Jarvis, A.P., 2010. Inventory of aquatic contaminant flux arising from historical metal mining in England and Wales. Sci. Total Environ. 408 (17), 3576–3583.*

3.3 Regulating services

3.3.1 Carbon

Given the considerable burden extractive and mineral processing industries have on global carbon emissions, efforts to mitigate these impacts have been developing in recent decades. Oberle et al. (2019) estimate that global iron and steel production is responsible for a quarter of global industrial energy use and emissions doubled between 2000 and 2015. Aluminum production is also a highly energy-intensive process (usually high-temperature alkali digestion of bauxite in the Bayer process) and another significant contributor to carbon dioxide emissions (~0.8% global CO_2 emissions; Carbon Trust, 2011). These high-temperature processes also give rise to a range of silicate-rich by-products, which, when weathered, sequester atmospheric carbon, usually in the process of forming stable carbonate minerals [Eq. (1); Gomes et al., 2016a,b].

$$(Ca,\ Mg)SiO_{3(s)} + CO_{2(g)} \rightarrow (Ca,\ Mg)CO_{3(s)} + SiO_{2(s)} \quad (1)$$

Crucially, the rates of mineral weathering of slags associated with iron and steel production are much higher than many nonanthropogenic silicate minerals (Renforth, 2019). As such, there has been much interest in evaluating the effectiveness of silicate-rich mineral wastes for mineral carbonation to offset some of the emissions associated with these carbon-intensive industries. Much of this experimentation takes place under conditions of high temperature, pressure, or CO_2 concentration, to produce very efficient silicate-to-carbonate conversion under what is termed "active" carbonation (i.e., manipulated conditions to enhance carbonation; Brück et al., 2018). However, under environmental disposal conditions, such scenarios are not feasible (e.g., Riley et al., 2020).

Recent investigations of the prospects for "passive" carbonation—where slag material emplaced in mounds weathers and carbonates unintentionally—have shown that the carbonation potential of deposited slags is far from realised in traditional disposal settings (Mayes et al., 2018; Pullin et al., 2019). Using a combination of techniques for assessing the composition and volume of slag mounds and integrating these with long term carbon balances in downstream waters and changes in mineralogy of slag with depth, it was estimated that less than 3% of the potential carbonation potential of a historical steel slag heap had been met since disposal approximately four decades previously (Mayes et al., 2018; Pullin et al., 2019). The factors driving this limited carbonation include armoring of slag by secondary precipitates, which reduces reactivity of silicate minerals (Hobson et al., 2017), particle size, and limited gas exchange into the heap (Pullin et al., 2019). As such, proactive approaches to encouraging slag weathering and subsequent carbonation should be considered as part of by-product processing and disposal chains. This could include comminution (i.e., grinding of material to increase reactivity where carbon balances are favorable) and weathering in shallow mounds to encourage ingress of atmospheric carbon dioxide. Analogous approaches are adopted by the alumina sector to manage the extreme alkalinity of bauxite processing residue. At some disposal areas, a process known as "mud-farming" is adopted, which uses amphibious amphirollers to overturn surface deposits to encourage reaction with atmospheric CO_2 (Finngean et al., 2018).

In both the steel slag and bauxite processing residue examples considered above, carbonation has complementary benefits for (1) sequestering atmospheric CO_2, (2) lowering alkalinity of the substrate (through reaction of hydroxide phases), which improves prospects for plant growth, and (3) potentially increases prospects for bulk reuse of material (e.g., with slags that are less prone to expansion during weathering of free lime: Gomes et al., 2016a,b). Recent assessment of the potential for alkaline wastes to contribute towards negative emissions technologies (i.e., processes that draw out atmospheric CO_2 and stabilise it) has suggested a carbon dioxide storage potential of 2.9–8.5 billion tonnes per year by 2100 (Renforth, 2019). This equates to emissions offsetting at current technological levels, but with efforts to improve efficiency and fuel sources in these carbon-intensive industries (e.g., IPCC, 2018), could push these systems towards negative emissions (Renforth, 2019). For example, the carbonation potential of steel slag is 368–620 $kg\,CO_2\,t^{-1}$, while current CO_2 emissions per tonne of slag produced are in the region 12,000 $kg\,CO_2\,t^{-1}$, but potentially as low as $<300\,kg\,CO_2\,t^{-1}$ slag with efficiency and fuel adjustments (e.g., hydrogen; Renforth, 2019).

Research interest in mineral carbonation of legacy wastes is not confined to those alkaline materials from high-temperature processes such as steel slags, combustion ashes, and alumina wastes (Brück et al., 2018). Many mine wastes are also rich in silicate minerals that can passively sequester atmospheric CO_2 under ambient environmental weathering conditions. Wilson et al. (2011) demonstrate the weathering products of kimberlitic mine tailings at the Diavik Diamond Mine, Northwest Territories, Canada, to include extensive nesquehonite [$MgCO_3 \cdot 3H_2O$] and Na- and Ca-carbonate minerals. Nearly 90% of the carbon in these secondary minerals was fingerprinted by radiocarbon and isotopic analyses to be from modern sources and the authors suggested a potential carbon uptake rate of 102–114 g C/m^2/year, which is double that of natural silicate weathering at the high latitude subarctic study site (Wilson et al., 2011). Similar limitations on carbon capture, as in the steel slag examples (Pullin et al., 2019), are apparent in field estimation of carbon uptake in mine wastes, such as coarse particle size limiting reactivity and limited diffusion of CO_2 into waste heaps (van Haren et al., 2017). As such, field testing and optimisation of methods that encourage CO_2 ingress, such as carefully engineered waste heap morphology, particle size manipulation, or wind-powered air pumps to ventilate heaps, should be seen as an environmental engineering priority for realising the potential carbonation benefits.

3.4 Underpinning and cultural services

3.4.1 Biodiversity

The benefits of abandoned waste sites for biodiversity have long been identified (e.g., Bradshaw and Chadwick, 1980). Through observation of spontaneous recovery and revegetation of abandoned waste tips, the importance of harsh and unusual substrates in promoting floristic diversity has been documented at a range of acid to alkaline sites. These cover a range of soil pH conditions from highly acidic (e.g., pyrite-rich mine wastes) to extremely alkaline (e.g., bauxite processing residue, lime waste), with associated gradients of nutrient and metal(loid) availability and variability in particle size distribution. At extremes of pH, nutrient availability and mobility of potentially phytotoxic elements typically increase (Batty and Hallberg, 2010), while fine particle size provides constraints to soil stability and also poses a risk of particulate matter pollution (Gelencsér et al., 2011). Table 2 provides a summary of some of the plant communities and associated fauna that have been documented colonising legacy waste sites. In many cases, these communities include otherwise declining habitats of conservation importance (e.g., calcareous grassland, which has declined with agricultural intensification) and even communities that receive formal conservation designation that are primarily anthropogenic. The best example of the latter is calaminarian grassland that consists of metallophyte (metal-tolerant) plants (e.g., spring sandwort: *Minuartia verna*; alpine pennycress: *Noccaea caerulescens*; moonwort: *Botrychium lunaria*) that colonise metal-rich spoil heaps and fluvial gravels that are protected under the European Union Habitats Directive. Beyond the substrates that can promote floristic biodiversity (and as is often assumed, though less frequently tested, insect biodiversity), mine buildings and structures can also be of conservation value. For example, mine adits and drainage structures form important roost sites for bats and may be incorporated into conservation designations (e.g., approximately 10% of the UK population of greater horseshoe bat, *Rhinolophus ferrumequinum*, is recorded within tunnels at a former Oolitic limestone mine in Wiltshire; Natural England, 1991). Current work also highlights the value of such recolonised legacy sites to breeding declining farmland bird species in the UK (e.g., Yellowhammer: *Emberiza citronella* and Linnets: *Carduelis cannabina*), considered to be on the UK red list (Eaton et al., 2015).

In some cases, proactive approaches to site restoration have explicitly engaged with ecological enhancement aims through assisted recovery and ecosystem "design." These are especially important on mined lands and wastes at risk of erosion, with extreme substrates, or where suitable species for recolonisation have low dispersal abilities (Baasch et al., 2012). Such approaches are particularly prevalent in the sand and gravel extractive industries where long-established guidance exists for promoting biodiversity postclosure, typically through the provision of mixed open water, reedbed, and marginal woodland habitat (Andrews and Kinsman, 1990). In many settings, the postclosure scenarios can be considered enhancement on previous

Table 2 Examples of communities of biological interest developing over extractive and processing industry wastes.

Type of extractive/processing site	Substrate	Communities/species of interest	Example sites	Example references
Coal and lignite mines	Pyrite-rich (FeS_2) coal spoil	Heathland, acid grassland, bog communities	Lusatian brown coal mines, Germany/Poland	Pietsch (1998)
Pb-Zn mines	Metal-rich waste rock and downstream gravels/fines	Calaminarian grassland	Goginan Pb mine, Wales; Gang Mine, England; Geul Valley, Netherlands	Baker et al. (2010)
Leblanc and Solvay waste	Calcareous, nutrient poor and alkaline (pH < 11)	Orchid-rich calcareous grassland	Darcy Lever, Greater Manchester, UK	Ash et al. (1994); Box (1999)
Steel slag	Calcareous, nutrient poor and alkaline (pH < 12)	Orchid-rich calcareous grassland into alkaline fen-swamp	Coatham Marsh, Cleveland, UK	Raper et al. (2015), Mayes et al. (2009a)
Lime waste	Calcareous, nutrient poor and alkaline (pH <12)	Orchid-rich calcareous grassland	Plumley Lime Beds, Cheshire, UK	Bradshaw and Chadwick (1980)
Sand and gravel extraction	Mosaic of open water, littoral habitat; *Phragmites australis* reedbed succeeding to alder-carr woodland	Reedbed, marginal wetland, soft sediments; Reed Bunting (*Emberiza schoeniclus*), Bittern (*Botaurus stellaris*), Lapwing (*Vanellus vanellus*)	Wykeham, North Yorks, UK; North Cave, Humberside, UK	Day et al. (2017)
Mine water, steel slag leachate passive treatment systems	Open water to littoral wetland	Reedbed, marginal wetland; Reed Bunting (*Emberiza schoeniclus*); Sedge Warbler (*Acrocephalus schoenobaenus*)	Coatham Marsh, Cleveland, UK; Woolley Mine Water Treatment Plant, South Yorks, UK	Unpublished data of authors

(preextraction) land use, which is often intensive agriculture in lowland settings, as demonstrated by Blaen et al. (2015) for sand and gravel extraction sites in Eastern England.

On more hostile substrates, a range of surface amendments can assist revegetation and the development of functional soils (Courtney and Harrington, 2012). Functioning, microbially diverse surface horizons are crucial for long term soil stability and can have an influence on pH and metal(loid) availability to significant depth, as demonstrated on alkaline bauxite residue deposits (Bray et al., 2017). Assisted recovery, through planting of appropriate native species, usually involving a mix of legumes (nitrogen-fixers), has been shown to both accelerate recolonisation of wastes and direct communities to those of higher conservation value (more diverse plant communities and fewer ruderals) when compared to reference treatments relying on spontaneous recolonisation (Ash et al., 1994; Baasch et al., 2012). As such, there are significant opportunities in restoration planning phases to direct postclosure mining and mineral wastes towards high conservation value plant communities. Where these incorporate existing priority or declining habitats at landscape scale, as is the case with calcareous grassland in many lime and steel working regions, then such efforts can complement broader strategic conservation goals that aim to reverse habitat loss and fragmentation (Wallis de Vries et al., 2002). Furthermore, ongoing low-intensity management by conservation agencies, serving as custodians of legacy sites, can also assist in minimising potentially damaging informal site uses (e.g., fly-tipping).

3.4.2 Culture, recreation, and leisure

The cultural significance of legacy mining and mineral processing sites is of great importance both to local communities and in documenting the industrial archaeology of regions. Often, the mine or processing facility was the hub of the community, and possibly the reason for a settlement being located in that place in the first instance, so even after closure waste heaps and the remnants of former site buildings are important in fostering community identity. The semicoke ash dumps associated with oil shale mining and processing in East-Viru County in northeast Estonia take on particular importance given the surrounding flat landscape. A series of semicoke ash mounds with relative heights over 100 m are not only the highest artificial landforms in the Baltic States, but also of great local importance as viewpoints in the landscape (Pae et al., 2005). Similar features are apparent across mining and mineral processing regions, which gain local cultural significance and affection including the "bings" of West Lothian in Scotland (spoil heaps associated with oil shale extraction; Heal and Salt, 1999), the Three Sisters coal spoil heaps in Wigan, Lancashire (known as the "Wigan Alps" by locals prior to restoration efforts), and Slag Bank in Barrow, Cumbria (a large steel waste dump to the north-west of the town which provides spectacular views over the adjacent Walney Island).

In many cases, former mine and processing site buildings receive formal built environment or cultural designations signifying their historical importance (Howard et al., 2015), which are often complementary to the natural environment designations discussed in Section 3.4.1. The industrial archaeological significance can be a driver for tourist interest, through documenting the history of the sites and integrating archaeology with other leisure activities (Crane et al., 2017). Such redevelopment initiatives can have significant social and economic regeneration benefits in postindustrial areas, as has been the case with the Magna Museum and Conference Centre in Rotherham, UK, at the site of a former steel mill (Alker and Stone, 2005). Formal built environment designations at former mining and processing sites range from national designations (e.g., Scheduled Ancient Monuments and Listed Buildings in the UK, e.g., Penallta Colliery, Hengoed, Wales) to United Nations World Heritage Site status (e.g., Cornish Tin and Copper Mining District, UK; Wallonia Mining Sites, Belgium; Zollverein Coal Mine Complex, Westphalia, Germany). Inventories of these structures have been included in some mine remediation exercises (e.g., Metal Mines Strategy for Wales; Johnston, 2004), but can often be overlooked in assessments of resource recovery or environmental remediation (Crane et al., 2017; Howard et al., 2015).

While the traditional endpoint for restoration efforts at mineral waste sites was either to return the land to productive agricultural use (for relatively benign materials) or landscaping (Bradshaw and Chadwick, 1980), a more diverse range of afteruses are now apparent, which hold varied cultural values. Leisure-based afteruses for legacy mining and processing wastes are commonplace in many locations and include mountain biking and hiking and are often integrated with biodiversity conservation efforts. Retail and housing developments are also commonplace where substrates are suitably stable, and risks associated with exposure to metal-rich materials are minimised (Pullin et al., 2019). This may involve extensive decontamination efforts, resource recovery hubs where secondary materials are recovered after plant closure to assist decontamination, or use of novel geoengineering techniques to minimise the risk of leachate generation from the wastes (Goss et al., 2006).

Legacy sites associated with extractive industries can also be both a source of inspiration and a canvas for artists. Former colliery spoil heaps in the Ruhr Valley, Germany have become a centerpiece for art installations along a heritage trail (Miccoli et al., 2014), while the waste materials themselves can be sculpted into artistic landscape features, as in the case of the *Northumberlandia* or *The Lady of the North* at Cramlington in Tyne and Wear, UK (Escobar, 2013). Visual artists also play a crucial role in documenting and disseminating information on the environmental hazards associated with legacy wastes to a diverse audience. The integration of compelling images with graphic design and scientific data provided a novel collaborative approach to communicating the impacts of extractive industries and associated petrochemicals in southern USA (Misrach and Orff, 2012). Edward Burtynsky has used a range of legacy wastes as his subject to visualise the vastness and otherworldliness of degraded landscapes associated with extractive industries, recently incorporating augmented reality approaches to aid scientific data presentation (Burtynsky and Baichwal, 2008). Similar integrating approaches have been adopted to communicate environmental impacts and narratives around polluting legacy waste sites (e.g., Scott et al., 2014) and provide an important means for communicating scientific information beyond typical academic audiences. As such, mineral waste disposal sites can provide a range of cultural values, sometimes inspired by environmental degradation that provides additional cultural legacies not encompassed in formal designations.

3.4.3 Science

Given the extreme geochemical conditions that mining and mineral wastes can produce, for example, with pH values that can range from the negative in acid mine drainage (Nordstrom et al., 2000) to pH $>$ 13 in bauxite processing residue (Mayes et al., 2011), there is considerable interest in extremophile communities at such sites. Researchers are interested in extremes for assessing communities and evolutionary processes as analogues for early Earth-like or extra-terrestrial environments (Tiago et al., 2004). Similarly, the extremely alkaline sites have been considered analogues for studying geochemical transport processes that may occur at nuclear disposal facilities (Duro et al., 2014) as well as carbonate-hosted hydrocarbon reservoirs (Bastianini et al., 2019). The tolerance of microbial communities to extreme conditions has revealed a range of novel metabolic and energy generating processes (e.g., Salah et al., 2019), which can also be exploited in industrial applications (Borkar, 2015). Microbial investigations under extreme conditions are also crucial to the development of biologically mediated remedial techniques for both acidic and alkaline wastes (Charles et al., 2019). The identification of acidophile sulfate-reducing bacteria has been key in many novel bioreactor technologies for remediating acid mine drainage (Neculita et al., 2007), while acidophilic microbial consortia have long been used for bioleaching of metals from low-grade ores and wastes (Norris et al., 2000), which can assist in resource recovery efforts when executed in a controlled manner. There has also been interest in translating such microbial consortia for bio-mining of other anthropogenic wastes, as demonstrated for municipal waste ashes using communities of *Acidothiobacillus thiooxidans* and *Acidothiobacillus ferrooxidans* isolated from acid mine drainage sites (Funari et al., 2019).

4. Conclusions—A way forward

Resource use in the 21st century needs to heed the lessons of the environmental legacies created by current and historical extractive industries to ensure that its footprint lies firmly within planetary boundaries. The extractive and processing sector has improved environmental practice in many areas in recent years, quite often in response to catastrophic events (e.g., Evans, 2016). However, the recent events in Brazil highlight that significant challenges remain to embed environmental and societal wellbeing within decision making across the sector. The traction of CE in national and international policy spheres (Deutz et al., 2017; Millar et al., 2019) offers a framework for extractive industries to reevaluate approaches to waste arisings, their legacies, and management, particularly where this is integrated with means of accounting for the multifaceted values/ESs that can arise at legacy sites. However, the CE discourse both in academic and policy-making circles is currently dominated by postconsumer wastes (Sapsford et al., 2019; Velenturf et al., 2019). Given the environmental impacts of mining and associated mineral processing, discussions of CE need to explicitly address these sectors. Our future societies will still produce waste materials from extractive industries that cannot immediately find productive use in a CE, either due to lack of demand at that particular moment in time, or the geographical disconnect between locations of waste production and potential demand. How we manage such materials and reap value from them is key to moving the extractive industries towards sustainability and resource circularity. The first step in this process is ensuring that mining and mineral processing activity moves beyond simple economic cost–benefit analysis (Bian et al., 2012) and takes account of the potentially extensive external (environmental and social) costs that the sector can produce. In this regard, there are considerable opportunities to learn from the mistakes of our industrial past, often from an era of lax environmental regulation. This can involve "leapfrogging" not just resource efficient technologies to industrialising countries as a means of decoupling resource use from environmental and social impacts (Oberle et al., 2019), but also remedial technologies and approaches to restoration used to manage the legacy issues, which have developed markedly in the last 3–4 decades and provided demonstrable ESs (Younger et al., 2002; Neculita et al., 2007; Jarvis et al., 2018). Such technologies may, however, need to be deployed at scales not previously seen before, as mine abandonment plans in China suggest (Lu et al., 2019), which provides further challenges for the sector.

Fig. 2 compares a current model of steel by-product use with an alternative model accounting for all potential environmental impacts and benefits. The example highlights the contribution of environmental engineering, not just in carefully designed passive treatment systems to manage legacy issues, but also reframing the disposal chain to realise the currently overlooked potential benefits of by-products for carbon sequestration and habitat creation. As with all idealised visions of future resource use, the constraints on implementation need to be considered. These have been reviewed by various workers (see Deutz et al., 2017; Velenturf and Purnell, 2017; Velenturf et al., 2019) and include:

- Governance and policy environments that are historically dominated by environmental regulation, which can stifle opportunities for innovative resource recovery;
- Inertia in some industrial sectors that means innovative practices are slow to be adopted, or long-term (multidecade) by-product agreements with downstream users that limit opportunities for alternative resource recovery options, as is the case with steel by-products in the UK (Deutz et al., 2017);

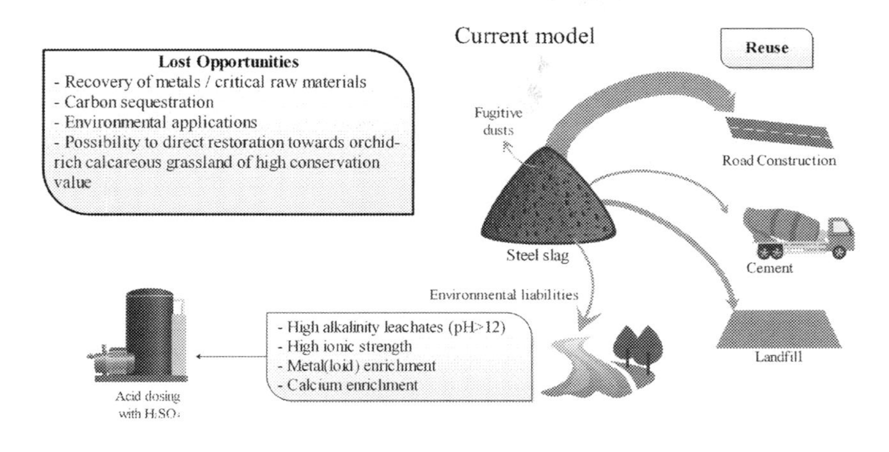

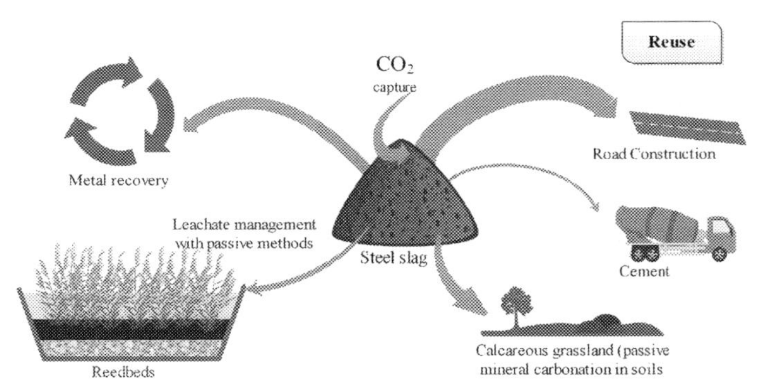

Fig. 2 Comparison between the current model of steel by-product use and an alternative model, accounting for all potential environmental impacts and benefits.

- Legacy issues where liability for sites is unclear or absent, which can provide financial barriers to resource recovery as well as concerns for companies about "accidentally" inheriting pollution legacies;
- Funding gaps for management of legacy sites that fall on public authorities (Jarvis and Mayes, 2012);
- Technology development gaps for a range of recovery technologies. Many recovery technologies have been demonstrated at laboratory scale, but full-scale demonstration is a key challenge for environmental engineers to permit both regulatory scrutiny and to foster uptake of novel technologies for material/metal recovery, carbon capture, and bioremediation.

Overcoming these potential obstacles requires multiagency approaches to formalise the benefits of legacy sites and manage enduring environmental impacts. There are excellent examples of initiatives that ally goals of resource recovery with environmental remediation at legacy sites, some operating at a national level (e.g., the UK Coal Authority's coal mine remediation program). It is no coincidence that such successes have strategic oversight in publically funded agencies, and clear lines of responsibility for legacy issues (Bailey et al., 2016). It is apparent that for some of the potential ESs that could be offered by mining and mineral wastes, such as passive mineral carbonation or biodiversity benefits, that long-term custody of the site is required and site managers need to acknowledge the protracted timescales over which benefits are potentially realised. As the technical basis for waste valorisation technologies and environmental engineering approaches to managing legacy issues improves further, embedding these in strategic planning across all phases of extractive industries from prospecting to long-term management of environmental legacies and site afteruses is essential for minimising the environmental footprint of the sector.

References

Alker, S., Stone, C., 2005. Tourism and leisure development on brownfield sites: an opportunity to enhance urban sustainability. Tour. Hospital. Plan. Dev. 2 (1), 27–38.

Allwood, J.M., Cullen, J.M., Milford, R.L., 2010. Options for achieving a 50% cut in industrial carbon emissions by 2050. Environ. Sci. Technol. 44 (6), 1888–1894.

Andrews, J., Kinsman, D., 1990. Gravel Pit Restoration for Wildlife: A Practical Manual. Royal Society for the Protection of Birds.

Arnold, M., Kangas, P., Makinen, A., Lakay, E., Isomaki, N., Laven, G., Gericke, M., Pajuniemi, P., Kaartinen, T., Wendling, L., 2019. Mine water as a resource: selective removal and recovery of trace antimony from mine-impacted water. Mine Water Environ. 38, 431–446.

Árvay, J., Hauptvogl, M., Jančo, I., Stanovič, R., Tomáš, J., Harangozo, Ľ., 2019. Impact of former mining activity to soil contamination by risk elements. Arch. Ecotoxicol. 1 (1), 1–6.

Ash, H.J., Gemmell, R.P., Bradshaw, A.D., 1994. The introduction of native plant species on industrial waste heaps: a test of immigration and other factors affecting primary succession. J. Appl. Ecol. 31, 74–84.

Baasch, A., Kirmer, A., Tischew, S., 2012. Nine years of vegetation development in a postmining site: effects of spontaneous and assisted site recovery. J. Appl. Ecol. 49 (1), 251–260.

Bailey, M.T., Gandy, C.J., Watson, I.A., Wyatt, L.M., Jarvis, A.P., 2016. Heat recovery potential of mine water treatment systems in Great Britain. Int. J. Coal Geol. 164, 77–84.

Baker, A.J., Ernst, W.H., van der Ent, A., Malaisse, F., Ginocchio, R., 2010. Metallophytes: the unique biological resource, its ecology and conservational status in Europe, Central Africa and Latin America. Ecol. Indus. Pollut. 18, 7–40.

Banks, D., 2012. An Introduction to Thermogeology: Ground Source Heating and Cooling. John Wiley & Sons.

Banks, D., Athresh, A., Al-Habaibeh, A., Burnside, N., 2019. Water from abandoned mines as a heat source: practical experiences of open-and closed-loop strategies, United Kingdom. Sust. Water Resour. Manag. 5 (1), 29–50.

Banks, D., Younger, P.L., Arnesen, R.T., Iversen, E.R., Banks, S.B., 1997. Mine-water chemistry: the good, the bad and the ugly. Environ. Geol. 32 (3), 157–174.

Bastianini, L., Rogerson, M., Mercedes-Martín, R., Prior, T.J., Cesar, E.A., Mayes, W.M., 2019. What causes carbonates to form "shrubby" morphologies? An Anthropocene limestone case study. Front. Earth Sci. 7, 236.

Batty, L.C., Hallberg, K.B. (Eds.), 2010. Ecology of Industrial Pollution. Cambridge University Press.

Bian, Z., Miao, X., Lei, S., Chen, S.E., Wang, W., Struthers, S., 2012. The challenges of reusing mining and mineral-processing wastes. Science 337 (6095), 702–703.

Blaen, P.J., Jia, L., Peh, K.S.H., Field, R.H., Balmford, A., MacDonald, M.A., Bradbury, R.B., 2015. Rapid assessment of ecosystem services provided by two mineral extraction sites restored for nature conservation in an agricultural landscape in eastern England. PLoS One 10 (4), e0121010.

Borkar, S., 2015. Alkaliphilic bacteria: diversity, physiology and industrial applications. In: Borkar, S. (Ed.), Bioprospects of Coastal Eubacteria. Springer, pp. 59–83.

Bowden, L.I., Jarvis, A.P., Younger, P.L., Johnson, K.L., 2009. Phosphorus removal from waste waters using basic oxygen steel slag. Environ. Sci. Technol. 43 (7), 2476–2481.

Box, J., 1999. Nature conservation and post-industrial landscapes. Ind. Archaeol. Rev. 21 (2), 137–146.

Bradshaw, A.D., Chadwick, M.J., 1980. The Restoration of Land: The Ecology and Reclamation of Derelict and Degraded Land. University of California Press.

Bray, A.W., Stewart, D.I., Courtney, R., Rout, S.P., Humphreys, P.N., Mayes, W.M., Burke, I.T., 2017. Sustained bauxite residue rehabilitation with gypsum and organic matter 16 years after initial treatment. Environ. Sci. Technol. 52 (1), 152–161.

Brück, F., Fröhlich, C., Mansfeldt, T., Weigand, H., 2018. A fast and simple method to monitor carbonation of MSWI bottom ash under static and dynamic conditions. Waste Manag. 78, 588–594.

Brundtland, G.H., 1987. Our common future—call for action. Environ. Conserv. 14 (4), 291–294.

Burtynsky, E., Baichwal, J., 2008. Manufactured Landscapes. British Film Institute.

Caballero Espejo, J., Messinger, M., Román-Dañobeytia, F., Ascorra, C., Fernandez, L., Silman, M., 2018. Deforestation and Forest degradation due to gold mining in the Peruvian Amazon: A 34-year perspective. Remote Sens. 10 (12), 1903.

Capasso, I., Lirer, S., Flora, A., Ferone, C., Cioffi, R., Caputo, D., Liguori, B., 2019. Reuse of mining waste as aggregates in fly ash-based geopolymers. J. Clean. Prod. 220, 65–73.

Carbon Trust, 2011. Aluminium. https://www.carbontrust.com/media/38366/ctc790-international-carbon-flows_-aluminium.pdf. (last accessed 10/8/19).

Charles, C.J., Rout, S.P., Wormald, R., Laws, A.P., Jackson, B.R., Boxall, S.A., Humphreys, P.N., 2019. In-situ biofilm formation in hyper alkaline environments. Geomicrobiol. J. 36 (5), 405–411.

Choi, Y., Song, J., 2017. Review of photovoltaic and wind power systems utilized in the mining industry. Renew. Sust. Energ. Rev. 75, 1386–1391.

Courtney, R., Harrington, T., 2012. Growth and nutrition of *Holcus lanatus* in bauxite residue amended with combinations of spent mushroom compost and gypsum. Land Degrad. Dev. 23 (2), 144–149.

Crane, R.A., Sinnett, D.E., Cleall, P.J., Sapsford, D.J., 2017. Physicochemical composition of wastes and co-located environmental designations at legacy mine sites in the south west of England and Wales: implications for their resource potential. Resour. Conserv. Recycl. 123, 117–134.

Day, G., Mayes, W.M., Wheeler, P.M., Hull, S.L., 2017. Can aggregate quarry silt lagoons provide resources for wading birds? Ecol. Eng. 105, 189–197.

Deutz, P., Baxter, H., Gibbs, D., Mayes, W.M., Gomes, H.I., 2017. Resource recovery and remediation of highly alkaline residues: a political-industrial ecology approach to building a circular economy. Geoforum 85, 336–344.

Díaz, S., Settele, J., Brondízio, E., Ngo, H., Guèze, M., Agard, J., Arneth, A., Balvanera, P., Brauman, K., Butchart, S., Chan, K., 2019. Summary for Policymakers of the Global Assessment Report on Biodiversity and Ecosystem Services of the Intergovernmental Science—Policy Platform on Biodiversity and Ecosystem Services. IPBES Secretariat.

Dobbie, K.E., Heal, K.V., Aumonier, J., Smith, K.A., Johnston, A., Younger, P.L., 2009. Evaluation of iron ochre from mine drainage treatment for removal of phosphorus from wastewater. Chemosphere 75 (6), 795–800.

Dunham, K.C., 1990. Geology of the Northern Pennine Orefield: Tyne to Stainmore. vol. 19 Stationery Office/Tso.

Duro, L., Bruno, J., Grivé, M., Montoya, V., Kienzler, B., Altmaier, M., Buckau, G., 2014. Redox processes in the safety case of deep geological repositories of radioactive wastes. Contribution of the European RECOSY collaborative project. Appl. Geochem. 49, 206–217.

Eaton, M.A., Aebischer, N.J., Brown, A.F., Hearn, R.D., Lock, L., Musgrove, A.J., Noble, D.G., Stroud, D.A., Gregory, R.D., 2015. Birds of conservation concern 4: the population status of birds in the United Kingdom, Channel Islands and Isle of Man. British Birds 108, 708–746.

Engels, F., 2005. The Condition of the Working Class in England. Routledge.

Escobar, P.G., 2013. Landscaped English beauty built with geosynthetics. Geosynthetics 31, 26–35.

European Commission, 2015. Closing the Loop—An EU Action Plan for the Circular Economy. COM/2015/0614 Final. Available at https://eur-lex.europa.eu/legal-content/EN/TXT/?uri=CELEX:52015DC0614 [last accessed 13/12/19].

European Commission, 2018. Critical Raw Materials. Available at: https://ec.europa.eu/growth/sectors/raw-materials/specific-interest/critical_en [last accessed 1/8/2019].

Euroslag. Statistics 2012. http://www.euroslag.com/products/ statistics/2012/ [last accessed 21/05/21].

Evans, K., 2016. The history, challenges, and new developments in the management and use of bauxite residue. J. Sust. Metal. 2 (4), 316–331.

Fakhri, M., Ahmadi, A., 2017. Recycling of RAP and steel slag aggregates into the warm mix asphalt: a performance evaluation. Constr. Build. Mater. 147, 630–638.

Finngean, G., O'Grady, A., Courtney, R., 2018. Plant assays and avoidance tests with collembola and earthworms demonstrate rehabilitation success in bauxite residue. Environ. Sci. Pollut. Res. 25 (3), 2157–2166.

Foulds, S.A., Macklin, M.G., Brewer, P.A., 2013. Agro-industrial alluvium in the swale catchment, northern England, as an event marker for the Anthropocene. The Holocene 23 (4), 587–602.

Funari, V., Gomes, H.I., Cappelletti, M., Fedi, S., Dinelli, E., Rogerson, M., Mayes, W.M., Rovere, M., 2019. Optimization routes for the bioleaching of MSWI fly and bottom ashes using microorganisms collected from a natural system. Waste Biomass Valorization 10, 1–10.

Galán-Arboledas, R.J., Álvarez de Diego, J., Dondi, M., Bueno, S., 2017. Energy, environmental and technical assessment for the incorporation of EAF stainless steel slag in ceramic building materials. J. Clean. Prod. 142, 1778–1788.

Galvão, J.L.B., Andrade, H.D., Brigolini, G.J., Peixoto, R.A.F., Mendes, J.C., 2018. Reuse of iron ore tailings from tailings dams as pigment for sustainable paints. J. Clean. Prod. 200, 412–422.

Gelencsér, A., Kováts, N., Turóczi, B., Rostási, Á., Hoffer, A., Imre, K., Nyirő-Kósa, I., Csákberényi-Malasics, D., Tóth, A., Czitrovszky, A., Nagy, A., 2011. The red mud accident in Ajka (Hungary): characterization and potential health effects of fugitive dust. Environ. Sci. Technol. 45 (4), 1608–1615.

Gomes, H.I., Funari, V., Mayes, W.M., Rogerson, M., Prior, T.J., 2018a. Recovery of Al, Cr and V from steel slag by bioleaching: batch and column experiments. J. Environ. Manag. 222, 30–36.

Gomes, H.I., Jones, A., Rogerson, M., Greenway, G.M., Lisbona, D.F., Burke, I.T., Mayes, W.M., 2017. Removal and recovery of vanadium from alkaline steel slag leachates with anion exchange resins. J. Environ. Manag. 187, 384–392.

Gomes, H.I., Jones, A., Rogerson, M., Burke, I.T., Mayes, W.M., 2016a. Vanadium removal and recovery from bauxite residue leachates by ion exchange. Environ. Sci. Pollut. Res. 23 (22), 23034–23042.

Gomes, H.I., Mayes, W.M., Baxter, H.A., Jarvis, A.P., Burke, I.T., Stewart, D.I., Rogerson, M., 2018b. Options for managing alkaline steel slag leachate: a life cycle assessment. J. Clean. Prod. 202, 401–412.

Gomes, H.I., Mayes, W.M., Rogerson, M., Stewart, D.I., Burke, I.T., 2016b. Alkaline residues and the environment: a review of impacts, management practices and opportunities. J. Clean. Prod. 112, 3571–3582.

Gomes, H.I., Rogerson, M., Courtney, R., Mayes, W.M., 2019. Integrating remediation and resource recovery of industrial alkaline wastes: case studies of steel and alumina industry residues. In: Resource Recovery From Wastes. The Royal Society of Chemistry, pp. 168–191.

Goss, S., Kane, G., Street, G., 2006. The Eco-Park: Green Nirvana or White Elephant? Clean Environment Management Centre, University of Teesside.

Heal, K.V., Salt, C.A., 1999. Treatment of acidic metal-rich drainage from reclaimed ironstone mine spoil. Water Sci. Technol. 39 (12), 141–148.

Hobson, A.J., Stewart, D.I., Bray, A.W., Mortimer, R.J., Mayes, W.M., Rogerson, M., Burke, I.T., 2017. Mechanism of vanadium leaching during surface weathering of basic oxygen furnace steel slag blocks: a microfocus X-ray absorption spectroscopy and electron microscopy study. Environ. Sci. Technol. 51 (14), 7823–7830.
Howard, A.J., Kincey, M., Carey, C., 2015. Preserving the legacy of historic metal-mining industries in light of the water framework directive and future environmental change in mainland Britain: challenges for the heritage community. Historic Environ. Policy Pract. 6 (1), 3–15.
Hudson-Edwards, K., 2016. Tackling mine wastes. Science 352 (6283), 288–290.
IPCC, 2018. Global warming of 1.5 °C. In: An IPCC Special Report on the Impacts of Global Warming of 1.5°C above Pre-Industrial Levels and Related Global Greenhouse Gas Emission Pathways, in the Context of Strengthening the Global Response to the Threat of Climate Change, Sustainable Development, and Efforts to Eradicate Poverty. Intergovernmental Panel on Climate Change, Incheon.
Jarvis, A.P., Mayes, W.M., 2012. Prioritisation of abandoned non-coal mine impacts on the environment. SC030136/R2. The national picture. In: Environment Agency Report SC030136. vol. 14.
Jarvis, A.P., Davis, J.E., Orme, P.H., Potter, H.A., Gandy, C.J., 2018. Predicting the benefits of mine water treatment under varying hydrological conditions using a synoptic mass balance approach. Environ. Sci. Technol. 53 (2), 702–709.
Johnston, D., 2004. A metal mines strategy for Wales. In: Proc IMWA Symp., Mine Water.
Kim, E., Spooren, J., Broos, K., Nielsen, P., Horckmans, L., Geurts, R., Vrancken, K.C., Quaghebeur, M., 2016. Valorization of stainless steel slag by selective chromium recovery and subsequent carbonation of the matrix material. J. Clean. Prod. 117, 221–228.
Leblanc, M., Morales, J.A., Borrego, J., Elbaz-Poulichet, F., 2000. 4,500-Year-old mining pollution in southwestern Spain: long-term implications for modern mining pollution. Econ. Geol. 95 (3), 655–662.
Lu, P., Zhou, L., Cheng, S., Zhu, X., Yuan, T., Chen, D., Feng, Q., 2019. Main challenges of closed/abandoned coal mine resource utilization in China. Energ. Source A 43, 1–9.
MacArthur, E., 2013. Towards the circular economy. J. Ind. Ecol. 2, 23–44.
Mathieux, F., Ardente, F., Bobba, S., Nuss, P., Blengini, G.A., Dias, P.A., Blagoeva, D., de Matos, C.T., Wittmer, D., Pavel, C. and Hamor, T., 2017. Critical Raw Materials and the Circular Economy. Publ. Off. Eur. Union. https://doi. org/10.2760/378123.
Marglin, S.A., 2008. The Dismal Science: How Thinking Like an Economist Undermines Community. Harvard University Press, Cambridge, MA.
Mayes, W.M., Batty, L.C., Younger, P.L., Jarvis, A.P., Kõiv, M., Vohla, C., Mander, U., 2009a. Wetland treatment at extremes of pH: a review. Sci. Total Environ. 407 (13), 3944–3957.
Mayes, W.M., Jarvis, A.P., Burke, I.T., Walton, M., Feigl, V., Klebercz, O., Gruiz, K., 2011. Dispersal and attenuation of trace contaminants downstream of the Ajka bauxite residue (red mud) depository failure, Hungary. Environ. Sci. Technol. 45 (12), 5147–5155.
Mayes, W.M., Potter, H.A., Jarvis, A.P., 2009b. Novel approach to zinc removal from circum-neutral mine waters using pelletised recovered hydrous ferric oxide. J. Hazard. Mater. 162 (1), 512–520.
Mayes, W.M., Potter, H.A.B., Jarvis, A.P., 2010. Inventory of aquatic contaminant flux arising from historical metal mining in England and Wales. Sci. Total Environ. 408 (17), 3576–3583.
Mayes, W.M., Riley, A.L., Gomes, H.I., Brabham, P., Hamlyn, J., Pullin, H., Renforth, P., 2018. Atmospheric CO_2 sequestration in iron and steel slag: Consett, county Durham, United Kingdom. Environ. Sci. Technol. 52 (14), 7892–7900.
Miccoli, S., Finucci, F., Murro, R., 2014. Criteria and procedures for regional environmental regeneration: a European strategic project. In: Applied Mechanics and Materials. vol. 675. Trans Tech Publications, pp. 401–405.
Milbrandt, A.R., Heimiller, D.M., Perry, A.D., Field, C.B., 2014. Renewable energy potential on marginal lands in the United States. Renew. Sust. Energ. Rev. 29, 473–481.
Millar, N., McLaughlin, E., Börger, T., 2019. The circular economy: swings and roundabouts? Ecol. Econ. 158, 11–19.
Mirazimi, S.M., Abbasalipour, Z., Rashchi, F., 2015. Vanadium removal from LD converter slag using bacteria and fungi. J. Environ. Manag. 153, 144–151.
Misrach, R., Orff, K., 2012. Petrochemical America. Aperture.

Muravyov, M., 2019. Bioprocessing of mine waste: effects of process conditions. Chem. Pap. 73, 3075–3083.
Muravyov, M.I., Fomchenko, N.V., 2018. Biohydrometallurgical treatment of old flotation tailings of sulfide ores containing non-nonferrous metals and gold. Miner. Eng. 122, 267–276.
Natural England, 1991. Box Mine SSSI Designation. Available at https://designatedsites.naturalengland.org.uk/PDFsForWeb/Citation/1005600.pdf. (last accessed 24.07.19).
Neculita, C.M., Zagury, G.J., Bussière, B., 2007. Passive treatment of acid mine drainage in bioreactors using sulfate-reducing bacteria. J. Environ. Qual. 36 (1), 1–16.
Nikolic, I., Drincic, A., Djurovic, D., Karanovic, L., Radmilovic, V.V., Radmilovic, V.R., 2016. Kinetics of electric arc furnace slag leaching in alkaline solutions. Constr. Build. Mater. 108, 1–9.
Ning, P., Lin, X., Wang, X., Cao, H., 2016. High-efficient extraction of vanadium and its application in the utilization of the chromium-bearing vanadium slag. Chem. Eng. J. 301, 132–138.
Nordstrom, D.K., Alpers, C.N., Ptacek, C.J., Blowes, D.W., 2000. Negative pH and extremely acidic mine waters from Iron Mountain, California. Environ. Sci. Technol. 34 (2), 254–258.
Norris, P.R., Burton, N.P., Foulis, N.A., 2000. Acidophiles in bioreactor mineral processing. Extremophiles 4 (2), 71–76.
Nriagu, J.O., Pacyna, J.M., 1988. Quantitative assessment of worldwide contamination of air, water and soils by trace metals. Nature 333 (6169), 134–139.
Oberle, B., Bringezu, S., Hatfield-Dodds, S., Hellweg, S., Schandl, H., Clement, J., Cabernard, L., Che, N., Chen, D., Droz-Georget, H., Ekins, P., 2019. Global Resources Outlook 2019: Natural Resources for the Future We Want. United Nations Environment Programme.
Pae, T., Luud, A., Sepp, M., 2005. Artificial mountains in north-East Estonia: monumental dumps of ash and semi-coke. Oil Shale 22 (3), 333.
Palumbo-Roe, B., Klinck, B., Banks, V., Quigley, S., 2009. Prediction of the long-term performance of abandoned lead zinc mine tailings in a Welsh catchment. J. Geochem. Explor. 100 (2–3), 169–181.
Pietsch, W., 1998. Colonization and development of vegetation in mining lakes of the Lusatian lignite area depending on water genesis. In: Acidic Mining Lakes. Springer, Berlin, Heidelberg, pp. 169–193.
Pullin, H., Bray, A.W., Burke, I.T., Muir, D., Sapsford, D., Mayes, W.M., Renforth, P., 2019. The atmospheric carbon capture performance of legacy iron and steel waste. Environ. Sci. Technol. 53, 9502–9511.
Purnell, P., Velenturf, A.P.M., Marshall, R., 2019. New governance for circular economy: policy, regulation and market contexts for resource recovery from waste. Resource Recovery from Wastes 63, 395.
Raper, E., Davies, S., Perkins, B., Lamb, H., Hermanson, M., Soares, A., Stephenson, T., 2015. Ecological conditions of ponds situated on blast furnace slag deposits located in south Gare site of special scientific interest (SSSI), Teesside, UK. Environ. Geochem. Health 37 (3), 545–556.
Raworth, K., 2017. A doughnut for the Anthropocene: humanity's compass in the 21st century. Lancet Planet. Health 1 (2), e48–e49.
Renforth, P., 2019. The negative emission potential of alkaline materials. Nat. Commun. *10* (1), 1401.
Riley, A.L., MacDonald, J.M., Burke, I.T., Renforth, P., Jarvis, A.P., Hudson-Edwards, K.A., McKie, J., Mayes, W.M., 2020. Legacy iron and steel wastes in the UK: extent, resource potential, and management futures. J. Geochem. Explor. 219, 106630.
Rivers Pollution Commission, 1874. The best means of preventing the pollution of rivers. In: Fifth Report of the Commissioners for 1868. vol. 1. HMSO, London.
Romero-Garcia, A., Iglesias-Gonzalez, N., Romero, R., Lorenzo-Tallafigo, J., Mazuelos, A., Carranza, F., 2019. Valorisation of a flotation tailing by bioleaching and brine leaching, fostering environmental protection and sustainable development. J. Clean. Prod. 233, 573–581.
Salah, Z.B., Charles, C.J., Humphreys, P.N., Laws, A.P., Rout, S.P., 2019. Genomic insights into a novel, alkalitolerant nitrogen fixing bacteria, Azonexus sp. strain ZS02. J. Genom. 7, 1.
Salem, J., Amonkar, Y., Maennling, N., Lall, U., Bonnafous, L., Thakkar, K., 2018. An analysis of Peru: is water driving mining conflicts. Res. Policy. https://doi.org/10.1016/j.resourpol.2018.09.010.
Sapsford, D.J., Crane, R.A., Sinnett, D., 2019. An exploration of key concepts in application of in situ processes for recovery of resources from high-volume industrial and mine wastes. In: Resource Recovery From Wastes: Towards a Circular Economy. Royal Society of Chemistry, pp. 141–167.

Sarfo, P., Das, A., Wyss, G., Young, C., 2017. Recovery of metal values from copper slag and reuse of residual secondary slag. Waste Manag. 70, 272–281.
Schröter, M., van der Zanden, E.H., van Oudenhoven, A.P., Remme, R.P., Serna-Chavez, H.M., De Groot, R.S., Opdam, P., 2014. Ecosystem services as a contested concept: a synthesis of critique and counter-arguments. Conserv. Lett. 7 (6), 514–523.
Scott, C., Mayes, W., Redman, V., 2014. Mynydd Parys & Afon Goch. Environmental Resistance Press, ISBN: 9780992983215.
Sethurajan, M., Van Hullebusch, E.D., Nancharaiah, Y.V., 2018. Biotechnology in the management and resource recovery from metal bearing solid wastes: recent advances. J. Environ. Manag. 211, 138–153.
Song, J., Choi, Y., Yoon, S.H., 2015. Analysis of photovoltaic potential at abandoned mine promotion districts in Korea. Geosyst. Eng. 18 (3), 168–172.
Sonter, L.J., Herrera, D., Barrett, D.J., Galford, G.L., Moran, C.J., Soares-Filho, B.S., 2017. Mining drives extensive deforestation in the Brazilian Amazon. Nat. Commun. 8 (1), 1013.
Tepanosyan, G., Sahakyan, L., Belyaeva, O., Asmaryan, S., Saghatelyan, A., 2018. Continuous impact of mining activities on soil heavy metals levels and human health. Sci. Total Environ. 639, 900–909.
Tiago, I., Chung, A.P., Veríssimo, A., 2004. Bacterial diversity in a nonsaline alkaline environment: heterotrophic aerobic populations. Appl. Environ. Microbiol. 70 (12), 7378–7387.
United Kingdom National Ecosystem Assessment, 2005. UKNEA Conceptual Framework. Available at http://uknea.unep-wcmc.org/About/ConceptualFramework/tabid/61/Default.aspx. (last accessed 02.10.20).
van Haren, J., Dontsova, K., Barron-Gafford, G.A., Troch, P.A., Chorover, J., Delong, S.B., Breshears, D.-D., Huxman, T.E., Pelletier, J.D., Saleska, S.R., Zeng, X., 2017. CO_2 diffusion into pore spaces limits weathering rate of an experimental basalt landscape. Geology 45 (3), 203–206.
Velenturf, A., Purnell, P., 2017. Resource recovery from waste: restoring the balance between resource scarcity and waste overload. Sustain. For. 9 (9), 1603.
Velenturf, A.P., Archer, S.A., Gomes, H.I., Christgen, B., Lag-Brotons, A.J., Purnell, P., 2019. Circular economy and the matter of integrated resources. Sci. Total Environ. https://doi.org/10.1016/j.scitotenv.2019.06.449.
Verhoeven, R., Willems, E., Harcouët-Menou, V., De Boever, E., Hiddes, L., Op't Veld, P., Demollin, E., 2014. Minewater 2.0 project in Heerlen the Netherlands: transformation of a geothermal mine water pilot project into a full scale hybrid sustainable energy infrastructure for heating and cooling. Energy Procedia 46, 58–67.
Viers, J., Dupré, B., Gaillardet, J., 2009. Chemical composition of suspended sediments in world rivers: new insights from a new database. Sci. Total Environ. 407 (2), 853–868.
Wallis de Vries, M.F., Poschlod, P., Willems, J.H., 2002. Challenges for the Conservation of Calcareous Grasslands in Northwestern Europe: Integrating the Requirements of Flora and Fauna.
Wilson, S.A., Dipple, G.M., Power, I.M., Barker, S.L., Fallon, S.J., Southam, G., 2011. Subarctic weathering of mineral wastes provides a sink for atmospheric CO_2. Environ. Sci. Technol. 45 (18), 7727–7736.
Younger, P.L., Banwart, S.A., Hedin, R.S., 2002. Mine Water: Hydrology, Pollution, Remediation. vol. 5 Springer Science & Business Media.
Zijlstra, J.J.P., Dessì, R., Peretti, R., Zucca, A., 2010. Treatment of percolate from metal sulfide mine tailings with a permeable reactive barrier of transformed red mud. Water Environ. Res. 82 (4), 319–327.

CHAPTER 23

"Closing two loops"—The importance of energy recovery in the "closing the loop" approach

Tihomir Tomić and Daniel Rolph Schneider
Department of Energy, Power Engineering and Environment, Faculty of Mechanical Engineering and Naval Architecture, University of Zagreb, Zagreb, Croatia

1. Introduction

Over the past century, Europe has developed an economy capable of generating enormous wealth for its population. Since 1995, EU countries have increased industrial production by 31% (European Commission. Eurostat, 2018). This growth comes with an increase in raw materials consumption. Over time, Europe's raw material needs outgrew its own production from natural sites and the EU began to rely on imported raw materials and energy. Today, up to 63% of EU raw materials needs are covered through import (Fig. 1).

This has positioned the EU as the world's second-largest importer with an import value of 1711 billion Euros in 2016 (European Commission, 2017) and with imports much higher than exports (Fig. 2). Excessive dependence on raw materials imports calls into question security of supply and becomes a significant, not only economic, but also political and security issue.

Even though dependences on the imports of various raw materials represent problems that need to be approached systematically from many different angles, in the end, they need to be addressed as one problem, and the overall efficiency of measures that are implemented to solve this problem needs to be analyzed. As it will be shown later in this chapter, this is not the case when the EU legislation framework is looked upon, neither in the previous analyses of this problem, where energy and material dependences are addressed separately. Furthermore, in this chapter, the importance of waste and waste recovery in solving this problem is identified, as well as an approach that gives a holistic view of this problem, which is further elaborated in the section on methodology. The presented approach is used for calculation and discussed in the results section, on the basis of which, the conclusions presented in the final section are drawn.

Circular Economy and Sustainability, Volume 1
https://doi.org/10.1016/B978-0-12-819817-9.00021-1

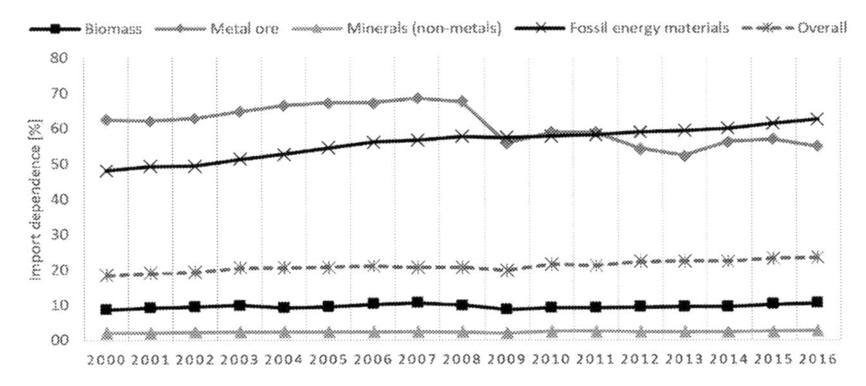

Fig. 1 Import dependency by main material category, EU-28 (European Commission, 2018a). *(Credit: Author.)*

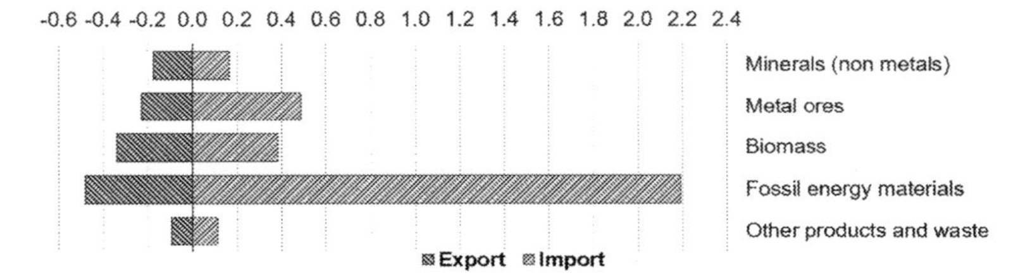

Fig. 2 Import/export per material category, EU-28, 2016 (tons per capita) (European Commission, 2018b). *(Credit: Author.)*

1.1 Legislative framework

For healthy industrial development, it is necessary to enable industry access to needed materials on time and at a reasonable price. Europe has recognized these problems and is trying to address them through a range of legislative frameworks, plans, and strategies—on the material side through the Raw Materials Initiative, European Innovation Partnership on Raw Materials, Resource efficient Europe, Europe 2020, and Roadmap to a Resource Efficient Europe, and on the energy side through the Green Book Towards a European Strategy for the Security of Energy Supply, 2020 Climate and Energy Package, 2030 Climate and Energy Framework, Roadmap for Moving to a Low-carbon Economy in 2050, Energy Roadmap 2050, Roadmap to a Single European Transport Area, etc.

To address the material dependency problem, within Roadmap to a Resource Efficient Europe, as part of the Europe 2020 strategy, the circular economy principle has been proposed as the best concept to guide economic transformation. It sets goals and means to transform the current economy, which is based on intensive use of resources, into a new model based on efficient resource use where waste is reintroduced back into the production processes. As this strategy emphasizes the importance of material recovery of waste, its objectives are complementary with European waste management legislation.

The new EU Action Plan for the Circular Economy, as a part of the wider EU Circular Economy Package, builds on the Waste Disposal Directive and the Waste Framework Directive and incorporates waste-related legislative changes with a goal to encourage a European transition to a circular economy. The circular economy introduces the concept of "closing the loop" of the life cycle of products and introduces measures which cover the entire life cycle of raw materials/materials/products, from extraction, production, and use through disposal and waste management to the market of secondary raw materials, recovery, and reuse. Closing of the loop between the end of the product life cycle and its production enables the circulation of resources, materials, and products within the EU economy and thus keeps its energy, material, and economic value within the economy for as long as possible. These measures reduce the utilization of resources. In particular, the revised waste proposals include higher targets for packaging recycling as well as municipal waste disposal reduction, which would benefit both the economy and the environment. How much this is important in the EU context is shown through waste quantities. In 2014, overall EU waste production was around 2500 million tons, of which around 1325 million tons were not reused or recovered and thus a large amount of resources were lost for EU economy (European Parliament, 2018).

1.2 The connection between problems

As can be seen, the approach to solving the single problem of sustainability of the EU economy is, in the legislative sense, divided into two separate approaches (one to solve the problem of material dependence, and the other to solve the energy dependence problem), while waste recovery is the only identified way to address both problems.

Thus, the Mineral Resources Initiative stresses the importance of recycling, while the European Innovation Partnership on Raw Materials seeks to promote Europe as a world leader in recycling. Likewise, Resource Efficient Europe emphasizes recycling as an important factor in relieving the pressure on the raw material supply system as well as in reducing energy consumption and greenhouse gas emissions throughout the production chain. Roadmap to a Resource Efficient Europe also seeks to increase resource efficiency and break the link between increasing energy and material consumption and the environmental impact of economic growth.

On the other hand, the transformation of the energy sector based on decarbonization is outlined through the 2020 Climate and Energy Package, 2030 Climate and Energy Framework, Roadmap for Moving to a Low-carbon Economy in 2050, Energy Roadmap 2050, and the White Paper on Transport, but also through the Climate-neutral Europe by 2050 strategy, which covers almost all EU policies and is in line with the aim of the Paris Agreement. In this respect, not only do the targets cover the waste management sector, but also the use of bio-waste and biogas as a means of reducing transport emissions and emphasizes the importance of switching from fossil fuels. In this context, energy recovery of waste found its place in the development of the EU energy sector. This is also emphasized through the Heat Roadmap Europe (Connolly et al., 2012, 2013) where waste is classified as the primary source of heat in centralized heat systems.

1.3 Approach to the problem

There are a number of papers that put emphasis on “closing the loop” on the material side and which analyze the influence of material recovery (Niero et al., 2016; Niero and Olsen, 2016; Huysman et al., 2017), including a developed circular economy model (George et al., 2015), which is in accordance with EU legislation. While material recovery of waste reduces the use of primary raw materials, energy produced from waste replaces the energy carriers from primary energy sources, leading to a partial transition of energy consumption to energy-from-waste (EfW). Produced energy can be used to meet the energy needs of the waste management system itself, which is a step further in the concept of closing the loop. Due to the importance of energy recovery of waste, its energy potential has been analyzed in a number of papers (Rajaeifar et al., 2017; Fernández-Nava et al., 2014; Bovea et al., 2010; Bueno et al., 2015; Giugliano et al., 2011; Cherubini et al., 2008; Ouda et al., 2016; Dzene et al., 2016), but in the majority of these analyses, only energy the side is considered.

While a number of papers indirectly compared energy and material recovery systems, especially from the life cycle assessment (LCA) point of view, a very few analyzed the interaction of material and energy recovery. The problem with the analysis of interaction between those systems, and of making direct comparisons, is in the difference of products, i.e., materials and energy. LCA analysis gained on importance due to position of the EC, which incorporated into the Waste Framework Directive that potential deviations from the waste hierarchy must be justified through considerations encompassing whole life cycle (where the LCA is a standardized scientific method for assessing impact at the life cycle level (ISO, 2006a, 2006b)) and declared the LCA as “the best framework for assessing the potential environmental impacts of products” (Commission of the European Communities, 2003). Despite this, in many cases, simpler and more practical forms of analysis should be used (Petrov, 2007). A simplified LCA approach is used in many sectors (construction (Zabalza Bribián et al., 2009), car manufacturing (Danilecki et al., 2017), solar systems (Beccali et al., 2016)) and is considered an important tool for overcoming LCA complexity concerns and difficulties in understanding, and for presentation of the results. This is especially important to ensure understanding from broader groups of people, for decision-makers that require rapid product comparisons or identification of possible areas for performance improvement, as well as in the planning sector, where it is used to compare scenarios.

All of these approaches to LCA include energy indicators. Energy indicators are used to simplify analyses and have been used as an indicator of environmental impact in a wide range of activities such as energy production (Huijbregts et al., 2010), material production (Arvidsson et al., 2012), transportation, and waste treatment (Scipioni et al., 2013). Furthermore, their use in waste management is important, as direct LCA comparison, with respect to material and energy recovery, should be avoided in favor of other LCA-based indicators such as cumulative energy demand (CED) (Bueno et al., 2015), which is also used for life cycle impact assessments (LCIA). This conclusion is based on the properties of the CED method, which integrates energy and material flows, as well as related environmental impacts, into a single value, thus enabling simultaneous assessment and comparison of life cycles of material and energy flows that are not directly comparable or interchangeable.

From the beginning, primary energy (PE) consumption has been used in LCA studies as one of the key indicators (Frischknecht et al., 2015). The PE consumption indicator, also called the cumulative energy demand (CED) or life cycle embodied energy (LEE), is also part of the engineering guidelines (VDI (VDI, 1997) and SIA (SIA, 2010)) and construction standards (EN 15804, 15978 and 15643-2 (CEN, 2011, 2012a, 2012b)). The International Organization for Standardization classifies fossil energy depletion as an impact category in ISO/TR 14047 (ISO, 2012), although an energy indicator is not required according to its LCA standards. The CED is used to represent the total PE consumption in the considered production chain, and takes into account all background processes that indirectly participate in the production, thus being an energy indicator for evaluating the results of LCA (Rohrlich et al., 2000) that is quantitative and includes all energy flows that affect the life cycle (Huijbregts et al., 2006). CED correlates with more complex LCIA methods that give single score results (such as eco-indicator, ecological footprint, climate footprint, cumulative exergy extraction in the natural environment, and ecological scarcity) (Arvidsson et al., 2012) and is intermediary for environmental impact assessment (Arvidsson et al., 2012; Mert et al., 2017), making it an appropriate decision-making tool (Rohrlich et al., 2000).

CED is one way to calculate the embodied energy (EE) of products. The EE represents the sum of the total PE consumed throughout the life cycle of the analyzed product (Zero Waste Europe, 2017) and shows its sustainability, through sustainability of its production, by reducing the energy and material consumption for its production to the PE equivalent. EE includes the consumed energy as well as the energy of the raw materials. As an indicator, EE is most commonly used for analyses and comparisons of sustainability in the construction sector (Dixit, 2017; Vukotic et al., 2010), but is also used to evaluate the sustainability of products (Penny et al., 2013) and energy sources (Agostinho and Siche, 2014). As it only accounts for energy flows, material flows that are taken directly from the environment are indirectly taken into account as energy that needs to be consumed for their use (e.g., for extraction, pumping, processing, conditioning, etc.).

Based on everything previously said, the hypothesis of this chapter can be defined as: an energy analysis approach, which seems to be the most appropriate, can be used to show the influence of energy and material recovery of waste on sustainability of waste management as well as a whole economy from two points of view—one which shows the influence on overall sustainability, and another that shows the influence on the closing the loop agenda. While material recovery is a recognized approach from the EU legislation and energy recovery is not the preferred solution, they should not be analyzed separately as they complement each other and form a synergy effect. So, using energy recovery can boost quantity of resources returned to the economy as well as boost sustainability of recycled materials. This way, by "closing two loops", one from the material and the second from the energy side, overall sustainability of the waste management/recovery system and the whole economy can be boosted even further.

2. Methodology

In order to elaborate on a given hypothesis, it was necessary to design scenarios on the basis of which this approach can be demonstrated. Due to the importance of cities in the European context, the scenarios are based on a real city of a younger EU member state (with a low level of separate waste collection of 13.5% and whose municipal waste management system does not meet the objectives of European legislation), while the considered systems are modeled using an LCA-based modeling approach (through Ecoinvent (Ecoinvent Centre, 2016) and USLCI (NREL, 2016), literature data, and actual conditions). By choice of this case study, projection of the situation in accordance with European legislation, and analysis of multiple scenarios, a wide range of cities with more or less advanced municipal waste management/recovery systems is described.

Scenarios describe the current situation as well as four alternatives for the development of a waste management system. Fig. 3 shows the existing waste management system and all major waste streams, which track the flows within the waste treatment process (dashed lines), as well as energy flows that visualize energy consumption (solid lines). In addition to the internal flows, the connection of the city, through the waste collection system, as well as with the most significant material and energy markets, is visualized. Next to internal waste streams, energy flows, and product flows (material and energy), material consumption flows (consumption of natural resources, market materials, as well as services such as disposal of special waste categories) could not be graphically shown because of their sheer number, however, their impacts are included in the results in the form of the total PE consumption equivalent.

Fig. 3 illustrates the generally accepted principle of closing the loop on the material side, where the market for materials is linked back with the corresponding waste streams

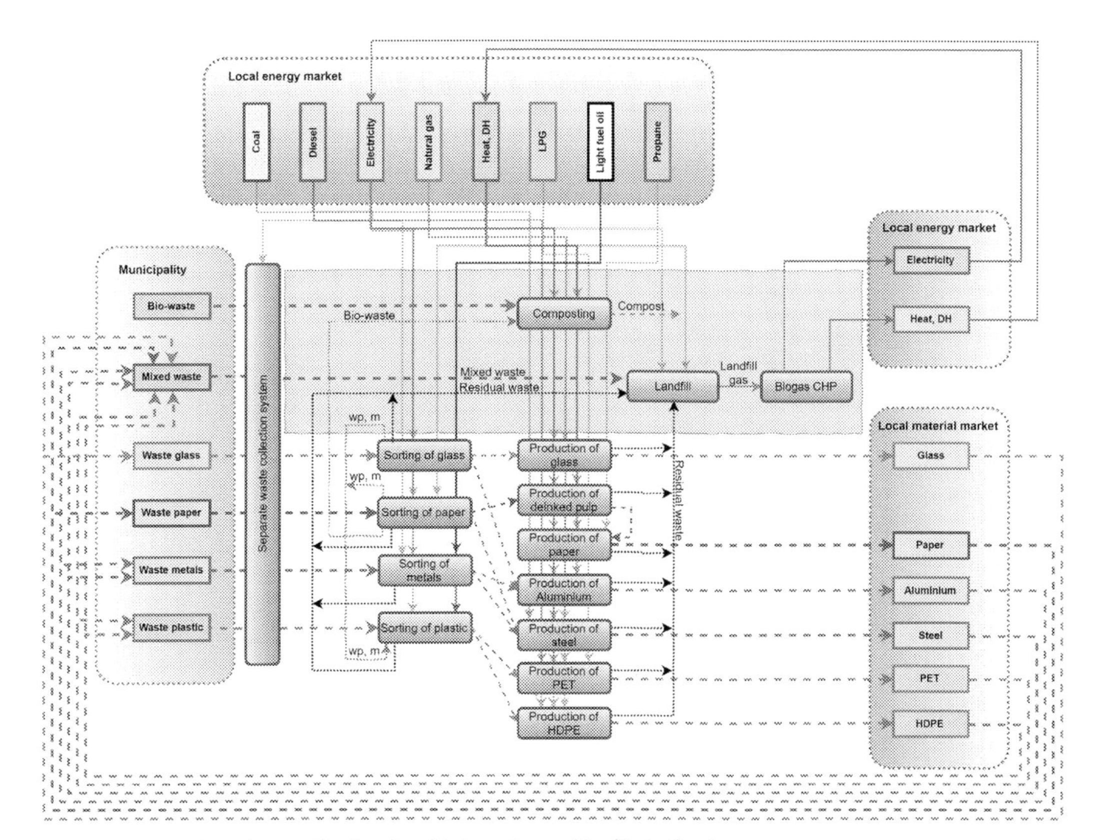

Fig. 3 Existing waste management system with visualized interactions. *(Credit: Author.)*

in the municipality (dashed outer lines), since, by closing the loop, the produced materials will reenter the waste management system (partly as separately collected, and partly as mixed waste). On the other hand, closing the loop can also be achieved from the energy side—the produced energy vectors are sold on energy markets where they reduce the consumption of primary energy carriers. Therefore, the energy market in which energy vectors are sold is linked to the energy supply market (solid outer lines).

This "closing two loops" effect is evaluated on the basis of considered systems by reducing all input and output material and energy flows to PE, via a CED indicator, and calculating the share of produced PE (in the form of material and energy products) compared to PE that entered the system in the form of waste materials, i.e., by calculating the primary energy return index (PERIndex) (Tomić and Schneider, 2017). On the other hand, by closing the loop on the energy side, energy recovery can (partially) fulfill the energy needs of the entire waste management/recovery chain, and thus increase the sustainability of the produced (recycled) materials, which can be shown through changes in the EE of recycled materials, i.e., by calculating the return of embodied energy (REE) factor (Tomić and Schneider, 2018).

Alternative scenarios differ from the existing, landfill-based, system by the technologies included in the energy recovery chain, and the part of the system that changes according to alternatives is highlighted in Fig. 3. At the same time, the material recovery chain is the same in all scenarios. Therefore, alternative systems/scenarios are defined by a different configuration of the (mainly) energy recovery chain (i.e., the highlighted part) as shown in Fig. 4, which is responsible for treatment of residual waste and bio-waste.

Analyzed scenarios consist of the same recycling technologies, with the exception of composting, which is in some scenarios replaced by anaerobic digestion (AD), which will prove to be a more sustainable option. Regarding waste separation, scenarios differ in the presence or lack of a mixed waste separation facility (MBT). Analyzed energy recovery technologies include landfill with landfill gas collection system (in the Existing system), incineration of mixed waste in a waste-to-energy (WtE) plant (Mass burn scenario), thermal treatment of residual waste in the cement plant (Material recovery scenario), refuse-derived fuel (RDF) from MBT plant incineration in the WtE plant (Energy recovery scenario), and the AD of bio-waste (Energy recovery and Material recovery with AD scenarios).

Input flows are based on characteristic values of the existing system and the projected situation in 2020 and 2030. The projections were made using the LCA-IWM model (Boer et al., 2005), which takes into account a wide range of socioeconomic indicators, while the boundary conditions are set by the EU legislation—prognosis represents legislatively defined situation in corresponding years. The quantities of waste collected and their shares are shown in Table 1.

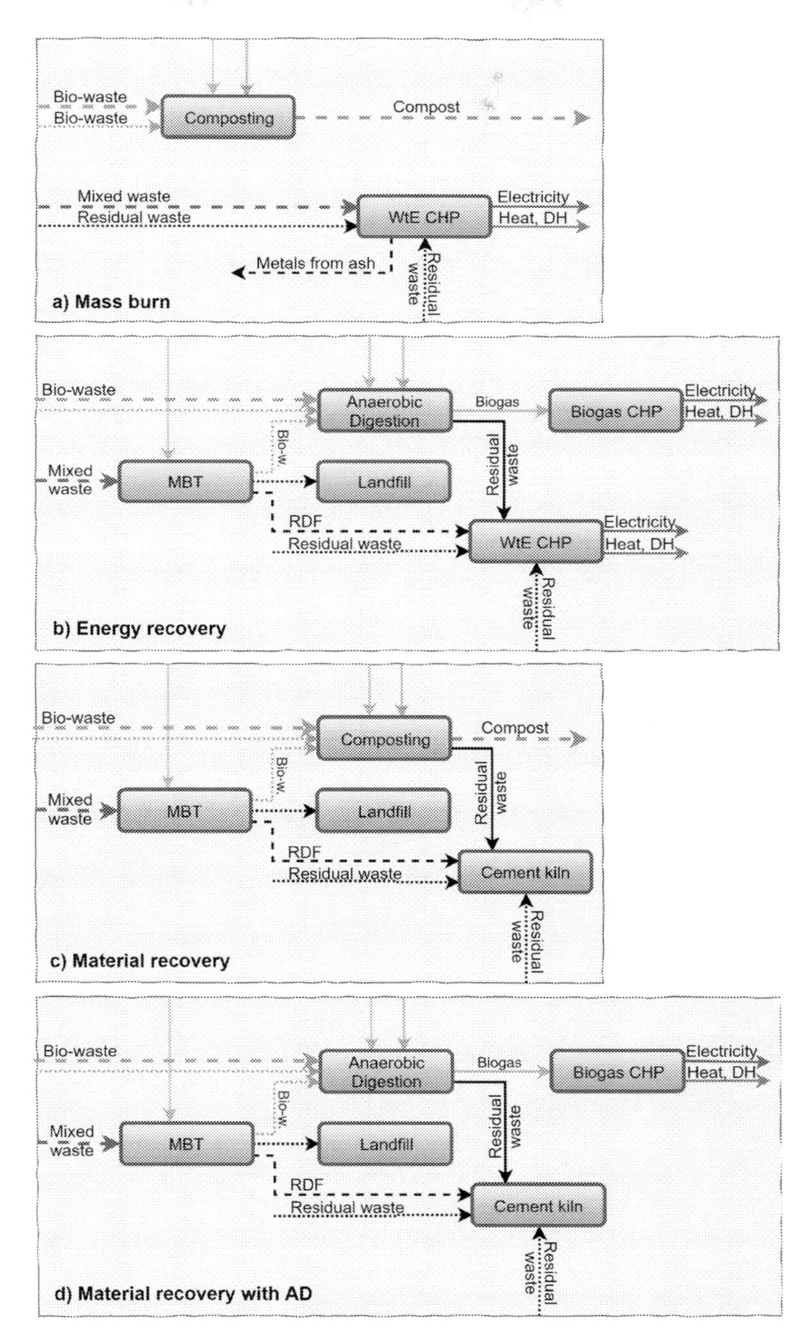

Fig. 4 Alternative systems/scenarios. *(Credit: Author.)*

Table 1 Collected waste.

		Paper	Glass	Fe	Al	PE	PET	Garden waste	Kitchen waste	Mixed waste
Today	Quantity (t)	1447	818	447	53	381	1.314	27.935	3.773	232.587
	Share (%)	0.54	0.30	0.17	0.02	0.14	0.49	10.39	1.40	86.54
2020	Quantity (t)	59.700	5500	1.520	180	8.565	29.535	28.630	78.570	108.300
	Share (%)	18.63	1.72	0.47	0.06	2.67	9.22	8.93	24.51	33.79
2030	Quantity (t)	82.200	9.300	2.146	254	14.837	51.163	28.808	104.792	123.600
	Share (%)	19.71	2.23	0.51	0.06	3.56	12	6.91	25.12	29.63

3. Results and discussion

From the input waste flows and the LCI data for the technologies in the analyzed chain, the energy consumption of each technology, and thus the entire system, can be calculated. The consumption of each scenario by energy vectors is shown in Table 2.

In order to allow comparison of energy consumption, consumption by energy vector has been reduced to the PE equivalents, by multiplying calculated energy consumptions with CED factors for the corresponding energy vectors (Fig. 5). From the results, it is evident that the majority of the systems' total energy needs are met through electricity and heat, thus justifying biogas transformation in cogeneration plants. With an increase in primary separation, the energy consumption of the system increases due to the higher energy consumption in the collection and separation, as well as material and energy recovery. In addition, the consumption is higher for systems with more separation and recovery technologies integrated.

The increase in energy consumption corresponds to an increase in material (Table 3) and energy recovery (Table 4). The production of materials from the recycling plants as well as the energy from the energy recovery plants was calculated from previously calculated material and energy flows and the corresponding LCI data for all technologies in the production chain.

While material production varies according to scenarios, paper production differs very little between scenarios due to inability of its secondary separation. Generated differences in Material recovery, Material recovery with AD, and Energy recovery scenarios are mainly the result of the integration of a MBT plant. As the Existing system and Mass burn scenarios do not have an integrated MBT plant, the amount of produced materials is the same with the exception of metals, as they are subsequently extracted from the ash produced in the mass burn scenario. The impact of MBT can also be seen in the Material recovery, Material recovery with AD, and Energy recovery scenarios, where the difference is noticeable only for materials that are not always separated (plastics in the Energy recovery scenario were used as part of the RDF) and for compost, as the bio-waste in the Energy recovery and Material recovery with AD scenarios is used in the AD plant.

Even though all scenarios have integrated material recovery and meet EU separation/recycling goals, there is always some waste that can be used in energy recovery. So, in the Existing system, most of the energy is produced through landfill gas collection, while in the following two scenarios it is done through the WtE plant. In the Energy recovery and Material recovery with AD scenarios, the energy production from biogas is the largest and, together with the WtE production (in the scenario Energy recovery), leads to the highest energy production. The biogas yield is higher only in the Existing system with the current volume of landfilling.

Presented data do not include data for the Material recovery scenario, because it does not have direct energy production, although residual waste is recovered via external

Table 2 Energy consumption.

	Existing today	Existing 2020	Existing 2030	Mass burn today	Mass burn 2020	Mass burn 2030	Mat. today	Mat. 2020	Mat. 2030	Energy today	Energy 2020	Energy 2030	Mat. AD today	Mat. AD 2020	Mat. AD 2030
Light fuel oil (kg)	68	231	326	68	231	326	381	427	486	381	427	485	381	427	486
Heavy fuel oil (t)	29	192	325	29	192	325	292	365	486	292	365	486	292	365	486
Diesel (t)	1200	4440	6938	1200	4440	6938	5707	7177	9561	1425	4456	6888	5565	7001	9040
Natural gas ('000 m3)	163	2064	3532	206	2094	3554	3074	3904	5241	1044	2641	4064	3074	3904	5241
Propane (GJ)	106	4346	5984	106	4346	5984	107	4347	5985	107	4347	5985	107	4347	5985
Coal (t)	6	20	28	29	36	40	32	36	41	32	36	41	32	36	41
Heat (TJ)	22	798	1099	50	814	1118	22	799	1100	135	930	1264	94	889	1212
Elect. (GWh)	14	121	171	10	120	170	34	135	184	16	125	175	37	138	188
LPG (t)	0	6	11	0	6	11	9	12	16	0	6	11	9	12	16

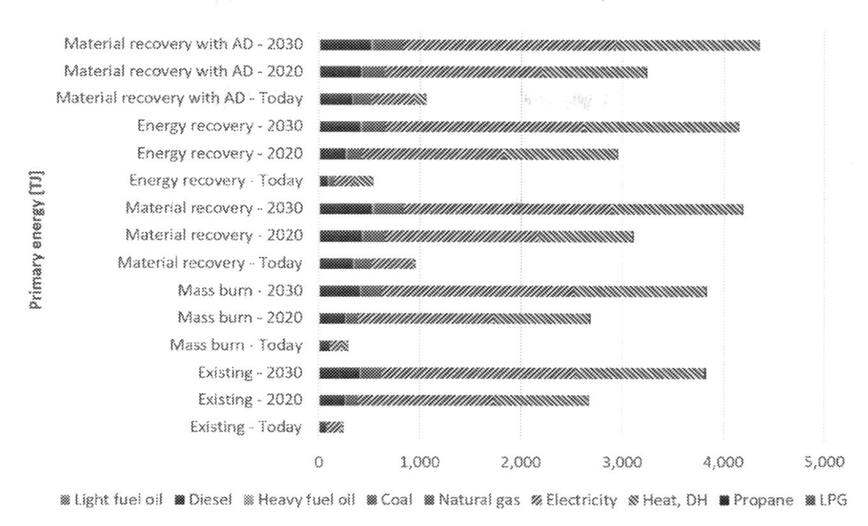

Fig. 5 Energy consumption—PE equivalent. *(Credit: Author.)*

technology—the cement kiln. This way, in the material recovery and material recovery with AD scenarios, primary fuel consumption is replaced by waste (using a method of equivalent delivered energy). Thus, 78,080/46,660/55,440 tons of waste from the MBT plant replaces 30,652/14,516/19,101 tons of coal in the considered timeframes, respectively, for today, 2020, and 2030.

3.1 Sustainability analysis—Primary energy recovery analysis

To calculate the recovery of PE, all input/output flows that intersect the system boundary are reduced to PE equivalents. The recycled materials were valued as substitutes for primary materials from the local market, i.e., their PE equivalent, and the same applies to energy vectors whose production was converted to PE on the basis of LCI data for local energy mix. As for the compost and compost-like materials from AD, since there is no alternative production of these products, their production was evaluated by their chemical composition (amount of nitrogen (N), phosphorus (P), and potassium (K) in the material) (ISWA, 2015), and considered as substitutes for fertilizers, i.e., their EE equivalent (Gilbert, 2009). In addition to material and energy flows, special waste categories that cross the boundary of analysis were reduced to the PE consumption of technology for their disposal/treatment. Summarizing the obtained results by technologies, the total PE recovery was calculated and the results are shown in Fig. 6.

As can be seen, there are significant differences in the values of the PE recovery indicators between the Existing system and alternative scenarios, between which the differences are relatively small. The material recovery with AD scenario recovers PE through material recovery and energy recovery of bio-waste, substitution of coal with RDF in the

Table 3 Material production.

	Paper (t)	Glass (t)	Steel (t)	Alu (t)	PE (t)	PET (t)	Compost (t)
Existing system—Today	1942	1211	406	56	254	811	17,122
Existing system—2020	80,001	8148	1415	204	5747	18,327	57,910
Existing system—2030	110,161	13,778	1997	296	9926	31,652	72,175
Mass burn—Today	1942	1211	5472	275	254	811	17,122
Mass burn—2020	80,001	8148	2603	358	5747	18,327	57,910
Mass burn—2030	110,161	13,778	2864	408	9926	31,652	72,175
Material recovery—Today	1978	12,375	2269	333	8469	27,007	52,841
Material recovery—2020	80,025	15,481	2580	379	10,858	34,624	65,491
Material recovery—2030	110,183	20,578	2943	440	14,689	46,840	81,894
Energy recovery—Today	1978	12,375	2269	333	258	824	0
Energy recovery—2020	80,025	15,481	2580	379	5750	18,335	0
Energy recovery—2030	110,183	20,578	2943	440	9928	31,660	0
Material recovery with AD—Today	1978	12,375	2269	333	8469	27,007	52,841
Material recovery with AD—2020	80,025	15,481	2580	379	10,858	34,624	65,491
Material recovery with AD—2030	110,183	20,578	2943	440	14,689	46,840	81,894

cement kiln, and also values digestate as a substitute for fertilizers, thus achieving higher PE recovery than other scenarios. The other three alternative scenarios (mass burn, material recovery, and energy recovery) show similar values of PE recovery, among which the energy recovery scenario shows the highest recovery.

By dividing the PE recovery factor by the overall EE of materials contained in municipal waste that enter the analyzed system, the annual PERIndex can be calculated (Fig. 7).

The PERIndex represents the share of energy contained in the collected waste (entering the analyzed system) that is recovered and returned to the economy through material and energy recovery, and represents a more understandable sustainability indicator for systems and products. This index also shows some other trends. While the PERIndex is increasing over the years, a faster increase can be observed in the first period, which

Table 4 Energy production.

Source	Product	Exist. today	Exist. 2020	Exist. 2030	Mass burn today	Mass burn 2020	Mass burn 2030	Energy today	Energy 2020	Energy 2030	Mat. AD today	Mat. AD 2020	Mat. AD 2030
Biogas	Electricity (GWh)	0	0	0	0	0	0	32.6	41.9	53.1	32.6	41.9	53.1
	Heat (TJ)	0	0	0	0	0	0	202.0	259.3	328.4	202.0	259.3	328.4
Landfill gas	Electricity (GWh)	37.8	15.8	21.6	0	0	0	0	0	0	0	0	0
	Heat (TJ)	234.1	97.5	135.2	0	0	0	0	0	0	0	0	0
WtE	Electricity (GWh)	0	0	0	119.1	71.8	81.9	111.2	67.8	77.1	0	0	0
	Heat (TJ)	0	0	0	1908.4	1150.2	1312.8	1780.0	998.1	1106.3	0	0	0

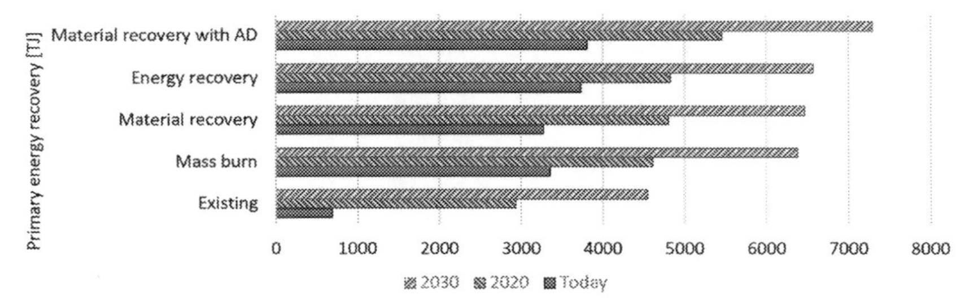

Fig. 6 Primary energy recovery. *(Credit: Author.)*

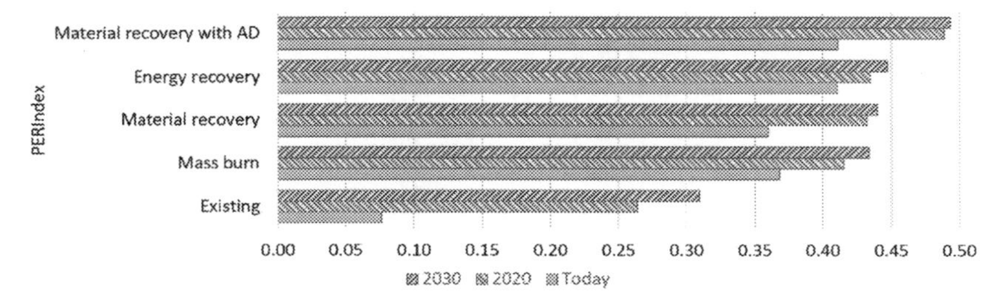

Fig. 7 PERIndex values. *(Credit: Author.)*

is even more pronounced in scenarios that put emphasis on material recovery and those without integrated secondary waste separation. This increase is the result of primary waste separation and is not a result of changes in the waste management system, as this increase is also present in the Existing system, which describes the situation without any changes in the system. It shows that if slow progress in primary waste separation can be expected, investment in energy recovery is a more sensible decision from the sustainability standpoint.

In later years, a convergence of the indices can be observed, with a diminishing increase as primary separation increases (from 2020 to 2030). This defines a certain maximum, after which further investment in primary waste separation is no longer profitable and more concrete changes to the municipal waste management system need to be made in order to further increase the sustainability of the system, such as the integration of new technologies, which would boost sustainability results. This can be seen in the example of the material recovery scenario, where exchange of composting with AD facility significantly boosts PERIndex.

PERIndex shows that energy recovery of waste boosts the overall sustainability of waste management/recovery systems by increasing share of recovered resources that are returned to the economy. Energy recovery builds upon results of material recovery and, in the same system configuration, boosts recovery share from 44%, and almost reaching 50% of recovered resources invested in production of materials, which, at the end of their useful lifetime, become a part of municipal solid waste.

3.2 Closing the loop analysis—Analysis of the reduction in embodied energy

The previously shown approach shows comparable results with the CED approach, which is widely used in other fields, and adds a twist that makes it more useful for conducting analysis in waste management. Nevertheless, it is a system-by-system approach that yields system-level results whose analysis and interpretation can lead to understanding the events within the system.

While PERIndex represents a good tool for sustainability analysis and comparison of the systems, it does not take into account connections between energy and material recovery and the mutual influence of one on the other. This can be done through the calculation of changes in the EE of the produced recycled materials. To do this, it is necessary to analyze the possible (partial) covering of energy needs of the system, by individual energy vectors, through EfW production, which can be done by comparing data on energy production and consumption by energy vectors (Tables 2 and 4).

While the EE of the primary materials (produced from primary raw materials and with existing energy mix) was taken from the CED data for the considered materials, the calculation of the EE of the recycled (secondary) materials was made on the basis of systems' models and calculated data on energy and material consumption and production (Tables 2–4).

Recycled materials have a lower value of EE compared to primary materials due to the use of already formed materials (waste materials) as input raw materials, which eliminates the extraction phase of natural raw materials, shortens the production chain, and reduces energy consumption as well as waste generation during the processing/production process. A comparison of the EE of primary and recycled materials with the use of existing energy mix (for the current situation) is presented in the first two columns in Table 5.

This difference is only generated through the material recovery of waste, while the waste management, separation, and material recovery chain was powered by energy vectors with the usual local energy mix. When the production of EfW is integrated into the system, it (partially) covers the energy needs of the analyzed system (in this case for electricity and heat), which changes the energy intensity of the system. These changes must be taken into account.

The potential for additional reductions in EE was calculated using EfW in the production of recycled materials. This is achieved by reducing the consumption of PE in the material waste recovery chain due to use of energy produced from waste, i.e., by subtracting produced energy through energy recovery technologies (Table 4) from the total system consumption (Table 2), by individual energy vectors. In this way, the EE of

Table 5 Embodied energy—example of an existing situation.

	Embodied (primary) energy (MJ/kg)		
Material	**Primary material**	**Recycled materials with an existing energy mix**	**Recycled materials with EfW**
Paper	46.66	30.80	5.91
Glass	18.64	10.93	9.15
Fe	24.44	15.91	9.80
Al	174.74	15.64	9.45
PE	77.28	14.63	6.87
PET	79.09	24.98	11.93

Table 6 REE index values.

	Timeframe	Reduction in EE of recycled materials due to material recovery (%)	Additional reduction in EE of recycled materials due to energy recovery (%)	The total reduction in EE of recycled materials—REE index (%)
Existing	Today	73.18	52.95	87.38
	2020	73.46	30.74	81.62
	2030	73.55	19.16	78.61
Mass burn	Today	73.23	53.17	87.46
	2020	73.46	37.10	83.31
	2030	73.55	32.81	82.23
Energy recovery	Today	62.59	67.12	87.70
	2020	69.08	58.58	87.20
	2030	69.27	56.82	86.73
Material recovery	Today	62.74	0.00	62.74
	2020	69.16	0.00	69.16
	2030	69.32	0.00	69.32
Material recovery with AD	Today	62.74	56.56	83.82
	2020	69.16	41.40	81.93
	2030	69.32	16.37	74.35

recycled materials is further reduced, as shown in the example of the existing situation and the data given in the third column of Table 5. Based on calculated EE, REE index is calculated, which shows an overall decrease in EE of produced material in comparison with primary material in a situation where the usual energy mix is used for powering the analyzed system or when EfW is used for (partially) covering its energy needs (Table 6).

Results of REE index show that the use of secondary raw materials (separated municipal waste) in production reduces the average energy of the produced (recycled) materials by 73%–74% in the Existing system and Mass burn scenarios and 63%–69% in the Energy recovery and Material recovery scenarios, depending on the timeframe, compared to the production of primary materials. These reductions confirm the thesis that material recovery of waste and waste materials can contribute to increasing the sustainability of the produced secondary (recycled) materials and is in line with EU legislation. At the same time, this is only one aspect, because integration of EfW, in the same scenarios/calculations, leads to an additional reduction of the embodied energy by 19%–53% in the Existing system, 33%–53% in the Mass burn scenario, 57%–67% in the case of the Energy recovery scenario, and 16%–57% in the Material recovery with AD scenario, depending on the timeframe. Greater reductions can be observed in the timeframe "Today", which is the result of full coverage of electrical and thermal demands due to the relatively low energy consumption that is a result of a relatively small production of secondary materials.

REE index results show that energy recovery of waste can even boost the sustainability of material recovery by making its products more resource-efficient. These results can help in understanding energy recovery and show its opponents that energy recovery is not competing technology to material recovery and that it has its role even in the material-oriented understanding of sustainability.

4. Conclusion

Although there is one distinctive problem of sustainability of economic development that is triggered by raw material scarcity of the European Union, and resulting dependence on imports, the ways in which the EU tackles this problem is divided into two separate approaches that seek to solve the problems of energy and material import dependence independently of each other. The only contact area that can be identified in both approaches is the area of waste management where waste, through its material and energy recovery, can be used to alleviate both problems. Although it connects these two approaches, this division is also expressed in the field of waste management, where the concept of a circular economy, while recognizing energy recovery, places emphasis on material recovery and gives it precedence.

Therefore, in most papers that analyze waste management/recovery systems, the emphasis is put on recycling and reducing the use of raw materials. Energy flows are neglected or observed only through the reduction of energy consumption and the possibility of switching to renewable energy sources, while energy recovery is considered only as one of the analyzed options of waste treatment, separate from material recovery. This is not the case when looking at the entire waste management/recovery system, where waste can be used as raw material for energy and material production and these two approaches can complement each other.

This can be directly deduced from the results, which shows that avoiding energy recovery leads to worse results—the Material recovery scenario provides higher PE recovery only to the Existing system, having better PERIndex results than the Mass burn and Energy recovery scenarios, while by REE, it shows the worst result. However, these results would be even worse if the energy recovery of the waste, which is integrated through thermal treatment in a cement kiln, was completely eliminated from this scenario. Therefore, although in this scenario more secondary materials are produced, the materials have higher embodied energy, i.e., their production is less sustainable. This situation changes when a dedicated energy-recovery facility is added to the scenario based on material recovery: the AD plant.

Results show that a holistic view of this problem, which at the same time looks at the issue of sustainability from the material and energy side, shows the best results. By this, a new link between energy and material waste recovery has been identified, through an energy feedback loop that enables the sustainability of manufactured (recycled) materials to be increased by reducing the use of primary energy sources, thus defining a more sustainable "closing two loops" principle in waste management. This approach should be more acceptable to a wider range of people by showing them that material and energy recovery of waste complement each other and can lead to a more sustainable economy.

In addition, the role of energy recovery in the circular economy and the accepted principle of closing the loop is clearly defined. While at one level it helps to increase the recovery of primary energy and assists material recovery in reduction of the environmental impact, as well as the dependence on raw material imports, at the second level it helps to close the loop in the circular economy and promotes symbiosis among industries, directly increasing the sustainability of material recovery and the resulting products. This approach reexamines the EU's current position on the understanding of "closing the loop" and the circular economy itself, where the focus should be put more on the sustainability of the overall economy and, within these frameworks, establishing the link between the objectives of separate legislation frameworks seeking to address material and energy dependence on imports.

Using this united framework for further sustainable development of the EU economy, the pursuit of overall (simultaneous energy and material) sustainability of the economy can be established, and the phrase "material recovery" could get another, wider, meaning as produced/recovered, more sustainable, materials are the product of combined material and energy recovery.

References

Agostinho, F., Siche, R., 2014. Hidden costs of a typical embodied energy analysis: Brazilian sugarcane ethanol as a case study. Biomass Bioenergy 71, 69–83.

Arvidsson, R., Fransson, K., Fröling, M., Svanström, M., Molander, S., 2012. Energy use indicators in energy and life cycle assessments of biofuels: review and recommendations. J. Clean. Prod. 31, 54–61.

Beccali, M., Cellura, M., Longo, S., Guarino, F., 2016. Solar heating and cooling systems versus conventional systems assisted by photovoltaic: application of a simplified LCA tool. Sol. Energy Mater. Sol. Cells 156, 92–100.

Boer, E.D., Boer, J.D., Jager, J., 2005. Waste Management Planning and Optimisation (LCA IWM). Obidem-Verlag, Stuttgart.

Bovea, M.D., Ibáñez-Forés, V., Gallardo, A., Colomer-Mendoza, F.J., 2010. Environmental assessment of alternative municipal solid waste management strategies. A Spanish case study. Waste Manag. 30, 2383–2395.

Bueno, G., Latasa, I., Lozano, P.J., 2015. Comparative LCA of two approaches with different emphasis on energy or material recovery for a municipal solid waste management system in Gipuzkoa. Renew. Sust. Energ. Rev. 51, 449–459.

CEN, 2011. CSN EN 15978. Sustainability of Construction Works—Assessment of Environmental Performance of Buildings—Calculation Method.

CEN, 2012a. BS EN 15804: Sustainability of Construction Works–Environmental Product Declarations–Core Rules for the Product Category of Construction Products.

CEN, 2012b. Sustainability of Construction Works—Assessment of Buildings.

Cherubini, F., Bargigli, S., Ulgiati, S., 2008. Life cycle assessment of urban waste management: energy performances and environmental impacts. The case of Rome, Italy. Waste Manag. 28, 2552–2564.

Commission of the European Communities, 2003. Integrated Product Policy—Building on Environmental Life-Cycle Thinking—COM(2003) 302 Final. Available from https://eur-lex.europa.eu/LexUriServ/LexUriServ.do?uri=COM:2003:0302:FIN:en:PDF.

Connolly, D., Mathiesen, B.V., Østergaard, P.A., Møller, B., Nielsen, S., Lund, H., et al., 2012. Heat Roadmap Europe 1: First Pre-Study for the EU27. Aalborg University, Aalborg, Denmark.

Connolly, D., Mathiesen, B.V., Østergaard, P.A., Møller, B., Nielsen, S., Lund, H., et al., 2013. Heat Roadmap Europe 2: Second Pre-Study for the EU27. Aalborg University, Aalborg, Denmark.

Danilecki, K., Mrozik, M., Smurawski, P., 2017. Changes in the environmental profile of a popular passenger car over the last 30 years—results of a simplified LCA study. J. Clean. Prod. 141, 208–218.

Dixit, M.K., 2017. Life cycle embodied energy analysis of residential buildings: a review of literature to investigate embodied energy parameters. Renew. Sust. Energ. Rev. 79, 390–413.

Dzene, I., Barisa, A., Rosa, M., Dobraja, K., 2016. A conceptual methodology for waste-to-biomethane assessment in an urban environment. Energy Procedia 95, 3–10.

Ecoinvent Centre, 2016. Ecoinvent database v3.2. Inf. Syst. Sustain. Dev.

European Commission, 2017. Eurostat. Energy, Transport and Environment Indicators 2017. Available from https://ec.europa.eu/eurostat/documents/3217494/8435375/KS-DK-17-001-EN-N.pdf/18d1ecfd-acd8-4390-ade6-e1f858d746da.

European Commission, 2018a. Eurostat. Material Flow Accounts (online data codes: env_ac_mfa). Available from: http://ec.europa.eu/eurostat.

European Commission, 2018b. Eurostat. Material Flow Accounts Per Capita (online data codes: env_ac_mfa, demo_gind). Available from: http://ec.europa.eu/eurostat.

European Commission. Eurostat, 2018. Production in Industry (online data codes: sts_inpr_m). Available from: http://ec.europa.eu/eurostat.

European Parliament, 2018. Waste Management in the EU: Infographic with Facts and Figures. Available from: http://www.europarl.europa.eu/news/en/headlines/society/20180328STO00751/eu-waste-management-infographic-with-facts-and-figures.

Fernández-Nava, Y., del Río, J., Rodríguez-Iglesias, J., Castrillín, L., Marañón, E., 2014. Life cycle assessment of different municipal solid waste management options: a case study of Asturias (Spain). J. Clean. Prod. 81, 178–189.

Frischknecht, R., Wyss, F., Knöpfel, S.B., Lützkendorf, T., Balouktsi, M., 2015. Cumulative energy demand in LCA: the energy harvested approach. Int. J. Life Cycle Assess. 20, 957–969.

George, D.A.R., Lin, B.C.-A., Chen, Y., 2015. A circular economy model of economic growth. Environ. Model. Softw. 73, 60–63.

Gilbert, J.C., 2009. Comparison and Analysis of Energy Consumption in Typical Iowa Swine Finishing Systems. M.Sc. Thesis, Iowa State University. Available from: https://lib.dr.iastate.edu/cgi/viewcontent.cgi?referer=https://www.google.com/&httpsredir=1&article=2019&context=etd.
Giugliano, M., Cernuschi, S., Grosso, M., Rigamonti, L., 2011. Material and energy recovery in integrated waste management systems. An evaluation based on life cycle assessment. Waste Manag. 31, 2092–2101.
Huijbregts, M.A.J., Rombouts, L.J.A., Hellweg, S., Frischknecht, R., Hendriks, A.J., Van De Meent, D., et al., 2006. Is cumulative fossil energy demand a useful indicator for the environmental performance of products? Environ. Sci. Technol. 40, 641–648.
Huijbregts, M.A.J., Hellweg, S., Frischknecht, R., Hendriks, H.W.M., Hungerbühler, K., Hendriks, A.J., 2010. Cumulative energy demand as predictor for the environmental burden of commodity production. Environ. Sci. Technol. 44, 2189–2196.
Huysman, S., De Schaepmeester, J., Ragaert, K., Dewulf, J., De Meester, S., 2017. Performance indicators for a circular economy: a case study on post-industrial plastic waste. Resour. Conserv. Recycl. 120, 46–54. Available from.
ISO, 2006a. ISO 14040:2006—Environmental Management—Life Cycle Assessment—Principles and Framework. CEN (European Committee for Standardisation), Brussels.
ISO, 2006b. ISO 14044:2006—Environmental Management—Life Cycle Assessment—Requirements and Guidelines. CEN (European Committee for Standardisation), Brussels.
ISO, 2012. ISO/TR 14047:2012—Environmental Management—Life Cycle Assessment—Illustrative Examples on How to Apply ISO 14044 to Impact Assessment Situations.
ISWA, 2015. Circular Economy: Carbon, Nutrients and Soil. Vienna, Austria. Available from: https://www.iswa.org/fileadmin/galleries/Task_Forces/Task_Force_Report_4.pdf.
Mert, G., Linke, B.S., Aurich, J.C., 2017. Analysing the cumulative energy demand of product-service systems for wind turbines. Proc. CIRP 59, 214–219.
Niero, M., Olsen, S.I., 2016. Circular economy: to be or not to be in a closed product loop? A life cycle assessment of aluminium cans with inclusion of alloying elements. Resour. Conserv. Recycl. 114, 18–31.
Niero, M., Negrelli, A.J., Hoffmeyer, S.B., Olsen, S.I., Birkved, M., 2016. Closing the loop for aluminum cans: life cycle assessment of progression in cradle-to-cradle certification levels. J. Clean. Prod. 126, 352–362.
NREL, 2016. USLCI Database. Available from http://www.nrel.gov/lci/.
Ouda, O.K.M., Raza, S.A., Nizami, A.S., Rehan, M., Al-Waked, R., Korres, N.E., 2016. Waste to energy potential: a case study of Saudi Arabia. Renew. Sust. Energ. Rev. 61, 328–340.
Penny, T., Collins, M., Aumônier, S., Ramchurn, K., Thiele, T., 2013. Embodied energy as an indicator for environmental impacts—a case study for fire sprinkler systems. In: Håkansson, A., Höjer, M., Howlett, R.J., Jain, L.C. (Eds.), Sustain. Energy Build. Smart Innov. Syst. Technol. Springer, Berlin, Heidelberg, pp. 555–565.
Petrov, R.L., 2007. Original method for car life cycle assessment (LCA) and its application to LADA cars. In: 2007 World Congr.
Rajaeifar, M.A., Ghanavati, H., Dashti, B.B., Heijungs, R., Aghbashlo, M., Tabatabaei, M., 2017. Electricity generation and GHG emission reduction potentials through different municipal solid waste management technologies: a comparative review. Renew. Sust. Energ. Rev. 79, 414–439.
Rohrlich, M., Mistry, M., Martens, P.N., Buntenbach, S., Ruhrberg, M., Dienhart, M., et al., 2000. A method to calculate the cumulative energy demand (CED) of lignite extraction. Int. J. Life Cycle Assess. 5, 369–373.
Scipioni, A., Niero, M., Mazzi, A., Manzardo, A., Piubello, S., 2013. Significance of the use of non-renewable fossil CED as proxy indicator for screening LCA in the beverage packaging sector. Int. J. Life Cycle Assess. 18, 673–682.
SIA, 2010. 2032 Graue Energie von Gebäuden.
Tomić, T., Schneider, D.R., 2017. Municipal solid waste system analysis through energy consumption and return approach. J. Environ. Manag. 203, 973–987.
Tomić, T., Schneider, D.R., 2018. The role of energy from waste in circular economy and closing the loop concept—energy analysis approach. Renew. Sust. Energ. Rev. 98, 268–287.

VDI, 1997. In: Ingenieure, V.D. (Ed.), Cumulative Energy Demand—Terms, Definitions, Methods of Calculation. VDI-Richtlinien 4600.

Vukotic, L., Fenner, R., Symons, K., 2010. Assessing embodied energy of building structural elements. Proc. Inst. Civ. Eng. Eng. Sustain. 163, 147–158.

Zabalza Bribián, I., Aranda Usón, A., Scarpellini, S., 2009. Life cycle assessment in buildings: state-of-the-art and simplified LCA methodology as a complement for building certification. Build. Environ. 44, 2510–2520.

Zero Waste Europe, 2017. Embodied Energy: A Driver for the Circular Economy? Available from https://zerowasteeurope.eu/downloads/embodied-energy-a-driver-for-the-circular-economy.

CHAPTER 24

Investigation of the sustainable waste transportation in urban and rural municipalities—Key environmental parameters of the collection vehicles use

Piotr Nowakowski[a], Krzysztof Szwarc[b], and Mariusz Wala[c]
[a]Faculty of Transport, Silesian University of Technology, Katowice, Poland
[b]Institute of Computer Science, University of Silesia in Katowice, Sosnowiec, Poland
[c]PST Transgór S.A, Rybnik, Poland

1. Introduction and literature review

A circular economy concept has been introduced in many countries. The main principle of this approach is closing the loop of the flow of materials in the economy and focuses on reusing, recycling, and recovering various categories of products and materials (Andersen, 2007; Korhonen et al., 2018). Landfilling should be reduced as much as possible. This environmentally friendly approach requires intensified effort both from legislative authorities on the global, national, and local levels, and from society, obliged to fulfill the law and to be responsible in everyday activities that have a negative impact on the natural environment by the disposal of waste and end-of-life products (Parajuly et al., 2020; Zhang and Klenosky, 2016). One of the most important management systems in the circular economy is solid waste management. The key role of solid waste management is to provide all necessary resources for collection, handling, transportation, and recovery of the waste stream generated in society and other economic activities including services, trade, and manufacturing (Hoornweg and Bhada-Tata, 2012). Waste management depends on legislation and two key players in the system—society and waste collection companies (Knickmeyer, 2020). The hierarchy, main tasks for the key players, and potentially negative implications for human health and the natural environment in waste management are illustrated in Fig. 1.

The composition of waste and volume of the waste stream has big variations depending on the economy, development level, gross domestic product, population, climate zone, type of housing, etc. The fate of the collected waste can be dumping, through

Circular Economy and Sustainability, Volume 1
https://doi.org/10.1016/B978-0-12-819817-9.00026-0

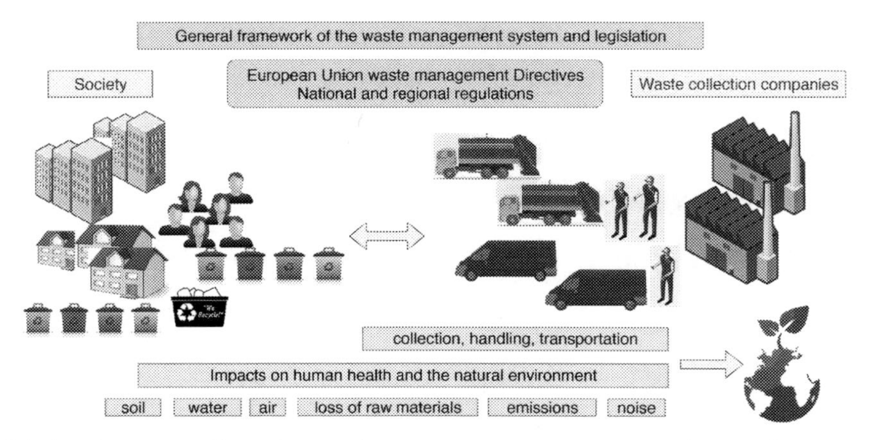

Fig. 1 Relations between policymakers, society, waste management companies, and the environment. *(Source: Graphic created by the authors.)*

sanitary landfilling, and different scenarios of recovery, including waste-to-energy, as a form of recovery of energy and recycling (Hoornweg and Bhada-Tata, 2012; Pires et al., 2011).

In the case when solid waste management is conducted improperly, numerous negative impacts on the natural environment may occur (Ayomoh et al., 2008). These can be from the behavior of residents, disposing of waste in green or inhabited areas (Fig. 2), or storing excessive waste in residential premises (Nowakowski, 2019; Timlett and Williams, 2011). When the waste stream escapes legal collection schemes and further processing in waste treatment facilities, the secondary raw materials are lost. Some categories of waste containing hazardous substances possess a significant threat to the natural environment and human health. The most dangerous components from waste electrical and electronic equipment (WEEE) must be collected at the disassembly stage and neutralized (Bogaert et al., 2008; European Commission, 2012). Another category of hazardous waste is portable batteries (European Parliament, 2006). The collection and transportation of waste require using a fleet of vehicles ranging from a typical van to specialized waste collection trucks (Bilitewski et al., 1996). The direct influences on the natural environment by waste transportation are the vehicles' exhaust emissions (Dotoli and Epicoco, 2017; Maimoun et al., 2013).

Waste collection planning should take into consideration key economic and environmental factors. In the majority of cases, waste collection is conducted in agreement with local authorities from municipalities. The collection companies can operate as a public or private entity, but cost reduction is one of the main factors for collection planning. Therefore, planning waste collection frequency, selection of vehicles, and assigning staff must be rational and fulfil the demands of residents, who will expect high-quality service

Fig. 2 Examples of illegal disposal of waste electrical equipment (refrigerator) and plastics (garden furniture). *(Source: Authors' private archive.)*

provided by the waste collection company. Waste collection protects the natural environment because residents are discouraged to dispose of any category of waste illegally. The design of an efficient waste collection and transportation framework using the best available methods can result in a significant reduction in secondary raw material loss and pollution from hazardous elements and substances contained in various categories of waste. At the same time, reduction of exhaust emissions brings a reduction of air pollution.

Depending on its complexity, the waste collection process can be considered as an instance of the traveling salesman problem (the simplest approach), vehicle routing problem (VRP; and its variants), or arc routing problem (ARP). Indicated problems belong to the class of NP-hard problems, making it impossible to solve their larger instances accurately and forcing the use of approximate approaches in real applications.

Among approximate methods of solving the vehicle routing problem with time windows (VRPTW)—a problem that models on-demand e-waste collection perfectly—special attention should be paid to metaheuristics, which are commonly used in the process of solving NP-hard problems. The literature on the subject describes, among others, a new adaptive memetic algorithm, which dynamically adapted itself according to the current state of the search, reducing the need for a time-consuming tuning process (Nalepa and Blocho, 2016). An interesting approach to solve time-constrained vehicle routing problems (i.e., VRPTW) has been proposed by Vidal et al. (2013a), who have developed a hybrid genetic search with adaptive diversity control algorithm, which effectively combines the exploration capabilities of genetic algorithms with improvement procedures based on local search.

The use of metaheuristics to solve other VRP variants is also noteworthy, which can be a canvas for the development of specialized models meeting the requirements of WEEE collection. Bell and McMullen (2004) proposed some modifications to the ant colony optimization (ACO) algorithm, used firstly to solve the traveling salesman problem to allow the search of the multiple routes of the VRP. Cattaruzza et al. (2014) described a hybrid genetic algorithm for the multitrip vehicle routing problem, while Vidal et al. (2012) have proposed a hybrid genetic algorithm for three VRP variants—the multidepot VRP, the periodic VRP, and the multidepot periodic VRP with capacitated vehicles and constrained route duration. Furthermore, a survey of heuristics and metaheuristics for multiattribute vehicle routing problems has been prepared by Vidal et al. (2013b), and an adaptation of other known metaheuristic methods (i.e., particle swarm optimization) to solve different VRP variants was analyzed by Marinakis et al. (2018); in total, around 100 papers were found.

Solid waste collection can be modeled using different ARP variants. Among articles dealing with these issues, some valuable results are presented by Ghiani et al. (2005), who used ARP to model waste collection in Southern Italy. Bautista et al. (2008) solved an urban waste collection problem in Spain using ants heuristics. Among the interesting approaches to solving the capacitated variant of the problem, the use of tabu search (TS) (Brandão and Eglese, 2008) and an improved ant colony optimization-based algorithm (Santos et al., 2010) stand out.

This chapter focuses on various waste streams transportation issues. The methodology applied in this study guides the reader through a generalized view of the separated waste streams and the statistics in the European Union, then a more detailed insight is presented for Poland. We investigate variations of the collection and transportation of waste for rural communities and urban municipalities. As the main motive for our study is an environmental perspective on use of vehicles, some important impacts on the natural environment are discussed. They include improper disposal of waste, vehicle emissions, and noise. For a case study illustrating the possible application of artificial intelligence algorithms, we included several examples of separate waste collections: bulky waste, waste electrical and electronic equipment, and separated waste collection (plastics, paper, and glass). To solve the optimization problem and minimize environmental impacts we used three approximate algorithms. As a result, we optimized vehicles' travel routes for waste collections, including various restrictions. The chapter includes examples of solving the selected optimization problems. The case study analyzes six optimization tasks for bulky waste collection (three instances), a bulky waste collection that requires a return to the company base after driving 20 km (which can simulate the application of electric vehicles for waste collection), separate waste collection, and an extended instance of e-waste mobile collection with time windows. For each problem, we verified the effectiveness of individual algorithms by analyzing the obtained objective function value—route length and the number of points/streets served. The results show the necessity of applying artificial intelligence algorithms in supporting various categories of waste collection and minimizing the environmental impacts of collection vehicles.

2. Waste categories and waste treatment in the European Union

In the European Union, the average individual produces about 500 kg of waste per year. Approximately 40% of it is reused or recycled. In several countries in the European Union, more than 70% of the waste stream is landfilled (Eurostat, 2019b,c). The objectives set in European legislation and directives focus on improvement in solid waste management, like limiting landfilling, reducing the number of landfills, promoting education to improve individuals' behavior in the disposal of waste, and encouraging innovation in the collection, treatment, and recycling of waste (Williams, 2013). In addition, remanufacturing, reusing, and repairing are important activities for the elimination of waste, and using products for a longer time has benefits for the environment (Laustsen, 2007; Wolf and Gutowski, 2013). Improved waste management systems can reduce the negative impacts on human health and the environment, such as soil, water, and air pollution. Efficient waste collection systems also reduce littering and disposal of waste in green areas.

Waste management policy has been shifted to redefine waste hierarchy and management methods, so that the priority is prevention, then reusing, recycling, and recovering. The least preferred option is landfilling and incineration without energy recovery. The objectives for waste policy in the EU focus on the reduction of waste generated, maximizing reusing and recycling, and phasing out landfilling (European Commission, 2013).

The priorities of the European Union indicate the main goals to be achieved in solid waste management but significant differences waste treatment methods between individual countries still exist, as shown in Fig. 3. In many countries, landfilling is still one of the most common waste treatment methods, whereas others, like Denmark, Germany, Netherlands, Ireland, Finland, and Sweden, have reduced the use of landfilling to a minimum. Significant differences also exist in the waste stream ratio in kg/capita. In this case,

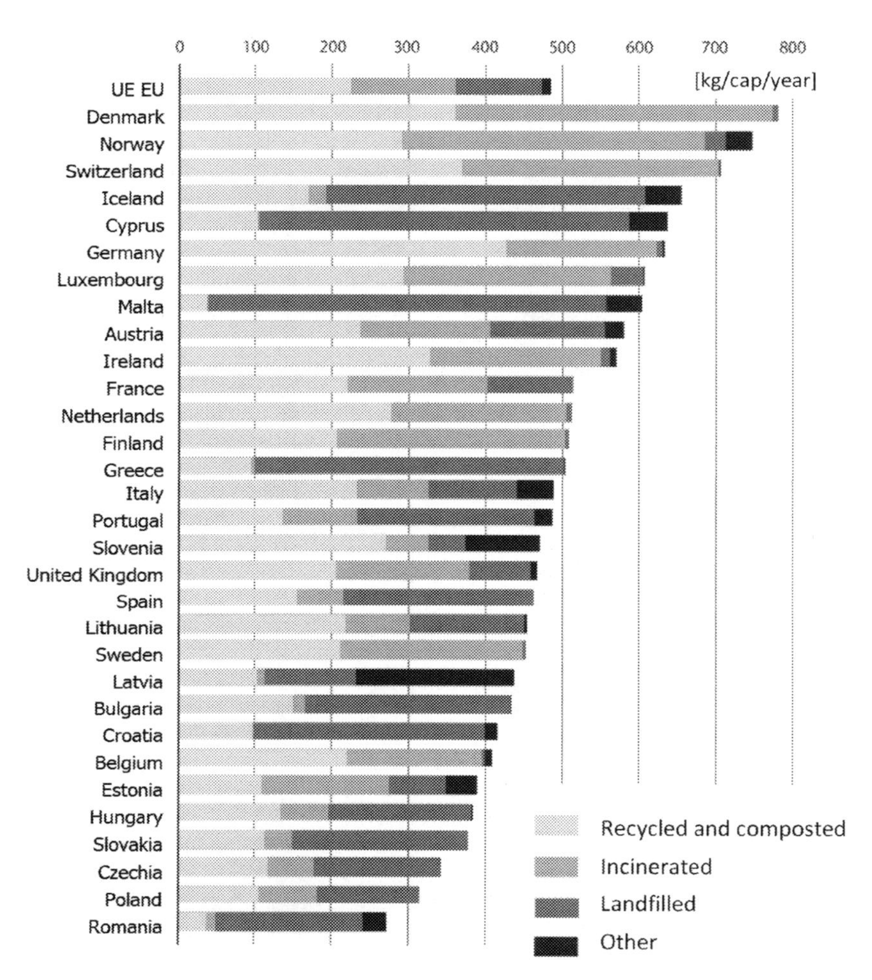

Fig. 3 Waste treatment methods in the European Union in 2015. *(Source: Eurostat.)*

some richer countries can produce a much bigger waste stream compared to countries with much lower purchasing power parity (PPP) but with larger populations.

The larger amount of waste generated by some members of the European Union requires more frequent collections or assignment by the waste collection companies of a larger vehicle fleet. Several categories of waste are hazardous for human health and the natural environment. These categories are generated from households in much lower volumes compared to separated or mixed solid waste. However, they contain a significant proportion of substances dangerous for living species and humans. These categories are waste electrical and electronic equipment and portable batteries. Dangerous substances present in products from these categories include mercury, cadmium, lead, lithium,

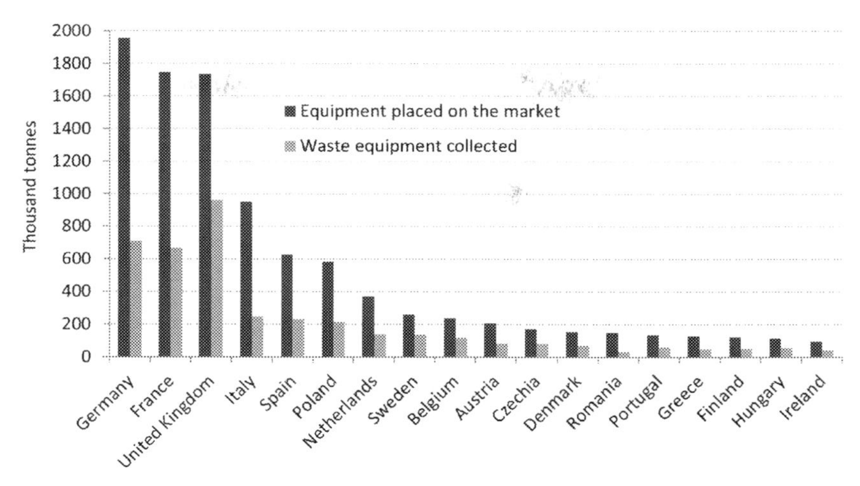

Fig. 4 Waste electrical and electronic equipment placed on the market and collected in the European Union member statess with over 100,000 tonnes of equipment placed on the market in 2017. *(Source: Eurostat.)*

polychlorinated biphenyls, polybrominated biphenyls, and some other substances (Inglezakis and Moustakas, 2015).

Electrical and electronic equipment varies in shape and size. The general requirements and recommendations considering collection and treatment for this category of waste have been defined in the Waste Electrical and Electronic Equipment European Directive (European Commission, 2012). In this case, the number of possible places where waste electrical and electronic equipment can be disposed of is much higher. The collection is provided by supermarkets, municipal collection centers, events to increase knowledge concerning protection of the natural environment by individuals. Therefore, the variation of waste collection methods is higher compared to other municipal waste categories. The total mass of electrical and electronic equipment placed on the market and collected for selected member states of the European Union is presented in Fig. 4 (Eurostat, 2019a).

More complex is the collection of waste portable batteries. A place where they can be deposited can be in any shop, supermarket, institution, or municipal collection center, depending on local regulations, and schools and other public or private institutions can participate in collection schemes to promote environmentally friendly behavior. However, the tasks of the collection companies are more demanding. They have to use waste collection vehicles even to collect a small amount of waste. Fig. 5 shows recent data considering portable batteries placed on the market and waste batteries collected. The average collection rate is 50% for the EU member states with developed waste management systems. It is much lower than expected because, although the dimensions of a single product like AA, AAA, or button-type batteries are small, the substances contained

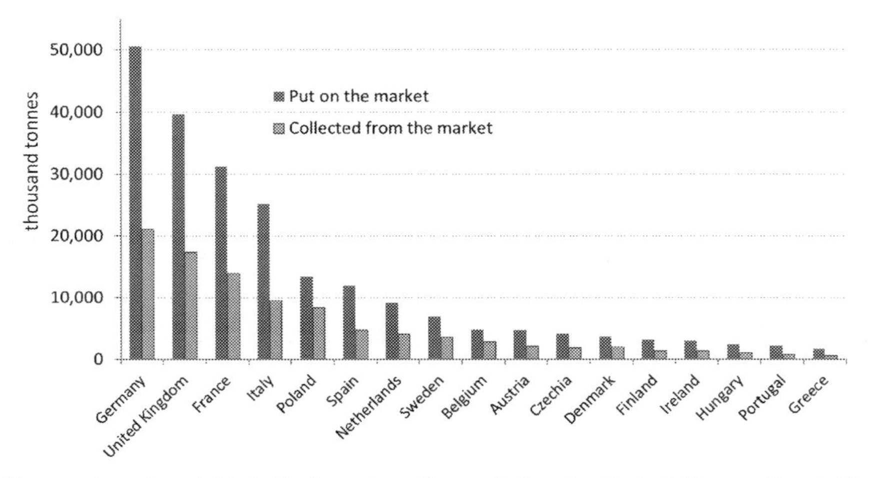

Fig. 5 The number of portable batteries put on the market and collected. *(Source: Eurostat.)*

in batteries can be harmful both for human health and the environment (Eurostat, 2019d). The small dimensions of portable batteries encourage the public to dispose of these products with general waste; therefore, more information and education campaigns need to be conducted for improving public consciousness of the proper disposal of this category of waste.

3. The main indicators of solid waste stream in Poland

The population of Poland is 38 million inhabitants (Statistics Poland, 2020). In 2005, Poland joined the European Union and, therefore, its legislation had to comply with EU regulations as applied to waste categories and waste management. Each municipality had to establish a local municipal waste management system. Depending on the municipality, public or private companies participate in the waste collection. Various collection schemes apply to different categories of waste. The collection of mixed solid waste and separated waste is the most frequent. The separated waste categories are divided into subcategories like paper, glass, plastics, and metal (Journal of Laws of The Republic of Poland, 2013). Biodegradable waste usually has varied frequency depending on the season. More collections are required in spring, or autumn, due to clearing, moving, and pruning in green areas. Furthermore, in summer, some areas require frequent waste collections due to fast decomposition and odors coming from that category of waste. Other categories of waste, including bulky waste, construction and demolition, and waste electrical and electronic equipment, are conducted as a curbside collection or on-demand collection.

Local administration encourages individuals to participate in selective waste collection from households. Waste collection fees are reduced for residents participating in

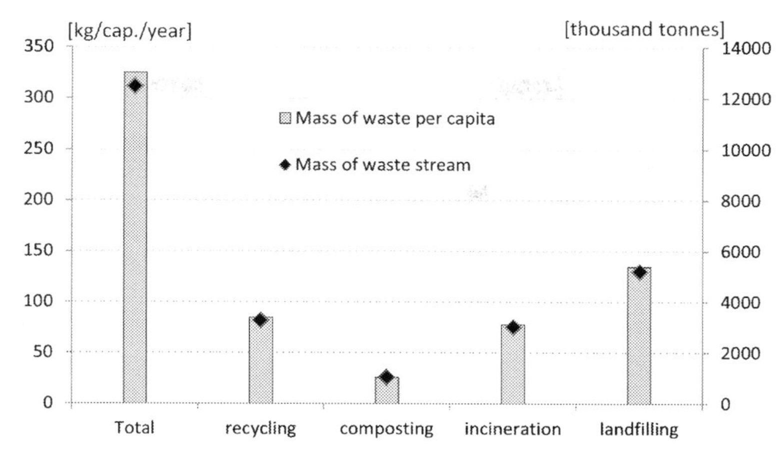

Fig. 6 Treatment of waste in Poland in 2018. *(From Statistics Poland, 2019a. Environment 2019 - Ochrona środowiska 2019, 190.)*

selective collection schemes. The volume and mass from the selective collection of waste have been increasing year by year, but landfilling is still one of the most common methods of waste treatment in Poland. The waste treatment methods in 2018 in Poland are shown in Fig. 6 (Statistics Poland, 2019a).

In Poland in 2018, the average mass per capita of waste from selective collection was 94 kg, including biodegradable waste 26 kg, mixed packaging waste 15 kg, bulky waste 14 kg, glass 13 kg, plastics 9 kg, and paper and cardboard 7 kg. Fig. 7 shows the distribution of the waste streams in the voivodships (regional administrative districts) in Poland (Statistics Poland, 2019a).

The categorization of waste has a major impact on the selection of the collection method, containers, bags, collection vehicles, and assignment of staff. Each municipality has to assign waste collection schemes and general waste management systems individually.

4. Schedules and types of waste collection in municipalities

Each method of waste collection requires the assignment of vehicles. On the one hand, it may seem obvious to assign a classical waste truck with compactor for a selected area; on the other hand, in some cases, such a vehicle is not suitable to drive through narrow roads within suburbs with blocks of flats or other residential collection areas that are difficult to access. In such a case, a lower-capacity vehicle must be used. As a result, the total collection costs, traveling distances, and exhaust emissions may increase. Waste collection planners use a fixed schedule for mixed municipal waste or separated waste collections.

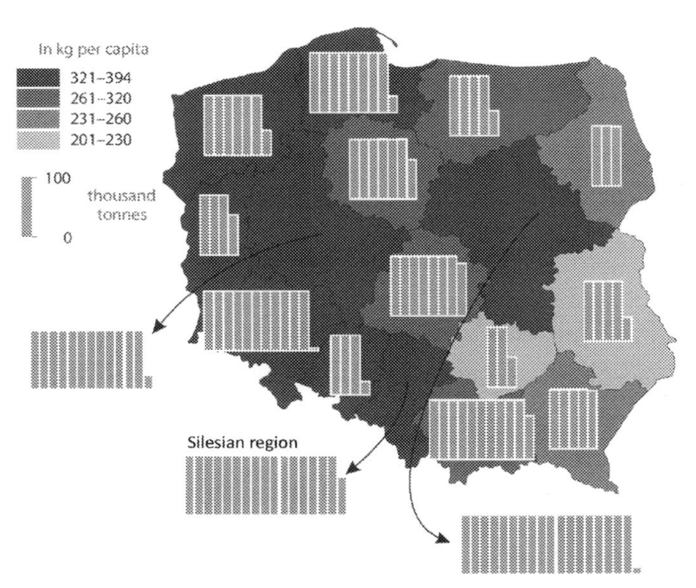

Fig. 7 Generated municipal waste in different regions (voivodships) in Poland in 2018. *(From Statistics Poland, 2019a. Environment 2019 - Ochrona środowiska 2019, 190.)*

A similar collection method applies to the collection of bulky waste or special waste like electrical and electronic equipment. In such a case, the route is similar, but collection made only once a month, a quarter, or even once per year (Bilitewski et al., 1996; Ghiani et al., 2014; Nowakowski et al., 2017; Zbib and Wøhlk, 2019).

Other categories of waste like construction and demolition waste, large WEEE appliances, and bulky waste can be collected as an on-demand waste collection (Król et al., 2016). An individual calls a collection service or uses an online form and requests the waste pick-up. After the required number of calls is reached, a collection company prepares a plan, including vehicles, staff, and routing (Nowakowski et al., 2018). Another method for the collection of waste is container collection, especially in densely populated areas. It is the most suitable for the collection of separated waste, but is often applied for the selection of placed containers ad hoc in various locations, as was investigated for waste electrical and electronic equipment (Nowakowski and Mrówczyńska, 2018). Finally, an individual can use his/her own vehicle and transport any category of waste to a municipal or regional collection center. In such a case, the exhaust emissions depend on a type of vehicle in possession of the individual, route length, and the total number of residents using this method. Table 1 shows the relations between waste collection companies (B—business) and individuals (C—consumers). Other columns indicate the common collection methods of the main categories of waste from households, indicating environmental impacts. The main negative impact on the environment is the emissions of the

Table 1 Specification of waste collection.

Type of waste	Relation (waste collection company/individual)	Method of collection	Vehicles	Environmental impacts
Mixed Solid waste	B2C	Scheduled	Waste collection truck	Improper disposal—loss of materials; vehicle emissions
Separated	B2C	Scheduled	Waste collection truck, lorry, van	Improper disposal—loss of materials; vehicle emissions; exceeded allowed noise level
Biodegradable	B2C	Scheduled	Waste collection truck	Vehicle emissions
WEEE	B2C; C2B	Scheduled; on-demand; flexible	Lorry, van	Improper disposal—loss of materials; contamination by hazardous substances of the natural environment; vehicle emissions; contamination by pollutants (e.g., mercury vapor from compact fluorescent lamps, oil from cooling equipment compressors, etc.)
Portable batteries	C2B	Flexible	Van	Improper disposal—loss of materials, contamination by hazardous substances of the natural environment; vehicle emissions, contamination by elements with negative impact (cadmium, lithium, etc.) during improperly conducted collection

Continued

Table 1 Specification of waste collection—cont'd

Type of waste	Relation (waste collection company/individual)	Method of collection	Vehicles	Environmental impacts
Construction and demolition waste	B2C	On-demand; flexible	Hook container truck, gantry truck	Destruction of green areas; vehicle emissions
Bulky waste	B2C	Scheduled; on-demand;	Garbage truck, lorry, van	Improper disposal—loss of materials

collection vehicles (Demir et al., 2014; Helmers et al., 2019; Maimoun et al., 2013; Reşitoğlu et al., 2015).

Employees and residents may be exposed to high levels of noise during waste collection activities and, as an effect, they risk damage to their hearing. Therefore, exposure to noise should be controlled at source using engineering means and hazardous noise should be avoided and eliminated (Department of Environment Regulation, 2014; HSE, 2005). The exceeded noise level occurs mainly in waste loading activities. The highest noise levels were registered in the unloading of separated waste from large bins or containers for glass fraction. Other activities of using hydraulic equipment and unloading the waste are usually within the allowed limits for employees. The results of broad research conducted in the United Kingdom show the typical level of noise exposure might vary depending upon the particular circumstances. In the majority of cases, they were within the range of 71–85 dBA. However, in unloading activities, especially for collection vehicles with side-loading, it exceeded 85 dBA and in some cases reached up to 95 dBA (Jobling and WRAP, 2012).

5. Waste collections in urban and rural municipalities in Poland

The case study focuses on the analysis of the waste stream in selected municipalities in Poland. In this research, the data for the analysis were collected for a town, Wodzisław Śląski; a rural municipality, Marklowice (both in the Silesian region, see Fig. 8); and a big city, Warsaw—the capital of Poland. Wodzisław Śląski is a town located in the south of Poland. The population in 2015 was about 49,000 inhabitants. The households consist approximately of 4500 single-family houses and 16,500 flats in blocks. Marklowice is a rural municipality in the Silesian region and belongs to the broader area of the Wodzisław Śląski municipality. The population of Marklowice is approximately 5500 inhabitants. A great majority of the population live in single-family houses (Statistics Katowice,

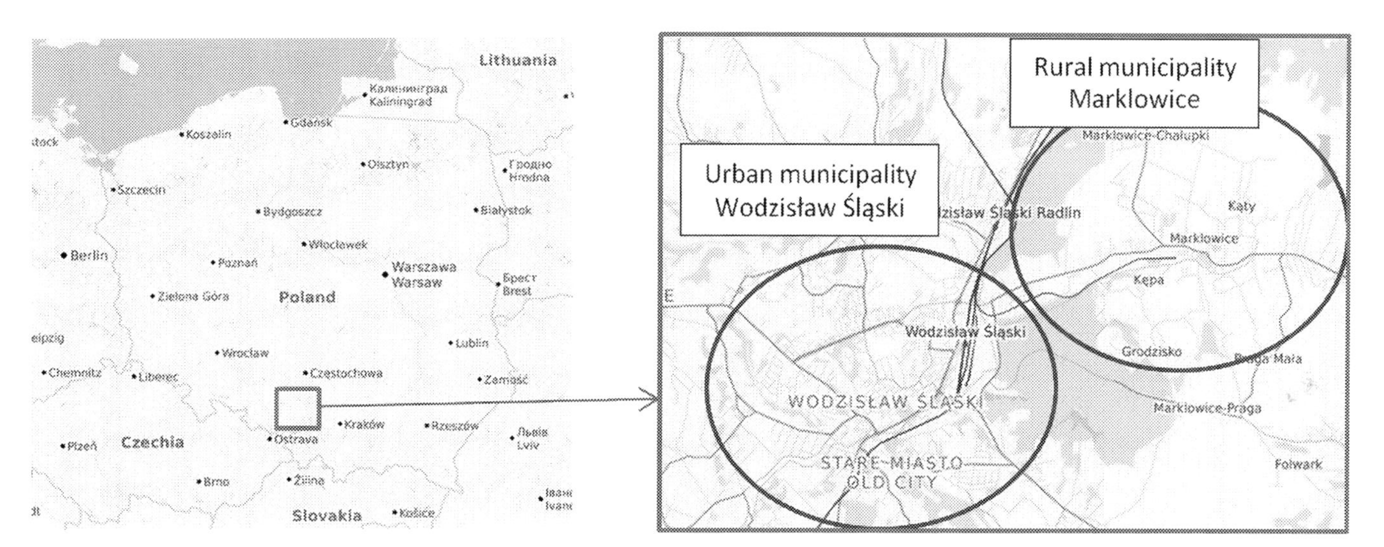

Fig. 8 Localization of two municipalities in the Silesian region for a case study. *(Source: Graphic created by authors; map: openstreetmap.org.)*

2016). The collection of waste for these two municipalities—the urban municipality, Wodzisław Śląski, and the rural municipality, Marklowice—is operated by the Transgór S.A. company. The on-demand WEEE collection is presented for Warsaw. It is the capital of Poland with a population of 1.8 million inhabitants and including a greater metropolitan area of 3.1 million residents (World Population Review, 2020). The number of single-family houses is 263 thousand and there are 1.2 million flats in multistory buildings. The total waste stream generated in 2018 was 560 thousand tones including 20% from selective collection (Statistics Poland, 2019b).

The methods and results of waste collection for the two municipalities are intended to show a practical approach in waste collection planning and further, to propose some routing optimization algorithms for improvement. It will focus on route optimization for on-demand bulky waste collection and selective collection for plastics, glass, and paper. The case study of Warsaw is intended to show WEEE collection planning for a large area including 460 waste collection requests. For route planning and selection of vehicles for the collection, the optimization results are presented in the next section of this chapter.

Waste collection schedules in rural municipalities and urban municipalities commonly applied in waste management in Poland are presented in Table 2.

The results of the collection of various categories of waste for municipalities in Silesian region—Marklowice and Wodzisław Śląski—in 2019 are shown in Table 3.

The average mass of the collected waste per capita in 2019 for various categories of waste is presented in Fig. 9. The results of the total mass of collected waste are similar to

Table 2 Waste collection schedules in rural and urban municipalities in the Silesian region in Poland.

	Schedule of the waste collection	
Category of waste	**Rural municipalities**	**Urban municipalities**
Mixed municipal solid waste	Once per month or twice per month depending on season	Once per month or twice per month for single-family houses, Multistory buildings up to three times per week
Separated waste (paper, plastics, glass, metals)	Once per month	Once per month or twice per month
Bulky waste	Once per year or twice per year	Once per year or twice per year as curbside collection, possibility of on-demand service
Biodegradable waste	Once per month or twice per month depending on season	At least twice a month spring, summer; rest of the year, once per month
Construction and demolition waste	Curbside collection or on-demand collection	On-demand collection or transportation of the waste by a resident to municipal collection center

Table 3 Total mass of the collected categories of waste for two Silesian municipalities in 2019.

Category of waste	Marklowice (tones/year)	Wodzisław Śląski (tones/year)
Mixed fraction	518	12,012
Biodegradable	262	1306
Glass	154	633
Plastics	187	705
Paper	51	350
Bulky waste	67	912
Other fractions	404	347

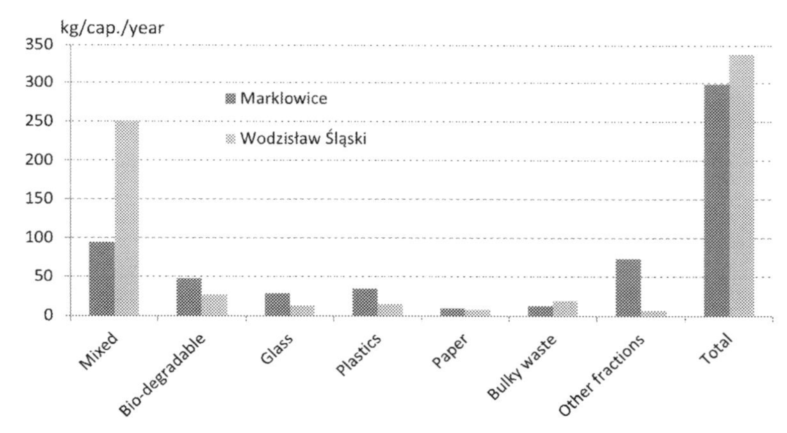

Fig. 9 Collection of various categories of waste per capita in the rural municipality (Marklowice) and urban municipality (Wodzisław Śląski) from the Silesian region in Poland.

the average for Poland. It is significant that the urban municipality residents tend to dispose of the majority of household waste as a mixed category. As a result, potential secondary raw materials can go to landfills, or they would require separation in the transfer station. This step is necessary for retrieving all recyclable fractions from the mixed waste stream.

The waste collection company Transgór S.A. operates a fleet of vehicles suitable to collect all categories of waste. Fig. 10 shows a practical approach to loading bulky waste and separated waste. Heavy-duty waste compacting trucks are mainly used for mixed waste and biodegradable waste. Other vehicles are equipped with a special cage for securing the load, which is essential for the transportation of bulky waste (Fig. 10—left) and separated waste in bags (Fig. 10—middle). The types of vehicles used for the collection are shown in Table 4 and all are diesel-engine vehicles.

Fig. 10 Collection of bulky waste, separated waste in bags, and heavy-duty 3 axes waste compacting truck.

Table 4 Average distances travelled by collection vehicles from the Transgór S.A. fleet in 2019.

Type of vehicle	Route length per vehicle in 2019 (km)
Waste compacting lorry—loading capacity 3.5 t	33,310
Waste compacting truck—2 axes loading capacity 8 t	27,018
Waste compacting truck—2 axes loading capacity 7.5 t	21,478
Heavy duty 3 axes waste compacting truck I	21,687
Heavy duty 3 axes waste compacting truck II	21,997
Heavy duty 3 axes waste compacting truck III	22,881
Heavy duty 3 axes waste compacting truck IV	23,484
Large commercial vehicle—extended body van I	29,888
Large commercial vehicle—extended body van II	27,195
Large commercial vehicle—extended body van III	33,128
Large commercial vehicle—extended body van IV	27,553

Table 5 Average emissions level per vehicle of the Transgór S.A. waste collection company in 2019.

Vehicle type	Annual average route length (km)	Emission factors (kg/year)		
		NO_x	PM	CO_2
Waste collection truck—3 axes compacting vehicle	22,500	252.4	4.67	25,080
Large commercial vehicle—extended body van	29,500	34.5	3.79	8690

The average exhaust emissions depending on type of vehicle are shown in Table 5. They include common components that contribute to smog, particulate matter (PM), nitrogen oxides (NO_x), and greenhouse gas—carbon dioxide (CO_2).

Depending on the category of waste and schedule, the Transgór S.A. company selects vans, lorries, or specialized waste compacting garbage trucks. The operation of these vehicles is the main contributor to exhausts emissions. It depends on the type and age of vehicle. Waste collection planning and vehicle routing should be supported by artificial intelligence algorithms to minimize emissions. The next section of this chapter includes routing optimization for six tasks including on-demand bulky waste collection for various numbers of collection requests from residents, separate collection of plastics, glass, and paper (for Wodzisław Śląski and Marklowice), and WEEE collection planning for Warsaw. Various methods of e-waste collection are used in Warsaw. However, one of the most innovative methods, acknowledged by the public, is requesting pick-up of waste equipment by on-line form or phone call. The collection planning including assignment of staff and vehicle routing requires the application of artificial intelligence algorithms. In this study, we have used a set of data for e-waste collections including 460 real-world requests from the residents of Warsaw. Each resident required a collection of mainly medium or large home appliances suitable for resident time windows. The collection company had to assign a suitable number of vehicles and create a collection plan for each vehicle.

6. Supporting waste collections by artificial intelligence algorithms—A case study for municipalities in the Silesian region of Poland

A case study includes the application of several heuristics approaches for supporting the collection on-demand of bulky waste and selective waste collection for urban municipalities—Warsaw and Wodzisław Śląski—and the rural community—Marklowice. The purpose of vehicle route optimization in collection planning is the reduction of vehicle emissions and increasing the amount of collected waste. For typical bulky waste, it discourages the public from illegal dumping of waste in green spaces, and for e-waste, it protects the natural environment against pollution from hazardous substances—some examples of illegal disposal of waste are presented in Fig. 2 of this chapter.

To determine the possibility of using heuristic and metaheuristic algorithms to increase the effectiveness of waste collection, three methods were implemented: greedy algorithm (GrA), simulated annealing (SA), and TS. GrA assumes taking the best decision from the local perspective. This means that in our implementation algorithm, it selects another analyzed vehicle based on the cost of its usage in such a way that the cheapest available vehicle is selected. In addition, the decision concerning the sequence of waste collection points is based on the distance from the last visited node—the nearest visited point is selected. SA is the local search method, which allows approving a worse solution than the one which is currently considered, in this way omitting the local optimum. The described passage is executed based on the parameter referred to as temperature. Nowakowski et al. demonstrate the significant effect of the algorithm applied when planning the routes of mobile e-waste collection (Nowakowski et al., 2018). TS, like SA, is a method of local search, enabling bypassing of the local optimum by moving to a solution described with a worse value of an objective function. The process is carried out using the so-called tabu period, which prevents repeated execution of the same movement in a set period. Both TS and SA are based on the initial solution determined by GrA. In this way, it has been ensured that the solution developed employing these methods will not be worse than the one created by GrA.

The following values of algorithm parameters are used in this chapter (further described in the paper by Ghiani et al. (2005); they were selected based on a paper by Nowakowski et al. (2018), and experimental research conducted): cooling rate $\alpha = 0.98$, length of epoch $L = 100$, the likelihood of moving to a worse solution in the first iteration $P_0 = 0.3$, the maximum number of iterations without improving solution for SA = 100, tabu period = 10, the maximum number of iterations without improving solution for TS = 1000, and $\zeta = 0.05$. The neighborhood for TS and SA is determined in the same way as in Nowakowski et al. (2018): it is created by removing and planning new travel routes for one or two vehicles participating in the collection.

The effectiveness of applying specific algorithms is determined by utilizing objective function f:

$$\text{minimize} f = v_f + c_f + t_f + p_f,$$

where:

v_f is the parameter describing the value of collected waste—determined as the product of quotient of the value obtained in the basic and analyzed solution and 1/2 of p_p weight;

c_f is the parameter describing the cost of conducting collection—determined as the product of quotient of the costs incurred in the basic and analyzed solution and 1/2 of p_p weight;

t_f is the parameter describing the period of conducting waste collection—determined as the product of quotient of the period of conducting collection in the basic and analyzed solution and 1/2 of p_t weight; and

p_f is the parameter describing the number of served points—determined as the product of quotient of the number of served points in the basic and analyzed solution and 1/2 of p_g weight.

The following weight values were selected: profit significance $p_p = 0.02$, time significance $t_p = 0.01$, and significance of the number of served points $p_g = 0.97$. The base solution is determined by GrA. The research was conducted on the "test bed" consisting of six tasks that characterize various types of waste collection, as presented in Table 6. The visualization of analyzed streets for the P5 case is shown in Fig. 11 (properly located vertices force the passage along a given edge) and the visualization of collection points for case P6 is shown in Fig. 12.

Table 6 Characteristics of waste collection tasks in the case study.

Name of the case	Number of waste collection points/ streets	Collection type and category
P1	14 waste collection points	Bulky waste collection
P2	7 waste collection points	Bulky waste collection
P3	7 waste collection points	Bulky waste collection
P4	14 waste collection points	Bulky waste collection with the requirement to return to the company's base after driving 20 km
P5	Collection along 5 streets	Separate waste collection
P6	460 waste collection points	*E*-waste collection with time windows

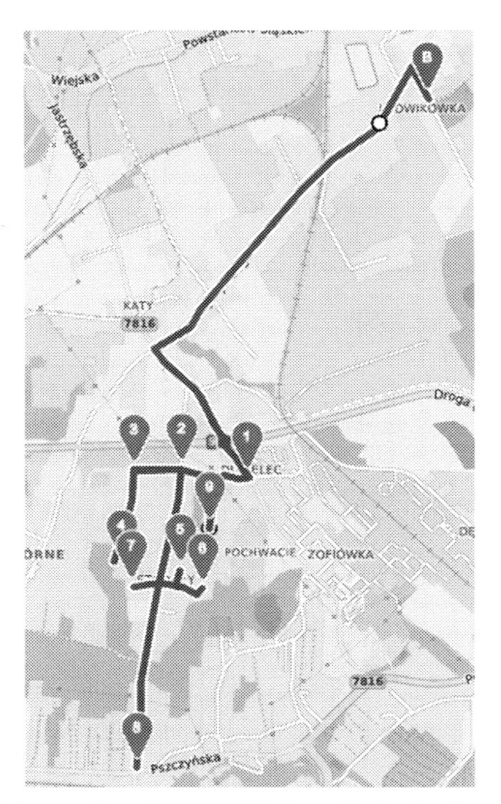

Fig. 11 Visualization of the routing for case P5 from Table 6. *(Source of map: https://maps.openrouteservice.org.)*

It has been assumed that all routing for waste collection tasks in the cases P1–P5 may be served in any time frame by a single vehicle from a waste collection company. In case P6, additionally, available time windows, the weight of the reported equipment, its potential recycling value, and its volume were included. Moreover, any number of vehicles, characterized by the same costs, capacity, and payload, could be assigned for each collection plan.

The results for route optimization for case P1 are presented in Table 7. The application of both SA and TS enabled the reduction of the route length in comparison with the solution determined by GrA. The waste collection vehicle travel routes constructed by GrA, SA, and TS are presented in Figs. 13–15, respectively.

Fig. 16 indicates a possible reduction of vehicles' emissions. Shortening the route has a direct impact on lowering emissions as the driving cycle for the waste collection is similar in urban or rural municipalities. It is defined by the speed limit in an urban area and it includes a cycle: "start," "go," and residential access (Keller and Wüthrich, 2014).

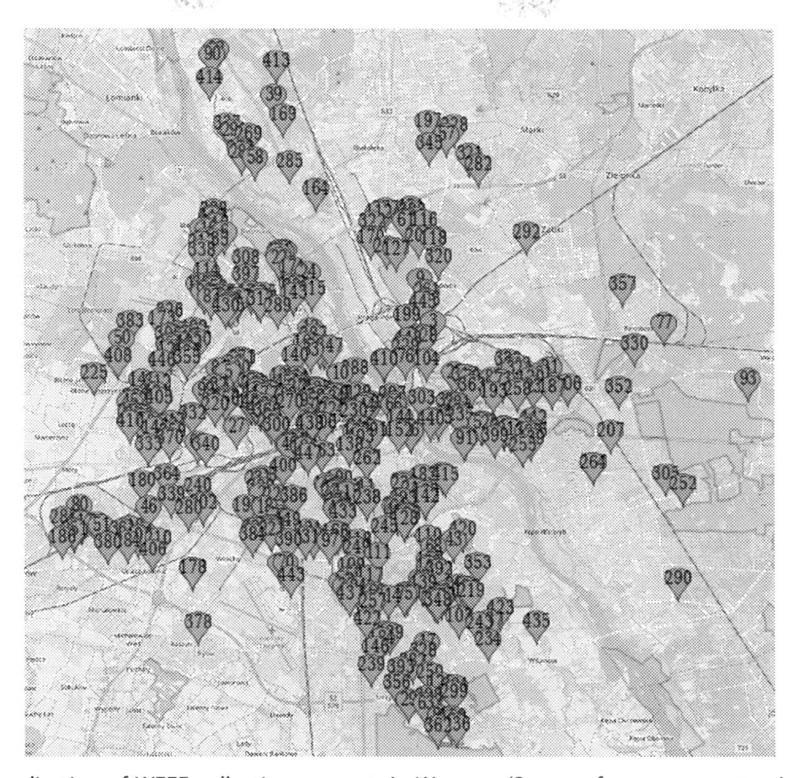

Fig. 12 Localization of WEEE collection requests in Warsaw. *(Source of map: mapcustomizer.com.)*

Table 7 Summary of algorithm results for the P1 case.

Parameter/algorithm	GrA	SA	TS
f	1	0.998746654	0.998768574
Route distance (km)	47.88	41.73	42.54
Served points	14	14	14

The results for case P2 are presented in Table 8. According to them, the application of metaheuristics (SA and TS) allowed reduction of the travel route again, while maintaining the same number of applications, which demonstrates the possibility to reduce the collection costs by using the appropriate optimization methods.

For case P3, all analyzed methods determined the same travel plan (the results are presented in Table 9). This means that the greedy approach proved to be effective for the location of reception points occurring in the task.

The effectiveness of particular methods for the task in which the vehicle needed to return to base after driving 20 km and then to continue collection is presented in

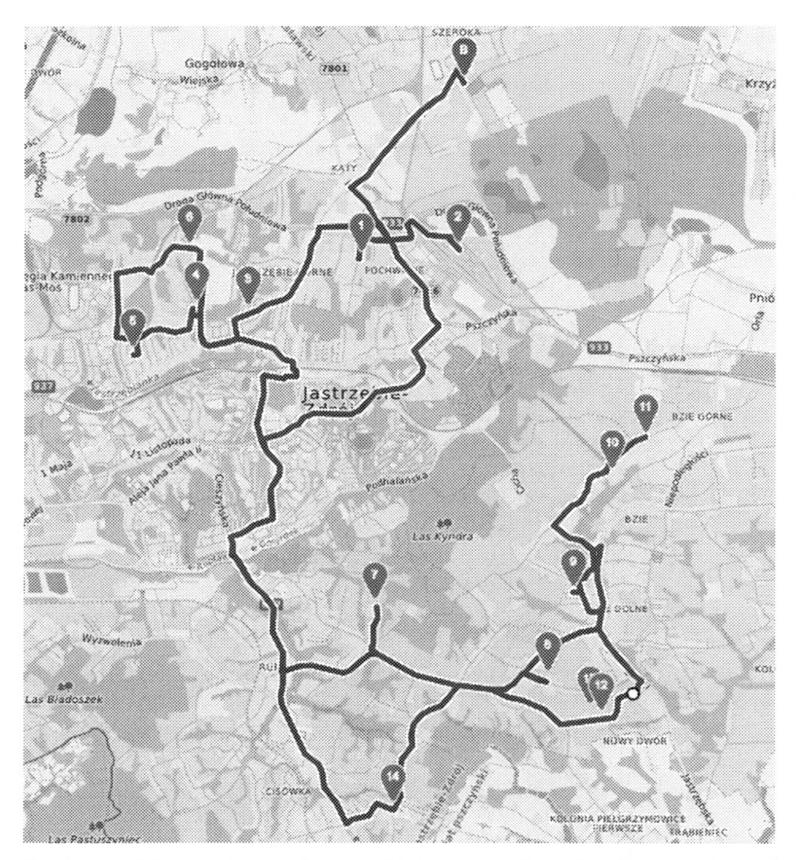

Fig. 13 A collection vehicle travel route for case P1 determined by GrA. *(Source of map: https://maps.openrouteservice.org.)*

Table 10. In line with the summary of results, metaheuristics demonstrated significant effectiveness again, thus allowing for significant reduction of the length of travel route.

Possible reduction of vehicles' emissions for case P4 including 14 waste collection points is shown in Fig. 17. In this case it was possible to limit emissions by 15% using the SA algorithm for route optimization and by 14% using the TS algorithm.

The results for case P5 are presented in Table 6. In line with these results, a collection that requires driving along five streets (instead of visiting waste collection points only) may also be optimized by using the appropriate metaheuristics. A possible reduction of emissions is shown in Fig. 18, indicating a 5% reduction for both SA and TS algorithms (Table 11).

The results for a relatively big optimization task, which included 460 WEEE collection points, are presented in Table 7 (an example of a travel route for one vehicle is shown in Fig. 6). Significant efficiency of the greedy approach was noticed, proving the

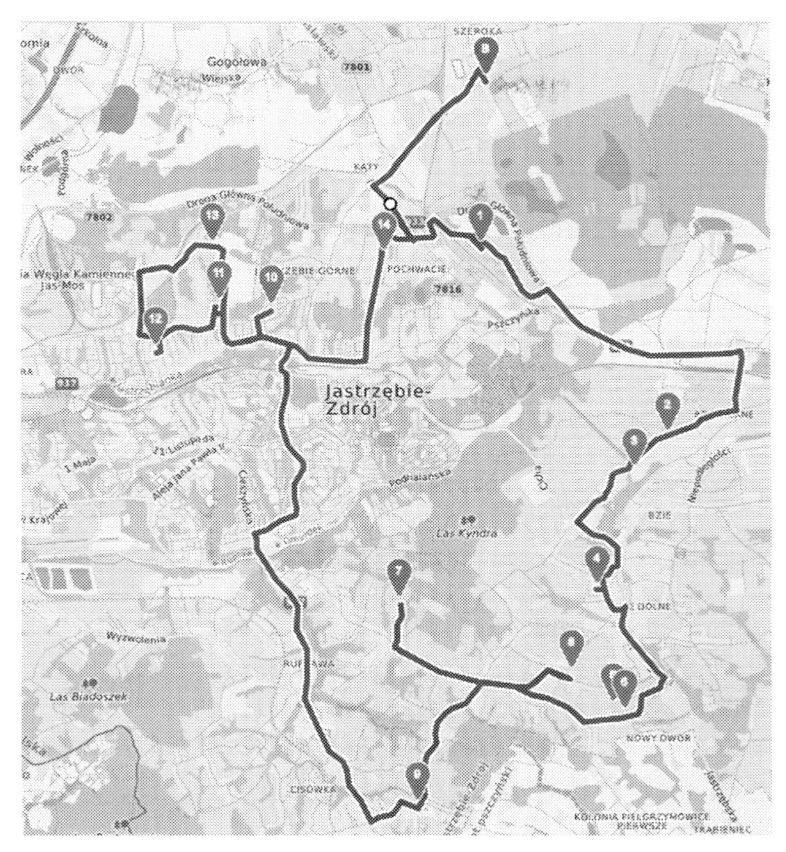

Fig. 14 A collection vehicle travel route for case P1 determined by SA. *(Source of map: https://maps.openrouteservice.org.)*

ineffective determination of the neighborhood for SA and TS in the case when multiple collection points are present (Table 12). TS slightly improved the result obtained by GrA, because one collection point was not served, which reduced the route length for this case (Fig. 19).

For case P6 (Table 6)—collection on-demand of e-waste—the purpose of collection planning was to assign the number of vehicles for the requests coming from inhabitants of Warsaw. The on-demand collection method can be efficient only if the vehicle routing problem with time windows is supported by artificial intelligence algorithms. This method allows direct handling of end-of-life equipment for the collection company's staff and discourages individuals from illegal disposal of waste equipment. Therefore, it produces an improvement in collecting secondary raw materials from WEEE and protecting the environment against illegal disposal.

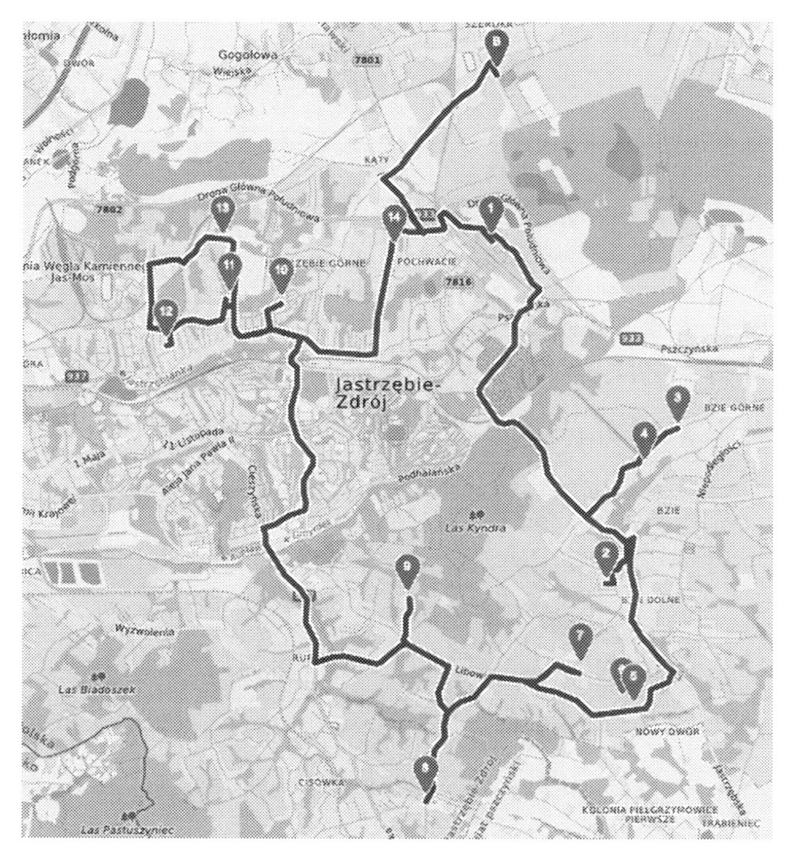

Fig. 15 A collection vehicle travel route for case P1 determined by TS. *(Source of map: https://maps.openrouteservice.org.)*

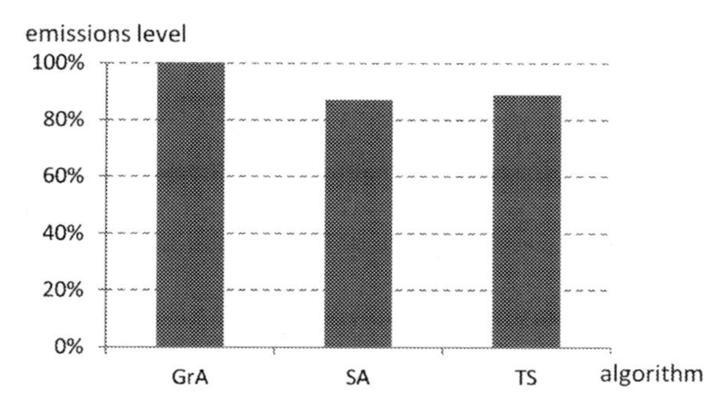

Fig. 16 Possible reduction of vehicle emissions for case P1—reduction by SA algorithm 13% and by TS algorithm 11%. *(Source: Authors.)*

Table 8 Summary of algorithm results for the P2 case.

Parameter/algorithm	GrA	SA	TS
f	1	0.999596	0.999596
Route distance (km)	23.84	23.3	23.3
Served points	7	7	7

Table 9 Summary of algorithm results for the P3 case.

Parameter/algorithm	GrA	SA	TS
f	1	1	1
Route distance (km)	23.26	23.26	23.26
Served points	7	7	7

Table 10 Summary of algorithm results for the P4 case.

Parameter/algorithm	GrA	SA	TS
f	1	0.998746654	0.998768574
Route distance (km)	52.65	44.75	45.07
Served points	14	14	14

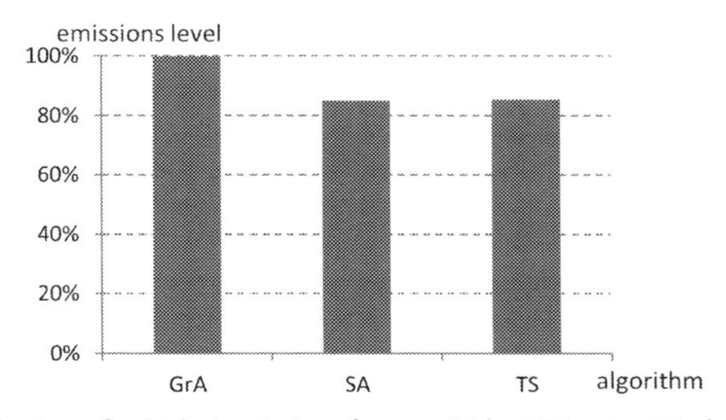

Fig. 17 Possible reduction of vehicles' emissions for case P4 by 15% using SA algorithm, and by 14% using TS algorithm. *(Source: Authors.)*

7. Discussion and conclusions

One of the fundamental activities in the circular economy is waste collection. The legislation in the European Union and the member states' national regulations impose a gradual increase in the collection and recycling rate for selected waste categories. Some categories of waste contain valuable secondary raw materials but some waste streams from households also contain hazardous substances like WEEE, medical waste, or portable

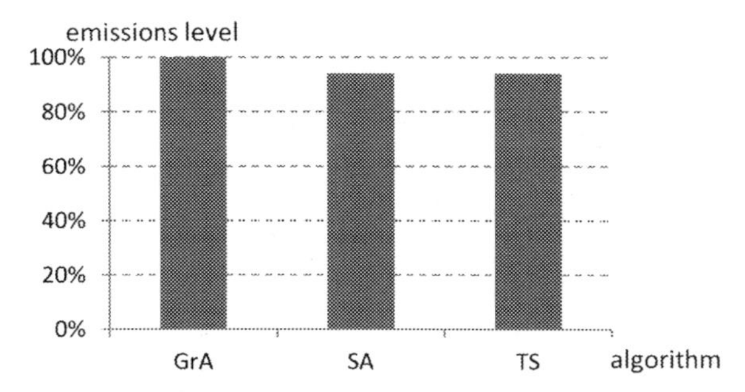

Fig. 18 Possible reduction of emissions for case P4—possible 5% reduction by SA and TS algorithms. *(Source: Authors.)*

Table 11 Summary of algorithm results for the P5 case.

Parameter/algorithm	GrA	SA	TS
f	1	0.999535	0.999535
Route distance (km)	10.03	9.44	9.44

Table 12 Summary of algorithm results for the P6 case.

Parameter/algorithm	GrA	SA	TS
f	1	1	0.990853
Used vehicles	13	13	13
Route distance (km)	1478.042	1478.042	1456.882
Served points	460	460	459

batteries. This chapter indicated some issues to be considered concurrently to meet the requirements of the circular economy: to maximize collection for all categories of waste and to prevent illegal or inappropriate waste disposal. Additionally, novel methods, such as artificial intelligence, for an individual category of waste can increase the efficiency of waste collection. At the same time, it is crucial to minimize negative health and environmental impacts during the transportation of waste (Kalmykova et al., 2018).

This chapter provides an insight into waste generation in the European Union and provides details of the methods, features, and results of the waste collection in urban and rural municipalities in Poland. Some negative factors accompanying waste collection and transportation are collection vehicles' exhaust emissions and noise generated during the collection and loading of some categories of waste. New studies on vehicle routing problems take into consideration pollution and emissions. Several approaches to minimize vehicle emissions discussed pollution routing problems and greenhouse gas routing problems (Bektaş et al., 2016; Bektaş and Laporte, 2011; Pradenas et al., 2013).

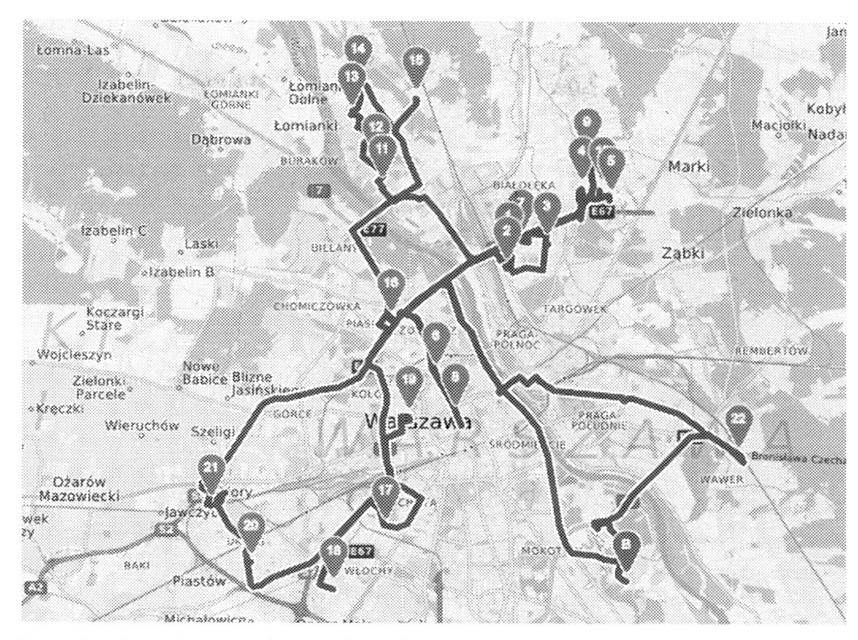

Fig. 19 Example of travel route of one vehicle for case P6 from Table 6. *(Source of map: https://maps.openrouteservice.org.)*

Regardless of the type of modeling of vehicle routing, different approaches must be applied for various categories of waste. For separated waste collection, the companies need to apply vehicle route optimization using the arc routing problem, combining collecting two or more fractions, and thus, limiting emissions. The results of this study show the potential of limitation of vehicles' exhaust emissions. They are presented in examples of on-demand bulky waste collections. As for a low number of waste pick-up requests, it is not necessary to apply artificial algorithms. Our results show the growing number of waste collection requests show a possible reduction of route length. A potential reduction of emissions can be within 5%–15% per route. This should encourage waste collection companies to apply the software for routing optimization. It also applies to waste collection planning along streets (ARP instance).

Other categories of waste like WEEE require a different approach in collection. It is most convenient for residents to request a waste pick-up by the collection company. This category of waste can be collected directly from households, as an on-demand service. Some recent studies indicate growing interest worldwide for such services (Nowakowski et al., 2020; Sun et al., 2020; Xue et al., 2019). In the case study presented in this chapter, we show an efficient online on-demand collection system (Nowakowski et al., 2018) operation for a large city—Warsaw. This method indicates benefits not only by a reduction of vehicle emissions for optimized routes but also increases the number of collected items containing hazardous substances. This way, the task of dealing with 460 requests was solved for a large city and required the use of 13 vehicles. Solving such tasks can only be efficiently conducted with the support of artificial intelligence algorithms.

Finally, the environmentally conscious behavior of the residents is a key factor in collecting secondary raw materials. It ranges from commonly used materials like plastics, metal, or paper from the separated waste collection to more valuable sources, such as waste electrical and electronic equipment. As was indicated in this study, illegal dumping or disposal brings harm to the natural environment starting from visual pollution of the countryside and ending by contamination of soil, water, and air. This issue requires educational and information campaigns addressed to residents of all age groups. Currently, the main activities in education focus on the youngest age groups. At the same time, waste collection companies must take responsibility for convenience and adapt to the requirements and volume of the waste stream collections. Novel methods of waste collection like the on-demand collection presented in this study have the potential to be applied widely. For larger tasks, they need to be supported by artificial intelligence algorithms. Future studies should extend to categories of waste such as portable batteries and pharmaceutical waste.

References

Andersen, M.S., 2007. An introductory note on the environmental economics of the circular economy. Sustain. Sci. 2, 133–140.

Ayomoh, M.K.O., Oke, S.A., Adedeji, W.O., Charles-Owaba, O.E., 2008. An approach to tackling the environmental and health impacts of municipal solid waste disposal in developing countries. J. Environ. Manag. 88, 108–114. https://doi.org/10.1016/j.jenvman.2007.01.040.

Bautista, J., Fernández, E., Pereira, J., 2008. Solving an urban waste collection problem using ants heuristics. Comput. Oper. Res. 35, 3020–3033. https://doi.org/10.1016/j.cor.2007.01.029.

Bektaş, T., Laporte, G., 2011. The pollution-routing problem. Transp. Res. B Methodol. 45, 1232–1250.

Bektaş, T., Demir, E., Laporte, G., 2016. Green vehicle routing. In: Green Transportation Logistics. Springer, pp. 243–265.

Bell, J.E., McMullen, P.R., 2004. Ant colony optimization techniques for the vehicle routing problem. Adv. Eng. Inform. 18, 41–48. https://doi.org/10.1016/j.aei.2004.07.001.

Bilitewski, B., Härdtle, G., Marek, K., 1996. Waste Management. U.S. Government Printing Office.

Bogaert, S., Van Acoleyen, M., Van Tomme, I., De Smet, L., Fleet, D., Salado, R., 2008. Study on RoHS and WEEE Directives. ARCADIS & RPA for the DG ENTR, European Commission.

Brandão, J., Eglese, R., 2008. A deterministic tabu search algorithm for the capacitated arc routing problem. Comput. Oper. Res. 35, 1112–1126. https://doi.org/10.1016/j.cor.2006.07.007.

Cattaruzza, D., Absi, N., Feillet, D., Vidal, T., 2014. A memetic algorithm for the multi trip vehicle routing problem. Eur. J. Oper. Res. 236, 833–848. https://doi.org/10.1016/j.ejor.2013.06.012.

Demir, E., Bektaş, T., Laporte, G., 2014. A review of recent research on green road freight transportation. Eur. J. Oper. Res. 237, 775–793. https://doi.org/10.1016/j.ejor.2013.12.033.

Department of Environment Regulation, 2014. Guide to Management of Noise from Waste Collection and Other Works. Department of Environment Regulation - Australia. [WWW Document]. URL https://www.der.wa.gov.au/images/documents/your-environment/noise/Guide_to_management_of_noise_from_waste_collection_and_other_works-DRAFT.pdf. (accessed 5.1.20).

Dotoli, M., Epicoco, N., 2017. A vehicle routing technique for hazardous waste collection. IFAC-PapersOnLine 50, 9694–9699. https://doi.org/10.1016/j.ifacol.2017.08.2051.

European Commission, 2012. Directive 2012/19/EU of the European Parliament and of the Council of 4 July 2012 on Waste Electrical and Electronic Equipment (WEEE)Text with EEA Relevance. LexUriServ.do.
European Commission, 2013. Environment Action Programme to 2020. [WWW Document]. URL https://eur-lex.europa.eu/legal-content/EN/TXT/?uri=CELEX:32013D1386. (accessed 5.1.20).
European Parliament, 2006. Directive 2006/66/EC of the European Parliament and of the Council of 6 September 2006 on Batteries and Accumulators and Waste Batteries and Accumulators and Repealing Directive 91/157/EEC (Text with EEA Relevance). [WWW Document]. URL https://eur-lex.europa.eu/legal-content/EN/TXT/?uri=CELEX%3A32006L0066. (accessed 4.1.20).
Eurostat, 2019a. Eurostat—WEEE Data. [WWW Document]. URL http://appsso.eurostat.ec.europa.eu/nui/submitViewTableAction.do. (accessed 11.18.19).
Eurostat, 2019b. Recycling Rate of Municipal Waste. [WWW Document]. URL https://ec.europa.eu/eurostat/databrowser/view/sdg_11_60/default/table?lang=en. (accessed 3.27.20).
Eurostat, 2019c. Waste Generated by Households by Year and Waste Category [WWW Document]. Waste generated by households by year and waste category. URL https://ec.europa.eu/eurostat/web/environment/waste/main-tables. (accessed 10.7.19).
Eurostat, 2019d. Waste Statistics—Recycling of Batteries and Accumulators.
Ghiani, G., Guerriero, F., Improta, G., Musmanno, R., 2005. Waste collection in southern Italy: solution of a real-life arc routing problem. Int. Trans. Oper. Res. 12, 135–144. https://doi.org/10.1111/j.1475-3995.2005.00494.x.
Ghiani, G., Manni, A., Manni, E., Toraldo, M., 2014. The impact of an efficient collection sites location on the zoning phase in municipal solid waste management. Waste Manag. 34, 1949–1956. https://doi.org/10.1016/j.wasman.2014.05.026.
Helmers, E., Leitão, J., Tietge, U., Butler, T., 2019. CO2-equivalent emissions from European passenger vehicles in the years 1995–2015 based on real-world use: Assessing the climate benefit of the European "diesel boom". Atmos. Environ. 198, 122–132. https://doi.org/10.1016/j.atmosenv.2018.10.039.
Hoornweg, D., Bhada-Tata, P., 2012. What a Waste: A Global Review of Solid Waste Management. World Bank, Washington, DC.
HSE, 2005. Noise in the Waste Management and Recycling Industry. Health and Safety Executive. [WWW Document]. URL https://www.hse.gov.uk/waste/noise.htm. (accessed 5.1.20).
Inglezakis, V.J., Moustakas, K., 2015. Household hazardous waste management: a review. J. Environ. Manag. 150, 310–321. https://doi.org/10.1016/j.jenvman.2014.11.021.
Jobling, B., WRAP, 2012. Noise Exposure in Glass Collections for Glass Recycling WRAP's Vision is a World Without Waste, Where Resources are used Sustainably. The Waste and Resources Action Programme, p. 76. Available at: https://wrap.org.uk/sites/default/files/2021-03/noise-exposure-in-glass-collections-for-recycling.pdf. (Accessed 18 November 2019).
Journal of Laws of The Republic of Poland, 2013. Ustawa z dnia 14 grudnia 2012 r. o odpadach. Ustawa z dnia 14 grudnia 2012 r. o odpadach [in Polish: ACT of 14 December 2012 on wastes Dz.U. 2013 item. 21]. Republic of Poland.
Kalmykova, Y., Sadagopan, M., Rosado, L., 2018. Circular economy—from review of theories and practices to development of implementation tools. Resour. Conserv. Recycl. 135, 190–201. https://doi.org/10.1016/j.resconrec.2017.10.034.
Keller, M., Wüthrich, P., 2014. Handbook Emission Factors for Road Transport 3.1 / 3.2 Quick Reference. INFRAS - Forschung und Beratung, Bern.
Knickmeyer, D., 2020. Social factors influencing household waste separation: a literature review on good practices to improve the recycling performance of urban areas. J. Clean. Prod. 245, 118605. https://doi.org/10.1016/j.jclepro.2019.118605.
Korhonen, J., Honkasalo, A., Seppälä, J., 2018. Circular economy: the concept and its limitations. Ecol. Econ. 143, 37–46.
Król, A., Nowakowski, P., Mrówczyńska, B., 2016. How to improve WEEE management? Novel approach in mobile collection with application of artificial intelligence. Waste Manag. 50, 222–233.
Laustsen, G., 2007. Reduce–recycle–reuse: guidelines for promoting perioperative waste management. AORN J. 85, 717–728. https://doi.org/10.1016/S0001-2092(07)60146-X.
Maimoun, M.A., Reinhart, D.R., Gammoh, F.T., McCauley Bush, P., 2013. Emissions from US waste collection vehicles. Waste Manag. 33, 1079–1089. https://doi.org/10.1016/j.wasman.2012.12.021.

Marinakis, Y., Marinaki, M., Migdalas, A., 2018. Particle swarm optimization for the vehicle routing problem: a survey and a comparative analysis. In: Martí, R., Panos, P., Resende, M.G.C. (Eds.), Handbook of Heuristics. Springer, Cham, pp. 1163–1196.

Nalepa, J., Blocho, M., 2016. Adaptive memetic algorithm for minimizing distance in the vehicle routing problem with time windows. Soft. Comput. 20, 2309–2327. https://doi.org/10.1007/s00500-015-1642-4.

Nowakowski, P., 2019. Investigating the reasons for storage of WEEE by residents—a potential for removal from households. Waste Manag. 87, 192–203. https://doi.org/10.1016/j.wasman.2019.02.008.

Nowakowski, P., Mrówczyńska, B., 2018. Towards sustainable WEEE collection and transportation methods in circular economy—comparative study for rural and urban settlements. Resour. Conserv. Recycl. 135, 93–107. https://doi.org/10.1016/j.resconrec.2017.12.016.

Nowakowski, P., Król, A., Mrówczyńska, B., 2017. Supporting mobile WEEE collection on demand: a method for multi-criteria vehicle routing, loading and cost optimisation. Waste Manag. 69, 377–392. https://doi.org/10.1016/j.wasman.2017.07.045.

Nowakowski, P., Szwarc, K., Boryczka, U., 2018. Vehicle route planning in e-waste mobile collection on demand supported by artificial intelligence algorithms. Transp. Res. Part D: Transp. Environ. 63, 1–22. https://doi.org/10.1016/j.trd.2018.04.007.

Nowakowski, P., Szwarc, K., Boryczka, U., 2020. Combining an artificial intelligence algorithm and a novel vehicle for sustainable e-waste collection. Sci. Total Environ., 138726. Article in press https://doi.org/10.1016/j.scitotenv.2020.138726.

Parajuly, K., Fitzpatrick, C., Muldoon, O., Kuehr, R., 2020. Behavioral change for the circular economy: a review with focus on electronic waste management in the EU. Resour. Conserv. Recycl. 6, 100035. https://doi.org/10.1016/j.rcrx.2020.100035.

Pires, A., Martinho, G., Chang, N.-B., 2011. Solid waste management in European countries: a review of systems analysis techniques. J. Environ. Manag. 92, 1033–1050. https://doi.org/10.1016/j.jenvman.2010.11.024.

Pradenas, L., Oportus, B., Parada, V., 2013. Mitigation of greenhouse gas emissions in vehicle routing problems with backhauling. Expert Syst. Appl. 40, 2985–2991.

Reşitoğlu, İ.A., Altinişik, K., Keskin, A., 2015. The pollutant emissions from diesel-engine vehicles and exhaust aftertreatment systems. Clean Techn. Environ. Policy 17, 15–27. https://doi.org/10.1007/s10098-014-0793-9.

Santos, L., Coutinho-Rodrigues, J., Current, J.R., 2010. An improved ant colony optimization based algorithm for the capacitated arc routing problem. Transp. Res. B Methodol. 44, 246–266. https://doi.org/10.1016/j.trb.2009.07.004.

Statistics Katowice, 2016. Urząd Statystyczny w Katowicach. Statystyczne Vademecum Samorządowca - Gmina miejska, Wodzisław Śląski (in polish: statistical Office in Katowice - statistical Vademecum of the local government - urban commune - Wodzisław Śląski – 2016).

Statistics Poland, 2019a. Environment 2019—Ochrona środowiska 2019. Statistics Poland, Warsaw, p. 190.

Statistics Poland, 2019b. Municipal Waste Management in Mazowieckie voivodship (Poland) in 2018. Statistics Poland, Warsaw. In Polish "Gospodarka odpadami komunalnymi w województwie mazowieckim - 2018.".

Statistics Poland, 2020. Population. Size and Structure and Vital Statistics in Poland by Territorial Divison. As of December 31, 2019. Statistics Poland, Warsaw.

Sun, Q., Wang, C., Zhou, Y., Zuo, L., Tang, J., 2020. Dominant platform capability, symbiotic strategy and the construction of "Internet + WEEE collection" business ecosystem:a comparative study of two typical cases in China. J. Clean. Prod. 254, 120074. https://doi.org/10.1016/j.jclepro.2020.120074.

Timlett, R., Williams, I.D., 2011. The ISB model (infrastructure, service, behaviour): a tool for waste practitioners. Waste Manag. 31, 1381–1392. https://doi.org/10.1016/j.wasman.2010.12.010.

Vidal, T., Crainic, T.G., Gendreau, M., Lahrichi, N., Rei, W., 2012. A hybrid genetic algorithm for multidepot and periodic vehicle routing problems. Oper. Res. 60, 611–624. https://doi.org/10.1287/opre.1120.1048.

Vidal, T., Crainic, T.G., Gendreau, M., Prins, C., 2013a. A hybrid genetic algorithm with adaptive diversity management for a large class of vehicle routing problems with time-windows. Comput. Oper. Res. 40, 475–489. https://doi.org/10.1016/j.cor.2012.07.018.

Vidal, T., Crainic, T.G., Gendreau, M., Prins, C., 2013b. Heuristics for multi-attribute vehicle routing problems: a survey and synthesis. Eur. J. Oper. Res. 231, 1–21. https://doi.org/10.1016/j.ejor.2013.02.053.

Williams, P.T., 2013. Waste Treatment and Disposal. Wiley.

Wolf, M.I., Gutowski, T.G., 2013. EOL product remanufacturing and reuse economics: consumer side and producer side. In: Proceedings of the International Symposium on Sustainable Systems and Technologies.

World Population Review, 2020. Warsaw Population 2020 (Demographics, Maps, Graphs). [WWW Document]. URL https://worldpopulationreview.com/world-cities/warsaw-population/. (accessed 4.1.20).

Xue, Y., Wen, Z., Bressers, H., Ai, N., 2019. Can intelligent collection integrate informal sector for urban resource recycling in China? J. Clean. Prod. 208, 307–315. https://doi.org/10.1016/j.jclepro.2018.10.155.

Zbib, H., Wøhlk, S., 2019. A comparison of the transport requirements of different curbside waste collection systems in Denmark. Waste Manag. 87, 21–32. https://doi.org/10.1016/j.wasman.2019.01.037.

Zhang, L., Klenosky, D.B., 2016. Residents' perceptions and attitudes toward waste treatment facility sites and their possible conversion: a literature review. Urban For. Urban Green. 20, 32–42. https://doi.org/10.1016/j.ufug.2016.07.016.

CHAPTER 25

New age zero waste sustainable apparel industry: Design practices, innovative approaches, and technological intervention

Indranil Saha and Deepak John Mathew
Department of Design, Indian Institute of Technology Hyderabad, India

1. Introduction

The apparel industry caters to the need for clothing, which is recognized as being one of the basic physiological human needs (Maslow, 1943). Activities involved in apparel design and production are controlled by core human skills of design with a huge variety of raw materials, product types, manufacture technologies, variable manufacture volumes, retail markets, and brands. Firms range from small individual businesses to multinationals. Apparel supply chains are usually universal, with raw materials being obtained from many different countries. This multiplicity is challenging to match in any other industry (Carr and Latham, 2008). The past two decades have seen massive changes in the structure of the apparel industry. This industry is continuously innovating with its design skillset, product development, procurement, logistics, and supply chain management. Domestic apparel production is also growing by catering to niche markets using professional skills (Carr and Latham, 2008). A growing number of nations, including India, are taking advantage of the apparel industry for economic development (Bheda et al., 2003; Dicken, 1998; Dickerson, 1995). Over the years, apparel design and manufacture technology have been gradually improved and there is qualitative and quantitative growth in the Indian apparel industry. The objective of this chapter is to explore the various aspects of the new age apparel industry and its innovative sustainable practices, which contribute towards a circular economic system. After the ongoing introduction section, which sets out the background of the chapter, a short narrative on the methodology is provided. The chapter is then categorized into sections to discuss the following. Firstly, it establishes the typical nature of the apparel industry and its adverse effects on the environment and society. Secondly, it reviews different methods of sustainable apparel design, manufacturing, and consumption procedures in order to minimize the harmful effects of the industry. It also discusses the ethical concerns associated with

Circular Economy and Sustainability, Volume 1
https://doi.org/10.1016/B978-0-12-819817-9.00035-1

both apparel firms and consumers. Finally, it assesses various zero waste design practices in the field of apparel design at different levels of its creation. The chapter ends with a closing note on the overall conception of the contribution of the whole study.

The current study implemented a systematic method of selecting and analyzing the existing literature (Johnsen et al., 2017) to reduce researcher bias concerning the addition or elimination of studies. Three important steps, planning, conducting, and reporting, were carried out to obtain information during the review process. The keywords were identified to suit the objective of the study. It was recognized that the topic of the circular economy in apparel and fashion industry was mainly introduced early in the last decade and it has received growing interest in the research field. Additionally, related search terms in the sphere of sustainability, including apparel design, manufacturing, and consumption perspectives, were also covered. The terms were thoroughly checked in the titles, abstracts, and the findings of the articles listed through a popular research search engine. This initial list was further reduced in terms of domain-specific relevance in their full texts. The studies that mainly focused on domains other than the textile, apparel, or fashion industries were excluded. For instance, domain-specific articles that focused primarily on engineering, mathematics, chemistry, energy, etc. were excluded to have a more precise group of articles for consideration. Finally, 75 studies were read thoroughly to evaluate them as eligible for the study. The following sections present the key findings of the analysis.

2. Features and ecological challenges in the apparel industry

Apparel design and production are often characterized by high competitiveness as well as high demand for product quality (Scott, 2006; Hassler, 2003; Forza and Vinelli, 2000). Carr and Latham (2008) have identified three special features that drive this extremely diverse industry. Firstly, the apparel industry requires a quick response as it integrates products ranging from high fashion luxury exclusives to mass-produced commodities (Carr and Latham, 2008). The nature of versatility and receptiveness towards the market demand calls for continuous technological innovation. Secondly, the apparel industry is highly labor intensive, which typically requires low fixed capital investment (Perry and Towers, 2012; Abernathy et al., 2006). The reason for this is recognized as being the simplicity of the essential process in apparel production, sewing, that controls the output of an apparel factory, irrespective of the size of the businesses. The sewing process makes up about 30% of the total cost of a garment (Perry and Towers, 2012). Thirdly, apparel design and production have industrialized supply chains globally. In the absence of a domestic supplier or producer, the crucial role of technology is to cater to the methods of handling the aspects of product development efficiently.

The apparel industry is going through a dynamic shift driven by the changing perception of consumers' purchase behavior (Nayak and Padhye, 2015). Consumer preference and demographic changeability are some of the significant challenges in the apparel industry. Modern consumers are seeking more multidimensional and intricate styles (Forward, 2003) due to increased responsiveness and access to existing information on the products and designs. On the other hand, the apparel supply chain is also multifaceted and challenging due to global outsourcing, lengthier lead times, and shorter seasons. The bricks-and-mortar apparel retailers are facing competition from online retailers, which continues to destabilize the apparel sector. As far as design and production methods are concerned, various jobs within the industry are performed by production personnel who execute cutting, sewing, finishing, and other processes in an assembly line. Regardless of the technical progressions, one of the leading global issues in the apparel industry is increasing labor costs. Apparel companies that deal with the same product categories sell similar clothing items to similar consumer segments or store types to survive in the market, resulting in stiff competition throughout every season.

Apart from the above-mentioned challenges, the most alarming concern faced by this industry is the ecological challenges. The apparel industry produces incredibly adverse effects on the environment. The global trend in the apparel business is shifting towards the use of ecofriendly materials and fewer chemical finishes (Fletcher, 2014). However, excessive usage of water, contamination from chemical treatments utilized in dyeing and preparation, the discarding of large quantities of used and unsold stock through burning or landfill deposits, and poor labor policies make clothing one of the highest impact industries in the world (Henninger et al., 2016; Pal and Gander, 2018). The nonprofit organization Global Fashion Agenda and global management consulting firm Boston Consulting Group forecast that worldwide apparel consumption will increase by 63%, from 62 million tons in 2017 to 102 million tons in 2030—the equivalent of 500 billion T-shirts. They also indicate that only 20% of clothing is picked up for recycling and reuse, with the rest typically ending up in landfills (Boston Consulting Group, Global Fashion Agenda, 2017). To minimize the environmental and social impact, the demand from consumers for apparel products that are designed, manufactured, and distributed sustainably is increasing (Nayak and Padhye, 2015). The apparel industry is the second most polluting industry globally (Henninger et al., 2016). Repeatedly, the apparel industry has been put under the spotlight for negative reasons, such as landfill caused by discarded apparel, pollution in water-bodies, labor rights issues, antifur campaigns, etc. (Henninger et al., 2017). In 2013, more than 1000 people were killed and 2500 people were injured in the tragedy of the Rana Plaza factory collapse in Bangladesh (Ryding et al., 2014). To reduce manufacturing costs, many apparel companies take advantage of a lack of environmental awareness and not-so-strict environmental supervisory systems in developing countries (Nagurney and Yu, 2012). Fashion brands like C&A, Benetton, and Adidas are accused of developing unsustainable supply chains (Shen et al., 2012).

Even though these negative instances dominate the apparel industry, recent years have seen a change. Both designers and retailers are promoting sustainable initiatives.

Many apparel companies are giving importance to sustainability in their business and they are also integrating green practices into their supply chain. For instance, Marks & Spencer started the initiative *Shwopping* (an initiative by M&S with Oxfam to help reduce unwanted clothing items going to landfill), which appeared as a new trend in 2012 (M&S. Ask Marks &, 2016), or the *Close the Loop* recycled apparel collection by H&M (Witte, 2016). Simultaneously, most of the newly established apparel firms are considering the aspects of sustainability as a form of environment, social, and economic element rather than looking at it as an add-on component. The emergence and development of new design practices, which effectively and efficiently generate products that are long-lasting and fit for restoration and recycling, are results of this paradigm shift (Todeschini et al., 2017). However, while discussing sustainability in the apparel industry, some considerations concerning ethics and consumers' perspective need to be explored to get a more comprehensive picture of this growing concern.

3. Sustainable apparel design, production, and consumption

Sustainable apparel is a concept that brings together a circular economy and clothing. In this day and age, it is essential to form a promise for the future with the aid of this pair (Clark, 2008). The circular economy offers several possibilities by establishing frameworks and stepping stones for a resilient, sustainable system for innovative product design, process, and business models (Webster, 2017). This feature of the circular economy is consistent with the definition of "sustainable development," i.e., "development that meets the needs of the present without compromising the ability of future generations to meet their own needs" (WCED, 1987).

The real concern in sustainable apparel design is the target of greater and long term change. Instead of defying changes in clothing habits, benefitting on a practical level from sustainable design viewpoints is more constructive. Cradle to cradle, a design philosophy established by McDonough and Braungart (2009), suggests that the lifecycle of materials should be the essential consideration instead of merely preserving natural resources (McDonough and Braungart, 2009). Instead of discarding the products, their material should be reused. Similar principles are being followed by designing for recycling (DFR) and designing for disassembly (DFD), which indicates that recycling should be integrated with the design process (Fletcher, 2008). The functional design concept described by Papanek (1984) requires that design must be in line with human needs to bring together utility, aesthetics, associations, and system (tools, materials, and processes) (Papanek, 1984). The concept of slow fashion is less defined than cradle to cradle or functional design; however, it has its origins in the slow food movement and slow design theory in the background. While the word "slow" signifies aspects like appreciating local resources, transparent production systems, and sustainable products, redesigning and recreating are identified as elements of slow fashion and ways of saving resources (Clark, 2008).

While exploring the ecological aspects of fibers, the typical considerations are the consumption of energy and the pollution produced by fiber processing. Whether the fiber is made from renewable resources, is fully biodegradable, or if it could be recycled or reused (Chen and Burns, 2006), all of these need to be taken into account. The concept of zero waste design includes many different approaches that all aim to eliminate material waste. The idea of zero waste apparel is deeply rooted; for example, Indian dhotis, saris, and the traditional Japanese kimonos all use one whole piece of fabric without wasting any of it. In some modern styles, the fabric is consumed by utilizing rectangular and triangular patterns that fit together, similar to a jigsaw puzzle (Gwilt and Rissanen, 2011). In addition to that, the technology of 3D printing entirely eradicates any possibility of waste that may occur in apparel production, by printing the particular size of the piece that is needed for the final product. Hence, while reducing the cost of production, this technology is also capable of minimizing the environmental impact by eliminating the misuse of materials during manufacturing (Valtas and Sun, 2016). Handcrafting apparel can also be considered as a way of sustainable production as hand-made manufacture saves energy; however, it is not feasible sustainable production in mass-market and contemporary manufacturing (Aakko and Koskennurmi-Sivonen, 2013).

Throughout the production phase, as well as the usage and disposal of apparel, energy efficiency is a crucial factor for consideration. Laundering, which typically consists of washing, drying, and ironing, is essentially one of the most vital actions to give attention to in the whole lifecycle of sustainable apparel. Frequent laundering of clothing results in more energy consumption than what was consumed in the process of making it. The contemporary cheap products of "fast-fashion" make self-mending of apparel look wasteful and unappealing, as in most of the cases, buying new apparel is more economical and time-saving than mending existing items. Clothing on rent is anticipated as a possible way to achieve sustainable clothing consumption. Allwood et al. (2006) mentioned that renting apparel offers the right to wear more clothing and minimizes the need to buy clothes only to wear once on special occasions or solely for the wish for a change (Allwood et al., 2015).

Sustainability has formerly been defined as an uncertain concept (Markusen, 2003) that lacks a precise definition. Therefore, sustainability can be and is described in multi-dimensional ways and can include an economic, social, and/or environmental approach. To restate this fact further, sustainability is an innately comprehended concept that is construed in a very subjective manner and hence, it is dependent upon the person and/or context (Henninger et al., 2016; Gunder, 2006). The phenomenon of sustainability has been observed since the 1960s, when consumers started becoming more conscious about the environmental impact of their consumption patterns (McCormick, 1995; Peattie, 1995). In the context of the apparel industry, sustainability issues are crucial (Battaglia et al., 2014); however, sustainable apparel is also not formally defined. Apparel products are produced and consumed; hence, the designer, the producer, and the wearer are all accountable for their choices. The more methodically the different elements of sustainability are practiced, the nearer the commonly acknowledged principles of sustainable apparel are achieved.

4. Ethical and consumption-based concerns of sustainable apparel

The concept of ethics and sustainability in the apparel industry, specifically in the fast fashion (Mcneill and Moore, 2015) context, is a complex concept and appears to be an oxymoron, as the desire to acquire fashion products is never truly satisfied. Traditionally, sustainable fashion refers to: "Fashionable clothes that incorporate fair trade principles with sweatshop-free labour conditions while not harming the environment or workers by using biodegradable and organic cotton"(Joergens, 2006). While this definition is limited to cotton production, the concept has been developed to be a more adaptable term: "Clothing, which incorporates one or more aspects of social or environmental sustainability, such as Fair Trade manufacturing or fabric containing organic-grown raw material"(Goworek et al., 2012). Apparel companies are being pressured to lessen the environmental impact of their actions and to develop sustainable fashion businesses (Chan and Wong, 2012; Gupta and Hodges, 2012; Shen et al., 2012), and this highlights a further distinction between environmentally responsible business (ERB) and socially responsible business (SRB). The environmental impact comprises of the chemicals and insecticides used in plantations that contaminate the environment during natural fiber production,and discarding of textiles and apparels in landfill that harm the environment (Bianchi and Birtwistle, 2010). To reduce this impact, the apparel industry has embraced small-scale moral retailers and has valued the launch of sustainable collections by big retailers (Goworek et al., 2012). Sustainable fashion can range from apparel made from organic fibers to recycling apparel and textiles and transforming them into new clothing designs (Goworek et al., 2012; Shen et al., 2012). In modern times, there has been a rising awareness of sustainable apparel production and consumption (Goworek et al., 2012). Diverse expressions like "ethical fashion," "eco-fashion," "fashion with a conscience," or "green fashion" are evidence of this growing interest. Although big retailers like H&M show dedication towards ethics and sustainability within their ethical collection and in their annual sustainability report (Witte, 2016), the universal view of sustainability within the apparel industry is not so positive. Besides natural resources concerns, increasingly, negligence of companies towards public health and security (Ryding et al., 2014) is observed to be a major issue.

Consideration of sustainability from the consumers' perspective leads to an emphasis on ethical consumption (Henninger et al., 2017). Ethical use includes a series of consumer behaviors, for example, animal welfare, sustainability, ecology, or human rights that are driven by motivations and moral obligations (Carrigan and Attalla, 2001). Sustainability in a clothing context is related to the idea of eco-fashion (Chan and Wong,

2012). However, awareness of sustainable fashion does not always show a relationship with buying intention; hence, it does not relate to an increase in eco-fashion sales (Joergens, 2006). Apart from eco-awareness, the prominence of the brand image is an additional variable concerning sustainable apparel consumption (Ryding et al., 2014). Some consumers are willing to buy sustainable apparel if there is no perceived risk associated with the extra price paid or there is no gap in quality (Gupta and Hodges, 2012; Carrigan and Attalla, 2001). Conventionally, consumers engaged in ethical behavior only if there were something that could have a health impact rather than unethical work practices (Joergens, 2006). Researchers provide a comprehensive foundation that classified four different types of ethical behavior comprising: (i) caring and ethical—those who act on ethical concerns (and who are a minority); (ii) confused and uncertain—consumers who are interested in ethical purchase but confused about the conflicting communications around commercial ethical behavior; (iii) cynical and disinterested—who have a lack of belief about businesses being truthfully ethical and will buy ethically if there is no effect on price, brand choice, and accessibility; and (iv) oblivious—consumers whose ethical concerns are not reflected in their buying choices (Carrigan and Attalla, 2001). Consumer psychological literature includes ample research work under ethical segments; however, very little work exists in recent research into fashion typologies, which has potential to add significant understanding in this area (Henninger et al., 2017). In line with this, ethical apparel brands should deliver more information in their marketing activities and endorse the benefits of sustainable consumption for consumers to make better ethical decisions (Chan and Wong, 2012; Gupta and Hodges, 2012; Carrigan and Attalla, 2001). This importance of source and clarity is a comparatively new call for action and fashion vocabulary and communications are being infused with more evidence. Hence, it is becoming a common trend for an apparel brand to make some effort at presenting reports on sustainability or including an environmentally oriented collection. Sustainability is a present-day concern and this term is often used in regards to pretty much every industry, such as food, development, fashion, and economics, etc. The sustainable movement has made a number of approaches in many diverse categories of fashion and apparel is certainly one of them. Designers need to ensure that they communicate the importance of sustainable products for the longevity of the planet (Devaki et al., 2017).

5. Zero waste design practices in apparel design

The concept of zero waste design is gaining popularity as a sustainable way of designing and manufacturing products (Carrico and Kim, 2014). Apart from fashion, the zero waste method is being explored in other manufacturing areas, for example, Caterpillar, DuPont, Proctor & Gamble, and Subaru are just a few of the companies that claim to have zero-waste-to-landfill plans. Anastas and Zimmerman (2003) listed one of the

12 green engineering principles as "It is better to prevent waste than to treat or clean up waste after it is formed" in their article (Anastas and Zimmerman, 2003). While sustainability can be a consequence of such design practices, in the apparel industry, zero waste design can also be an extremely ingenious challenge. For instance, zero waste patternmaking, which not only enables the designer to ideate creatively about shapes of the patterns within the garment, but may also demand innovative methods of sewing, closures, and seam finishes. Designers play an essential role in a rounded approach to creating apparel, considering aesthetics and function at the same time. Zero waste apparel design is an idea that attempts to eliminate any wasted fabric or material from the production of clothing by conceptualizing designs and production methods that consume 100% of a length of fabric or material so that there are no remainders. The average waste of fabric in the most common way of apparel production—cut and sew method—is 15% (Carrico and Kim, 2014; Townsend and Mills, 2013). Historically, the reason for the waste is that pattern pieces for most apparel products have asymmetrical shapes that do not seamlessly join with each other like jigsaw puzzle pieces. Although the apparel industry employs different approaches for lessening that waste, such as utilizing software to plot the pattern placements for cutting or cutting different sizes and styles together, allowing better probabilities for pattern pieces of interconnecting optimally, these methods do not entirely eradicate the wastage of fabric that occurs on the cutting-room floor. Researchers have also argued that if it is possible to eliminate the process of cutting from apparel production entirely with the aid of 3D printing technology, it will result in zero material waste (Valtas and Sun, 2016).

5.1 Zero waste apparel design: Historical and contemporary

The zero waste approach has been present in clothing habits since ancient times. Patterns traced from ancient clothing show that less fabric was wasted in the course of developing fashionable apparel. Ethnic clothing and traditional dress, such as the Indian Dhoti and Saree and the Greek Ionic Chiton, are instances of zero waste design concepts. As "cloth was clothing itself" (Rudofsky, 1947), they are pieces of fabric draped around the body without being cut or stitched, which results in zero wastage of fabric. Fig. 1 shows an Indian woman in a Saree.

Similarly, all pieces of the Japanese Kimono, another zero waste traditional clothing, are constrained to the width and length, generating no waste in the cutting process (Rudofsky, 1947). Manufacturing of textiles and garments was a much lengthier process during the Pre-Industrial Revolution period; hence, fabrics were considered a valuable resource. During that period, apparel producers aimed to utilize every piece of cut fabric, which often resulted in 100% utilization (Burnham, 1973). For centuries, the tailors and dressmakers of Europe were successful in eliminating fabric wastage during pattern development and cutting process (Saeidi and Wimberley, 2018). Before the 1800s, the width

Fig. 1 Ms. Lahama Bose, dancer in a Saree, Kolkata, India, September 15, 2018.

and cost of fabric used to determine the way of cutting. Patterns were typically joined closely together and had minimal cut-offs to utilize fabric resourcefully (Hill and Bucknell, 1987).

Post the Industrial Revolution, as new technologies accelerated textile production, some fabrics became very cheap and the cost associated with wastage of that fabric became irrelevant (Schneider, 1989). Claire McCardell, a fashion designer of the 1940s, used to get two rectangular fabric pieces stitched together in her designs, allowing fitting with the help of elastic bands either under the bust, waist, or at the shoulder, thus, contributing to zero waste apparel design (Rudofsky, 1947). The initial period of the 21st century has seen the rise of zero waste fashion designers who have initiated completely eradicating or decreasing waste in their designs using different methods. One of the means of achieving zero waste design is to recycle or to reuse the fabric scraps as an embellishment, patchwork, or applique in the garment. An American sustainable fashion company, Alabama Chanin, established by Natalie Chanin, makes all their garments either handcrafted by local artisans or with recycled and organic fabrics, and waste fabric during the cutting process is used as embellishments (Saeidi and Wimberley, 2018). Mark Liu, a fashion and textile designer, used a unique cutting technique to save 15% of the material on each garment. He cut the edges of the negative space of the fabric with a laser cutter and converted them as decorative external seams (Rissanen, 2013a). An Indian company, Doodlage Retail, founded by Kriti Tula, uses leftover fabric, a biodegradable fabric made from corn and banana fibers, and utilizes every last bit of fabric, at times by making stuffing or small accessories (YKA, 2016), thus contributing to the zero waste fashion movement.

The zero waste approach indicates that the designer's knowledge of the fabric and design dimension should be able to create a zero waste garment. Rissanen (2013b) in his research paper, mentioned that the width of fabric is the most crucial concern in zero waste design. It is impossible to design a zero waste apparel design unless the fabric width is known, as fabric width is the area to create the design. McQuillan (2011) has categorized different methods to achieve zero waste fashion design through pattern making, which includes tessellation and jigsaw puzzle within the static area of the width of the fabric.

5.2 Basic zero waste design approaches through pattern making

Tessellation methods can fill the width and length of the fabric by repeating one shape or motif. Depending upon the tessellated motifs, there is a possibility of having wasted areas that are not incorporated in the design; however, those areas are mostly placed along the selvedge of the fabric. To overcome the problem, McQuillan (2011) has tried using mathematical elements called fractals, which have random shapes, to lessen or remove the waste at the selvedge. But, this method would require significant mathematical expertise and calculation. Another possible solution would be applying smaller tessellated shapes as they easily come close together at the selvedge of the fabric (McQuillan, 2011).

The jigsaw method requires proficiency in pattern cutting to control the pattern pieces to join together without any wastage of fabric (McQuillan, 2011). Zero waste fashion designer Mark Liu (2010) describes the technicality of zero waste design as "it involves fitting all the flat pieces of your clothing pattern like a jigsaw puzzle so no fabric is wasted." In contrast to the tessellation method, the jigsaw method may use a range of different pattern shapes (Liu, 2010). McQuillan (2011) emphasizes two factors as determinants to proceed with her design process: "the width of the fabric" and "the fixed area." McQuillan aims to eradicate waste rather than to use less fabric, so, she does not limit the design to a specific length. Therefore, a comprehensive and clear plan is needed for designing zero waste apparel, which includes both technical details and the final appearance of the attire. While these approaches for zero waste design call for more time than traditional fashion design, the advantage is in the reduction of the environmental footprint by decreasing waste of resources (Saeidi and Wimberley, 2018).

5.3 Zero waste design approaches through creative pattern making

Creative methodology in pattern making is an initial point to remove fabric waste (Rissanen and McQuillan, 2016). In the 2D form of pattern making, it is challenging to consider the shape of the pattern pieces, how to join them together, and the fit of the garment all at once. As observed by (Rissanen, 2008), 3D form draping techniques may resolve the complications of using a 2D flat pattern method. Transformational reconstruction is a pioneering pattern-making method by Shingo Sato, a Japanese designer, who manipulates pattern pieces in 3D rather than 2D. In this method, pattern cutting is combined with the design process as desired design lines are drawn on a readily test-fitted toile on a dress form to make the pattern. Traditional fitting techniques, such as waistline seams and darts, can be removed by transforming them into the seam lines of the drawn design lines. To use pattern pieces in 2D, the toile is detached from the dress form, cut separately along the design lines, and then spread flat on top of the fabric as pattern pieces. Rissanen and McQuillan (2016) state that processes that include both pattern making and pattern cutting as an integral part of the design process distinguish zero waste apparel design from the conventional form.

5.4 Zero waste apparel design: Fabric production

5.4.1 *Fully fashioned knitted fabrics*

Fully fashioned is a frequently used term in knitted apparel production. The garment pieces are knitted separately and then stitched or linked with each other (Black, 2002). In its most accurate form, a fully fashioned garment production includes no cutting and thus generates little or no material waste. Domestic hand knitting is a familiar example where the front, back, and sleeves are knitted separately and then seamed together to make a sweater. According to BTTG (1999), 2% nylon yarn wastage occurs in the knitting of a blouse. Nevertheless, a lot of knitwear is a combination of fully fashioned and cut and sew, as arranging a knitting machine for each new style can be expensive (British Textile Technology Group, 1999).

5.4.2 *Seamless knitting*

Seamless knitting is a technically upgraded version of fully fashioned knitted fabrics. In this process, a machine knits a finished garment to its 3D shape; thus, it eradicates sewing from the apparel production process (Choi and Powell, 2005; Black, 2002). While this technology has only been made accessible for complicated apparel designs since the mid-1990s, typically, knitted gloves and socks have been manufactured in the industry using seamless knitting for decades. Allwood et al. (2006) mentioned that this method should be applied to produce more styles frequently to get rid of material waste. In this process, the wastage is reduced since the knitting machine produces finished garments from yarn without cutting and sewing (Allwood et al., 2015).

5.4.3 Nonwoven fabrics

Nonwoven fabrics are engineered fibrous assemblies that are important to the functional performance of products used in diverse medical, consumer, and industrial applications as part of daily life. Traditionally, the nonwovens industry has progressed within different areas of the polymer, paper, and textile industries, and today it has a distinct identity. (Mao and Russell, 2015). Nonwoven fabrics are widely used in the production of both disposable and durable apparel, such as synthetic leather fabrics, shoe linings, protective clothing, garment linings, interlinings, waddings, etc. The Woolmark Company and Canesis Network (AgResearch) developed nonwoven fabrics appropriate for woolen apparel and the process was essentially based on carding and mechanical bonding. A variety of fabrics with diverse colors, patterns, and textures for clothing were developed for applications in fashion using nonwoven fabrics. Sturdy sleeveless jackets manufactured from mechanically bonded nonwoven fabric have been available for commercial use for business wear in Australia for a long time (Mao and Russell, 2015). Manel Torres, a Ph.D. researcher in association with Fabrican Ltd., produced fiber that can be sprayed directly on the wearers' body to create a garment (Figure 4), demonstrating a zero waste fully fashioned method using nonwoven fabric (Rissanen, 2013b).

5.4.4 Woven fabrics

Woven fabrics are typically produced in rectangular lengths through looms. However, technologies for weaving shaped pieces other than the rectangular form of fabric are also present in the industry. TI Rissanen (2013) mentioned that hand-weaving can create shaped woven pieces. The Roman toga was woven into half-oval form rather than being cut (Tarrant, 1994). A German company, Shape 3, produces fully fashioned, 3D woven products for the medical industry using industrial equipment. From an apparel design perspective, Indian company August uses "DPOL" technology that manufactures ready-to-stitch shaped woven garment components finished with selvedge, contributing to the zero waste method using woven fabric (August, 2009).

As more design practitioners explore the area of zero waste apparel design, this field of design has possibilities to grow in the future, both from a sustainability viewpoint and from facing new creative methodological challenges. However, zero waste design includes some obvious complications if it is to be applied to the mass manufacturing level. Issues like pattern grading and creating size sets are not addressed with the methods of tessellation and jigsaw, as they do not allow any room for modification of the pattern pieces. Additionally, to eradicate wastage of scrap fabric, the surplus fabric may stay within the garment needlessly. This extra fabric may remain in the clothing in the form of broader facings, excessive dart (folded and sewn fabric to take in easily and provide shape to a garment) and pleat (a fold in a garment's fabric, held by stitching the top or side) intake, or ornamental shirring or smocking (Carrico and Kim, 2014). Although these elements provide decorative components to apparel design, they may result in the consumption of excessive fabric. For example, a pleated bib (the rectangular panel that runs up the front, doubling the fabric) for a tuxedo shirt adds aesthetic value, which calls for additional fabric in both conventional and zero waste garments. To eliminate these obstacles, the possibility of integrating other technologies with zero waste design concepts is required.

5.5 Technological intervention: 3D printed apparel

Additive manufacturing or three-dimensional (3D) printing is a popular aid in visualization and prototype testing, typically in the fields of aircraft, biomedical implants, tissue engineering, and medical implants (Yeong et al., 2004; Thomas et al., 1996). Gradually, additive manufacturing is being applied in the creation of artistic pieces as well as customized fashion products like apparel and accessories (e.g., footwear, sunglasses, watches, jewelry) allowing mass customization (Vanderploeg et al., 2017; Mpofu et al., 2014; Yap and Yeong, 2014; Ferreira et al., 2012). 3D printing of apparel consist of several steps, including product design, creation of a model in 3D modeling software, dividing the 3D model into a sequence of 2D slices, and then printing the series of 2D slices layer by layer into a 3D product (Perry, 2017; Valtas and Sun, 2016). From a design perspective, additive manufacturing facilitates designers to construct apparel design with more freedom by allowing them to explore formerly inaccessible technical restrictions. This advantage is particularly relevant for the apparel industry, which integrates the elements of design and creativity.

Furthermore, high-quality end products and a new platform for effective customization have been offered by additive manufacturing to match the diverse market demands of custom-made jewelry and clothing (Ferreira et al., 2012). At the same time, it also provides consumers with possibilities to create precisely what they want (Tarmy, 2016; Kuhn and Minuzzi, 2015). Hence, 3D printed apparel is projected to be the future of fashion. Even though, in comparison to apparel, direct and indirect additive manufactured accessories have more material choices, in the past few years, many eminent fashion shows have showcased 3D printed garments, for example, haute couture from Iris Van Herpen, a Dutch fashion designer, who is one of the innovators to embrace additive manufacturing in fashion apparel, and a black netted gown by Francis Bitonti, formed by selective laser sintering (SLS). Consequently, 3D printed garments have the potential to widely alter the apparel market as well as consumers' lives (Tarmy, 2016).

By and large, sustainability includes three standard dimensions: social, economic, and environmental. All these dimensions are supposed to be considered concurrently and equally in sustainable decision-making. While evaluating the impact on the sustainability of any production process—specifically, an additive manufacturing process—the entire life cycle of the final product and its raw materials are to be considered. There are several ecofriendly features of additive manufacturing (Mani et al., 2014), as follows:

i. The nature of the additive manufacturing processes (constructing layer by layer and placing material only where needed) results in a smaller amount of waste when compared to mass production processes such as cutting and sewing (where lots of raw material is scrapped as cut or trimmed fabrics, for instance, in the case of apparel production).
ii. Additive manufacturing processes do not need any specific tools or equipment. Thus it eliminates the need for manufacturing and disposal of these at the end of their life cycle.
iii. Additive manufacturing processes are capable of generating durable, functionally lightweight parts, which have significantly less impact in their life cycle sustainability (e.g., reduced weight of particular aerospace components causes reduced requirement for energy to move them around).
iv. Additive manufacturing processes use raw material only when it is required. Hence, it decreases the need for keeping large amounts of raw material within the supply chain, warehousing, and transportation.
v. Additive manufacturing processes are generally more material-efficient when compared with traditional machining and casting.
vi. Additive manufacturing processes are also capable of generating optimized geometries with near-perfect (compared with wrought material) strength-to-weight ratios, which is essential for lightweight structures such as airplanes.
vii. There is less impact of the generated part over its complete life cycle, which results in a lower carbon footprint, less embodied energy, and an improved economic model.
viii. Additive manufacturing processes can make demand-based spare parts, decreasing or eliminating the need for inventory and transportation of these spare parts over vast distances.

The sustainable advantages of additive manufacturing derived from analysis and interpretation of prior literature have been extracted for particular applications. While the perception of sustainability caters to the environmental, societal, and economic aspects, the consumption of the product and material life cycles indicate that the ecological dimensions of sustainability have appeared to be most important in the discussed literature. Although a few aspects of social sustainability have also emerged, some elements like employment and the distribution of labor, health and safety, ethics, quality of life, creativity, and self-expression are not included in the existing literature (Ford and Despeisse, 2016). This dearth of literature calls for further thorough investigations into the environmental, societal, and economic sustainability of additive manufacturing that take forward and validate the previous research work (Kohtala, 2015; Huang et al., 2013).

The findings of this chapter suggest that a number of research studies have been carried out to theorize varied dimensions of sustainable design and development practices in the apparel industry from the zero waste perspective. Most of the zero waste-oriented studies have been focused on textile manufacturing and pattern making/cutting methodologies. At the same time, there has been a rapid increase of technological innovation, such as, additive manufacturing in apparel design and development practices in order to overcome hurdles within the industry. Hence, the emerging zero waste methodology to address the sustainability issue using technological innovations is important in the perspective of apparel design. The findings of the study provide contemporary apparel design experts and consumers with directions for applying zero waste methods in mainstream practice to reinvent apparel product development, concentrating on constructive benefits towards businesses, society, consumers, and the overall ecosystem.

References

Aakko, M., Koskennurmi-Sivonen, R., 2013. Designing sustainable fashion: possibilities and challenges. Res. J. Text. Appar. 17 (1), 13–22. [Internet]. Available from: https://doi.org/10.1108/RJTA-17-01-2013-B002.

Abernathy, F.H., Volpe, A., Weil, D., 2006. The future of the apparel and textile industries: prospects and choices for public and private actors. Environ. Plan. A 38 (12), 2207–2232.

Allwood, J.M., Laursen, S.E., De Rodriguez, C.M., Bocken, N.M.P., 2006. Well Dressed. Present Futur Sustain Cloth Text United Kingdom. University of Cambridge, p. 1.

Allwood, J.M., Laursen, S.E., de Rodriguez, C.M., Bocken, N.M.P., 2015. Well dressed?: The present and future sustainability of clothing and textiles in the United Kingdom. J. Home Econ. Inst. Aust. 22 (1), 42.

Anastas, P.T., Zimmerman, J.B., 2003. Peer Reviewed: Design through the 12 Principles of Green Engineering. ACS Publications.

August, 2009. August: Background. [Internet]. [cited 2018 Sep 17]. Available from: http://august.synthasite.com/faqs.php.

Battaglia, M., Testa, F., Bianchi, L., Iraldo, F., Frey, M., 2014. Corporate social responsibility and competitiveness within SMEs of the fashion industry: evidence from Italy and France. Sustainability 6 (2), 872–893.

Bheda, R., Narag, A.S., Singla, M.L., 2003. Apparel manufacturing: a strategy for productivity improvement. J. Fash. Mark. Manag. 7 (1), 12–22.

Bianchi, C., Birtwistle, G., 2010. Sell, give away, or donate: an exploratory study of fashion clothing disposal behaviour in two countries. Int. Rev. Retail Distrib. Consum. Res. 20 (3), 353–368.

Black, S., 2002. Knitwear in Fashion. Thames & Hudson, London.

Boston Consulting Group, Global Fashion Agenda, 2017. Pulse of the Fashion Industry 2017., p. 139. Available from: http://globalfashionagenda.com/wp-content/uploads/2017/05/Pulse-of-the-Fashion-Industry_2017.pdf.

British Textile Technology Group, 1999. Report 4: Textile Mass Balance and Product Life Cycles.

Burnham, D.K., 1973. Cut My Cote. Textile Department, Royal Ontario Museum.

Carr, H., Latham, B., 2008. In: Tyler, D.J. (Ed.), Carr and Latham's Technology of Clothing Manufacture, fourth ed. Blackwell Publishing Ltd, Oxford. [Internet]. Available from: www.blackwellpublishing.com.

Carrico, M., Kim, V., 2014. Expanding zero-waste design practices: a discussion paper. Int. J. Fash. Des. Technol. Educ. 7 (1), 58–64.

Carrigan, M., Attalla, A., 2001. The myth of the ethical consumer–do ethics matter in purchase behaviour? J. Consum. Mark. 18 (7), 560–578.

Chan, T., Wong, C.W.Y., 2012. The consumption side of sustainable fashion supply chain: understanding fashion consumer eco-fashion consumption decision. J. Fash. Mark. Manag. 16 (2), 193–215.

Chen, H.L., Burns, L.D., 2006. Environmental analysis of textile products. Cloth. Text. Res. J. 24 (3), 248–261.

Choi, W., Powell, N.B., 2005. Three dimensional seamless garment knitting on V-bed flat knitting machines. J. Text. Apparel Technol. Manag. 4 (3), 1–33.

Clark, H., 2008. SLOW + FASHION—an oxymoron—or a promise for the. Fash. Theory J. Dress Body Cult. 7419 (January), 427–446.

Devaki, E., Suganya, M., Sreelakshmi, S., 2017. Sustainable fashion—a review. In: Parthiban, M., Srikrishnan, M.R., Kandhavadivu, P. (Eds.), Sustainability in Fashion and Apparels: Challenges and Solutions. Woodhead Publishing India Pvt. Ltd, New Delhi, pp. 73–82 (96A).

Dicken, P., 1998. Global Shift. Paul Chapman, London. Google Sch.

Dickerson, K.G., 1995. Textiles and Apparel in the Global Economy. Prentice Hall, Englewood Cliffs.

Ferreira, T., Almeida, H.A., Bártolo, P.J., Campbell, I., 2012. Additive manufacturing in jewellery design. In: ASME 2012 11th Biennial Conference on Engineering Systems Design and Analysis. American Society of Mechanical Engineers, pp. 187–194.

Fletcher, K., 2008. Sustainable Fashion and Clothing. Design Journeys. Basım Routledge, London.

Fletcher, K., 2014. Sustainable Fashion and Textiles: Design Journeys, second ed. Routledge New York, NY, New York.

Ford, S., Despeisse, M., 2016. Additive manufacturing and sustainability: an exploratory study of the advantages and challenges. J. Clean. Prod. 137, 1573–1587. [Internet]. Available from https://doi.org/10.1016/j.jclepro.2016.04.150.

Forward, R., 2003. Twenty Trends for 2010: Retailing in an Age of Uncertainty. OH Retail Forw, Columbus, p. 8.

Forza, C., Vinelli, A., 2000. Time compression in production and distribution within the textile-apparel chain. Integr. Manuf. Syst. 11 (2), 138–146.

Goworek, H., Fisher, T., Cooper, T., Woodward, S., Hiller, A., 2012. The sustainable clothing market: an evaluation of potential strategies for UK retailers. Int. J. Retail Distrib. Manag. 40 (12), 935–955.

Gunder, M., 2006. Sustainability: Planning's saving grace or road to perdition? J. Plan. Educ. Res. 26 (2), 208–221.

Gupta, M., Hodges, N., 2012. Corporate social responsibility in the apparel industry: an exploration of Indian consumers' perceptions and expectations. J. Fash. Mark. Manag. 16 (2), 216–233.

Gwilt, A., Rissanen, T., 2011. Shaping Sustainable Fashion [Internet], Shaping Sustainable Fashion: Changing the Way We Make and Use Clothes. pp. 1–192. Available from: http://www.scopus.com/inward/record.url?eid=2-s2.0-84920443014&partnerID=tZOtx3y1.

Hassler, M., 2003. The global clothing production system: commodity chains and business networks. Global Netw. 3 (4), 513–531.

Henninger, C.E., Alevizou, P.J., Oates, C.J., 2016. What is sustainable fashion? J. Fash. Mark. Manag. 20 (4), 400–416.

Henninger, C.E., Ryding, D., Alevizou, P.J., Goworek, H., 2017. Sustainability in Fashion: A Cradle to Upcycle Approach. Springer. pp. 1–270.

Hill, M.H., Bucknell, P.A., 1987. The Evolution of Fashion: Pattern and Cut from 1066 to 1930. Gardners Books.

Huang, S.H., Liu, P., Mokasdar, A., Hou, L., 2013. Additive manufacturing and its societal impact: a literature review. Int. J. Adv. Manuf. Technol. 67 (5–8), 1191–1203.

Joergens, C., 2006. Ethical fashion: myth or future trend? J. Fash. Mark. Manag. 10 (3), 360–371. [Internet]. Available from: https://doi.org/10.1108/13612020610679321.

Johnsen, T.E., Miemczyk, J., Howard, M., 2017. A systematic literature review of sustainable purchasing and supply research: theoretical perspectives and opportunities for IMP-based research. Ind. Mark. Manag. 61, 130–143.

Kohtala, C., 2015. Addressing sustainability in research on distributed production: an integrated literature review. J. Clean. Prod. 106, 654–668.

Kuhn, R., Minuzzi, R.F.B., 2015. The 3D printing's panorama in fashion design. Moda Doc Museu. Meme Des. 11 (1), 1–12.

Liu, M., 2010. What is Zero-Waste Fashion (and Why Does It Matter)? Ecouterre. [Internet]. [cited 2018 Sep 16]. Available from: https://inhabitat.com/ecouterre/what-is-zero-waste-fashion-and-why-does-it-matter/.

M&S. Ask Marks &, 2016. Spencer—How Do I Find Out About Shwopping? [Internet]. [cited 2018 Sep 16]. Available from http://help.marksandspencer.com/support/company-website/shwopping.

Mani, M., Lyons, K.W., Gupta, S.K., 2014. Sustainability characterization for additive manufacturing. J. Res. Natl. Inst. Stand. Technol. 119, 419.

Mao, N., Russell, S.J., 2015. Fibre to fabric: nonwoven fabrics (Internet). In: Textiles and Fashion: Materials, Design and Technology. Elsevier Ltd, pp. 307–335. Available from https://doi.org/10.1016/B978-1-84569-931-4.00013-1.
Markusen, A., 2003. Fuzzy concepts, scanty evidence, policy distance: the case for rigour and policy relevance in critical regional studies. Reg. Stud. 37 (6–7), 701–717.
Maslow, A.H., 1943. A theory of human motivation. Psychol. Rev. 50 (4), 370–396.
McCormick, J., 1995. Environmental policy and the European Union. Contrib. Polit. Sci. 355, 37–50.
McDonough, W., Braungart, M., 2009. Cradle to Cradle : Remaking the Way We Make Things, second ed. Vintage, London: London.
Mcneill, L., Moore, R., 2015. Sustainable fashion consumption and the fast fashion conundrum: fashionable consumers and attitudes to sustainability in clothing choice. Int. J. Consum. Stud. 39 (3), 212–222.
McQuillan, H., 2011. Zero-waste design practice: strategies and risk taking for garment design. In: Shaping Sustainable Fashion: Changing the Way We Make and Use Clothes, vol. 83, p. 97.
Mpofu, T.P., Mawere, C., Mukosera, M., 2014. The impact and application of 3D printing technology. Int. J. Sci. Res. 3 (6).
Nagurney, A., Yu, M., 2012. Sustainable fashion supply chain management under oligopolistic competition and brand differentiation. Int. J. Prod. Econ. 135 (2), 532–540. [internet]. Available from https://doi.org/10.1016/j.ijpe.2011.02.015.
Nayak, R., Padhye, R., 2015. Introduction: the apparel Industry. In: Rajkishore, N., Padhte, R. (Eds.), Garnment Manufacturing Technology. First. Woodhead Publishing, Cambridge, pp. 1–469.
Pal, R., Gander, J., 2018. Modelling environmental value: an examination of sustainable business models within the fashion industry. J. Clean. Prod. 184, 251–263.
Papanek, V., 1984. Design for the Real World: Human Ecology and Social Change. Academy Chicago.
Peattie, K., 1995. Environmental Marketing Management: Meeting the Green Challenge. Financial Times Management.
Perry, A., 2017. Factors comprehensively influencing acceptance of 3D-printed apparel. J. Fash. Mark. Manag. 21 (2), 219–234.
Perry, P., Towers, N., 2012. Fashioning a socially responsible garment supply chain: a qualitative exploration of corporate social responsibility in Sri Lankan export garment manufacturers. In: Fashion Supply Chain Management: Industry and Business Analysis. IGI Global, pp. 327–330.
Rissanen, T.I., 2008. Creating fashion without the creation of fabric waste. In: Suatainable Fash why Now? A Conversat about issues, Pract possibilities.
Rissanen, T., 2013a. Zero-Waste Fashion Design: A Study at the Intersection of Cloth, Fashion Design and Pattern Cutting (Doctoral dissertation). p. 240.
Rissanen, T., 2013b. Zero-Waste Fashion Design : A Study at the Intersection of Cloth, Fashion Design and Pattern Cutting. [cited 2018 Sep 16]; Available from: https://opus.lib.uts.edu.au/handle/10453/23384.
Rissanen, T., McQuillan, H., 2016. Zero Waste Fashion Design. vol. 57 Bloomsbury Publishing.
Rudofsky, B., 1947. Are Clothes Modern?: An Essay on Contemporary Apparel. Chicago Paul Theobald.
Ryding, D., Navrozidou, A., Carey, R., 2014. The impact of eco-fashion strategies on male shoppers' perceptions of brand image and loyalty. Int. J. Bus. Glob. 13 (2), 173–196.
Saeidi, E., Wimberley, V.S., 2018. Precious cut: exploring creative pattern cutting and draping for zero-waste design. Int. J. Fash. Des. Technol. Educ. 11 (2), 243–253. [Internet]. Available from https://doi.org/10.1080/17543266.2017.1389997.
Schneider, J., 1989. Rumpelstiltskin's bargain: folklore and the merchant capitalist intensification of linen manufacture in early modern Europe. Cloth Hum Exp, pp. 177–213.
Scott, A.J., 2006. The changing global geography of low-technology, labor-intensive industry: clothing, footwear, and furniture. World Dev. 34 (9), 1517–1536.
Shen, B., Wang, Y., Lo, C.K.Y., Shum, M., 2012. The impact of ethical fashion on consumer purchase behavior. J. Fash. Mark. Manag. 16 (2), 234–245. [Internet]. Available from: https://doi.org/10.1108/13612021211222842.
Tarmy, J., 2016. The Future of Fashion Is 3D Printing Clothes at Home. Bloomberg. [Internet]. [cited 2018 Sep 1]. Available from: https://www.bloomberg.com/news/articles/2016-04-15/3d-printing-is-poised-to-bring-haute-couture-into-the-home.

Tarrant, N., 1994. The Development of Costume. Routledge, London.
Thomas, C.L., Gaffney, T.M., Kaza, S., Lee, C.H., 1996. Rapid prototyping of large scale aerospace structures. In: Aerospace Applications Conference, 1996 Proceedings, 1996 IEEE. IEEE, pp. 219–230.
Todeschini, B.V., Cortimiglia, M.N., Callegaro-de-Menezes, D., Ghezzi, A., 2017. Innovative and sustainable business models in the fashion industry: entrepreneurial drivers, opportunities, and challenges. Bus. Horiz. 60 (6), 759–770. [Internet]. Available from https://doi.org/10.1016/j.bushor.2017.07.003.
Townsend, K., Mills, F., 2013. Mastering zero: how the pursuit of less waste leads to more creative pattern cutting. Int. J. Fash. Des. Technol. Educ. 6 (2), 104–111.
Valtas, A., Sun, D., 2016. 3D printing for garments production: an exploratory study. J. Fash. Technol. Text. Eng. 04 (03). [Internet]. Available from: http://www.scitechnol.com/peer-review/3d-printing-for-garments-production-an-exploratory-study-BISE.php?article_id=5467.
Vanderploeg, A., Lee, S.-E., Mamp, M., 2017. The application of 3D printing technology in the fashion industry. Int. J. Fash. Des. Technol. Educ. 10 (2), 170–179. [Internet]. Available from: https://doi.org/10.1080/17543266.2016.1223355.
WCED, 1987. Report of the World Commision on Environement and Development: Our Common Future. Oxford Pap. [Internet]. Report of:400. Available from http://www.un-documents.net/wced-ocf.htm.
Webster, K., 2017. The Circular Economy: A Wealth of Flows. Ellen MacArthur Foundation Publishing.
Witte, E., 2016. H &M Close the Loop Recycled Clothing Fashion Line. [Internet]. [cited 2018 Sep 16]. Available from: https://www.refinery29.com/2016/09/124862/hm-close-the-loop-recycled-clothing.
Yap, Y.L., Yeong, W.Y., 2014. Additive manufacture of fashion and jewellery products: a mini review: this paper provides an insight into the future of 3D printing industries for fashion and jewellery products. Virtual Phys. Prototyp. 9 (3), 195–201.
Yeong, W.-Y., Chua, C.-K., Leong, K.-F., Chandrasekaran, M., 2004. Rapid prototyping in tissue engineering: challenges and potential. Trends Biotechnol. 22 (12), 643–652.
YKA, 2016. Sustainable Entrepreneurship In Fashion: Doodlage Founder Explains. [Internet]. [cited 2018 Sep 16]. Available from: https://www.youthkiawaaz.com/2017/01/doodlage-turns-waste-to-fashion/.

CHAPTER 26

A conceptual and empirical study into the process and emerging patterns enabling the transition to a circular economy: Evidence from the Dutch dairy sector

Hilde Engels and Jan Jonker
Department of Strategy, Institute for Management Research, Radboud University Nijmegen, Nijmegen, The Netherlands

1. Introduction: Circular economy from a transitional perspective

In 2015, 195 countries signed the Paris Climate Agreement (United Nations Framework Convention on Climate Change, 2015) to combat climate change. The implementation of this climate agreement requires a far-reaching change of societal arrangements. The linear economy is the current dominant industrial model of "take-make-dispose" or "end-of-life," based on a process of extraction of resources, followed by industrial production and use, and ended by the deposition of waste (Ellen MacArthur Foundation, 2013). Recently, the ambition to change towards a circular economy has gained increased attention from academics, politicians (European Commission, 2015, 2019), and practitioners as it is viewed as a promising paradigm of organizing a sustainable future.

Circular economy (CE) holds the ability to achieve prosperity for all, while realizing a more balanced system of production and consumption. It aims to reduce the need for materials, reuse, recycle, and restore these, thus enabling the retention of their value as long as possible in the loop (Blomsma and Brennan, 2017; Ellen MacArthur Foundation, 2013; Geissdoerfer et al., 2017; Jonker et al., 2018; Murray et al., 2017; Stahel, 1982). The advantages of a CE are the reduction of costs through the whole value chain because less resources are needed for production activities, resulting in reduced emission in the value chain and life cycle (Van Buren et al., 2016). However, CE requires a different way of organizing society (Jonker et al., 2018) as it is argued to be based on (1) the cooperation between partners in loops and networks, (2) leading to cooperative or collective value creation, and (3) value that is understood not only as economic values, but also ecological and social values. Organizing the CE in various functional subsystems

https://doi.org/10.1016/B978-0-12-819817-9.00019-3

might deliver a valuable and impactful contribution to implementing the Paris Climate Agreement, leading to a drastic reduction of GHG-emissions.

Kirchherr et al. (2017) also identified multiple values when they conceptualized an integral definition of CE based on their research of 114 published definitions of CE. We use a slightly adapted version for the operationalization of CE in our research: a circular economy describes an economic system that is based on business models which replace the 'end-of-life' concept with reducing, alternatively reusing, recycling and recovering materials in production/distribution and consumption processes, thus operating at the micro level (innovation structures developing technologies, networks), meso level (regime structures and rules, cooperating networks) and macro level (long term societal trends), with the aim to accomplish sustainable development, which implies creating environmental quality, economic prosperity and social equity, to the benefit of current and future generations. (Adapted from Kirchherr et al., 2017, p. 224–225). Here we use this definition as a starting point to investigate one subsystem, namely food production and consumption. We need the broadness of this definition because the functioning of the food system includes more than only an approach of resources and materials, often referred to as the 4 Rs of reduce, reuse, recycle, and recover. For the production of food, a favorable environment is needed in which, i.e., water supply, soil quality, and biodiversity regarding soil and further ecosystem services play a key role. Hence, environmental quality must be a part of this definition.

The Dutch food system, which includes the structure, culture, and practices "involved in feeding the population: growing, harvesting, processing, packaging, transporting, marketing, consuming, and disposing food and food-related items, including the inputs needed and outputs generated at each of these steps" (Hollander et al., 2017, p. 17) still functions as a mainly linearly organized, complex system. The dairy sector poses important societal challenges including GHG-emissions, nitrate emissions, animal welfare, and biodiversity loss. The Dutch Minister of Agriculture, Nature, and Food Quality has stated that "the way we produce our food today .... is not sustainable .... while the future of our food provision can only be secured by switching to a circular agriculture" (Ministry of Agriculture, Nature and Food Quality, 2018, p. 5). The first outline of a more circular food system in the Netherlands was presented recently (Scholten et al., 2018). It is now politically acknowledged that the change to circular food production means a fundamental shift in the Dutch agriculture system, which is denominated as a "transition" (Scholten et al., 2018).

Fundamental societal changes are explored by the field of transition studies, which is rapidly growing (Chappin and Ligtvoet, 2014). A "transition" can be defined as a radical structural transformation of a societal system (or of subsystems), which entails mutual changes in the fundamental societal structures related to economic, cultural, technological, ecological, and institutional arrangements at different societal levels (Rotmans et al., 2001). Within transition studies, strategic niche management (SNM) is a conceptual framework concerning "the creation, development and controlled phase-out of protected spaces (called "niches") for the development and use of promising technologies by means of experimentation" (Kemp et al., 1998, p. 186). The intention of SNM is to guide the adoption of new technologies in their early stages in order to ensure the creation of path dependency in the direction of sustainability and eventually enhance a regime shift (Kemp, 1994; Kemp et al., 1998; Rip and Kemp, 1998). A "regime" can be seen as a rule-set that describes how the established order of society is organized with regard to institutional arrangements (Markard and Truffer, 2008). The regime is the primary factor for the stability of existing large-scale systems. Innovative activities are directed towards incremental improvements within the rule-set (Geels, 2002), and the regime is, simultaneously, the primary source of resistance to radical change because of the regime actors' vested interests (Unruh, 2000). The regime functions as the selection environment for technological development in a specific field and forms a significant barrier to the diffusion of radical innovations. From this perspective, a transition may be regarded as a regime shift (Geels, 2002). The multilevel perspective (MLP) appears to be an appropriate framework for the analysis of the dynamics of a transition (Geels, 2002). The MLP considers "niches" to be the micro level of spaces but also opportunities for radical innovations; and the "sociotechnical regime" to be the meso level of institutionalized dominant structures. Both of them are embedded in a broader "landscape," seen as the macro level of autonomous, exogenous context. A transition can be better understood by combining and thus reinforcing interaction of processes on these three levels.

This chapter aims to contribute to the knowledge of transitional processes by analyzing the extant literature on niche development and subsequently applying the findings to a case study of circular economy in the subsystem of agrifood.

2. Understanding transitional processes by applying a multiphase approach

Understanding transitional processes is imperative for guidance that aims to enhance the acceleration of transition towards sustainability. This section explains the multiphase approach as a conceptual framework for the understanding of the process of niche development.

In order to develop this understanding, a systematic literature review was conducted on peer-reviewed papers that were selected from the Web of Science Core Collection by searching for "strategic niche management" and "transition" and "niche development" in the period of 1998 to 2019. This search produced 143 papers, which were examined to determine their contribution to the exploration of the process of niche development. The dominant findings of this research are described in this section.

2.1 Taking a multiphase approach to transitions

Although the multiphase approach was one of the first concepts elaborated on in transition studies, its exploration has been limited. Four dynamic phases were explained (Loorbach, 2007; Rotmans et al., 2001):

(1) "Predevelopment," in which an increasing political and societal sense of urgency is being felt whereby societal pressure is building up;

(2) "Take-off," in which the process of change begins to build up and is fueled by the shared goals and agendas of political and private institutions;

(3) "Acceleration," also referred to as "breakthrough," in which rapid structural changes become apparent from the transformations in societal subsystems; and.

(4) "Stabilization," in which the rate of societal changes decreases, and a new dynamic equilibrium is established.

The process of niche development must be placed in this transitional process and can be divided into four congruent phases (Geels, 2005; Geels and Kemp, 2007).

Fig. 1 depicts these four phases of niche development in an uncomplicated and ideal-typical pathway of emerging niche innovations advancing a regime, thus adopting a niche-to-regime process (Schot and Geels, 2008). However, their categorization in phases is not always clear because one phase follows from another, and there is some

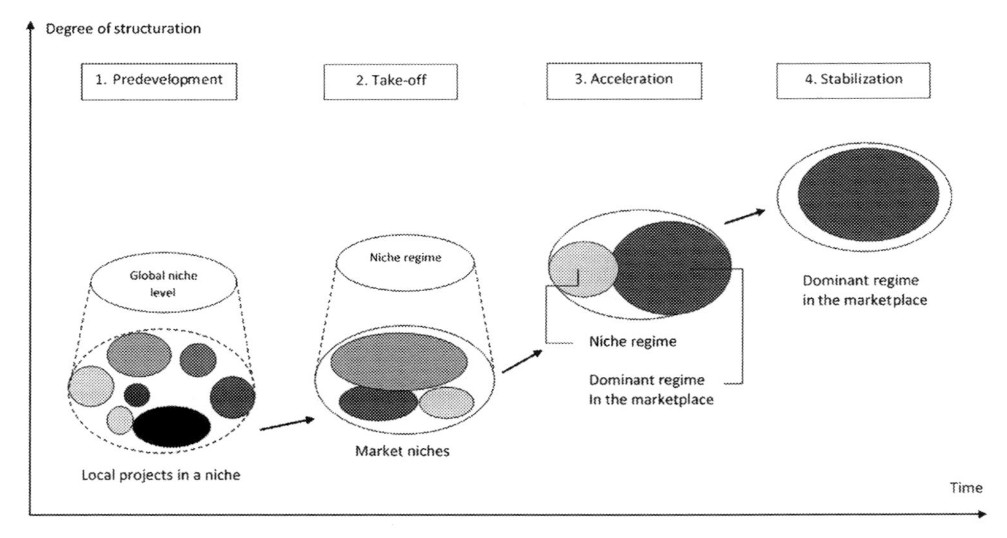

Fig. 1 Four phases of niche development in a simple and ideal-typical niche-to-regime pathway. *(Adapted from Geels, F.W., Raven, R., 2006. Non-linearity and expectations in niche-development trajectories: ups and downs in Dutch biogas development (1973–2003). Technol. Anal. Strateg. Manag. 18 (3–4), 375–392; Schot, J.W., Geels, F.W., 2008. Strategic niche management and sustainable innovation journeys: theory, findings, research agenda, and policy. Technol. Anal. Strateg. Manag. 20 (5), 537–554.)*

overlap between the phases. Yet, each phase of niche development has its own organizational characteristic activities (as is summarized in Table 1):

(1) The first phase—predevelopment—can be characterized by experimentation in a protected environment (Kemp et al., 1998, 2001; Smith and Raven, 2012). The focus of the radical innovation lies on three internal processes: articulation of expectations and vision, building social networks, and learning. From a variety of local projects, a global niche level develops (Geels and Raven, 2006). At this level knowledge (Beers et al., 2019) and resources (Geels and Deuten, 2006) are exchanged. An intermediary actor (i.e., a user-group, branch organization, or supporter) may take different roles: coordinating for gaining support (Smith and Raven, 2012), or inspiring, advancing, or linking projects (Martiskainen and Kivimaa, 2018).

(2) The second phase—take-off—is related to the early diffusion of a new technology, the growth and specialization of its innovation structure. Experiments contribute to transitions by deepening (focused learning), broadening (application in different contexts), and scaling up (dissemination) (Van den Bosch and Rotmans, 2008). Learning processes lead to a "dominant design" of niche structure (Geels, 2005). A niche regime is constituted as a new coherent set of stable rules and routines. When the market niche scales up and grows beyond the initially protected space (of which the protection phases out), the market niche is ready for a breakthrough in full market competition and may challenge the existing regime (Geels and Kemp, 2007). We define a "dominant design" as the dominant set of rules of how practices are performed, structure is organized, and culture guides these practices and structures (based on Rotmans and Loorbach, 2010).

(3) The third phase—acceleration—can be characterized by a wide diffusion of the radical innovation structure. The niche regime in which it is embedded fully interacts with the established regime. The opportunity for the niche regime to advance depends on the interplay of three processes (Geels, 2014; Turnheim and Geels, 2012, 2013): (a) pressures from the economic and sociopolitical environments; (b) destabilization of the regime caused by internal problems; a and b provide windows of opportunity but further diffusion of the niche depends on (c) the problem-solving potential of the innovation or its potential to fulfil societal needs in general. Different transitional paths are conceptualized (De Haan and Rotmans, 2011) and empirically identified (Geels and Schot, 2007). Destabilization of regimes follows five stages, ending in dissolution of the regime (Turnheim and Geels, 2012).

(4) In the fourth phase—stabilization—a new dynamically stable equilibrium is established. The continued refinement of the "dominant design" leads to incremental improvements within the rule set of the dominant regime. The mutual, interdependent, co-evolutionary development of structure, culture, and practices leads to a lock-in situation of an inert marketplace (Unruh, 2000, 2002), whereby the regime is subject to new challenges from the environment as well as from within.

Table 1 Four phases of niche development and their characteristic activities; constructed from transition literature.

(1) Predevelopment characteristics	(2) Take-off characteristics	(3) Acceleration characteristics	(4) Stabilization characteristics
• Space: niches. • Experimentation. • Internal processes: (1) articulation of expectations and visions, (2) building of social networks, (3) learning. The interaction of these processes lead to more articulated shared rules. • Protection processes induced by external forces: shielding, nurturing, empowerment. • Generation of a global niche level from local projects: sharing knowledge, creating resources, guiding activities in local projects. Intermediary actors at the global niche level play an important role.	• Space: market niches. • First transactions. • Further technical specialization. • Stabilization of learning processes leading to a dominant design of the innovation structure. • Internal processes: (1) networks become larger, (2) rules become more stable, (3) a niche regime is constituted as a new coherent set of rules and routines, created by the mutual development of the new technology, user preferences, infrastructure, industry structure, scientific knowledge, policy, and cultural meaning. • Protection processes by external forces phase-out.	• Space: competitive marketplace. • Full interaction of niche regime with the existing dominant regime. Success of breakthrough depends on pressures from the economic and sociopolitical environments in combination with the destabilizing tensions from within the regime and the problem-solving potential of the radical innovation. • Five most common paths of regime change in which radical innovations play a role: (1) reconfiguration; (2) substitution; (3) backlash; (4) teleological path; (5) lock-in path.	• Space: inert marketplace. • New dynamically stable equilibrium. • Co-evolutionary development of structure, culture, practices of the dominant regime leads to a lock-in state. • The regime is subject to new challenges from the environment and from within.

2.2 Conclusions from our literature review

We conducted a systematic literature review to identify contributions on niche development of transitions and applied the multiphase approach to find the characteristics of four phases. Based on this research, a remarkable observation is that it seems as if only thorough empirical research has been conducted in the first (predevelopment) and third (acceleration) phases, while research in the second phase (take-off) and fourth phase (stabilization) seems to be unbalanced, yet each phase is primordial in the process of a transition. In particular, the second—take-off—phase, when the innovation structure is developing its "market niche," seems most underexposed. One of the gaps we identified in the take-off phase concerns the dynamic process to move from (multiple) market niches with a bundle of various innovations to a dominant market niche. This is crucial because the market niche is ready for a breakthrough into full market competition once it has constituted a niche regime and its learning processes have led to a "dominant design" of niche structure that can be deployed. We performed a case study to test the findings of this analysis.

3. Circular economy of Dutch dairy in a transitional perspective

3.1 Methodology and data sources

The change to circular food production implies a transition in Dutch agriculture. We conducted a case study to explore how circular economy of the Dutch food system emerges, especially in the dairy sector, thus taking a research strategy of qualitative research in a case setting (Yin, 2016). The research focuses on a process of development and thus requires process research. Process research explains outcomes as the result of the contribution of a sequence of events (Geels and Schot, 2010; Van de Ven and Engleman, 2004). Process studies adopt a narrative event-driven approach in aiming to explain how a process unfolds.

We focused our research on a positive case regarding CE, in this research understood as a food network with a track record of demonstrated efforts to applying circular economy practices during at least the last 10 years. We selected the dairy sector because of its socioeconomic significance: the dairy sector is the largest agricultural sector of the Netherlands with regard to economics (1.2% contribution to the Dutch economy in 2014) and land-use (28% of Dutch surface area) (Statistics Netherlands, 2018). First, we reviewed archival sources to investigate relevant historical events. Additionally, we interviewed seven key informants of actors of the dairy network: a dairy farming company, a dairy production company, a retail chain, a financial institution, an NGO, and two environmental agencies, in the period of February until May 2019. The interviewed informants were (sustainability) directors of their companies. We operationalized CE using the 17 dimensions given by Kirchherr et al. (2017). The gathered data contain events in their context as the unit of observation at the micro, meso, or macro level. We used the multi-phase approach to structure the results of the case study.

3.2 Results of the case study of Dutch dairy

In this section, the transitional phases are presented showing how the Dutch food system and the embedded network of dairy developed. To understand the context of today, we start with a sketch of the history of Dutch agriculture from secondary data; starting from 1945 as the post-war period marked the beginning of the rationalization of Dutch agriculture. We end by presenting the last 10 years from primary data.

3.2.1 1945–1957 Never hunger again

Before World War II, farms were circularly organized at a company scale or at a local scale when they practiced mixed farming, and 70% of the farms were smaller than 5 ha (van Merriënboer, 2006). The farmers owned some cattle and a piece of land for growing crops, while the manure was applied to fertilize their land. In 1945, after the severe "hunger winters" of World War II, the first priority was to restore food provision for the Dutch people. In 1948, the (post-)war food shortages were solved, primarily as the result of a decisively acting government. The Minister of Agriculture, Mansholt, firmly believed in stimulating exports, which in his view needed a policy of rationalization and liberalization (van Merriënboer, 2006). He initiated the modernization of the agricultural system by facilitating large-scale farming. He introduced import duties and export restitutions. At the same time, he protected the farmers by assessing guaranteed minimum prices for main agricultural products. He developed his governmental policies in consultation with the trade and industry, as in his view, the market was essential for the further development of the sector. The government invested in the development of strong knowledge institutions.

3.2.2 1958–1971 The roll out

In 1958, the European Economic Community was founded by West Germany, France, Belgium, Luxemburg, Italy, and The Netherlands. Mansholt believed in integration of agriculture at a European scale and he became the first European Commissioner of Agriculture, which he stayed until 1973. In this position, he was able to roll out his ideas of specialization, intensification, and scaling-up, to the detriment of small family enterprises (van Merriënboer, 2006). Specialized dairy farming and intensive livestock farming grew and simultaneously, the agricultural production, which was supported by mechanization and education. In 1959, import duties were decreased which fueled European Community trade. The food system that was created now functioned at a European scale. It could have been circular, but for two main reasons was not. First, imports from countries overseas were admitted, which led to many imports of cheap cattle feed; second, the manure-food cycle was interrupted: the manure stayed in the country of dairy production, while exports of dairy products grew rapidly.

3.2.3 1972–2014 Continued growth

By selective breeding and applying concentrated feed, the milk production per cow continued to increase (from 2.890 L in 1950, to 5.000 in 1980, and 8.200 in 2016) (Statistics Netherlands, 2017). The Common Agricultural Policy guaranteed minimum prices, which created European excesses of milk, butter, cheese, and other products. Yet it was difficult to change this policy, as farmers seemed to have become dependent on these subsidies. Finally, in 1984, a European milk quota was agreed upon to control the produced amounts of milk in the European Community. In 1972 "The limits to growth" was published (Meadows et al., 1972). For the first time, scientists showed that the depletion of resources could mean an end to growth. Problems rose related to dairy farming: "acid rain" (1979), "mad cow disease" (1993), and foot-and-mouth disease (2001), influencing public opinion on intensive farming. Regulations and laws were enacted to limit the manure supply and also the cattle population, but the industrial and farmers' lobby stayed strong. From 2006, Europe granted permission for "derogation," an exemption of the regulatory norms on nitrates. Meanwhile, NGOs put pressure on animal welfare. In 2006, the Party for Animal Welfare entered the Second Chamber of Parliament, in the same year that the FAO report "Livestock's long shadow" was published (FAO, 2006). However, the majority of Dutch politicians prioritized the economic position over the state of animal welfare, public health, or the natural environment. The European Commission announced, in 2008, the end of the milk quota in 2015. The Dutch government and dairy sector are content to be a top-exporter of the world year after year (Statistics Netherlands, 2019).

3.2.4 2015–2019 In turmoil

From the interviews conducted, 2015 appeared to be a turning point in the mindset of participants in the dairy network. This was the year the European milk quota ended. Dutch farmers called it "liberation day," taking the opportunity for growth. Some dairy producers had enlarged their capacity to be able to process an increased amount of milk. But the consequence of this new growth was that the European regulatory phosphate levels were exceeded (Ministerie van Landbouw, Natuur en Voedselkwalteit, 2019) and the derogation was threatened. The government rapidly came up with new laws and regulations and the number of cows decreased by 160,000 (10%) so as to achieve the agreed phosphate levels. At the "landscape level," extreme rainfall and droughts created awareness of the consequences of climate change. Research has shown a 75% decrease of insects in the countryside during the last 30 years (Hallmann et al., 2017; Kleijn et al., 2018). There is a growing sense of urgency in Dutch society that a transition of the agrifood system is needed, but the agricultural lobby is strong, recently having led to protests and demonstrations.

In 2018, the Dutch Minister of Agriculture took the initiative for developing a vision that should lead to circular agriculture, with the involvement of industry, farmers, nature conservation associations, NGOs, provincial governments, and societal organizations (Ministry of Agriculture, Nature and Food Quality, 2018). The transformation of the food system has the aim to achieve a transition to a sustainable, more resilient, and circular form of food production, which will contribute to climate goals as well as biodiversity goals (Scholten et al., 2018). Scholten proposes a "circular agriculture" assuming:

- minimal losses of resources of biomass through the closure of loops of resources and biomass within the agricultural system;
- constraining the use of land and resources to what is necessary to produce food for human consumption;
- promoting the reuse of food waste as feed for cattle, and subsequently only exceptionally may land be used for production of feed for cattle;
- promoting the use of organic fertilizers instead of chemical fertilizers.

The indirect consequence of this new circular system is a reduction of amounts of livestock, which is believed to alleviate the pressure on multiple sustainability goals.

3.2.5 Results from the interviews

In the interviews we held with participants in the dairy network, we were confronted with the diverse interpretations of the concept of circular economy applied to agrifood. All respondents referred in abstract terms to CE ("keeping nutrients in cycles"); reduction of resources and recycling of food waste were the most familiar elements of CE. The dairy farmer has a partnership with an arable farmer to dispose of his cattle's manure but does not yet receive crops as feed in return. There was a sense that circularity is not always necessarily sustainable: efficiency does not always lead to an integral sustainable mode. A hot issue is the decreasing environmental quality. All interviewees recognize the importance of business models that enhance a better environmental quality while having an earning capacity for the farmer (enhancing a greener environment can only exist within the boundaries of "the cycle of the wallet," as the farmer calls it). A partnership between the farmer, the dairy producer, and the retail organization has the aim to enhance the ecological quality while offering economic returns for the farmer and showing transparency of resource and product flows to consumers. The interviewees agree that collaboration with chain partners is necessary to achieve a sustainable, circular economy. This collaboration is argued to be most promising if activated at a regional scale.

3.3 Case analysis

Our case study shows the development of the Dutch agrifood system through the lens of the four transitional phases: predevelopment, take-off, acceleration, and stabilization.

3.3.1 *Predevelopment and take-off phase: 1945–1957*

Since World War II, the development of the agricultural sector took off that was initiated by the Dutch government. Minister Mansholt, had a strong articulated vision of a modern agriculture (i.e., low-cost, large-scale, intensive), which provided the new "dominant design" of the regime (characteristic of the take-off phase). This vision was shaped in dialogue with relevant stakeholders and effectively guided the development. The government facilitated resources to organize research and education, which generated a shared level of knowledge (a global niche level). The government protected the developing new regime from the (European and world) market with tax exemptions and income subsidies. These are characteristics of a predevelopment and take-off phase.

Regarding CE: the by origins circularly organized sector was transformed into a linear one, valuing economic and social dimensions, but undervaluing the ecological dimension.

3.3.2 *Acceleration phase: 1958–1971*

The new regime developed a strong community, including farmers, supporting political parties, strong industrial food and feed partners, financial institutions, and knowledge institutions. The Dutch agricultural economy was ready for competition in the European (and world) marketplace. However, the protection measures stayed, yet shifted from the national scale to EEC scale. Under EEC regulations, stable rule structures developed.

Regarding CE: the political-economic environment favored the development of linearly organized large-scale intensive farming as the dominant design and allowed overproduction, which was degraded to feed or even waste.

3.3.3 *Stabilization phase: 1972–2014*

The sector developed a stable structure (i.e., regulations and resources), culture (i.e., paradigms, norms, values), and practices. Although system flaws were disclosed, the regime of the food system remained strong. Incremental improvements were applied within the rule-set to develop further to a low-cost, large-scale, intensive form of farming. Although innovations appeared, such as organic farming, the cohesion of the regime acted as a strong source of resistance to radical change.

Regarding CE: the European food system performed as a linear system, with large amounts of external inputs (i.e., chemical fertilizers and feed), generating large amounts of losses and waste, including manure, which interrupts the mineral cycles.

3.3.4 Destabilization: 2015–2019

The regime is subject to challenges from the environment and from within.

Regarding CE: again, a minister put forward a vision for a new organizational design of the regime: circular agriculture. Dutch society has entered a period of turmoil: (1) characterized as a trial of strength between different partners of the food chain, whereby farmers are demanding an equal position and resist on room for new growth; (2) Dutch national politicians who challenge the force of EU legislation; (3) meanwhile radical innovators are experimenting with new technologies and organization models of food production.

Referring to the interviews, most respondents prefer to view a circular economy from an integral system level, taking into account the effects on the domains of economic prosperity, as well as social equity and environmental quality. Overall there is a strong sense that circular economy is a means to achieve sustainability through these three domains, which are connected to the Sustainable Development Goals of the United Nations (2017). Collaboration with partners in the network is seen as necessary to achieve a CE. This cooperation must take into account the different interests and motivations of the various partners; some take interest in more earnings, others in enhancing environmental or social quality.

4. Conclusions and discussion

4.1 Practical implications

The transition of the agrifood system of The Netherlands after World War II was fueled by a vision of focused development to large-scale intensive agriculture driven by lowest costs enabled by large-scale mechanization. Successive national governments of the post-war period facilitated the transformation of the agrifood sector by strong and consistent institutional support at a national scale; further transformation was institutionalized at the European scale. As a result, a new institutional level of legislative authority was added, which might function as an anchor for agreed legislation, for example for environmental issues, when national politicians keep prioritizing economic values.

The Dutch dairy network now seems to be at the threshold of returning to a circular economy, initiated by the government. But today, the societal context is far more complex than in the post-war period. Although the aims and principals of this circular agricultural are clear, the concept has to be substantiated in concrete measures for the whole sector and society without losing sight of the consequences for individual farmers. One of the key questions for the design of this new circular agrifood regime is, at which scale is this system to be organized: regional, national, European, global? Our interviewees suggest organizing a CE at a regional scale. In the societal context of today, we expect a step-by-step transformation to a sustainable, circular economy of the agrifood system, following a teleological path: at the "landscape level" forced by destabilizing, tightening legislation in favor of environmental and social aspects; at the "regime level" pressured by weak economic results of farmers, foresighted regime players and stronger voices of NGO's; at the micro level opening up windows of opportunity for path-breaking innovative entrepreneurial niches. These interacting forces might create an adaptation of the regime over time, meaning it might take a decade to see some real changes.

4.2 Theoretical implications

Taking a multiphase approach to transitions, we constructed a four phase transitional process and added key characteristics to each phase based on theoretical considerations (Table 1). We then applied this multiphase approach to the case study of the Dutch dairy sector. Based on our research and interviews we found that most characteristics of each phase were confirmed.

However, the transition of the dairy sector is in some ways different from the typical pathway provided by strategic niche management, which focuses on emerging niche innovations advancing a regime (niche-to-regime). Neither a radical innovation, nor new entrants or outsiders played a key role in the first phases of this transition. Instead, a top-down governmental vision guided the development of the regime of the dairy sector. The government acted as a collaborative incumbent of this regime and facilitated effectively the institutionalization of new rules. Of course, this development also created room for co-evolutionary technological innovations, including pharmaceuticals, industrial feed, chemical fertilizers, and various forms of equipment, but these were not the drivers of regime development. Thus, this pathway reflects *a regime-to-niche process* as opposed to the typical *niche-to-regime process of Strategic Niche Management*. As a consequence of the strong articulated vision, which provided the new "dominant design" of the regime, open-ended experimentation was replaced by a focused development. As a result, the early phases of predevelopment and take-off could take place in a relatively short period: in fact the phase of 1945–1957 could be categorized as a predevelopment and take-off phase in one.

On the surface, partners in the dairy network judge collaboration and cooperation as necessary in order to achieve a CE. This collaboration has impact on two different levels. First, there is a clear wish to change the nature of collaboration within the value chain. Not only with the intention to optimize resource flows and prevent waste, but also to equalize the power balance in the chain, where farmers believe themselves to be in a minor position. Second, collaboration with actors of other sectors is seen as necessary, i.e., the dairy farmer and arable farmer for the exchange of manure and feed. Also, cooperation between partners of the production chain and other network partners is appreciated, but must lead to value capturing, taking into account the different interests of partners. In this respect, value is not only seen as monetary value, but also as recognition, esteem, and giving legitimacy. These findings correspond with other publications (Jonker et al., 2018).

Based on the depicted research we suggest further research to elaborate at least three fundamental aspects of the emerging transition in the dairy-sector: (a) trans-sectoral concepts of collaboration based on closed loops, (b) investigation into (new) business models underpinning the viability of certain transitions, and (c) stable institutional and regulatory support (top-down) to enhance the required transitions. In order to contribute to the transitions, the nature of the research had best be action- and involvement-based. Given the growing urgency of the matter, shaping transitions on a trial-and-error basis is the most promising way to move towards a more sustainable society and economy.

Acknowledgments

The authors would like to thank the anonymous reviewer for the valuable comments on an earlier version of this chapter.

We have no declarations of interest to report. This research did not receive any specific grant from funding agencies in the public, commercial, or not-for-profit sectors.

References

Beers, P.J., Turner, J.A., Rijswijk, K., Williams, T., Barnard, T., Beechener, S., 2019. Learning or evaluating? Towards a negotiation-of-meaning approach to learning in transition governance. Technol. Forecast. Soc. Chang. 145, 229–239.

Blomsma, F., Brennan, G., 2017. The emergence of circular economy: a new framing around prolonging resource productivity. J. Ind. Ecol. 21 (3), 603–614.

Chappin, E.J.L., Ligtvoet, A., 2014. Transition and transformation: a bibliometric analysis of two scientific networks researching socio-technical change. Renew. Sust. Energ. Rev. 30, 715–723.

De Haan, F.J., Rotmans, J., 2011. Patterns in transitions: understanding complex chains of change. Technol. Forecast. Soc. Chang. 78 (1), 90–102.

Ellen MacArthur Foundation, 2013. Towards the Circular Economy. Economic and Business Rationale for an Accelerated Transition. vol. 1 Ellen MacArthur Foundation. In: 596/09-08-2013.

European Commission, 2015. Closing the Loop—An EU Action Plan for the Circular Economy. Retrieved from https://eur-lex.europa.eu/resource.html?uri=cellar:8a8ef5e8-99a0-11e5-b3b7-01aa75ed71a1.0012.02/DOC_1&format=PDF.

European Commission, 2019. The European Green Deal. Retrieved from https://eur-lex.europa.eu/resource.html?uri=cellar:b828d165-1c22-11ea-8c1f-01aa75ed71a1.0002.02/DOC_1&format=PDF.

FAO, 2006. Livestock's Long Shadow. FAO, Rome.

Geels, F.W., 2002. Technological transitions as evolutionary reconfiguration processes: a multi-level perspective and a case-study. Res. Policy 31 (8–9), 1257–1274.

Geels, F.W., 2005. Processes and patterns in transitions and system innovations: refining the co-evolutionary multi-level perspective. Technol. Forecast. Soc. Chang. 72 (6), 681–696.

Geels, F.W., 2014. Reconceptualising the co-evolution of firms-in-industries and their environments: developing an inter-disciplinary triple embeddedness framework. Res. Policy 43 (2), 261–277.

Geels, F.W., Deuten, J.J., 2006. Local and global dynamics in technological development: a socio-cognitive perspective on knowledge flows and lessons from reinforced concrete. Sci. Public Policy 33 (4), 265–275.

Geels, F.W., Kemp, R., 2007. Dynamics in socio-technical systems: typology of change processes and contrasting case studies. Technol. Soc. 29 (4), 441–455.

Geels, F.W., Raven, R., 2006. Non-linearity and expectations in niche-development trajectories: ups and downs in dutch biogas development (1973–2003). Tech. Anal. Strat. Manag. 18 (3–4), 375–392.
Geels, F.W., Schot, J., 2007. Typology of sociotechnical transition pathways. Res. Policy 36 (3), 399–417.
Geels, F.W., Schot, J., 2010. Reflections: process theory, causality and narrative explanation. In: Transitions to Sustainable Development: New Directions in the Study of Long Term Transformative Change, pp. 93–101, https://doi.org/10.1016/j.mechmachtheory.2009.10.001.
Geissdoerfer, M., Savaget, P., Bocken, N.M.P., Hultink, E.J., 2017. The circular economy—a new sustainability paradigm? J. Clean. Prod. 143, 757–768.
Hallmann, C.A., Sorg, M., Jongejans, E., Siepel, H., Hofland, N., Schwan, H., et al., 2017. More than 75 percent decline over 27 years in total flying insect biomass in protected areas. PLoS One 12 (10), 1–21.
Hollander, A., Temme, E.H.M., Zijp, M.C., 2017. The Environmental Sustainability of the Dutch Diet. National Institute for Public Health and the Environment, Bilthoven.
Jonker, J., Kothman, I., Faber, N., Navarro, N.M., 2018. Circulair Organiseren. Available at www.circulairebusinessmodellen.nl.
Kemp, R., 1994. Technology and the transition to environmental sustainability: the problem of technological regime shifts. Futures 26, 1023–1046.
Kemp, R., Schot, J.W., Hoogma, R., 1998. Regime shifts to sustainability through processes of niche formation: the approach of strategic niche management. Tech. Anal. Strat. Manag. 10 (2), 175–198.
Kemp, R.P.M., Rip, A., Schot, J.W., 2001. Constructing transition paths through the management of niches. In: Garud, R., Karnoe, P. (Eds.), Path Dependence and Creation. Psychology Press, New York, pp. 269–299.
Kirchherr, J., Reike, D., Hekkert, M., 2017. Conceptualizing the circular economy: an analysis of 114 definitions. Resour. Conserv. Recycl. 127, 221–232.
Kleijn, D., Bink, R.J., Ter Braak, C.J.F., Van Grunsven, R., Ozinga, W.A., Roessink, I., Zeegers, T., et al., 2018. Achteruitgang van insectenpopulaties in Nederland: trends, oorzaken en kennislacunes. Wageningen Environmental Research, Wageningen. No. 2871.
Loorbach, D., 2007. Transition Management: New Mode of Governance for Sustainable Development. Erasmus University Rotterdam, https://doi.org/10.3141/2013-09.
Markard, J., Truffer, B., 2008. Technological innovation systems and the multi-level perspective: towards an integrated framework. Res. Policy 37 (4), 596–615.
Martiskainen, M., Kivimaa, P., 2018. Creating innovative zero carbon homes in the United Kingdom—intermediaries and champions in building projects. Environ. Innov. Soc. Trans. 26, 15–31.
Meadows, D.H., Meadows, D.L., Randers, J., Behrens, W.W., 1972. The limits to growth. In: A Report for THE CLUB OF ROME'S Project on the Predicament of Mankind, second ed., https://doi.org/10.4324/9780429493744.
Ministerie van Landbouw, Natuur en Voedselkwalteit, 2019. Realizing the Phosphate Allowances System. Totstandkoming fosfaatrechtenstelsel, Dutch. https://www.rijksoverheid.nl/onderwerpen/mest/fosfaatrechten/totstandkoming-fosfaatrechtenstelsel.
Ministry of Agriculture, Nature and Food Quality, 2018. Agriculture, Nature and Food: Valuable and Interconnected (in Dutch). Ministry of Agriculture, Nature and Food Quality, The Hague.
Murray, A., Skene, K., Haynes, K., 2017. The circular economy: an interdisciplinary exploration of the concept and application in a global context. J. Bus. Ethics 140 (3), 369–380.
Rip, A., Kemp, R., 1998. Technological change. In: Human Choice and Climate Change. vol. 2, pp. 327–399, https://doi.org/10.1007/BF02887432.
Rotmans, J., Loorbach, D., 2010. Towards a better understanding of transitions and their governance: A systemic and reflexive approach. In: Grin, J., Rotmans, J., Schot, J. (Eds.), Transitions to Sustainable Development: New Directions in the Study of Long-Term Transformative Change. Routledge, New York, London, pp. 103–220.
Rotmans, J., Kemp, R., Van Asselt, M., 2001. More evolution than revolution: transition management in public policy. Foresight 3 (1), 15–31.
Scholten, M., Bianchi, F., De Boer, I., Conijn, S., Dijkstra, J., Doorn, V., Termeer, K., 2018. Technical Briefing Circular Food Production. vol. 31 Wageningen.

Schot, J.W., Geels, F.W., 2008. Strategic niche management and sustainable innovation journeys: theory, findings, research agenda, and policy. Tech. Anal. Strat. Manag. 20 (5), 537–554.

Smith, A., Raven, R., 2012. What is protective space? Reconsidering niches in transitions to sustainability. Res. Policy 41 (6), 1025–1036.

Stahel, W.R., 1982. The product-life factor. In: Orr, S.G. (Ed.), An Inquiry into the Nature of Sustainable Societies: The Role of the Private Sector. HARC Houston Area Research Center, pp. 72–96.

Statistics Netherlands, 2017. Larger Dairy Farms and More Milk (Dutch: Grotere melkveebedrijven en meer melk). https://www.cbs.nl/nl-nl/nieuws/2017/18/grotere-melkveebedrijven-en-meer-melk.

Statistics Netherlands, 2018. Dutch Agricutural Exports 2017 (De Nederlandse landbouwexport 2017). Retrieved from https://opendata.cbs.nl/#/CBS/nl/dataset/80780ned/table?dl=6135.

Statistics Netherlands, 2019. Agricultural exports exceed 90 billion euros in 2018 (Dutch: Landbouwexport ruim 90 miljard euro in 2018). https://www.cbs.nl/nl-nl/nieuws/2019/03/landbouwexport-ruim-90-miljard-euro-in-2018.

Turnheim, B., Geels, F.W., 2012. Regime destabilisation as the flipside of energy transitions: lessons from the history of the British coal industry (1913–1997). Energ Policy 50, 35–49.

Turnheim, B., Geels, F.W., 2013. The destabilisation of existing regimes: confronting a multi-dimensional framework with a case study of the British coal industry (1913-1967). Res. Policy 42 (10), 1749–1767.

United Nations, 2017. The Sustainable Development Goals Report. United Nations Publications, https://doi.org/10.18356/3405d09f-en.

United Nations Framework Convention on Climate Change, 2015. Adoption of the Paris agreement. Proposal by the president. In: Conference of the Parties. Twenty-First Session. vol. 21932, p. 32. https://doi.org/FCCC/CP/2015/L.9/Rev.1.

Unruh, G.C., 2000. Understanding carbon lock-in. Energ Policy 28, 817–830.

Unruh, G.C., 2002. Escaping carbon lock-in. Energ Policy 30 (4), 317–325.

Van Buren, N., Demmers, M., Van der Heijden, R., Witlox, F., 2016. Towards a circular economy: the role of Dutch logistics industries and governments. Sustainability 8 (7), 1–17.

Van de Ven, A.H., Engleman, R.M., 2004. Event- and outcome-driven explanations of entrepreneurship. J. Bus. Ventur. 19 (3), 343–358.

Van den Bosch, S., Rotmans, J., 2008. Deepening, Broadening and Scaling up. A Framework for Steering Transition Experiments. Knowledge Centre for Sustainable System Innovations and Transitions (KCT).

van Merriënboer, J., 2006. Mansholt. Boom Amsterdam, Nijmegen.

Yin, R.K., 2016. Qualitative Research from Start to Finish, second ed. The Guilford Press, New York.

CHAPTER 27

The contemporary research on circular economy in industry

Victor Fukumoto and Alexandre Meira de Vasconcelos
Federal University of Mato Grosso do Sul, Brazil

1. Introduction

In the early days of the industrial revolution, new manufacturing methods and abundant resources at low cost boosted industrial production, but the current situation is different (Lieder and Rashid, 2016). The predictions for the next few decades are a large growth in population and consumers, resulting in an increase of at least a third in the global demand for and consumption of resources such as oil, coal, and iron (EMF, 2013). In 2010, consumption has exceeded the planet's regenerative capacity by 50%; i.e., the current linear system (take-make-dispose) is increasingly physically impossible for the biosphere to maintain (Bonciu, 2014). In addition to the environment limitation, the linear production system has become economically infeasible because the volatility of resource prices exposes companies to risks (EMF, 2013).

Scientific reports present circular economy as an economic and research strategy to promote development without compromising future generations (Ghisellini et al., 2016; Yuan et al., 2006). Circular economy is a strategy focused on environmental protection, pollution prevention, and sustainable development (Hu et al., 2011). The concept focuses on regeneration, which seeks the maximization of resource efficiency, minimizes the production of waste, or utilizes waste in new processes, so that outputs become inputs for other industries (Bonciu, 2014). The concept of circular economy is extremely broad and, therefore, there is a wide variety of scientific investigations that follow different schools of thought as to the objectives and forms of deployment (Zink and Geyer, 2017).

The growth of scientific literature on circular economy associated with industrial production is a trend, whereas the "take-make-dispose" production model little has changed since the industrial revolution (EMF, 2013). Modern companies have realized that linear consumption increases the exposure to risk, due to volatility of and increases in prices of natural resources. In addition, population growth and increasing urbanization move resource extraction to more distant locations with higher logistics costs and increase environmental costs associated with its breakdown. Regarding the prices of resources, despite the slight fall of the price of commodities in 2011, the percentage growth is higher than

Circular Economy and Sustainability, Volume 1
https://doi.org/10.1016/B978-0-12-819817-9.00023-5

global economic production and this growth has the potential to affect several markets (Dobbs et al., 2014).

The limitations of the linear system of consumption account for the growth of scientific literature on circular economy. It is inferred that this phenomenon arises from the official manifesto of the European Commission about the radical change of perception on the use of resources, in July 2014, which introduced the circular production as part of the challenge for Europe 2020 (Bonciu, 2014). In recent years, the regions that most published on circular economy in the industry were Europe and Asia, which also have the greatest demand for natural resources and/or lower availability of them.

The research, in general, have a multi or transdisciplinary approach to the concept of circular economy, and address issues such as cleaner production, sustainability, and the environment in public or private industrial and services corporations. Transdisciplinarity is a way of integrating disciplines without rigid boundaries among them. The researchers' recommendation for discussion on circular economy in corporations to stimulate the removal, replacement, or rational use of natural resources can be highlighted. The association of circular economy with sustainable development was noted, but there is a tendency to change the theme of sustainability to circular economy because the latter directs more clearly the instrumentalization of organizations for change and signals the opportunity for greater financial gains, which normally appeals to more investors.

The researchers intended to understand ways of socioeconomic development with less industrial assault on the environment. Studies were more aligned with economic development, due to its importance in a market society, and were designed to relate it with the circular economy. The last stage of transition to the circular economy occurs when the motivation is the prospect of gathering economic benefits by individual economic actors (Lieder and Rashid, 2016). The shift from linear to circular economy involves hybrid strategies of integrating new business models and design, to close material and energy cycles systemically from a social, economic, and environmental point of view. This discussion is still theoretical and lacks real-world case studies (Bocken et al., 2016).

The plurality of approaches and theoretical affiliations is a problem for the consolidation of the concept in the academic environment and it is necessary, first, to organize and explain what has already been published to evidence whether there are priority approaches of researchers. Therefore, the objective of this exploratory study is, through content analysis, to determine conceptual categories, related to the circular economy in industry, present in articles from indexed scientific journals to extract knowledge about the approaches prioritized by researchers.

2. Methodological procedures

For this study, the scientific literature on circular economy in industry was analyzed using the content analysis technique with the support of Iramuteq software (*Interface de R pour* *lês Analyses Multidimensionelles de Textes et de Questionnaires*). Scientific papers published until 2018 on circular economy and industry were selected in the Scopus database, which is one of the largest multidisciplinary databases. For the selection of scientific papers, searches were performed in the metadata "title," "summary," and "keywords," using the search string "CIRCULAR ECONOMY" AND INDUSTR*. The use of characters followed by asterisk aims to identify metadata in the singular or plural (industry or industries, for example) and/or variations using the same radical (industrial, or industrialization, or industrialization, for example).

We chose to analyze the abstracts of the articles, because they are, a priori, representations of the scientific work content, which were grouped into a single file with 1537 abstracts of papers published between 2001 and 2018, called textual corpus. The descending hierarchical classification (DHC) method shown in Fig. 1, one of the features of Iramuteq, was used to sort the words and generate an analytical model of the word classes that was explained based on the reference literature and the textual corpus. It was possible to evaluate through the DHC the specific vocabulary used by the authors to characterize the circular economy in industry and, through the construction of the text, understand in what context the words were used.

3. Results and discussion

It was found that there are two main factors of interest ("waste management" and "production management," subdivided into "business management" and "resources management"), confirming the growing association between circular economy and development, integrating socioenvironmental management with economic and financial gains. The DHC dendrogram shows the vocabulary present in the classes and the most important words of each. Vocabulary is presented in a factorial plan with three well defined factors, without overlapping words.

3.1 Business management

Business management is the most expressive factor (50.93%) and is included in production management. This class shows studies related to modernization of technology and processes and to a servitization trend of business models, based on circular economy paradigm and guidance about deploying sustainable business models applied in industry.

The introduction of servitized business models (BMs), when the use or the function of a product is sold instead of the product itself, has been acknowledged as one possible enabling factor of the CE paradigm into companies. The literature has pointed out the potential benefits that companies may gather from servitized BMs, such as strengthening customer relations, creating higher barriers for competitors, and generating new and resilient revenue streams (Bressanelli et al., 2018).

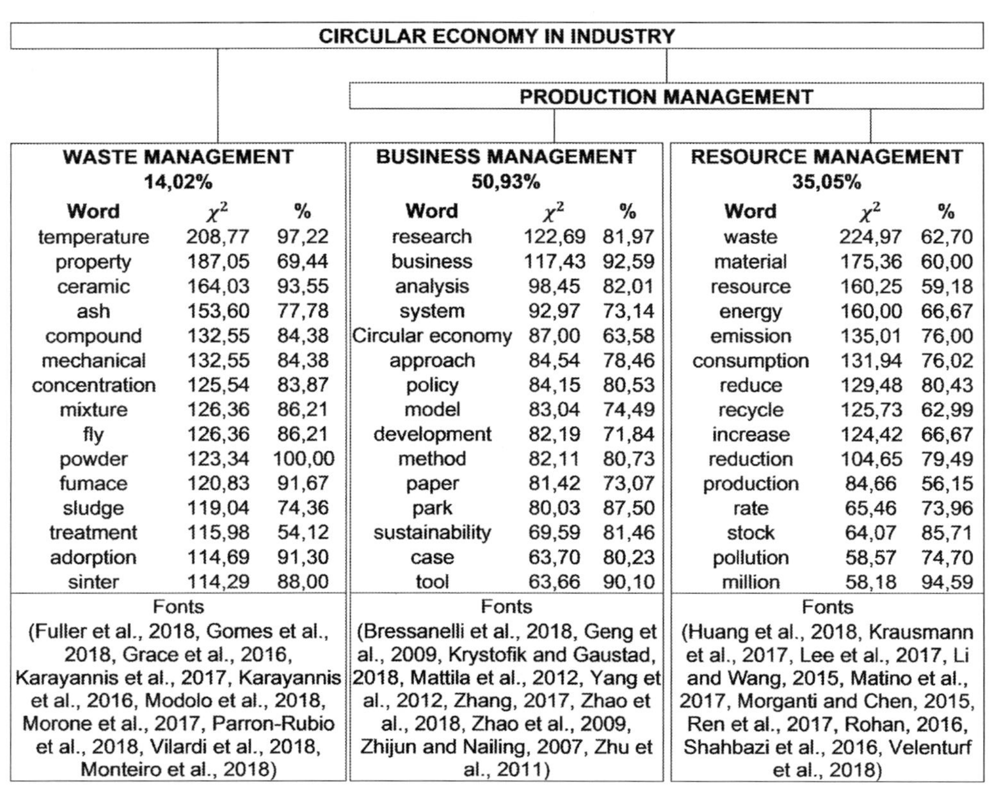

CIRCULAR ECONOMY IN INDUSTRY

PRODUCTION MANAGEMENT

WASTE MANAGEMENT 14,02%

Word	χ^2	%
temperature	208,77	97,22
property	187,05	69,44
ceramic	164,03	93,55
ash	153,60	77,78
compound	132,55	84,38
mechanical	132,55	84,38
concentration	125,54	83,87
mixture	126,36	86,21
fly	126,36	86,21
powder	123,34	100,00
fumace	120,83	91,67
sludge	119,04	74,36
treatment	115,98	54,12
adorption	114,69	91,30
sinter	114,29	88,00

Fonts
(Fuller et al., 2018, Gomes et al., 2018, Grace et al., 2016, Karayannis et al., 2017, Karayannis et al., 2016, Modolo et al., 2018, Morone et al., 2017, Parron-Rubio et al., 2018, Vilardi et al., 2018, Monteiro et al., 2018)

BUSINESS MANAGEMENT 50,93%

Word	χ^2	%
research	122,69	81,97
business	117,43	92,59
analysis	98,45	82,01
system	92,97	73,14
Circular economy	87,00	63,58
approach	84,54	78,46
policy	84,15	80,53
model	83,04	74,49
development	82,19	71,84
method	82,11	80,73
paper	81,42	73,07
park	80,03	87,50
sustainability	69,59	81,46
case	63,70	80,23
tool	63,66	90,10

Fonts
(Bressanelli et al., 2018, Geng et al., 2009, Krystofik and Gaustad, 2018, Mattila et al., 2012, Yang et al., 2012, Zhang, 2017, Zhao et al., 2018, Zhao et al., 2009, Zhijun and Nailing, 2007, Zhu et al., 2011)

RESOURCE MANAGEMENT 35,05%

Word	χ^2	%
waste	224,97	62,70
material	175,36	60,00
resource	160,25	59,18
energy	160,00	66,67
emission	135,01	76,00
consumption	131,94	76,02
reduce	129,48	80,43
recycle	125,73	62,99
increase	124,42	66,67
reduction	104,65	79,49
production	84,66	56,15
rate	65,46	73,96
stock	64,07	85,71
pollution	58,57	74,70
million	58,18	94,59

Fonts
(Huang et al., 2018, Krausmann et al., 2017, Lee et al., 2017, Li and Wang, 2015, Matino et al., 2017, Morganti and Chen, 2015, Ren et al., 2017, Rohan, 2016, Shahbazi et al., 2016, Velenturf et al., 2018)

Fig. 1 Dendrogram of the DHC (Fuller et al., 2018; Gomes et al., 2018; Grace et al., 2016; Karayannis et al., 2017; Karayannis et al., 2016; Modolo et al., 2018; Morone et al., 2017; Parron-Rubio et al., 2018; Vilardi et al., 2018; Monteiro et al., 2018; Bressanelli et al., 2018; Geng et al., 2009; Krystofik and Gaustad, 2018; Mattila et al., 2012; Yang et al., 2012; Zhang, 2017; Zhao et al., 2018; Zhao et al., 2009; Zhijun and Nailing, 2007; Zhu et al., 2011; Huang et al., 2018; Krausmann et al., 2017; Lee et al., 2017; Li and Wang, 2015; Matino et al., 2017; Morganti and Chen, 2015; Ren et al., 2017; Rohan, 2016; Shahbazi et al., 2016; Velenturf et al., 2018). *(Source: Author (2019).)*

In business management, servitization is an alternative to encourage businesses to adopt the circular economy and migrate from selling products to supplying services (Bocken et al., 2016). To change, it is necessary to transform the value proposition and the business model, generating greater competitiveness and sustainability simultaneously. However, the change is complex, both for companies and customers (Tukker, 2004; Adrodegari and Saccani, 2017). Between the pure sale of products and the pure provision of services there is a spectrum with different options in which products and services are combined to different degrees (Tukker, 2004). From the point of view of the ownership, it can be replaced by other business models such as leasing contracts, shared savings, or substitution of products for services (Michelini and Razzoli, 2004). In this way, the business model can be focused on the product, the use, or the result (Tukker, 2004). This classification is inspired by the three values of the circular economy from the Ellen MacArthur Foundation (EMF, 2013): increasing resource efficiency, increasing lifetime, and closing the cycle.

In business management with a product-oriented model, the focus is to deliver tangible value to the end customer. Therefore, the value is an embedded attribute in the product that will be delivered. The right to property is transferred to the client, and what is left to companies are the services of product enhancement and restoration of their functionality, such as maintenance, repairs, and extended warranties, which are considered noncore business, leading to outsourcing (Kujala et al., 2010). The life cycle of the product is not the company's responsibility and the revenue stream is represented by the value of the original sale. Thus, business management focused on product does not encourage companies to achieve goals based on the principles of circular economy. The solutions offered rarely increase resource efficiency and the increase in lifetime is not interesting for the company that produces the product (Tukker, 2015).

In management with the use-oriented business model, the customer does not buy the product, but pays to have access (Reim et al., 2015). The property rights to the product belong to the company and, consequently, the responsibility for the life cycle, as well as the maintenance, repairs, and control (Kujala et al., 2010). This business model requires that companies design their products with reusable elements and enable improvements to the customer (Adrodegari and Saccani, 2017; Reim et al., 2015). However, it is a complex business model with high internal cost that requires advanced services of monitoring and remote diagnosis, advanced training, and predictive maintenance (Adrodegari and Saccani, 2017). Despite the high initial capital and long payback period, the use-oriented management model is more promising to achieve the objectives of the circular economy, since it contributes to increases in resource efficiency and the closure of the cycle through maintenance and return contracts. Nevertheless, there are questions as to the increase in life cycle, as the lack of ownership could bring less careful use by the client and greater wear and tear of the product.

In management with the result-oriented business model, the customer does not purchase the product or system, but pays according to the result set contractually in terms of use performance (Tukker, 2004). This business model requires an advanced network of services with technicians constantly interacting with customers for assistance and monitoring of the use of products, because measuring the results is a high complexity task (Tukker, 2015; Reim et al., 2015). In addition, the company is responsible for all product life cycle costs, which creates greater incentive to follow circular economy principles, as the company minimizes life cycle and operational costs. The revenue stream is based on the result of the use, encourages the conservation of resources and the collection of products to generate cycle closure (Bressanelli et al., 2018; Tukker, 2015).

It was observed that the circular economy approximates to the performance economy, which is an approach in which the economy takes place in cycles with positive impacts on job creation, resource reduction, waste prevention, and economic competitiveness. The objectives of the performance economy are the extension of a product's life cycle, reconditioning of processes, and waste prevention. Therefore, the sale of services in place of products is intrinsic to the performance economy (Stahel, 2010).

3.2 Resources management

The resources management factor is inserted in production management and comprises studies aimed at reducing waste, optimizing the use of resources and material flow. Digital technologies facilitate the transition to circular economy through innovative designs, concept dissemination, increased value perception, shared use, extended life, reform, remanufacturing, and recycling, in closed cycles of materials and energy (Bressanelli et al., 2018).

This chapter uses the first Circular Economy Park for the coal industry to discuss the strategies that China's coal enterprises have been using to shrink their carbon footprints. Through the use of two circulated economic industrial chains involving coal-electricity-building materials and the coal-chemical industry to reduce, reuse, and recycle resources, the Tashan Circular Economy Park has achieved significant energy conservation and reduction of CO_2 emissions (Li and Wang, 2015).

The life cycle theory was used to analyze different production technologies, and it was found that comprehensive waste use to produce sulfoaluminate clinker reduces resource consumption, global warming, chemical oxygen demand, particulate matter, primary energy consumption, acidification, eutrophication, and solid waste generation (Ren et al., 2017).

There are several definitions of circular economy and, in resources management, we highlight the definition around the paradigm of the 3Rs (reduction, reuse, and recycling). The reduction consists of innovating processes, products, and business models to reduce the use of nonrenewable raw materials and replace them with renewable, biodegradable, or recyclable resources (Goyal et al., 2016). Reuse is the search for the extension of the product's life cycle and a conduit for unused material and disposables into a new economic cycle. Through this approach, the reuse of waste allows the reduction of material that would "leak" from the linear model, and the maximization of economic value. The concepts that are closest to the topics covered in this category are regenerative design, cradle to cradle, and industrial ecology.

Companies should be the first to deploy the circular economy on a micro scale, which would then be developed at the industrial parks level. From there, the circular economy must be managed at the macro level of cities and regions in a growing scale of deployment. The materials cycle must be closed in an ecological chain in the internal environment of organizations, minimizing environmental impact and maximizing resources.

The economy of resources and energy happens in production and the decrease of the effects of the whole life cycle from raw material extraction to the product's discard (Zhijun and Nailing, 2007).

Regenerative design is a continuous self-renewal structure of energy and material supply. It means replacing the linear production flow system with cyclic flows in its operations, such as the replacement of conventional energy with solar and raw material consumption through recycling and reuse. The principles are operational integration with the natural and social processes; minimal use of fossil fuels; minimal use of nonrenewable resources; use of renewable resources within their recyclability; and disposal within the environment's capacity to receive them without damage (Lyle, 1996). The "cradle to cradle" design concept addresses safe and productive processes, which use the organic logic of nature as a model to develop industrial material flows by a technical metabolism. This train of thought is designed to eliminate the concept of waste, because designs of products with reusable materials or with biological metabolism safe for the environment maximizes the use of renewable energy and manages the use of water to protect ecosystems (EMF, 2013).

With respect to industrial ecology, the Ellen MacArthur Foundation (EMF, 2013) defines it as the study of flows of materials and energy in industry, which is observed in a systemic and concentrated manner in human operations within the industrial ecosystem. The systemic vision requires that processes follow local ecological restrictions, eliminating the concept of by-products (EMF, 2013). Another definition compares the industrial ecology with the biologic one, in which organisms consume water, sunlight, and minerals and are consumed by others that produce residues. These residues—food for some organisms—are converted into minerals that supply the primary organisms, thus closing the cycle. In this way, industrial ecology aims to form this network of dependent processes, like the biological concept, which, although not perfect, generates many gains in resources management (Socolow et al., 1997).

3.2.1 Waste management

The waste management factor has received the least coverage in scientific publication on circular economy in industry (14.02%) and highlights the treatment and utilization of industrial waste. It discusses concepts like the resource management factor, but the scientific literature is more specifically focused on waste. The research has an empirical approach, applied with concepts of physics, chemistry, and biology, while in the management of resources, researches are more theoretical.

> *[...] one of the main problems in the iron and steel industry is waste generation and byproducts that must be properly processed or reused to promote environmental sustainability. One of these byproducts is steel slag. The cement substitution with slag strategy achieves two goals: raw materials consumption reduction and waste management. Parron-Rubio et al. (2018)*

> *[...] the suitability of potential media by quantifying their adsorption potential across a variety of common drinking water contaminants. The media investigated were fly ash, Bayer residue,*

ground granular blast furnace slag, coconut shell, tea/coffee waste, rice husk, crushed concrete, masonry waste, and wood waste. There is a potential for the use of these media in the water treatment sector [...]. Grace et al. (2016)

Reverse logistics combined with the selection and processing of recycled materials is a way for companies in developing countries to create value, alongside policies for reducing, recycling, and reusing resources from different sources (Goyal et al., 2016). A considerable quantity of natural resources have become concentrated in manufactured products that are, or have been, in use, such as structures, machinery, and other durable goods. Much of this stock in use has become obsolete throughout the 20th century and has resulted in a considerable environmental liability, becoming waste at its end-of-life. The dynamics of these materials should be better understood, since recycling has not contributed to the closure of the materials cycle, evidenced by the global recycling percentage of plastics, metals, and biomass, which represent only 12% of the total end-of-life (Krausmann et al., 2017).

The presence of the "blue economy" principles in academic papers was highlighted, with physics, chemistry, and biology applied for innovation in renewable and sustainable practices (Pauli, 2010). This is an open-source movement that gathers concrete case studies on the use of waste for inputs of new cash flows (EMF, 2013). The principles of this train of thought resemble those from industrial ecology, with solutions determined by the local environment, limited by physical and ecological features (Pauli, 2010).

4. General considerations

Independent of classes, it was possible to understand trends in scientific production. Regarding the content, it was found that the African continent, Europe, and South America use closer vocabularies, however there is no exacerbated differentiation among continents in general.

[...] the world is neither globally successful in remanufacturing and reuse of products nor recycling of waste materials this requires a combination of circular economy management systems business models and novel technologies; there are contrasting views in literature regarding models for a circular economy [...]. Mativenga et al. (2017)

[...] the discovery and improvement of processes for effective recycling of polyethylene terephthalate pet packages meet high priority and short term demands for the implementation of a circular economy; one of the most interesting approaches is enzyme catalyzed depolymerization which to date has been exclusively reported as hydrolysis reactions [...]. Castro and Carniel (2017)

[...] in Europe of E-waste and of packaging waste have already been recycled or reused in some cases this article analyzes the challenges of implementing EPR and provides useful insights for what has worked well and what challenges remain [...]. Kunz et al. (2018)

Chronologically, the research on "circular economy in industry" has used different language in the past 3 years (2016, 2017, and 2018), which indicates differences in features among them. At the same time, research with an empirical aspect is supported by physical and biochemical concepts applied in industry. They also dealt with more sustainable and competitive business models, practices, and methods. The focus of academia nowadays is not simply to conserve and limit; the goal is to innovate from the form of consumption to modify the linear system.

> *[…] it considers the recycling of scrap into new steel as closed material loop recycling and thus recycling steel scrap avoids the production of primary steel; the methodology developed shows that for every kg of steel scrap that is recycled at the end of the products life […]. Broadbent (2016)*

> *[…] human made material stocks accumulating in buildings infrastructure and machinery play a crucial but underappreciated role in shaping the use of material and energy resources building maintaining and in particular operating in use stocks of materials require raw materials and energy […]. Krausmann et al. (2017)*

> *[…] sustainable economy dimensions and sustainable business goals sustainable business goals and sustainable technology business innovation principles and sustainable technology business innovation principles and technology and business models […]. Jaksic (2017)*

Research on circular economy in industry has increased every year and the trend is for this growth to continue. Despite this increase, studies are still at an early stage and limited to certain countries, especially China, which also holds the predominance of authors with more publications. Excepting Europe and Asia, it is important that researchers from peripheral regions and in possession of natural resources, such as the Americas and Africa, contribute to the evolution of the theme with greater production. It was observed that there was a predominance of subjects on business management with the innovation of business models, processes, and practices based on economic performance. However, this language came into prominence only in 2018, suggesting more interest from academics on this issue and that industrial ecology, recycling, and optimization of resources are declining when it comes to scientific research.

The detachment of research from waste and resources management was noticed, that is, research on the development of cleaner technologies and more efficient resources. The focus of academia was directed to innovation in industry as a business model, with solutions to the root of the problems that may affect the environment. Research seeks to transform the paradigm of pure products sale and also, tangibility. This is at an early stage, mainly with theoretical approaches, but with proposed benefits to the environment and the opening of new markets, making them economically attractive. The empirical research on this subject is still an area to be considered by researchers.

References

Adrodegari, F., Saccani, N., 2017. Business models for the service transformation of industrial firms. Serv. Ind. J. 37 (1), 57–83.

Bocken, N.M.P., de Pauw, I., Bakker, C.A., van der Grinten, B., 2016. Product design and business model strategies for a circular economy. J. Ind. Prod. Eng. 33 (5), 308–320.

Bonciu, F., 2014. The European economy: from a linear to a circular economy. Rom. J. Eur. Aff. 14 (4), 78–91.

Bressanelli, G., Adrodegari, F., Perona, M., Saccani, N., 2018. Exploring how usage-focused business models enable circular economy through digital technologies. Sustainability 10 (3), 639.

Broadbent, C., 2016. Steel's recyclability: demonstrating the benefits of recycling steel to achieve a circular economy. Int. J. Life Cycle Assess. 21 (11), 1658–1665.

Castro, A.M.D., Carniel, A., 2017. A novel process for poly(ethylene terephthalate) depolymerization via enzyme-catalyzed glycolysis. Biochem. Eng. J. 124, 64–68.

Dobbs, R., Oppenheim, J., Thompson, F., Mareels, S., Nyquist, S., Sanghvi, S., 2014. Resource Revolution: Tracking Global Commodity Markets: Trends Survey 2013. McKinsey Global Institute, London.

EMF, 2013. Towards the Circular Economy, Economic and Business Rationale for an Accelerated Transition. Ellen MacArthur Foundation, Cowes, UK. Available from: https://www.ellenmacarthurfoundation.org/publications/towards-a-circular-economy-business-rationale-for-an-accelerated-transition.

Fuller, A., Maier, J., Karampinis, E., Kalivodova, J., Grammelis, P., Kakaras, E., et al., 2018. Fly ash formation and characteristics from (co-) combustion of an herbaceous biomass and a Greek lignite (low-rank coal) in a pulverized fuel pilot-scale test facility. Energies 11 (6), 1581.

Geng, Y., Mitchell, B., Zhu, Q., 2009. Teaching industrial ecology at Dalian University of Technology: toward improving overall eco-efficiency. J. Ind. Ecol. 13 (6), 978–989.

Ghisellini, P., Cialani, C., Ulgiati, S., 2016. A review on circular economy: the expected transition to a balanced interplay of environmental and economic systems. J. Clean. Prod. 114, 11–32.

Gomes, H.I., Funari, V., Mayes, W.M., Rogerson, M., Prior, T.J., 2018. Recovery of Al, Cr and V from steel slag by bioleaching: batch and column experiments. J. Environ. Manage. 222, 30–36.

Goyal, S., Esposito, M., Kapoor, A., 2016. Circular economy business models in developing economies: lessons from india on reduce, recycle, and reuse paradigms. Thunderbird Int. Bus. Rev. 60 (5), 729–740.

Grace, M.A., Clifford, E., Healy, M.G., 2016. The potential for the use of waste products from a variety of sectors in water treatment processes. J. Clean. Prod. 137, 788–802.

Hu, J., Xiao, Z., Zhou, R., Deng, W., Wang, M., Ma, S., 2011. Ecological utilization of leather tannery waste with circular economy model. J. Clean. Prod. 19 (2–3), 221–228.

Huang, B., Wang, X., Kua, H., Geng, Y., Bleischwitz, R., Ren, J., 2018. Construction and demolition waste management in China through the 3R principle. Resour. Conserv. Recycl. 129, 36–44.

Jaksic, M.L., 2017. Sustainable innovation of technology and business models: rethinking business strategy. South East Eur. J. Econ. 14 (2), 127–139.

Karayannis, V., Ntampegliotis, K., Lamprakopoulos, S., Kasiteropoulou, D., Papapolymerou, G., Spiliotis, X., 2016. Development of extruded and sintered clay bricks with beneficial use of industrial "scrap-soil" as admixture. Rev. Rom. Mater. 46 (4), 523–529.

Karayannis, V., Moutsatsou, A., Domopoulou, A., Katsika, E., Drossou, C., Baklavaridis, A., 2017. Fired ceramics 100% from lignite fly ash and waste glass cullet mixtures. J. Build. Eng. 14, 1–6.

Krausmann, F., Wiedenhofer, D., Lauk, C., Haas, W., Tanikawa, H., Fishman, T., et al., 2017. Global socioeconomic material stocks rise 23-fold over the 20th century and require half of annual resource use. Proc. Natl. Acad. Sci. U. S. A. 114 (8), 1880–1885.

Krystofik, M., Gaustad, G., 2018. Tying product reuse into tying arrangements to achieve competitive advantage and environmental improvement. Resour. Conserv. Recycl. 135, 235–245.

Kujala, S., Artto, K., Aaltonen, P., Turkulainen, V., 2010. Business models in project-based firms—towards a typology of solution-specific business models. Int. J. Proj. Manage. 28 (2), 96–106.

Kunz, N., Mayers, K., Van Wassenhove, L.N., 2018. Stakeholder views on extended producer responsibility and the circular economy. Calif. Manage. Rev. 60 (3), 45–70.

Lee, R., Wolfersdorf, C., Keller, F., Meyer, B., 2017. Towards a closed carbon cycle and achieving a circular economy for carbonaceous resources. Oil Gas Eur. Mag. 43 (2), 76–80.

Li, J., Wang, F., 2015. How China's coal enterprise shrinks carbon emissions: a case study of Tashan Circular Economy Park. Energy Sources 37 (19), 2123–2130.

Lieder, M., Rashid, A., 2016. Towards circular economy implementation: a comprehensive review in context of manufacturing industry. J. Clean. Prod. 115, 36–51.

Lyle, J.T., 1996. Regenerative Design for Sustainable Development. John Wiley & Sons.

Matino, I., Colla, V., Baragiola, S., 2017. Quantification of energy and environmental impacts in uncommon electric steelmaking scenarios to improve process sustainability. Appl. Energy 207, 543–552.
Mativenga, P.T., Agwa-Ejon, J., Mbohwa, C., Sultan, A.A.M., Shuaib, N.A., 2017. Circular economy ownership models: a view from South Africa industry. Procedia Manuf. 8, 284–291.
Mattila, T., Lehtoranta, S., Sokka, L., Melanen, M., Nissinen, A., 2012. Methodological aspects of applying life cycle assessment to industrial symbioses. J. Ind. Ecol. 16 (1), 51–60.
Michelini, R., Razzoli, R., 2004. Product-service for environmental safeguard: a metrics to sustainability. Resour. Conserv. Recycl. 42 (1), 83–98.
Modolo, R., Senff, L., Ferreira, V., Tarelho, L., Moraes, C., 2018. Fly ash from biomass combustion as replacement raw material and its influence on the mortars durability. J. Mater. Cycles Waste Manage. 20 (2), 1006–1015.
Monteiro, M., Matos, E., Ramos, R., Campos, I., Valente, L.M., 2018. A blend of land animal fats can replace up to 75% fish oil without affecting growth and nutrient utilization of European seabass. Aquaculture 487, 22–31.
Morganti, P., Chen, H.D., 2015. From the circular economy to a green economy. Note 1. Chitin nanofibrils as natural by-products to manage the human and environment ecosystems. J. Appl. Cosmetol. 33 (3–4), 101–113.
Morone, M., Costa, G., Georgakopoulos, E., Manovic, V., Stendardo, S., Baciocchi, R., 2017. Granulation–carbonation treatment of alkali activated steel slag for secondary aggregates production. Waste Biomass Valoriz. 8 (5), 1381–1391.
Parron-Rubio, M., Perez-García, F., Gonzalez-Herrera, A., Rubio-Cintas, M., 2018. Concrete properties comparison when substituting a 25% cement with slag from different provenances. Materials 11 (6), 1029.
Pauli, G.A., 2010. The blue economy: 10 years, 100 innovations, 100 million jobs. Japan Spotlight, 14–17. Jan/Feb (NGOs—Key Players in Fighting Global Environmental Challenges).
Reim, W., Parida, V., Örtqvist, D., 2015. Product–service systems (PSS) business models and tactics—a systematic literature review. J. Clean. Prod. 97, 61–75.
Ren, C., Wang, W., Mao, Y., Yuan, X., Song, Z., Sun, J., et al., 2017. Comparative life cycle assessment of sulfoaluminate clinker production derived from industrial solid wastes and conventional raw materials. J. Clean. Prod. 167, 1314–1324.
Rohan, M., 2016. Cement and concrete industry integral part of the circular economy. Rev. Rom. Mater. 46 (3), 253–258.
Shahbazi, S., Wiktorsson, M., Kurdve, M., Jonsson, C., Bjelkemyr, M., 2016. Material efficiency in manufacturing: Swedish evidence on potential, barriers and strategies. J. Clean. Prod. 127, 438–450.
Socolow, R., Andrews, C., Berkhout, F., Thomas, V., 1997. Industrial Ecology and Global Change. Cambridge University Press.
Stahel, W., 2010. The Performance Economy. Springer.
Tukker, A., 2004. Eight types of product–service system: eight ways to sustainability? Experiences from SusProNet. Bus. Strateg. Environ. 13 (4), 246–260.
Tukker, A., 2015. Product services for a resource-efficient and circular economy—a review. J. Clean. Prod. 97, 76–91.
Velenturf, A., Purnell, P., Tregent, M., Ferguson, J., Holmes, A., 2018. Co-producing a vision and approach for the transition towards a circular economy: perspectives from government partners. Sustainability 10 (5), 1401.
Vilardi, G., Di Palma, L., Verdone, N., 2018. Heavy metals adsorption by banana peels micro-powder: equilibrium modeling by non-linear models. Chin. J. Chem. Eng. 26 (3), 455–464.
Yang, C.-M., Cao, D.-G., Zhang, B., 2012. Regional coal industry evaluation in perspective of its supportiveness of circular economy. Res. J. Appl. Sci. Eng. Technol. 4 (5), 521–525.
Yuan, Z., Bi, J., Moriguichi, Y., 2006. The circular economy: a new development strategy in China. J. Ind. Ecol. 10 (1–2), 4–8.
Zhang, W., 2017. Construction and stability studies on industrial chain network of circular economy of organic chemical industry. Chem. Eng. Trans. 62, 1507–1512.
Zhao, S., Sheng, X., Qian, Y., Zhang, Y., Zhang, J., 2009. Environmental performance management system for enterprises in industrial parks based on CCR model. China Environ. Sci. 29 (11), 1227–1232.

Zhao, H., Guo, S., Zhao, H., 2018. Comprehensive benefit evaluation of eco-industrial parks by employing the best-worst method based on circular economy and sustainability. Environ. Dev. Sustain. 20 (3), 1229–1253.

Zhijun, F., Nailing, Y., 2007. Putting a circular economy into practice in China. Sustain. Sci. 2 (1), 95–101.

Zhu, B., Hong, L., Yu, Y., Chen, D., Jiang, D., Hu, S., 2011. DEU-based comparative analysis of natural resource utilization in different countries. J. Tsinghua Univ. Sci. Technol. 51 (4), 482–487.

Zink, T., Geyer, R., 2017. Circular economy rebound. J. Ind. Ecol. 21 (3), 593–602.

CHAPTER 28

The role of collaborative leadership in the circular economy

Nermin Kişi
Zonguldak Bülent Ecevit University, Zonguldak, Turkey

1. Introduction

A rapidly increasing global population, rising resource consumption, and adverse environmental effects have led enterprises to alternative growth models (Wautelet, 2018). In spite of the fact that current economic models are largely based on a linear concept consisting of mass production, mass consumption, and mass destruction, a global consensus has been reached on the necessity to transit from these wasteful models to sustainable models (Sillanpaa and Ncibi, 2019). As a result of this, circular economy (CE), which has emerged as a new industrial paradigm, has recently aroused interest as the way of transition toward a more sustainable future (Hanumante et al., 2019). The CE approach describes a renewable economy that intends to design, use, and sustain products at the highest level of utility (Ionescu et al., 2017). In addition, CE provides a viable option to transform linear economy models and obtain better sustainable benefits (Lei and Yi, 2004). Transitioning from linear to CE business models is crucial in terms of avoiding both the continuous increase of global resource use and the related environmental and socioeconomic issues (Wilts, 2017). That is to say, it contributes to unification of economic activities and environmental welfare in a sustainable manner (Murray et al., 2015).

However, its implementation poses serious challenges in different areas including technical and financial resources, qualified personnel, information management systems, technology, consumers, leadership, etc. (Ormazabal et al., 2018). Therefore, it is clear that one of the necessary cornerstones in the successful implementation of circular processes is an effective leadership approach to knowing how to overcome the many challenges in CE transitions. Regarding the leadership factor, the key question this chapter aims to answer is: What kind of leadership is needed for companies that desire to transition to a CE? With this in mind, the chapter attempts to fill a gap in the literature by focusing on leadership themes for the CE. It suggests a collaborative leadership approach in moving to a CE system. The main purpose of this chapter is to explore the basic functions of collaborative leadership in the CE. The chapter is structured as follows. First, a brief theoretical foundation on CE challenges is presented. Next, the importance of the leadership factor in CE is discussed. Then, collaborative leadership is described as an

Circular Economy and Sustainability, Volume 1
https://doi.org/10.1016/B978-0-12-819817-9.00011-9

appropriate leadership style for circular businesses. This is followed by a discussion of the findings. The chapter concludes with its contributions, its limitations, and its insights for future research.

2. Theoretical framing of the challenges of circular economy

Although CE has been of considerable interest from both theoretical and practical standpoints, rather less attention has been paid to investigating the challenges of CE implementation. Nevertheless, some researchers have argued that many challenges or barriers slow down the practice of CE. The findings that emerge from these studies can be summarized as follows.

A number of researches have been conducted on the challenges of CE in small and medium enterprises (SMEs). Rizos et al. (2015) determine the main barriers to implement circular business models of SMEs by analyzing two case studies. Some of these barriers are summarized as poor environmental consciousness, financial constraints, inadequate government encouragement and ineffective legislation, lack of knowledge on CE, administrative costs, inadequate technical expertise, and reluctance of suppliers and customers to adopt a greener supply chain. They frequently emphasize the importance of financial barriers in their studies. In contrast to this study, Kirchherr et al. (2018), who classify the CE barriers as cultural, regulatory, market, and technological, claim that the core barriers are related to cultural factors. In their survey studies on SMEs in regions of Navarra and the Basque Country of Spain, Ormazabal et al. (2018) investigate the degree CE models are applied in SMEs, the willingness to work with other companies, and their barriers and opportunities in practice. They classify CE barriers in two clusters. These are hard barriers (e.g., shortage of technical and financial resources, inadequate technology, and insufficient information management systems for monitoring) and human-centered barriers (e.g., problems of leadership and low level of consumer interest in environmental issues). In particular, Singh and Giacosa (2019) stress that cognitive bias of consumers about CE practices is a substantial part of the human-centered CE barrier. They put forward that consumer bias causes challenges in development of CE practices, as CE-based systems do not fulfill consumer needs and expectations.

Several researchers have paid special attention to the country-centered challenges of the transition to CE. For example, Geng and Doberstein (2008), who emphasize that the CE approach presents a feasible way for governments in achieving more sustainable results, classify barriers and challenges in developing the CE in China in three clusters: policy, technology, and public participation. Similarly, Su et al. (2013) identify challenges that may prevent progress in implementation of the circular strategies in China in detail. According to them, these challenges arise from paucity of reliable knowledge, lack of high technology development, shortage of economic incentives, weak implementation of laws, ineffective managerial and leadership behaviors, weak public consciousness, and absence of systematic performance evaluation systems. In another study performed in China, Naustdalslid (2014) ascertains the constraints for implementing circular strategies as an inadequate set of government indicators, inadequate supporting policies, and lack of public participation. In their studies in Russia, Fedotkina et al. (2019) emphasize that Russian companies have started to adopt CE business models, and that sustainable management of waste in the country is a fundamental prerequisite for the practice of CE. They state that some barriers to the waste management system are related to infrastructure, technology, market, and cultural factors.

Some researchers discuss the CE challenges or barriers obtained from literature analysis. A bibliometric research done by Galvaoa et al. (2018) categorizes the major challenges in performing CE actions found in the literature as technological, political, economic, administrative, indicator-specific, customer-centered, and social barriers. Likewise, Govindan and Hasanagic (2018) analyze drivers, challenges, and practices of CE systems by using a systematic literature review. Based on this review, they outline eight types of barriers: government-oriented, economic, technologic, structural, managerial, cultural and social, knowledge and skill, and market issues. They also point out that managerial refers to poor leadership and management of CE.

To conclude, a set of micro and macro challenges of CE is presented in previous studies conducted by using case studies, survey studies, and conceptual studies. These practical challenges can be summarized as social, political, financial, economic, technological, cultural, and managerial barriers. Nonetheless, the content of managerial barriers in CE research has not been sufficiently addressed. To close this knowledge gap, the issue of which type of leadership can be built into CE-based systems must be researched. Therefore, the following sections of this chapter concentrate on the leadership aspect of CE and propose a leadership approach for CE practices.

3. The importance of leadership in circular economy

The circular economy comprises various stages of value chains such as innovation and investment, production, secondary raw material use, consumption, and waste management (Holstein et al., 2019). Implementing and advancing a CE system requires a wide range of stakeholders, long-term processes, and close coordination and cooperation between actors in all the value chains (Lei and Yi, 2004). A multistakeholder collaboration with different actors ensures the most effective way for transitioning to CE. That is why it is crucial that all stakeholders consisting of designers, corporate strategists, and marketing experts are equipped with the necessary skills to generate potential solutions for the abovementioned CE challenges by thinking with a systems approach. In particular, leaders in circular projects have a prominent role in setting strategies in order to achieve the CE goal (Ellen MacArthur Foundation, 2018). The significance of leaders has also been documented in driving change and forwarding sustainability strategies (Bernon et al., 2018).

Overall, leaders, who are acknowledged as the key element to implementing circular strategies (Bolger and Doyon, 2019), are supposed to effectively respond to complex environmental demands through collaborative network approaches in developing products and services for CE systems (Brown et al., 2018). They should handle issues of how individuals interact with others to tackle complex problems in the workplace and what are the potential effects of collaboration on the businesses (Sirman, 2008). Thanks to these activities, on the one hand they can create a cultural change where positive working relationships based on mutual trust can thrive (Penhall and Gram, 2008), on the other hand, they can encourage higher level collaborations in sustainable business practices (Gosling et al., 2016). Thereby, success or failure of CE leaders is closely associated with their skills to manage collaboratively in today's fast-paced business world (Sirman, 2008). To this respect, a concept of leadership covering governments, business, and civil society is extremely important to facilitate the transition to CE business models (PACE, 2019). What is more, in the wake of transformations in national and organizational culture worldwide, self-directed and team-oriented approaches have become a more common management style (Sirman, 2008). There is no doubt that new leadership styles are needed to respond to the changing circumstances and create more efficient models in regards to sustainability and circularity (Earley, 2016). In this chapter, a new leadership style called collaborative leadership is proposed for the successful implementations of CE.

4. The key factor to success in circular economy: Collaborative leadership

Moving toward CE practices requires a major change that spreads throughout the system and includes stakeholders. This destructive transition underlines the need to reconsider existing ways of working (Ritzen and Sandström, 2017). For instance, new generation leaders are expected to adopt flexible and collaborative approaches instead of traditional top-down leadership in order to adapt to internal and external environment. Because, leadership with a collaborative approach has been recognized as a catalyst to effectively achieve cultural change (Otter and Paxton, 2017; Davenport and Mattson, 2018), it has also become a source of competitive advantage, especially in networked, team-based, and partnership-oriented business environments (Lowry, 2016). The term "collaborative leadership" has been defined by different authors in different ways. According to Forman (2014), collaborative leadership is an impact relation that provides reliance, faith, and loyalty among leaders and team members, who aim to get a permanent or transforming change reflecting their shared target, collective vision, and common purposes; in another definition, Jameson (2007) states that it refers to a teamwork process that requires power-sharing, authority-sharing, knowledge, and responsibility.

Knowing how to build effective partnerships through collaboration is an important leadership skill to improve business success in an organization. Collaboration unites the knowledge, experience, and creativity of leadership team members (Stowell and Mead, 2017). Thus, collaborative leadership requires openness to learning from different perspectives, effective communication, shared power, shared responsibility, flexibility, etc. (Chrislip, 2002; Lawrence, 2017). It also includes group decisions, sharing information, promoting the generation of new ideas, accountable and self-selected teams, attitudes of expecting success but allowing mistakes, and matching skills and interests to responsibilities. Collaborative leaders find the right talent and trust them, create an open workplace environment, and free them to analyze and investigate. Accordingly, they create and maintain cultures that foster innovation (Pixton, n.d.). Furthermore, the task of achieving value from diversity is the essence of collaborative leadership (Archer and Cameron, 2008). Making innovation for sustainable businesses depends on leaders' ability to think together and collaborate across nations, institutions, and cultures. To attain this, leaders should harmonize different organizational cultures within the company and should encourage cooperation among the multiple stakeholders (Kuenkel, 2016). With a collaborative mindset, managers may create an inclusive environment that energizes teams, releases creativity, and enables working together (Goman, 2014).

Collaboration is absolutely necessary for integration of social, environmental, and economic sustainability in the all stages of product life cycles (Brown et al., 2018). Many companies are aware of the requirement for internal and external collaboration to ensure sustainable development. In addition, they know that communicating with others is vital to address sustainability challenges, shape their social environment, and explore new market opportunities (Kiron et al., 2015). One of the primary functions of collaborative leaders in an organization is to create networks, taking into account all stakeholders, numerous relationships, and interdependence. Another principle of collaborative leadership is distributing power and delegating decision-making authority to team members. This principle makes companies more resilient and increases their performance (Stoner, 2017).

5. Discussion

The focus of this chapter is about which leadership style may be required to make a success of CE. Within this scope, the author of the chapter proposes the collaborative leadership style for circular business practices. Undoubtedly, it can be argued that collaborative leadership skills of managers are at the heart of CE, as overcoming various challenges caused by circularity can be daunting without a collaborative mindset. As such, it can be useful to understand the key collaborative leadership roles in relation to CE to implement CE business processes efficiently. Collaborative leaders mainly generate value-added and effective solutions for their organizations through three actions:

facilitate sustainability, promote innovation, and build resilience. These functions are well-suited for fulfilling the requirements of the CE systems.

To begin with, the globalization of the world economy causes economic, environmental, and social challenges in many areas. In order to compete effectively in such an environment, companies try to reduce unnecessary use of resources by keeping items in the cycle as long as possible (Mattos and Albuquerque, 2018). As a consequence of this, they tend toward circular business models, which are accepted as a way to provide solutions for the problems of global sustainability with better resource management (Velenturf et al., 2019). On the other hand, it is a noteworthy fact that adopting new leadership styles to facilitate sustainability practices is a fundamental success factor in the implementation of circular models. This fact raises the question of which leadership activities can lead CE executives to success. In general, leaders should participate in broader networks to understand which relationships are vital for long-term sustainable success. Hence, collaborative leaders can be considered as one of the most effective leadership styles to ensure sustainability because of having collaborative networks containing multiple stakeholders. As mentioned earlier, communicating with team members on a professional level can enhance sustainability awareness in systems and explore many paths of achieving sustainable success. Thus, leaders with a collaborative mindset can facilitate the movement toward CE thanks to learning from others and strategic networking skills (Archer and Cameron, 2008).

Moreover, moving toward a CE contains a variety of challenges to capabilities, networks, and business processes for companies. Radical and systemic innovations are required to address these barriers in a rapidly changing world (Antikainen and Valkokari, 2016). Companies also need to develop new business trends in cooperation with their partners to provide sustainable competitive advantage. Briefly, innovation is an essential factor for sustainability in circular businesses (Bucea-Manea-Tonis, 2017). Given the requirement of the innovation factor, it is also important that leaders should have new sets of capabilities by transforming themselves, their partners, and their organization in order to strengthen innovation activities within the company (Smet et al., 2018). In this regard, collaborative leaders think about how they can develop an innovation culture in their organization and how they can use the different skills and experience of their partners. As a result, they release the creative potential of partners by removing boundaries among departments and enabling them to work together. To put it simply, collaborative leaders can promote innovation activities of organizations to fulfill the requirements of CE approaches (Archer and Cameron, 2008).

Another issue to consider in moving to a CE-based system is that it requires organizational transformation in building resilience through diversity and bringing together stakeholders who operate different functions in the system (Ellen MacArthur Foundation, 2013; ICAO, 2019). Collaborative leaders take an indispensable role in managing organizational transformations. They are conscious of understanding and assessing the risks and opportunities associated with management quality, workflow systems, processes, and organizational structures within the changing business ecosystem (Bennet and Bennet, 2003). In order to solve problems with collaboration and to create common visions, they know how to bring team members together, how to help them develop trust, and how to design constructive processes (Chrislip, 2002). So as to produce new insights, they move away from the hierarchical control system and create conditions whereby people throughout the organization can share information. Thereby, they can help to create a resilient organizational structure that can collaborate well both internally and externally (Archer and Cameron, 2008). In this way, they enable faster and more effective response to market conditions (Hays, 1999).

All these functions considered, it could be supposed that the collaborative leadership style facilitates the transition to CE based systems and increases the success of CE business models.

6. Conclusion

Implementation of CE faces a wide variety of social, political, economic, technological, cultural, and managerial challenges. Leaders have a significant impact on shaping CE models by tackling specifically managerial challenges. Therefore, this strong relationship between successful practices of CE and adopted leadership styles should be indicated explicitly. To the best knowledge of the author, this chapter is the first to address the leadership issue for CE practices. It claims that collaborative leadership lies at the heart of the success of CE. In this context, it emphasizes the three collaborative leadership functions that can support to enhance the chance of success of CE practices: facilitate sustainability, promote innovation, and build resilience.

This chapter adds to current CE research by offering an opportunity to advance our knowledge of effective leadership styles in circular policy implementations. Specifically, it provides an understanding of the influence of collaborative leadership in CE. In terms of the practical implications, the findings of this chapter could guide leaders as to how they can improve existing skills of management while moving to CE. That is why it is expected that research results will be a source for researchers and practitioners interested in CE. However, the conceptual considerations presented in this chapter have some limitations. One of the limitations is the lack of quantitative research. Another limitation is regarding lack of detailed comparative assessments with other leadership approaches. Consequently, more research needs to be conducted on leadership in CE in the future. First, circular projects of use of collaborative leadership styles require a comprehensive discussion with case studies. Secondly, the impacts of other leadership approaches on CE can be examined. Thirdly, an in-depth discussion of the managerial barriers of CE is a potential area for further research. For instance, research can be carried out to analyze cross-country or cross-sectoral differences with respect to managerial challenges of CE. Fourthly, studies can also be conducted on stakeholders' expectations from managerial roles in CE-based systems.

References

Antikainen, M., Valkokari, K., 2016. A framework for sustainable circular business model innovation. Technol. Innov. Manag. Rev. 6 (7), 5–12.

Archer, D., Cameron, A., 2008. Collaborative Leadership: How to Succeed in an Interconnected World. Butterworth-Heinemann, Burlington, MA.

Bennet, A., Bennet, D., 2003. Organizational Survival in the New World: The Intelligent Complex Adaptive System. Butterworth-Heineniann, Burlington, MA.

Bernon, M., Tjahjono, B., Ripanti, E.F., 2018. Aligning retail reverse logistics practice with circular economy values: an exploratory framework. Prod. Plan. Control 29 (6), 483–497.

Bolger, K., Doyon, A., 2019. Circular cities: exploring local government strategies to facilitate a circular economy. Eur. Plan. Stud. 27 (11), 2184–2205.

Brown, P., Bocken, N., Balkenende, R., 2018. Towards understanding collaboration within circular business models. In: Moratis, L., Melissen, F., Idowu, S.O. (Eds.), Sustainable Business Models: Principles, Promise and Practice. Springer International Publishing, Cham, Switzerland, pp. 169–201.

Bucea-Manea-Tonis, R., 2017. Assumptions on innovation into a circular economy. J. Appl. Econ. Sci. 5 (51), 1319–1327.

Chrislip, D.D., 2002. The Collaborative Leadership Fieldbook: A Guide for Citizens and Civic Leaders. Jossey-Bass, San Francisco, CA.

Davenport, A., Mattson, K., 2018. Collaborative leadership as a catalyst for change. Knowl. Quest 46 (3), 14–21.

Earley, R., 2016. Whole circles: models for academic textile design research leadership in the circular economy. In: Earley, R., Goldsworthy, K. (Eds.), Proceeding of a Mistra Future Fashion Conference on Textile Design and the Circular Economy, Chelsea College of Arts & Tate Britain, London, 23–24 November 2016, pp. 68–80.

Ellen MacArthur Foundation, 2013. Towards the Circular Economy: Business Rationale for an Accelerated Transition. Ellen Macarthur Foundation, Cowes, UK.

Ellen MacArthur Foundation, 2018. The Circular Economy Opportunity for Urban & Industrial Innovation in China. Ellen MacArthur Foundation, Cowes, UK.

Fedotkina, O., Gorbashko, E., Vatolkina, N., 2019. Circular economy in Russia: drivers and barriers for waste management development. Sustainability 11, 1–21.

Forman, D., 2014. Introduction. In: Forman, D., Jones, M., Thistlethwaite, J. (Eds.), Leadership Development for Interprofessional Education and Collaborative Practice. Palgrave Macmillan, London, pp. 1–13.

Galvaoa, G.D.A., Nadae, J., Clemente, D.H., Chinen, G., Carvalhoa, M.M., 2018. Circular economy: overview of barriers, in: Sakao, T., Lindahl, M., Liu, Y., Dalhammar C. (Eds.), Proceeding of 10th CIRP Conference on Industrial Product-Service Systems, IPS2 2018. Linköping, Sweden, 29-31 May 2018. Procedia CIRP 73, 79–85.

Geng, Y., Doberstein, B., 2008. Developing the circular economy in China: challenges and opportunities for achieving "leapfrog development". Int. J. Sustain. Dev. World Ecol. 15 (3), 231–239.

Goman, C.K., 2014. Collaborative leadership: 8 tips to build collaboration in your team or organization. Leadersh. Excell. 31 (4), 35.

Gosling, J., Jia, F., Gong, Y., Brown, S., 2016. The role of supply chain leadership in the learning of sustainable practice: toward an integrated framework. J. Clean. Prod. 137, 1458–1469.

Govindan, K., Hasanagic, M., 2018. A systematic review on drivers, barriers, and practices towards circular economy: a supply chain perspective. Int. J. Prod. Res. 56 (1–2), 278–311.

Hanumante, N.C., Shastri, Y., Hoadley, A., 2019. Assessment of circular economy for global sustainability using an integrated model. Resour. Conserv. Recycl. 151, 1–13.

Hays, S., 1999. Our future requires collaborative leadership. Workforce 78 (12), 30.

Holstein, F., Salvatori, G., Böhme, K., 2019. Local and regional strategies accelerating the transition to a balanced circular economy. Spatial Foresight Brief 11, 1–10.

ICAO, 2019. ICAO Environmental Report 2019: Aviation and Environment. ICAO Publications, Canada.

Ionescu, C.A., Coman, M.D., Lixandru, M., Groza, D., 2017. Business model in circular economy. Valahian J. Econ. Stud. 8 (22), 101–108 (4).

Jameson, J., 2007. Investigating Collaborative Leadership for Communities of Practice in Learning and Skills. The Centre for Excellence in Leadership (CEL), Lancaster.

Kirchherr, J., Piscicelli, L., Bour, R., Kostense-Smit, E., Muller, J., Huibrechtse-Truijens, A., Hekkert, M., 2018. Barriers to the circular economy: evidence from the European Union (EU). Ecol. Econ. 150, 264–272.

Kiron, D., Kruschwitz, N., Haanaes, K., Reeves, M., Fuisz-Kehrbach, S.-K., Kell, G., 2015. Joining forces: collaboration and leadership for sustainability. MIT Sloan Manag. Rev. 2015, 1–31.

Kuenkel, P., 2016. The Art of Leading Collectively: Co-Creating a Sustainable, Socially Just Future. Chelsea Green Publishing, White River Junction, Vermont.

Lawrence, R.L., 2017. Understanding collaborative leadership in theory and practice. New Dir. Adult Contin. Educ. 156, 89–96.

Lei, S., Yi, Q., 2004. Strategy and mechanism study for promotion of circular economy in China. Chin. J. Popul. Resour. Environ. 2 (1), 5–8.

Lowry, C., 2016. Are You a Collaborative Leader? https://asentiv.com/are-you-a-collaborative-leader/. (Accessed 20 November 2019).

Mattos, C.A., Albuquerque, T.L.M., 2018. Enabling factors and strategies for the transition toward a circular economy (CE). Sustainability 10, 1–18.

Murray, A., Skene, K., Haynes, K., 2015. The circular economy: an interdisciplinary exploration of the concept and application in a global context. J. Bus. Ethics 140, 369–380.

Naustdalslid, J., 2014. Circular economy in China-the environmental dimension of the harmonious society. Int. J. Sustain. Dev. World Ecol. 21 (4), 303–313.

Ormazabal, M., Prieto-Sandoval, V., Puga-Leal, R., Jaca, C., 2018. Circular economy in Spanish SMEs: challenges and opportunities. J. Clean. Prod. 185, 157–167.

Otter, K., Paxton, D., 2017. A journey into collaborative leadership: moving toward innovation and adaptability. New Dir. Adult Contin. Educ. 156, 33–42.

PACE (Platform for Accelerating the Circular Economy), 2019. Leadership. https://pacecircular.org/. (Accessed 20 November 2019).

Penhall, S., Gram, R., 2008. Collaboration and Competition: A Leadership Challenge. The Centre for Excellence in Leadership (CEL), London.

Pixton, P., n.d. Collaborative Leadership: An Overview. http://leadit.databasemonth.com/it-business-management/Collaborative-Leadership-An-Overview (Accessed 20 November 2019).

Ritzen, S., Sandström, G.Ö., 2017. Barriers to the circular economy-integration of perspectives and domains, in: McAloone, T.C., Pigosso, D.C.A., Mortensen, N.H., Shimomura, Y. (Eds.), Proceeding of the 9th CIRP IPSS Conference: Circular Perspectives on Product/Service-Systems, IPSS2017, Kongens Lyngby, Denmark, 19–21 June 2017. Procedia CIRP 64, 7–12.

Rizos, V., Behrens, A., Kafyeke, T., Hirschnitz-Garbers, M., Ioannou, A., 2015. The Circular Economy: Barriers and Opportunities for SMEs. Centre for European Policy Studies, Brussels, Belgium, pp. 1–22.

Sillanpaa, M., Ncibi, C., 2019. Circular economy: here and now. In: Sillanpaa, M., Ncibi, C. (Eds.), The Circular Economy Case Studies about the Transition from the Linear Economy. Academic Press, London, United Kingdom, pp. 37–68.

Singh, P., Giacosa, E., 2019. Cognitive biases of consumers as barriers in transition towards circular economy. Manag. Decis. 57 (4), 921–936.

Sirman, R., 2008. Collaborative leadership- a sound solution to complex problems. Employ. Relat. Today 35 (2), 31–42.

Smet, A.D., Lurie, M., George, A.S., 2018. Leading Agile Transformation: The New Capabilities Leaders Need to Build 21st-Century Organizations. McKinsey & Company, Houston.

Stoner, J.L., 2017. 8 Principles of Collaborative Leadership. Seapoint Center for Collaborative Leadership, Berkeley, CA.

Stowell, S.J., Mead, S., 2017. What Does Collaborative Leadership Look Like? https://trainingindustry.com/articles/leadership/what-does-collaborative-leadership-look-like/. (Accessed 20 November 2019).

Su, B., Heshmati, A., Geng, Y., Yu, X., 2013. A review of the circular economy in China: moving from rhetoric to implementation. J. Clean. Prod. 42, 215–227.

Velenturf, A.P.M., Archer, S.A., Gomes, H.I., Christgen, B., Lag-Brotons, A.J., Purnell, P., 2019. Circular economy and the matter of integrated resources. Sci. Total Environ. 689, 963–969.

Wautelet, T., 2018. Exploring the Role of Independent Retailers in the Circular Economy: A Case Study Approach (Unpublished Master's Thesis). Eufom European University for Economics and Management, Luxembourg.

Wilts, H., 2017. Key challenges for transformations towards a circular economy – the status quo in Germany. Int. J. Waste Resour. 7 (1), 1–5.

CHAPTER 29

Issues, interventions, and innovations in the cement industry: A comparative trajectory analysis of eco-cement transitions in the Netherlands, China, and Japan

Serdar Türkeli[a], Beijia Huang[b], Satoshi Ohnishi[c], and René Kemp[a]
[a]United Nations University-MERIT, Maastricht University, Maastricht, The Netherlands
[b]School of Environment and Architecture, University of Shanghai for Science and Technology, Shanghai, China
[c]Tokyo University of Science, Faculty of Science and Technology, Department of Industrial Administration, Noda-shi, Chiba-ken, Japan

JEL: L74, O33, O38, O52, P16, Q55.

1. Introduction

Total CO_2 emissions from the cement industry are estimated at 8% of global CO_2 emissions (Andrew, 2018). Regarding specific political geographies in particular, by 1982, China had become the world's largest producer of cement. Originally a technology and equipment importer, China became the largest supplier of cement equipment. Since the mid-1990s, cement production in China has grown by almost 12 times, such that 73% of the global growth in cement production occurred in China (van Oss, 2017). In 2016, globally, 4.6 billion metric tonnes of cement was produced, of which 2.41 billion metric tonnes were produced in China (over half of the world's total cement production in 1 year). In contrast, in 2016, Japan accounted for 1.2% of the global cement production volume. While China produced 57% of the world's cement in 2016, its emissions were 52% of the total emissions registered globally, as a consequence of its clinker ratio being less than 0.60, which is below the world average (Andrew, 2018). On the other hand, 7 of the 10 most polluted cities in China are located in Hebei Province, which is a major regional center for the cement industry (Reuters, 2018).

In this chapter, we analyze three political geographies, namely, China, Japan, and the Netherlands, which differ by orders of magnitude in their cement emissions. Especially, we analyze the slow, ongoing transition from ordinary Portland cement (OPC) to eco-cement and associated transition trajectories in these different contexts with respect to our research questions focusing on the dynamics of: (1) the constellations of issue actors and

Circular Economy and Sustainability, Volume 1
https://doi.org/10.1016/B978-0-12-819817-9.00027-2

networks, (2) intervention capacity and capabilities of institutions and society, and (3) scientific and industrial advancements for alternatives to traditional OPC.

This chapter proceeds as follows. Section 2 provides materials (Section 2.1, theoretical and conceptual background of co-evolution analysis between studies of technological change and policy change) (Kemp et al., 2017) and methods (Section 2.2, data and methods) used in this chapter. In Section 3, we analyze the results and discuss these three contexts in-depth. In Section 4, we provide concluding remarks and future research directions.

2. Materials and methods

2.1 Theoretical background for a co-evolutionary analysis of transitions

Policy change is a topic of policy scholars and technical change is a topic of innovation research (Van den Belt and Rip, 1987; Nelson, 1994; Kemp, 1994). The incrementalist model of policy change offered by Lindblom (1959, 1979) is extended by: (1) the punctuated equilibrium model of Baumgartner and Jones (1991, 2002); (2) the technical/rational model (Bridgman and Davis, 2003); (3) models of power and interest groups (Weiss, 1983); (4) policy networks (Rhodes, 1990); (5) advocacy coalition framework (Sabatier, 1988, 1993); and (6) multiple streams framework (Kingdon, 1984, 1995).

A limitation of those models is that the interaction of policy with techno-economic developments is poorly considered, because of the focus on the problem definitions, policy agendas, and coalitions between policy actors. In our chapter, we give special attention to the techno-economic dimension in terms of the material consequences of policy actions, namely cement choices, and in terms of the degree to which policy is based on eco-cement possibilities as perceived by capable agencies interacting with innovation actors. We combine the problems-politics-policy streams model of Kingdon (1984, 1995) with the three subsystems identified in the SUSTIME project (Zundel et al., 2005): a sociocultural subsystem of problem frames, a politico-administrative subsystem of politics and policy choices, and a techno-economic subsystem of innovation and research activities in old and new technologies and products.

This framework helps us investigate the ways in which the actors of these subsystems influence the streams. We attribute an important role to techno-economic assessments performed by capable agencies with the help of domain experts on policy acts. In an earlier paper of two of the authors of this chapter, this framework is used to study the co-evolutionary dynamics between policy choices, societal issues discussed in the media, policy and politics, and technical advances in old and new cement technologies in the Netherlands in-depth (Kemp et al., 2017). A graphical representation of the framework used in this chapter is given in Fig. 1.

The streams model is based on the assumption that successful innovation policy for transitions or path changes requires a combination of windows and favorable conditions

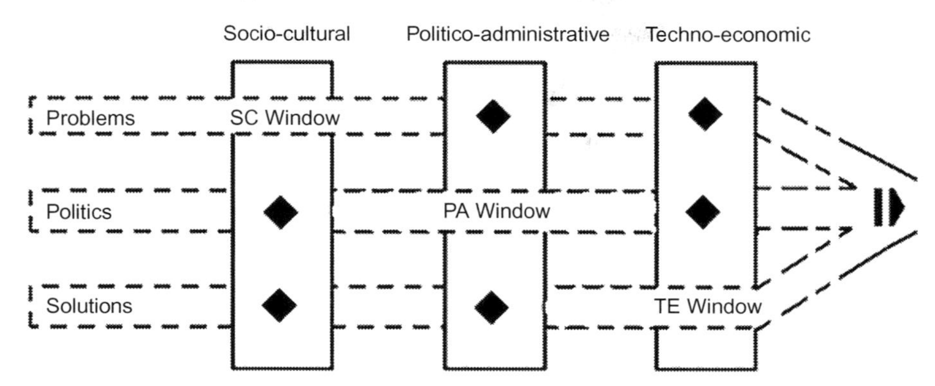

Fig. 1 Multiple streams framework and policy windows. *SC*, socio-cultural window; *PA*, politico-administrative window; *TE*, techno-economic window.

(Zundel et al., 2005), where policy windows are opened by the convergence of problems, politics, and policy streams. This convergence, which necessitates the alignment of capable actors with certain capacities, sparks the design and implementation of certain public policies, and this, in turn, forms the basis for further policy actions. In general, the convergence of these three streams depends on the perceived need for change among critical policy actors (including businesses and NGOs), capable agencies, and available solutions (real or perceived).

Keeping the policy window open, utilized, and updated requires capacities and capabilities of all relevant actors operating in different subsystems, or at the boundaries of these subsystems. Dimensions marked with an asterisk (*) in Table 1 are also scrutinized in Kemp et al. (2017).

For the case of eco-cement transition in the Netherlands, an earlier co-evolutionary analysis revealed six key dimensions: (1) standards and regulations, as shapers of cement choices; (2) the emission trading systems; (3) circular economy policies (co-incineration of waste, the recycling of concrete); (4) research and innovation in new types of cement (industrial innovation, scientific applied research); (5) civil society actions; and finally (6) the market demand for eco-cement (dynamics of take up) (Kemp et al., 2017). In this chapter, those aspects will also be examined for China and Japan in a comparative way, with special attention to: (1) the techno-economic window, (2) the politico-administrative window, and (3) the sociocultural window. These windows refer to material elements and symbolic ones leading actors to do something in terms of making pleas for change (because of the specific issues), policy action, and investments in alternative types of cement.

Consequently, we concentrate on the issues (problems/problematization by actors), the interventions (scales and levels of political-administrative interventions), and the innovations (in terms of actor constellations/networks and technologies). In doing so,

Table 1 Issues, interventions, and innovations.

Streams	Objects of analysis
Issues Problems and problematization (with respect to incumbent products and emerging technologies)	**Traditional cement:** Problems/ problematization associated with traditional cement (e.g., environmental, economic, social, political problems) by different actors (e.g., in different subsystems: politicians, administration, companies, civil society actions*, international organizations/ commitments) **Alternative cements:** Problems/ problematization about traditional cement alternatives (e.g., industrial and scientific development of new types of cement: eco/ green/sustainable*; status and prospects of market demand*)
Interventions (Politico-administrative) Scale and level sophistication of political-administrative interventions	Relatively **smaller scale** interventions (e.g., introducing or amending standards, regulations*, financing of alternative cements R&D) Relatively **larger scale** interventions: (e.g., emission trading systems*, industrial restructuring, policy mixes (e.g., with circular economy policies*), demand side policies, mission-oriented policies)
Innovations (Technology) Solutions	Processes and actors behind **old cement products** and **technologies** Processes and actors behind **innovative cement** solutions* in research and development, demonstration, and utilization

Aspects with * are scrutinized in Kemp et al. (2017) in depth.

we seek to reveal the actors and their interactions among different streams throughout sociocultural, politico-administrative, and techno-economic domains in the Netherlands, China, and Japan for the comparative co-evolutionary analysis of eco-cement transition trajectories.

2.2 Case selection, data, and method

Three political geographies, namely, China, Japan, and the Netherlands, differ by orders of magnitude in their cement emissions. Fig. 2 provides the trends related to cement production-based CO_2 emissions in China, Japan, and the Netherlands from 1980 to 2015 (Liu et al., 2015; Olivier et al., 2016; Andrew, 2018). While the figures for the Netherlands were in the 1.0–2.0 Mt CO_2 band in the 1980s and decreased throughout

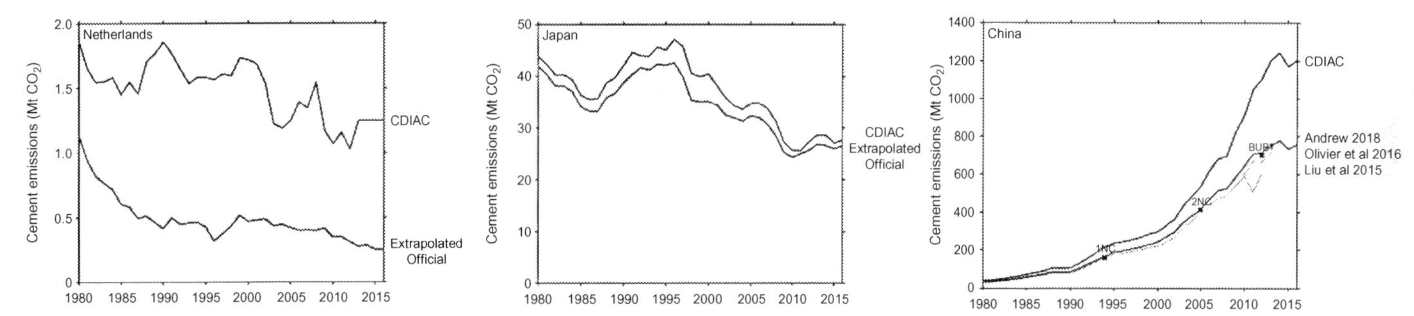

Fig. 2 Cement emissions in the Netherlands, Japan and China, Mt CO_2 (1980–2015). *(Source: Andrew, R.M., 2018. Global CO_2 emissions from cement production. Earth Syst. Sci. Data 10(1), 195.; Olivier, J.G.J., Janssens-Maenhout, G., Muntean, M., Peters, J.A.H.W., 2016. Trends in Global CO2 Emissions: 2016 Report, PBL Netherlands Environmental Assessment Agency, Ispra/European Commission, Joint Research Centre, The Hague, Available from: http://edgar.jrc.ec.europa.eu/; Liu, Z., Guan, D., Wei, W., Davis, S. J., Ciais, P., Bai, J., Peng, S., Zhang, Q., Hubacek, K., Marland, G., Andres, R.J., Crawford Brown, D., Lin, J., Zhao, H., Hong, C., Boden, T.A., Feng, K., Peters, G.P., Xi, F., Liu, J., Li, Y., Zhao, Y., Zeng, N., He, K., 2015. Reduced carbon emission estimates from fossil fuel combustion and cement production in China, Nature 524, 335–338. https://doi.org/10.1038/nature14677, 2015;* CDIAC, *Carbon Dioxide Emission Analysis Center.)*

the following years to the 0.5–1.5 Mt CO_2 band, the initial condition of cement emissions in Japan was around 40–45 Mt CO_2 band in the 1980s, and reduced to 25–30 Mt CO_2 band in the 2010s, roughly 20 times larger than the cement production-based CO_2 levels in the Netherlands. Yet for China, we observe 200 Mt CO_2 in 1995 and an increase to 600–800 Mt CO_2 in 2015, with respect to different analysis approaches (for details see Liu et al., 2015; Olivier et al., 2016; Andrew, 2018), which is again roughly 20 times larger than the Japanese cement-based CO_2 emissions.

The methodology used in this chapter involves gathering of data and information through a number of primary and secondary sources in these three political geographies. Informed by six key dimensions listed in Table 1, a number of face-to-face, open-ended interviews with key experts in the cement industry are undertaken in each political geography under study. The experts are selected for having specific information on, and a thorough knowledge of specificities of the cement sector in each context. For reasons of confidentially, their names are not disclosed. Thus, in the chapter, interviews serve as the primary source of data.

Considering the number of interviews, Guest et al. (2006) indicates that 6–12 expert interviews as sufficient in cases of a rather homogenous sample, and Romney et al. (1986) considers sample sizes as small as four experts to suffice if the experts are highly competent in the domain of inquiry. Considering the specificity of the eco-cement topic, we fulfill these quality criteria. Where interviews were not available, we used secondary data sources, such as official reports and newspaper articles from internationally recognized sources, thus, all limitations relevant to the use of secondary data in the published literature are also applicable for this chapter.

3. Analysis and discussions

3.1 Issues

3.1.1 Traditional cement in China, the Netherlands, and Japan

In China, problems/issues related to traditional cement were brought forward by a powerful techno-economic actor, the chairman of China National Building Material Group (NBMG) Co. Ltd. The chairman indicated that the Mark 32.5 cement, which is of low strength and takes up 60% of the market, should be eliminated, while most countries around the world use Mark 42.5 cement and above, and interventions relating to upgrading the cement standards via regulations, with the aim of eliminating outdated capacity for low-standard cement, which takes up too much of the output are called for (October 10, 2016, Monday, China Daily European Edition). In this respect, the chairman acted as a boundary policy entrepreneur indicating a need to merge different streams, e.g., issue stream (Mark 32.5 cement), policy (standards and regulations), and politics (consolidation of Chinese cement industry), to open up a policy window for initiating the change.

Another organization operated at the boundary of politico-administrative and techno-economic subsystems is the Chinese Cement Association (CCA), which assists the Chinese government in formulating strategic plans for industrial development, industrial policies, regulatory policies, industrial codes and standards, and implementation assistance. The president of the Chinese Cement Association (CCA) voiced similar concerns to the chairman of the NBMG in stating that "the industry is mired in low efficiency and chaos because most cement factories are small and scattered, lacking in high technologies and management expertise. There are too many cement enterprises, which makes it very hard to utilize resources effectively" (October 10, 2016 Monday, China Daily European Edition). Following the opening up of a policy window, policy measures targeting the standards and industrial structure have been taken to replace Mark 32.5 in China, for example, China's current universal cement product standard GB175-2007 "Universal Portland Cement" and its No. 1 and No. 2 Amendments aiming at removing overcapacity in the cement industry and adjusting its structure. No. 2 Amendment especially provides institutional grounds for the de-compounding of Mark 32.5 Portland cement as of December 1, 2015, but retaining the composite 32.5R Portland cement. Yet, due to the retention of composite 32.5R cement, the actual conversion of the original composite 32.5 Portland cement into composite 32.5R Portland cement by business, in actual implementation, has not been achieved. This situation stresses the importance of policy responsiveness from the actors of techno-economic subsystem to policy windows.

The Netherlands is characterized by different cement formulations, institutional and social issue networks, and actor profiles. The main cement type produced is the CEM III standard blast furnace slag (BFS), constituting almost 70% of the current market (Kemp et al., 2017). In terms of regulations concerning the use of blended cements in the Netherlands, these regulations have historically evolved from initially prohibiting the use of slag in cement from 1912, to being continuously more liberal in allowing the use of the slag upon prior consent by the client and the contractor (1930, 1940, and 1952 regulations), with a 1984 regulation explicitly recommending the use of slag cement in aggressive environments (Kemp et al., 2017). For alternative cement, a long-time policy window existed in the Netherlands, yet unlike China, we also observe rising civil society actions, in the form of protests by local residents and environmental activists, to excavation activities and the co-combustion of waste. The interaction of the sociocultural window (especially the citizen protests) with the bad economies of the main cement company (ENCI) and adverse market conditions in the techno-economic subsystem influenced the policy and technology trajectory of eco-cement transition. A covenant was signed between the ENCI company, the municipality of Maastricht, the Limburg government, nature organizations, and local resident groups that excavation would be stopped in 2018, with the area obtaining a new destination as an area for water recreation and nature protection (NRC, 2009).

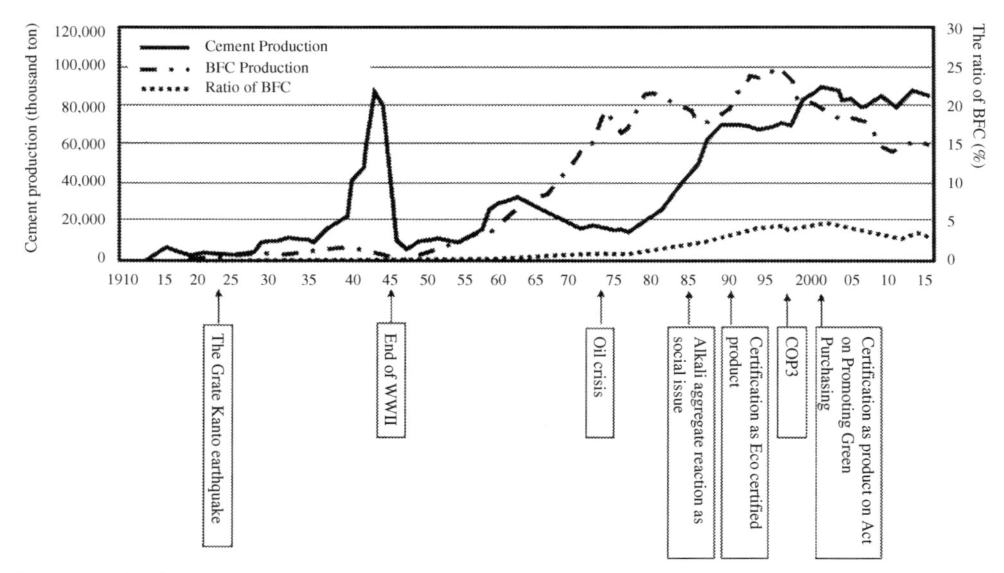

Fig. 3 Trend of cement production and the ratio of Blast Furnace Cement (BFC) in Japan.

Quoting Kemp et al. (2017) on this:

> *The closure of the quarry and clinker production is a result of various factors: A major role though was played by the civil society from both the Netherlands and Belgium, which was pointing to the environmental damages caused by mining limestone, in terms of air pollution, odour, dust and noise nuisance, as well as soil contamination and decline of groundwater levels and the quality of the Belgian drinking water, and the endangered flora and fauna (Dagblad de Limburger, 2003a, b, 2004a, b, c; Het Belang van Limburg, 2003). The opposition, led by ENCI Stop and Sint Pietersberg Adembenemend environmental organisations, peaked when ENCI Maastricht applied for the extension of mining concession till 2030 and possibly beyond and also for the allowance for deeper mining (Dagblad de Limburger, 2007, de Trompetter, 2007, Het Belang van Limburg, 2004; Enci Stop, 2001).*

In Japan, the role of decisive social pressure for promoting eco-cement and opening up of a policy window emanated from a completely different reason, which was the shortage of land reclaimed with blast furnace slag and fly ash, which was used in industrial parks with iron/steel and power factories and final disposal sites. For actors in the politico-administrative subsystem, the Japanese cement industry was the most promising candidate to accept blast furnace slag and fly ash as materials, since the result of R&D activities have standardized eco-cement from blast furnace and fly ash since the 1970s. Fig. 3 shows the trend of cement production in Japan and the ratio of blast furnace cement (BFC) since the 1910s. In the era during WWII, the demand of traditional cement, especially for military use, had been increased, though the lack of clinker production capacity had been a concern. The production amounts of BFC with proven technology, originally from

Germany, had rapidly spread. Based on R&D activities in the 1970s, the ratio of BFC had increased from the 1980s, promoted by public concern about the alkali-aggregate reaction in OPC, certification as an Eco-certified product, and certification through the Act on Promoting Green Purchasing. Subsequently, the ratio of BFC has been stable at 20% (Fig. 3).

What broadened the policy window in Japan was incineration ash from municipal solid waste (MSW), which was also targeted to be utilized as material to produce cement and to reduce the final disposal amounts as a relief measure for energy-intensive industries (EIIs), which had struggled with the recession due to the collapse of a bubble economy and absolute changes in the industrial structure. Additionally, cement industries in Japan were large candidates to accept many kinds of wastes from both municipal and industrial activities. Eco town programs from 1997 to 2006 had played an important role in investing in facilities through a subsidy system (Van Berkel et al., 2009). Consequently, in Japan, 35,220 thousand tonnes/year of OPC and 9450 thousand tonnes/year of eco-cement were produced in 2014, representing 70.8% and 21.6%, respectively, in total cement production in the same year.

3.1.2 Issues related to eco-cement in Japan, the Netherlands, and China

Dewald and Achternbosch, in their chapter titled *Why more sustainable cements failed so far? Disruptive innovations and their barriers in a basic industry*, indicate that demand-oriented knowledge formation and the reliance on existing prescriptive standards impede progress toward more sustainable alternatives to conventional cements. The authors state that the implementation of new materials in cement requires new types of collaboration between R&D and market actors, a combination of synthetic knowledge with analytic knowledge, and redefinition of standards and norms (Dewald and Achternbosch, 2016). Likewise, a close cooperation of incumbent actors along the construction value chain is pointed out as a precondition for the success of disruptive innovations in cement. After WWII, Japanese industries had recognized that the quality of BFC had been low, thus, the distribution of BFC had become diminished. In line with the arguments of Dewald and Achternbosch (2016), in Japan, eco-cement was developed through the cooperation of both public and private institutions, which helped a solutions stream to merge. The New Energy and Industrial Technology Development Organization (NEDO) of Japan, operating at the boundary of politico-administrative subsystem and techno-economic subsystem, together with private companies, developed the technology for the manufacture of eco-cement (Lopez, 2008). A blended cement BFC was developed by Taiheiyo Cement Corporation in Japan. Yet not every attempt for change was effective, although the policy window is open. The carbon prices of the ETS program failed to have much impact on eco-cement production. In an effort to support eco-cement, the Manufacturing Industries Bureau (METI) conducted a survey on diffusion of blended cement as eco-cement, with special attention on how to promote BFC in 2016. The survey revealed that big general construction companies often used BFC for retaining structure, abutment, and revetment in the public sector and underground joist in the private sector. The reasons why the companies employed BFC were firstly that specifications of structures demanded the utilization of BFC, especially in public sector construction, and secondly, that BFC mitigates alkali-aggregate reactions. CASBEE (Comprehensive Assessment System for Built Environment Efficiency), which is a method for evaluating and rating the environmental performance of buildings and the built environment authorized by the Ministry of Land, Infrastructure, Transport, and Tourism, Japan, played an important role to promote using BFC. Technological problems were identified, such as rapid neutralization rate, low durability, and chapters in the report on diffusion policy of energy saving manufacturing process in cement industry and discussions on diffusion of blended cement by METI.

Issues related to eco-cement in the Netherlands revealed several political economy elements of industrial research. One example is that The Heidelberg Group (of which ENCI is a part) studies alkali-activated cement in a covert way (Source: interviewee). Cement actors also managed to restrict the investigation of the costs of the ways to make the concrete chain more sustainable to those options in which the sector is commercially interested, showing the influence of the industry on the technology and policy options for consideration. The use of carbon capture and sequestering and utilization (CCSU) and the use of less concrete in construction works were excluded from investigations for strategic reasons (Van der Vooren et al., 2015, p. 20). Companies are worried that the inclusion of those options may lead to policies in support of those options; they did not want those options to appear in the policy window for cement-based carbon reductions (Kemp et al., 2017).

In China, several issues related to overcapacity, different energy consumption levels among different enterprises, energy consumption of a few cement enterprises that reached internationally advanced levels, ordinary cement intensity levels, helped problems, politics, and solutions streams to merge and this led to an opening up of a policy window that resulted in the cancelation of Marks 32.5 and 32.5R, causing the elimination of shaft kiln cement. For compounding sludge into cement, a process was first developed at Lafarge cement plants in France that is currently being used in Japan; and for China, the Deputy Secretary General of China Cement Association indicated that "such an undertaking is rarely seen in Chinese cement plants and Lafarge's innovation has set a good example of a circular economy system in China" in commenting on the Chairman and CEO of Lafarge Group's announcement that "Lafarge is willing to bring its technology and experiences into China" (Global Green Cement Market, 2016). Yet, for the case of China, there is no explicit definition and standard about green cement, eco-cement, or healthy cement.

Table 2 Issues and main actors of streams, subsystems, and policy windows.

Issues	Context	PA	PA/TE	TE	TE/SC	SC	SC/PA
Traditional cement	**China**	*	*	*			
	The Netherlands		*		*		*
	Japan	*	*				*
Alternative cements	**China**	*	*	*			
	The Netherlands		*		*	*	*
	Japan	*	*	*			*

* denotes significant influence; *PA*, politico-administrative actors; *TE*, techno-economic actors; *SC*, sociocultural actors; *PA/TE*, politico-administrative and techno-economic interface/boundary; *TE/SC*, techno-economic and sociocultural interface/boundary; *SC/PA*, sociocultural and politico-administrative interface/boundary.

Table 2 summarizes the different contexts and leading actors for the issues related to traditional cement and alternative cement that helped in opening up and stabilizing a policy window.

3.2 Interventions in China, Japan, and the Netherlands

The Chinese government, as a politico-administrative actor, has made several efforts on tightening regulations and policies in the cement sector in order to reduce industry-driven pollution. Two initiatives launched in China in 2006 are the "Cement Industry Development Policy," which aims to abandon backward production capacity, and "Special Planning for Cement Industry Development," the regulation that has led to the closure and merger of many small cement plants, which were causing serious environmental pollution and resource destruction. These policies put significant pressures on cement companies to upgrade their technology and invest in advanced and cleaner production equipment. Results are expected to have positive impacts on environmental efficiency in the sector. At provincial level, Guangdong had started allotting "388Mt of carbon emission quotas to selected enterprises," according to the Provincial Development and Reform Commission. Initially, 242 companies from the cement, power, iron and steel, and petrochemical industries were included in the quota allocation. Through the scheme, it is intended to cap CO_2 emissions at 350 Mt for 2013 (Global Cement, 2013).

Meanwhile, at international level, the US-China Climate Change Working Group held a special event in Beijing to celebrate the achievements of private sector partnerships to advance carbon capture, use, and storage, and develop green data centers, reduce hydrofluorocarbons (HFCs) from refrigeration and air conditioning, and promote "green cement" (IISD, 2014). These achievements underscore the critical role of the private sector and other nongovernmental actors to test and deploy innovative solutions to climate change and other pressing global challenges (US Department of State (July 9, 2014)). Following the publication of "Policy Analysis of China's Cement Industry" in

2015, the central government reinforced the promotion of supply-side reform at the end of 2015 by making elimination of outdated capacity a top priority, removing "zombie enterprises," and promoting industrial reorganization (Global Cement, 2016, 2017). The Vice Minister of Industry and Information Technology, said that "dealing with 'zombie companies' is the key to solving the excess capacity in China" (China Daily, 2016a,b). Likewise, the Minister of Industry and Information Technology stated at a State Council Information Office (SCIO) press conference that in "stabilizing industry growth and readjusting structure that they will vigorously deal with 'zombie enterprises'" (SCIO, 2016). Adding further force to this development, the China Cement Association has recently asked the Ministry of Industry and Information Technology to speed up the consolidation process in the local cement industry. According to documents reported as seen by the South China Morning Post, the cement body wants the Ministry to consolidate at least 60% of the country's cement production capacity into 10 producers by 2020. The association made its proposals in July 2016 and has since chased the Ministry for a response (Global Cement, 2016). The utilization of policy windows is thus being challenged in the Chinese context for an eco-cement transition due to the misalignment of capabilities in cement industry and industrial structure.

In Japan, two oil shocks triggered the diffusion of BFC for reasons of reducing energy costs in the cement production process. Another trigger was the research reports on the effect of mitigation of alkali-aggregate reaction by utilizing BF as cement material (Kobayashi et al., 1985). Those triggers helped R&D activities to standardize eco-cement in the1970s. After the Rio Summit in 1992, BFC was recognized as a low GHG emission product. Although the policy window was open after this, nonetheless, the use of BFC had reached a celling in the early 1990s. At the time that the Japanese government developed a recycling system, called the Sound-Material Cycle Society, BFC was certified by the Act on Promoting Green Purchasing in 2001. General construction companies had developed technologies with low GHG emission BFC and implementation methods for private buildings and public infrastructure (Kobayashi et al., 1985). In Japan, in order to better compete, one of the strategies of Japanese energy-intensive industries (EIIs) was to accept industrial waste, and the Japanese government supported them to develop new business models by introducing a comprehensive political framework including strict regulations, free waste disposal, and subsidies.

Companies in Japan engaged in environmental management systems, such as ISO14000, in the late 1990s, and the waste management issue was a high priority in promoting their environmentally friendly activities to the Japanese public. Industrial sectors also started to ask for the treatment and recycling of waste by the cement industry (e.g., waste as fuel) due to the fact that the cement industry can treat various types of industrial wastes safely and properly with huge capacity. On the other hand, although several municipalities in Japan enacted the ETS program and several cement companies joined the program and these initiatives do not seem to be effective to promote eco-cement

Table 3 Alternative Cement Standards in Japan and the Netherlands.

Japan Industrial Standard	The Netherlands/EU Standards
JIS R 5210: OPC (5 types with compressive strength varying from 22.5 to 42.5 N/mm^2, 28 days curing time, since 1950).	**CEM I:** Portland cement with a maximum of 5% other materials.
JIS R 5211: Blast Furnace Slag Cement (3 types with SCM percentage varying from 5–30 wt% to 60–70 wt%, since 1950).	**CEM III:** blast furnace/Portland cement mixture in 3 classes: A, B, and C; whereby CEM III/A contains the least (40%) and CEM III/C contains the most (90%) slag.
JIS R 5212: Silica Cement (3 types with SCM content percentage varying from 5–10 wt% to 20–30 wt%, since 1950).	**CEM IV:** Pozzolana cement varieties (pozzolana is a natural or artificial material containing silica in a reactive form).
JIS R 5213: Fly Ash Cement (3 types with SCM content percentage varying from 3 ~ 10 wt% to 20–30 wt%, since 1960).	**CEM II**: all kind of hybrids of Portland cement with for example slate, fly ash, slag, etc., minimum 65% Portland cement.
JIS R 5214: Eco-Cement (since 2002).	**CEM V:** composite cements, with mixtures of Portland cement, slag, and pozzolana.

CEM I Portland cement (>95% clinker); CEM II Portland-composite cement (65%–94% clinker); CEM III Blast furnace cement (5%–64% clinker); CEM IV Pozzolanic cement (45%–89% clinker); CEM V Composite cement (20%–64% clinker). For the case of China, there is no definition and standard about green cement, eco-cement, and healthy cement.

production via voluntary action plan (VAP), Japanese industry has claimed that VAP can play a more effective role in solving environmental issues for the Sound-Material Cycling Society (this term is similar in concept to the circular economy in Japan) in addition to a low carbon society. Stakeholders in Japan must consider not only recycling technology but also the social framework represented by the Sound-Material Cycling Society.

The number of reports and books retrievable using "Sound Material Recycling Society" keyword has also increased since the end of the 1990s, and is second only to the number of publications concerning recycling technology. The Eco-Town program guaranteed implementation of a recycling plant via financial support for plant construction and enabled the development of niche technologies through collaboration between national and local stakeholders. As a result, local and national governments decided to implement the Eco-Town project, including eco-cement production. In Japan, a Japan Industrial Standard (JIS) has defined "Eco-Cement" as cement with high incineration ash and sewage sludge content. Here, the terminology "Eco-Cement" is a specified definition of JIS to be distinguished from the general word "Eco-cement." As a recent development, a tradeoff has emerged between promotion of BFC with low carbon emissions and waste management businesses to produce eco-cement by clinker (Table 3).

According to the Organization for Applied Scientific Research (TNO), the Netherlands have almost a century of experience in the use of ground granulated blast furnace slag (GGBS) cements with high slag content, which is comparable to the current CEM III/B (66%–80% slag) standard as defined in EN 197-1 (2011) for major infrastructure, including marine concrete (TNO, 2014). In the Netherlands, the penetration of OPC (CEM I) has been relatively low, with a market share of below 25% between 1990 and 2007, while instead, the main cement type produced is the CEM III standard blast furnace slag (BFS), constituting almost 70% of the current market share (CEMBUREAU, 2013a,b).

In the Netherlands, relevant policies for eco-cement include the carbon emission allowances for the cement industry and a voluntary green deal. The green deal is a covenant about making the concrete chain more sustainable, in which the Dutch government and seven sectors involved in the production and use of concrete promise to take action on policy agendas such as climate policy and the resource efficiency roadmap, and for Dutch actors, the closure of the ENCI quarry and the green deal to make concrete more sustainable, and circular economy policies (MVO Nederland, 2016). In the Netherlands, policymakers are involved in the innovation trajectory of eco-cement in multiple ways to widen and deepen the policy window and the utilization of it by building regulations, sector policies, waste policies, and science and innovation policies. In this sense, political economy aspects of regulation and innovation in the cement industry can be exemplified as the cooperative approach of waste authorities with regard to reuse of waste, absence of policies to put an effective price on CO_2 emissions from cement production, together with the specificities of the cement market, and bans on the disposal of fly ash and sewage sludge resulting in the use of those materials either as a supplementary cementitious material or a fuel. Demand for green cement is presently growing but meets with several obstacles (see Kemp et al., 2017).

3.3 Innovations in the Netherlands, China, and Japan

In the Netherlands, most innovations are historically based on the substitution of Portland clinker with blast furnace slag and fly ash, and the substitution of dried sewage sludge and bone meal for fossil fuels in the incineration process. Within the techno-economic subsystem, two other contemporary innovations are Via Verde concrete and ASCEM cement. Via Verde concrete is used for concrete foot paths, bicycle paths, parking places, and secondary roads (Kemp et al., 2017). It was developed by a special partnership of BAM Wegen, IntroVation, and ORCEM en Van Nieuwpoort, with support of the Province Utrecht. The combined use of fly ash and furnace ash helps to achieve a CO_2 reduction of 80% compared to OPC, according to the partnership's sustainability manager (Kemp et al., 2017). The Netherlands is also aligned and active in international scientific and applied cement research in European projects (Kemp et al., 2017). The science and research agenda has its own autonomy but is influenced by policy agendas at the boundary of politico-administrative and techno-economic subsystems, such as climate policy and the resource efficiency roadmap, and for Dutch actors, the closure of the

ENCI quarry and the green deal to make concrete more sustainable, and circular economy policies (MVO Nederland, 2016)..

Policy windows were opened for innovations in China after the China Cement Association formulated the 13th Five-Year Development Plan for Cement Industry, in which the key points of technological innovations mainly included the following five aspects:

(1) Cement kiln co-processing technology and equipment, complete set of cement preparation technology and equipment, cement-based composite materials research and development, and intelligent manufacturing technology and equipment packages;

(2) Development of new types of cement with high performance and low CO_2 emissions, such as high belite cement;

(3) Continuing to accelerate the integration of R&D and technology research on high-performance firing system technology (the second generation of new dry-process technology);

(4) Research and development of fine dust (PM2.5) particle capture, deep flue gas desulfurization, high efficiency denitrification capture and its integrated collaborative control technology and equipment; and

(5) R&D of technologies to control hexavalent chromium and mercury in raw materials, control technology of hexavalent chromium and mercury in research and development of production process, research and development of emission control capture technology, and equipment for end emission control.

In China, three R&D projects independently undertaken by China 20MCC Construction Group (20MCC Group) have already passed through the authentication by MCC Group and the research achievements were rated to be of international advanced level. The R&D projects included Key Construction Technology for Large-sized Eco-Cement Production Facility (MCC, 2016), in which, the world's most advanced production process of "one kiln and four mills" is employed for the first time. Key technologies have been developed for civil works, steel structures, and electromechanical construction works. In total, 14 items of patent have been approved, and a patent group has been formed, which enjoys the prospect of wide application (MCC, 2016). Along with alternative solutions, such as the sludge usage project, a waste heat recovery (WHR) project has also begun at the Nanshan Cement Plant. The 82 million yuan project can meet 30% of the total electricity consumption of the Nanshan plant. It is predicted that the investment will be paid off within 3 years: "The market price of electricity is 0.5 yuan per kWh, but our cost to produce a kWh of electricity is merely 0.1 yuan"—Deputy Secretary General of China Cement Association (China Daily November 3, 2008). Thus, in China, alignments of capable actors of scientific and technological innovations with new businesses coincide with policy developments in various fields. Cleaner production, recycling, and alternative cement are seen as three effective approaches for the sustainable development of the cement industry in China (Shen et al., 2017). Thus, effective approaches to reduce the environmental impacts and economic cost of the cement industry in China also include promoting clean cement production technology, decreasing the consumption of limestone and energy use, increasing the energy recovery rate, and optimizing transport distance (Yang et al., 2017). According to Gao et al. (2017), energy efficiency improvements will be the main driver of emission reductions. However, the exponential growth of Chinese in-use cement stock in this sense underlines the need for extending the lifetime and reducing cement intensity of buildings and infrastructures to be able to realize dematerialization in China (Cao et al., 2017).

Chairman and CEO of Yulong Eco-Materials Ltd. (a company which controls nearly 51% of the brick market and 30% of the concrete market in Pingdingshan, according to company data) summarizes the interaction between policies, policymakers, and innovators in the cement industry in China as:

> *Yulong Company manufactures eco-friendly fly-ash bricks and concrete, will benefit from China's increased focus on anti-pollution initiatives. The growing emphasis on anti-pollution policy and regulations means there will be a very big market for our business. The government has also expressed that it's going to increase the speed of urbanization, which will give the whole construction market a big stimulus, too .... There was a lot of construction waste and a lot of materials: crushed stone, sand, etc. I was just thinking of how to reutilize these things .... Fly-ash bricks are eco-friendly bricks because of the fly ash, which is the residual from burning coal. Developers tend to buy eco-friendly products, so there's a general trend. This business doesn't have any bottleneck in terms of environmental policy, so basically the government has encouraged us. The company is building up its business model with the goal of expanding the recycling technology to other cities already using the production model as a prototype. New production plants that will handle the recycling of construction waste and increased brick production also are in the works and set to open soon. Elected officials from cities in Henan province and elsewhere already have visited the facilities in Pingdingshan. The provincial level government in Henan thinks this is a very important project. It is a problem in every city in China, so there's going to be a very huge market for us to grow.*
>
> ***Freifelder (2015)***

In Japan, research activity on recycling and eco-cement dates back to 1985. Since the fiscal year 1985, the Japanese government has invested over 3.5 trillion JPY through Grants-in-Aid for Research and Science, resulting in the funding of 505 research projects involving recycling technologies and systems. Since the 1990s, the focus has shifted from innovative technology development to socially relevant technology research, into the planning, design, and evaluation of recycling networks in order to increase their efficiency and effectiveness. A research project with a collaborative team of general construction companies, cement production companies, and a university had started, from 2008 to 2013, to develop ECM (energy, CO_2 minimum) cement, which achieved more than 60% CO_2 emission reduction compared to OPC. This project was one of the innovation points from the technological aspect.

As a result of certification and R&D, BFC has been stably used. BFC has occupied 92.4% in total eco-cement production and almost BFC was classified as B-type BFC with 30–60 wt% BF as a by-product in 2014. Alignment of actors in techno-economic subsystem also occurred in a novel research project entitled "Development of Fundamental Technology of Innovative Cement Production Process," which received 720 million JPY from the New Energy and Industrial Technology Development Organization (NEDO) between 2010 and 2015. An important finding is that the recent issue that has promoted eco-cement in Japan is climate change mitigation to achieve Paris Agreement targets rather than the circular economy, as a factor keeping the policy window open and utilized in Japan. Utilization for cement companies is that they have a trade-off between GHG reduction through promotion of eco-cement, especially BFC, to reduce the production of clinker and waste management business with a treatment fee to accept wastes such as sludge, waste tires, etc. as a material of clinker. BFC has been the dominant eco-cement in Japan, and eco-cement is around 1% of total cement production. BFC is mainly distributed by by-product exchange among steel industries and cement industries with public support to stimulate the demand of BFC through circular economy policy and climate change mitigation policy. On the other hand, eco-cement has been developed mainly by circular economy policy. Hence, we distinguish between BFC and eco-cement in the Japanese context.

4. Concluding remarks and future research directions

Although the continuing transitions in Dutch, Japanese, and Chinese contexts from ordinary Portland cement (OPC) to eco-cement follow different conjunctural causation patterns and temporalities due to "multiple stream determinants" emanating from varying degrees of (i) intervention capacities and capabilities of actors in politico-administrative and sociocultural domains, (ii) socio-technical alignment of the policy stream to techno-economic and scientific possibilities, and (iii) the nature of market demand with regard to both traditional and alternative cements, a common target in each context is to reduce the contribution of this basic industry in overall carbon dioxide (CO_2) emissions, with circular economy policies playing an increasingly significant role in each trajectory.

Regarding the intervention capacities and capabilities of actors in the politico-administrative domain, our findings indicate that although a policy window can be opened by the lead of politico-administrative actors, as took place in China, in order to intervene in the issues emanating from traditional cement, the cement industry, and its industrial organization, socio-technical alignment of the policy stream to techno-economic and scientific possibilities is also needed for an effective utilization, widening, and deepening opened policy window. We observed this relatively more effective socio-technical alignment of actors in Japan and the Netherlands for the utilization of opened policy windows. However, the Netherlands demonstrates a more diverse set of

Table 4 Interventions, their scales, and levels.

Interventions	Context	Large scale	Multi-level
Traditional cement	**China**	Circular economy policies, industrial restructuring, emission trading system (ETS)	Guangdong ETS, national level policies, US-China Climate Change Working Group
	The Netherlands	Emission trading system, circular economy policies, green deal	Local/province level Limburg (ENCI), national level, EU standards
	Japan	Comprehensive political framework including strict regulations and subsidies, circular economy policies, low carbon policy, ETS (very limited), voluntary action plan, eco-town	Eco-town, municipalities, national level (e.g., Sound-Material Cycling Society), Kyoto Protocol

bottom-up, sociocultural capacities, and capabilities of actors and organizations with respect to its historical market development for alternative cements, current application niches, market dynamics, and the issue of ownership, even in terms of protests, if needed, against traditional cement and industry, by society, compared to the Japanese context (Table 2).

We also argue that socio-technical alignments play an important role regarding the scale, level, and effectiveness of interventions. Both in China and Japan, several large-scale interventions (e.g., ETS design and implementation, voluntary action plan, and working group vision strategies (see Table 4)) fell short and were relatively less or not as effective as they were intended to be in the design stage. In this respect, the socio-technical alignment of actors from politico-administrative, techno-economic, and socio-cultural domains in the Netherlands can be argued to be supporting the widening and deepening of the policy window and utilization of it, e.g., via large-scale interventions (e.g., a dynamic emission trading system and green deal (see Table 4)). As for innovations, circular economy policies are an important pull factor in each context, where the Netherlands and Japan come forward with respect to their research and development (R&D) and science-based alternative cement development activities compared to China.

As concluding remarks on the eco-cement trajectories in China, Japan, and the Netherlands, we would like to emphasize, with respect to our findings, that:

- Eco-cement transitions are both introduced (opening up the policy window) and carried out (socio-technical alignment) by different issue (and solution) actors and their networks in different institutional contexts, which together shape the structuration process of ongoing eco-cement transition trajectories in each context.

- In this respect, as theoretical implications, we can state that while the multiple stream framework of Kingdon (1984, 1995) provides explanatory insights for identifying the conditions for opening up policy windows, Zundel et al. (2005) provides complementary and useful insights for detecting the conditions through which effective utilization of such open policy windows take place via configurations of different degrees of socio-technical alignments.
- In this regard, co-evolutionary (policy and technology) analysis of transitions (Kemp et al., 2017) come forward as a relevant framework to study the interactions between policy (Kingdon, 1984, 1995) and technology domains (Zundel et al., 2005).
- For policy implications, we highlight that the scales and the levels of interventions (ranging from regulatory amendments and technical standards to industrial restructuring, ETS) can be integrated more to create multilevel (e.g., geographical deepening) and multidomain (e.g., cross-sectoral widening) synergies to utilize open policy windows more effectively. This is performed by *ex ante* (e.g., policy design) and *ex post* (e.g., policy implementation) socio-technical alignments of actors with capabilities.
- Regarding the content and targets of such policies and technologies, not only an industrial policy response in terms of industrial reorganization and cleaner production, but also circular economy policies targeting the use of waste-as-fuel, recycling (especially recycling of concrete, construction, and demolition waste), and emergence of commercially viable alternative cements, come forward as the main co-evolutionary drivers of an eco-cement transition in all contexts.

Finally, we would like to emphasize the importance of future research directions targeting especially the role of circular economy policies for technology transitions, recycling of demolition and construction waste, economic and social determinants of production, and utilization of alternative cements in increasing the utilization rate of open policy windows in each context from a socio-technical transition perspective. Analysis of multiscalar interventions, especially, future research in studying the conditions for widening and deepening the policy windows of emerging opportunities should not be overlooked for studying circular economy and sustainability transitions.

Acknowledgment

This chapter was prepared as a part of the SINCERE (Sino-European Circular Economy and Resource Efficiency) project, and received funding from NWO for the Dutch research (467-14-154), grants from the National Natural Science Foundation of China (Nos. 71461137008; 71690241; 71403170; No 71974129) and the UK Economic and Social Science Research Council (ES/L015838/1). This chapter also benefited from valuable comments provided by members of SINCERE consortium and the reviewers and the Environment Research and Technology Development Fund (3-1709) of the Environmental Restoration and Conservation Agency of Japan.

References

Andrew, R.M., 2018. Global CO_2 emissions from cement production. Earth Syst. Sci. Data 10 (1), 195.

Baumgartner, F.R., Jones, B.D., 1991. Agendas and Instability in American Politics. University of Chicago Press, Chicago.

Baumgartner, F.R., Jones, B.D., 2002. Policy Dynamics. University of Chicago Press, Chicago.

Bridgman, P., Davis, G., 2003. What use is a policy cycle? Plenty, if the aim is clear. Aust. J. Public Admin. 62, 98–102.

Cao, Z., Shen, L., Liu, L., Zhao, J., Zhong, S., Kong, H., Sun, Y., 2017. Estimating the in-use cement stock in China: 1920–2013. Resour. Conserv. Recycl. 122, 21–31.

CEMBUREAU, 2013a. Cements for a Low-Carbon Europe. Accessed from: http://www.cembureau.be/sites/default/files/documents/Cement%20for%20low-carbon%20Europe%20through%20clinker%20substitution.pdf.

CEMBUREAU, 2013b. The Role of Cement in the 2050 Low Carbon Economy. Accessed from: http://lowcarboneconomy.cembureau.eu/uploads/Modules/MCMedias/1380546575335/cembureau- - - full-report.pdf.

China Daily, November 3, 2008. Green Cement Expected to Reduce Pollution. Accessed from: http://www.china.org.cn/environment/news/2008-11/03/content_16703540.htm.

China Daily, 2016, October 10. Concentrating Cement Capacities. Accessed from: http://www.chinadaily.com.cn/bizchina/2016-10/10/content_27010011.htm.

China Daily, 2016, May 21. 345 'Zombie' Enterprises to be Cleaned up. Accessed from: https://www.chinadaily.com.cn/china/2016-05/21/content_25398030.htm.

Dagblad de Limburger, 2007. Bijna 10.000 Keer 'Nee' Tegen Plannen ENCI. Retrieved from. http://www.6212hp.nl/SES-archief/krant/overz_div23.htm#top. (Accessed 24 April 2016).

Dagblad de Limburger, 2003a. ENCI: Nieuwe Installatie voor Verbranden Slib. Retrieved from. http://www.6212hp.nl/SES-archief/krant/overz_div14.htm#top. (Accessed 24 April 2016).

Dagblad de Limburger, 2003b. Toename Klachten over ENCI. Retrieved from. http://www.6212 hp.nl/SES-archief/krant/overz_div10.htm. (Accessed 24 April 2016).

Dagblad de Limburger, 2004a. Belgen Vrezen Vuil water. Retrieved from. http://www.6212hp.nl/SES-archief/krant/overz_div15.htm#top. (Accessed 24 April 2016).

Dagblad de Limburger, 2004b. Bezwaren tegen ENCI-Vergunning. Retrieved from. http://www.6212hp.nl/SES-archief/krant/overz_div15.htm#top. (Accessed 24 April 2016).

Dagblad de Limburger, 2004c. Milieufederatie Steunt Verzet tegen ENCI. Retrieved from. http://www.6212hp.nl/SES-archief/krant/overz_div15.htm#top. (Accessed 24 April 2016).

de Trompetter, 2007. Iniatiatief Tegen Vergunning ENCI. Retrieved from. http://www.6212hp.nl/SES-archief/krant/overz_div23.htm#top. (Accessed 24 April 2016).

Dewald, U., Achternbosch, M., 2016. Why more sustainable cements failed so far? Disruptive innovations and their barriers in a basic industry. Environ. Innov. Soc. Transit. 19, 15–30.

Enci Stop, 2001. Waarom stichting EnciStop? Retrieved from. http://www.encistop.nl/. (Accessed 24 April 2016).

Freifelder, J., 2015. China's eco-friendly companies stand to gain: CEO, China Daily USA. Updated: 19 March 2015 11:23. Accessed from http://www.chinadaily.com.cn/world/2015-03/19/content_19854791.htm.

Gao, T., Shen, L., Shen, M., Liu, L., Chen, F., Gao, L., 2017. Evolution and projection of CO_2 emissions for China's cement industry from 1980 to 2020. Renew. Sust. Energ. Rev. 74, 522–537.

Global Cement, 2013. Lessons From the Europe ETS for the Chinese Cement Industry. Accessed from: https://www.globalcement.com/news/item/2129-lessons-from-the-europe-ets-for-the-chinese-cement-industry.

Global Cement, 2016. China Cement Association asks Government to Speed up Sector Consolidation. Accessed from: https://www.globalcement.com/news/item/5330-china-cement-association-asks-government-to-speed-up-sector-consolidation.

Global Cement, 2017. China Cement Association. Accessed from: http://www.globalcement.com/news/itemlist/tag/China%20Cement%20Association.

Global Green Cement Market, 2016. Global Cement Market Report 2018-2023: Largest Players are Lafarge Holcim, Anhui Conch, Jidong Development and Heidelberg Cement. Accessed from: https://markets.businessinsider.com/news/stocks/global-cement-market-report-2018-2023-largest-players-are-lafarge-holcim-anhui-conch-jidong-development-and-heidelberg-cement-1022162659.

Guest, G., Bunce, A., Johnson, L., 2006. How many interviews are enough? An experiment with data saturation variability. Field Methods 18 (1), 59–82.

Het Belang van Limburg, 2003. Jonge Milieu-Activisten Protesteren tegen ENCI. Retrieved from. http://www.6212hp.nl/SES-archief/krant/overz_div14.htm#top. (Accessed 24 April 2016).

Het Belang van Limburg, 2004. ENCI Graaft Observant af en Wil Voorraden tot 2042. Retrieved from. http://www.6212hp.nl/SES-archief/krant/overz_div15.htm#top. (Accessed 24 April 2016).

IISD, 2014. US, China to Cooperate on Energy and Forests to Tackle Climate Change. http://sdg.iisd.org/news/us-china-to-cooperate-on-energy-and-forests-to-tackle-climate-change/.

Kemp, R., 1994. Technology and the transition to environmental sustainability. The problem of technological regime shifts. Futures 26 (10), 1023–1046.

Kemp, R., Barteková, E., Türkeli, S., 2017. The innovation trajectory of eco-cement in the Netherlands: a co-evolution analysis. IEEP 14 (3), 409–429.

Kingdon, J.W., 1984. Agendas, Alternatives, and Public Policies. 45 Little, Brown, Boston, pp. 16528–36169.

Kingdon, J.W., 1995. Agenda, Alternatives and Public Policy. Little, Brown, Boston.

Kobayashi, A.S., Hawkins, N.M., Barker, D.B., Liaw, B.M., 1985. Fracture process zone of concrete. In: Application of Fracture Mechanics to Cementitious Composites. Springer, Dordrecht, pp. 25–50.

Lindblom, C.E., 1959. The science of 'muddling through'. Public Adm. Rev. 19, 79–88.

Lindblom, C.E., 1979. Still muddling, not yet through. Public Adm. Rev. 39, 517–526.

Liu, Z., Guan, D., Wei, W., Davis, S.J., Ciais, P., Bai, J., Peng, S., Zhang, Q., Hubacek, K., Marland, G., Andres, R.J., Crawford Brown, D., Lin, J., Zhao, H., Hong, C., Boden, T.A., Feng, K., Peters, G.P., Xi, F., Liu, J., Li, Y., Zhao, Y., Zeng, N., He, K., 2015. Reduced carbon emission estimates from fossil fuel combustion and cement production in China. Nature 524 (335–338), 2015. https://doi.org/10.1038/nature14677.

Lopez, R.A., 2008. Progress in Sustainable Development Research. Nova Publishers.

MCC, 2016. Many MCC Scientific and Technological Achievements Win Technical Innovation Achievement Awards at 2nd China Construction Engineering Construction Technology Awards. http://www.mcc.com.cn/mccen/culture/_360288/405353/index.html.

MVO Nederland, 2016. MVO Netwerk Beton. Retrieved 11 April 2016, from http://mvonederland.nl/mvo-netwerk-beton.

Nelson, R.R., 1994. The co-evolution of technology, industrial structure, and supporting institutions. Ind. Corp. Chang. 3, 47–63.

NRC, 2009. ENCI sluit groeve in. Accessed from: https://www.nrc.nl/nieuws/2009/11/05/enci-sluit-groeve-in-2018-11807581-a1074678.

Olivier, J.G.J., Janssens-Maenhout, G., Muntean, M., Peters, J.A.H.W., 2016. Trends in Global CO_2 Emissions: 2016 Report. PBL Netherlands Environmental Assessment Agency, Ispra/European Commission, Joint Research Centre, The Hague. Available from: http://edgar.jrc.ec.europa.eu/.

Reuters, January 18, 2018. Cities in China's Hebei Province Still Top List of Smoggiest Places. Accessed from: https://in.reuters.com/article/us-china-pollution/cities-in-chinas-hebei-province-still-top-list-of-smoggiest-places-idINKBN1F70CA.

Rhodes, R.A., 1990. Policy networks a British perspective. J. Theor. Polit. 2 (3), 293–317.

Romney, A.K., Weller, S.C., Batchelder, W.H., 1986. Culture as consensus: a theory of culture and informant accuracy. Am. Anthropol. 88 (2), 313–338.

Sabatier, P.A., 1988. An advocacy coalition framework of policy change and the role of policy-oriented learning therein. Policy Sci. 21 (2–3), 129–168.

Sabatier, P.A., 1993. Policy change over a decade or more. In: Sabatier, P., Jenkins-Smith, H. (Eds.), Policy Change and Learning: An Advocacy Coalition Approach. Westview Press, Boulder, CO, pp. 13–39.

SCIO, 2016. SCIO Press Conference on Stabilizing Industry Growth and Readjusting. Accessed from: https://www.scio.gov.cn/32618/Document/1469891/1469891.htm.

Shen, W., Liu, Y., Yan, B., Wang, J., He, P., Zhou, C., Ding, Q., 2017. Cement industry of China: driving force, environment impact and sustainable development. Renew. Sust. Energ. Rev. 75, 618–628.

TNO, 2014. Slag cement concrete-the Dutch experience. http://www.vegvesen.no/_attachment/588406/binary/940831?fast_title=Nr+270+Slag+cement+concrete-+the+Dutch+experience.pdf.

Van Berkel, R., Fujita, T., Hashimoto, S., Geng, Y., 2009. Industrial and urban symbiosis in Japan: analysis of the Eco-Town program 1997–2006. J. Environ. Manag. 90 (3), 1544–1556.

Van den Belt, H., Rip, A., 1987. The Nelson-Winter-Dosi model and synthetic dye chemistry. In: Bijker, W.E., Hughes, T.P., Pinch, T. (Eds.), The Social Construction of Technological Systems: New Directions in the Sociology and History of Technology. MIT Press, Cambridge, pp. 135–158.

Van der Vooren, A., Reudink, M., Hanemaaijer, A., 2015. Eco-innovatie in gevestigde productieketens. Een analyse van de beton- en de glastuinbouwketen. PBL Planbureau voor de Leefomgeving.

van Oss, H.G., 2017. Cement. In: (USGS) United States Geological Survey (Ed.), 2014 Minerals Yearbook. Available from: https://minerals.usgs.gov/minerals/pubs/commodity/cement/index.html.

Weiss, C.H., 1983. Ideology, interests, and information. In: Ethics, the Social Sciences, and Policy Analysis. Springer, US, pp. 213–245.

Yang, D., Fan, L., Shi, F., Liu, Q., Wang, Y., 2017. Comparative study of cement manufacturing with different strength grades using the coupled LCA and partial LCC methods—a case study in China. Resour. Conserv. Recycl. 119, 60–68.

Zundel, S., Erdmann, G., Kemp, R., Nill, J., Sartorius, C., 2005. Conclusions-a time-strategic ecological innovation policy. In: Sartorius, C., Zundel, S. (Eds.), Time Strategies, Innovation, and Environmental Policy. Edward Elgar, Cheltenham, pp. 322–348.

CHAPTER 30

The potential for a circular economy in the nonroad mobile machinery industry—The case of Linde Material Handling GmbH

Sergey Makaryan[a,b], Holger Hoppe[a], and Karen Fortuin[b]
[a]Sustainability Management Department, Linde Material Handling GmbH, Aschaffenburg, Germany
[b]Environmental Systems Analysis Group, Wageningen University, Wageningen, Netherlands

1. Introduction

To balance future resource needs with current resource consumption, a future-oriented economic model is needed. The current linear or "take-make-waste" economic model follows the principle of extracting resources or nutrients (take), converting these into products (make), which are then disposed of after use (waste/disposal) (Circle Economy, 2018). The global community has established policies, projects, and platforms to develop a circular economy (CE) (European Commission, 2018; WEF, 2017). A CE aims to decouple the growth imperative from the environmental constraints (Ghisellini et al., 2016). More specifically, this means modifying the linear economy in such a way that it becomes a loop of resources (i.e., materials, energy) (Circle Economy, 2018). Hereby, waste and excess resources from production are turned into new materials and thus, products, closing the "loop" and eliminating the element of waste.

Generally, private companies also need to apply CE (Antikainen et al., 2017). Key drivers for companies to apply CE into their business are increasing resource efficiency, thus reducing resource risks as well as price volatility, and complying with government regulation. Moreover, increasing pressure from customers and the public is "pushing" companies to move towards more circular solutions (McIntyre and Ortiz, 2016). Therefore, decoupling the use of natural resources from production process is key for private companies (Ghisellini et al., 2016). Pioneering companies are already putting CE policies into practice, and in doing so, developing their own approaches and experiencing challenges (McIntyre and Ortiz, 2016).

Linde Material Handling GmbH (LMH) is an industrial company in the nonroad mobile machinery (NRMM) sector and is one of the world's largest manufacturers of forklift trucks and warehouse equipment. Furthermore, the company offers intralogistics

Circular Economy and Sustainability, Volume 1
https://doi.org/10.1016/B978-0-12-819817-9.00006-5

services such as, fleet management, driver assistance systems, and many more (Linde, 2020). Due to the recent above average growth, LMH (KION Group, 2016a) faced a clear disparity between the need to minimize environmental impact and to meet the demand from the business side. Implementing the concept of CE was considered to provide a solution to this issue.

1.1 Key principles of circular economy

The definition of CE used in this study is taken from (Geissdoerfer et al., 2017):

> *[W]e define the circular economy as a regenerative system in which resource input and waste, emission, and energy leakage are minimised by slowing, closing, and narrowing material and energy loops. This can be achieved through long-lasting design, maintenance, repair, reuse, remanufacturing, refurbishing, and recycling.*

This definition is accompanied by four key principles of CE (Table 1).

The first principle, "circular design," implies that the whole product "journey" with multiple use phases (2nd, 3rd loop) and the end-of-life (EoL) must be considered during product design (EMF, 2017). Crucial success factors for circular design are appropriate material selection and standardized components. Furthermore, products are "designed-to-last," can be sorted/separated easily at their EoL, and are fit for reuse (EMF, n.d.-a). The second principle, "reduction," is applied to lower environmental impacts and reduce costs for companies in the long term. It dictates that resource use, e.g., "the input of primary energy, raw material and waste," should be minimized (Ghisellini et al., 2016). Crucial success factors materialize in simpler, lighter, and less packaged products (Ghisellini et al., 2016) while monetizing efficiency gains through a higher price for the product to compensate for the rebound effect (Freire González, 2010). The third principle, "reuse," dictates that components of previously manufactured products are to be taken after the EoL (with minor adjustments) and assembled into new products (Ghisellini et al., 2016). This principle offers economic and environmental benefits because of the avoidance of using virgin materials (Ghisellini et al., 2016), thus involving

Table 1 Key principles of CE guiding this study (EMF, n.d.-b; Ghisellini et al., 2016).

Circular design	During (new) product design, the whole product life cycle with a second and third loop and the end-of-life (EoL) must be considered (EMF, 2017)
Reduction	The use of resources (product design, manufacturing, waste, etc.) should be minimized (Ghisellini et al., 2016)
Reuse	Components of previously manufactured products are to be taken after the EoL and assembled into new products (Ghisellini et al., 2016)
Recycling	Components are broken down into raw materials, recovered from waste or at EoL, and recovered as secondary raw materials for further use (Ghisellini et al., 2016)

lower extraction costs, stable material availability, and less energy use (Castellani et al., 2015) as compared to the production of new components. This reuse principle is often overlooked or hampered in practice as crucial success factors are often not fulfilled due to practical limitations such as underdeveloped take-back mechanisms, lack of appropriate design, and the use of toxic materials (Berndtsson et al., 2017; Ghisellini et al., 2016).

The last principle, "recycling," refers to the "reprocessing of [...] materials" (Ghisellini et al., 2016) where components are broken down into raw materials, recovered from waste or at EoL, and used as secondary raw materials for further applications. The "recycling" principle is predominantly used in practice, whereas the "reuse" principle might offer more potential benefits (Ghisellini et al., 2016). Crucial success of recycling results in retention of material quality, thus avoiding a "down-cycle," appropriate design, and choice of materials (Berndtsson et al., 2017).

This study identifies CE implementation on a micro level, i.e., the business possibilities for LMH in the NRMM sector. It aims to answer the following research question: What are the business, environmental, and technological opportunities and challenges to implementing CE principles at LMH?

2. Methodology

The study was conducted through literature review, the use of internal documents, and in-person interviews with representatives from LMH. For the literature review, search engines such as Scopus.com, sciencedirect.com, springerprofessionals.de, and the Wiley online library were used to search for publications combining the search terms "circular economy," "industry," and "sustainability." The results were then filtered according to languages: English and German review articles were prioritized. After an initial scan of abstracts, a total of 55 articles were selected for literature review. Additional articles were considered from the original sources of the articles where these seemed appropriate to provide context. Further, gray literature in the form of official publications from NGOs, company websites, and global policy reports (i.e., from UN, EU) were obtained for specific literature review on CE, since certain NGOs (e.g., EMF) are thought-leaders on this topic and, furthermore, provide a practical context for potential implementation. Lastly, to gain specific insights into the LMH company, multiple internal sources (e.g., intranet) were reviewed, as well as public documents such as annual reports and sustainability reports. Furthermore, to assess the company internal structures, a total of 16 internal, in-depth interviews were performed. The companies' interviewees were selected based on their position (Head of, Senior Director etc.) in the company and their assumed specific knowledge on certain topics (e.g., used trucks).

The internal interviews were conducted by the first author using a topic-specific questionnaire (business, environment, technology) as a guideline. Interviewees were, however, encouraged to state their opinions freely, allowing for more flexibility and possible insights that might not have been covered by a rigid adherence to the questionnaire. Moreover, to introduce the CE topic to the interviewees, an infographic was created to explain the concept of CE and connect the topic to the company. The interviews were recorded and written down with high-fidelity and anonymized after the analysis. The analysis consisted of summarizing answers from different interviewees on the same specific topics (e.g., summary business aspects) and the answers provided set up subquestions to offer a robust content basis through a common topic-wise overlap.

The scope of the study was limited to the main LMH production site in Aschaffenburg, Germany, and excluded all the other sites. Studying all the other sites was not feasible within the timeframe of this study. The in-depth knowledge gained at the Aschaffenburg plant was assumed to serve as a model for other KION Group products and sites. More specifically, the study focused on the main products of LMH, which are electric and internal combustion (IC) counterbalance forklift trucks (e.g., E16 and H30, respectively). Moreover, this study considered CE principles on a business (i.e., micro) level. Hence, *meso* (i.e., cities/regions) and macro (i.e., countries) level applications as outlined by Ghisellini et al. (2016), were not researched.

3. Results

Below, business, environmental, and technological aspects of CE are outlined. Firstly, the findings from literature are presented. Secondly, the current situation at LMH is described. Thirdly, opportunities and challenges are identified, based on the comparison of the findings from the literature review and the situation at LMH, and the interview results.

3.1 Circular business models—A literature review

Various circular business models (CBMs) were found in literature, with the central themes being value creation and retention. The most relevant ones for this study are related to incorporating service offerings into a goods selling company (i.e., servitization). They are described below.

3.1.1 Product-service-system

In the PSS business model, the company remains the ownership of a product and creates value by offering customers access to this product. Because the company remains the owner, the total costs of product ownership are optimized, automatically shifting the focus in product development to longevity, reliability, and reusability (Lacy and Rutqvist, 2015). For the customer, additional value is created by focusing more on customer orientation and precise performance. PSS are often seen as the preferred business model/practice to develop a CE (Aminoff and Kettunen, 2016; Sousa-Zomer et al., 2017; Tukker, 2015). However, for a PSS to be successful, careful design and planning is required, since a PSS itself does not inherently increase business profits or sustainability. If applied correctly, the results from PSS correspond to the CE principles reduce (fewer products), reuse (take-back and refurbishment), and eventually, recycling.

3.1.2 Product life extension and after-sales services

Another relevant business model is the "follow up" strategy in the form of ASS (after-sales services, sometimes called life cycle services). ASS prolongs the use cycles of products and, ultimately, their life cycle, by repairing, upgrading, and reselling them. This business model ensures close contact with the customer, and as such, might contribute to achieving a seamless transition into more circular business models (Simons, 2017). Generally, "servitization" of a business model can be applied to achieve financial advantages. These advantages include: (i) higher margins as compared to a product sale (5–20 times higher), (ii) more secure and reliable revenue streams through fee-based long-term arrangements, and (iii) more value creation through services and retaining product ownership as a manufacturer (DLL, 2017).

3.1.3 Resource recovery and reverse logistics

Bringing used products, materials, and if possible, energy back to a valuable state to sell/remarket is another CBM. This CBM is enabled by various recovery methods, such as reuse, refurbishment, remanufacturing, or eventually, recycling (Accenture, 2014).

Ultimately, the supply chain and thus the reverse supply chain, govern the take-back of products to the manufacturer, to allow for a recovery of the products/materials. Therefore, value chains in the CE need to be considered carefully, since the added reverse supply chain (reverse logistics) is an inherent element in embedding individual efforts of one company into larger circular business operations (Belvedere and Grando, 2017).

3.2 Business model at LMH

The main findings concerning the business model(s) at LMH are summarized in Fig. 1.

The business model at LMH has developed from a linear sales model towards a more service-oriented business and is currently still in a transformation phase, with the aim to become an intralogistics solutions provider, increasing the number of services.

After production and manufacturing, the trucks are distributed to the customers either through a direct sale or through a rental contract, with vastly different profit margins (lower and higher, respectively). However, the most attractive offerings, by profit margin, are the ASS. These offerings are often interconnected and difficult to separate but in sum, they clearly represent a PSS. LMH offers leasing and short-term rental (STR) contracts. After the (first) use phase, the products with rental contracts are taken back and, depending on the configuration and condition, are remanufactured in specialized centers (under the label "approved used trucks") for a second life to be sold to end customers. Trucks that do not qualify according to the rigorous quality standards are sold

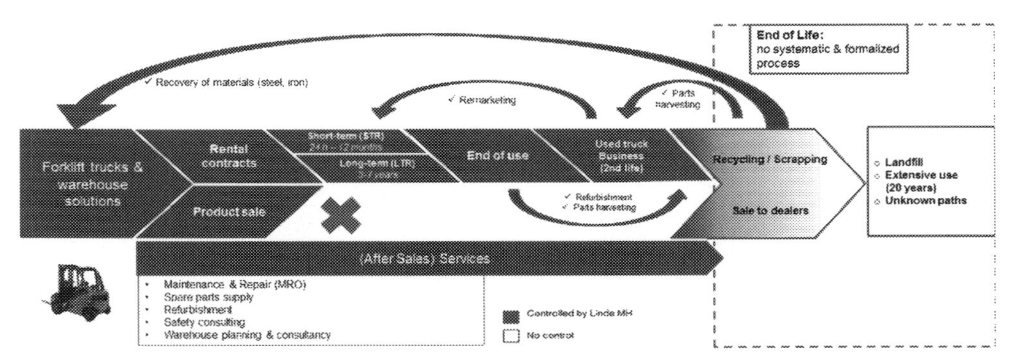

Fig. 1 Business model for Linde MH trucks. *Credit: own depiction, based on interviews.*

to dealers or, for some extremely damaged trucks, parts harvesting, scrapping, and recycling is applied. The used trucks re-enter the market with rental contracts, as product sales, or are remarketed directly to dealers. At end-of-life, trucks are commonly scrapped and recycled or sold to dealers who export them to other markets. Since there is no formalized and systematic process, as well as very little to no control over the EoL, LMHs' products might end up in landfills or be disassembled for spare parts or take other unknown paths. To ensure a longer product use and minimal downtime, LMH offers services such as repairs and maintenance (Linde, 2017b) and spare parts supply. Furthermore, product modifications and upgrades can be applied to suit a different need with the same core product, making the trucks more suited for an intensive use for various logistical tasks. To prevent larger damages and thus avoid diminishing the trucks' value and use, LMH suggests routine check-ups for wear and tear and depending on the contract, schedules maintenance for intensively used trucks (Linde, 2017b). Digital services such as "Linde connect" allow the driver to gather data on use patterns, identify errors, report them to customer service, access and track the trucks, and analyze the truck's performance (Linde, 2018a). Moreover, consulting services in energy management improve battery efficiency and ensure an optimal energy use with minimal dissipation and complement the ASS offering.

The service business seems to be more resilient in terms of revenue over time (brandeins.de, 2013; KION Group, 2013, 2014, 2015a, 2016b, 2017). However, today, services are not even a competitive advantage anymore, but a right to play, according to internal interviews.

3.3 Business opportunities and challenges

Following from the results in literature and the findings at LMH, opportunities for more circular business activities arise. LMH has already started multiple services and solutions, thus the development of these offerings is key. Developing the business into more service-based (asset/fleet management, energy management) solutions, breaking down internal "silos," and using a more common and global approach to business with high-tech digital tools can unlock more circular opportunities such as "pay per moving pallet capacity" or "capacity in tons/hour" schemes. According to interviews, future stricter financial regulations support such schemes. Therefore, streamlining the customer's balance sheet by leasing offerings is a clear advantage, as leasing costs do not appear in the balance sheet but the profit and loss statement. The achievement of a remanufacturing-like process and possibly certification can establish more trust in the recovered products. With the use of appropriate KPIs (key performance indicators), the continuous improvement of refurbishment and recovery methods in general can be driven forward, to better target the end-customer and circumvent independent dealers as well as (unnecessary) scrapping of trucks and parts. An industrialized remanufacturing process would increase competitiveness. Social benefits arise from remanufacturing due to the jobs created in the whole value chain, from take-back to quality inspections and sales management (Linde, 2017a). Eventually, LMH might find a large opportunity in "energy management" (i.e., distribution of batteries throughout the fleet). Lastly, "reuse" potential might be realized in a highly controlled STR, when a 3rd lifecycle and battery reuse are implemented. Enabling reverse supply chains through digital services (e.g., Linde connect) could be developed into fully fledged telematics services.

Currently, LMH is in a process of transformation process-wise and culturally, due to a higher servitization and digitalization of the business, and additional highly qualified employees are needed. Potential chances regarding servitization are expected from new and unfamiliar business concepts (e.g., rental of Li-ion batteries). To really drive the circularization of LMHs' business model, a more cross-functional coordination is needed internally. Tackling business challenges towards greater circularity demands comprehensive and standardized digital tools.

Challenges of the remanufacturing program are increasing efficiency and ensuring premium quality to stay competitive, while keeping an appropriate amount of trucks coming back (remanufacturing.eu, 2016). Elevating recovery processes closer to "remanufacturing" is something of a business challenge, as the used trucks would pose a threat to new trucks (lower price and same warranty), which are still the core business, thus threatening to cause a certain "cannibalization." The issue of control, whether internal or external, emerged throughout the interviews. For LMH, it is very challenging to control the nonconsolidated, third-party dealerships who have a lot of customer interaction (e.g., servicing and take-back of the trucks). Without this control, implementing closed-loop systems and thus a more circular business model will remain difficult. Ultimately, support and governance from upper management are needed to instill a long-term "circular" vision, which currently seems to be lacking throughout the organization.

3.4 Environmental aspects in literature review

In this study, the relevant industry is manufacturing, where CE has been met with insights from the established fields such as "industrial ecology" (Lieder and Rashid, 2016) and, at a micro level, "eco-innovation," focusing on resource efficiency and eco-design (de Jesus et al., 2018). These influences allow for a systemic view on sustainability in manufacturing and embed production and industrial systems into the network of material and energy flows.

Environmental impacts of the relevant products in this study are very diverse and related to the powertrain of the trucks (Lajunen et al., 2018), which is hard to change. In this study the environmental impact focus lies on resource and material use: steel, iron, lithium, cobalt (Li-ion batteries (LIB)), specifically. The use of materials such as lithium and cobalt will increase in the future in the NRMM sector as manufacturers move towards electric and hybrid vehicles (machinedesign.com, 2017). The environmental impacts of iron can be indicated by the amount of kg CO_2 equiv. produced, which is a common indicator used in life cycle assessments (LCA). The impacts of iron (Fe) and steel in 2006 were 2.4 $GtCO_2$ equiv. per year (Allwood et al., 2010; Nuss and Eckelman, 2014), divided by the total amount of crude steel in 2005 (IISI, 2005), resulting approximately in 2.1 kg CO_2 equiv./kg for iron (cradle-to-gate). This is similar to the value provided by Voet et al. (2013), which is at 2.4 kg CO_2 equiv./kg for iron. Furthermore, Voet et al. (2013) find one of the highest impacts of metals (also iron) to be in mining and metal extraction, which create a vast amount of waste rock and tailings (finely crushed ore residue) that need to be disposed of.

The use of Li-ion technology, offering a significant efficiency gain compared to conventional lead-acid batteries, is a clean technology due to the lack of exhaust emissions. However, the production of batteries does emit considerable greenhouse gas and the emissions during battery use hinge upon the electricity supplied during charging (renewable vs. fossil). Finally, the EoL of batteries, if not conducted appropriately, can lead to tremendous impacts (Ellingsen et al., 2013). In their cradle-to-gate LCA study, Ellingsen et al. (2013) studied, i.a. lithium (nickel$_x$cobalt$_y$manganese$_z$)dioxide batteries. The results show that for one 253 kg Li-ion battery containing 5 kg of lithium, the global warming potential (GWP) is at 4.6 tons of CO_2 equiv. (18.2 kg CO_2 equiv./kg). The largest contributor (51%) to the GWP of the battery is the use of electricity from nonrenewable sources such as coal (Ellingsen et al., 2013) for the battery cell manufacture. As the content of lithium in the batteries is rather low ~2% (Ellingsen et al., 2013) to ~4 (Langkau and Tercero Espinoza, 2018), the lithium will most likely not be recycled. Ellingsen et al. (2013) state that it is not economically sensible to recover lithium due to the low prices for lithium ore. This has changed however, in comparison to the price assumption in 2012—lithium prices doubled in the time period between 2012 to 2017 (adjusted for inflation) (metalary.com, 2018). For cobalt (Co), the demand and thus the impacts in the future are dependent on which type of Li-ion battery will prevail. While the exact GWP of cobalt (and lithium) could not be found in literature, as results are often provided for the whole battery pack, environmental impacts are characterized by Richa et al. (2017) as potentially toxic due to leaching into the soil, groundwater, and surface water if not properly disposed of.

3.5 Environmental aspects at LMH

Since most of the GHG emissions related to counterbalance forklift trucks stem from the use phase (Linde Material Handling, 2014), LMHs' main leverage point is changing the powertrain to become more efficient or using alternative technologies. For example, in the IC-trucks' lifecycle the use phase makes up over 80% of the GHG emissions (Linde Material Handling, 2014), thus if circular approaches would be taken towards keeping materials in the "loop," the material impacts would be rather low (~20%). Moreover, assuming new trucks are becoming more efficient, it might even be a disadvantage to keep the "outdated" truck models in circulation, as they would be more impactful than new trucks.

The emission profile of electric trucks is highly dependent on how the customers power the trucks (Linde Material Handling, 2014). If renewable energy is used during the use phase, lowering the impact, additional circular approaches could be a possible solution to reduce GHG emissions from the material use and production of the trucks.

Streamlining efforts at LMH are continuously undertaken during new developments, e.g., more electric trucks, substituting IC-trucks, as *E*-trucks need less maintenance and require fewer secondary materials. Furthermore, the use of "common parts" (modular parts used in IC- as well as electrical vehicles) reduces the complexity (managing components, production lines, assembly) in the trucks (KION Group, 2015b).

A precise calculation of the potential CO_2 equiv. was not possible due to a lack of data. Yet it is evident that the extension of the use phase is the largest leverage point for reducing the impact from production and manufacturing and is accomplished through offering used trucks and refurbishment options as much as possible.

LMH trucks are made up of steel and iron, which, as materials, have a high recyclability. Additionally, main components, such as the truck body, the counterbalance and axles, as well as the mast are suitable for recycling. The LCA conducted by LMH proves that recycling can "compensate for up to 50% of the environmental impacts in the production phase" (Linde Material Handling, 2014). Generally, interviewees stressed the longevity of LMH products, outlasting many use cycles, even when these use cycles are not always controlled by LMH. Further, LMH technicians have experienced, trucks in the "field" being operated significantly past the limit they were designed for.

3.6 Environmental opportunities and challenges

From an environmental point of view, the opportunities lie in the products' readiness for repairs, maintenance, and modularity. Moreover, a systematic take-back process in a plug-and-play style recovery of the product (remanufacturing, etc.) could result in a decrease of emissions from production and, depending on the thoroughness and degree of implementation, also on waste creation. To a certain degree, these aspects are already implemented at LMH, however formal processes, such as design for recycling, circularity, disassembly, etc., known from literature are not fully formalized at LMH. Only when a switch to renewable energy (use phase) is made, will reducing environmental impacts through material conservation (i.e., circular approaches) offer significant savings on an individual product basis. Many interviewees suggested a reduction of the total use of trucks in customer fleets (i.e., delivering results through services, digitalization, and minimal hardware use) to reduce overall emissions. After the push towards LIB technology, the reuse and extension of life through repairs is a leverage point to balance out the initially high emissions during the production of batteries. By distributing emissions over a possibly long life cycle, ideally involving multiple use cycles (second, maybe third), this balance is fortified. A similar approach is assumed for lead-acid batteries, which have been used thus far and have an established EoL approach, consisting of take-back and recycling. To advance efforts of reuse, a refurbishment center was set up and processes are under continuous development, allowing for up to 1500 trucks to be refurbished annually and creating 40 jobs at the facility (Linde, 2018b).

The main challenges found at LMH regarding product-related environmental aspects are to systematically include them in the product design and development process (e.g., setting specific long-term targets and KPIs). An LCA was conducted to identify the environmental impacts of key products. The implications of this assessment for product design are currently implemented into the relevant processes. Additionally, as LMH loses its control over the trucks after the direct sale, enforcing environmentally beneficial refurbishment often depends on the negotiation skill of the individual manager at LMH (truck take-back for refurbishment and remarketing) and the individual dealers.

Challenges from recycling are seen due to the relatively low (total) environmental savings compared to primary material extraction. Especially, pitfalls such as an abundance of low-quality material or an insufficient supply chain coordination hinder efficient use of recycled materials. As LMH does not have its own recycling facility, the company is dependent on public governance and the infrastructure that is offered (public or private).

3.7 Technological aspects in literature

Throughout the literature, the main aim of CE has been attributed to preserving value by extending product lifetime (e.g., upgradeability (hardware) and updateability (software) (Ghosh et al., 2017)), by reuse and eventually closing the loop through various recovery methods after that life cycle (den Hollander et al., 2017; EMF, n.d.-c, 2013), ideally from the onset by circular design. Even though some emission of waste cannot be avoided, the value retention in products would be so vast that a small amount of waste could be handled (den Hollander et al., 2017). Ultimately, the lifetime of a product is composed of multiple "use cycles," where, at the end of every use cycle, a recovery method restores the product value, reversing the value loss and thus preventing obsolescence. However, the preservation of value is not always evident, as the quality increase in the product or component follows from the higher work and time investment. Once the product is recycled, product value (i.e., embodied energy—"emergy") can be destroyed, yet material recovery is still possible (King et al., 2006).

To simplify understanding of the differences, refurbishment and remanufacturing can be considered as an example. Refurbishment is superior to simple recovery methods such as repairing and maintenance, due to formalizing the take-back process (reverse logistics). Moreover, through refurbishment, the quality of a product or its components will be considerably improved (Leino et al., 2016; Reike et al., 2018). King et al. (2006) mention that after the process of refurbishment/reconditioning, a limited manufacturer's warranty is issued, which boosts trust in the product (EMF, n.d.-b). The refurbished products are considered "almost new" since these products are usually taken back as used products after the first lifecycle (e.g., after a lease contract). The more advanced recovery method of remanufacturing dictates that the whole product is disassembled "down to the nuts and bolts" (Caterpillar Inc., 2009), cleaned, checked, and reassembled (King et al., 2006), whereas refurbishment results in disassembly at a modular level (Yoruk, 2004). Furthermore, after remanufacturing, original equipment manufacturers (OEM) offer a full warranty, since the remanufactured products are not differentiated from new products built from virgin raw materials. Lastly, CE is a relatively data-driven concept due to its iterative and "feedback-rich nature" (EMF, 2016), thus digital concepts (Internet of things (IoT), telematics) complement and support CE implementation. Better (circular) management of products during use is enabled through knowledge of their location, condition, and availability, hence transforming products into "intelligent assets" (EMF, 2016).

3.8 Technological aspects at LMH

According to company interviews, the most impactful but also hardest to implement CE principle is circular design. The R&D department at LMH is responsible for product design in coordination with relevant departments and suppliers. The steering of design and engineering criteria for LMH trucks is coordinated through a specifications sheet where all technical and otherwise relevant requirements for a product are stated. The trucks are designed to be ready for repair, maintenance, and serviceability on a component level.

Moreover, the trucks must fulfill a range of criteria, from stringent safety requirements to material constraints, cost, customers' specifications, "manufacturability," and ergonomics for the driver, as well as production and service technicians. New designs at LMH undergo an iterative design cycle, alternating between incremental and major changes to be implemented. LMH forklifts "require significantly fewer components than conventional forklift trucks, need less-frequent oil changes and stand out from sector competitors" (KION Group, 2018). With the current development of the new

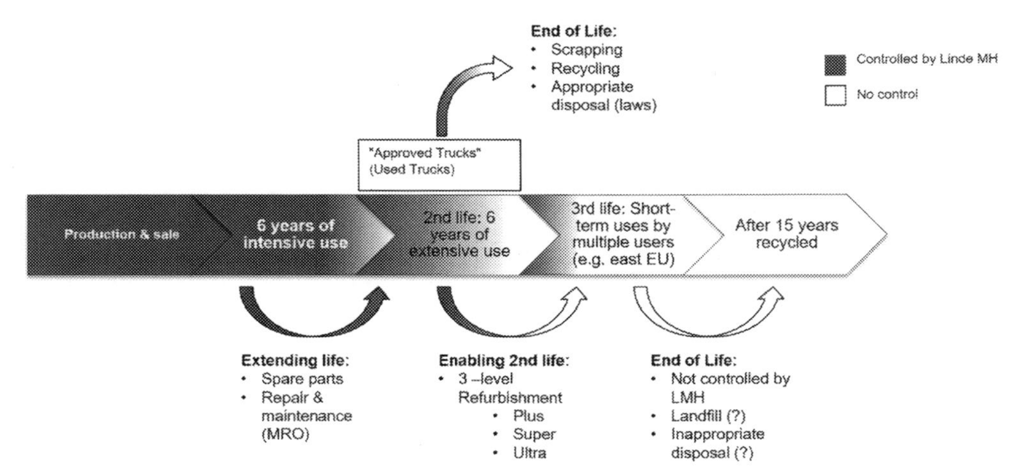

Fig. 2 A common technical lifecycle of Linde MH products. *Credits: own depiction, based on interviews.*

generation of trucks, in the 120 × and 125 × series, the company is further radically reducing complexity and the number of individual components, increasing the number of common (standardized) parts.

Key findings concerning technological aspects at LMH are summarized in (Fig. 2).

The current (counterbalance) trucks are designed to last about 20,000 use hours in intensive (i.e., 24 h, in three 8-h shifts) applications, often more use phases are possible; in field applications, engineers have seen trucks in (extensive) use from 1980 or trucks with more than 40,000 h. After the production and sale, the first lifecycle of the trucks begins. By offering various product life-extending services (spare parts provision, MRO, refurbishment, used trucks) a second or even third life is enabled. Furthermore, updates and upgrades for hard- and software keep the trucks up to date. According to interviews, the on-board software currently cannot be updated as new electronics and cables would need to be built-in. However, in the future series, an appropriate electronics infrastructure is in place and the "over the air" updates are planned.

Due to limited ownership and the geographical distribution, no internal closed loops are established and EoL is not completely controlled by LMH. For example, the materials needed to produce the counterweight in an LMH-owned foundry, such as steel and iron and aggregates (e.g., magnesium, Mg) are acquired from a local recycling company. Only for prototypes from tests of new trucks does the material go directly to the foundry.

3.9 Technological opportunities and challenges

The largest opportunities regarding CE principles implementation at LMH lie in the development of the new series 12xx, which will be introduced starting in 2020. The specification requirements for this series are already very comprehensive and aim to cover sustainability aspects. Circular design criteria are already implemented to a certain degree, such as modularity, disassembly, recyclability, reparability, and the push for more common parts, reducing complexity as well as the need for special tools and resources. Moreover, opportunities are expected from digitalization, especially in the field of preventive and predictive maintenance, which will be enabled through additional sensors and electronics in the trucks. Through more efficient maintenance efforts, service intervals can be prolonged, thus extending the use cycles of products. Predictive maintenance for telematics offers a significant potential for LMH. Through existing infrastructures such as "Linde connect," data collection and analytics of the trucks is improved. Ultimately, the thorough and holistic implementation of modularity, disassembly, and other principles that already exist at LMH combined with keeping the goal of closing (material) loops offers great potential, as the concepts are already known, but must be followed more thoroughly, especially with a more "sustainability/circularity" mindset.

Key challenges for implementing circular design from a technological aspect lie mainly in the specific requirements a truck must fulfill regarding safety and customer demands. In addition, the following drivers need to be considered:

- Regulatory requirements (e.g., Eco Design, REACH);
- Systematic integration of environmental aspects (i.e., targets and KPIs) in R&D processes;
- Uncertainty of technological changes and market availability; and
- Economic constraints (e.g., for the realization of part refurbishment).

There are further developments to make disassembly and refurbishment faster and easier demanded from the sales and service department, however, the engineers are cautious, as third parties not using LMH's OEM specifications can exploit such developments, possibly resulting in safety risks and ultimately, in accidents. Often these challenges are specific to the German market, due to the independent dealership structure, which is not as present in other countries, where dealerships are partly consolidated, allowing for more control by LMH.

Upon investigating whether material flows (loops) that LMH does have control over could be closed or at least tightened, interviewees acknowledged that such suggestions were not yet considered in-depth. Additionally, due to a limited control over EoL, the realization of the trucks high recyclability cannot be ensured. This challenge stems from the lack of control after certain use cycles (Figs. 1 and 2).

4. Discussion and recommendations

When critically assessing the results of this study, the following discussion points need to be addressed.

4.1 Research scope

The "reclassification of materials" and "renewable energy" principles were not studied in detail, whereas these two principles are important to consider for a successful implementation of a CE. Renewable energy acts as the underlying energy source for production, operation, and recovery of products. Reclassification of materials governs the EoL of materials and the appropriate flow of materials on a more global scale. The omission of the principle of "reclassification of materials" does not, however, strongly affect the results, as this principle is more global and concerned with technical material flows over multiple life cycles, which include multiple companies. A deeper investigation could have led to more insights into vast material flows, for which data might not even be present. In the case of the "renewable energy" principle, the omission does influences the results to some extent. In literature, many methods to lower environmental impacts are related to the reduction of energy or at least emissions from energy needed to manufacture goods such as LIB (Ahmadi et al., 2014; Richa et al., 2017), permanent magnets (Jin et al., 2016), or recovery and recycling of products (ALBA Group, 2017; Gutowski et al., 2011). For LMH, energy management is becoming ever more important, especially with the shift towards more e-trucks. Since LMH has little control over the use phase, where energy use is the largest environmental factor, further investigations would not offer actionable recommendations.

In this research, rubber and plastics used in trucks were excluded. These materials might be financially important, however, they only represent a small amount by weight (<10%) and were left out due to time limitations.

4.2 Research methodology

Very little literature on the NRMM sector exists and the only comprehensive study that covers NRMM/HDOR and CE is by Saidani et al. (2017). To get robust results, literature on CE and gray literature were comprehensively reviewed.

Moreover, in the internal interviews (company employees), the majority of interviewees were mainly from the business perspective, as these interviewees agreed to meet, thus the results in the technological and environmental perspective possibly lack the robustness of the business perspective. To compensate for that lack of robustness, gray literature (e.g., sustainability report for LMH) was considered.

As the interviews were conducted with internal experts, some biases of the interviewees cannot be ruled out. In particular, since the interviews were on a topic that was new to most of the participants, one might assume that they would have tried to provide answers which they thought fit best (e.g., according to the availability heuristic or to the belief bias).

However, the robustness of the results is verified by the in-depth interviews with multiple interviewees, who independently arrived at the same statements and generally had minimal inconsistencies. Moreover, internal documents from LMH and KION Group support many of the findings. Furthermore, assumptions based on personal expertise, company knowledge and insights gathered during the interviews corroborate the research.

4.3 Results

The results are also confirmed by Saidani et al. (2017), who found industry-specific challenges, such as "lack of transparency in the EoL," "lack of control in the value chain," "long lifecycles of trucks," and a certain "displacement effect," as described by Korhonen et al. (2018). Moreover, the business model at LMH can be seen as an example of the PSS identified by Tukker (2015) and it displays similar challenges, such as "finding successful use cases and monetizing Circular Business model," and "Cannibalization of products through services." The use of after-sales services as described by Simons (2017) is well developed at LMH, especially considering the recovery methods and additional services (e.g., preventive and predictive maintenance) are underway with the push towards digitalization. The "lock-in" and "path dependency" challenges described by Korhonen et al. (2018) were not found at LMH to an extent which would hinder a successful CE implementation. Yet, some challenges regarding the governance and management as described by Korhonen et al. (2018) were found and are described as a lack of cross-functional communication and collaboration at LMH. All important elements for the technical implementation of the CE as echoed throughout the literature (Bakker et al., 2014; Bocken et al., 2016; den Hollander et al., 2017; EMF, 2017) could be identified at LMH.

Lastly, one aspect which was not found in literature is a business-related aspect, specifically the importance of KPI setting. Setting appropriate KPIs was suggested during the interviews and represents an effective way to incentivize employees and departments to align their operation towards closing material loops and retaining value of products.

4.4 Recommendations

It is still unclear whether the implementation of CE principles in individual businesses (micro level) can be realized by small incremental steps, thus by making marginal gains over time and developing mechanisms "organically" (e.g., as in the case of LMH with the used truck and rental schemes), as opposed to the radical "paradigm shift" mentioned in literature (Ritzén and Sandström, 2017). This gap would require more research on an academic and applied level. Moreover, understanding the importance of KPIs and goals in driving the implementation of CE in companies should be further investigated.

Implications of this study for LMH's top management are to acknowledge and unify the current activities in a systemic approach to achieve a successful CE implementation. On an operational level, a first step would be implementing CE principles in the internal product design process and ultimately, move towards results-oriented product-service-systems such as asset, fleet, and energy management.

5. Conclusion

This study aimed to identify business, environmental, and technological opportunities and challenges to implement CE principles at Linde Material Handling.

Generally, implementing CE principles at LMH seems feasible, as key mechanisms and solutions are already in place. The biggest potential at LMH for the reduction of environmental impacts lies in the conservation of primary resources through reuse and recovery methods and the deliberate procurement and use of secondary and tertiary materials from trusted suppliers, without compromising product safety and quality.

From an environmental perspective, the unintended cross-regional recycling and scrap trading mechanism offers the potential for connecting material recovery and enabling tighter closed loops for materials. Moreover, the long lifetime of the products and the development of a modular vehicle series with a higher number of common parts present a major opportunity in line with implementing circular elements. Furthermore, through fleet- and asset management the number of overall trucks in use could be reduced, leading to a higher use phase for the remaining products, amplifying the importance of fuel and electricity consumption.

The current lead-acid batteries are in a certain recycling and recovery loop and documenting the impacts would greatly help create momentum around environmentally conscious activities. This approach should be applied to LIB, when possible.

The environmental challenges are mainly, overcoming the current gaps between EoL and new products to fully establish closed-loop systems or if possible, open-loops systems with key suppliers.

From a business model perspective, LMH is offering purchase and leasing contracts and a wide-ranging service portfolio, which will be enhanced by digital tools, thus resulting in a PSS and streamlining the use of products to focusing more on services. Moreover, the rental and leasing schemes are a key linkage to reverse logistics and recovery operations as these mechanisms enable a take-back of products (e.g., asset return). Furthermore, a high-quality remanufacturing and battery take-back represent the first closed-loop systems. More resilience in the business and the supply chain can be achieved through modular vehicles, which can be disassembled, reused or recovered, therefore reducing material costs—key to LMHs costs structure. Additional revenues can be gained from the optimization of the used truck and rental schemes, as whole products and even individual components can be valorized multiple times. The used truck and rental businesses, accompanied by services, represent a substantial part of the revenue at LMH. Lastly, recycled material flows could be studied closely to enable material reuse (i.e., steel and other metals from scrap) in the foundry, thus creating a closed loop.

Challenges from a business point of view are congruent with findings from literature; especially that recycling is often "overused," instead of exploring other recovery methods thoroughly. These challenges ultimately stem from the lack of control in the value chain and the lack of digital tools to precisely monitor material flows.

The systematic implementation of CE principles at LMH needs to encompass a strong orientation towards circular design approaches with heavy use of telematics and analytics, which are strong pillars to build a cross-functional and collaborative CE implementation on a micro level. Setting long-term goals would create more flexibility and allow for a cross-functional distribution of spare parts, leading to parts harvesting and redistribution. Lastly, the opportunities in recycling are small, but comparatively easy to implement. Investigating how the scrapped steel and iron can be directed to the foundry could be a "low-hanging fruit" option to effectively implement more circularity in the organization. The main technological challenges in implementing circular design criteria are extensive technological and future regulatory requirements. Challenges such as efficient parts harvesting and distribution remain. Furthermore, there are still safety considerations in reusing a material too often, due to material fatigue and inappropriate refurbishment by third parties.

Ultimately, the implementation of CE principles at LMH is already in progress, even if it is not verbalized as such. Important mechanisms and solutions are in place. Yet key challenges such as realigning the internal organization by instilling a certain awareness for "circular thinking" and retaining enough control over the supply chain to steer material flows for reverse logistics remain.

References

Journal publication

Ahmadi, L., Fowler, M., Young, S.B., Fraser, R.A., Gaffney, B., Walker, S.B., 2014. Energy efficiency of Li-ion battery packs re-used in stationary power applications. Sustain. Energy Technol. Assess. 8, 9–17. https://doi.org/10.1016/j.seta.2014.06.006.

Allwood, J.M., Cullen, J.M., Milford, R.L., 2010. Options for achieving a 50% cut in industrial carbon emissions by 2050. Environ. Sci. Technol. 44 (6), 1888–1894. https://doi.org/10.1021/es902909k.

Aminoff, A., Kettunen, O., 2016. Sustainable supply chain management in a circular economy—towards supply circles. In: Setchi, R., Howlett, R.J., Liu, Y., Theobald, P. (Eds.), Sustainable Design and Manufacturing 2016. 52. Springer International Publishing, pp. 61–72, https://doi.org/10.1007/978-3-319-32098-4_6.

Antikainen, M., Aminoff, A., Kettunen, O., Sundqvist-Andberg, H., Paloheimo, H., 2017. Circular economy business model innovation process—case study. In: Campana, G., Howlett, R.J., Setchi, R., Cimatti, B. (Eds.), Sustainable Design and Manufacturing 2017. vol. 68. Springer International Publishing, pp. 546–555, https://doi.org/10.1007/978-3-319-57078-5_52.

Bakker, C., Wang, F., Huisman, J., Den, H., 2014. Products that go round: exploring product life extension through design. J. Clean. Prod. 69, 10–16 (Scopus) https://doi.org/10.1016/j.jclepro.2014.01.028.

Bocken, N.M.P., Pauw, I.D., Bakker, C., Grinten, B.V.D., 2016. Product design and business model strategies for a circular economy. J. Ind. Prod. Eng. 33 (5), 308–320. https://doi.org/10.1080/21681015.2016.1172124.

Castellani, V., Sala, S., Mirabella, N., 2015. Beyond the throwaway society: a life cycle-based assessment of the environmental benefit of reuse: LCA of second-hand shops. Integr. Environ. Assess. Manag. 11 (3), 373–382. https://doi.org/10.1002/ieam.1614.

de Jesus, A., Antunes, P., Santos, R., Mendonça, S., 2018. Eco-innovation in the transition to a circular economy: an analytical literature review. J. Clean. Prod. 172, 2999–3018. https://doi.org/10.1016/j.jclepro.2017.11.111.

den Hollander, M.C., Bakker, C.A., Hultink, E.J., 2017. Product design in a circular economy: development of a typology of key concepts and terms. J. Ind. Ecol. 21, 517–525. https://doi.org/10.1111/jiec.12610.

Ellingsen, L.A.-W., Majeau-Bettez, G., Singh, B., Srivastava, A.K., Valøen, L.O., Strømman, A.H., 2013. Life cycle assessment of a lithium-ion battery vehicle pack. J. Ind. Ecol. 18 (1), 113–124. https://doi.org/10.1111/jiec.12072.

Freire González, J., 2010. Empirical evidence of direct rebound effect in Catalonia. Energy Policy 38 (5), 2309–2314. https://doi.org/10.1016/j.enpol.2009.12.018.

Geissdoerfer, M., Savaget, P., Bocken, N.M.P., Hultink, E.J., 2017. The circular economy—a new sustainability paradigm? J. Clean. Prod. 143, 757–768. https://doi.org/10.1016/j.jclepro.2016.12.048.

Ghisellini, P., Cialani, C., Ulgiati, S., 2016. A review on circular economy: the expected transition to a balanced interplay of environmental and economic systems. J. Clean. Prod. 114 (Suppl C), 11–32. https://doi.org/10.1016/j.jclepro.2015.09.007.

Jin, H., Afiuny, P., McIntyre, T., Yih, Y., Sutherland, J.W., 2016. Comparative life cycle assessment of NdFeB magnets: virgin production versus magnet-to-magnet recycling. Proc. CIRP 48, 45–50. https://doi.org/10.1016/j.procir.2016.03.013.

King, A.M., Burgess, S.C., Ijomah, W., McMahon, C.A., 2006. Reducing waste: repair, recondition, remanufacture or recycle? Sustain. Dev. 14 (4), 257–267. https://doi.org/10.1002/sd.271.

Korhonen, J., Honkasalo, A., Seppälä, J., 2018. Circular economy: the concept and its limitations. Ecol. Econ. 143, 37–46. https://doi.org/10.1016/j.ecolecon.2017.06.041.

Lacy, P., Rutqvist, J., 2015. In: Lacy, P., Rutqvist, J. (Eds.), Waste to Wealth. Palgrave Macmillan UK, p. 8.

Lajunen, A., Sainio, P., Laurila, L., Pippuri-Mäkeläinen, J., Tammi, K., Lajunen, A., Sainio, P., Laurila, L., Pippuri-Mäkeläinen, J., Tammi, K., 2018. Overview of powertrain electrification and future scenarios for non-road mobile machinery. Energies 11 (5), 1184. https://doi.org/10.3390/en11051184.

Langkau, S., Tercero Espinoza, L.A., 2018. Technological change and metal demand over time: what can we learn from the past? Sustain. Mater. Technol. 16, 54–59. https://doi.org/10.1016/j.susmat.2018.02.001.

Leino, M., Pekkarinen, J., Soukka, R., 2016. The role of laser additive manufacturing methods of metals in repair, refurbishment and remanufacturing—enabling circular economy. Phys. Proc. 83, 752–760. https://doi.org/10.1016/j.phpro.2016.08.077.

Lieder, M., Rashid, A., 2016. Towards circular economy implementation: a comprehensive review in context of manufacturing industry. J. Clean. Prod. 115, 36–51. https://doi.org/10.1016/j.jclepro.2015.12.042.

McIntyre, K., Ortiz, J.A., 2016. Multinational corporations and the circular economy: how Hewlett Packard scales innovation and technology in its global supply chain. In: Clift, R., Druckman, A. (Eds.), Taking Stock of Industrial Ecology. Springer International Publishing, pp. 317–330, https://doi.org/10.1007/978-3-319-20571-7_17.

Nuss, P., Eckelman, M.J., 2014. Life cycle assessment of metals: a scientific synthesis. PLoS One 9 (7). https://doi.org/10.1371/journal.pone.0101298.

Reike, D., Vermeulen, W.J.V., Witjes, S., 2018. The circular economy: new or refurbished as CE 3.0?—exploring controversies in the conceptualization of the circular economy through a focus on history and resource value retention options. Resour. Conserv. Recycl. 135, 246–264. https://doi.org/10.1016/j.resconrec.2017.08.027.

Richa, K., Babbitt, C.W., Gaustad, G., 2017. Eco-efficiency analysis of a lithium-ion battery waste hierarchy inspired by circular economy: LIB waste hierarchy inspired by circular economy. J. Ind. Ecol. 21 (3), 715–730. https://doi.org/10.1111/jiec.12607.

Ritzén, S., Sandström, G.Ö., 2017. Barriers to the circular economy—integration of perspectives and domains. Proc. CIRP 64, 7–12. https://doi.org/10.1016/j.procir.2017.03.005.

Saidani, M., Yannou, B., Leroy, Y., Cluzel, F., 2017. Heavy vehicles on the road towards the circular economy: analysis and comparison with the automotive industry. Resour. Conserv. Recycl. https://doi.org/10.1016/j.resconrec.2017.06.017.

Simons, M., 2017. Comparing Industrial Cluster Cases to Define Upgrade Business Models for a Circular Economy. Springer International Publishing, pp. 327–356, https://doi.org/10.1007/978-3-319-45438-2_17.

Sousa-Zomer, T.T., Magalhães, L., Zancul, E., Cauchick-Miguel, P.A., 2017. Exploring the challenges for circular business implementation in manufacturing companies: an empirical investigation of a pay-per-use service provider. Resour. Conserv. Recycl. https://doi.org/10.1016/j.resconrec.2017.10.033.

Tukker, A., 2015. Product services for a resource-efficient and circular economy—a review. J. Clean. Prod. 97, 76–91. https://doi.org/10.1016/j.jclepro.2013.11.049.
Voet, E.V.D., United Nations Environment Programme, & Working Group on the Global Metal Flows, 2013. Environmental Risks and Challenges of Anthropogenic Metals Flows and Cycles. Report 3.

Book

Accenture, 2014. Circular Advantage. Accenture Strategy.
Belvedere, V., Grando, A., 2017, March. Sustainable Operations and Supply Chain Management. Wiley. Com. https://www.wiley.com/en-us/Sustainable+Operations+and+Supply+Chain+Management-p-9781119284956.
Yoruk, S., 2004. Some Strategic Problems in Remanufacturing and Refurbishing (Thesis). University of Florida.

Chapter in an edited book

Berndtsson, M., Drake, L., Hellstrand, S., 2017. Is circular economy a magic bullet? In: Dahlquist, E., Hellstrand, S. (Eds.), Natural Resources Available Today and in the Future. Springer International Publishing, pp. 281–295, https://doi.org/10.1007/978-3-319-54263-8_11.

Reference to a website

ALBA Group, 2017. The results of the study. https://www.resources-saved.com/index.php?id=31&L=3.
brandeins.de, 2013. Auf eigene Rechnung—Brand eins online. https://www.brandeins.de/magazine/brand-eins-wirtschaftsmagazin/2013/besitz/auf-eigene-rechnung.
Caterpillar Inc, 2009. Cat Reman Products for Engines. Caterpillar Dealer, Puckett Machinery-Flowood Location. http://www.puckettmachinery.com/parts/cat-reman-products.
Circle Economy, 2018, January. The CIRCULARITY GAP Report. https://docs.wixstatic.com/ugd/ad6e59_733a71635ad946bc9902dbdc52217018.pdf.
DLL. (2017). Servitized-business-models.pdf.
EMF, 2013. Towards the circular economy—vol. 1. J. Ind. Ecol., 23–44.
EMF, 2016. Intelligent Assets: Unlocking the Circular Economy Potential. The Ellen MacArthur Foundation and World Economic Forum as part of Project MainStream. https://www.ellenmacarthurfoundation.org/publications/intelligent-assets.
EMF, 2017, January. New Circular Design Guide. The Ellen MacArthur Foundation and IDEO, Davos. https://www.ellenmacarthurfoundation.org/news/new-circular-design-guide-launched.
EMF. (n.d.-a). Building Blocks of a Circular Economy—Circular Economy Design & Circular Economy Business Models. Retrieved 23 February 2018, from https://www.ellenmacarthurfoundation.org/circular-economy/building-blocks.
EMF. (n.d.-b). How refurbisment can work, even when safety and performance matter the most. Retrieved 4 June 2018, from https://www.ellenmacarthurfoundation.org/case-studies/how-refurbishing-can-work-even-when-safety-and-performance-matter-the-most.
EMF. (n.d.-c). What is a Circular Economy?|Ellen MacArthur Foundation. Retrieved 13 February 2018, from https://www.ellenmacarthurfoundation.org/circular-economy.
European Commission, 2018. Circular Economy Strategy—Environment—European Commission. http://ec.europa.eu/environment/circular-economy/index_en.htm.
Ghosh, S., Eckerle, K., Morrison, H., 2017. Turning Waste into Value With Your Supply Chain. Accenture Strategy, p. 13.
Gutowski, T.G., Sahni, S., Boustani, A., Graves, S.C., 2011. Remanufacturing and energy savings. Environ. Sci. Technol. 45 (10), 4540–4547. https://doi.org/10.1021/es102598b.
IISI, 2005. Steel Statistical Yearbook 2006. International Iron and Steel Institute. https://www.google.com/url?sa=t&rct=j&q=&esrc=s&source=web&cd=3&ved=0ahUKEwig6p-YmLXcAhUB6qQKHcbdCvEQFggzMAI&url=https%3A%2F%2Fwww.worldsteel.org%2Fen%2Fdam%2Fjcr%3Ae20c1da7-ed4f-4429-8195-ade85c39e38a%2FSteel%2Bstatistical%2Byearbook%2B2006.pdf&usg=AOvVaw06Nr7wYw_2UQCgcKz3SISc.

KION Group, 2013. KION GROUP AG Annual Report 2013—Revenue. http://reports.kiongroup.com/2013/ar/management-report/financial-position-and-financial-performance/business-situation-and-financial-performance/revenue.html.

KION Group, 2014. KION GROUP AG Annual Report 2014—Revenue. http://reports.kiongro2up.com/2014/ar/management-report/financial-position-and-financial-performance/business-situation-and-financial-performance/revenue.html.

KION Group, 2015a. KION GROUP AG Annual Report 2015—Revenue. http://reports.kiongroup.com/2015/ar/management-report/financial-position-and-financial-performance/business-situation-and-financial-performance/revenue.html.

KION Group, 2015b. https://reports.kiongroup.com/2015/ar/management-report/non-financial-performance-indicators/research-and-development.html.

KION Group, 2016a. Annual Report 2016|KION GROUP AG-Market and Influencing Factors. http://reports.kiongroup.com/2016/ar/management-report/fundamentals-of-the-kion-group/company-profile/market-and-influencing-factors.html.

KION Group, 2016b. KION GROUP AG Annual Report 2016—Revenue. http://reports.kiongroup.com/2016/ar/management-report/financial-position-and-financial-performance/business-situation-and-financial-performance-of-the-kion-group/revenue.html.

KION Group, 2017. KION GROUP AG Annual Report 2017—Revenue. http://reports.kiongroup.com/2017/ar/management-report/financial-position-and-financial-performance/business-situation-and-financial-performance-of-the-kion-group/revenue.html.

KION Group, 2018. Available at: https://reports.kiongroup.com/2017/sr/servicepages/downloads/files/entire_kiongroup_sr17.pdf.

Linde Material Handling, 2014. Available at: http://bericht.linde-mh.de/nachhaltigkeitsbericht/2014/produkte/oekobilanzen.html.

Linde, M.H., 2017, November. Linde opens new Remanufacturing Center for Used Trucks in the Czech Republic!. https://www.linde-mh.com/en/technical/News-Detail_18902.html.

Linde, M.H., 2017, Decemer 21. Wartung und Reparatur. https://www.linde-mh.de/de/Service/Wartung-Reparatur/.

Linde, M.H., 2018, April 17. Nachrüstung & Zubehör. https://www.linde-mh.de/de/Service/Nachruestung-Zubehoer/.

Linde, M.H., 2018, June 25. Linde opens New Remanufacturing Center for Used Trucks in the Czech Republic!. https://www.linde-mh.com/en/technical/News-Detail_18902.html.

Linde, M.H., 2020, February 25. About us—Linde Material Handling. https://www.linde-mh.com/en/About-us/Company/.

machinedesign.com, 2017, November 14. Are Off-Highway Equipment and Heavy-Duty Trucks Going All-Electric Soon? Machine Design. https://www.machinedesign.com/mechanical/are-highway-equipment-and-heavy-duty-trucks-going-all-electric-soon.

metalary.com, 2018. Lithium Price. Metalary. https://www.metalary.com/lithium-price/.

remanufacturing.eu, 2016. Linde Forklift Trucks—Refurbishment. https://www.google.com/url?sa=t&rct=j&q=&esrc=s&source=web&cd=4&ved=0ahUKEwjjrMzapZnbAhUJwBQKHYYzBuAQFghJMAM&url=http%3A%2F%2Fwww.remanufacturing.eu%2Fstudies%2Fa3de0ef53fbba9caac85.pdf&usg=AOvVaw1G-3hqga9KC-5daHsz96o2.

WEF, 2017. Circular Economy. World Economic Forum. https://www.weforum.org/projects/circular-economy/.

CHAPTER 31

VALUABLE—Transition of automotive supply chain to the circular economy

Alberto Fernandez Minguela[a], Robin Foster[a], Alistair Ho[a], Emma Goosey[b], and Juyeon Park[c]
[a]HSSMI Limited, London, United Kingdom
[b]Envaqua Research Limited, Warwick, United Kingdom
[c]National Physical Laboratory (NPL), Electrochemistry Group, Teddington, United Kingdom

Abbreviations

ATF	Authorized treatment facility
CRM	Critical raw material
EU	European Union
EV	Electric vehicle
EoL	End of life
ELEV	End of life electric vehicle
GHG	Greenhouse gas
GBA	Global Battery Alliance
GWP	Global warming potential
IAB	Industrial Advisory Board
ICE	Internal combustion engine
LIB	Lithium ion battery
SBSG	Sustainable Batteries Steering Group
SDG	Sustainable Development Goals
SoH	State of health
UK	United Kingdom
USA	United States of America
VALUABLE	Value Chain and Battery Lifecycle Exploitation

1. Introduction

This chapter is based on the outputs from the VALUABLE (Value Chain and Battery Lifecycle Exploitation) project. The project is comprised of partners across the EV traction battery value chain in the United Kingdom (UK). These are, in alphabetical order: Aceleron Ltd., Aspire Engineering Ltd., Envaqua Research Ltd., HSSMI Ltd., National Physical Laboratory (NPL) Electrochemistry and Metrology teams, University College London (UCL) Electrochemical Innovation Lab (EIL), and Tevva Ltd. This project, led by HSSMI Ltd., is co-funded by Innovate UK (project number: 104182) (Innovate UK, 2017), the innovation agency of UK Research and Innovation (UKRI), as part of a

Circular Economy and Sustainability, Volume 1
https://doi.org/10.1016/B978-0-12-819817-9.00008-9

project cohort funded under the Faraday Battery Challenge (The Faraday Institution, 2017). Fig. 1 describes the project setting, vision, and benefits, in more detail.

The project is examining legal and regulatory responsibilities, as well as developing new commercially available metrology and test processes. These will be integrated with closed loop supply chain concepts for the recycling, reuse, and remanufacture of automotive LIBs. This embraces a vision of enhancing the added supply chain value of EV batteries within the UK, seeking to reduce life cycle impacts and optimize battery design for EoL. The project is not only supported by representation from the supply chain, but by an industrial advisory board (IAB), including stakeholders from the EV batteries value chain, from automotive OEMs (original equipment manufacturers) to recyclers, bringing a holistic approach, supporting the transition to a circular economy, as set forth in the project vision.

2. A cleaner future—A waste and resource challenge

To provide further background to the reader, in this section we describe the current trend in EV battery growth and the challenges of waste management and resource security brought forward by the push towards more sustainable road transport.

2.1 A solution for sustainable transport—A challenge of battery waste

It is generally acknowledged that the trend in the automotive industry is to supply more EVs onto the road networks, to meet current and future CO_2 emissions legislation. As with ICE (internal combustion engine) vehicles, there is a need to reduce the amount of EoL (end of life) waste arising from end of life electric vehicles (ELEVs), thereby avoiding landfill, with materials ultimately reentering supply chains through recovery.

The European Union (EU) has set targets for reducing GHG emissions under a series of progressive targets, for 2020 and 2030. To achieve these targets, all market sectors need to contribute to the low-carbon transition, according to their technological and economic potential. The deployment of battery powered vehicles is seen as a promising approach for early decarbonization of road transport.

According to the Global Battery Alliance (GBA) 2030 vision report (Global Battery Alliance, 2019) commissioned from McKinsey and SystemIQ, "batteries of any kind could enable a 30% reduction in the carbon emissions of transport and power sectors, provide access to electricity for 600 million people who currently have no access, and create 10 million safe and sustainable jobs around the world."

For EVs, the production (and disposal) of the traction battery—the one that powers the electric motors in an EV—constitutes a significant part of the full life cycle socioeconomic and environmental impacts of each vehicle, where the battery pack accounts for up to 40% of the production costs (HSSMI, 2019). With the take-up of EVs anticipated to accelerate in the coming years, minimization of these impacts will be increasingly important. Back in 2014, Foster et al. stated that "the number of lithium-ion batteries

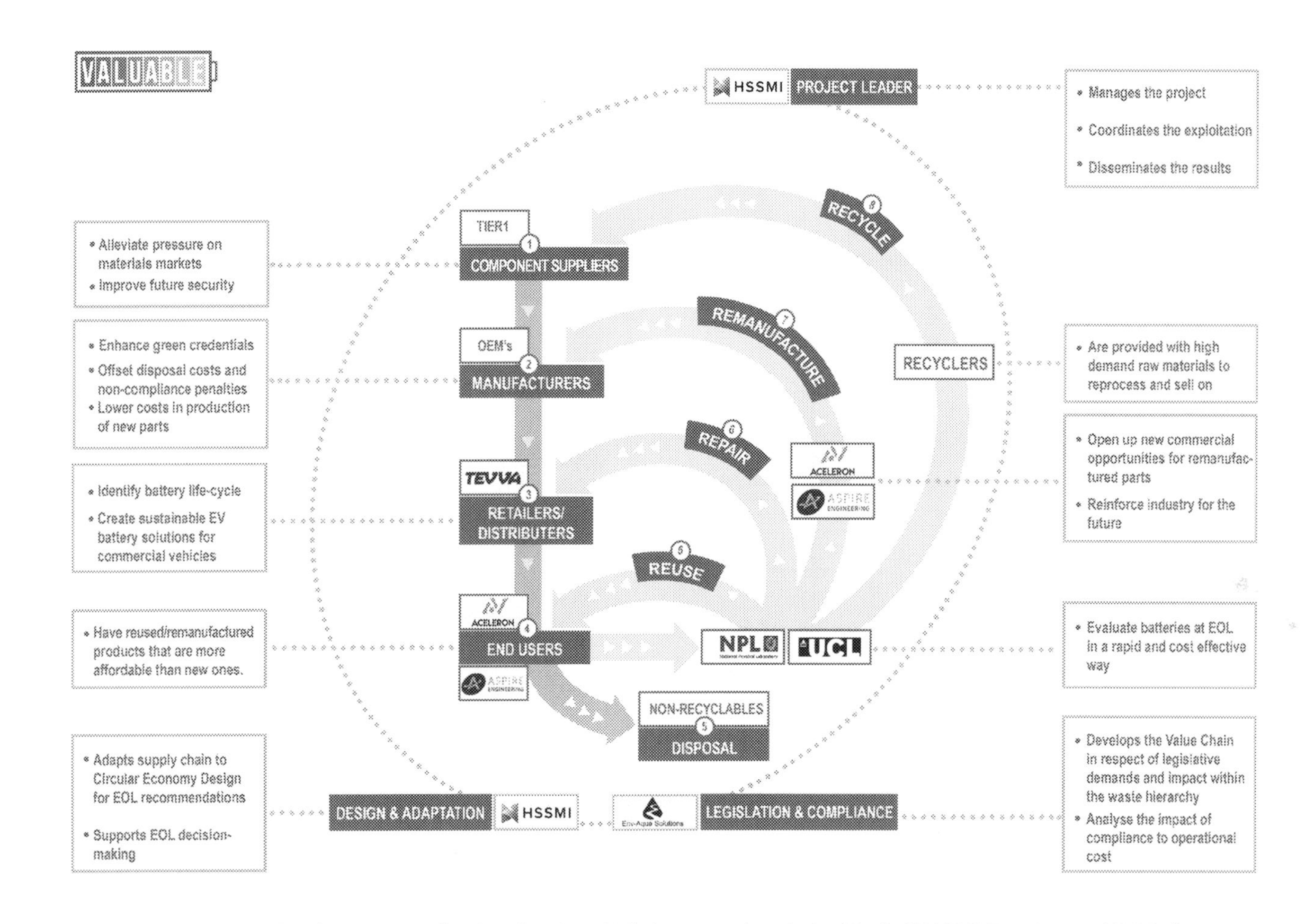

Fig. 1 Project VALUABLE—Circular economy for the electric vehicle battery value chain. *(Credit: VALUABLE project—HSSMI Ltd).*

becoming available annually for remanufacturing, recycling and repurposing is likely to exceed 3,000,000 between 2029 and 2032" (Foster et al., 2014). Fast forward to 2018, Drabik and Rizos estimate the quantity of EV batteries reaching EoL as 1,163,500 in 2030; 2,596,100 in 2035; and up to 5,380,000 in 2040. This assumes that an average EV battery will last for 8 years.

Common recovery processes, such as recycling and reuse, divert these EoL batteries from landfill or incineration. In the EU, neither landfill nor incineration is permitted under current regulations. EV batteries are included in the group "industrial batteries" under the Battery Directive (2006/66/EC). Given the growth trajectory of the EV market, the cumulative burden of EoL batteries requiring disposal will be substantial, while the available capacity to responsibly dispose of these batteries is limited. For instance, as of 2020, there is currently no UK facility for EoL treatment of EV batteries at scale. Therefore, all EoL batteries arising from EV production, such as those with defects, prototype products, or batteries from ELEVs, are currently exported to China or South Korea and, to a lesser extent, to mainland Europe (Melin, 2018), where they are disposed of, typically through recycling or reuse in other applications.

2.2 An increased demand for batteries—A critical raw materials resource challenge

In the UK, the challenge presented by greater production of batteries extends beyond a waste problem to the resourcing of essential critical raw materials (CRMs), such as cobalt, nickel, or lithium, and componentry (Drabik and Rizos, 2018).

The global fleet of EVs on the roads reached 8 million in 2019 (EV-volumes, 2019). In 2018, over 2 million EVs were sold worldwide and according to BloombergNEF (2019) this trend is only expected to continue, growing by up to 56 million in 2040, with China as the biggest EV market. Europe is also rapidly developing, pulling ahead as the second largest EV market, up from third in 2018, when it was behind the USA. The reasons behind this surge in European EV sales are a stricter regulatory framework, consumer incentives, and increased pressure on fossil fuel powered vehicles driven by the GHG emissions targets set by the EU for more sustainable road transport. To support this growth in the EV market, the production of batteries for EVs is expected to reach 1.6 TWh per year globally by 2028, up from 297 GWh per year in 2018 (The Faraday Institution, 2019).

In order to meet such demand, a substantial increase in the supply of CRMs and components to produce these batteries will be required. According to the Faraday Institution, the global demand for cobalt, lithium, and nickel will increase 2.3, 3.9, and 0.9 times, respectively (Faraday Institution, 2020).

The UK is following a similar trend in EV market growth and battery demand, so will see a similar increase in demand for CRMs and componentry.

2.3 VALUABLE industrial advisory board and the UK battery value chain

From the outset of the VALUABLE project in 2018, the IAB has met six times. The global shift towards a more sustainable transport sector and associated battery demand has been discussed at length, along with the overarching challenges these are bringing. Initially, the aim of this discussion was to understand the challenges of the UK battery value chain, as shown in Fig. 2, examining how the IAB members and their extensive supply chain could tackle them.

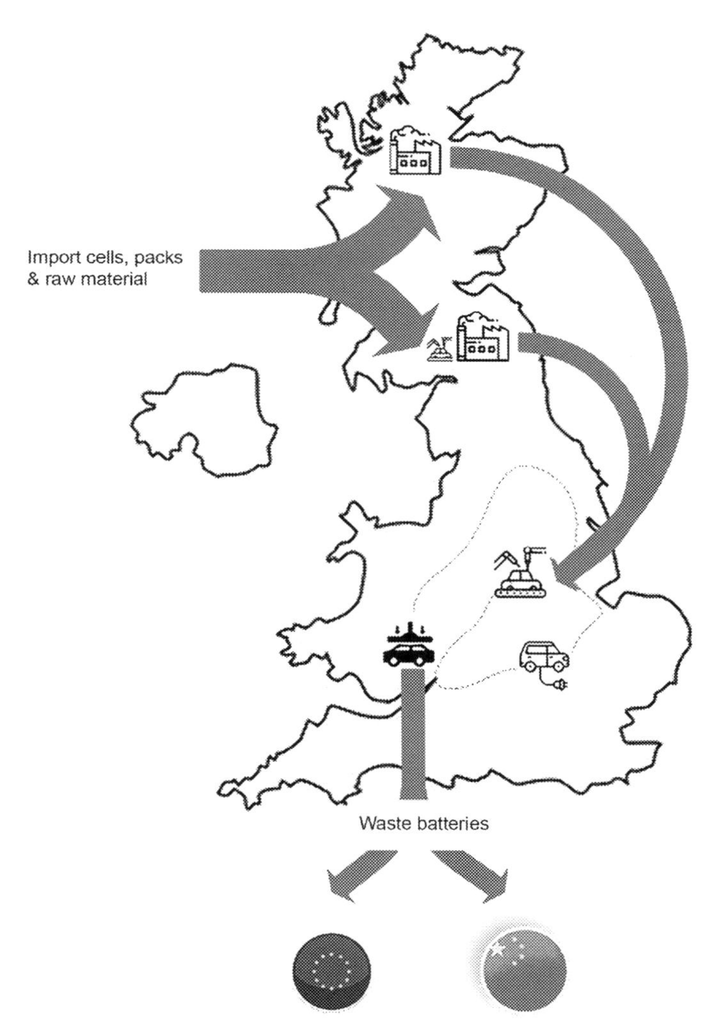

Fig. 2 Current linear economy in the UK battery value chain. *(Credit: VALUABLE project—HSSMI Ltd—Alberto F. Minguela).*

Fig. 2 highlights the UK's dependence on overseas countries to support the global trend towards a zero emission transport sector, posing a significant risk to the same automotive supply chain that is undertaking this shift. The UK automotive industry initially pays a premium to import the required supply of CRMs and essential components to make the batteries. This is followed by the manufacturing and assembly process, which starts adding value. Then the EVs are sold to the customer, subsequently used, and then at EoL of either the vehicle or battery, the battery is taken out and recovered. It is then sent for responsible regulatory disposal through a network of authorized treatment facilities (ATFs). At this point, the UK's value chain story ends, with the export of EoL batteries for recovery through recycling, mostly in mainland Europe, China, or South Korea (HSSMI, 2019).

The present situation, with the high value battery product leaving the automotive OEM hands, produces a serious loss to their balance sheet. As of 2020, only one UK-based EV manufacturer has successfully demonstrated a recovery strategy to retain the value of its batteries, through intelligent grading for the purpose of second life applications as energy storage (Element Energy, 2020).

Project partners also took the opportunity to raise awareness among the IAB members of the economic and environmental potential of embracing the concepts and vision proposed by a circular battery value chain and the benefits of the research undertaken in project activities. Furthermore, we worked together to reach consensus on what collective actions can be taken to successfully embrace these concepts and overcome the challenges identified in the UK automotive supply chain.

3. Why a circular economy?

The only step in the waste hierarchy (Fig. 3) which does not require more energy or use more resources is the initial step: "reduce." Once a product is manufactured, it begins to move down the waste hierarchy, each step employing more energy and encompassing more losses (energy and resources) directly and/or indirectly until finally, a small amount of the energy (and maybe resources) may be recovered through "energy from waste" treatment processes. Thus, when treating EoL products, the higher up the waste hierarchy the product can remain, the greater the overall energy savings that can be achieved. This process of reusing a product or article (part of a product) forms a circular economy, whereby items are recirculated back into the use phase. The more cycles that can be achieved, the greater the savings that are achieved by preventing the need for new and primary materials.

As a strategy for sustainability, this circular movement of goods helps retain embedded energy (the energy used to manufacture a product or article) and replaces the need for primary raw materials and new products.

For batteries, the waste hierarchy and the creation of a circular economy is important, as batteries contain a plethora of materials, including many which are listed as CRMs.

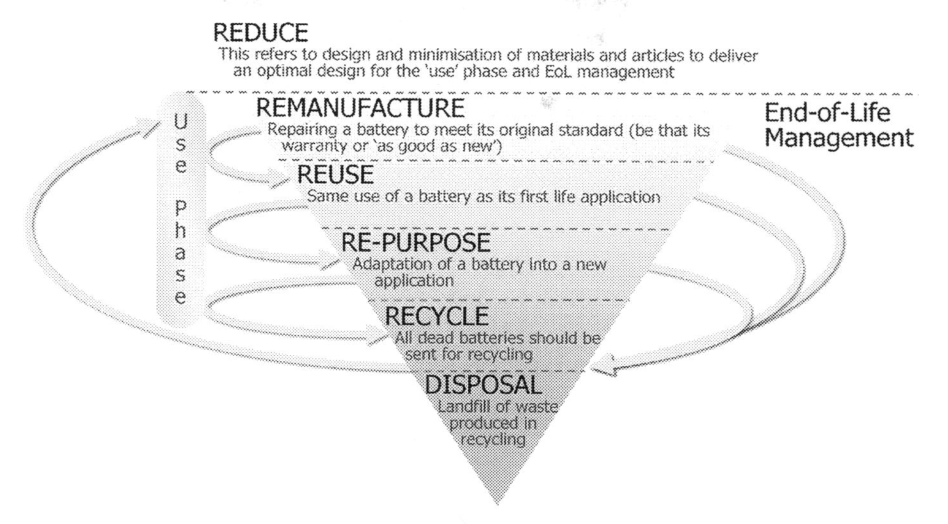

Fig. 3 The waste hierarchy, including concepts of the circular economy. *(Credit: Emma Goosey (Adapted from DEFRA)).*

As the world moves towards greener energy sources, batteries are considered to offer a practical solution to renewable energy storage, which in the past, has not been viable (due to energy density issues). In addition, the movement to battery electric vehicles (BEVs) is hailed as a turning point for urban air pollution, health, and climate change. Yet, while batteries are helping to reduce environmental impacts from energy and automotive industries, in order to produce a battery, a significant amount of energy and resources are used and the same is true for recovery of their metals at EoL. Mining and refining of primary or secondary metals are impactful activities. Battery production is also a considerable contributor to environmental impacts, with manufacturing plant productivity and the electricity grid mix being the major parameters effecting emissions (Philippot et al., 2019). This is leading to battery manufacturing creating serious environmental impacts. To minimize these impacts, use of the circular economy to extend the lifetime of batteries and their cells is being developed.

The circular economy is being demonstrated as one of the most important steps in creating sustainability for the battery industry. Batteries are a great example of what a circular economy can achieve. Producing a automotive traction LIB creates 61–106 kg CO_2 equiv. GWP impacts (Emilsson and Dahllof, 2019). During their "use" phase, their impacts can be minimized by using renewable energy. However, recycling batteries for material recovery is also highly impactful and the current recycling state of the art is a balance between material recovery and material footprint. The circular economy aims to mitigate the impacts of new battery production by extending their operational lifetime (remanufacturing and reuse), repurposing EV batteries, and by using closed loop recycling of the materials to mitigate raw material demands.

Typically, at end of first use, an EV battery can maintain 70%–80% state of charge. This capacity is insufficient for EV operation, but perfectly suitable for a multitude of other applications. Thus, to directly recycle EV batteries squanders their residual capacity. The circular economy offers an opportunity to utilize batteries through their cascading state of charge, without compromising the quality of the product they are used within.

3.1 Environmental drivers—Sustainable future power

During the late 20th century, environmental impacts from energy production and fossil fuel use were having marked effects on the environment. In 1990, the United Nations noted that if the world continued its current trajectory with CO_2 emissions, a possible 2°C warming could occur within 35 years (by 2025). In response to this, the Paris Agreement on Climate Change was implemented to bring nations together, to share the burden and responsibility of tackling actions to minimize climate change effects. The objective of this agreement has been to limit global warming to 1.5°C, with participating nations setting individual targets and ambitions to reduce their impacts. Many countries identified one of the leading causes for global climate change to be the burning of fossil fuels. As such, many have established electricity production and transportation as their primary targets. However, as it stands, the simplicity of individual nations counting their own emissions causes discrepancies between manufacturing and service economies, where goods made for the benefit of other nations remain the burden of the producer and, in turn, no responsibility over consumption and excess is placed on the end users. With this in mind, environmental concerns are being overtaken by sustainability concerns.

Developing from early environmental observation, the United Nations established 17 sustainable development goals (SDGs) (adopted in 2015), in which they aim to deliver sustainable guidelines and globally comparative tools to quantify how, nationally, we live and operate our lives. These SDGs directly address climate change through the three pillars of sustainable development—environmental, social, and economic—which strive to help the world operate within its means, and without causing irreversible change.

These multinational initiatives have been drivers for nations to minimize climate change effects, leading to countries establishing renewable energy sources and low carbon operations. In 2015, approximately 18% of electricity globally was generated from renewable sources (World Bank, 2018). Energy consumption has risen as populations have grown, along with average use per capita, but fossil fuel electricity production per capita in the UK dropped by $>$60% between 1990 and 2016 (Ritchie and Roser, 2019), due in part to energy efficiency initiatives (e.g., EU Energy Efficiency Directive 2012/27/EU and amending Directive 2018/2002), with renewable energy developing on the back of these international agreements.

As batteries have begun to flood the market and be hailed as an environmentally beneficial change, their sustainability has come into focus. While their utility can help reduce fossil fuel consumption, their production has been noted to be associated with child labor, modern day slavery, substandard workplace health and safety, and substantial environmental degradation, as well as GHG emissions.

In response to this, the GBA was launched in 2017, as battery technology was welcomed as a critical enabler of the fourth industrial revolution by the World Economic Forum. This public-private coalition was created in direct response to the global growth in battery demand and the identification that this changing face of energy supply has multiple sustainability challenges. Their aim was to address batteries as a global technology, linking the entire supply chain and establishing sustainable operations throughout. Resulting from this, the GBA seeks to tackle battery product development as a global project.

3.2 Social drivers—Regulation

EV batteries are governed by a breadth of regulations with regards to their EoL treatment. In particular, the End of Life Vehicles Directive (ELV 2017/2096/EC) and the Batteries and Accumulators Directive (2006/66/EU) are two overarching regulations, which are in place to assure that 95% of an EoL vehicle is recycled, and that its batteries are collected and recycled, making incineration and landfill illegal. The aim of these regulations is to minimize the environmental impacts of "waste" and prevent environmental damage.

While these regulations concentrate on what happens at EoL, more recently, regulatory developments have begun to concentrate on the supply chain. When operated with renewable energy, batteries offer a shift in technology that will help reduce the environmental impacts of transportation and electricity production (energy storage). But still, in 2020, materials used in these batteries are sourced from mining operations that abuse human rights and are conducted in environmentally impacting ways. Cobalt used in batteries is sourced predominantly from the Democratic Republic of Congo (DRC). Mining operations in this country are reported to include severe human rights violations and significant environmental damage. As a move to prevent companies from overlooking atrocities that happen along their supply chains, regulations like the UK's Modern Day Slavery Act 2015 or California's Transparency in Supply Chains Act 2010 make large companies responsible for ensuring their supply chains reflect the ethics and human rights of the company and the country they operate in. This, in turn, has impacts upon EoL treatment of batteries and the metals they contain. By fostering closed loop recycling and use of secondary raw materials, companies can better control their supply chains.

There is a developing responsibility to account for all emissions produced in the creation of a product. In the case of batteries, much of the production is conducted in countries relying heavily on fossil fuel energy. Thus, there can be a shift in the environmental burdens to other regions. Often, these regions are those with lower emissions controls and regulations. It has been recognized that shifting the burden does not contribute positively to global climate change efforts. Large companies are trying to implement their company ethics throughout their supply chains, to establish a level of quality in their products that is reflected by all the components and materials they contain.

3.3 Economic drivers—Competitiveness, critical materials

Referring to the waste hierarchy (Fig. 3), as a product moves down through the EoL management scenarios, economic losses are incurred. These are both directly and indirectly through the need to conduct repairs or changes and subsequently, through state of charge and state of health (SoH) testing. However, in these scenarios, if the battery can remain higher up in the waste hierarchy, it can command a higher price—with the stimulation of second life markets through careful quality assurance and quality control, the batteries are worth more than the sum of their materials—and when the battery capacity eventually drops, recycling can be used to recover the metals. Currently, over 95% of the metals can be recovered.

EV powertrains are a potential source of multiple CRMs. For regions with no natural supplies, governments and industries are looking to these products, at their end of lives, as potential reserves for urban mining. Within a power train, materials like cobalt, niobium, platinum group metals, and rare earth elements, which are all CRMs, plus other raw materials such as nickel and lithium, are important metals for EU manufacturing. As a result, EV components at EoL are considered to be valuable sources of these important materials and are in demand by recyclers.

3.4 Technological drivers—First life extension

Whilst the proportion of EV cost accounted for by the battery is expected to reduce, from 57% in 2015 to 20% in 2025 (Bullard, 2019), it still represents a significant expenditure if it needs to be replaced. This is especially relevant when automotive OEMs are providing warranty schemes that cover up to 8 years or 100,000 miles (Whatcar, 2019), meaning they will cover battery replacement costs. Efforts to extend the life of the battery and its constituent parts must, therefore, be made to limit financial implications to the OEM.

Repair is the first port of call for extending battery life. What Car's reliability study reported that most faults with electric cars are associated with electrical systems, with only 4% related to battery-specific issues (Whatcar, 2019). This means that replacement of the whole battery pack system is likely to be unnecessary, as the most expensive component, the lithium-ion cells, will still be functional. Accessibility to components and quick diagnosis is fundamental to quick and cost-effective repair.

Remanufacture is common practice for supporting warranty schemes for internal combustion engine (ICE) vehicles and is described as creating a product as good as or better than the original (Oakdene Hollins, 2016). Remanufacturing is conducted by harvesting parts from faulty vehicle components and rebuilding them back to the manufacturer's specifications, using a combination of harvested and new parts. The cost saving for an ICE powertrain can be 20%–80% (MCT Reman, 2016); similar savings could be seen for EVs.

There are, however, challenges specific to battery packs, the first being that lithium-ion cells degrade over their lifetime. This means that remanufactured batteries that utilize used cells cannot meet the definition of a remanufactured product. To achieve a similar system for EVs to that with ICE vehicles, OEMs may have to consider what is expected from a warranty scheme, e.g., replacing faulty batteries with one of the same age-related performance.

4. Barriers—What is in the way?

While there are clearly potential benefits to reusing EV batteries, there are also barriers that need to be addressed in order for this to become a more widely accepted practice.

Key barriers include health and safety concerns, regulation, technical challenges, and perceptions of reused products. For example, due to the potential for reputational damage, there is a risk that there may be a reluctance from EV manufacturers to let others use their batteries for second life applications (HSSMI, 2019).

The Commission for Environmental Cooperation notes that refurbishment is "not considered practical" and "is too dangerous" due to the flammable and explosive nature of the active material, the toxicity of the gaseous emissions in such an event, and because batteries are capable of discharging in excess of 200 V (CEC, 2015). Some of these health and safety issues could be overcome through improvements in the initial battery design and manufacturing process. For example, batteries could be designed in such a way that they can be accessed easily in an expert manner and the components easily replaced without any risks. However, at present there is currently no standardized process for battery packs, making it difficult to do this (Harper et al., 2019).

The lack of a clear regulatory position regarding reused EV batteries also presents a challenge. This lack of regulation was highlighted by RECHARGE, noting that for reuse to be successful, regulations will be essential to deliver reliable and durable reused batteries. However, they also state that this is likely to prove challenging, given that so many factors can impact on SoH (which in turn affects consistency of performance) and that these can only be controlled by the user (Recharge, 2014).

The lack of regulation also raises liability concerns. The battery manufacturer designs and manufactures them for a specific purpose and may not anticipate the battery then being remanufactured, for a different purpose, in the future. If the repurposed battery then malfunctions in its new environment, it is unclear at present where liability would lie. As a result, EV battery manufacturers may wish to discourage repurposing (Elkind, 2014).

4.1 Uncertainties

New battery chemistries are being developed that may well overtake lithium-ion as the primary choice of battery for electrical vehicles in the future—particularly if they prove to be less hazardous and have greater capacity. This could result in the current generation of LIBs being unfit for reuse, simply because technology has moved on (Hill et al., 2019). However, it may also take a long time for new advanced chemistries to fully enter the market—typically, at least 10 years (Wu et al., 2020). It is then imperative to develop a flexible battery management system that is adaptable to different cell chemistries (Podias et al., 2018).

The number of EV batteries believed to be reaching EoL are estimations based on EV sales and product lifetime. The business case for recovering the EoL batteries is usually built on the assumption of seizing a share of the total volume of the forecast EoL available. However, businesses keen to recover these batteries do not know when and where the batteries will reach EoL (HSSMI, 2019).

4.2 Economic costs

Stakeholders gathered together by Ricardo Energy and Environment (Hill et al., 2019) highlighted that repurposing of batteries could drive up costs, as transport can be up to or greater than 50% of the overall recycling cost. Another stakeholder suggested that these high costs may be due to the UN Dangerous Good Regulation, meaning that expensive packaging is needed in some cases. Additionally, the costs associated with recovery of batteries exceed the raw material value, adding further difficulties to the economic case.

Recycling LIBs is technologically challenging. Lithium-ion is not a single chemistry design, but rather a family of designs, each requiring different recycling technologies. This adds complexities to EoL management costs. In addition, these batteries contain blended materials that are not designed for ease of disassembly to recycle. Most technologies for recycling of LIBs take the approach of burning each battery and its components in a kiln to recover the most valuable materials, thus, the residual value is lost rather than recycled (Harper et al., 2019) making it less attractive economically.

4.3 Collection

Current collection services and infrastructure may be out of date or inadequate to meet future supply needs. For rechargeable batteries in EVs, stakeholders outlined that collection infrastructure is steadily improving, but still requires more efficient sorting processes to separate the batteries based on their chemistries (Hill et al., 2019).

The development of improved collection systems and recycling facilities may be constrained by lack of a reliable supply of batteries or related products, making the investment in and development of recycling infrastructure not economically justifiable (HSSMI,

2019). This unstable supply is influenced by the variable direction of market development and the resulting heterogeneity of products.

4.4 Complexity of product and not knowing what you have

LIBs consist of a complex combination of active ingredients and chemistries and can be arranged in either single or multiple electrochemical cells connected in various configurations, which presents a huge challenge in recycling them. As LIB chemistries become more specialized, they include a larger number of elements, increasing the cost of recycling. The wide variety of mixed LIB chemistries reaching recycling streams is already generating further challenges in recycling processes (Harper et al., 2019).

Furthermore, the diversity of cathode chemistries presents an additional challenge, since they may require divergent conditions for an efficient recycling process. Thus, treating them simultaneously may increase the required resources (e.g., energy and time consumed by a sorting mechanism) or may have a negative impact on the quality of recovered components (e.g., incompatibilities among cathode chemistries or processing conditions). Current developments in battery chemistry may also introduce additional elements in the future (sodium, silicon) (Hill et al., 2019). On the other hand, consensus between battery manufacturers as to the use of a standardized LIB labelling system (e.g., based on the cathode chemistry) might allow automatic sorting operations (HSSMI, 2019). In this way, resources would be better used to obtain a suitable feed for efficient processing.

5. Enablers—Clear the way

5.1 Design for end of life, recovery, and repair

While a clear business case for a circular battery value chain must be developed, little consideration is being made during the battery design stage to enable it. The latest and next generation of batteries currently utilize production methods, such as welding and adhesive bonding, that make disassembly difficult. The methods are selected to address the challenging operational requirements of a high-performance battery, e.g., high voltages, currents, temperatures, and ingress of road dirt.

Existing strategies at ATFs and recyclers rely on speed—quick processing of the depreciated assets is critical to provide a profit, due to a narrow scope for margin (DEFRA, 2011). The ATF team will quickly depollute the vehicle, removing the catalytic converter, lead acid battery, fluids, and any valuable items before compacting it for recycling.

When considering an electric vehicle arriving at an ATF, the process becomes more complex and—importantly—slow. The battery of a BEV represents a significant proportion of the volume and weight of the car, not only reducing the potential steel scrap value, but often requiring specialist lifts. BEV batteries will also be rated at voltages upwards of 300 V, requiring further training, high voltage tooling, and insurance, all considerations not required for the relatively inert ICE vehicles.

Furthermore, value chain stakeholders immediately after ATFs have similar problems. Second life battery manufacturers and remanufacturers that require access to battery modules are often left with complex procedures to disassemble in a safe, nondestructive way, eating into their processing time and profits.

The final issue lies with the recyclers. Current procedures utilize shredders and furnaces to access the valuable battery materials. Treating the battery pack as a single unit for recycle severely impacts material recovery and, therefore, the value of packs for recycling. For this reason, EoL and waste batteries are considered a liability, which manufacturers must pay recyclers to remove.

Regardless of which business case appears best for batteries post vehicle, design for disassembly is a fundamental enabler from servicing to recycling.

One feature of the battery pack that can be challenging during disassembly is electrical joints, specifically cell to cell, cell to bus bar, and module to module type connections. The number of joints can range from hundreds to thousands in one pack, which means quality is key.

When selecting and designing electrical connectors there are many parameters that affect contact reliability. These factors can be separated into two categories: performance factors and design technology factors (Braunovic et al., 2006). Performance factors as shown in Fig. 4 are concerned with the conditions the system operates at and the environment in which it does this.

Design technology factors shown in Fig. 5 are concerned with specific design elements and decisions. While the designer may be limited on which parameters they can change, it helps to aid in highlighting where influence is possible and to indicate issues early.

Naturally, the best joining methods to aid in disassembly are reversible, mechanical joints like compression fit, or bolting. There is however a trade-off between consistency of good electrical connections and ease of disconnection. For high-power applications it is unlikely that bolted joints will be able reproduce the same contact area and conductivity throughout the pack. This can lead to monitoring, balancing, heating, and motion (vibration) issues. Resistance or laser welding would provide a higher level of consistency but means the connections have been made for life.

This leads to the conclusion that, while design for disassembly is a highly important concept, strategies for its implementation may differ from pack to pack. High-power packs may have removable module interconnect bus bars to allow for servicing and second life at module level but commit further disassembly and separation operations at cell level to shredding prior to recycling. Equally a low-power pack will be able to utilize reversible joints like bolting, providing greater flexibility and options for cell and module reuse.

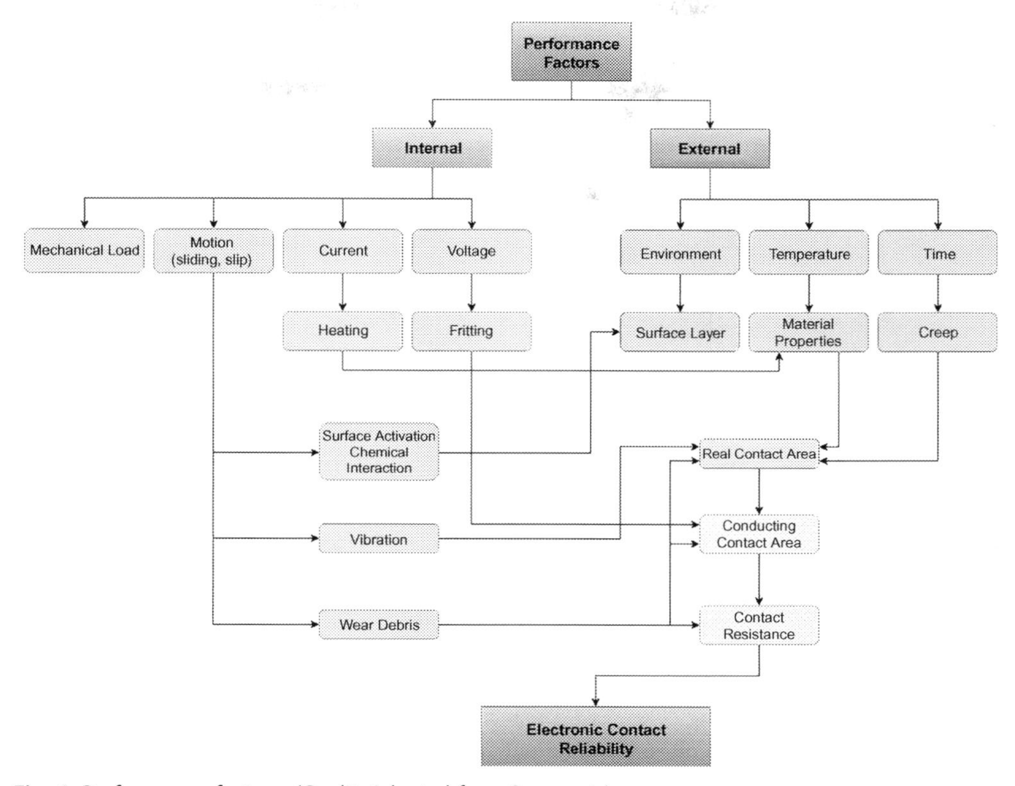

Fig. 4 Performance factors. *(Credit: Adapted from Braunovic).*

Considering the impact battery pack manufacturing methods have on the future opportunities that a battery has is fundamental to building and enabling a circular economy. The automotive OEMs that adopt this principal will not only benefit the value chain but will help to reduce costs and improve customer experience, through lower cost servicing and reduced or even no compliance scheme fees.

5.2 Rapid testing for decision making

Various challenges remain before direct recycling of automotive battery materials can become economically viable (e.g., recycling cost of approx. $10/kWh is still higher than the expected market value of the reclaimed material). Reuse is another solution capable of absorbing a large quantity of used batteries, as electric vehicle batteries typically retain approximately 80% of their initial capacity after removal from a car. The cost of second life batteries is around $50/kWh, which is also attractive compared to $200–300/kWh for a new build. This will remain competitive until the price of new batteries falls below $90/kWh (Capgemini, 2019).

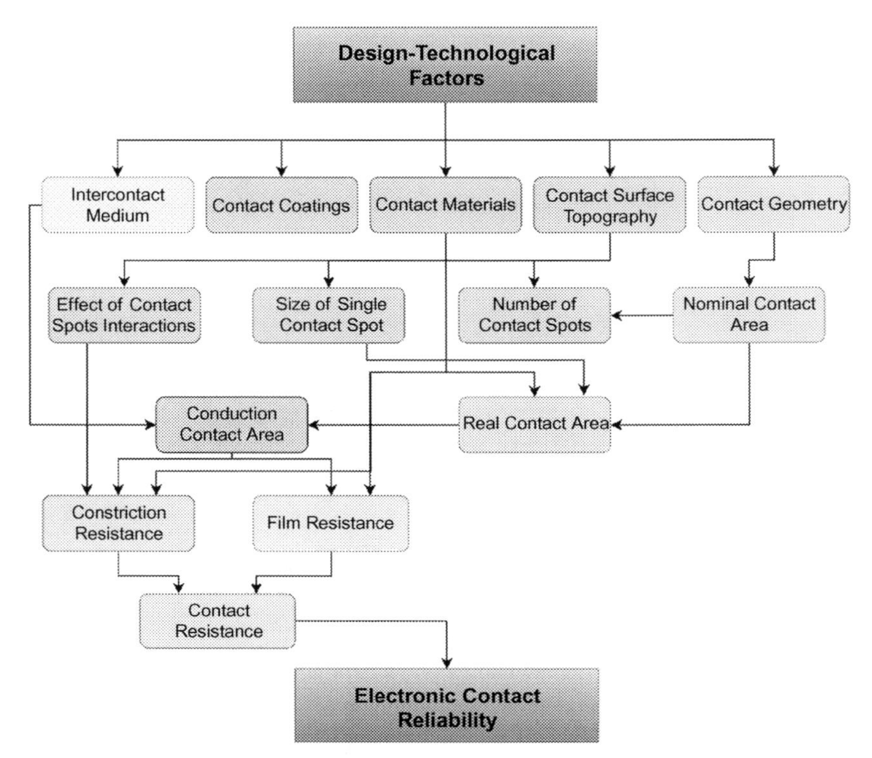

Fig. 5 Design technology factors. *(Credit: Adapted from Braunovic).*

For direct reuse of battery packs in other vehicles, there is no need to dismantle and reassemble cells/modules. This approach offers a potentially attractive business opportunity to automotive companies as it is easier and cheaper than other types of repurposing. The effectiveness of direct reuse depends critically on accurate knowledge of first life usage, making it easier for automotive manufacturers to match the second life requirements and expectations, however, there is generally a lack of readily available data on first life usage. In addition, the inflexible requirement in direct reuse, to retain the original pack volume and weight, and the inability to optimize capacity or power capability for the second use application limit the range of possible direct reuse applications.

In order to expand the market for large-scale repurposing, stationary use could be promising, in which used batteries are refurbished and integrated into large energy storage systems for various grid applications. In this case, the battery SoH would have to be determined efficiently and accurately in order to assemble a new, homogeneous second life battery pack. The characterization process must be fast and robust, in order to provide a competitive advantage compared to a fresh battery, especially when considering the economics of the process.

The performance of LIBs degrades by capacity fade and impedance increase. In general, these two factors are used to define SoH; however, there is no clear consensus on a single definition of SoH. Typically, SoH is evaluated as the ratio of measured capacity to the capacity of a fresh cell, under the same experimental conditions. Capacity measurement entails a full charge-discharge cycle at a relatively low current (e.g., C/3 with respect to nominal C-rate,[a] corresponding to approximately 6.5 h/cycle including setup time) for sufficient accuracy. Thus, although knowledge of the SoH is essential for business, this traditional charge-discharge methodology is not suitable for used battery testing, as it is time-consuming and has high labor costs. The battery replacement in first life, i.e., the choice of when the battery is to be removed from automotive use, may be straightforward, based on available performance criteria. However, establishing a lifespan warranty for second use service is challenging, as lifespan is highly sensitive to the second life duty cycle and will depend upon compliance with defined operating conditions, considering environmental factors such as climate and battery thermal management.

New techniques for SoH estimation have been explored in recent years (Quinard et al., 2019). Adaptive methods such as Kalman filters and neural networks can be very accurate but are computationally expensive and complex to implement. Internal resistance measurement using impedance can be conducted quickly, but a major concern remains how best to interpret data in terms of trends for ohmic, capacitive, and inductive behavior according to the applied frequency. Differential analysis (dQ/dV where Q: charge stored, V: voltage) can detect degradation mechanisms, combining the advantages of both adaptive and empirical techniques, namely accuracy and low computational effort. Hence, incremental capacity analysis is gaining increasing traction in the battery community. However, no single measurement technique is yet capable of providing a clear and unambiguous health assessment.

To overcome the current technology limitations outlined above, a novel testing methodology for battery diagnostics, analogous to the medical screening of a patient (triage), will need developing to support second life battery decision-making. Such a methodology would determine the health of the battery using multiple noninvasive/rapid "fingerprinting" techniques, such as thermal, mechanical, dimensional, acoustic, and X-ray measurements, either coupled with battery testing or as stand-alone techniques. Development of standard protocols for used battery testing is also needed. To approach these goals, an understanding of battery degradation based on fundamental scientific principles and data disclosure from the automotive industry will be essential, as such collaboration and knowledge of first life data will accelerate the process of battery performance assessment for second life. Collaborative development of understanding will also benefit

[a] C-rate is a measure of how quickly a battery is charged or discharged and is defined relative to the current required to fully charge/discharge the battery in 1 h. For example, charging a battery to capacity from the fully discharged state at a C-rate of C/3 will take 3 h.

automotive manufacturers, since the feedback from second use will improve first life battery manufacturing and lifespan, further stimulating the circular economy.

5.3 Global passport for batteries

Previous sections described design and testing strategies to support recovery strategies under the umbrella of the circular economy. Following this trend of enablers, it is the information that is provided when receiving an EoL battery that can truly enable the best decision. At each stage of the battery value chain, from raw material extraction to recycling, many operations and events occur, some of them adding materials value in the form of manufacturing and assembly, but also damage and degradation, affecting the battery SoH, described in previous sections. However, this information about the events in a battery life cycle is either not captured, or if recorded, not shared across the supply chain stakeholders. This is particularly true when the stakeholders are not in a close relationship, such as joint ventures, business agreements, and parts of the same network, to mention some examples (HSSMI, 2019). For instance, automotive OEMs do capture vast amounts of data in the battery management system that is rarely utilized, and this could provide vital information to predict maintenance through data analysis and tell the user the best time to substitute a battery before it breaks (Hill et al., 2019).

To overcome this challenge of information sharing across the battery value chain, the material passport concept has been proposed as a solution by the GBA (Global Battery Alliance, 2019). This concept aims to provide products with an individual tag or documentation that brings together all the pertinent information to support its recovery at EoL. It also brings benefits to all the stakeholders in the value chain, as the information held by this tag can provide understanding of the potential value and operations required to handle the product.

For the specific case of EV batteries, this passport could provide relevant information to further enable the transition to a circular value chain, allowing stakeholders to make the right decision based on the information stored about the battery. It would enable an understanding, for instance, of the use patterns and provenance of materials. The latter does support a socially fair transition towards sustainable energy. As mentioned previously, human rights are often disregarded in the CRM mining industry, as in the case of cobalt in the Democratic Republic of Congo (DRC) (Tsurukawa et al., 2011). This passport would highlight the provenance of battery materials, encouraging suppliers of CRM to support local communities and respect human rights in places like the DRC.

Furthermore, if a distributed ledger or blockchain technology is coupled with this concept, issues with data privacy and confidentiality could be overcome. Only providing the most relevant information to each stakeholder in the battery value chain would enable the best circular economy decision to be taken in the waste hierarchy, retaining the maximum available value possible at all times. The VALUABLE project, as a participant of the circular economy working group of the GBA, is a supporter of this global passport for EV batteries and will continue working alongside other stakeholders to demonstrate and validate this solution.

6. A circular vision for the UK battery value chain

Within the scope of the VALUABLE project, we have had the opportunity to access influencing groups of stakeholders sharing a similar vision, such as the Faraday Institution, the GBA, RELIB, EPriME, and the European Battery Alliance, and make connections with large automotive OEMs, such as Ford and Jaguar Land Rover.

Within the framework of the Faraday Institution and backed by the Faraday Battery Challenge from Innovate UK, the project lead Alberto Fernandez Minguela became:

- Chair of the Industrial Advisory Board of the VALUABLE project
- Friend of the GBA part of the World Economic Forum
- Founding member of the Sustainable Batteries Steering Group in the UK
- Member of the 100 CE Network of the Ellen MacArthur Foundation

In each of these positions, the VALUABLE project presented its vision for a circular economy for LIBs (Fig. 6).

One of the many contributions that the partners in the VALUABLE project provided was towards the GBA 2030 vision report (Global Battery Alliance, 2019). The report supports moving from a linear to a circular value chain, improving both the environmental and the economic footprint of batteries, by obtaining more information from in use batteries, and thereby harvesting EoL value. This enables a reduction in the GHG intensity of the value chain, while creating additional economic value.

As discussed in Section 2, a circular battery value chain offers many environmental, social, and economic benefits. It will, however, not be achieved without an active shift from the current developmental trajectory. This requires coordinated and immediate actions by companies, investors, and policymakers, in consultation with all stakeholders.

This report resulted in a set of recommendations for stakeholders of the wider battery value chain community to take specific actions and, while overcoming the described challenges, capitalize on the opportunities a circular economy for LIBs can bring.

In parallel to contributing to this report, the Sustainable Batteries Steering Group (SBSG) was developed. Driven by the VALUABLE project and supported by its IAB, the SBSG brought together policy-makers, the Environment Agency, the Department of Environment and Rural Affairs (DEFRA), industry, and the research community, to act on the recommendations being proposed by the Faraday Institution and its funding of academic led projects, such as RELIB and the VALUABLE project.

In the VALUABLE project, HSSMI conducted a thorough analysis of the value lost in CRMs from exporting EoL batteries. This was based on ELEV forecasts for the UK and provided a deeper understanding of the business case behind setting up a facility to treat

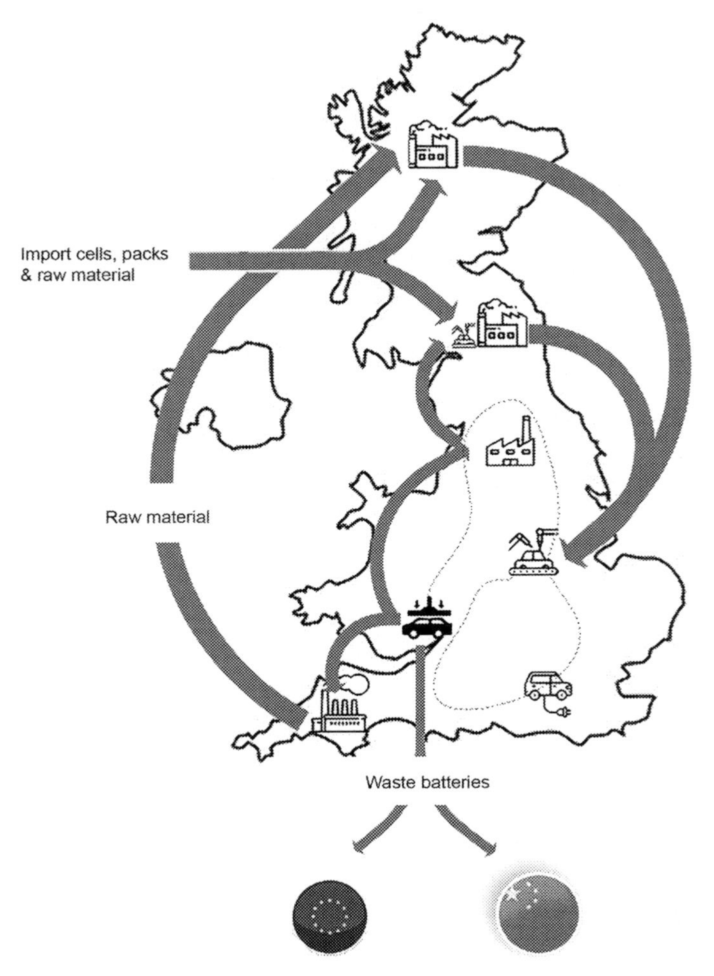

Fig. 6 A circular vision for the UK battery value chain. *(Credit: VALUABLE project—HSSMI Ltd.—Alberto F. Minguela).*

EoL batteries in the UK to recover their embedded value. This analysis was combined with a study of the necessary processes and equipment requirements for such a facility by Aspire Engineering.

In conclusion, a transition of the automotive supply chain to the circular economy is possible. Achieving this will bring environmental, social, and economic benefits; however, as reported in this chapter, they do come with many challenges at all levels. Nonetheless, the extensive research and investment in innovation expands each day the literature for this field. Governments, industry, and other stakeholders of the automotive and battery value chains do back the transition to a circular economy. Furthermore, most stakeholders are keen to seek guidance from the research and innovation community. The pool of knowledge that is being created will serve society in taking further steps on the transition towards a more sustainable and cleaner future.

References

BloombergNEF, 2019. Electric Vehicle Outlook 2019. Available from: https://about.bnef.com/electric-vehicle-outlook/#toc-viewreport. (Accessed 6 April 2020).

Braunovic, M., Myshkin, N., Konchits, V., 2006. Electrical Contacts: Fundamentals, Applications and Technology. CRC Press.

Bullard, N., 2019. Electric Car Price Tag Shrinks Along With Battery Cost. https://www.bloomberg.com/opinion/articles/2019-04-12/electric-vehicle-battery-shrinks-and-so-does-the-total-cost. (Accessed 12 December 2019).

Capgemini, 2019. Second Life Batteries: A Sustainable Business Opportunity, not a Conundrum. Available from: https://www.capgemini.com/2019/04/second-life-batteries-a-sustainable-business-opportunity-not-a-conundrum/. (Accessed 4 April 2020).

CEC, 2015. Environmentally Sound Management of End-of-Life Batteries. Commission for Environmental Cooperation, Montreal, Canada.

DEFRA, 2011. Depolluting End-of-Life Vehicles – Guidance for Authorised Treatment Facilities. https://assets.publishing.service.gov.uk/government/uploads/system/uploads/attachment_data/file/31736/11-528-depolluting-end-of-life-vehicles-guidance.pdf. (Accessed 2 November 2019).

Drabik, E., Rizos, V., 2018. Prospects for Electric Vehicle Batteries in a Circular Economy. Centre for European Policy Studies (CEPS).

Element Energy, 2020. The UK Energy Storage Lab (UKESL) Project. http://www.element-energy.co.uk/wordpress/wp-content/uploads/2020/01/UKESL-Non-technical-Public-Report_2020.pdf. (Accessed 25 March 2020).

Elkind, E. N., 2014. Reuse and Repower. How to save Money and Clean the Grid with Second-Life Electric Vehicle Batteries, Berkeley, California, September.

Emilsson, E., Dahllof, L., 2019. Lithium-ion Vehicle Battery Production. Status 2019 on Energy Use, CO_2 Emissions, Use of Metals, Products Environmental Footprint, and Recycling. www.ivl.se/download/18.14d7b12e16e3c5c36271070/1574923989017/C444.pdf. (Accessed 11 December 2019).

EV-volumes, 2019. EV-volumes.com. https://www.ev-volumes.com/country/total-world-plug-in-vehicle-volumes/. (Accessed April 2020).

Faraday Institution, 2020. Faraday Insights. https://faraday.ac.uk/wp-content/uploads/2020/04/Faraday_Insights_6_WEB.pdf. (Accessed April 2020).

Foster, M., Isely, P., Standridge, C.R., Hasan, M.M., 2014. Feasibility assessment of remanufacturing, repurposing, and recycling of end of life vehicle application lithium-ion batteries. J. Ind. Eng. Manag. 7 (3), 698–715.

Global Battery Alliance, 2019. A Vision for a Sustainable Battery Value Chain in 2030 Unlocking the Full Potential to Power Sustainable Development and Climate Change Mitigation. World Economic Forum, Geneva.

Harper, G., et al., 2019. Recycling lithium-ion batteries from electric vehicles. Nature 575 (7781), 75–86.

Hill, N., Clarke, D., Blair, L., Menadue, H., 2019. Circular Economy Perspectives for the Management of Batteries used in Electric Vehicles Final Project Report. Ricardo Energy, Joint Research Centre (JRC), EU Commission.

HSSMI, 2019. Industrial Advisory Board – Consensus. VALUABLE Project, London, UK.

Innovate UK, 2017. ISCF Winners. Faraday Battery Challenge – HSSMI. https://www.youtube.com/watch?v=2bddKDAaKHY.

MCT Reman, 2016. Remanufacturing. https://www.mctreman.com/Remanufacturing. (Accessed 2 November 2019).

Melin, H.E., 2018. The Lithium-Ion Battery End-of-Life Market 2018-2025. Circular Energy Storage, United Kingdom.

Oakdene Hollins, 2016. What is Remanufacturing? https://www.remanufacturing.eu/about-remanufacturing.php. (Accessed 5 November 2019).
Philippot, M., et al., 2019. Eco-efficiency of a lithium-ion battery for electric vehicles: influences of manufacturing country and commodity prices on GHG emissions and costs. Batteries 23, 5.
Podias, A., et al., 2018. Sustainability assessment of second use applications of automotive batteries: ageing of li-ion battery cells in automotive and grid-scale applications. World Electr. Veh. 9, 24.
Quinard, H., Redondo-Iglesias, E., Pelissier, S., Venet, P., 2019. Fast electrical characterizations of high-energy second life lithium-ion batteries for embedded and stationary applications. Batteries 33, 5.
Recharge, 2014. Re-use and Second Use of Rechargeable Batteries. European Association for Advanced Rechargeable Batteries (RECHARGE). https://rechargebatteries.org/wp-content/uploads/2014/04/RECHARGE-Information-Paper-on-Re-use-and-second-use-October-2014-v.14.pdf. (Accessed 5 November 2019).
Ritchie, H., Roser, M., 2019. Fossil Fuels. ourworldindata.org/fossilfuels. (Accessed 11 December 2019).
The Faraday Institution, 2017. Faraday Battery Challenge. https://faraday.ac.uk/post-with-video/. (Accessed 5 April 2020).
The Faraday Institution, 2019. Faraday Insights. https://faraday.ac.uk/wp-content/uploads/2019/08/Faraday_Insights-2_FINAL.pdf. (Accessed 4 April 2020).
Tsurukawa, N., Prakash, S., Manhart, A., 2011. Social Impacts of Cobalt Mining in Katanga. Oko-Institut e. V., Democratic Republic of Congo, Freiburg.
Whatcar, 2019. How Long Do Electric Car Batteries Last For? https://www.whatcar.com/advice/owning/how-long-do-electric-car-batteries-last-for/n18117. (Accessed 7 December 2019).
World Bank, 2018. Renewable Energy Consumption (% of Total Final Energy Consumption). https://data.worldbank.org/indicator/. (Accessed 11 December 2019).
Wu, F., Maier, J., Yu, Y., 2020. Guidelines and trends for next-generation rechargeable lithium and lithium-ion batteries. Chem. Soc. Rev. 49 (5), 1569–1614.

CHAPTER 32

Circular economy in the cosmetics industry: An assessment of sustainability reporting

O. Mikroni, G. Fountoulakis, P. Vouros, and K.I. Evangelinos
Department of Environment, University of the Aegean, Mytilini, Greece

1. Introduction

Current levels of production and consumption are unsustainable, and the circular economy is a response to the degradation of natural resources and the environment. This economic model focuses on minimizing the resources required in production processes, making use of renewable resources, reusing products, and producing energy from waste.

In this chapter, the concept of the circular economy will be discussed and assessed. Specifically, the implementation of circular economy practices on manufacturing industry will be evaluated through the assessment of corporate social responsibility reports of cosmetic companies for the year 2017. The aim of this study is to highlight the policy of each company regarding circular economy strategies including aspects such as prevention, recycling, and material reuse.

2. Circular economy and sustainable development

Nowadays, the circular economy theory is used with increasing frequency, aiming to highlight practices that will contribute to sustainable development and, therefore, to human prosperity. At the U.S. General Assembly called Our Shared Future in 1987, also known as the Brundtland Report, sustainable development was first defined (Geissdoerfer et al., 2017). Over the years, the definition of sustainable development was reshaped (Broman and Karl-Henrik, 2017) to encompass certain principles that are the core of the strategy. Ensuring conditions in which society can function and evolve is considered to be the principal criterion for achieving this. Inevitably, circular economy is linked with, and also refers to the definition of, sustainable development, as society and economic prosperity are associated with the protection of the environment (Mavroeidis and Papaioannou, 2005).

A definition of circular economy was devised by Kirchherr et al. (2017), according to whom, the circular economy is defined as:

Circular Economy and Sustainability, Volume 1
https://doi.org/10.1016/B978-0-12-819817-9.00030-2

a financial system which replaces the concept "end of life cycle" by reducing, alternatively reusing, recycling and reacclimating materials in production/distribution and consumption procedures. It works at a micro-level (products, companies, consumers), medium-level (eco-industrial parks) and macro-level (city, region, nation and beyond) with a view to achieving sustainable development, creating, at the same time, environmental quality, economic prosperity and social justice, in the interest of today's and future generations.

It is worth mentioning that in the past, the linear financial model was the dominant one, according to which, a product was produced in order to be consumed and afterwards disposed of: "production-consumption-disposal" (Stahel, 2016). Once the products had fulfilled the purpose they were made for, such as material intended for food production or consumer goods, they were disposed of. This practice resulted in ecological disaster due to pollution (Mavridou and Vasilaki, 2019).

Additionally, overpopulation and the ever-increasing need for more consumption inevitably led to a demand for goods, quite often in excessive amounts and consequently resulted in the depletion of natural resources and environmental degradation. Thus, the linear model focusing on extraction of raw materials, production of products, use of products, and waste in landfill is not considered a sustainable solution and another alternative is needed in order to turn products identified as waste into raw material. Products must be designed to remain within a material circle and sustain their added value, with, ideally, no or minimal waste. The need for new resources, necessitating financial and environmental cost, will be reduced (Mavridou and Vasilaki, 2019).

The circular economy represents a new production and consumption model and stresses the importance of the reuse, renewal, and recycling of materials and products and repair in order to extend their use (Proedros, 2019). Thus, while the circular economy is considered to be an evolution of recycling, it is not the same. Specifically, in recycling a product already used is destroyed and a new product is created, while in the circular economy, the product is designed to be reusable (Mavridou and Vasilaki, 2019).

However, moving to a circular economy presupposes the effective involvement of every individual and policymaking by manufacturers. The principal objective of the circular economy is to preserve the value of products, materials, and resources for as long as possible, as well as the minimization of waste production (Eurostat, 2018).

The European Union has already taken steps towards implementation of the circular economy and the statistical office, Eurostat, is monitoring its progress and the different stages of development in the EU, indicating which policies are successful and which need more action. The main purpose of business planning is the development of a viable, cost-efficient, and competitive economy (Eurostat, 2018). Some key strategies for the circular economy in the EU are to extend reuse and recycling (Bocken et al., 2016).

Not only does the environment benefit from this model, but also the economic and social life of the public benefits; with its innovative measures, the need for new competitive products, new business opportunities and markets, and employment opportunities follow (Mavridou and Vasilaki, 2019).

3. Survey methodology

In this research, in order to find the reports of the European companies producing cosmetics in 2017, the GRI (global reporting initiative) database was used. The literature reveals that the global reporting initiative is a database of sustainable development, which monitors and stores corporate social responsibility (CSR) reports. From this global database, policy makers and researchers can identify CSR reports from around the world (GRI, 2020).

The companies produce CSR reports on the basis of those guidelines identified by the application guides of the GRI. The objectives of CSR reports are to enable each company to declare their complete internal processes. By making use of the report guides and the indicators of the GRI, CSR reports gain credibility and become standards on how a report is produced. Finally, by making use of the relevant guides of the GRI, companies report their work in line with the guidelines. Consequently, the disclosure and communication of issues related to corporate accountability is improving transparency and accountability, benefiting both the company and the stakeholders (Hedberg and Von Malmborg, 2003).

The GRI's purpose is to provide a guide to companies interested in reporting their work externally, but from the results of these reports, the most important fact is that by reviewing the directives and collecting the data needed, the management achieves a useful overview of the company.

The search engine of GRI database allows identification of companies that are fully compliant with GRI and those that do not meet all the requirements set out by the GRI standards (GRI, 2020). For the purpose of this report, meeting the GRI database requirements was chosen, meaning that non-GRI reports were omitted. An additional eligibility criterion for the companies chosen was to be classified according to the GRI database, as large companies, or multinational enterprises (MNEs).

Finally, aiming for easily accessible information, reports written in Greek and English were selected. The five companies that met the prerequisites mentioned are listed in Table 1.

In this research, all the GRI indicators linked to recycling and reuse were recorded. Specifically, recycling is defined and implemented as the reintroduction of materials considered as waste in the production process, and is a major preventive and alternative cost-effective measure of managing hazardous, environmentally damaging materials. The safe recovery of materials (e.g., plastic, paper, etc.) depends on the qualitative and quantitative characteristics of the waste, on how companies obtain recyclable

Table 1 Companies included in the research.

Company name	Company size	Country of origin
GlaxoSmithKline (GSK)	MNE	United Kingdom
Merck	MNE	Germany
Pfizer	Large	Greece
Vianex	Large	Greece
Weleda	Large	Switzerland

material, etc. Through recycling, waste reduction and energy and raw material savings are achieved (Mousiopoulos, 1998).

Furthermore, according to the Greek law 4042/2012, the term reuse is defined as "every operation through which the products or ingredients that are not waste are reused for the same purpose they were initially designed" (OGN 24/A/13.02.2012). These indicators were chosen, as the purpose of this paper is to present the practices used by cosmetic companies related to reuse and recycling.

4. Companies

4.1 Glaxo Smith Kline (GSK)

This company produced 137,000 t of waste in 2016, which represents 4% less waste than in 2015 and 23% less than the reference year, 2010. Of this waste, 73% was incinerated for energy production or was recycled, while 5% ended up in landfill sites. It is worth mentioning that approximately 70% of these company chains have achieved zero waste by the redefinition of materials that could have been wasted. The term redefinition refers to the procedure in which an item with a specific use is repurposed for an alternative use. The company does not clearly mention the material recycled, though part of the 2016 report mentions products made of plastic.

The company takes part in the "Complete the Cycle" program, which collaborates with patients and pharmacists, aiming to recycle ventilators. A further step made by the company is to convert some of the products consisting of inactive plastic ingredients. Finally, their cooperation with Company Shop, a food and products distribution company in the UK, enables them to convert 23 t of products close to expiry date, so as to be sold on the retail market. In the present report of GSK, repossession of product-packaging is noted, but no provision for their products' biodegradation into the environment is made.

4.2 Merck

Through cooperation with Triumvirate Environmental, Merck have achieved the recycling of around 450 t of waste per year. They changed the labelling of their packaging, from plastic to paper and plastic, resulting in 61% less greenhouse gas emissions. They use Titripac, a plastic package with a fit-in tap, which offers a more environmentally friendly solution.

The company separates different kinds of waste, achieving the recycling of plastic of around 5 t/year. In addition, over 20,000 containers with expanded polystyrene (EPS) are reused every year, resulting in waste reduction. Clients return the empty containers to the company for cleaning and reuse. According to the company's report, the aim is to maximize their waste reduction while limiting, at the same time, raw material usage. When waste recycling is not possible, nonrecyclable material is disposed of in environmentally friendly ways.

It should be emphasized that Merck's report mentions that their products are subject to safety evaluations from the first stage of production, to assess and avoid the hazardous disposal of their chemicals, thereby avoiding impact on the environment or human health. The company's policy is to prohibit the use of toxic acids and mutagenic and hazardous substances. The performance materials sector produce materials that can be used in order to manufacture viable and environmentally compatible products, such as liquid crystals, which enhance screen resolution.

The company's policy follows national and international laws and the EU's statute for environmental protection. In particular, the raw materials for manufacturing cosmetic products meet the directives for cosmetics, such as Fair Production Practices (EFFCI GMP). This includes, for example, the gradual abolition of plastic microbeads, which are contained in skin care products, by 2020. The existence of these plastic microbeads in cosmetics and other products has been found to be particularly harmful for aquatic and terrestrial ecosystems.

Finally, since consumer demand for natural cosmetics has increased, the company aims to manufacture products by using raw materials that meet Cosmos Ecocert standards for natural and organic cosmetics.

4.3 Pfizer

This company aims to comply with any applicable law relating to the protection of the environment, while also supporting programs related to environmental sustainability. Additionally, recycling paper, batteries, electronical devices, plastic, ink, and aluminum is mentioned in the company report. In 2016, the company recycled 2600 kg of paper.

According to the Greenhouse Gas Protocol, Pfizer was involved in reducing their carbon footprint in 2016 as a result of their successful paper recycling and waste reduction program. Measurements for 2016 indicate greenhouse gas emission levels were 0%, both for paper and waste. One of the company's achievements concerning the delivery of products with high toxicity (cytotoxic) is that adaptations to packaging have been made to avoid leakage in case of accident.

The company's future objectives are:

- Informing employees about environmentally related issues and recycling
- Constant annual reports in relation to waste
- Carbon footprint measurements and reduction.

4.4 Vianex

In this company, recycling of many materials takes place, such as paper, glass, batteries, metallic objects (e.g., equipment), wooden objects (e.g., packaging and pallets), aluminum containers, plastic containers, and computers recycling. To motivate and raise awareness among employees, recycling bins have been introduced. The employees of this corporation recycle plastic bottle tops to exchange them for wheelchairs.

Paper use leads to the largest amounts of waste; Vianex highlights paper recycling in their report. Tables 2–4 detail paper and carton purchasing and recycling within the company's three plants for 2014, 2015, and 2016.

From Table 4, plants A, B, and C included both paper and cartons in their recycling operations. Relevant data are presented for those years that data were available.

Electronic device recycling reached a total of 1640 kg from 2014 to 2016, and battery recycling is presented in Table 5.

Table 2 Printing paper purchase.

2014	2015	2016
18,200 kg	18,900 kg	19,400 kg

Table 3 Cartons used.

2014	2015	2016
19,500 kg	42,500 kg	48,400 kg

Table 4 Paper and carton recycling.

Plant	2014	2015	2016
A	21,635 kg	21,955 kg	10,210 kg
B	–	8980 kg	17,240 kg
C	–	4660 kg	6800 kg
Total	*21,635 kg*	*35,595 kg*	*34,250 kg*

Table 5 Battery recycling.

2014	2015	2016
35 kg	49 kg	63.5 kg

The company's future objectives are:

- Further efforts to reduce paper consumption
- Increased recycling in all facilities.

4.5 Weleda

Weleda aims to create recyclable packages for their products in order to effectively maintain a balance between material savings and recycling. In addition, during the production of pharmaceutical products, the materials no longer usable in production are separated and given to a nonprofit organization, Offcut. Offcut is an organization that stores the materials and sells them to secondhand stores for artists. In 2015 and 2016, for example, they sold preused pharmaceutical bottles with lids. In Weleda, they focus on the optimum separation of recyclable and reusable waste, to prevent this material ending up in incinerators.

The company's objective for the year 2016 was to use recyclable and more environmentally friendly packaging for their natural and organic products. The achievements for the year were: (a) the amount of recycled material for glass bottles increased by 85%, (b) an external study about bioplastic, and (c) increased cooperation with environmentally friendly strategic platforms regarding packaging. The objective for 2017 was to maintain and continue the objectives of 2016.

In the middle of 2016, the company used green glass bottles, of which 85% were recycled. The supply of these glass bottles increased by 176 t and thus, the recycling of this material from 37.3% to 48.7%.

Reusable waste was reduced in 2015, total recycled waste reached 59% (814 t), and the waste used for thermal application reached 37% (500 t). In Table 6, the amount of waste and the percentage of recycled material in packaging is analyzed.

The percentage of the waste recycled from 2015 to 2016 appeared to have decreased by 2%. At the same time, however, the proportion of product packaging recycling seems to have increased by 11%.

Table 6 Weleda waste recycling analysis.

	2015	2016
Percentage of recycled waste	98%	96%
Recycling proportion of product packaging	38%	49%

Table 7 Natural waste and its reuse in tonnes.

	2014	2015	2016
Natural waste	738	751	722
Reuse	41	37	33

Table 8 Hazardous waste in tonnes.

	2014	2015	2016
Hazardous waste	3	4	1

It is important to highlight that, while the company does not clearly state the kinds of natural waste, their separation from hazardous/toxic waste leads to the conclusion that this waste does not cause major environmental damage.

Tables 7 and 8 demonstrate that more waste was produced in 2015. It is important to emphasize that the company reuses natural waste, showing a more sensitized attitude and practice towards the environment.

5. Conclusion

Having examined the CSR reports of the companies analyzed above, it was concluded that these European companies do not all completely meet the practices of the circular economy as they do not all reuse material. All companies recycle their products and packaging, however, the procedure could have been more successful.

Some of the companies produce their products in such a way that they can be recycled and reused. Specifically, GSK recycle and reclaim products and also convert them into products of a different use than the initial one. In cooperation with other companies, they convert products close to expiry date and promote them to be sold on the retail market. Merck follows recycling and reuse practices for their products and produces materials which are in turn used in manufacturing viable products. Their primary priority over recent years is producing cosmetics that do not contain nonbiodegradable plastic microbeads.

Pfizer comply with the applicable laws related to environmental protection and although they make a clear reference to recycling in their report, this is not the case with reusing. However, it should be noted that specific provision is made for the delivery of toxic products. Vianex recycles as does Weleda, except that they reuse products as well. In cooperation with other companies they resell them to second-hand stores for artists in order not to dispose of excess materials. There are numerous weaknesses related to informing and raising awareness among the public and employees on environmentally related issues and material reuse. In relation to recycling, the results achieved are rather satisfying.

Studies have shown that in order to implement circular economy principles, there should be legislative provisions that not only promote the proper functioning of the company, but aim also to safeguard the environment. It is imperative, at both the European and national level, for countries to safeguard the practices of recycling and reuse by companies, showing in practice their contribution to society. The reinforcement of a financial model that will respect and defend the environment will ensure the prosperity and viability of our planet.

References

Bocken, N.M.P., DePauw, I., Bakker, C., Vander Grinten, B., 2016. Product design and business model strategies for a circular economy. J. Ind. Prod. 33 (35), 308–320.

Broman, G.I., Karl-Henrik, R., 2017. A framework for strategic sustainable development. J. Clean. Prod. 140, 17–31.

Eurostat, 2018. How is the EU Progressing Towards the Circular Economy? Retrieved from https://ec.europa.eu/eurostat/documents/2995521/8587408/8-16012018-AP-EN.pdf/aaaaf8f4-75f4-4879-8fea-6b2c27ffa1a2.

Geissdoerfer, M., Savaget, P., Bocken, N.M.P., Hultink, E.J., 2017. The circular economy—a new sustainability paradigm? J. Clean. Prod. 143, 757–768.

Global Reporting Initiative (GRI), 2020. Sustainability Disclosure Database: Data Legend. Global Reporting Initiative (GRI), Amsterdam, pp. 1–15.

Hedberg, C.J., Von Malmborg, F., 2003. The global reporting initiative and corporate sustainability reporting in Swedish companies. Corp. Soc. Responsib. Environ. Manag. 10, 153–164.

Kirchherr, J., Reike, D., Hekkert, M., 2017. Conceptualizing the circular economy: an analysis of 114 definitions. Resour. Conserv. Recycl. 127, 221–232.

Mavridou, D., Vasilaki, B., 2019. Tribute. Circular economy. A new economic model of sustainable development. Innov. Res. Technol. (115), 12–17.

Mavroeidis, H., Papaioannou, M., 2005. Sustainable Development: International and European Developments and Perspectives. Ministry of Environment, Physical Planning and Public Works, Athens.

Mousiopoulos, N., 1998. Recycling. Ziti Publications, Thessaloniki.

Proedros, M., 2019. Issue note. Innov. Res. Technol. (115), 1.

Stahel, W.R., 2016. Circular economy. Nature 531, 435–438.

CHAPTER 33

Company perspectives on sustainable circular economy development in the South Karelia and Kymenlaakso regions and in the publishing sector in Finland

R. Husgafvel[c], L. Linkosalmi[a], D. Sakaguchi[b], and M. Hughes[c]
[a]Wood Material Technology Research Group, Department of Bioproducts and Biosystems, School of Chemical Technology, Aalto University, Espoo, Finland
[b]Clean Technologies Research Group, Department of Bioproducts and Biosystems, School of Chemical Technology, Aalto University, Espoo, Finland
[c]Faculty of Health Science, Department of Human Care Engineering, Nihon Fukushi University, Aichi, Japan

1. Introduction and background

This study addresses sustainable circular economy (CE) development in the Etelä-Karjala (South Karelia) and Kymenlaakso regions and in the publishing sector in Finland. The study was conducted from the company perspective encompassing topics such as drivers, barriers, and potential for cross-border collaboration. Special emphasis is placed on the identification of necessary sustainability considerations, supportive drivers, and potential barriers perceived by companies in the studied regions and sector that need to be addressed in the future to promote the creation of enabling conditions for CE development. A similar regional and sectoral approach was applied in our previous research (Husgafvel et al., 2017, 2018a,b).

The CE is expected to create a significant amount of economic, social, and environmental benefits in Finland. The main goal is to maximize the circulation of products, materials, and component and the associated value in the economy, with strong emphasis on systemic change (Sitra, 2016). CE development at the EU level (EC, 2015) is associated with, e.g., the global sustainable development agenda, gaining sustainable competitive advantages at the European level, and development of a sustainable, low carbon, and resource efficient economy. Development of new business opportunities and innovative production and consumption approaches are also essential (EC, 2015). Currently, the EU is putting more emphasis, e.g., on the recycling of plastics and the analysis of practical, legal, or technical problems associated with recycled materials (EC, 2017). The recycling of plastic has been rapidly implemented and there are different technological flows of recycling plastics, for instance, from material and chemical recycling to thermal recycling to achieve material circulation (Hamba and Ida, 2018). The latest developments also

Circular Economy and Sustainability, Volume 1
https://doi.org/10.1016/B978-0-12-819817-9.00013-2

include reparability requirements for electronic devices, which are connected to the Ecodesign Directive and its requirements for continuous improvement of reparability, recyclability, and sustainability of devices (YLE, 2019), a ban on disposable plastics, and new requirements for recycling of plastic bottles (HS, 2019). The main goal is set at minimizing the environmental impact and social cost, which will contribute to a sustainable material life cycle (Saito et al., 2018). In addition to plastics and electronic devices, a recycling system for components of automobiles has been developed (ERCC, 2015; Fujii, 2017), to which the current main barriers are cost efficiency and the processing of various materials used in each component (Onoda, 2019).

In its essence, CE aims at keeping products, materials, and component in circulation and at their highest utility level at all times, including both biological and technical materials and cycles. It is about a new economic model that seeks to decouple economic growth from the consumption of finite resources, based on principles that highlight better management of natural capital, optimization of resource yields through circularity, and system-level effectiveness (Ellen MacArthur Foundation, 2015).

In Finland, CE requires practical actions and incentives, and changing consumer trends and demands are already important drivers. The associated new solutions based on sustainable use of natural resources can create profitable business. The key focus areas encompass sustainable products, recycling, and changes from product design to consumption toward products that remain in the cycle (HS, 2017). Currently, there is a lot of focus on innovations that promote the replacement of, e.g., plastics with wood, such as the use of ionic liquids to modify wood material (HS, 2018). For example, the recycling of plastics creates multiple challenges such as sorting, cleaning, and the need for new kinds of package design. A lot of plastic is still burned and is not within the recycling cycle (HS Kuukausiliite, 2018). Plastics production and the burning of plastic waste both cause carbon dioxide emissions and thus have an impact on climate change (Loimu, 2018).

Witjesa and Lozano (2016) noted that there is still limited research on key CE aspects (e.g., bridging production and consumption activities and resources derived from waste materials). Promotion of CE requires integrated approaches encompassing focus on long-term system change, overcoming barriers, and involvement of multiple public and private actors (Van Buren et al., 2016). Addressing the practical challenges associated with CE requires transdisciplinary research efforts and focus on, for example, sustainability and life cycle considerations (Sauvé et al., 2016).

Further research in the field of CE is needed on, for example, (1) industrial associations and (2) business associations and their small and medium-sized enterprise (SME) members including their role in both promoting industrial symbiosis and closing of material loops. Further research is also needed on business opportunities provided by CE for SMEs, best SME practices, and on societal actions that focus on CE as a major contributor to long-term sustainability (Ormazabal et al., 2018). Heshmati (2015) defined that CE is a sustainable development strategy driven by principles of reducing, reusing, and recycling materials, energy, and waste and noted that literature on the relationship between entrepreneurship and CE is in its infant stage.

Lakatos et al. (2016) noted that the adoption of new consumption patterns to promote CE development depends strongly on appropriate incentives and benefits. Moreno et al. (2016) suggested that more focus is needed on design of new circular business models in practice in addition to studies on new business models and technical/biological cycles. Assessment of social and environmental impacts of new product, service, or business models is also essential (Moreno et al., 2016). Lieder and Rashid (2016) recognized that the implementation of the CE concept requires support from all stakeholders and appropriate new business models, social awareness, and multidisciplinary industrial-level approaches.

Hall et al. (2010) reviewed the role of entrepreneurship in sustainable development and they noted that the emergence and legitimization of sustainable development within business and policy will change the rules of the game. They suggested further research should focus on complexities created by sustainability considerations (advantages and disadvantages for and evolution of industry and new ventures) and the conditions in which (1) entrepreneurs would pursue sustainable ventures and be able to generate economic growth and advance social and environmental goals, (2) entrepreneurship creates welfare including all externalities, and (3) public policy positively influences sustainable entrepreneurship (Hall et al., 2010).

CE can play a role in addressing many global challenges. For example, Potočnik (2017) highlighted many challenges such as climate change, environmental degradation, urbanization, social inequality, population growth, and the issues associated with the level of wellbeing globally (production and consumption). He also suggested that there is a need for increased resource efficiency and more responsible use of resources including supportive market and public policy efforts driven by political will. Implementation of SDGs (cf. UN, 2019) requires leadership, system change, and improved global governance (Potočnik, 2017). In the field of energy, utilizing renewable energy could significantly contribute to the reduction of CO_2 emissions, as well as environmental load (Shimizu et al., 2018). Large-scale adoption of renewable energy systems can lead to the establishment of added-value systems with circular supply chain management (Akiba, 2019).

It is increasingly important to use less resources (e.g., materials and energy) to support economic growth and use them wisely to reduce environmental impact (Steiner, 2017). Circularity in production and consumption supported by new multistakeholder partnerships (including broader public-private partnerships of much larger scale) can contribute to the achievement of SDGs, and in many cases, requires sustainable resource management (Noronha, 2017).

The EU encourages all regions and stakeholders to take part in measures to support transition toward a CE including focus on the associated systemic change aspects (EC, 2015). In this context, the engagement of stakeholders encompasses, for example, sharing of best practices among regions, voluntary business approaches, development of new skills, and public-private partnerships. It is also essential to take into account the important role of SMEs and consider the social implications of activities. Measurement of progress calls for attention to, e.g., a set of reliable indicators. The EC also promotes efficient use of bio-based resources including guidance on best practices on the cascading use of biomass and research on integrated bio-refineries that could process both biomass and bio-waste. Other related aims comprise application of extended producer responsibility and focus on environmentally sound options (EC, 2015).

At the European level, the recycling of construction and demolition waste is a major focus area encompassing, e.g., development of indicators to assess environmental performance of buildings taking into account full life cycle considerations (EC, 2015). Other key goals comprise focus on product design and production processes, recycling of plastics (whole value chain and life cycle focus), recovery of critical raw materials, addressing food waste issues, creation of markets for secondary raw materials (waste to resources), cross-border circulation of secondary raw materials, and enhanced recycling efforts in waste management (EC, 2015).

In Finland, the selected focus areas (Sitra, 2016) comprise forest-based and technical loops, joint actions, a sustainable food system, and transport and logistics. The role of the state is to facilitate the creation of an enabling platform for CE, taking into account both domestic and global markets (e.g., addressing incentives, potential barriers, and taxation aspects). Collaboration between the private and public sectors and enhancement of the roles of education and research in the advancement of CE are among the focus areas. Increased use of renewable natural resources and development toward a carbon-neutral and waste-free society as well as advancement of services, new consumption models, and sharing in the economy are also highly relevant in this context (Sitra, 2016). In addition, the recycling of plastics should be a major focus area (Loimu, 2018).

2. Material and methods

2.1 Aims of the study

This study aimed at assessing company perspectives on CE development and challenges in the South Karelia and Kymenlaakso regions and in the publishing sector in Finland. The idea was to identify potential drivers and barriers as well as to address associated sustainability and business aspects. The chosen target group consists of regional and sectoral companies in Finland. The approach includes elements of both current status and future outlook. The intention was that the obtained results could support informed decision-making regionally and sectorally as well as in both private and public sectors. The set-up of the study and its objectives are linked to the overall CE development framework at the EU and national levels and to key issues as identified by previous research and writings.

2.2 Methodology

The applied methodology in the two regional case studies and in the sectoral study was a questionnaire survey (Patten, 2011; Hirsjärvi et al., 2007). Survey is an information collection method that can be applied to describe, explain, and compare, e.g., knowledge and preferences (Fink, 2009). The chosen research approach was an online survey, which means it was a self-administered questionnaire (Fink, 2009), and was carried out in phases (Hirsjärvi et al., 2007). It began with choosing of the topic and themes (CE and sustainability aspects). The chosen approach is qualitative and descriptive. The questionnaire was structured and formal and consisted of multiple choice closed questions (Hirsjärvi et al., 2007; Gillham, 2007). Sudman and Bradbrun (1982) noted that the questionnaire should be professionally designed and look easy to answer, with clear questions. Fink (2009) noted that survey data analysis uses qualitative and statistical methods to describe, interpret, and compare preferences, attitudes, values, and behavior of respondents. The qualitative research approach allows appropriate selection of research subjects and objects in accordance with the research goals (Hirsjärvi et al., 2015).

2.3 Survey and questionnaire analysis

The survey was implemented via email and the key term CE was defined in the beginning of the survey (cf. Fink, 2009; Hirsjärvi et al., 2007). The survey aimed at describing the essential features of sustainable CE development as perceived by companies. The self-administered questionnaire was short and simple and the purpose of the survey was explained in the accompanied email (cf. Fink, 2009). Research design refers to strategy/plan/logic of the research and it forms a basis for the research specification (Oppenheim, 1997). The logic of this research was based on sustainable CE considerations and the closed questions were simple and specific. Gillham (2007) noted that the benefits of questionnaires include, e.g., short implementation time, capturing a lot of information from a lot of people, and easy analysis.

Construction of effective questions often requires both expertise and creativity supported by review of, e.g., problems and opportunities (Saris and Gallhofer, 2014). The online survey was designed for companies and the selected themes and answering options of the questionnaire surveys were based on (1) literature review, (2) previous studies (e.g., Husgafvel et al., 2017, 2018a,b; Lakatos et al., 2016; Lieder and Rashid, 2016; Moreno et al., 2016; Sauvé et al., 2016; Witjesa and Lozano, 2016), (3) national and EU level CE goals (EC, 2017, 2016; SITRA, 2016, 2014), and (4) some elements of sustainability assessment (Husgafvel et al., 2017). The face to face approach can be used to check

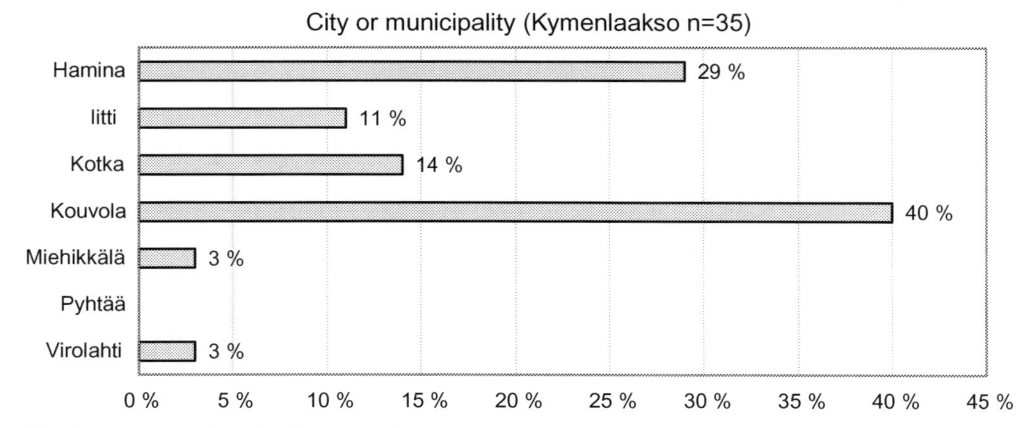

Fig. 1 City or municipality (Kymenlaakso).

the quality of the questionnaire (Saris and Gallhofer, 2014). Therefore, the chosen themes and associated sets of answering options were also assessed based on face to face assessment (peers).

The chosen themes comprised the following: (1) drivers and opportunities of CE, (2) barriers to CE, (3) future CE aspects, (4) ways in which CE can improve economic, social, and environmental sustainability, (5) how the CE operational environment can be advanced, (6) the technological measures that can advance CE operational environment, (7) how public procurement can advance CE, (8) how cross-border CE can be advanced, and (9) the reasons that could lead to company investments in CE. Each theme encompassed a set of answering options based on the chosen research approach.

Questionnaires were sent to companies in the Etelä-Karjala (South Karelia) and the Kymenlaakso regions and in the publishing sector. The response rate was 12% for the South Karelia region (24 companies responded), 11% for the Kymenlaakso region (35 companies responded) and 4.7% for the publishing sector (17 companies responded). The locations (cities or municipalities for the regional companies and the provinces for the sectoral companies) of the responding companies are presented in Figs. 1–3. The sizes of the companies are presented in Fig. 4 and the fields of business of the regional companies are presented in Fig. 5. The operational levels of the companies are presented in Fig. 6.

The answering options of the survey were: "please select 1–3 options." The survey questionnaires links were sent directly to the personal addresses of the company representatives through email and the respondents included multiple sized companies and staff from various positions (Figs. 1–4) in each survey study. The results are presented as figures based on themes, answers to the surveys, and the number of responding companies with the idea to identify and indicate the most preferred options in each theme.

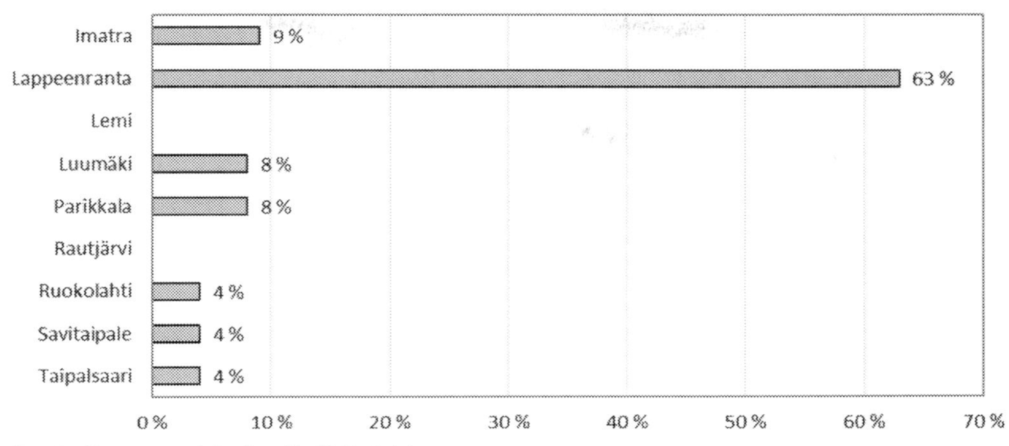

Fig. 2 City or municipality (Etelä-Karjala).

Closed-end questions include all possible responses as specified by the researchers (Peterson, 2000). Closed questions are easy to process and compare, do not require much time, and are good for testing specific hypotheses (Oppenheim, 1997). The theoretical framework of sustainable engineering (Abraham, 2006; Allen and Shonnard, 2012; Allenby, 2012; Graedel and Allenby, 2010; Rosen, 2012) was taken into account in this study, including economic, social, and environmental sustainability considerations.

3. Results

The results of the regional studies indicate that the main regional drivers and opportunities of CE encompass sustainable, long-lasting, and fixable products, increased recycling/reuse of products, components, and materials, and use of wastes as raw material. The key drivers and opportunities in the publishing sector include new services (including substitution of products), new products, and wastes as raw material (Fig. 7). The main regional barriers to CE include profitability, lack of information (for example supply/product chain), lack of communication and cooperation, and lack of seeing "the big picture." Sectorally, the main barriers are profitability, lack of information (for example supply/product chain), and competition/competitive activities (Fig. 8).

Regionally, the main future CE aspects include sustainability and long lifespan of products, components, and materials, holistic system-level thinking, use of renewable energy, completely new products, components, and materials, and replacement of plastics with renewable raw materials. Completely new products, components, and materials, sustainability, and long lifespan of products, components, and materials, and holistic system-level thinking were the main future CE aspects in the publishing sector

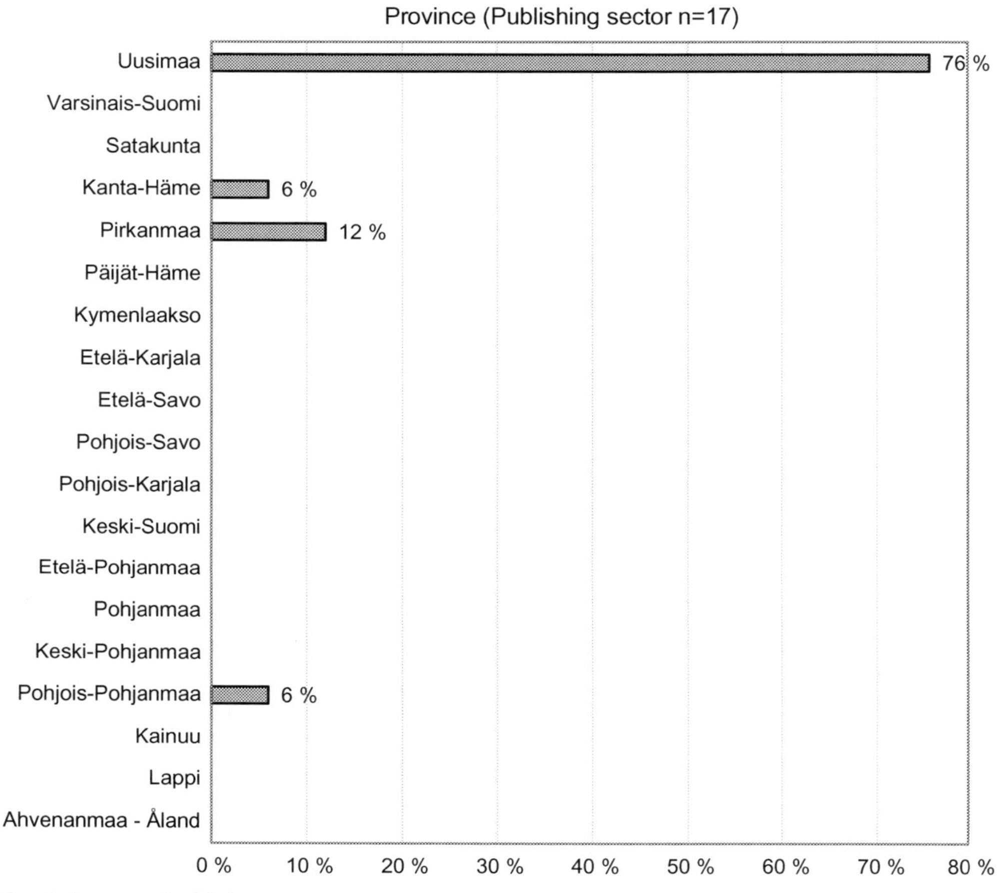

Fig. 3 Province (publishing sector).

(Fig. 9). In the studied regions, CE can improve economic, social, and environmental sustainability in the following areas: recycling and reuse, renewable energy and raw materials, supply chain management, and staff skills and training. In the publishing sector, those areas are supply chain management, recycling and reuse, and staff skills and training (Fig. 10).

Based on the company responses in the regional studies, the CE operational environment can be advanced through the following governance measures: research and development subsidies, taxation, guidance, and networks. In the sectoral study, those measures are taxation, research and development subsidies, and guidance (Fig. 11). Regionally, CE operational environment can be advanced through the following business measures: profitable products and services innovated by forerunners, use of renewable raw materials, use of renewable energy, new business models, and marketing and market creation.

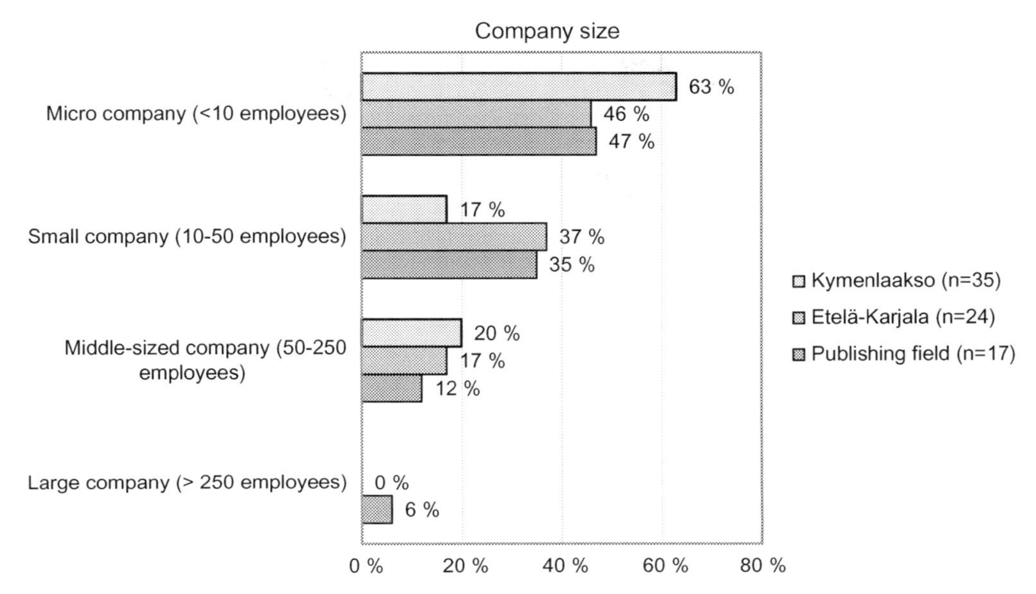

Fig. 4 Company size.

Sectorally, those measures are new business models, profitable products and services innovated by forerunners, and use of renewable raw materials (Fig. 12).

The capability to handle all materials in recycling, product design, replacement of plastics with renewable raw materials, and product fixability are the technological measures that can advance CE operational environment in the studied regions. The capability to handle all materials in recycling, product design, and technological change of the whole supply chain are the technological measures that can advance CE sectorally (Fig. 13). In the studied regions, public procurement can advance CE through the following measures: acknowledgment of the whole product chain and life cycle, guidance, replacement of plastics with renewable raw materials in all possible procurement, legislation, and obligatory recycling of products, components, and materials. Sectorally, those measures are the acknowledgment of the whole product chain and life cycle, obligatory recycling of products, components, and materials, and legislation (Fig. 14).

Cross-border CE can be advanced regionally through the following measures: company cooperation, agreements between neighboring countries, logistical restructuring (e.g., border traffic), and reassessment of the whole supply chain. Sectorally, those measures are company cooperation, agreements between neighboring countries, and reassessment of the whole supply chain (Fig. 15). The results of the sectoral studies indicate that the reasons that could lead to company investments in CE encompass better grasp of sustainability thinking (economic, social, and environmental), better grasp of life cycle thinking, use of renewable energy and raw materials, significant improvement of resource

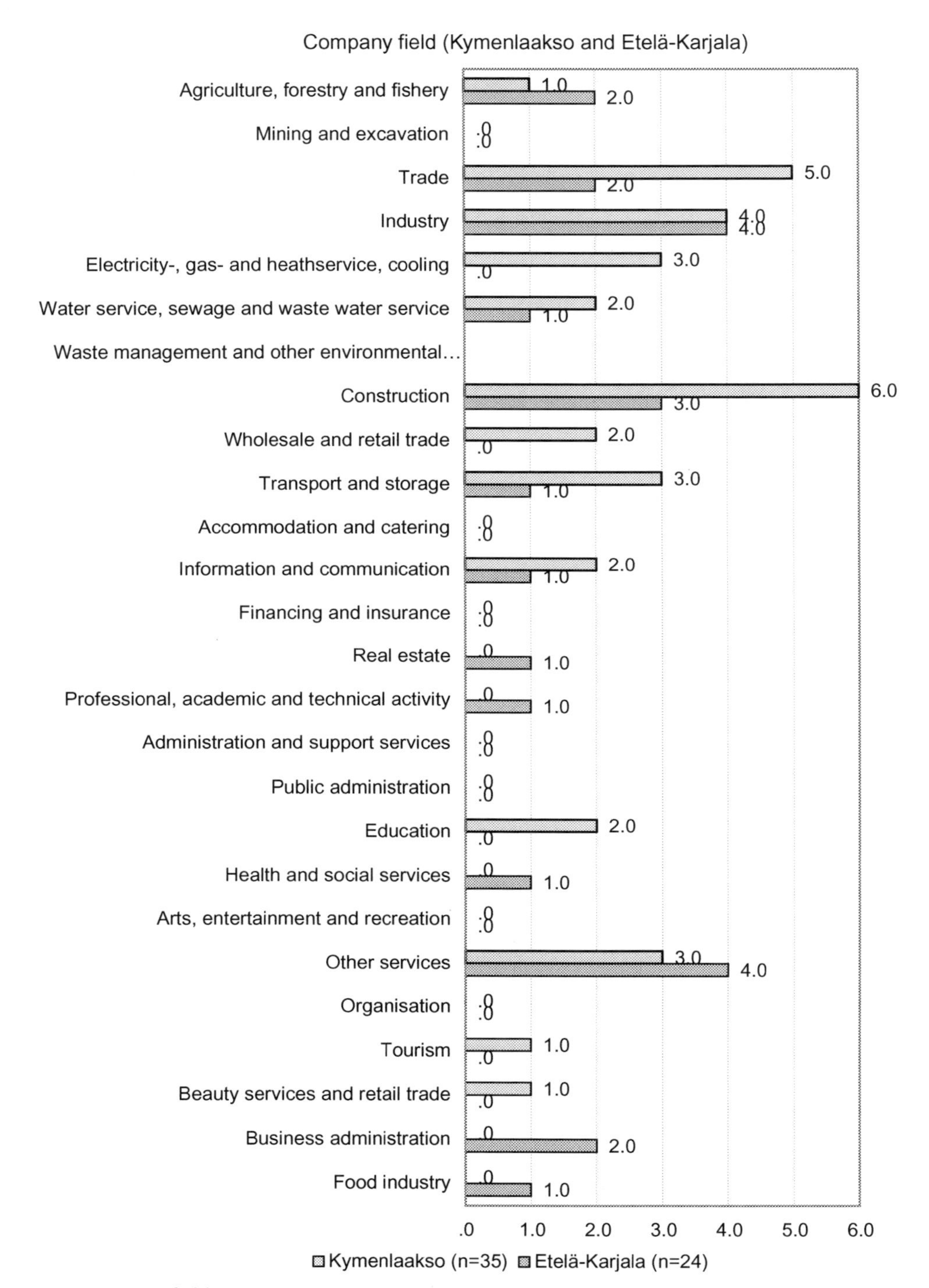

Fig. 5 Company fields.

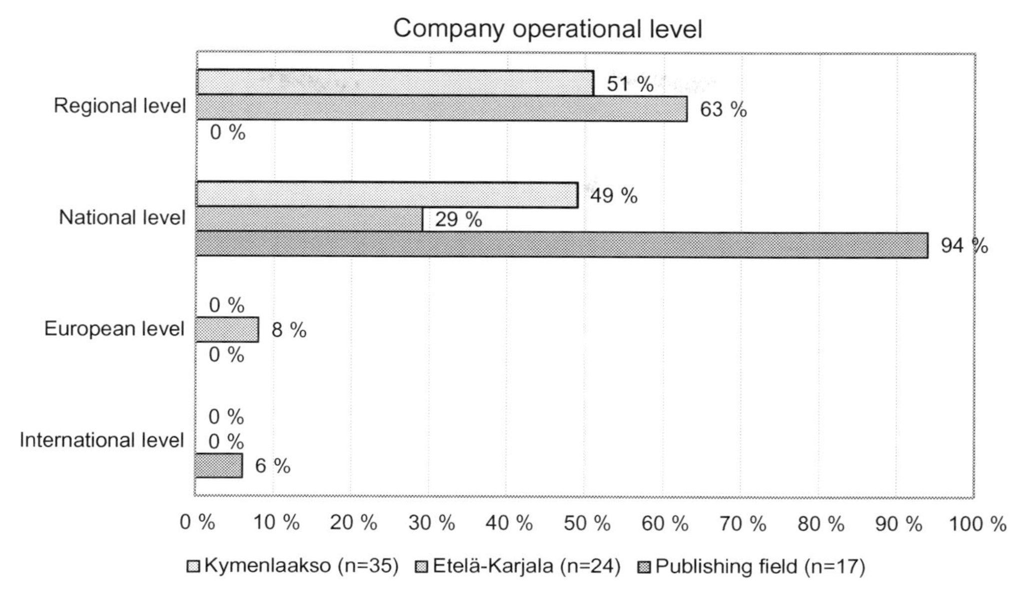

Fig. 6 Company operational levels.

efficiency, and awareness and competence building. The results of the sectoral study indicate that those reasons include better grasp of life cycle thinking, better grasp of sustainability thinking (economic, social, and environmental), better management of the whole supply/product chain, and significant improvement of resource efficiency (Fig. 16).

4. Discussion

The results of this study indicate that the main regional drivers and opportunities of CE encompass sustainable, long-lasting, and fixable products, increased recycling/reuse of products, components, and materials, and wastes as raw material, whereas the drivers and opportunities in the publishing sector include new services (including substitution of products), new products, and wastes as raw material. Cohen and Winn (2007) noticed that it is important to focus on entrepreneurial innovations in addressing market imperfections and address possibilities to support more sustainable markets. Dean and McMullen (2007) recognized that sustainable and environmental entrepreneurship can help to resolve environmental problems associated with global socio-economic systems through discovery, evaluation, and exploitation of opportunities created by market failures. Small firms and new entrants typically stimulate the transformation of industry toward sustainability in the early stages and this trend is followed by sustainability initiatives made by larger industry (Hockerts and Wüstenhagen, 2010). Hockerts and Wüstenhagen (2010) suggested that the interaction between small firms and industry

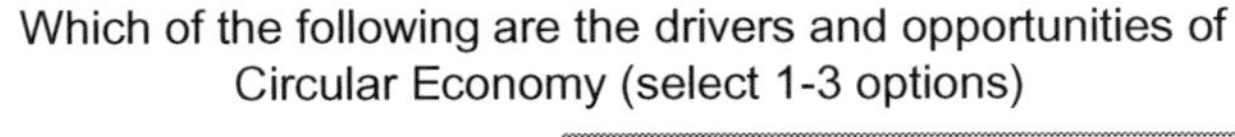

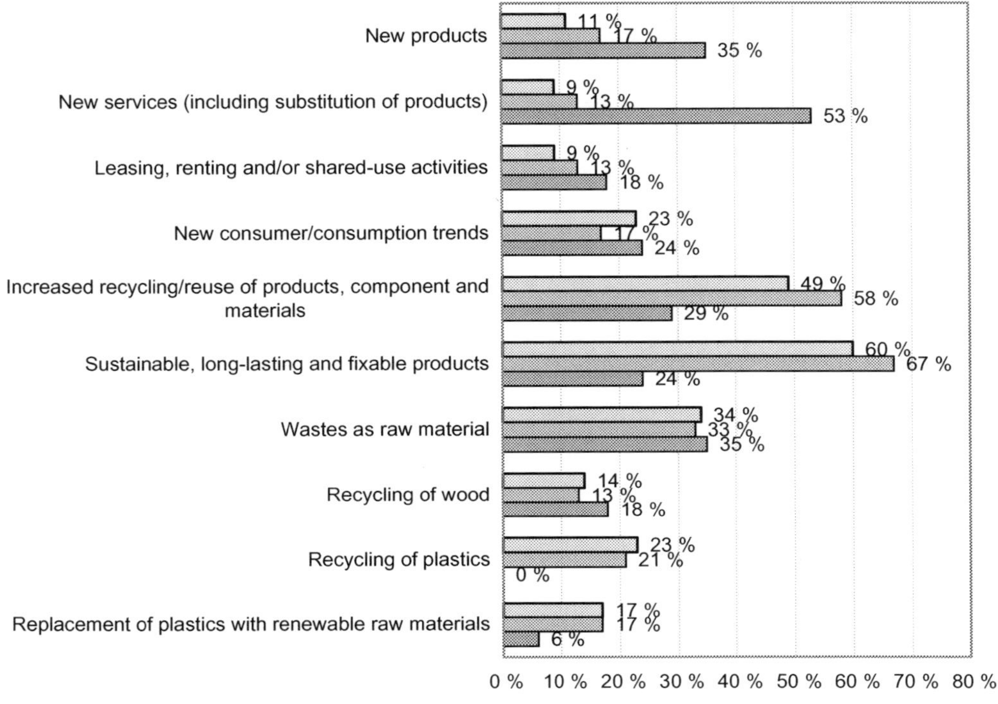

Fig. 7 Drivers and opportunities of circular economy.

is essential for sustainable entrepreneurship and the sustainable transformation of industries. They noticed that interesting focus areas comprise, e.g., the optimal allocation of roles between firms and industry for simultaneous achievement of high economic, social, and environmental performance. SMEs play an important role in promoting CE and new sustainable solutions to local needs (Khosla, 2017). A proactive approach to socially responsible buying can (1) help to avoid stakeholder pressures and negative publicity, (2) be an expression of corporate values, and (3) create business benefits (Maignan et al., 2002). Baumgartner (2009) studied sustainability in the mining industry and concluded that it is necessary to show the business case for sustainable development (demonstrate the value of sustainability values) including making it measurable and a part of controlling processes. It should also be linked to shareholder value and the sustainability performance of managers should be measured. Senior managers need to value the importance of and be committed to sustainable development (Baumgartner, 2009).

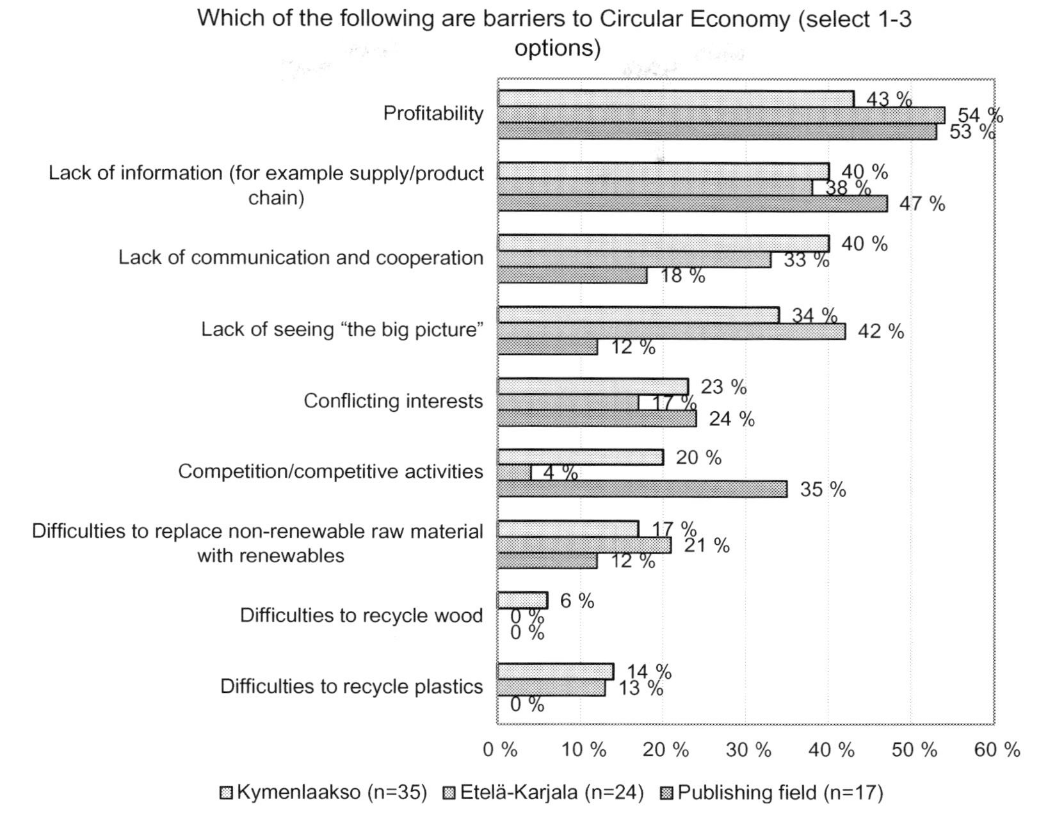

Fig. 8 Barriers to circular economy.

The results indicate that the main regional barriers to CE include profitability, lack of information (for example supply/product chain), lack of communication and cooperation, and lack of seeing "the big picture," whereas the main sectoral barriers are profitability, lack of information (for example supply/product chain), and competition/competitive activities. Van Buren et al. (2016) recognized that integrated and systemic approaches, including engagement of both public and private actors and measures to overcome institutional and economic barriers, are needed to advance CE development. Identification of barriers to change toward more sustainable activities, products, and services and planning of organizational changes toward more holistic sustainability efforts can help to incorporate and institutionalize corporate sustainability (Lozano, 2013). Kirchherra et al. (2018) noted that CE implementation in the EU has been limited and there is a need for significant efforts to maintain a specific CE-focused momentum apart from the overall sustainable development framework. Cultural barriers (e.g., lack of consumer awareness and interest) and indecisive company culture are the main barriers to

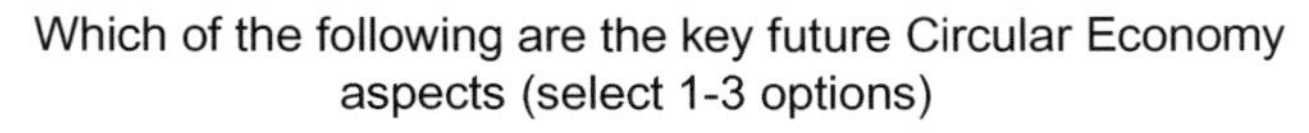

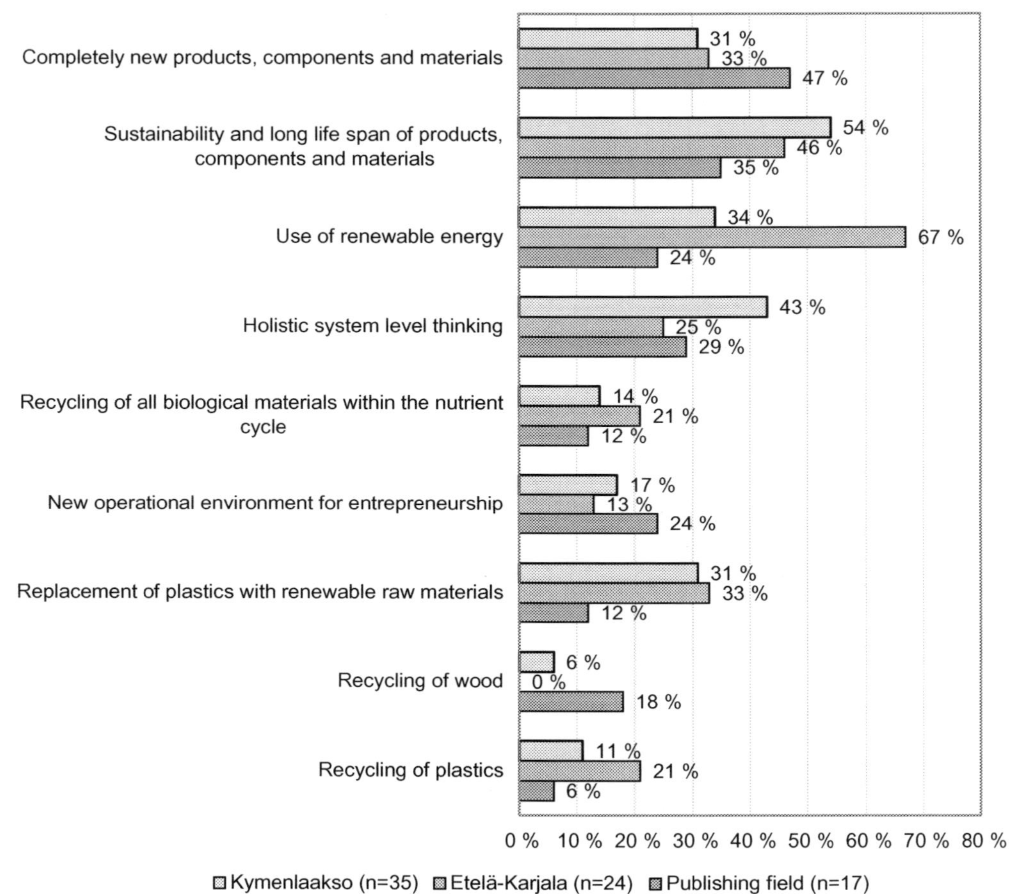

Fig. 9 Future circular economy aspects.

CE implementation. They also noted the role of market barriers and associated failed governmental interventions in this context in addition to difficulties related to CE business models (e.g., high initial investment costs and competitiveness against linear products). New ideas could encompass, for example, policies that promote circular products and associated reduced value added taxation (Kirchherra et al., 2018). There are specific studies on CE-related activities from the wood sector and, for example, cascading use[a] of wood is limited by multiple barriers such as use of toxic preservatives (pollutant materials

[a] Refers to efficient utilization of resources by using residues and recycled materials for material use to extend total biomass availability within a given system (EC, 2016).

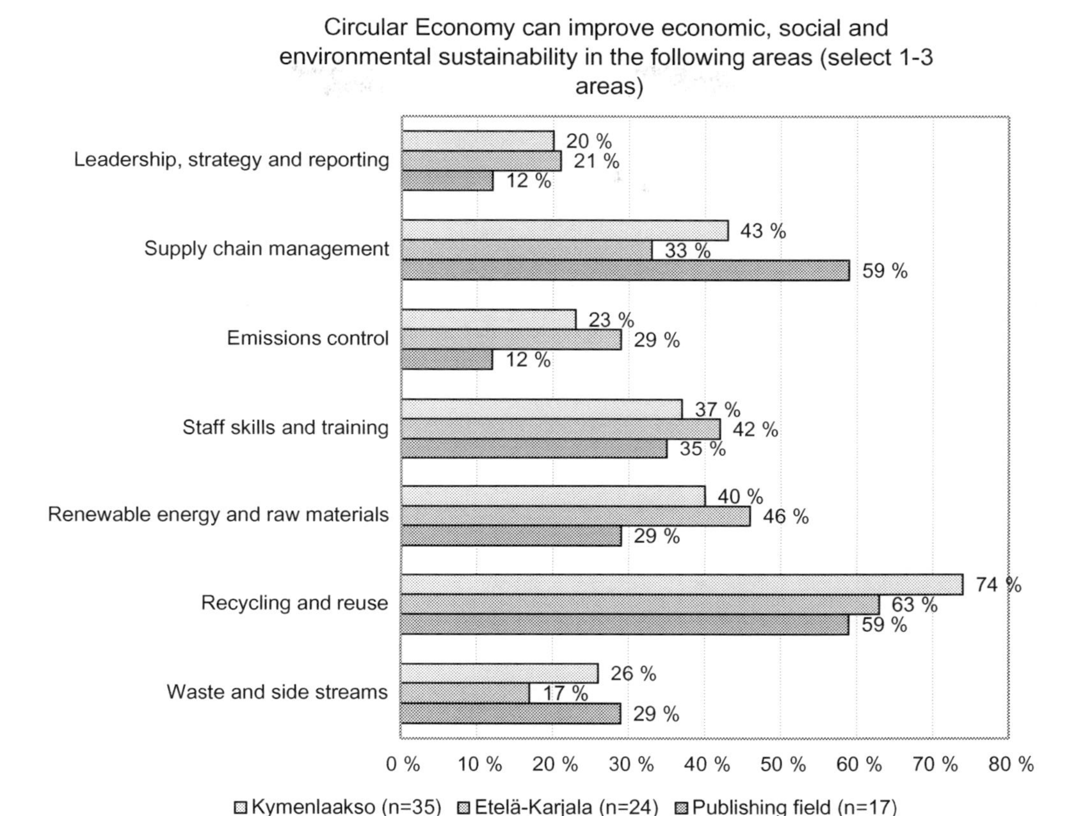

Fig. 10 Circular economy and sustainability.

in the wood), lack of sorting, recovery targets, wood classes, recycling technologies, and market incentives. Moreover, there is a lack of cost-effective cleaning and sorting methods and limited markets and commercial potential (EC, 2016). According to the EC (2016), there is a need to address various market, governance, and technological barriers to cascading use of wood. Additionally, there is a need for integrated approaches to material and energy use of wood and biomass and for a mix of coordinated approaches suited to specific local conditions (EC, 2016). The role of public institutions is significant and currently they form a major barrier to CE development. Policy instruments, financial stimulation, and technological modernization are essential in this context (Ormazabal et al., 2018).

The results indicate that the main regional future CE aspects include sustainability and long lifespan of products, components, and materials, holistic system-level thinking, use of renewable energy, completely new products, components, and materials, and replacement of plastics with renewable raw materials. Completely new products, components,

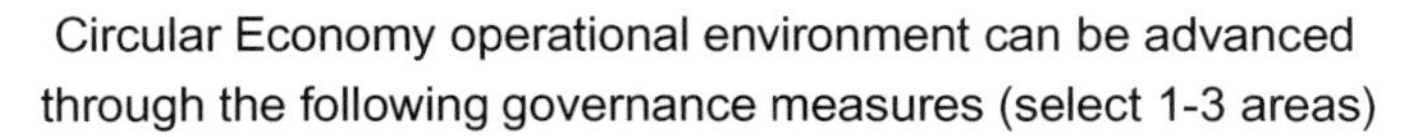

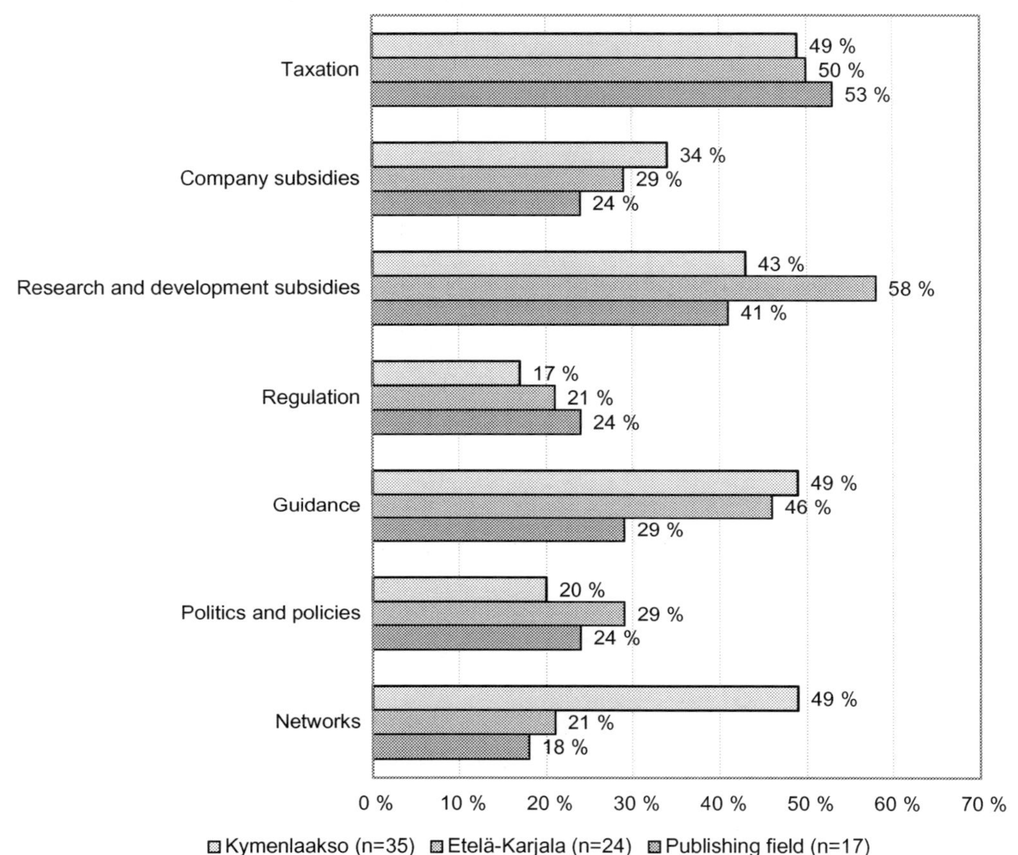

Fig. 11 Circular economy and governance.

and materials, sustainability and long lifespan of products, components, and materials, and holistic system-level thinking are the main future CE aspects in the publishing sector. Lieder and Rashid (2016) noted that the transition toward CE requires involvement of all actors of the society and joint creation of collaboration and exchange activities taking into account appropriate economic return on investments. Ghiselli et al. (2016) drew similar conclusions about the need for the involvement of and collaboration between all actors of society. Mylan et al. (2016) noted that CE is about moving toward a more sustainable society and concluded that the role of domestic consumption and associated practices (e.g., social value) should be better taken into account in this development.

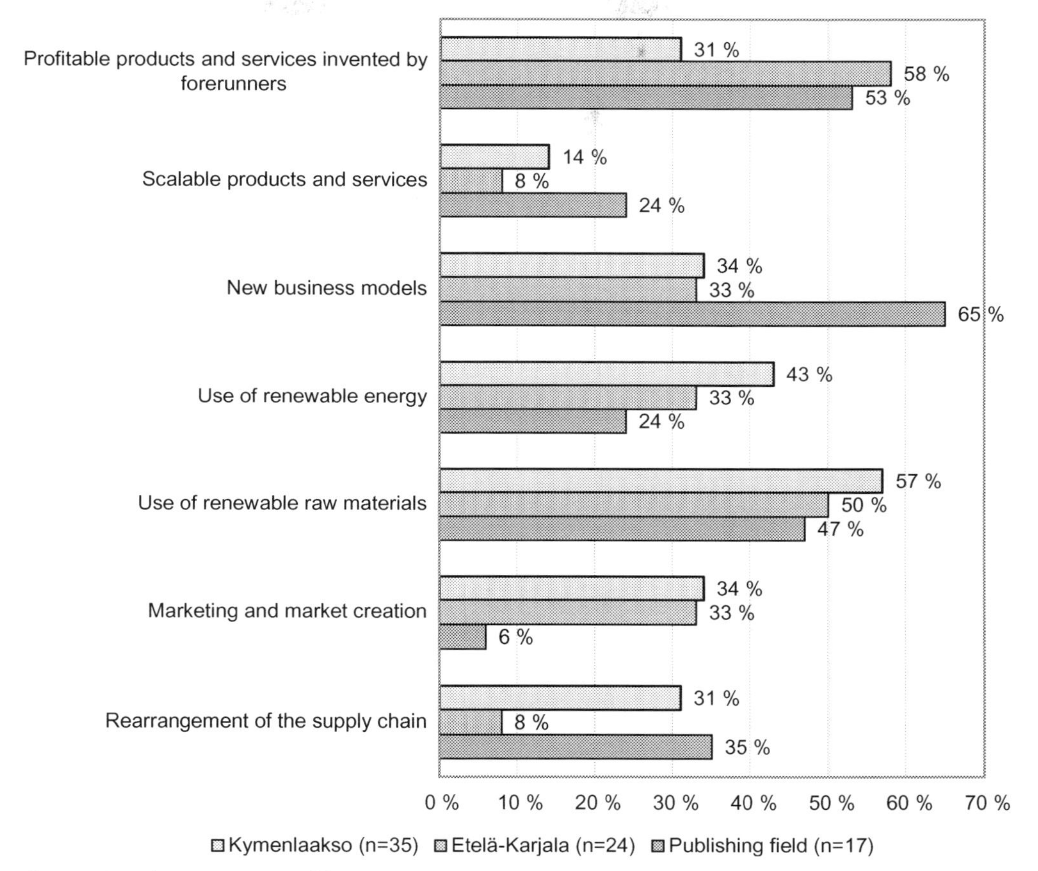

Fig. 12 Circular economy and business.

Baumgartner (2009) recognized that a precondition for a sustainable corporation is that sustainability is embedded in the organizational culture and leadership.

The results indicate that recycling and reuse, renewable energy and raw materials, supply chain management, and staff skills and training are the areas in which CE can improve economic, social, and environmental sustainability regionally, whereas in the publishing sector, those areas are supply chain management, recycling and reuse, and staff skills and training. Witjesa and Lozano (2016) noted that both CE and sustainability goals could be promoted through more service-oriented systems and enhanced collaboration in procurement and supply practices. Moving toward sustainable industrial development with focus on closed loop solutions is an important goal in the context of CE development (Heshmati, 2015). Companies can incorporate social responsibility criteria into

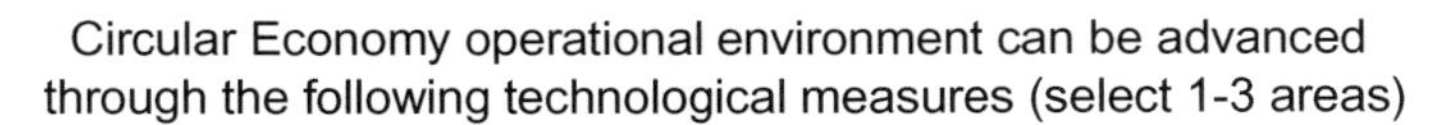

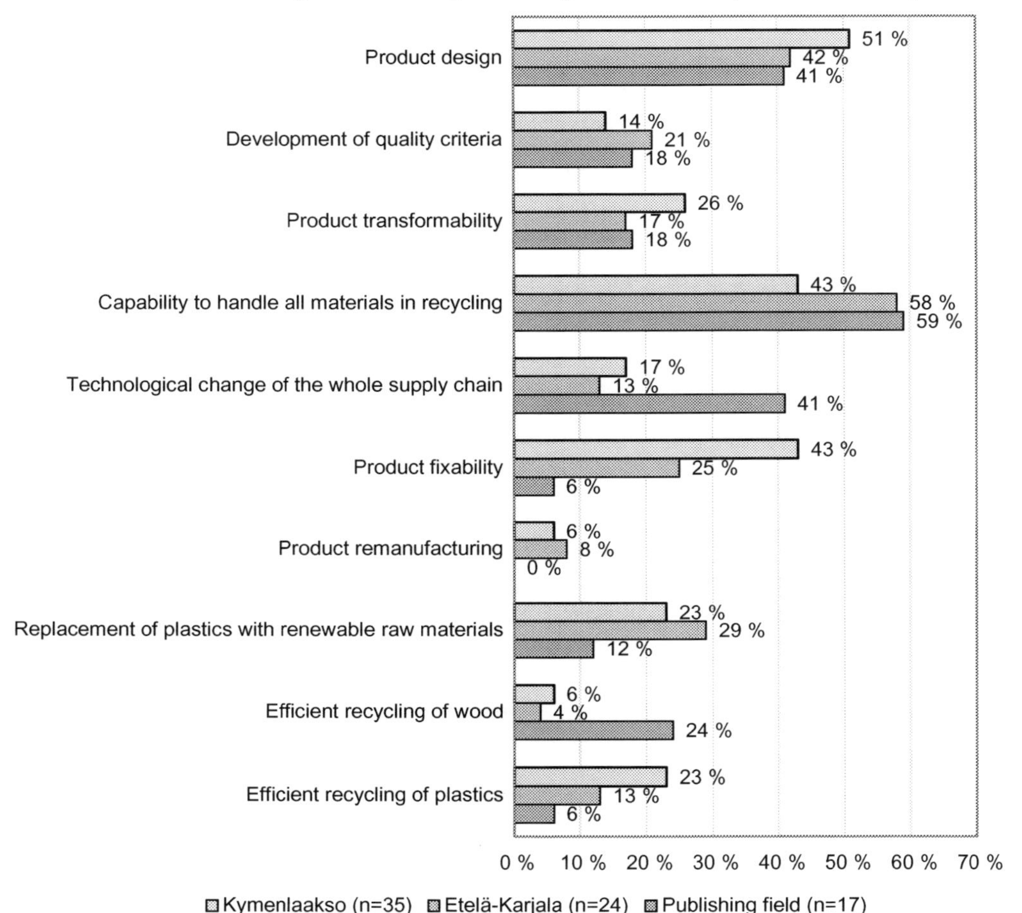

Fig. 13 Circular economy and technology.

their purchasing decisions through, e.g., assessment of stakeholder pressures, purchasing policies and principles based on organizational values, estimation of business benefits and costs, a proactive socially responsible buying strategy, and implementation and marketing of responsible purchasing practices (Maignan et al., 2002). In previous studies, it has been recognized that social and environmental criteria are becoming more important in the context of procurement processes and overall supply management to address sustainable development goals (Preuss, 2009; Srivastava, 2007). CE should create economic, environmental, and social benefits with due focus on local activities and impacts (Guohui and Yunfeng, 2012).

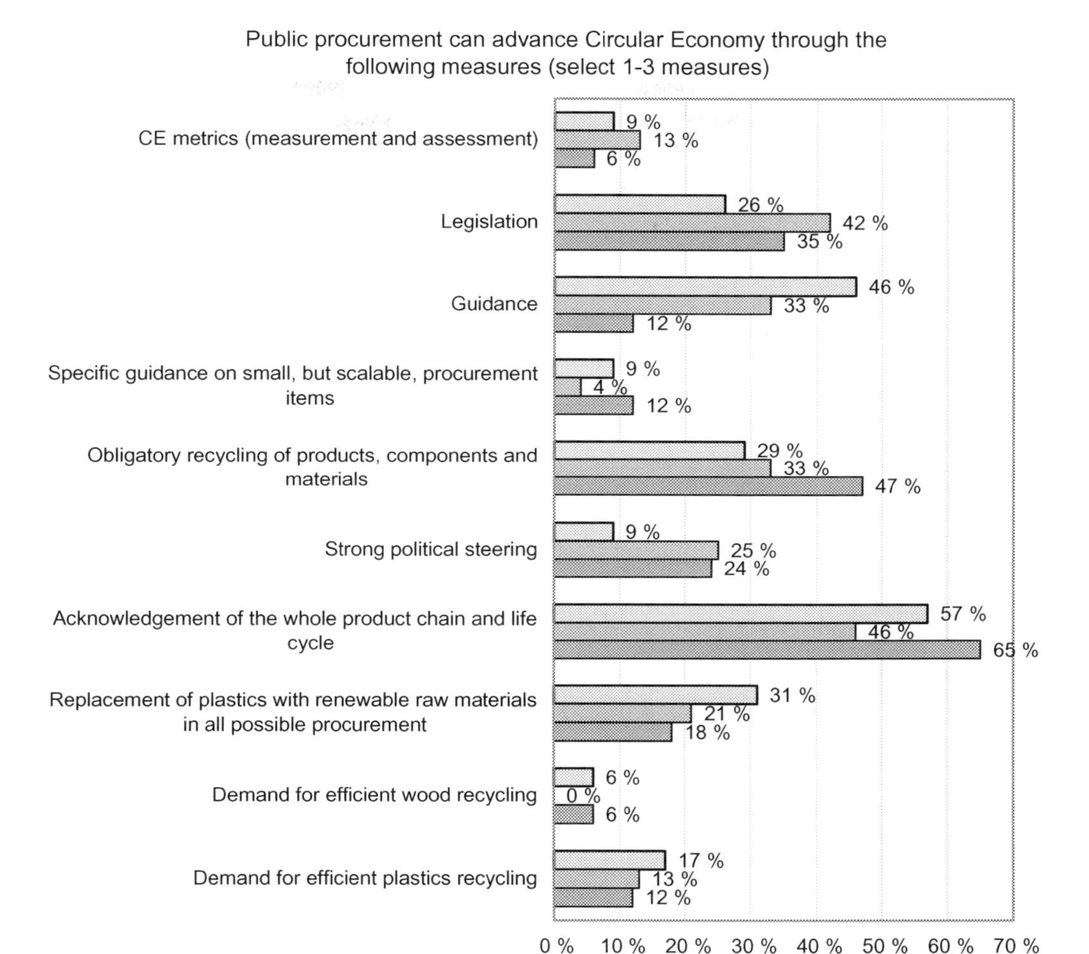

Fig. 14 Circular economy and public procurement.

The results indicate that research and development subsidies, taxation, guidance, and networks are the governance measures through which regional CE operational environment can be advanced, whereas taxation, research and development subsidies, and guidance are those measures in the publishing sector. Liu and Bai (2014) recognized that there was a large gap between the awareness of firms about developing CE and their actual behavior in that field. There is lack of enthusiasm among firms to adopt CE principles. Moreover, knowledge and awareness are not sufficient preconditions and therefore, policy makers should use regulations and incentives to support the building of modern corporate governance systems, which could overcome barriers and engage firms more

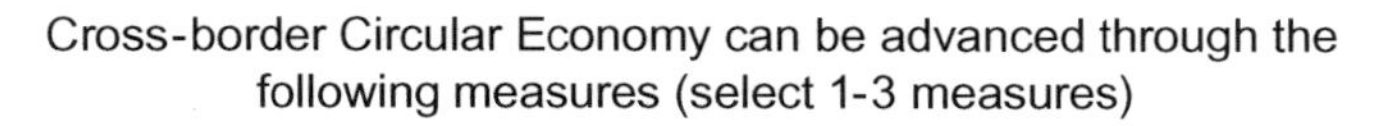

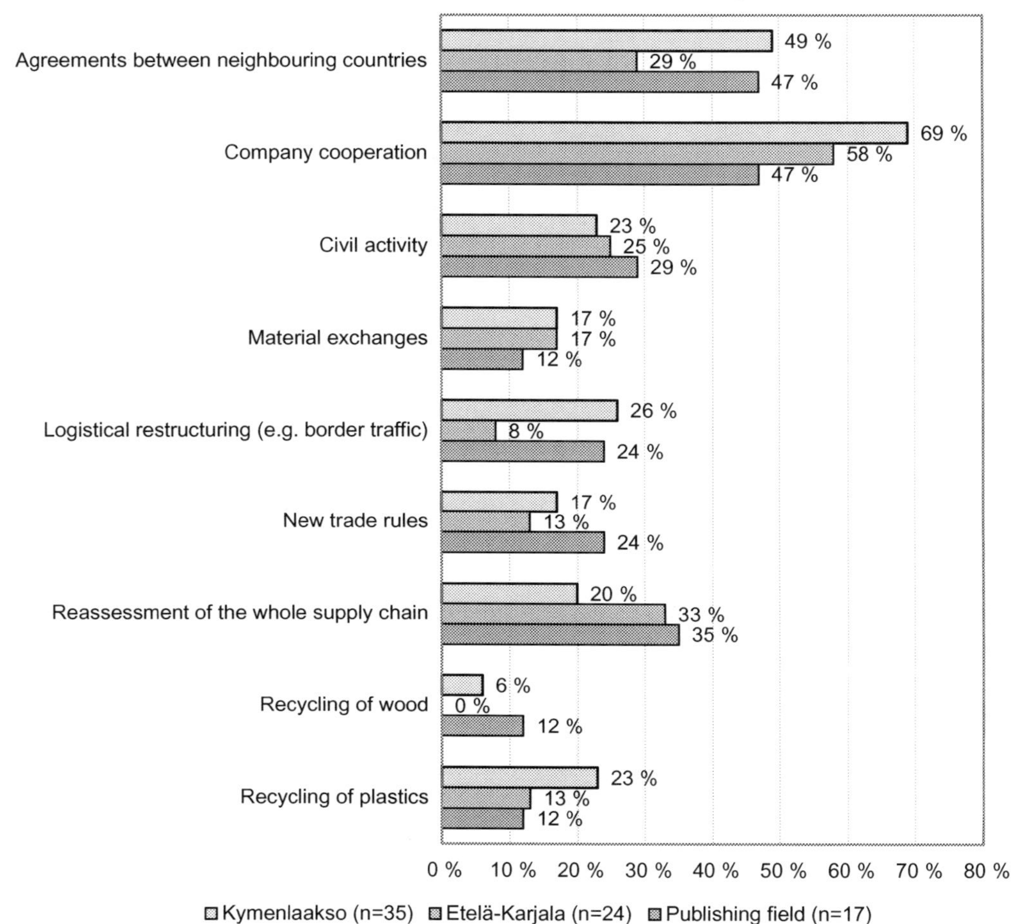

Fig. 15 Cross-border circular economy.

effectively. The behavior of firms was not improved by environmental pressure from market competition and customers. Thus, stricter regulatory market mechanisms and promotion of green consumption could be helpful (Liu and Bai, 2014). Guohui and Yunfeng (2012) recognized that CE needs to play a key role soon in sustainable economic development and environmental protection in West China jointly with the development of a legal system for CE including environmental law enforcement and much stronger environmental protection awareness. Kirchherr et al. (2017) noted that the definition of CE is often related to recycling, reduction, and reuse activities, whereas the systemic perspective is not often mentioned. They added that CE must be understood as a

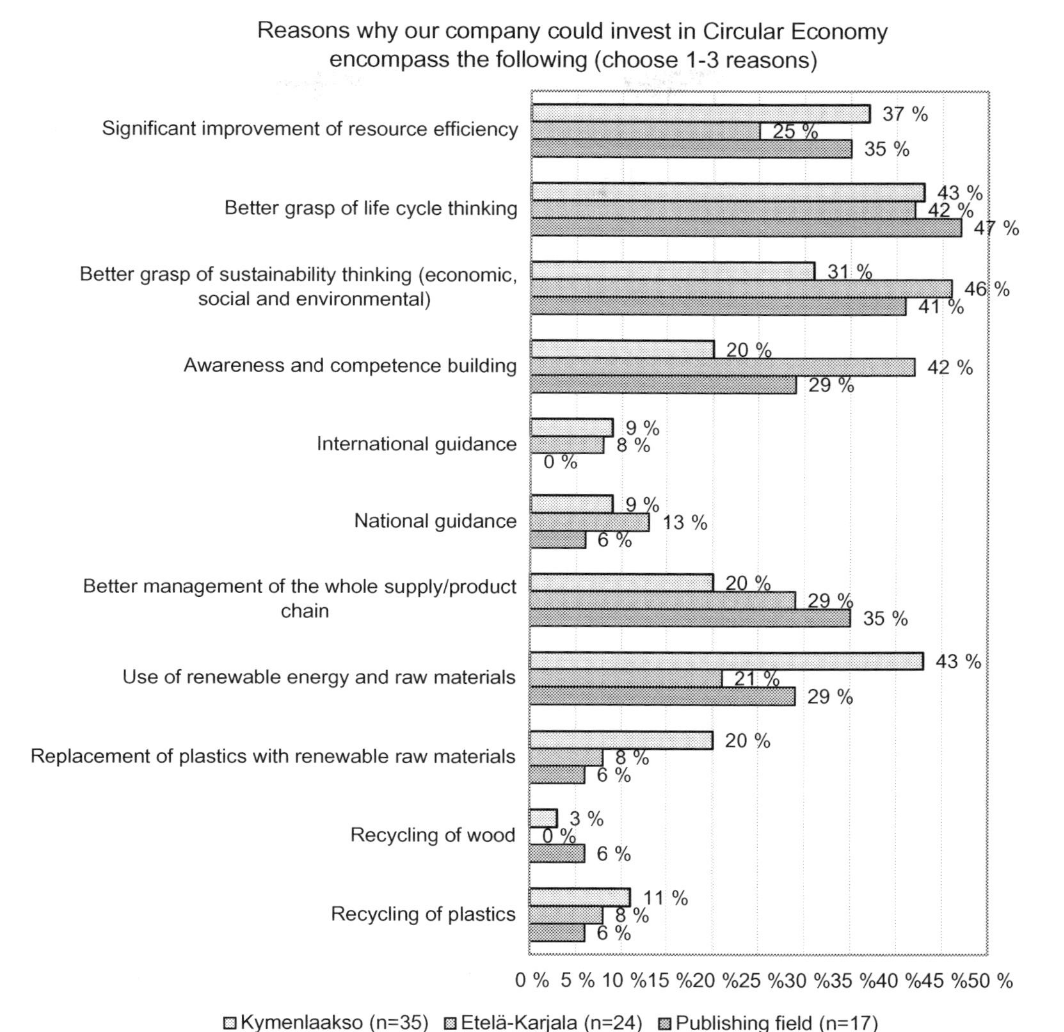

Fig. 16 Circular economy investments.

fundamental systemic change. There are explicit linkages between CE definitions and sustainable development, including economic prosperity as the main CE aim and environmental quality as the second aim. However, the impacts of CE on social equity and future generations are almost never mentioned. Business models and consumers are not often mentioned as enablers of CE. Future research on CE should focus on the neglected dimensions, such as the consumer perspective (Kirchherr et al., 2017).

The results indicate that profitable products and services innovated by forerunners, use of renewable raw materials, use of renewable energy, new business models, and marketing and market creation are the business measures through which regional CE operational environment can be advanced, whereas those measures are new business models, profitable products and services innovated by forerunners, and use of renewable raw materials in the publishing sector. Sauvé et al. (2016) noted that the practical challenges associated with CE encompass provision of economic incentives to close production loops and identification of appropriate ways to promote the manufacturing of more durable goods taking into account life cycle approaches and the internalization of full environmental costs. According to Lakatos et al. (2016), consumers are aware of the importance of CE business models for both the economy and the environment. However, their results indicated that the adoption of new consumption patterns specific to CE business models would require direct or indirect incentives and benefits in addition to awareness-raising and educational campaigns. The organization design of sustainability-driven entrepreneurs is typically guided by perpetual use of resources, qualitative management (decision criteria based on the quality of expected outcomes and processes), strategic identification of satisfactory outcomes of multiple objectives, multiple benefits from each operational activity, and consideration of the worthiness of various stakeholders in benefit sharing (Parrish, 2010).

Gerholdt (2017) noted that circular SMEs can be supported through (1) creation of pilots, case studies, and toolkits, (2) development of voluntary circular metrics to measure and communication progress, (3) access to capital using alternative channels, and (4) appropriate policy frameworks that support innovation and competitiveness. According to Srivastava (2007), environmentally sound choices need to be integrated into supply chain management practices and research. Lozano (2013) noted that there is limited focus on organizational systems and softer issues (e.g., values, visions, policies, and change in management practices) in the efforts to address corporate sustainability using various tools and initiatives. The existing potential for sustainable entrepreneurship should be encouraged and supported (Kuckertz and Wagner, 2010). Lozano (2012) noted that various voluntary tools, initiatives, and approaches have been developed to engage corporations with sustainability and that corporate leaders and employees have increasingly recognized their role in contributing to sustainability. However, there has been limited focus on how these initiatives support embedding sustainability into company systems (e.g., strategy, management, operations, procurement, and processes) and how they contribute to different dimensions of sustainability. The challenge for corporate leaders in addressing sustainability (all dimensions) within company systems is to understand the structure of their companies, the core competences of their companies, and the contexts of their operations (Lozano, 2012).

According to Huhtala (2015), it is important to critically address the market-based drivers of recycling (e.g., prices and costs), the implications of substitution of virgin raw materials by waste materials, and market failures and the underlying reasons for suboptimal behavior patterns by companies and consumers in the context of CE development and planning of associated public steering measures. Corporate governance system could support CE development through better disclosure, stakeholder empowerment, penalties for failure, and stronger boards of directors. (Liu and Bai, 2014). Lozano (2012) concluded that different corporate sustainability initiatives have both advantages regarding their scope and focus for the sustainability dimensions and disadvantages regarding addressing the broadness and complexity of sustainability. Procurement, organizational systems, and marketing are the least addressed parts of company systems in terms of sustainability (Lozano, 2012). Ormazabal et al. (2018) studied the roles of small and medium-sized enterprises in the implementation of CE and they concluded that companies focus on profits and competitiveness and do not typically commit themselves to environmental issues. In addition, companies are worried about their images and prefer savings, e.g., through resource reutilization. Currently, companies do not usually share infrastructure, but there is potential for that practice (Ormazabal et al., 2018).

The results indicate that the capability to handle all materials in recycling, product design, replacement of plastics with renewable raw materials, and product fixability are the technological measures that can advance regional CE operational environment. The capability to handle all materials in recycling, product design, and technological change of the whole supply chain are the technological measures that can advance CE in the publishing sector. Moreno et al. (2016) noted that the promotion of CE requires product design for closed loops taking into account holistic implications and revenue generation aspects. Singh and Ordoñez (2016) studied waste management and product design in the context of CE and they concluded that current product design is facing new social, economic, and environmental challenges. George et al. (2015) noted that the improvement of environmental quality requires an increase in the recycling ratio. For example, the use of ionic liquids to modify wood requires that the liquids are recyclable, nontoxic, chemically stable, and effective (Helsingin Sanomat, 2018). Tähkälä (2017) noted that there is no utilization capacity for recovered impregnated wood in Finland even though the recycling system is efficient and operational. Therefore, more focus is needed on, e.g., the development of a district heating plant (thermal energy), supportive legislative environment, and on environmental aspects of public procurement such as waste management and recycling fees (Tähkälä, 2017). For example, Sakaguchi et al. (2017) noted that the enhancement of the wood cascading potential in the Finnish construction sector requires focus on the development of the primary design, jointing systems, and demolition method, reuse of complete elements (or replacement of damaged parts), and consideration of the recovered length and condition through evaluation, regulation, and grading systems to generate more detailed information. Cascading wood is one of the main approaches to promote resource efficiency in the establishment of a sustainable system (Sakaguchi et al., 2017).

The results of the regional studies indicate that public procurement can advance CE through acknowledgment of the whole product chain and life cycle, guidance, replacement of plastics with renewable raw materials in all possible procurement, legislation, and obligatory recycling of products, components, and materials. In the publishing sector, public procurement can advance CE through the acknowledgment of the whole product chain and life cycle, obligatory recycling of products, components, and materials, and legislation. Pöyhönen (2017) noted that there is a significant lack of CE knowledge and expertise in the public procurement sector and also pointed out the associated need for training and collaboration supported by dissemination of good practices and capacity-building through pilot experiments. Public procurement adds up to a significant amount of consumption and thus it is imperative to promote CE principles and procurement in that sector and to harness all possibilities toward that end (Pöyhönen, 2017). Genovese et al. (2017) noted that the integration of CE principles within supply chain management can provide environmental advantages. CE development involves the creation of self-sustaining production systems with cyclical use of materials and there are benefits from enhanced emphasis on sustainable supply chain management (Genovese et al., 2017).

Public procurement plays a major role in the advancement of CE (EC, 2015). Firms that are involved in public procurement have a more legally oriented approach to corporate social responsibility and less voluntary orientation (Snider et al., 2013). According to Bratt et al. (2013), public procurement has potential to promote corporate sustainability and to steer the decisions of both procurers and producers toward more sustainable direction. Expert bodies that are responsible for the development of procurement criteria play a major role. More focus is needed on clearer definition of sustainability objectives, broader impact perspective, and more clear and solid short- and long-term process and criteria objectives (Bratt et al., 2013). Brammer and Walker (2011) noticed that sustainable public procurement is encouraged internationally and is also expected to contribute to the advancement of sustainability in the private sector. Most public sector organizations are using some sustainability criteria in their procurement, but there is significant regional variation in sustainable procurement practices. Environmental aspects are typically well addressed, but there is variation in how well other aspects, such as purchasing from diverse suppliers, support of human rights, and safe practices in the supply chain are taken into account (Brammer and Walker, 2011).

However, Brammer and Walker (2011) noted that it is important to remember that different countries have different interpretations of the concept and that there is no single right approach. Leadership is a significant factor (senior managers need to be engaged to realize implementation by the purchasing team) and the financial concerns are the main barrier (resistance to paying more to buy sustainably). Public sector organizations are more likely to implement sustainable procurement provided there is a supportive governmental policy and legislation environment. Shared learning about sustainability practices across regions is likely to benefit all (Brammer and Walker, 2011). Good examples of the promotion of CE in public procurement (Pöyhönen, 2017) encompass, for example, material selection that supports wood construction (renewable material), use of renewable energy choices, reuse of natural rock materials, innovative use of empty buildings and spaces (that are in good condition in terms of health and safety), new construction for multiple purposes, creation of an enabling environment for locally produced food including vegetables and fish, and strong focus on early childhood, preschool, and primary education using new learning methods (Pöyhönen, 2017). The establishment of platforms such as Eco Value Interchange (EVI) and mobility sharing will lead to the linkage of industry, enterprise, and users through environmental approaches, and the platforms result in raising the awareness of users to environmental protection and also the CE (Kato, 2019).

The results of the regional studies indicate that cross-border CE can be advanced through company cooperation, agreements between neighboring countries, logistical restructuring (e.g., border traffic), and reassessment of the whole supply chain, whereas in the publishing sector, those measures are company cooperation, agreements between neighboring countries, and reassessment of the whole supply chain. No previous studies could be identified that addressed this issue.

The results of the sectoral studies indicate that better grasp of sustainability thinking (economic, social, and environmental), better grasp of life cycle thinking, use of renewable energy and raw materials, significant improvement of resource efficiency, and awareness and competence building are the elements that could lead to company investments into CE. In the publishing sector, those elements encompass better grasp of life cycle thinking, better grasp of sustainability thinking (economic, social, and environmental), better management of the whole supply/product chain, and significant improvement of resource efficiency. Cohen and Winn (2007) noted that sustainable entrepreneurship could generate both entrepreneurial rents and improvement of local and global social and environmental conditions. Parrish (2010) noted that entrepreneurship is increasingly expected to contribute to sustainable development and defined that the logic behind sustainable entrepreneurship is about using human and natural resources in a way that maintains and enhances the quality of their functioning in the long-run. Various voluntary corporate tools and initiatives to address sustainability have been developed because corporations and their leaders have recognized their responsibility and role with respect to the environment and society (Lozano, 2013).

Future research should focus on, for example, the sustainability aspects of CE (all economic, environmental, and social benefits and impacts), capacity and awareness building, continuous improvement of sustainability, repairability and recyclability of products, components, and materials, design for sustainability and circularity, skills and competence, and whole supply chains. One really interesting future research focus area is CE metrics and its qualitative and quantitative dimensions (similarly to sustainability metrics). Research on metrics is particularly important due to the fact that both informed decision-making and operational management need to be supported by assessment and measurement. Research on CE should also cover public-private interactions, collaboration, and networks. It is important to study both public and private sector drivers, barriers, and dynamics.

Hockerts and Wüstenhagen (2010) noted that research should focus on comparative studies of sustainable entrepreneurial initiatives in both small and large firms including specific challenges in their efforts to create more broad and deep impact. Cohen and Winn (2007) suggested that further research should focus on sustainable entrepreneurship and the relationship between entrepreneurial opportunities and market imperfections. Snider et al. (2013) noted that more research is needed on how public procurement affects corporate social responsibility in companies dealing with state and local governments. They also noticed that studies could inform policy-makers about the influences of their decisions on firms. Preuss (2009) encouraged further research on sustainable supply chain management for the public sector and noted that existing local practices encompassed the roles of small local businesses (economic), voluntary organizations (social), and replacement of hazardous materials in products and services (environmental). Local authorities have also applied sustainability risk assessments and dissemination of sustainability information (Preuss, 2009). Further research is needed on, for example, environmental performance measurement, best practices, and transfer of green technology (Srivastava, 2007). Further research is needed on the ways to develop organizational culture toward sustainable development and on the potential of sustainable development to be a starter for organizational change processes (Baumgartner, 2009). There is little empirical research on the awareness and behavior of firms about developing CE (Liu and Bai, 2014). Kirchherra et al. (2018) suggested that future research should address CE barriers in business models and in various sectors. According to Lozano (2012) further research is needed on embedding sustainability into company systems including use of a combination of initiatives to address all dimensions of sustainability (environmental, economic, social, and time, i.e., intra- and intergenerational equity). The design and management of sustainable business models has received insufficient research focus and there is a need to consider the development of a research agenda to address the creation of sustainable value through business models for sustainable innovation (Boons and Lüdeke-Freund, 2013).

5. Conclusions

The results of the regional studies and the sectoral study suggest, for example, that sustainability and long lifespan of products, components, and materials are important for the future development of CE. Sustainable, long-lasting, and fixable products, and new services and products are among the important drivers and opportunities. Regionally and sectorally, profitability and lack of information are major barriers. CE can improve regional and sectoral economic, social, and environmental sustainability through, for example, supply chain management and recycling and reuse. Overall, the CE operational environment can be advanced through governance measures such as research and development subsidies, taxation, and guidance. The business measures that can advance CE operational environment regionally and sectorally encompass, for example, profitable products and services innovated by forerunners, new business models, and use of renewable raw materials. The capability to handle all materials in recycling and product design are among the technological measures that can advance CE operational environment both regionally and sectorally.

Sectorally and regionally, public procurement can advance CE through, for example, acknowledgment of the whole product chain and life cycle, legislation, and obligatory recycling of products, components, and materials. Cross-border CE can be advanced regionally and sectorally through, for example, company cooperation, agreements between neighboring countries, and reassessment of the whole supply chain. The results of the sectoral studies and the sectoral study suggest that the reasons that could lead to company investments in CE encompass, for example, better grasp of sustainability thinking (economic, social, and environmental), better grasp of life cycle thinking, and significant improvement of resource efficiency. Overall, the promotion of CE requires both private and public sector measures including due focus on sustainability and circular economy management and assessment. For example, product sustainability should be considered early in the design phase, including life cycle thinking. Future research could focus on, for example, company measures to advance CE, including sustainability considerations, life cycle thinking, and renewable raw materials and energy. In addition, interesting focus areas encompass public procurement (for example, whole product chain and life cycle approach), product design and recycling of all materials, new business models, taxation, research and development, CE networks, sustainable products, and measures to overcome barriers such as profitability and lack of information.

Acknowledgments

This study was supported by the Kymenlaakso Regional Fund of the Finnish Cultural Foundation, the Carelian Culture Foundation, and the Waldemar von Frenckell Foundation.

References

Abraham, G. (Ed.), 2006. Sustainability Science and Engineering: Defining Principles. Elsevier, London, UK.

Akiba, T., 2019. Large-scale hydrogen energy system using renewable energy. J. Jpn. Weld. Soc. 88 (1), 50–55. https://doi.org/10.2207/jjws.88.50 (in Japanese).

Allen, D.T., Shonnard, D.R., 2012. Sustainable Engineering. Concepts, Design, and Cases Studies. Prentice Hall, Upper Saddle River, NJ. 223 p.

Allenby, B.R., 2012. The Theory and Practice of Sustainable Engineering. Pearson Education Limited, England, UK.

Baumgartner, R.J., 2009. Organizational culture and leadership: preconditions for the development of a sustainable corporation. Sustain. Dev. 17, 102–113. https://doi.org/10.1002/sd.405.

Boons, F., Lüdeke-Freund, F., 2013. Business models for sustainable innovation: state-of-the-art and steps towards a research agenda. J. Clean. Prod. 45, 9–19. https://doi.org/10.1016/j.jclepro.2012.07.007.
Brammer, S., Walker, H., 2011. Sustainable procurement in the public sector: an international comparative study. Int. J. Oper. Prod. Manag. 31 (4), 452–476. https://doi.org/10.1108/01443571111119551.
Bratt, C., Hallstedt, S., Robèrt, K.H., Broman, G., Oldmark, J., 2013. Assessment of criteria development for public procurement from a strategic sustainability perspective. J. Clean. Prod. 52, 309–316. https://doi.org/10.1016/j.jclepro.2013.02.007.
Cohen, B., Winn, M.I., 2007. Market imperfections, opportunity and sustainable entrepreneurship. J. Bus. Ventur. 22, 29–49. https://doi.org/10.1016/j.jbusvent.2004.12.001.
Dean, T.J., McMullen, J.S., 2007. Toward a theory of sustainable entrepreneurship: reducing environmental degradation through entrepreneurial action. J. Bus. Ventur. 22, 50–76. https://doi.org/10.1016/j.jbusvent.2005.09.003.
EC, 2015. Communication from the Commission to the European Parliament, the Council, the European Economic and Social Committee and the Committee of the Regions. Closing the Loop—An EU Action Plan for the CE (COM(2015)614 Final). http://eur-lex.europa.eu/legal-content/EN/TXT/?uri=CELEX:52015DC0614. (Accessed 15 January 2020).
EC, 2016. CASCADES. Study on the Optimized Cascading Use of Wood. https://publications.europa.eu/en/publication-detail/-/publication/04c3a181-4e3d-11e6-89bd-01aa75ed71a1 (Accessed 27 May 2018) doi:https://doi.org/10.2873/827106.
EC, 2017. Report from the Commission to the European Parliament, the Council, the European Economic and Social Committee and the Committee of the Regions on the Implementation of the CE Action Plan. Brussels (26.1.2017 COM(2017) 33 Final). https://op.europa.eu/en/publication-detail/-/publication/391fd22b-e3ae-11e6-ad7c-01aa75ed71a1/language-en/format-PDF (Accessed 17 January 2020).
Ellen MacArthur Foundation, 2015. Towards a Circular Economy: Business Rational for an Accelerated Transition. https://www.ellenmacarthurfoundation.org/assets/downloads/TCE_Ellen-MacArthur-Foundation_9-Dec-2015.pdf. (Accessed 15 January 2020).
ERCC, 2015. Report of evaluation and review of the enforcement status of automobile recycling system. Gov. Environ. Spec. Rep. 50 (5), 57–92 (in Japanese) https://www.env.go.jp/council/03recycle/y033-43/mat03_2.pdf. (Accessed 22 April 2019).
Fink, A., 2009. How to Conduct Surveys: A Step-by-Step Guide, fourth ed. Sage. 125 p.
Fujii, K., 2017. Report of Current Status of Automobile Recycling, ARC Report (RS-1021). pp. 1–20. https://www.asahi-kasei.co.jp/arc/service/pdf/1021.pdf. (Accessed 21 January 2020 (in Japanese).
Genovese, A., Acquaye, A.A., Figueroa, A., koh, L., 2017. Sustainable supply chain management and the transition towards a CE. Omega 66, 244–257. https://doi.org/10.1016/j.omega.2015.05.015.
George, D.A.R., Lin, B.C., Chen, Y., 2015. A CE model of economic growth. Environ. Model. Softw. 73, 60–63. https://doi.org/10.1016/j.envsoft.2015.06.014.
Gerholdt, J., 2017. Supporting growth of circular SMEs. In: Senior Director, CE and Sustainability Programs, U.S. Chamber of Commerce Foundation. World CE Forum, June 5th, Helsinki, Finland. https://www.slideshare.net/WorldCircularEconomyForum/jennifer-gerholdt. (Accessed 28 January 2020).
Ghiselli, P., Cialani, C., Ulgiati, S., 2016. A review on CE: the expected transition to a balanced interplay of environmental and economic systems. J. Clean. Prod. 114, 11–32.
Gillham, B., 2007. Developing a Questionnaire, second ed. Continuum International Publishing Group. 112 p.
Graedel, T.E., Allenby, B.R., 2010. Industrial Ecology and Sustainable Engineering, International ed. Pearson Education Inc., Upper Saddle River, Prentice Hall.
Guohui, S., Yunfeng, L., 2012. The effect of reinforcing the concept of CE in West China environmental protection and economic development. Procedia Environ. Sci. 12, 785–792. https://doi.org/10.1016/j.proenv.2012.01.349.
Hall, J.K., Daneke, G.A., Lenox, M.J., 2010. Sustainable development and entrepreneurship: past contributions and future directions. J. Bus. Ventur. 25, 439–448. https://doi.org/10.1016/j.jbusvent.2010.01.002.
Hamba, M., Ida, H., 2018. Current status and business trends for plastics recycling in Japan. Mater. Cycles Waste Manag. Res. 29 (2), 99–107 (in Japanese) https://doi.org/10.3985/mcwmr.29.99.

Heshmati, A., 2015. A Review of the CE and its Implementation (Discussion Paper No. 9611). Institute for the Study of Labor.
Hirsjärvi, H., Remes, P., Sajavaara, P., 2007. Tutki Ja Kirjoita. 13. Painos. Tammi, Helsinki. 448 p.
Hirsjärvi, S., Remes, P., Sajavaara, P., 2015. Tutki ja Kirjoita. 20. Painos. Tammi, Helsinki.
Hockerts, K., Wüstenhagen, R., 2010. Greening Goliaths versus emerging Davids—theorizing about the role of incumbents and new entrants in sustainable entrepreneurship. J. Bus. Ventur. 25, 481–492. https://doi.org/10.1016/j.jbusvent.2009.07.005.
HS, 2017. Kiertotalous kaipaa käytännön toimia. Pääkirjoitus (18.4.2017). Helsingin Sanomat.
HS, 2018. Puu on uusi muovi. B6-7 (3.5.2018). Helsingin Sanomat.
HS, 2019. Hyvästi muovipillit: EU-parlamentti hyväksyi kertakäyttömuoveja kieltävän direktiivin. Helsingin Sanomat. https://www.hs.fi/kotimaa/art-2000006050264.html?share=a0c6ea200f49513f771f842207f0f0b9.
HS Kuukausiliite, 2018. 5/2018. Ikuinen ongelma. 21–31 p.
Huhtala, A., 2015. Kansantalous on pian “kiertotalous”. Kansantaloudellinen aikakauskirja 4/2015.
Husgafvel, R., Poikela, K., Honkatukia, J., Dahl, O., 2017. Development and piloting of sustainability assessment metrics for arctic process industry in Finland—the biorefinery investment and slag processing service cases. Sustainability 9 (10). https://doi.org/10.3390/su9101693, 1693.
Husgafvel, R., Linkosalmi, L., Hughes, M., Kanerva, J., Dahl, O., 2018a. Forest sector circular economy development in Finland: a regional study on sustainability driven competitive advantage and an assessment of the potential for cascading recovered solid wood. J. Clean. Prod. 181, 483–497. https://doi.org/10.1016/j.jclepro.2017.12.176.
Husgafvel, R., Linkosalmi, L., Dahl, O., 2018b. Company perspectives on the development of the circular economy in the seafaring sector and the Kainuu region in Finland. J. Clean. Prod. 186, 673–681. https://doi.org/10.1016/j.jclepro.2018.03.138.
Kato, H., 2019. Contribution of eco policy CE and decarbonated society. Environ. Meet. 51, 50–55 (in Japanese).
Khosla, A., 2017. Meeting the SDGs needs a circular and inclusive economy. SMEs are the key. In: World CE Forum, June 5th, Helsinki, Finland. https://www.slideshare.net/worldcirculareconomyforum/ashok-khosla-world-circular-economy-forum-2017-helsinki-finland. (Accessed 28 January 2020).
Kirchherr, J., Reike, D., Hekkert, M., 2017. Conceptualizing the CE: an analysis of 114 definitions. Resour. Conserv. Recycl. 127, 221–232. https://doi.org/10.1016/j.resconrec.2017.09.005.
Kirchherra, J., Piscicellia, L., Boura, R., Kostense-Smitb, E., Mullerb, J., Huibrechtse-Truijensb, A., Hekkerta, M., 2018. Barriers to the CE: evidence from the European Union (EU). Ecol. Econ. 150, 264–272.
Kuckertz, A., Wagner, M., 2010. The influence of sustainability orientation on entrepreneurial intentions—investigating the role of business experience. J. Bus. Ventur. 25, 524–539. https://doi.org/10.1016/j.jbusvent.2009.09.001.
Lakatos, E.S., Dan, V., Cioca, L.I., Bacali, L., Ciobanu, A.M., 2016. How supportive are Romanian consumers of the CE concept: a survey. Sustainability 8, 789. https://doi.org/10.3390/su8080789.
Lieder, M., Rashid, A., 2016. Towards CE implementation: a comprehensive review in context of manufacturing industry. J. Clean. Prod. 115, 36–51.
Liu, Y., Bai, Y., 2014. An exploration of firms' awareness and behavior of developing CE: an empirical research in China. Resour. Conserv. Recycl. 87, 145–152. https://doi.org/10.1016/j.resconrec.2014.04.002.
Loimu, 2018. Maapallo ei saa hukkua muovijätteeseen! 3/2018. Luonnon-, ympäristä ja metsätieteilijöiden liitto Loimu ry.
Lozano, R., 2012. Towards better embedding sustainability into companies' systems: an analysis of voluntary corporate initiatives. J. Clean. Prod. 25, 14–26. https://doi.org/10.1016/j.jclepro.2011.11.060.
Lozano, R., 2013. Are companies planning their organisational changes for corporate sustainability? An analysis of three case studies on resistance to change and their strategies to overcome it. Corp. Soc. Responsib. Environ. Manag. 20, 275–295. https://doi.org/10.1002/csr.1290.
Maignan, I., Hillebrand, B., McAlister, D., 2002. Managing socially-responsible buying: how to integrate non-economic criteria into the purchasing process. Eur. Manag. J. 20 (6), 548–641. https://doi.org/10.1016/S0263-2373(02)00115-9.

Moreno, M., De los Rios, C., Rowe, Z., Charnley, F., 2016. A conceptual framework for circular design. Sustainability 8, 937. https://doi.org/10.3390/su8090937.
Mylan, J., Holmes, H., Paddock, J., 2016. Re-introducing consumption to the 'CE': a sociotechnical analysis of domestic food provisioning. Sustainability 8, 794. https://doi.org/10.3390/su8080794.
Noronha, L., 2017. Momentum for change—perspectives from the UN Environment. In: Director, UN Environment Program. World CE Forum, June 5th, Helsinki, Finland. https://media.sitra.fi/2017/02/08113108/momentum_of_change_UNEP_LN.pdf. (Accessed 19 January 2020).
Onoda, H., 2019. Current status and issues of efforts to improve the quality of automobile recycling system. Enermix 98 (1), 78–82 (in Japanese) 10.20550/jieenermix.98.2_9802anno_1.
Oppenheim, A.N., 1997. Questionnaire Design, Interviewing and Attitude Measurement. Pinter, London and Washington. 303 p.
Ormazabal, M., Prieto-Sandoval, V., Puga-Leal, R., Jaca, C., 2018. Circular economy in Spanish SMEs: challenges and opportunities. J. Clean. Prod. 185, 157–167. https://doi.org/10.1016/j.jclepro.2018.03.031.
Parrish, B.D., 2010. Sustainability-driven entrepreneurship: principles of organization design. J. Bus. Ventur. 25, 510–523. https://doi.org/10.1016/j.jbusvent.2009.05.005.
Patten, M.L., 2011. Questionnaire Research: A Practical Guide. Pyrczak Pub, Glendale CA.
Peterson, R.A., 2000. Constructing Effective Questionnaires., https://doi.org/10.4135/9781483349022.
Potočnik, J., 2017. Global use of natural resources—in crisis or not? In: Co-chair, UNEP International Resources Panel. World CE Forum, June 5th, Helsinki, Finland. https://www.slideshare.net/WorldCircularEconomyForum/janez-potonik-world-circular-economy-forum-2017-helsinki-finland. (Accessed 18 January 2020).
Pöyhönen, T., 2017. Kiertotaloutta edistävät julkiset hankinnat—Case: Kouvolan kaupunki. Ammattikorkeakoulun opinnäytetyö. Kestävä kehitys. Hämeen ammattikorkeakoulu (HAMK). 88 p.
Preuss, L., 2009. Addressing sustainable development through public procurement: the case of local government. Supply Chain Manag. 14 (3), 213–223. https://doi.org/10.1108/13598540910954557.
Rosen, M.A., 2012. Engineering sustainability: a technical approach to sustainability. Sustainability 4 (9), 2270–2292.
Saito, Kumagai, Y., Kameda, T., Yoshioka, T., 2018. Current issues and future prospects in plastic recycling. J. Mater. Cycles Waste Manag. Res. 29 (2), 152–162. https://doi.org/10.3985/mcwmr.29.152 (in Japanese).
Sakaguchi, D., Takano, A., Hughes, M., 2017. The potential for cascading wood from demolished buildings: potential flows and possible applications through a case study in Finland. Int. Wood Prod. J. 8 (4). https://doi.org/10.1080/20426445.2017.1389835.
Saris, W.E., Gallhofer, E.N., 2014. Design, Evaluation and Analysis of Questionnaires for Survey Research, second ed. John Wiley & Sons.
Sauvé, S., Bernard, S., Sloan, P., 2016. Environmental sciences, sustainable development and CE: alternative concepts for trans-disciplinary research. Environ. Dev. 17, 48–56. https://doi.org/10.1016/j.envdev.2015.09.002.
Shimizu, K., Hondo, H., Moriizumi, Y., 2018. CO_2 reduction potential of solar water heating systems on a municipal basis. J. Jpn. Inst. Energy 97 (6), 147–159 (in Japanese) https://doi.org/10.3775/jie.97.147.
Singh, J., Ordoñez, I., 2016. Resource recovery from post-consumer waste: important lessons for the upcoming CE. J. Clean. Prod. 134 (Part A), 342–353. https://doi.org/10.1016/j.jclepro.2015.12.020.
Sitra, 2014. Kiertotalouden mahdollisuudet Suomelle (CE Possibilities in Finland). Sitran selvityksiä n:o 84. 72 p https://www.sitra.fi/julkaisut/Selvityksi%C3%A4-sarja/Selvityksia84.pdf. (Accessed 12 January 2020).
Sitra, 2016. Leading the Cycle. Finnish Road Map to a CE 2016–2025. Sitra Studies 117. 56 p. http://www.sitra.fi/sites/default/files/sitra_leading_the_cycle_report.pdf. (Accessed 15 January 2020).
Snider, K.F., Halpern, B.H., Rendona, R.G., Kidalova, M.V., 2013. Corporate social responsibility and public procurement: how supplying government affects managerial orientations. J. Purch. Supply Manag. 19 (2), 63–72. https://doi.org/10.1016/j.pursup.2013.01.001.
Srivastava, S.K., 2007. Green supply-chain management: a state-of-the-art literature review. Int. J. Manag. Rev. 9 (1), 53–80. https://doi.org/10.1111/j.1468-2370.2007.00202.x.

Steiner, A., 2017. CE as means to achieve the UN sustainable development goals and reduce poverty. In: Director, Oxford Martin School. World CE Forum, June 5th, Helsinki, Finland. https://www.slideshare.net/WorldCircularEconomyForum/achim-steiner. (Accessed 12 January 2020).
Sudman, S., Bradbrun, N.M., 1982. Asking Questions: A Practical Guide to Questionnaire Design. The Jossey-Bass Series in Social and Behavioural Sciences, Jossey-Bass Publishers. 397 p.
Tähkälä, T., 2017. Kestopuu kiertotalouden ytimessä. Pääkirjoitus. 3/2017. PUUMIES. Puumiesten Liitto Ry.
UN, 2019. Sustainable Development Goals. The United Nations. https://sustainabledevelopment.un.org/?menu=1300. (Accessed 22 January 2020).
Van Buren, N., Demmers, M., Van der Heijden, R., Witlox, F., 2016. Towards a CE: the role of Dutch logistics industries and governments. Sustainability 8, 647. https://doi.org/10.3390/su8070647.
Witjesa, S., Lozano, R., 2016. Towards a more CE: proposing a framework linking sustainable public procurement and sustainable business models. Resour. Conserv. Recycl. 112, 37–44. https://doi.org/10.1016/j.resconrec.2016.04.015.
YLE, 2019. EU pidentää sähkölaitteiden käyttöikää - korjauskelvottomia kodinkoneita ei enää päästetä myyntiin. https://yle.fi/uutiset/3-10592132.

CHAPTER 34

Approaches to the circular economy in Armenia and Portugal: An overview

K.S. Winans[a], Irina Mkrtchyan[b], and João Pedro Moreira Gonçalves[c]
[a]UCD Industrial Ecology Program, Davis, CA, United States
[b]"ISSD - Innovative Solutions for Sustainable Development of Communities" NGO, Yerevan, Armenia
[c]Finca Equilibrium, Montijo, Portugal

1. Introduction

The emergence of the circular economy concept challenges the way we think about the linear "take-waste-dispose" production and economic models. The take-waste-dispose mode of using resources and materials contributes to economic uncertainty, excessive waste generation, environmental degradation, and various other unintended consequences (e.g., human health issues). It seems inevitable, at least in part, to move away from these linear models and paradigms, and to create more regenerative systems.

The circular economy is based on several schools of thought, like biomimicry (Merrill, 1982; Benyus, 1997) and cradle to cradle (McDonough and Braungart, 2002) that are inspired by observations of nature's processes. This chapter explores some examples of how the circular economy concept is implemented in Armenia and Portugal.

The organization of the chapter is as follows. In the first section, we briefly describe the methodology for developing this work and the regional context of Portugal and Armenia. Then, we describe the article's central themes (e.g., emerging circular economy policies) and a summary of the study findings for Armenia (in Section 5) and Portugal (in Section 6).

2. Methodology

Our first step in this study was to review current peer-reviewed and grey literature in Armenia and Portugal. We reviewed the literature to help define each country's regional context (e.g., historical, geographical, and political contexts), and current circularity practices using keywords such as circular economy and policy, recycling, reuse, remanufacture, and performance. Then, we summarized the findings into a set of thematic areas and interviewed select individuals through semistructured interviews for purposes of evaluating the literature review findings. Interviewees were selected based on criteria such as their area(s) of expertise (e.g., academic or vocational training) and

Circular Economy and Sustainability, Volume 1
https://doi.org/10.1016/B978-0-12-819817-9.00003-X

current occupation. The outcomes of the semistructured interviews were used to refine the observations described in Sections 5 and 6.

3. Regional context for Armenia and Portugal

When considering emerging circular economy initiatives, it is essential to consider location (geographic and political), financial constraints, supply chains, and the main products produced in each country (Winans et al., 2017). Each of these factors plays a role in the potential advancement of circularity within a regional, national, or international context (Winans et al., 2017). During the literature review and interviews, we used the regional context to help contextualize potential opportunities for cross-sectoral collaborations and cross-supply chain management. Meaning, we discussed some options for industries to exchange products and materials, and information. Some supply chains and industries (e.g., textiles industries) are more advanced in circular economy practices. Other sectors (e.g., animal husbandry industries) can learn from them through strategic information exchange (e.g., enabling mechanisms, circular economy business models, etc.).

3.1 Armenia

Armenia is geographically situated within the South Caucasus and the Eastern European region and has political ties to the European Union (EU). Armenia is at the nascent phase of developing a national circular economy effort. It has acquired external financial support for specific circularity initiatives from the EU and Sweden (in 2018/19) (LIFE, 2020).

Broadly speaking, the main industries/products in Armenia include mineral products (copper ore), foodstuffs (e.g., rolled tobacco, liquor), metals (e.g., aluminum foil, ferroalloys), precious metals (e.g., gold, diamonds), and textiles (nonknit clothes) (OEC, 2017; UN COMTRADE, 2018).

3.2 Portugal

Among European countries, Portugal is a leader in circularity initiatives. It has received financial support from the EU for various circular economy initiatives (EU, 2020).

Industries/products in Portugal mainly include machines (e.g., insulated wire) and other materials (e.g., knit t-shirts), metals (e.g., iron), and foodstuffs (e.g., wine, rolled tobacco), as well as transportation, plastics, and rubber (OEC, 2017; UN COMTRADE, 2018). Both countries are partially dependent on an agricultural economy (40% of Portugal's economy is based on agricultural production and 56% in Armenia) (FAO, 2019).

4. Thematic areas of research

We investigate two themes that are often intertwined in each of the study countries: socioeconomic relationships and emerging circular economy policies. In the following sections (Sections 4.1 and 4.2), we provide a brief description of the broader context and rationale for focusing on these thematic areas. Also, we examine some of the community-government and community-based relationships that developed in recent years in each country.

4.1 Socioeconomic relationships

Our analysis of socioeconomic relationships observed in this work is informed by previous work by Giddens (in Binder, 2007; Binder et al., 2013; Muhar et al., 2017). We observe interactions between factors such as agents, actions, material flows, stakeholder groups, and policy to develop conceptual models of the linkages between these factors, and opportunties for exchange of materials, products and information between industries. Although not explicitly addressed, through this work we begin to track consumer preferences and competition (individual, business, etc.) associated with circular economy practices in the study regions (e.g., to examine competitive advantage strategies and potential for cooperative competition and collaboration).

4.2 Emerging circular economy policies

Worldwide, there are currently many circular economy policies. Some of the first circular economy policies developed in China (in the 1990s) and in the EU member states (in 2015) (EU, 2019). Some of the main focuses of these efforts are using secondary materials through closing loops and materials cycling (upcycling and downcycling) and reducing single-use materials (e.g., plastics, wastes, food/beverage wastes, and other packaging materials) (e.g., WBCSD, 2019). Notably, today, many countries are making significant strides in developing strategic approaches to enhancing a circular economy within their respective countries (e.g., see the Finnish Roadmap to a Circular Economy (2016–2025) and the Netherlands Green Deals). In the last 5 years, many regional policies focusing on cross-sectoral collaborations and initiatives at multiple scales (e.g., initiatives that span the national, urban, and localized levels) have emerged (WBCSD, 2019).

5. An overview of circular economy efforts in Armenia

In Armenia, small businesses and community members are among the early adopters of the circular economy. It is common for small companies in Armenia to incorporate circularity practices like reusing and repurposing materials. They do so out of necessity to maintain their businesses' economic viability. These businesses and circularity initiatives in Armenia, in general, are often linked through community-government relationships.

5.1 Brief historical context

In Armenia in the 1920s, privately owned land was placed under state rule in the Armenian Soviet Socialist Republic. This period was near the beginning of the industrialization of Armenia. In 1991, Armenia declared independence from the Soviet Union. Around that time, the Soviet Union collapsed, and Armenia moved away from a planned economy and toward a privatized one.

One of the next major shifts in the Armenian political economy occurred in 2015, when Armenia entered into a Treaty on the Eurasian Economic Union with the Russian Federation and other neighboring nations. They then entered into an Enhanced Partnership Agreement with the European Union in 2017. Today, Russia remains one ofArmenia's main benefactors and the primary consumer of materials such as mined materials (e.g., copper ore) and dairy products. Armenia's ties to Europe also influence production (and outcomes produced) within the country and financial initiatives and programs. That said, in many ways, Armenians remain in an age of postcommunist abundance.

5.2 Circular economy-related recent and emerging policies

In 2018, Armenia's new government started developing action plans and modifications in regulations to make Armenia more sustainable. The sustainability efforts focus on Armenia's main challenges (e.g., emigration, ongoing border issues, and growing the economy). Among Armenia's most recent efforts to achieve their sustainability goals are circularity strategic research and related action plans.

Recently, the Prime Minister of Armenia created a working group to address materials management issues in Armenia. An advisor to the Prime Minister is currently supervising this working group. One of the action items of this group is a partnership—between the Armenian Government, AUA Acopian Center for the Environment, the Swedish LIFE Foundation, and Miljö & Avfallsbyrån AB—to implement a project to update and further reform Armenia's national solid waste policy based on the principles of the circular economy (e.g., considering value loops and redesign of materials and products cycling) (AUA, 2019). The working group's main aim and the research conducted by AUA are to study the amount and composition of municipal solid waste in Armenia's various cities. The report presents Armenia's legislative and institutional gaps and information about the current waste management system, public awareness, and other waste management (AUA, 2020). The types of waste examined included construction, industrial, electrical and electronic, agricultural and horticultural, medical, and automotive waste (AUA, 2020). In concept, the research will help determine the energy use value of the waste and various other factors that show the importance of adjusting production models and material cycling in Armenia. Unfortunately, there are currently significant gaps in the report data.

5.3 Community-government interactions

The Prime Minister's Office, Ministry of Environment, and municipal governments are starting to show interest in supporting community-based efforts. An example of a successful event organized by the community and supported by the government occurred in late 2019, referred to as "Circle Up Yerevan." The event showcased the early adopters of circular economy practices in Armenia. Six ongoing small business projects were presented during the event, from aquaponics to wine producers. During the event, circular economy experts provided hands-on workshops for a wide range of community members (e.g., a design workshop focused on designing out waste in production systems).

Other notable community-based initiatives in Armenia are led by organizations like the "ISSD" Innovative Solutions for Sustainable Development of Communities NGO (or ISSD) that cooperate with the United Nations Development Programme—Global Environmental Finance (UNDP-GEF) and other institutions. These groups are working on developing new action plans for promoting the circular economy throughout the country. For example, the ISSD collaborates with regional and state governments to conduct research and propose legislative changes that support systematically reducing, reusing, repurposing, and recycling materials, following the circular economy principles.

In 2021, ISSD became the first organization in the region to recycle plastic polyethylene terephthalate (PET) waste into synthetic winterizer (filler) and yarn, which is later used in the textile industry. The team has also developed a business to business (B2B) program to facilitate partnerships to reduce and sort waste and extend the use phase of different products (through materials and products exchange between companies). Currently, over 400 organizations have joined the efforts and are transforming the solid waste management industry in the country. In late 2020, following the success case developed and tested by ISSD, several municipalities (including the municipality of Yerevan) started installing waste sorting bins in public areas to make waste sorting accessible for the general public. In addition, the ISSD is currently organizing educational programs that help educate community, private, and public organization members and others about circularity practices (e.g., opportunities to enhance circularity in the home or workplace).

6. An overview of circular economy efforts in Portugal

Portugal's circular economy program's overall objective is to achieve "balanced and environmentally sustainable economic development" (e.g., through decarbonizing the economy and reducing the need for energy from third countries) (Eionet, 2019, p. 6). Portugal's circular action plan is advanced compared to other EU member states' circular economy action plans. Portugal's efforts include a wide range of outcomes (e.g., Eco. nomia), which is a reasonably well-developed knowledge-sharing platform that provides an effective means for information exchange between a wide range of actors and various circularity topics from across Portugal (e.g., from small community-organized events to large government-based initiatives). In Portugal, there are also many community-based initiatives, one of which we highlight in Section 6.3.

6.1 Brief historical context

From 1932 until 1974, Portugal had a corporative dictatorship (Pinto, 2015). In 1974, the country transitioned from this dictatorship to a democratic republic. The Portuguese Constitution was first adopted in 1976, and revised several times including in 1989, with significant revisions enabling privatization and a free-market system (CP, 2005). Following Portugal's integration into the European Economic Community (now the European Union) in 1986, the EU policies shaped Portugal and other EU member states' policies.

6.2 Circular economy-related recent and emerging policies

Within the context of the Action Plan for Circular Economy in Portugal 2017–2020, there are three different levels: macro, meso, and micro (COM, 2017, 2019):

- Macrolevel actions aim to "exchange of information, materials, and energy within society."
- Meso (or sectoral) actions or initiatives engage actors in the "value chain of sectors relevant to raising productivity and the efficient use of the country's resources, seizing the economic, social, and environmental benefits."
- Micro (regional/local) actions or initiatives are at the "provincial or local government, economic, and [engage] social actors that incorporate a historical financial aspect and emphasize this as the approach to social challenges."

Here, we highlight some of the language used in the microlevel policy action framework to provide context for discussing a unique circularity initiative in Portugal that we describe in Section 6.3.

Per the microlevel policy, actions may include an individual business or a farm in a low-density area: "Development of low-density areas [can act] as drivers of economic competitiveness, guaranteeing territorial balance and exploitation of the unique resources of each territory (Eionet, 2019, p. 6)." To date, top-down financial support has been required to support many of the microscale actions. Funders include the European Commission LIFE program.

6.3 Community-based circular economy initiatives

The example presented in the following is a unique microlevel agrarian circularity system in Portugal in the Setubal District. It is a small-scale permacircular initiative referred to as Finca Equilibrium.

The landscape is characterized by dry rangeland with native grasses like *Stipa gigantea*. The land was transformed by the people and agriculture intensification. Industries introduced grape plantations for wine production, with many people as a workforce.

Soils are sandy and low in organic matter. Some areas have heavy clays. Later, with the introduction of large-scale irrigation/electricity projects, reliance on diesel pumps and concentrated animal feeding operations increased. Subsequently, animal and crop industrial-scale operations increasingly transformed the region into an area heavily dependent on fuel for industrial farming and electricity for pumping water.

Within this region is a small, six-hectare area referred to as Finca Equilibrium. At this location, within the last century, there was a transition from range grassland managed with cattle to a grape plantation (in 1940). All the grape vines were uprooted 20 years ago. Most recently , starting in 2017, the land managers transitioned to ecological intensification with permaculture and agroforest systems.

Today, the land managers use a low-input forage farming system, with animals like cattle and chickens doing free range foraging and harvesting. The land managers also implemented agricultural practices that will help decarbonize the economy and use less resources. In general, the plan is to plant and manage several sets of tree systems that will, among other functions, increase the living biomass in the soil, trap and condense air moisture, and harvest sunlight and wind energy all year round. For example, the managers introduced leguminous trees to collect nitrogen and to feed cattle. They are also building sets of forested swale earthworks for storm water that flows through flash flood runoff. In the concave surfaces of these trenches, living biomass in the soil will continue to increase as natural mulches fall overtime. The living biomass can be harvested for fertilization and increase carbon levels of annual crop gardens soils. Also, a form of concentrated carbon (similar to biochar) is produced by burning surplus wood in pits or ovens and fireplaces while cooking and boiling water. Consumables like drip tape are used and then repurposed for secondary and tertiary applications (e.g., as tying material). The land managers also import second-hand materials and products for reuse in the land dwellers living quarters (e.g., bathtubs, glass, etc.).

At the moment, the agrarian system is not visible in the region. It is at a very early stage of development. The families that work to develop this system aim to provide enough food primarily for their sustenance. Governance is by way of family cooperation. Soon, Finca Equilibrium social system will extend to community members through knowledge sharing and education activities, also in coordination with the local parish council (or parish-level government, since the commune is at the municipality level). In the future, it will likely resonate with local/regional past traditional practices of small family-scale intensive gardening.

7. A brief discussion: Observations from Portugal's unique permacircular system

The permacircular initiative referred to as Finca Equilibrium conceptually fits Portugal's action plan as a microlevel action or initiative. Finca Equilibrium employs the concept of regeneration. Here, we consider the definition of regenerative circular economy activities proposed by Morseletto (2020): "promotion of self-renewal capacity of the natural system to reactivate ecological processes damaged or over-exploited by human action." Studying the unique characteristics of regenerative systems (like the permacircular system at Finca Equilibrium) is critically essential for rebuilding ecosystems' capacity to regenerate. Scholars and practitioners need to work together to define frameworks (e.g., guidelines and procedures that enable and popularize regenerative circular economy activities in the ecosphere). These systems are likely to develop jointly with stewards of the land (i.e., the land managers and practitioners). They are unique and seemingly less present in current circular economy discussions.

Perhaps more common are systems that employ principles of restoration ecology. Of course, restorative circular economy activities are essential for accelerating the recovery of damaged ecosystems (e.g., Hobbs, 2018 in Morseletto, 2020). Still, honoring the uniqueness of regenerative systems and regenerative circular economy activities is critically important. The evolution of these systems is often informed directly by nature, as in the Finca Equilibrium system.

8. A brief discussion: Observations from Armenia's community-government interactions

Current circular economy initiatives in Armenia are primarily formed by linkages between government, researchers, and community members (e.g., the AUA associated initiative mentioned previously). In May 2021, the Ministry of Environment announced that together with partners from Lithuania, they will start recovering and recycling plastic and glass waste as a primary material input for new product production (e.g., in the textile industry). In addition, ISSD supports Armenia's transition toward a more circular economy. Such initiatives may help to build critical linkages and could develop into cross-sectoral collaborations.

9. Insights

We highlight some of the similarities and differences in Armenia and Portugal's current circular economy initiatives (Table 1). This list is not exhaustive. It provides a high-level overview of some of the factors observed in this work.

10. Conclusions

This chapter investigates vastly different communities that strive to shift from a linear economy toward a circular economy, considering emerging circular economy policies, socioeconomic relationships, community-government interactions in Armenia, and

Table 1 Some similarities and differences in current circular economy initiatives in Armenia and Portugal.

Similarities	Differences
Early adopters are small businesses and communities	Broader economic and political conditions
Private sector engagement/investment in new circular economy initiatives	Availability of circular economy action plans and government support at the state level
Ongoing development of community-based (bottom-up) circular economy initiatives	Knowledge sharing practices and knowledge sharing platforms
Material cycling (upcycling and downcycling) is a common practice (but still requires substantial research and support to become fully realized)	Development of cross-sectoral collaborations
Multistakeholder initiatives are common	Geographic location and related issues (e.g., border issues in Armenia that contribute to political and social instabilities)

a microlevel community-based initiative in Portugal. There is limited data available in Armenia on circular economy efforts. Most Armenian efforts focus on recycling (i.e., a last resort in most circular economy models because it is a lower efficiency option in circularity terms, often). However, in Armenia that is landlocked and has extreme landfill and waste issues, it is reasonable to consider recycling along with redesigning products and production processes to reduce wastes. In both Portugal and Armenia, the finance for circular economy initiatives comes from multiple sources (e.g., corporations, community members, NGO groups, government, and third party governments).

Financial investments in circularity initiatives must align with long-term circular economy goals. For example, the Eco.nomia information exchange platform, financed through conventional means in Portugal, contributes to an important long-term goal aligned with circular economy principles (i.e., a sharing economy). It contributes to a type of cultural practice, information exchange, and set of values aligned with the circular economy concept. The sharing economy focuses on sharing and enjoying what can be shared rather than privately consumed (Ostrom, 2000).

Still, the Portugal circular economy action plan could improve by embodying the language needed to support long-term outcomes aligned with the circular economy's core principles. For example, parts of Portugal's action plan and related documents focus on the "exploitation" of resources (Eionet, 2019, p. 6). It's critical to recall that the circular economy demands that we rethink terms like "exploitation" and "growth."

Moving forward, there is a clear need to develop a framework to help circular economy practitioners define the goals and scope and evaluate the outcomes of their circular economy efforts, using metrics based on the circular economy's core principles.

The framework needs to be adaptable to be able to account for various political, economic, social, and regionally specific environmental factors, long-term circular economy outcomes, and potential unintended consequences.

References

(AUA) Acopian Center for the Environment, 2020. Waste Governance in Armenia. American University of Armenia and Life Foundation. Retrieved from: https://ace.aua.am/files/2020/08/WGA-Report-Eng.pdf.

(AUA) American University of Armenia, 2019. Swedish-Armenian Cooperation Launches Waste Governance Policy Process for Armenia. Retrieved from: https://newsroom.aua.am/2018/12/24/swedish-armenian-cooperation-launches-waste-governance-policy-process-for-armenia.

(COM) Council of Minister's, 2017. Circular Economy Action Plan, Leading the Transition: Action for the Circular Economy in Portugal: 2017–2020. Retrieved from: https://circulareconomy.europa.eu/platform/sites/default/files/strategy_-_portuguese_action_plan_paec_en_version_3.pdf.

(COM) Council of Minister's, 2019. Report From the Commission to the European parliament, the Council, the European Economic and Social Committee and the Committee of the Regions: On the Implementation of the Circular Economy Action Plan. Retrieved from: https://eur-lex.europa.eu/legal-content/en/txt/?qid=1551871195772&uri=celex:52019dc0190.

(CP) Constitution Project, 2005. Portugal's Constitution of 1976 With Amendments Through 2005. Retrieved from: https://www.constituteproject.org/constitution/Portugal_2005.pdf.

(EU) European Union Commission, 2019. Report From the Commission to the European Parliament, the Council, the European Economic and Social Committee and the Committee of the Regions on the Implementation of the Circular Economy Action Plan. Retrieved from: https://ec.europa.eu/commission/sites/beta-political/files/report_implementation_circular_economy_action_plan.pdf.

(EU) European Union Commission, 2020. Retrieved from: https://ec.europa.eu/environment/circular-economy.

(FAO) Food and Agriculture Organization of the United Nations, 2019. Countries Profile: Portugal. Retrieved from: www.fao.org/countryprofiles/index/en/?iso3=PRT.

(LIFE) Life International Foundation for Ecology, 2020. Retrieved from: www.life.se.

(OEC) The Observatory of Economic Complexity, 2017. Retrieved from: www.oec.world/en/.

(UN COMTRADE) UN Comtrade Database, 2018. Retrieved from: www.comtrade.un.org/.

(WBCSD) World Business Council for Sustainable Development, 2019. Circular Transition Indicators: Proposed Metrics for Business, by Business. Retrieved from: https://docs.wbcsd.org/2019/07/WBCSD_Circular_Transition_Indicators_Proposed_metrics_for_business_by_business.pdf.

Benyus, J.M., 1997. Biomimicry: Innovation Inspired by Nature.

Binder, C., 2007. From material flow analysis to material flow management part II: the role of structural agent analysis. J. Clean. Prod. 15 (17), 1605–1617.

Binder, C.R., Hinkel, J., Bots, P.W.G., Pahl-Wostl, C., 2013. Comparison of frameworks for analyzing social-ecological systems. Ecol. Soc. 18 (4), 26. https://doi.org/10.5751/ES-05551-180426.

Eionet, 2019. Resource Efficiency and Circular Economy in Europe-Even More from Less: An Overview of Policies, Approaches and Targets of Portugal 2018. Retrieved from: https://www.eionet.europa.eu/etcs/etc-wmge/products/b-country-profile-portugal_finalised.pdf.

Hobbs, R.J., 2018. Restoration ecology's silver jubilee: innovation, debate, and creating a future for restoration ecology. Restor. Ecol. 26 (5), 801–805.

McDonough, W., Braungart, M., 2002. Design for the triple top line: new tools for sustainable commerce. Corp. Environ. Strateg. 9 (3), 251–258.

Merrill, C.L., 1982. Biomimicry of the Dioxygen Active Site in the Copper Proteins Hemocyanin and Cytochrome Oxidase: Part I: Copper (I) Complexes Which React Reversibly With Dioxygen and Serve to Mimic the Active Site Function of Hemocyanin. Part II: Mu-Imidazolato Binuclear Metalloporphyrin Complexes of Iron and Copper as Models for the Active Site Structure in Cytochrome Oxidase (Doctoral Dissertation). Rice University.

Morseletto, P., 2020. Restorative and regenerative: exploring the concepts in the circular economy. J. Ind. Ecol., 1–11. https://doi.org/10.1111/jiec.12987.

Muhar, A., Raymond, C., van den Born, R., Bauer, N., Böck, K., Braito, M., Buijs, A., Flint, C., de Groot, W., Ives, C., Mitrofanenko, T., Plieninger, T., Tucker, C., van Riper, C., 2017. A model integrating social-cultural concepts of nature into frameworks of interaction between social and natural systems. J. Environ. Plan. Manag. https://doi.org/10.1080/09640568.2017.1327424.

Ostrom, E., 2000. Collective action and the evolution of social norms. J. Econ. Perspect. 14 (3), 137–158.

Pinto, A.C., 2015. Corporatism and dictatorships in Portugal and Spain. Comparative perspectives. In: Bananen, Cola, Zeitgeschichte: Oliver Rathkolb Und Das Lange 20. Jahrhundert.

Winans, K., Kendall, A., Deng, H., 2017. The history and current applications of the circular economy concept. Renew. Sust. Energ. Rev. 68, 825–833.

Index

Note: Page numbers followed by *f* indicate figures and *t* indicate tables.

F

T

U

Printed in the United States
by Baker & Taylor Publisher Services